TRANSLATIONS

If f is a function and $c > 0$, th

the graph of $y = f(x) + c$ is the graph of $y = f(x)$ shifted *up* c units,
the graph of $y = f(x) - c$ is the graph of $y = f(x)$ shifted *down* c units,
the graph of $y = f(x + c)$ is the graph of $y = f(x)$ shifted c units to the *left*,
the graph of $y = f(x - c)$ is the graph of $y = f(x)$ shifted c units to the *right*.

INVERSE FUNCTIONS

If f is a one-to-one function, then f^{-1} is the function such that,

$$f(f^{-1}(x)) = x \text{ for all } x \text{ in the domain of } f^{-1} \text{ and}$$
$$f^{-1}(f(x)) = x \text{ for all } x \text{ in the domain of } f.$$

f^{-1} is called the inverse of f.

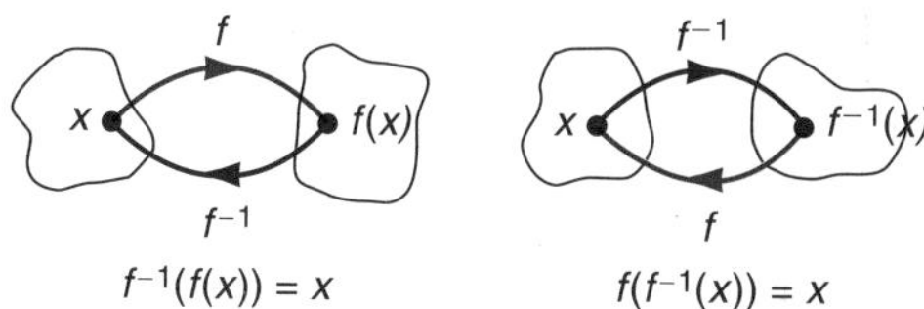

GRAPH OF A FUNCTION AND ITS INVERSE

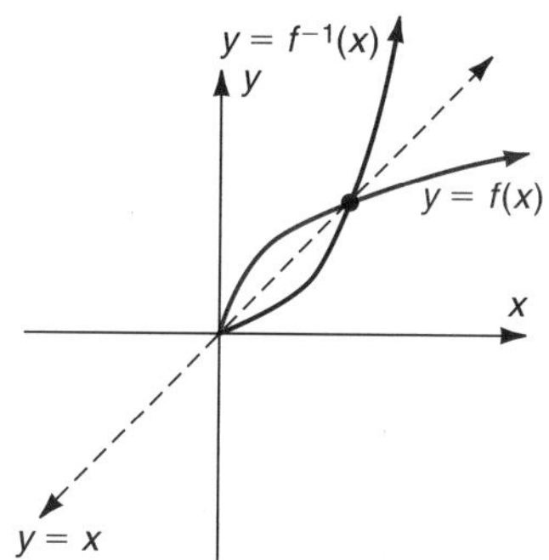

GRAPHS OF THE CONICS

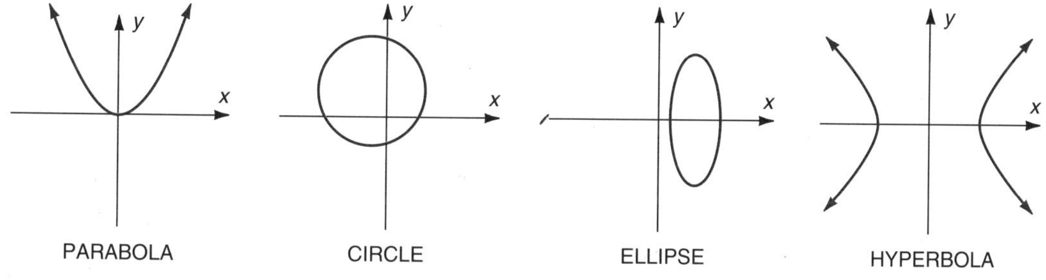

College Algebra

Linda L. Exley
Vincent K Smith
DeKalb College

PRENTICE HALL *Englewood Cliffs, New Jersey 07632*

Library of Congress Cataloging-in-Publication Data

Exley, Linda L.
College algebra / Linda L. Exley, Vincent K Smith.
p. cm.
Includes index.
ISBN 0-13-207754-X
1. Algebra. I. Smith, Vincent K. II. Title.
QA154.2.E85 1993 92-17537
512.9—dc20 CIP

Acquisitions Editor: Priscilla McGeehon
Editorial and Production Supervisor: Valerie Zaborski
Editor-in-Chief: Tim Bozik
Development Editor: Roberta Lewis
Marketing Manager: Paul Banks
Interior Design: Anne Bonanno and Lisa Domínguez
Cover Design: Bill McCloskey
Design Director: Florence Dara Silverman
Copy Editor: Carol Dean
Prepress Buyer: Paula Massenaro
Manufacturing Buyer: Lori Bulwin
Supplements Editor: Mary Hornby
Editorial Assistant: Marisol L. Torres

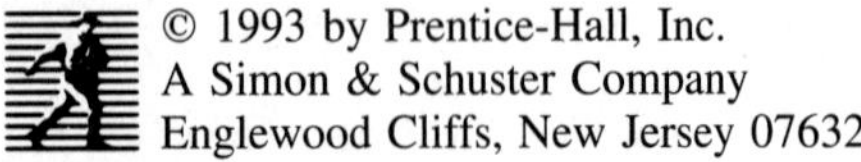

Printed in the United States of America

10 9 8 7 6 5 4 3 2 1

ISBN 0-13-207754-X

Prentice-Hall International (UK) Limited, *London*
Prentice-Hall of Australia Pty. Limited, *Sydney*
Prentice-Hall Canada Inc., *Toronto*
Prentice-Hall Hispanoamericana, S.A., *Mexico*
Prentice-Hall of India Private Limited, *New Delhi*
Prentice-Hall of Japan, Inc., *Tokyo*
Simon & Schuster Asia Pte. Ltd., *Singapore*
Editora Prentice-Hall do Brasil, Ltda., *Rio de Janeiro*

For Ken, Margaret,
and Irene

Contents

CHAPTER 3 GRAPHS AND FUNCTIONS 149

CHAPTER 4 WORKING WITH FUNCTIONS 225

CHAPTER 5 POLYNOMIAL EQUATIONS 306

CHAPTER 6 EXPONENTIAL AND LOGARITHMIC FUNCTIONS 351

Preface

Mathematics education in today's society holds many challenges. In order to ensure that students are properly prepared in college for technical careers, professional groups as well as government agencies are suggesting new ways to teach everything from arithmetic to linear algebra and beyond. Despite the advent of new technology, which provides new perspectives on learning math, students must nevertheless continue to learn the basics of college algebra and trigonometry in preparation for a calculus course.

In writing this text, we tried to maintain a balance between traditional topics and acknowledgement of the advantages of technology such as graphing calculators and computer-based symbolic processors. Although we have included boxes that specifically address the use of graphing calculators in the text, we do not assume their use. Any student, with or without a graphing calculator, should be able to follow the concepts introduced in our book.

After teaching college mathematics for many years, we have learned about students' level of preparedness for this course as well as the prerequisite knowledge they must have for the courses to follow. In writing this book, we wanted to address both ends of the spectrum—providing help to the students whose algebra skills are weak or rusty, and yet preparing them fully for a course in calculus.

REVIEWER'S CONFERENCE

Once a first draft of our manuscript was complete, we met with editors at Prentice Hall and with reviewers who teach this course at both four-year universities and two-year colleges. Over the course of two days, we turned every page of the manuscript and discussed ways in which to make it better—additional examples, a new slant on the exercise sets, a more precise statement. The reviewer's conference participants—Jim Newsom of Tidewater Community College, Lou Ann Mahaney of Tarrant County Junior College, Eldon Miller of the University of Mississippi, and Mary Yorke of Eastern Michigan University—each made a vital and individual contribution to the book.

DEVELOPMENT EDITOR AND ACCURACY CHECK

A full-time math development editor read through the manuscript and suggested how to incorporate these improvements, and established a high standard of quality for the final book. In order to ensure the accuracy of the final book, we worked carefully with the author of solutions manuals who simultaneously wrote the solutions and checked the answer key.

KEY CONTENT FEATURES

Examples, Exercises, Clarity of Writing

The most important features of any math text are clarity of writing, detail of examples, and wealth and variety of exercises. We think you will find that we have incorporated all of these characteristics into our books. As you look through it, you will note that it is written in a language that students can understand. Wordy statements and mathematical jargon are kept to a minimum. Although simple language is used and rigor is not stressed, correct mathematics is. Topics are illustrated with lots of examples containing detailed explanations. The exercise sets reflect students' varying abilities, and are designed to reinforce basic skills as well as challenge those who will go on to another course.

Graphing Calculators and Calculators

Handheld calculators have developed at an astonishing rate in the past few years. Excellent programmable scientific calculators are usually available for under $30. Since it is reasonable to expect every serious math student to own a scientific calculator, this book is written to be used in such an environment.

The more recent breakthrough of calculators that sell for under $100 and have exceptional graphing capability presents a stickier problem. Our approach in this textbook is to provide basic instruction (including keystrokes) for using the TI-81 and the Casio fx-7700G graphing calculators in a series of ''Graphing Calculator Boxes'' that are independent of the surrounding text. In some sections, there are also specially marked graphing calculator exercises. These boxes and exercises should provide help in getting students started with their graphing calculators, and to teach an appreciation of the power of the tool. (For students who do not own a graphing calculator, Prentice Hall's program X(PLORE) by David Meredith is an excellent substitute. The same information as in the graphing calculator boxes is available on a special tutorial program disk of X(PLORE).)

KEY PEDAGOGICAL FEATURES

The strength of this book lies in its problem sets. Each problem set is broken down into several parts, making selection for homework assignments easier.

Exercises

Warm-ups begin each problem set. They are keyed to examples, and provide basic drill and reinforcement of concepts.

Practice Problems are not keyed to examples, requiring students to make their own decisions about how to approach a problem. Because of this, Practice Problems are generally slightly more difficult than Warm-ups. The Practice Problems are presented in odd-even pairs. Many of the Practice Problems are color-coded, showing that they are Key Problems.

Key Problems are a representative group of problems which, if they are all assigned, ensure the instructor of covering all major concepts in the homework assignment.

Challenge Problems allow students to probe into natural extensions of the topics presented in the text, and prepare them for the types of problems they will confront in further courses.

In Your Own Words conclude each problem set, allowing students to use and enhance their understanding and writing abilities in the context of mathematics.

Chapter Introduction

Connections appear at the beginning of each chapter. They include several paragraphs which put upcoming material in its mathematical context, and show its connection to other chapters, other courses, real-life applications, and the historical development of mathematical ideas.

End-of-Chapter Material

Chapter Summaries can be valuable tools for students who are brushing up on skills or preparing for chapter or cumulative tests. Ours have been designed to provide only the most vital information from each chapter in a brief form. A glossary redefines key terms which were introduced in each chapter, followed by a list of important rules, procedures and formulas.

Review Problems provide a final opportunity to practice important skills. They are followed by a brief set of exercises labeled *Let's Not Forget,* which provide cumulative reinforcement of concepts students commonly forget through the course of the term.

SUPPLEMENTS

The following supplements are available to qualified adopters of the book:

For the Instructor

Instructor's Solutions Manual contains even-numbered solutions.

Testing will be available in several formats:

PHTestmanager testing (IBM and Macintosh) provides a bank of test items designed specifically for the book, in both free-response and multiple-choice form; they are fully editable for flexible use.

Test Item File contains a hard copy of test questions on PHTestmanager.

IPS Test/Algorithmic Testing (IBM) provides a number of algorithms or model problems for each chapter which can generate a large number of similar but not identical problems for tests or practice worksheets.

Syllabus with Instructor's Disk provides suggested syllabi and homework assignments, and an ASCII disk which allows customization of syllabi.

For the Student

Math Master Tutor software provides drill and practice, short tutorials, and quizzes on IBM and Macintosh formats.

Videotapes contain lectures introducing the graphing calculator for use in algebra (one set available with an adoption of 100 or more copies).

X(PLORE) Version 4.0, by David Meredith. A powerful, fully programmable, numeric mathematical processor for IBM-compatibles that evaluates expressions, graphs curves and surfaces, solves equations, and integrates and differentiates functions. Site license available free to adopters of the user's manual or book/disk. (Macintosh version in development for possible Fall 1993 use).

X(PLORE) Graphing Problems Tutorial Disk provides templates for graphing calculator problems in this book to be solved using X(PLORE).

Student's Solutions Manual contains solutions to odd-numbered problems.

REVIEWERS

We would like to thank the following reviewers, whose suggestions and comments were much appreciated:

Wayne Andrepont, *University of Southwestern Louisiana*
James D. Blackburn, *Tulsa Junior College*
Tom F. Davis, *Daytona Beach Community College*

William R. Fuller, *Purdue University*
Bill Gerson, *Prince Georges Community College*
Lou Ann Mahaney, *Tarrant County Junior College*
Eldon Miller, *University of Mississippi*
Jim Newsom, *Tidewater Community College*
Ian Walton, *Mission College*
Jon D. Weerts, *Triton College*
Mary Yorke, *Eastern Michigan University*

Finally, we would like to thank the staff at Prentice Hall for their support in this project: In particular, Valerie Zaborski, who is far too patient to be a production editor; designers Anne Bonanno and Lisa Domínguez; development editor Bobbie Lewis; and lastly, our editor and friend Priscilla McGeehon.

USING YOUR CALCULATOR

Since you will probably be expected to bring a scientific calculator to class and be able to use it, let's see if we can get you started on the right foot.

First, there are two major categories of calculators based on the logic they employ, those that use algebraic notation and those that use reverse polish notation (RPN). RPN machines are usually expensive and somewhat specialized. We have limited our instruction to machines that use algebraic notation. This is not meant to discourage you if you have an RPN calculator; they are very powerful and efficient and come with an excellent instruction manual.

It is important that you get to know your machine. You want it to be your friend and ally in the coming weeks. There are a few things you should find out about your calculator right away. Let's look at them.

ORDER TO PRESS KEYS

Within the class of machines that use algebraic notation are two categories. An example best illustrates the difference. Approximating the square root of 3 involves two important keys; the square root key and the key marked 3. Which do you press first? We will consider calculators that use the order

[3] [√‾]

to be our *basic calculator.* Most keystroke examples will use that logic. With this type of calculator, the display shows only numbers entered and the result of calculations.

The recent trend for slightly up-scale calculators and graphing calculators has been to use the logic

[√‾] [3] [ENTER] (or perhaps [EXE] instead of [ENTER])

for the operation. In the text, we will refer to these machines as *function-first* calculators. We will include keystrokes for function-first calculators when critical operations first occur. The display on these machines shows the expression being entered and the display looks like the way we would write the expression on paper. We enter keystrokes in the natural way. The function-first calculators have editing features that are useful.

Use the above example to find out if you have a basic calculator or a function-first calculator.

MODES

Your calculator has several "modes." Although you may not use some modes in this course, it is important that you know how to get out of a mode in case you accidentally get yourself into one. If your calculator has a statistical mode, it probably will interfere with some ordinary algebraic operations. Be sure you can get back to computational mode. Consult your instruction manual about modes.

THE KEYPAD

1. Find the location of the keys for [π], [(], [)], [y^x] or [x^y], and [1/x] or [x^{-1}].
2. Find out how to clear the screen. Such keys as [CLEAR], [AC], or [CL] are used.
3. If you look at your calculator keypad, you will notice that most keys have something written just above them. For example, you probably have a key that looks like,

Basic calculator:			Function-first calculator:		
√‾		x^2	√‾ I		x^2 H
[x^2]	or	[√‾]	[x^2]	or	[√‾]

The operation written *above the key* is called a second function. To use it, you first need to press a second function key. It is usually located near the upper, left-hand corner of the keypad. It is often color coded and is usually labeled

[2nd], [INV], or [SHIFT].

Function-first calculators have a letter *above* and *to the right* of most keys. It is accessed by the [ALPHA] key. The alpha keys are used to enter a letter of the alphabet. Letters are used to address storage locations, and as variables whose value is the contents of their storage location. We will indicate *second function keys* with a double box, such as [[x^2]].

ENTERING NEGATIVE NUMBERS

Find out how to enter a negative number such as −4. Most calculators will not let you use the subtraction key for negation. Look for a change sign key, usually

[+/−], [CHS], or [(−)]

and it may be a second function. Make sure you can do the addition problem $-4 + 4$ with the numbers in that order.

Basic logic: [4] [+/−] [+] [4] [=]

Function-first logic: [(−)] [4] [+] [4] [ENTER]

STORING AND RECALLING NUMBERS

Discover how to store a number and recall it.

Basic logic:

To store, look for a key like

STO, M in, x → M or →

and to recall, a key like

RCL, MR, or RM

To test yourself, store 5 in memory and then compute 4 + 5 by recalling 5 from memory.

5 STO *(Stores 5 in memory.)*

4 + RCL = *(Computes 4 + 5)*

Function-first logic:

The function-first calculators use letters to address storage locations. Since the letters can be used for the number in the storage locations, they are sometimes called *variables.*

The variables are accessed by the ALPHA key.

5 STO ALPHA A ENTER *(Stores 5 in variable* A.*)*

Now we can compute 4 + 5 by using the contents of storage location *A*.

4 + ALPHA A ENTER *(Computes 4 + 5.)*

Note: The STO key on TI-81 does not require using the ALPHA key.

In addition to the ideas already listed, *graphing calculators* and most *function-first calculators* have the capabilities below.

USING THE LAST ANSWER

Find an answer key, usually ANS and it may be a second function. The result of the last calculation made by the machine is stored in ANS. Press

4 X 5 + 5 ENTER

and then

√ ANS ENTER

You should get 5.

ENTERING MULTIPLICATION

See if your calculator will let you use $2x$ as shorthand for 2 times x. Try

[2] [√‾] [9] [ENTER]

If that doesn't yield 6, you will have to use

[2] [X] [√‾] [9] [ENTER]

Caution is required with some calculators calculating expressions such as $\frac{1}{3}\pi$. Enter [1] [÷] [3] [π] [ENTER]. The multiplication may be done before the division, which is not the intended order. This can be corrected by entering [1] [÷] [3] [X] [π] [ENTER] or [(] [1] [÷] [3] [)] [π] [ENTER]. $\left(\frac{1}{3}\pi \approx 1.0472\right)$

MENUS, STATUS LINES, EDITING, AND PROGRAMMING

Your calculator probably has menus and perhaps status lines. At least be sure you can get a menu off the screen in case you accidentally call it up. Many calculators have editing capabilities similar to a word processor. In addition, many of today's calculators have programming capabilities. Consult your instruction manual.

Of course, there is much more to learn about your calculator. But this should get you started. You will probably be a whiz by the end of the term.

BUYING A NEW CALCULATOR

If you need a new calculator for this course, the best advice we can give you is wait until the first day of class and ask your instructor for recommendations. If you need a calculator before you can speak to your instructor, consider the following points.

1. Decide if you want a graphing calculator. We recommend them highly. If you intend to continue on through the calculus sequence, a graphing calculator is almost a necessity. The only disadvantage is cost. If you shop around, an excellent graphing calculator can be found for under $80 at this writing.
2. If you decide against a graphing calculator, be sure to buy a scientific calculator. Somewhere on it will probably be written "scientific." In any event, it must have at least one memory and the following keys:

   [LOG], [LN], [SIN], [y^x] or [x^y], [1/x] or [x^{-1}], [(]

3. Do not hesitate to go to the math department of your school and corner any math professor for guidance about selection and best price.

Connections

Real-world **"Connections"** in the opening paragraphs of each chapter link the material to other chapters, other disciplines, math history, and current technology.

CHAPTER 4

Working with Functions

CONNECTIONS

In the last two sections of Chapter 3 we defined a function, explained the notation, and graphed functions utilizing our knowledge of the graph of an equation. An understanding of functions is essential for further study in mathematics. Those studying in preparation for calculus should realize that calculus is a *study* of functions. Graphing calculators are sometimes called *function machines* because they work with functions. In this chapter, we pursue the idea of function with emphasis on graphing the functions commonly used.

In Chapter 5 we continue the study of functions with emphasis on *polynomial equations.* In Chapter 6, we introduce the *exponential* function and its inverse, the *logarithmic* function.

An in depth look at concepts involved in graphing functions is important. Today, computers and graphing calculators can draw graphs of very complicated functions, freeing us from the drudgery of pencil and paper calculations. However, an understanding of the functions being graphed is necessary for precise use of such technology.

225

A Wealth of Examples

Both texts feature a wealth of well-chosen, detailed, worked examples. Each step is explained with careful attention to detail. These examples are then keyed to the **Warm-up** exercises at the end of each section.

EXAMPLE 2 Charles' Law says that when gas is heated at a constant pressure, the relationship between the temperature (in kelvin) of the gas and the volume of the gas is linear. The following information was recorded in a lab.

Temperature (kelvin)	*Volume (cubic centimeters)*
273	100
323	118.3

(a) Express the volume of the gas (V) as a function of temperature (T).

(b) Graph this function.

(c) As the temperature increases, what happens to the volume?

Solutions (a) Since the relationship is given to be *linear,* we can use the two points (273, 100) and (323, 118.3) and write an equation of the line. Notice that we are using temperature as the independent variable and volume as the dependent variable. Using the slope formula, we have

$$m = \frac{118.3 - 100}{323 - 273} = \frac{18.3}{50}$$

Now, the point-slope formula gives

$$V - 100 = \frac{18.3}{50}(T - 273)$$

$$V = 0.366T + 0.082$$

Using functional notation this becomes $V(T) = 0.366T + 0.082$.

(b) To graph the function, we plot the points (273, 100) and (323,118.3) as in Figure 4.4.

(c) Since the line is increasing, volume increases as temperature increases. □

V
V(T) = 0.366T + 0.082
(273, 100)
(323, 118.3)
T

Fig. 4.4

Quadratic Functions

> **Definition: Quadratic Functions**
>
> A function of the form $f(x) = ax^2 + bx + c$, where a, b, and c are real numbers and $a \neq 0$, is called a **quadratic function.** The graph of a quadratic function is a parabola.

Sample pages are reduced to 50% of original size.

PROBLEM SET 4.1

Warm-ups

In Problems 1 through 4, sketch the graph of each linear function. See Example 1.

1. $f(x) = 3x + 1$ **2.** $g(x) = -5x$ **3.** $h(x) = \frac{2}{3}x + 1$ **4.** $f(x) = -\frac{1}{2}$

For Problems 5 and 6, see Example 2.

5. Dr. Joan Rupp has determined that the amount (in grams) of lactic acid removed from the blood is a linear function of the amount (in liters) of oxygen consumed during slow recovery. Use the following data to express the amount of lactic acid (A) as a function of the amount of oxygen (x) consumed.

Lactic Acid (grams)	*Oxygen (liters)*
40	5
64	7

6. Heart rate (in beats per minute) is a linear function of the number of liters of oxygen consumed per minute. Express this relationship using the following data from one subject.

Heart Rate (beats per minute)	*Oxygen (liters per minute)*
142	2.0
125	1.6

In Problems 7 through 14, sketch the graph of each quadratic function. See Example 3.

7. $h(x) = 2x^2 - 1$ **8.** $f(x) = 1 - x^2$ **9.** $g(x) = (x + 2)^2$ **10.** $f(x) = -(x - 2)^2 + 3$
11. $g(x) = x^2 + 2x + 3$ **12.** $g(x) = x^2 + 4x - 1$ **13.** $h(x) = -2x^2 + 2x$ **14.** $f(x) = 3x^2 - 6x$

For Problems 15 and 16, see Example 4.

15. The profit on a certain Amazon River basin tour is given by $P(x) = -(x - 120)^2 + 8500$, where x represents the number of tour tickets sold. How many tickets must be sold to obtain a maximum profit? What will the maximum profit be?

16. The sum of two numbers is 20. What is the largest possible value for the product of the two numbers? What are the numbers when the product is largest?

In Problems 17 and 18, find the required value of each function. See Example 5.

17. $f(x) = 4 - 6x - 3x^2$. Find the maximum value. **18.** $g(t) = 5t^2 + 12t - 1$. Find the minimum value.

In Problems 19 through 26, sketch the graph of each absolute value function. See Examples 6 and 7.

19. $g(x) = |x + 4|$ **20.** $h(x) = |x - 3|$ **21.** $f(x) = |x - 1| - 2$ **22.** $g(x) = |x + 2| - 1$
23. $h(x) = |x| + 2$ **24.** $h(x) = |x| - 3$ **25.** $f(x) = 2 - |x + 3|$ **26.** $f(x) = 1 - |x|$

In Problems 27 through 32, sketch the graph of each square root function. See Example 8.

27. $f(x) = \sqrt{x - 2}$ **28.** $f(x) = \sqrt{x + 1}$ **29.** $f(x) = \sqrt{x} + 3$
30. $f(x) = \sqrt{x} - 1$ **31.** $f(x) = \sqrt{x - 1} + 2$ **32.** $f(x) = \sqrt{x + 2} - 1$

In Problems 33 through 36, use the graph shown in Figure 4.27 to graph each function. See Example 9.

33. $y = f(x) + 1$ **34.** $y = f(x + 2)$
35. $y = f(x + 1)$ **36.** $y = f(x + 2) - 1$

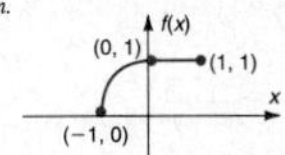

Fig. 4.27

Warm-ups

Each problem set is carefully designed to help students at all levels be successful. Exercises begin with **Warm-ups** which are specifically keyed to worked examples in the section.

Practice Exercises

Practice exercises that follow **Warm-ups** are mixed to provide decision-making opportunities.

In Problems 25 and 26, sketch the graph of each function. See Example 9.

25. $f(t) = \dfrac{t^2 - 9}{t + 3}$ **26.** $s(x) = \dfrac{x^2 - x - 12}{x - 4}$

Practice Exercises

In Problems 27 through 62, sketch the graph of each function.

27. $g(x) = \dfrac{1}{x^2}$ **28.** $f(x) = \dfrac{1}{(x-2)^2}$ **29.** $s(x) = \dfrac{-1}{x^2}$ **30.** $f(x) = \dfrac{-1}{(x-2)^2}$

31. $f(x) = \dfrac{4}{x+4}$ **32.** $g(x) = \dfrac{3}{x-2}$ **33.** $f(x) = \dfrac{-1}{x+4}$ **34.** $s(t) = \dfrac{-3}{t-2}$

35. $f(x) = \dfrac{x}{2x-1}$ **36.** $f(x) = \dfrac{2x}{x+2}$ **37.** $g(x) = \dfrac{x+3}{2x-3}$ **38.** $f(x) = \dfrac{2x-1}{3x+2}$

39. $h(t) = \dfrac{t}{2-t}$ **40.** $g(x) = \dfrac{3-2x}{x}$ **41.** $s(x) = \dfrac{x}{(2x-1)(x+1)}$ **42.** $f(x) = \dfrac{2x}{(3x+2)(x-1)}$

43. $f(x) = \dfrac{2-x}{x^2-1}$ **44.** $f(x) = \dfrac{2x+1}{x^2-9}$ **45.** $g(x) = \dfrac{x+4}{x^2}$ **46.** $s(t) = \dfrac{1-2t}{3t^2}$

47. $f(x) = \dfrac{x^2}{4x^2-1}$ **48.** $f(x) = \dfrac{(x+2)(x-3)}{(x+1)(2x-1)}$ **49.** $f(x) = \dfrac{x^2+x}{(x-1)^2}$ **50.** $f(x) = \dfrac{(x+2)^2}{(x+3)^2}$

51. $f(x) = \dfrac{-x}{x^2+1}$ **52.** $g(x) = \dfrac{(2x+1)(x-3)}{x^2+1}$ **53.** $s(x) = \dfrac{x^2+x+3}{x-1}$ **54.** $h(x) = \dfrac{2x^2-x-8}{x+2}$

55. $f(t) = \dfrac{2t^2+t-1}{2t-1}$ **56.** $g(x) = \dfrac{x^2-9}{x+3}$ **57.** $f(x) = \dfrac{x^2-2x-2}{x+1}$ **58.** $g(x) = \dfrac{2x^2+3x}{x+2}$

59. $p(t) = \dfrac{6t^2+t-2}{3t+2}$ **60.** $q(z) = \dfrac{3z^2+11z-4}{z+4}$ **61.** $g(x) = \dfrac{x^3-x^2}{x-1}$ **62.** $s(t) = \dfrac{t^3+1}{t+1}$

In Problem 63, graph the function f using the techniques of this section. In Problems 64 through 68, sketch each graph by reflecting or translating the graph of f found in Problem 63.

63. $f(x) = \dfrac{1}{x}$ **64.** $y = -\dfrac{1}{x}$ **65.** $y = \dfrac{1}{x} + 1$ **66.** $y = \dfrac{1}{x} - 2$

67. $y = \dfrac{1}{x+1}$ **68.** $y = \dfrac{1}{x-2}$

... *h graph by reflecting or translating the graph of*

... $= \dfrac{1}{x^2} + 2$ **71.** $y = \dfrac{1}{x^2} - 1$ **72.** $y = \dfrac{2}{x^2}$

...th a vertical asymptote ...e of $y = 3$.
... with the y-axis as a ...s as a vertical asymp-

75. Give a rule for a rational function with vertical asymptotes of $x = \pm 2$ and a horizontal asymptote of $y = 3$.

76. Give a rule for a rational function with no vertical asymptote and a horizontal asymptote of $y = 3$.

77. Boyle's Law states that pressure times volume is constant. That is, $PV = C$ or $P = C/V$. What would a sketch of the graph of this equation look like? If V increases, what happens to P? If P increases, what happens to V?

78. In manufacturing, unit cost is given by $U = T/n$, where T is the total cost and n is the number of units produced. If the total cost is made up of constant fixed costs F and variable costs of the form nV, we have $T = F + nV$ and we see that U is a function of n,

$$U(n) = \frac{F + nV}{n}$$

Sketch the graph of U taking F to be 20 and V to be 2. [For convenience in graphing, assume the domain of U is $[1, +\infty)$.]

Challenge Problems

79. Sketch the graph of $f(x) = \begin{cases} \dfrac{1}{x-1} & \text{if } x \neq 1 \\ 5 & \text{if } x = 1 \end{cases}$

80. Sketch the graph of $g(x) = \begin{cases} \dfrac{1}{x^2-1} & \text{if } |x| \neq 1 \\ 0 & \text{if } |x| = 1 \end{cases}$

81. In Problem 78, we *assumed* the domain of the function was $[1, +\infty)$ for ease in graphing. Read the problem again and determine the *actual* domain of U and draw the graph on the interval $[1, 5]$.

For Graphing Calculators

Graph each rational function and describe its behavior.

82. $y = \dfrac{x^2 + 2x + 2}{x^2 + x - 1}$ **83.** $y = \dfrac{3x^3 - 8x^2 - 5x - 22}{4x^3 + 16x^2 - 5x - 7}$

84. $y = \dfrac{x^4 - 3}{x^5 - 5}$ **85.** $y = \dfrac{x^4 - 3x^2 + x}{x^3 + 2x + 7}$

IN YOUR OWN WORDS . . .

86. Explain how to find the asymptotes of a rational function.

87. Explain how to find the x-intercepts and y-intercept of a rational function (if there are any).

4.4 ALGEBRA OF FUNCTIONS

If f and g are functions, we can make new functions by adding, subtracting, multiplying, and dividing the rules for f and g.

Definitions: Operations with Functions

$(f + g)(x) = f(x) + g(x)$	Sum function
$(f - g)(x) = f(x) - g(x)$	Difference function
$(fg)(x) = f(x) \cdot g(x)$	Product function
$\left(\dfrac{f}{g}\right)(x) = \dfrac{f(x)}{g(x)};\ g(x) \neq 0$	Quotient function

Sample pages are reduced to 50% of original size.

Challenge and In Your Own Words Problems

Each problem set also contains **Challenge Problems** to encourage students to probe the natural extension of the text's topics, as well as **In Your Own Words** problems that help students develop their writing skills.

Graphing Technology

Both texts feature graphing calculator content boxes which provide basic instruction for the TI-81 and the Casio fx-7700 G.

GRAPHING CALCULATOR BOX

Graphing Inverse Functions

A graphing calculator that has the capability to graph parametric equations can graph many inverse functions. Both the TI-81 and the CASIO fx-7700G have this capability. Suppose f is a one-to-one function whose rule is explicitly stated in terms of some independent variable such as x. We can state the function in equivalent form with the parametric equations, $\begin{cases} x = t \\ y = f(t) \end{cases}$. The graph of these parametric equations is the same as the graph of $y = f(x)$. Therefore, we can graph f^{-1} by graphing the parametric equations, $\begin{cases} x = f(t) \\ y = t \end{cases}$.

Let's graph the function $f(x) = \sqrt{x+2}$ and its inverse. First, we enter the function in parametric form: $\begin{cases} x = t \\ y = \sqrt{t+2} \end{cases}$.

TI-81

Press MODE, arrow down to the fourth line, select "Param" and ENTER. Press Y= and notice how the function memory has changed. Instead of functions Y1 through Y4, there are now three pairs; X1T, Y1T through X3T, Y3T. These are for graphing parametric equations. The cursor should be at X1T. Press X|T and notice that T has appeared. Arrow down to Y1T and enter the function $\sqrt{T+2}$ in the usual manner. Before we graph, press RANGE and notice the new entries, Tmin, Tmax, and Tstep. Set Tmin to -2 and Tmax to 10 and then set the window to $[-3, 6]$ for x and $[-3, 3]$ for y. Press GRAPH and you should see the graph of $f(x) = \sqrt{x+2}$. To graph the inverse, return to Y= and enter $\sqrt{T+2}$ in X2T and T in Y2T then regraph.

CASIO fx-7700G

Press MODE SHIFT x to select "PARAM" the parametric mode. Then call up the function memory F MEM and enter T as f1 and $\sqrt{T+2}$ as f2. Press RANGE to set the window at $[-3, 6]$ for x and $[-3, 3]$ for y. Press RANGE once again and we see the parametric range values. Set "min:" to -2 and "max:" to 10 and "ptch" to 0.1. Press RANGE once more to return to the computation screen. In parametric mode, when we press GRAPH we see,

...eive parametric equations for X and Y separated by a comma and terminated with ...ter, F2 1 , F2 2) EXE and see the graph. To graph the inverse, press ...ame keystrokes interchanging the 1 and 2.

Suppose M is the maximum load that can be applied to the beam in the figure at the bottom of page 300. What will the maximum load be if the supports are moved to be twice as far apart and the depth of the beam is doubled?

Challenge Problems

37. The maximum load on a column made of fine-grain hard maple varies jointly as the area of the cross section and the square of the least dimension of its cross section, and inversely as the square of its length. If such a column 25 feet long and 1 foot square in cross section can carry a maximum load of 28,149 pounds, what is the maximum load, to the nearest pound, that can be carried by a 20-foot-long column of the same material but circular in cross section with a diameter of 1.5 feet?

38. Kepler's Third Law of Planetary Motion states that the square of the time it takes a planet to make one revolution around the sun is directly proportional to the cube of the average distance of the planet from the sun. Assume the average distance of the earth from the sun is 93 million miles and it takes 365 days for the earth to make one orbit. If the average distance of Jupiter from the sun is 484 million miles, what is the length of the Jupiter year to the nearest earth-day?

IN YOUR OWN WORDS . . .

39. Compare direct, inverse, and joint variation.

40. Suppose A varies directly as the cube of k. Explain what happens to A if k is tripled.

CHAPTER SUMMARY

GLOSSARY

A **polynomial function of degree n:** A function of the form $f(x) = a_nx^n + a_{n-1}x^{n-1} + \cdots + a_1x + a_0$, where each coeffficient is a real number and n is a whole number. The given form is called **standard form.**

A **rational function:** A function of the form $f(x) = \frac{g(x)}{h(x)}$, where $g(x)$ and $h(x)$ are polynomials.

y **varies directly** as x: $y = kx$, where k is a nonzero constant.

y **varies inversely** as x: $y = \frac{k}{x}$, where k is a nonzero constant.

y **varies jointly** as w and x: $y = kwx$, where k is a nonzero constant.

LINEAR FUNCTIONS A function of the form $f(x) = mx + b$, where m and b are real numbers, is called a **linear function.**
The graph of a linear function is a line.

QUADRATIC FUNCTIONS A function of the form $f(x) = ax^2 + bx + c$; where a, b, and c are real numbers and $a \neq 0$, is called a **quadratic function.**
The graph of a quadratic function is a parabola.

Comprehensive Chapter Summary

Each chapter concludes with a detailed **Chapter Summary** which includes a **Glossary** of terms as well as important procedures and formulas.

Cumulative Review

Each chapter review exercise set concludes with **Let's Not Forget...** problems—cumulative exercises that provide regular review of difficult, but vital skills from previous chapters.

In Problems 69 and 70, find a partial fraction decomposition of each function.

69. $f(t) = \dfrac{t + 11}{(t - 1)(t + 3)}$

70. $g(x) = \dfrac{3x^2 + x - 6}{x^3 + 2x^2}$

In Problems 71 through 76, find $f^{-1}(x)$. Sketch f and f^{-1} on the same set of axes. Give the domain and range for both f and f^{-1}.

71. $f(x) = x + 2$

72. $f(x) = \dfrac{x}{x - 3}$

73. $f(x) = \sqrt{x - 3}$

74. $f(x) = 2 - x^2$; $x \geq 0$

75. A stone is thrown in a still pond making a circular ripple whose radius is increasing with time according to the function $r(t) = \frac{9}{2}t$, where t is seconds from the time of the splash and r is in feet. The function $A(r) = \pi r^2$ gives the area of a circle of radius r. Form the function $A \circ r$. What is $(A \circ r)(4)$? What is the meaning of $A \circ r$?

76. The daily cost in dollars of operating a television assembly line is given by the function

$$C(x) = 10x^{2/3} + 411;\ 1 \leq x \leq 100$$

where x is the number of television sets assembled in a day. Suppose the number of television sets assembled in h hours is given by the function $x(h) = 8h$; $0 \leq h \leq 12$. What does the function $C \circ x$ give? How much does it cost to operate the assembly line on a day when the assembly line operates 8 hours? What if it only operates 1 hour?

LET'S NOT FORGET . . .

77. Which form is more useful? $f(x) = \dfrac{x^3 - 4x}{x^2 - 2x - 15} = \dfrac{x(x - 2)(x + 2)}{(x - 5)(x + 3)}$

(a) If you are graphing the function f?
(b) If you are beginning a partial fraction decomposition?

78. Watch the role of negative signs. Graph the three functions.
(a) $f(x) = \sqrt{x}$
(b) $g(x) = -\sqrt{x}$
(c) $h(x) = \sqrt{-x}$.

79. From memory. Match each function in the left column with the best description of its graph in the right column.

1. $f(x) = 2x - 6$	**a.** A parabola opening downward
2. $g(x) = \sqrt{x - 3}$	**b.** A parabola opening upward
3. $h(x) = 2x^2 - 11x$	**c.** A line
4. $F(x) = \dfrac{x + 1}{x - 1}$	**d.** The upper half of a parabola
5. $G(x) = 13 - 4x - x^2$	**e.** Graph with a vertical asymptote

80. With a calculator. Given $f(t) = \dfrac{(3t - 2)(t - 1)^3}{0.12345(t^4 - 2)}$. Approximate $f(\pi)$ to five significant digits.

CHAPTER 1

Algebra Fundamentals

CONNECTIONS

No one really knows when humans began to organize counting and record keeping into an arithmetic. The Rhind Papyrus, dated approximately 1700 B.C., outlines some early Egyptian mathematics that included the solving of equations. The early Greeks developed arithmetic and geometry and laid the groundwork for most of our mathematics. Diophantus, a Greek mathematician who lived in the third century, introduced the idea of using symbols for numbers. This major step earned Diophantus the title "father of algebra."

At first, letters were used for positive numbers only. Gradually, the concept of a variable was extended to include all real numbers. It wasn't until the eighteenth century that complex numbers were accepted. In the nineteenth century, the idea of a variable was extended to let letters stand for sets, vectors, and matrices. This led to the development of vector algebra and linear algebra. As a consequence of this extension of the use of variables, mathematicians discovered that the properties of real numbers did not necessarily hold in other systems. Thus, they began to study properties and structure and developed the ideas of abstract algebra.

The algebra we study here is an extension of the algebra of Diophantus: arithmetic with letters standing for some numbers. This chapter is largely review. It should be taken seriously and used to sharpen skills learned earlier and to get "into the groove."

1.1 REAL NUMBERS

Sets

A set is a collection of objects. The objects are called **elements** or **members.** Often, we list the elements of a set and use capital letters as names for sets. For example, $N = \{1, 2, 3, \ldots\}$ names the set of counting numbers N. (The three dots, called an *ellipsis,* mean "continue on in the pattern that has been established.") If the number of elements in a set is a counting number, the set is called a **finite set.** If there are no elements in a set, the set is called the **empty set** or the **null set.** Otherwise a set is called an **infinite set.** The empty set is written as $\{\ \ \}$ or $\varnothing$.

The symbol $\in$ means **is an element of,** and the symbol $\notin$ means **is not an element of.** We could write $2 \in N$ to say "2 is an element of N," and $0 \notin N$ to say "0 is not an element of N."

Sometimes when defining a set, we do not list the elements. Instead we use a notation called **set builder notation.** $\{x \mid x \text{ is a vowel}\}$ describes the set {a, e, i, o, u} in set builder notation. We read the set builder notation as "the set of all x such that x is a vowel." The bar is read "such that." A colon can be used instead of the bar. For example, we could write $\{x : x \text{ is a day of the week}\}$ for the set {Sunday, Monday, Tuesday, Wednesday, Thursday, Friday, Saturday}.

Set A is a subset of set B if every element of A is also an element of B. The symbol $\subseteq$ means **is a subset of.** If B is the set of all the days of the week and A is the set containing only Monday and Tuesday, then A is a subset of B. We write this $A \subseteq B$ or

$$\{\text{Monday, Tuesday}\} \subseteq \{\text{Sunday, Monday, Tuesday, Wednesday, Thursday, Friday, Saturday}\}$$

Other subsets can be listed such as {Monday, Wednesday, Friday} and {Friday}.

Sets can be combined using the operations of **union** ($\cup$) and **intersection** ($\cap$). The union of two sets is the set of all elements that belong to one, or the other, or both of the two original sets.

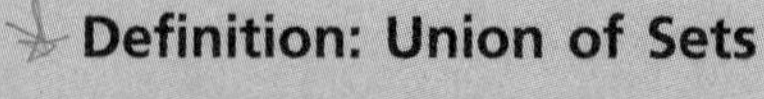

Definition: Union of Sets

$$A \cup B = \{x \mid x \in A \text{ or } x \in B\}$$

The intersection of two sets is the set of all elements that belong to *both* of the original sets.

Definition: Intersection of Sets

$$A \cap B = \{x \mid x \in A \text{ and } x \in B\}$$

EXAMPLE 1 Let $A = \{2, 4, 6\}$ and $B = \{1, 2, 3, 4, 5\}$.

(a) List all the subsets of A.

(b) Find $A \cap B$.

(c) Find $A \cup B$.

Solutions

(a) The subsets of A are $\varnothing, A, \{2\}, \{4\}, \{6\}, \{2, 4\}, \{2, 6\}, \{4, 6\}$. — The empty set and the set itself are always subsets of a given set.

(b) $A \cap B = \{2, 4\}$ — Because 2 and 4 belong to *both* A and B

(c) $A \cup B = \{1, 2, 3, 4, 5, 6\}$ — Because these elements belong to *either* A or B □

Real Numbers

We begin our study of algebra by studying the set of real numbers. The symbols used to name the numbers are called **numerals.** We start with the set $\{1, 2, 3, \ldots\}$, called the **natural numbers** or **counting numbers.** The **whole numbers** are the natural numbers with 0 added, and the set of **integers** is the whole numbers along with the negatives of the natural numbers.

A number line is helpful in visualizing the real numbers. Every point on the number line represents a real number, and every real number can be located on the number line. Draw a line, pick a starting point called the **origin,** and label it with the number 0. Choose a convenient unit of measure and represent the positive integers to the right of 0 and the negative integers to the left of 0. Fractions and decimals are located on the number line using the same unit of measure (Figure 1.1).

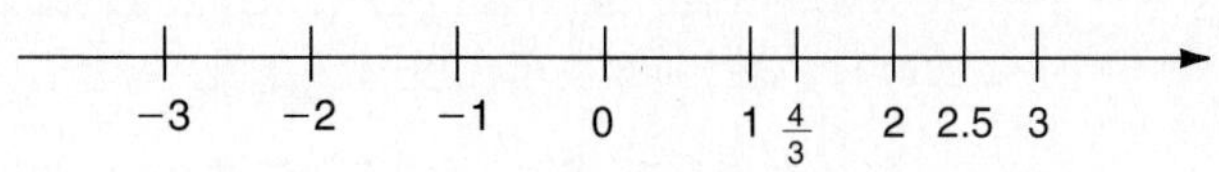

Fig. 1.1

The number associated with a point is called its **coordinate.** To **graph** a set of numbers means to locate them on a number line with a dot.

Natural Numbers

$$N = \{1, 2, 3, \ldots\}$$

Whole Numbers

$$W = \{0, 1, 2, 3, \ldots\}$$

Integers

$$Z = \{\ldots, -3, -2, -1, 0, 1, 2, 3, \ldots\}$$

Rational Numbers

$$Q = \left\{\frac{p}{q} \middle| p \text{ and } q \text{ are integers with } q \neq 0\right\}$$

Irrational Numbers

$$I = \{x \mid x \text{ is on the number line but is not a rational number}\}$$

Real Numbers

$$R = \{x \mid x \text{ is a rational or an irrational number}\}$$

Some examples of rational numbers are 7, −4, 3.56, and $\frac{8}{11}$. Rational numbers are often referred to as *fractions.* Notice that since $7 = \frac{7}{1}$, 7 is a rational number. Thus, all integers are rational numbers. Some examples of irrational numbers are $\sqrt{3}$, π, and $\sqrt[5]{13}$. Integers that end in 0, 2, 4, 6, or 8 are called **even** integers, while integers that end in 1, 3, 5, 7, or 9 are called **odd** integers.

Every real number has a decimal form. A rational number may be represented as the quotient of two integers and thus can be represented as a repeating decimal such as 1.3333 . . . for $\frac{4}{3}$. An irrational number cannot be written as a repeating decimal. $\sqrt[n]{x}$, where x is not a perfect nth power, cannot be written as a repeating decimal and is thus an irrational number. Figure 1.2 shows the relationships among subsets of real numbers.

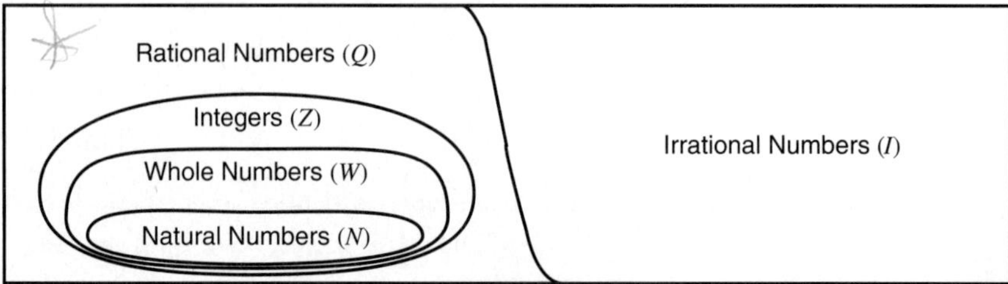

Fig. 1.2

The Ordering of Real Numbers

Graphing the real numbers p and q on a number line allows us to see which number is smaller by deciding which number lies to the left of the other on the number line. This idea is called **ordering.** If p is to the left of q on a number line, we write $p < q$ to indicate that p is less than q. Similarly, if p is to the right of q on a number line, we write $p > q$ to indicate that p is greater than q.

Ordering of Real Numbers

$p < q$ means that p is *less than* q. (p is to the left of q.)

$p > q$ means that p is *greater than* q. (p is to the right of q.)

$p \leq q$ means that p is *less than or equal to* q.

$p \geq q$ means that p is *greater than or equal to* q.

If p and q are real numbers, then p could be equal to q, p could be to the left of q, or p could be to the right of q; but *one and only one* of these conditions can be true. This idea is called the **Trichotomy Property.** (*Trichotomy* means ''splitting into three parts.'') Thus, $p = q$, or $p > q$, or $p < q$.

Looking at Figure 1.3, we can see that knowing that $p < q$ and that $q < r$ tells us that $p < r$. This idea is called the **Transitive Property** (Figure 1.3). From the same figure we see that $r > q$ and $q > p$ imply that $r > p$.

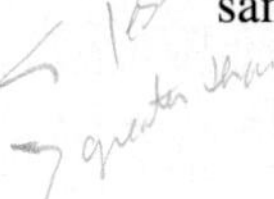

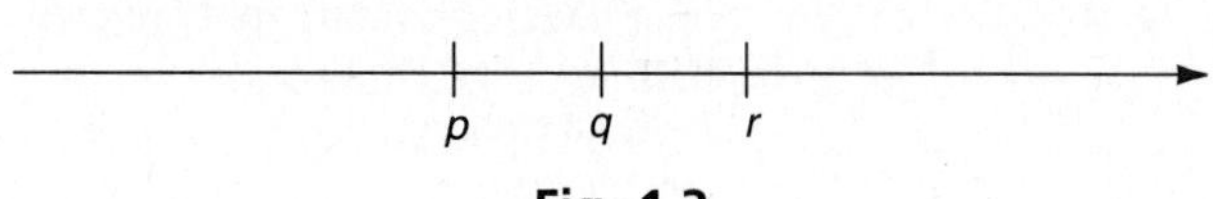

Fig. 1.3

Transitive Properties of Inequality

Let p, q, and r be real numbers.

If $p < q$ and $q < r$, then $p < r$.

If $r > q$ and $q > p$, then $r > p$.

Equality of real numbers also has the Transitive Property. We list it along with some other important properties of equality.

Properties of Equality

Let p, q, and r be real numbers.

1. $p = p$.	Reflexive
2. If $p = q$, then $q = p$.	Symmetric
3. If $p = q$ and $q = r$, then $p = r$.	Transitive

As a result of the properties of equality, we get the important Principle of Substitution.

Principle of Substitution

Let p and q be real numbers.

If $p = q$, then we may replace p with q.

Operations with Real Numbers

Arithmetic with real numbers uses the operations of addition, subtraction, multiplication, and division. Symbols are used to indicate which operation is to be performed.

When we express sums, differences, products, and quotients in symbols, we call them **expressions** or **algebraic expressions.** In a sum or difference, the numbers being added or subtracted are called **terms.** In a product, the numbers being multiplied are called **factors.**

Although we say that we use four operations, subtraction is defined in terms of addition, and division is defined in terms of multiplication.

Definition: Subtraction of Real Numbers

$$p - q = p + (-q)$$

To subtract q from p, we add the opposite (additive inverse) of q to p.

Definition: Division of Real Numbers

$$p \div q = p \cdot \frac{1}{q};\ q \neq 0$$

To divide p by q, we multiply p by the reciprocal (multiplicative inverse) of q. Of course, q must not be 0 since division by 0 is undefined.

Quotients and Products of Real Numbers with Zero

$$p \cdot 0 = 0 = 0 \cdot p$$

$$\frac{0}{p} = 0; \; p \neq 0$$

$$\frac{p}{0} \text{ is undefined.}$$

Often more than one operation is involved in an algebraic expression. The following rules determine which operation should be performed first. Grouping symbols such as parentheses ( ), brackets [ ], and braces { } are used to help make the order of operations clear.

Order of Operations

If grouping symbols are present, perform operations inside them, starting with the innermost symbol, in the following order.

1. Perform any exponentiations.
2. Perform all multiplications and divisions in order from left to right.
3. Perform all subtractions and additions from left to right.

If grouping symbols are not present, perform operations in the order given above.

EXAMPLE 2 Perform the operations indicated in each expression.

(a) $5(-3) - (-7)$ (b) $\sqrt{11 - 7} + (-3)^2 - \dfrac{8}{(-2)}$

Solutions (a)

$$\begin{aligned} 5(-3) - (-7) &= -15 - (-7) && \text{Multiply first.} \\ &= -15 + 7 && \text{Subtract.} \\ &= -8 \end{aligned}$$

(b)

$$\begin{aligned} \sqrt{11 - 7} + (-3)^2 - \frac{8}{(-2)} &= \sqrt{4} + 9 - (-4) \\ &= 2 + 9 + 4 \\ &= 15 \end{aligned}$$

□

EXAMPLE 3 Estimate the value of $\dfrac{2.337 - 12.034^2}{12.034 - 14.202\pi}$ and then approximate it to three decimal places.

Solution "Approximate" alerts us to a calculator computation. But first we are asked to estimate its value. Estimation is an important skill, particularly when working with a calculator. We wish to find a rough, "ballpark" evaluation of the expression. The numerator is about $2 - 12^2$ or -140, while the denominator is reasonably close to $12 - 14 \cdot 3$ or -30. Therefore, a rough guess for the entire expression is $\dfrac{-140}{-30}$ or about 4.

To approximate the value with a calculator, we must realize that the fraction bar acts as a grouping symbol just as a set of parentheses. Perhaps the best way to key in such a problem is to calculate the numerator and then divide by the *entire* denomina-tor using parentheses to hold the denominator together.

[2.337] [−] [12.034] [x^2] [=] [÷] [(] [12.034] [−] [14.202] [×] [π] [)] [=]

The display should show something like **4.372850819**.

The example requested three decimal places, so we write

$$\frac{2.337 - 12.034^2}{12.034 - 14.202\pi} \approx 4.373$$

Notice that this approximation is reasonably close to our estimate of 4. □

Properties of Real Numbers

Properties of Addition and Multiplication of Real Numbers

Closure	$p + q$ is a real number.
	pq is a real number.
Commutative	$p + q = q + p$
	$pq = qp$
Associative	$(p + q) + r = p + (q + r)$
	$(pq)r = p(qr)$
Identity	$p + 0 = p = 0 + p$
	$p \cdot 1 = p = 1 \cdot p$

Inverse	$p + (-p) = 0 = (-p) + p$
	$p \cdot \frac{1}{p} = 1 = \frac{1}{p} \cdot p;\ p \neq 0$
Distributive	$p(q + r) = pq + pr$

We can use this terminology to discuss the properties of real numbers. Zero is called the **additive identity.** The **multiplicative identity** is 1. $-p$ is called the **opposite** or **additive inverse** of p. For $p \neq 0$, $\frac{1}{p}$ is called the **reciprocal** or **multiplicative inverse** of p.

Absolute Value

A number line shows us that opposites such as 3 and -3 are both 3 units from zero. The measure of this distance is called the **absolute value** of 3 and -3. We write bars around the number to indicate absolute value. So, we see $|3| = |-3|$ (Figure 1.4).

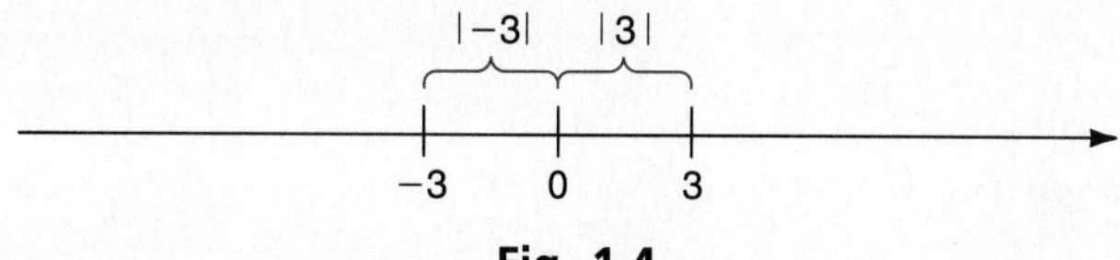

Fig. 1.4

Since absolute value is a distance, the absolute value of a real number is *never negative.* Since p and $-p$ are equally distant but on opposite sides of 0, we can see that $|p| = |-p|$. That is, the absolute value of opposites is the same.

Definition: Absolute Value

If p is a real number,

$$|p| = \begin{cases} p & \text{if } p \geq 0 \\ -p & \text{if } p < 0 \end{cases}$$

EXAMPLE 4 Suppose that r, s, t, and u are real numbers with $r > 0$ and $s < 0$. Evaluate the following absolute values if possible.

(a) $|r|$ (b) $|s|$ (c) $|t|$ (d) $|-u|$

Solutions (a) $|r| = r$ Because $r > 0$

(b) $|s| = -s$ Because s is negative, which means that $-s$ is positive

(c) $|t|$ Cannot be simplified because we do not know whether $t > 0$ or $t < 0$

(d) $|-u|$ Cannot be evaluated because we do not know whether $-u > 0$ or $-u < 0$; however, we could write $|-u| = |u|$. □

Students just beginning to study algebra are often confused by statements such as "$-s$ is positive." Just because there is a negative sign in $-s$ does not mean that $-s$ is a negative number. Remember that $-s$ represents the opposite of s. So, we cannot tell whether $-s$ is positive or negative unless we know whether s is positive or negative. If $s = -4$, notice that $-s = -(-4) = 4$. Hence, $-s$ is positive!

If p and q are the coordinates of two points on a number line, the distance between the two points is given by the value of $|p - q|$. Since $p - q$ and $q - p$ are opposites, we have $|p - q| = |q - p|$.

Distance Between Two Points on a Number Line

If p and q are coordinates of two points on a number line, the distance between them is

$$|p - q|$$

EXAMPLE 5 Find the distance between each pair of points whose coordinates are given. Write each answer without absolute values.

(a) -7 and 19 (b) π and $\sqrt{2}$ (c) x and y, if $x > y$ (d) s and t, if $t < s$

Solutions (a) The distance between the points whose coordinates are -7 and 19 is 26, since $|-7 - 19| = |-26| = 26$.

(b) The distance between the points whose coordinates are π and $\sqrt{2}$ is $\pi - \sqrt{2}$, since $\pi > \sqrt{2}$, $|\pi - \sqrt{2}| = \pi - \sqrt{2}$.

(c) If $x > y$, then $x - y > 0$.
$|x - y| = x - y$.
So the distance between these points whose coordinates are x and y is $x - y$.

(d) If $t < s$, then $s > t$ and $s - t > 0$.
$|t - s| = s - t$.
So the distance between these points whose coordinates are t and s is $s - t$.

□

Some Properties of Absolute Value

Let p and q be real numbers.

1. $\lvert p\rvert \geq 0$	**4.** $\lvert p\rvert = \lvert -p\rvert$
2. $\lvert p \cdot q\rvert = \lvert p\rvert \cdot \lvert q\rvert$	**5.** $\lvert p - q\rvert = \lvert q - p\rvert$
3. $\left\lvert \dfrac{p}{q} \right\rvert = \dfrac{\lvert p\rvert}{\lvert q\rvert}$; $q \neq 0$	**6.** $\lvert p\rvert^2 = p^2$

EXAMPLE 6 Use the properties of absolute value to simplify each expression if possible.

(a) $\lvert 3\pi\rvert$ (b) $\lvert -3x\rvert$ (c) $\left\lvert \dfrac{-\pi^2}{6} \right\rvert$

Solutions (a) $\lvert 3\pi\rvert = 3\pi$ since $3\pi > 0$.

(b) $\lvert -3x\rvert = \lvert -3\rvert \cdot \lvert x\rvert = 3\lvert x\rvert$ (We cannot simplify $\lvert x\rvert$ because we do not know if $x > 0$ or if $x < 0$.)

(c) $\left\lvert \dfrac{-\pi^2}{6} \right\rvert = \dfrac{\lvert -\pi^2\rvert}{\lvert 6\rvert} = \dfrac{\lvert \pi^2\rvert}{6} = \dfrac{\pi^2}{6}$ □

Proofs

From time to time we find it necessary to establish the truth of a statement. Such a process we call a **proof.** A proof consists of an argument that proceeds by logical steps from an agreed starting point and ends with conclusive evidence of the truth of the statement. The logical steps in the proof must consist of those things previously proven or valid assumptions.

EXAMPLE 7 Prove that the sum of two even numbers is an even number.

Solution Starting with two *even* numbers, say m and n, we must show conclusively that the number $m + n$ is *even.* What does it mean to say m is even? Since all even numbers are integers divisible by 2, there must be some integer p such that $m = 2 \cdot p$. Likewise, as n is also even, there must be an integer q such that $n = 2 \cdot q$. Therefore,

$m + n = 2p + 2q$	m and n are even.
$= 2(p + q)$	Distributive Property
$= 2k$; where k is an integer	The sum of two integers is an integer.

As the sum of m and n is 2 times an integer, it is divisible by 2 and therefore an even number. This completes the proof. □

GRAPHING CALCULATOR BOX

Arithmetic Operations and Editing Features

Read *Using Your Calculator* in the preface before reading this material. First, we examine ordinary arithmetic on a graphing calculator. The calculator must be in computational mode. Let's compute $\sqrt{7+9}+2^5-(-3)^2$.

TI-81

[2nd] [x²] [(] [7] [+] [9] [)] [+] [2] [^] [5] [−] [(] [(−)] [3] [)] [x²] [ENTER]

CASIO fx-7700G

[√‾] [(] [7] [+] [9] [)] [+] [2] [xʸ] [5] [−] [(] [−] [3] [)] [SHIFT] [√‾] [EXE]

The result is 27. Notice how to calculate powers on each calculator. Henceforth, we will indicate 2nd function keys with a double box. That is, keys such as [SHIFT] [√‾] (with x^2 above) on the Casio will be indicated by [[x²]] and [2nd] [x²] (with √‾ above) on the TI-81 will be shown as [[√‾]].

Editing

Graphing calculators have useful editing capabilities. Suppose we wanted the *fourth* power of 2 in the above example, instead of the *fifth*. We can edit the expression on the display to change the exponent to 4. We use the group of four arrowheads at the upper right corner of the keypad.

TI-81

To get into *edit mode,* press the *up arrow*. This causes the expression to be recopied with the cursor blinking at the end. We can now edit the expression. We use the left arrow to locate the cursor on the 5. Press [4] followed by [ENTER], and we have $\sqrt{7+9}+2^4-(-3)^2=11$.

CASIO fx-7700G

On the Casio, the *left arrow* or the *right arrow* puts us in edit mode. Otherwise the editing is the same as for the TI-81.

Editing can be done in a similar manner while entering an expression. Notice in the example that it was unnecessary to delete the 5 before entering the 4. Both calculators *overwrite* when editing. However, both have an *insert* key that allows insertion: [INS] on the TI and [[INS]] on the Casio. Suppose that when entering $\sqrt{7+9}$ we left off the first parenthesis. That is, we entered [√‾] [7] [+] [9] [)] and noticed the missing symbol before we finished entering the expression. We simply use the left arrow to place the cursor on the 7 and press the insert key followed by [(]; the display shows the correct expression, $\sqrt{\ }\,(7+9)$. Each graphing calculator also has a delete key, [DEL]. Experiment with these keys until you are comfortable with the editing capability of your machine.

PROBLEM SET 1.1

Warm-ups

For Problems 1 through 5, see Example 1.

1. Let $S = \{-2, 0, 2\}$ and $T = \{-4, -1, 0, 1, 3\}$. Find $S \cap T$ and $S \cup T$. List all the subsets of S.
2. Let $R = \{x \mid x \text{ is a month beginning with the letter } A\}$, and $W = \{x \mid x \text{ is a month with five letters in its name}\}$. Find $R \cap W$ and $R \cup W$.
3. Let $X = \{n + 1 \mid n \text{ is an even integer}\}$ and $Y = \{n + 1 \mid n \text{ is an odd integer}\}$. Find $X \cap Y$ and $X \cup Y$.
4. Let $A = \{(x, y) \mid x \in Z, \text{ and } y = 2x\}$ and $B = \{(x, y) \mid x \in N, \text{ and } y = 2x\}$. Find $A \cap B$ and $A \cup B$.
5. If M is a set, find $M \cap \varnothing$ and $M \cup \varnothing$.

In Problems 6 through 9, perform the indicated operations. See Example 2.

6. $-3^2 + \dfrac{15}{-3} - (-2)^2$
7. $\sqrt{27 + 22} - [5 + 2(11 - 15)]$
8. $\dfrac{1}{2} \div \left(\dfrac{2}{3} - \dfrac{1}{2}\right)$
9. $(12–17)^2 - (-37)$

In Problems 10 and 11, estimate and then approximate the value of the expression to three decimal places. See Example 3.

10. $\dfrac{503^2 - 272114}{523\pi + 708}$
11. $\dfrac{(1.00975)^2}{0.09331 - 0.75626}$

In Problems 12 through 15, simplify each expression if possible. Assume that $a > 0$ and $b < 0$. See Example 4.

12. $|a|$
13. $|-a|$
14. $|b|$
15. $|-b|$

In Problems 16 through 20, find the distance between each pair of points whose coordinates are given. Write the answer without absolute values. See Example 5.

16. -8 and -1
17. $\sqrt{5}$ and 3
18. $-\pi$ and 3.14
19. $-\sqrt{7}$ and $-\sqrt{11}$
20. c and d if $c < d$

In Problems 21 through 24, use the properties of absolute value to simplify each expression. See Example 6.

21. $\left|\dfrac{-8}{x}\right|$
22. $|-\sqrt{3}\pi|$
23. $\left|\dfrac{\pi}{-7}\right|$
24. $|-s|$

For Problem 25, see Example 7.

25. Prove that the sum of two odd numbers is even. (*Hint:* Odd numbers are integers that can be written in the form $2k + 1$, where k is an integer.)

Practice Exercises

In Problems 26 through 29, indicate which statements are true.

26. $W \subseteq I$
27. $I \subseteq R$
28. $Q \subseteq R$
29. $N \subseteq Z$

In Problems 30 through 41, perform the indicated operations.

30. $3 \cdot \left(-\dfrac{1}{2}\right)^2 - 2 \cdot \left(\dfrac{1}{2}\right) + 11$
31. $-3^2 + \dfrac{2}{3} \cdot \left(\dfrac{1}{2}\right)^2 - (-3)\left(\dfrac{1}{6}\right)\left(\dfrac{1}{2}\right)$

32. $14 \div 7 \cdot 2$

33. $2^2 - 4 \cdot 5$

34. $1.5 \div (2.75 - 3.5)$

35. $\frac{1}{8} \div \left(-\frac{1}{4}\right)^3$

36. $\sqrt{(-7)^2 - 4(3)^2}$

37. $\sqrt{-6^2 + (-6)^2}$

38. $\sqrt{3^2 + 4^2} - \left(3 + \frac{1}{3}\left(-\frac{2}{3} - 1\right)^2\right)$

39. $\sqrt{13^2 - (2 - 7)^2} - \frac{1}{2} - 1$

40. $-|15 - 12|^2 - |-11|$

41. $|-1^2 - 5^2| \cdot |-1|^3$

In Problems 42 through 51, rewrite the given expression using the indicated property.

42. $ax + ay$; Distributive Property

43. $3(a - b)$; Distributive Property

44. $-5 + 0$; identity

45. $x \cdot 1$; identity

46. $\frac{3}{2} \cdot \frac{2}{3}$; inverse

47. $\pi + (-\pi)$; inverse

48. $\frac{1}{2} \cdot [2(x + y)]$; Associative Property

49. $(2 + 3) + 4$; Associative Property

50. $5(x + y)$; Commutative Property (addition)

51. $5(x + y)$; Commutative Property (multiplication)

In Problems 52 through 67, use the properties of absolute value to write each expression without absolute values. Assume that x, y, and z are real numbers, with $x > 0$ and $y < 0$.

52. $|x|$

53. $|y|$

54. $|y^2|$

55. $|x^2|$

56. $\left|\frac{z^2}{-4}\right|$

57. $|-2z^2|$

58. $|8x^2|$

59. $\left|\frac{y^2}{-2}\right|$

60. $|x - y|$

61. $|y - x|$

62. $|-y|$

63. $|-x|$

64. $|\sqrt{7} - 3|$

65. $|\sqrt{5} - 1|$

66. $|x + 2|$

67. $|y - 1|$

68. Louise Blanchard went to Home Depot to buy enough pipe to go diagonally across a 40-foot square in her garden. Louise and the clerk calculated that she would need $40\sqrt{2}$ feet of pipe. Home Depot sells this pipe by the half-foot. Determine the least amount of pipe Louise should buy.

69. Thomas Wagoner is hosting a cookout for 17 neighbors. He is serving hamburgers $\left(\frac{1}{3}\text{ pounders}\right)$. In the Kroger store, he finds ground beef packages labeled 2.23, 2.14, 1.37, 2.33, and 1.39 pounds. What combination of packages should he buy to serve his neighbors one hamburger each and spend the least amount of money?

70. Suppose each page of the current edition of *National Geographic Magazine* weighs 0.07 ounce while the cover, glue, and other components weigh 1.08 ounces. How much does the completed magazine weigh, to the nearest hundredth of an ounce, if it contains 145 pages?

71. Ray wishes to fence a rectangular pasture 520 yards by 230 yards. He wants two gates in the fence that measure 3 yards each. If the fence costs \$2.80 a yard and each gate costs \$244, how much will it cost Ray to fence his field?

72. Tom Bigelow is making a circular pen of diameter 20 meters for use with his horses. His friends Sue and Pat, who teach mathematics, told him the circumference is π times the diameter. Smith Hardware sells fencing by the foot. What is the least amount of fencing that he should buy? (1 meter $\approx$ 3.28084 feet)

73. Each of the five salespeople of the Thompson Agragrow Company were given the same monthly sales quota. If Ken sold $\frac{2}{5}$ of his quota, June sold $\frac{3}{7}$ of her quota, Beth sold $\frac{4}{9}$ of her quota, Jim sold $\frac{5}{16}$ of his quota, and Harold sold $\frac{7}{20}$ of his quota, what was their order of finish in the monthly sales contest?

74. Suppose each page of a book weighs x ounces and the cover and components weigh y ounces. How much will a book containing p pages weigh?

75. Suppose Ray's pasture (see Problem 71) measures s yards by t yards and the gates measure 3 yards each. Further, suppose Ray wants n of the \$244 gates. If the fence costs C dollars per yard, write an expression for the total cost to fence the field.

76. Suppose $x > 6$. Write the following fractions in *increasing* order.

$$\frac{6}{x}, \frac{x}{6}, \frac{1}{x - 1}, \frac{x + 1}{6}, \frac{1}{x + 1}$$

77. Suppose $y < -1$. Write the following expressions in *increasing* order.

$$-y,\ y,\ |1 - y|,\ |1 + y|$$

78. Prove that the sum of an odd number and an even number is odd.

Challenge Problems

79. Investigate the relationship between $|p + q|$ and $|p| + |q|$ by substituting values for p and q into each expression. Make a conjecture about the relative ordering of $|p + q|$ and $|p| + |q|$.

80. Suppose $q > 0$. Investigate the relationship between q, q^2, and $\sqrt{q}$ by substituting values for q into each expression.

81. Prove the Absolute Value Property, $|p \cdot q| = |p| \cdot |q|$.

IN YOUR OWN WORDS . . .

82. Explain why $-x$ may or may not be negative.

83. Explain why $|x| = |-x|$.

1.2 INTEGER EXPONENTS

Multiplication of real numbers has been studied since ancient times. The product of identical factors, such as $5 \cdot 5 \cdot 5$, became common enough to warrant some sort of shorthand notation. Exponents were introduced in the seventeenth century to fill this need. For example, $6 \cdot 6 \cdot 6 \cdot 6$ was written 6^4 (read ''six to the fourth power''). The properties of this shorthand notation proved to be so powerful that exponents took on a life of their own. We begin with the definitions of integer exponents in this section and define rational exponents in Section 1.6. Irrational exponents are defined in calculus.

Integer Exponents

Definition: Natural Number Exponent

If x is a real number and n is a natural number, then

$$x^n = \underbrace{x \cdot x \cdot x \cdots x}_{n \text{ factors of } x}$$

Definition: Zero Exponent

If x is any nonzero number, then

$$x^0 = 1$$

Definition: Negative Integer Exponent

For x any nonzero real number and n a natural number,

$$x^{-n} = \frac{1}{x^n}$$

In the expression x^n, we call x the **base** and n the **exponent** or **power.**

EXAMPLE 1 Identify the base, perform the indicated multiplications where possible, and write each expression without zero or negative exponents.

(a) 2^4 (b) $(-2)^0$ (c) -2^{-4}
(d) $2x^{-4}$ (e) -2^0 (f) $(2x)^{-4}$

Solutions (a) 2 is the base, and 4 is the exponent. $2^4 = 2 \cdot 2 \cdot 2 \cdot 2 = 16$.

(b) -2 is the base. $(-2)^0 = 1$.

(c) 2 is the base. $-2^{-4} = -\dfrac{1}{2^4} = -\dfrac{1}{16}$.

(d) x is the base. $2x^{-4} = 2 \cdot \dfrac{1}{x^4} = \dfrac{2}{x^4}$.

(e) 2 is the base. $-2^0 = -1$.

(f) $2x$ is the base. $(2x)^{-4} = \dfrac{1}{(2x)^4} = \dfrac{1}{16x^4}$. □

Example 1 illustrates a useful result that is true for odd and even exponents.

Theorem: Odd and Even Powers

If p is a real number and n is an integer,

$$(-p)^n = \begin{cases} p^n & \text{if } n \text{ is even} \\ -p^n & \text{if } n \text{ is odd} \end{cases}$$

The proof of this theorem centers around even and odd integers and is left as an exercise.

The definition of natural number exponents led to properties that make exponents important in the study of algebra. The properties are summarized in the following table.

Properties of Exponents

If m and n are integers, and x and y are real numbers,

$x^m x^n = x^{m+n}$	Product with the same base
$(x^m)^n = x^{mn}$	Power of a power
$(xy)^n = x^n y^n$	Product to a power

$\left(\frac{x}{y}\right)^n = \frac{x^n}{y^n};\ y \neq 0$	Quotient to a power
$\frac{x^m}{x^n} = x^{m-n};\ x \neq 0$	Quotient with the same base

The first property tells us that we add the exponents in a product with the same base.

The middle three properties tell us when to *multiply* exponents. They are sometimes called *power rules* and are used in problems of the form

$$(\text{Expression})^{\text{power}}$$

We must be very careful in working problems of this type. The following example illustrates an important concept in working with an expression raised to a power.

EXAMPLE 2 Use the properties of exponents to write each expression in another form if possible.

(a) $(xy^4)^3$ (b) $(x + y^4)^3$ (c) $\left(\frac{s^4}{t^3}\right)^2$

Solutions (a) $(xy^4)^3$

This is an expression raised to a power. Notice that the expression is a **product.** It *can* be rewritten by multiplying exponents.

$$(xy^4)^3 = x^3y^{12}$$

(b) $(x + y^4)^3$

This also is an expression raised to a power. Notice that the expression is a **sum.** It *cannot* be written in another form by multiplying exponents.

$$\begin{aligned}(x + y^4)^3 &= (x + y^4)(x + y^4)(x + y^4)\\ &= x^3 + 3x^2y^4 + 3xy^8 + y^{12} \qquad \text{which is } \textit{not } x^3 + y^{12}\end{aligned}$$

We will study products like this in Section 1.3.

(c) $\left(\frac{s^4}{t^3}\right)^2$

This is another expression raised to a power. This time the expression is a **quotient.** Like a product, it *can* be written in another form by multiplying exponents.

$$\left(\frac{s^4}{t^3}\right)^2 = \frac{s^8}{t^6}$$

□

The point of Example 2 is that an expression to a power *may* be written in another form by multiplying exponents if the expression is a *product* or a *quotient.* It *cannot* be written in another form by multiplying exponents if the base is a *sum* or a *difference.* Remember that $(a + b)^n \neq a^n + b^n$.

Properties of Negative Exponents

For x any nonzero real number and n any natural number,

1. $\dfrac{1}{x^{-n}} = x^n$

2. If x^{-n} is a *factor* of the numerator (or denominator) of a fraction, the fraction may be rewritten with x^n as a factor of the denominator (or numerator).

3. $\left(\dfrac{p}{q}\right)^{-n} = \left(\dfrac{q}{p}\right)^n$; $p \neq 0$; $q \neq 0$

EXAMPLE 3 Write each of the following expressions without negative exponents. Assume all expressions are defined.

(a) $\dfrac{kx^{-3}y^2}{Cz^{-1}}$ (b) $\left(-\dfrac{sx^{-4}y^3}{tw^4z^{-7}}\right)^{-2}$ (c) $\dfrac{5^{-1} - 5^{-2}}{1 + 5^{-2}}$

Solutions (a) $\dfrac{kx^{-3}y^2}{Cz^{-1}} = \dfrac{ky^2z}{Cx^3}$

(b) $\left(-\dfrac{sx^{-4}y^3}{tw^4z^{-7}}\right)^{-2}$ is an expression to a power.

Since it is a *quotient,* the exponent properties allow us to multiply exponents.

$$\left(-\frac{sx^{-4}y^3}{tw^4z^{-7}}\right)^{-2} = \left(\frac{sx^{-4}y^3}{tw^4z^{-7}}\right)^{-2} \qquad \text{Even power}$$

$$= \frac{s^{-2}x^8y^{-6}}{t^{-2}w^{-8}z^{14}} \qquad \text{Multiplied exponents (quotient to a power and power of a power)}$$

$$= \frac{t^2w^8x^8}{s^2y^6z^{14}} \qquad \text{Properties of negative exponents}$$

(c) Notice in $\dfrac{5^{-1} - 5^{-2}}{1 + 5^{-2}}$ that neither the 5^{-1} nor the 5^{-2} is a *factor* of the numerator or denominator. The property of negative exponents applies only to factors in the numerator and the denominator. It is *important to realize that it does not apply here.* We must use the definition of negative exponent.

$$\frac{5^{-1}-5^{-2}}{1+5^{-2}} = \frac{\dfrac{1}{5}-\dfrac{1}{5^2}}{1+\dfrac{1}{5^2}} \qquad \text{Definition of negative exponent}$$

$$= \frac{\dfrac{1}{5}-\dfrac{1}{25}}{1+\dfrac{1}{25}} = \frac{\dfrac{5}{25}-\dfrac{1}{25}}{\dfrac{25}{25}+\dfrac{1}{25}} = \frac{\dfrac{4}{25}}{\dfrac{26}{25}} = \frac{4}{26} = \frac{2}{13}$$

□

Integer Exponents on a Calculator

The scientific calculator is a useful tool for approximating large or small powers of real numbers.

EXAMPLE 4 Approximate $(2.4)^{-7}$ to three significant digits.

Solution Enter the negative exponent with the [+/−] key.

With a scientific calculator: [2.4] [y^x] [7] [+/−] [=]

(The negative may have to be entered before the 7 on some calculators.)

By three significant digits, we mean to *include* the first three digits, *not beginning with a zero,* as we read the number from left to right. For example, the answer to

One significant digit	0.002
Two significant digits	0.0022
Three significant digits	0.00218
Four significant digits	0.002180

$(2.4)^{-7} \approx 0.00218$

□

Scientific Notation

In scientific notation we see positive and negative exponents in use. Multiply 123456 by 123456 on a calculator, and it may display an answer like

1.524138394^{10} or 1.524138394 10 or 1.524138394 E 10

Any one of these is short for $1.524138394 \times 10^{10}$, a large number written in scientific notation.

Divide 1 by 123456, and the calculator may display an answer like

8.10005184 $^{-06}$ or 8.10005184 −06 or 8.10005184 E − 06

These are each short for $8.10005184 \times 10^{-6}$, a small number written in **scientific notation.**

Each of these numbers is written as $c \times 10^p$, where $1 \leq c < 10$ and p is an integer.

Scientific notation is a convenient way to express very large and very small numbers. Two examples that occur in physics are Planck's constant,

$$h \approx 0.000000000000000000000000006625 \text{ erg-second}$$

and the speed of light in a vacuum,

$$c \approx 29900000000 \text{ centimeters per second}$$

Written in scientific notation, these become

$$h \approx 6.625 \times 10^{-27} \text{ erg-second and}$$

$$c \approx 2.99 \times 10^{10} \text{ centimeters per second}$$

These are both approximations, as the $\approx$ symbol indicates.

To Write a Number in Scientific Notation

1. Starting with the number in decimal format, move the decimal point until the number is between 1 and 10. Count the number of places moved.
2. Multiply the number formed in step 1 by 10 to the power equal to the number of decimal places moved. If the original number was between 0 and 1, the power of 10 is negative. If the original number was greater than 10, the power is positive.
3. This procedure is for a positive number. If the original number is negative, perform steps 1 and 2 on the absolute value of the original number and then attach a negative sign to the result.

EXAMPLE 5 Write each number in scientific notation.

(a) 93,000,000 (b) 0.000001554 (c) −254,000

Solutions (a) Think of 93,000,000 as 93000000.0 and then move the decimal point until the number is between 1 and 10.

$93{,}000{,}000 = 9.3 \times 10^7$

(b) $0.000001554 = 1.554 \times 10^{-6}$

(c) $-254{,}000 = -2.54 \times 10^5$

□

EXAMPLE 6 Write each number without exponents.

(a) 8.771×10^{14} (b) 3.2×10^{-13} (c) -9.99231×10^{-3}

Solutions To change scientific notation to standard decimal format, reverse the steps given above.

(a) $8.771 \times 10^{14} = 877{,}100{,}000{,}000{,}000$

(b) $3.2 \times 10^{-13} = 0.00000000000032$

(c) $-9.99231 \times 10^{-3} = -0.00999231$ □

To enter a number in a calculator in scientific notation we use the [EXP] or [EE] key.

EXAMPLE 7 Use a calculator to find the product of Planck's constant and the speed of light in a vacuum to three significant digits.

Solution Recall that Planck's constant in scientific notation is $h \approx 6.625 \times 10^{-27}$ erg-second and that the speed of light in a vacuum is $c \approx 2.99 \times 10^{10}$ centimeters per second. Therefore, we are to calculate

$$hc \approx (6.625 \times 10^{-27})(2.99 \times 10^{10})$$

We press the keys

[6.625] [EXP] [27] [+/−] [×] [2.99] [EXP] [10] [=]

and read 1.980875 −16 on the display. So,

$$hc \approx 1.98 \times 10^{-16}$$ □

GRAPHING CALCULATOR BOX

Using Memory

Graphing calculators have more than 20 usable memory locations, named with letters of the alphabet and accessed with the [ALPHA] key. Let's store 12 in memory location V.

TI-81

[12] [STO▷] [V] [ENTER]

Although the V is above the [6] key, the [ALPHA] key is not required on the TI because [STO▷] sets the keypad for an alphabetic entry.

CASIO fx-7700G

[12] [→] [ALPHA] [V] [EXE] (*Note:* [→] is the arrow key just under the [COS] key.)

Now 12 is stored in V. To see the contents of V, press [ALPHA] [V] [ENTER] (or [EXE]).

Let's evaluate $V^2 - V + 1$ if V is 12.

[ALPHA] [V] [x^2] [−] [ALPHA] [V] [+] [1] [ENTER] (or [EXE])

Since X is used so often as a variable, a special key will enter X. On the TI-81 it is [X|T], and on the Casio fx-7700G it is [X,θ,T]. The result of a computation can be stored in memory. Let's store $\frac{\pi}{3}$ in X.

TI-81

[π] [÷] [3] [STO▷] [X|T] [ENTER]

CASIO fx-7700G

[π] [÷] [3] [→] [X,θ,T] [EXE]

Evaluating Formulas for Multiple Variable Entries

Memory can be used when evaluating a formula for multiple variable entries. Let's calculate the volume of cones with radius and height given in the following table.

r	h
24.5	42.1
4.3	6.4

The formula for the volume of a cone is $V = \frac{1}{3}\pi r^2 h$. First, we store $\pi r^2 h \div 3$ in the *function memory,* using the [ALPHA] key for r and h.

TI-81

Press [Y =]. A menu appears with the cursor blinking after $:Y_1 =$. Enter the expression. Press [QUIT] to exit the [Y =] screen.

CASIO fx-7700G

Enter the expression and then press [F MEM] at the lower left of the keypad. Notice the function memory status line that appears at the bottom of the display. Just under the display are six keys, marked F1 through F6. They apply to the items on the status line. We wish to *store* in f_1, so we press [F1] [1]. The function memory is displayed with the expression in f_1.

Now we wish to evaluate the expression several times. Store the first numbers: 24.5 in r and 42.1 in h.

TI-81

Press [Y-VARS] for a listing of function names. Our function is Y_1, so we press [1]. Now, Y_1 is displayed. Press [ENTER], and the volume is calculated.

CASIO fx-7700G

The status line should still be showing. If not, press [F MEM]. Press [F2] [1] to recall the expression. Press [EXE].

With either calculator, the result 26463.2319 should be displayed. Enter the next numbers for r and for h and repeat the procedure (123.9211694).

PROBLEM SET 1.2

Warm-ups

In Problems 1 through 12, identify the base, perform the indicated multiplications where possible, and write each expression without zero or negative exponents. See Example 1.

1. -3^4 **2.** $(-3)^4$ **3.** $(2x)^3$ **4.** $-2x^3$

5. -5^0 **6.** $(-5)^0$ **7.** $\dfrac{(-3d^2)^0}{-2q^0}$ **8.** $\left(\dfrac{1}{2}P^3 - \dfrac{2}{3}Q^2\right)^0$

9. 5^{-2} **10.** -5^{-2} **11.** $(-5)^{-2}$ **12.** $-7x^{-1}z^{-2}$

In Problems 13 through 20, use the properties of exponents to write each expression in another form if possible. See Example 2.

13. $x^4 \cdot x^6$ **14.** $(2x)^6$ **15.** $\left(\dfrac{t}{3}\right)^3$ **16.** $\dfrac{(a-b)^8}{(a-b)^3}$

17. $(2^2x^3y^4)^3$ **18.** $(5x+3)^3$ **19.** $(k^3 - 2s^2)^2$ **20.** $\left(\dfrac{3t}{2s^2}\right)^4$

In Problems 21 through 26, write each expression without negative exponents. See Example 3.

21. $(2x^2y)^{-3}$ **22.** $\dfrac{2z^{-3}}{a^{-2}b}$ **23.** $\left(-\dfrac{3d^{-1}h^2k}{x^{-1}y^{-2}z^3}\right)^{-3}$ **24.** $\dfrac{3^{-1}+3^{-2}}{3^{-2}}$

25. $(x^2-3)^{-2}$ **26.** $(x^{-1}+2^{-1})^2$

In Problems 27 through 30, approximate the value of each expression to six significant digits. See Example 4.

27. 1.09330^9 **28.** $(\pi - \sqrt{17})^7$

29. $(\sqrt{11})^{-5}$ **30.** $\left(\dfrac{1}{\pi} - \pi\right)^{-3}$

In Problems 31 through 34, write each number in scientific notation. See Example 5.

31. 155,000 **32.** 0.0000008771 **33.** −0.00000254 **34.** −1,161,000,000

In Problems 35 through 38, write each number without an exponent. See Example 6.

35. 5.6473×10^{10} **36.** 9.92×10^{-5} **37.** -3.9022×10^{-8} **38.** -1.118×10^{11}

In Problems 39 through 42, approximate each to three significant digits. See Example 7. Use the following values:

$m_p \approx 1.6726 \times 10^{-27}$ Rest mass of a proton

$G \approx 6.6726 \times 10^{-11}$ Gravitational constant

$c \approx 2.9979 \times 10^{8}$ Speed of light in meters per second

39. cG **40.** $\dfrac{m_p}{G}$ **41.** $\dfrac{Gc}{m_p}$ **42.** $m_p c^2$

Practice Exercises

In Problems 43 through 66, use only the properties of exponents to rewrite each expression if possible. Answers should not contain zero or negative exponents.

43. $(-x)^4$ **44.** $(-k)^{10}$ **45.** $(-y)^{-5}$ **46.** $(-j)^{-15}$

47. $-2\left(-\dfrac{2}{3}\right)^{-2}$ **48.** $-\left(-\dfrac{3}{4}\right)^{-4}$ **49.** -4^2 **50.** -5^4

51. $-2x^0$ **52.** $-7w^0$ **53.** $-\left(\dfrac{x}{y^2}\right)^4$ **54.** $-\left(\dfrac{k^3}{M^2}\right)^{10}$

55. $(s^{-2} + t)^4$ **56.** $(m^3 - n^4)^{-2}$ **57.** $-(-n^2y)^0z^{-2}$ **58.** $-D^{-2}(-2k^3)^0$

59. $\left(\dfrac{2p}{q-1}\right)^2 \cdot \left(\dfrac{3p}{q-1}\right)^{-3}$ **60.** $\left(\dfrac{t+2}{3(t-2)}\right)^{-3} \cdot \left(\dfrac{t+2}{t-2}\right)^2$ **61.** $\dfrac{(-2k^3pz^{-1})^{-2}}{(k^{-1}pz^{-2})^2}$ **62.** $\left(\dfrac{-2r^{-2}st^3}{(4rs^{-1})^2t^{-2}}\right)^{-1}$

63. $(2^{-2} + 3^{-1})^{-1}$ **64.** $\dfrac{2^{-2} + 3^{-2}}{2^{-2} - 3^{-2}}$ **65.** $\dfrac{(2x-1)^{n+1}}{(2x-1)^n}$ **66.** $\dfrac{x^{2n+1}y^{2n-1}}{x^{2n+3}y^{2n}}$

67. Prove that $\left(\dfrac{p}{q}\right)^{-n} = \left(\dfrac{q}{p}\right)^n$, for any natural number n.

68. Prove that $(-p)^n = \begin{cases} p^n & \text{if } n \text{ is even} \\ -p^n & \text{if } n \text{ is odd} \end{cases}$ for any integer n.

69. The human race consumes energy at the rate of 10 billion kilowatts per year. If each person in the United States consumes 11 kilowatts per year and there are 240 million people in the United States, what percentage of energy is consumed in the United States?

70. In 1989 wind turbines supplied 1.5 million kilowatts of electricity to meet the needs of 750,000 Californians. How many kilowatts did the average Californian use?

71. A fusion device brings nuclei close enough together to fuse. If two deuterium nuclei come within 5 billionths of a millimeter of each other, fusion is likely to occur. If 1 inch $\approx$ 2.54 centimeters, approximate this distance in inches.

72. In a hydrogen molecule the nuclei of the two atoms stay 76 billionths of a millimeter apart. Approximately how many inches is this distance?

73. A light-year is a unit of measure used by astronomers to indicate the enormous distances between stars. One light-year is the distance light travels in 1 year. If light travels 186,280 miles in 1 second, how many miles is a light-year?

74. Proxima Centauri is the star nearest earth, other than our sun. If Proxima Centauri is 4.26 light-years distant, how far is it to Proxima Centauri in miles?

75. Claude Justice paid \$1 for one lottery ticket. Ten million tickets were sold. The prizes were a first prize of \$1,000,000, 5 second prizes of \$10,000, and 25 third prizes of \$1000 each. His expected winnings (E) are given by the formula $E = x_1p_1 + x_2p_2 + x_3p_3 + x_4p_4$, where each x_i is his net prize and p_i is his probability of winning that prize. Use the table below to calculate Claude's expected winnings.

Prize	*Net Prize*	*Probability of Winning*
1	\$999,999	$\dfrac{1}{10{,}000{,}000}$

2	\$9,999	$\frac{5}{10{,}000{,}000}$
3	\$999	$\frac{25}{10{,}000{,}000}$
None	−\$1	$\frac{9{,}999{,}961}{10{,}000{,}000}$

76. If P dollars is invested at an annual interest rate of r compounded n times a year, then the amount (A) accumulated after t years is given by the formula

$$A = P\left(1 + \frac{r}{n}\right)^{tn}$$

Approximate the amount of money accumulated after 5 years by investing \$100 at 11% if interest is compounded
(a) Annually ($n = 1$)
(b) Semiannually ($n = 2$)
(c) Quarterly ($n = 4$)
(d) Monthly ($n = 12$)
(e) Daily ($n = 365$)

77. Use the formula in Problem 76 and approximate the amount of money accumulated after 5 years by investing \$100,000 at 7.5% if interest is compounded as indicated in Problem 76.

Challenge Problems

78. Stacked exponents work from the top down. For example, 5^{2^3} is 5^8 whereas $(5^2)^3$ is 5^6. Experiment with different numbers for x and figure out what numbers will make $(5^x)^2$ less than 5^{x^2}.

79. If the world population is 5.2 billion, what will the world population be in 5 years if the growth rate is 1.7% each year?

80. If the world population is 5.2 billion, what will the world population be in 5 years if the growth rate is 1.5% each year?

IN YOUR OWN WORDS . . .

81. Explain the meaning of -5^2 and $(-5)^2$.

82. Explain the meaning of $x^2 + 16$ and $(x + 4)^2$.

83. Explain why scientific notation is useful.

1.3 POLYNOMIALS

The expression cx^n, where c is a nonzero real number and n is a whole number, is called a **monomial** in one variable. We call c the **coefficient,** x the **variable,** and n the **degree.** Some examples of monomials in one variable are $5x^4$, $-17z^2$, $\sqrt{13}t^{12}$, $\frac{2}{3}k$, and 11.

The expression cx^my^n, where c is a nonzero real number and m and n are nonnegative integers, is called a **monomial in two variables.** Its degree is $m + n$. For example, $-8p^4q^3$ is a monomial in two variables, p and q, with coefficient -8 and degree 7. Monomials with three or more variables may be constructed by the same scheme.

A **polynomial** is a finite sum of monomials. For example,

$$5x^4 + 6x^3z^2 + \sqrt{13}t^{12}$$

is a polynomial in three variables. The **degree of a polynomial** is the highest degree of the monomials that make it up. Consider the polynomial

$$6x^4 + (-2x^3) + 14x + (-3)$$

This is a fourth-degree polynomial in one variable. However, we would write it in the form

$$6x^4 - 2x^3 + 14x - 3$$

Notice that this polynomial is written so that the monomial of highest degree comes first and the degrees of the other monomials decrease in order. This is called **standard form.** The monomials that make up the polynomial are often called **terms.** The coefficient of the first term is called the **leading coefficient,** and the term with degree zero, if present, is called the **constant term.** Consider the polynomial in two variables

$$-x^7 + 4x^4y - 11x^3y^5 + 2y^6$$

Its degree is 8. Notice that the powers of x are in decreasing order. This is considered **standard form with respect to *x*.** The same polynomial written in standard form with respect to y is

$$2y^6 - 11x^3y^5 + 4x^4y - x^7$$

We have seen that polynomials with one term are called monomials. Polynomials with two terms are called **binomials,** and those with three terms are called **trinomials.** A general polynomial of degree n in the variable x can be written as

$$a_nx^n + a_{n-1}x^{n-1} + \cdots + a_1x + a_0;\ a_n \neq 0$$

For example, the fourth-degree polynomial

$$3x^4 - 2x^3 + x^2 + 7$$

can be written as

$$a_4x^4 + a_3x^3 + a_2x^2 + a_1x + a_0$$

where $a_4 = 3$, $a_3 = -2$, $a_2 = 1$, and $a_0 = 7$.

Evaluating Polynomials

To **evaluate** a polynomial is to find its value when given numbers are substituted for its variables.

EXAMPLE 1 Evaluate the following polynomial $x^2y - xy^2 + x^2y^2$ when x is 3 and y is -2.

Solution When x is 3 and y is -2,

$$\begin{aligned} x^2y - xy^2 + x^2y^2 &= (3)^2(-2) - 3(-2)^2 + (3)^2(-2)^2 \\ &= 9(-2) - 3 \cdot 4 + 9 \cdot 4 \\ &= -18 - 12 + 36 = 6 \end{aligned}$$

□

EXAMPLE 2 Evaluate the polynomial $x^5 - 3x^4 + 2x^2$, when x is 2.7113, to the nearest thousandth.

Solution This computation will use 2.7113 several times. We use the calculator memory to prevent having to key it in more than once.

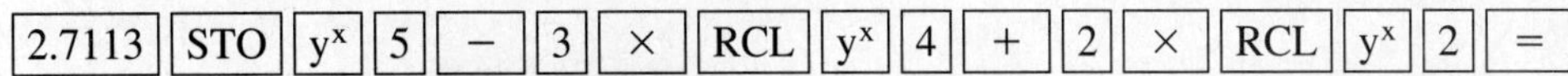

The display shows −.8988714204. The store and recall operations may be slightly more complicated on "superscientific" or graphing calculators that have many different memories. Consult the instruction manual. To the nearest thousandth,

$x^5 - 3x^4 + 2x^2 \approx -0.899$ when x is 2.7113. □

Addition and Subtraction of Polynomials

The associative, commutative, and distributive properties apply to polynomials just as to real numbers, since polynomials are expressions that represent real numbers. We can use these properties to add the polynomials $3x^2 + 5x - 4$ and $2x^2 - 3x - 6$.

$$\begin{aligned}(3x^2 + 5x - 4) + (2x^2 - 3x - 6) &= 3x^2 + 2x^2 + 5x - 3x - 4 - 6\\ &= (3 + 2)x^2 + (5 - 3)x - 10\\ &= 5x^2 + 2x - 10\end{aligned}$$

The first two steps in this problem are called **combining like terms. Like terms** are "alike" everywhere except for the coefficient. $7x^3$ and $5x^2$ are **unlike terms** because $3 \neq 2$.

We define the **zero polynomial** to be the number 0.

If P is a polynomial, $-P$ is called the **opposite** of P, and furthermore,

$$P + (-P) = 0$$

To subtract the polynomial Q from the polynomial P, we add the opposite of Q to P. That is,

$$P - Q = P + (-Q)$$

Multiplication of Polynomials

We multiply polynomials by using the properties of exponents and the properties of polynomials.

EXAMPLE 3 Find each product.

(a) $-3x^2y(5x^3z^2)$ (b) $6t^3(4t^2 + 7)$
(c) $(2x - 3y)(4x^2 + y^3)$ (d) $(3s + 2t)(s^2 - 4st + 5t^2)$

Solutions (a) $-3x^2y(5x^3z^2) = (-3)5x^2x^3yz^2 = -15x^5yz^2$

(b) $6t^3(4t^2 + 7) = 6t^3(4t^2) + 6t^3(7) = 24t^5 + 42t^3$

(c) $$\begin{aligned}(2x - 3y)(4x^2 + y^3) &= (2x - 3y)4x^2 + (2x - 3y)y^3 \\ &= 2x(4x^2) - 3y(4x^2) + 2x(y^3) - 3y(y^3) \\ &= 8x^3 - 12x^2y + 2xy^3 - 3y^4\end{aligned}$$

(d) $$\begin{aligned}(3s + 2t)(s^2 - 4st + 5t^2) &= (3s + 2t)s^2 - (3s + 2t)4st + (3s + 2t)5t^2 \\ &= 3s^3 + 2s^2t - 12s^2t - 8st^2 + 15st^2 + 10t^3 \\ &= 3s^3 - 10s^2t + 7st^2 + 10t^3\end{aligned}$$ □

Some multiplication patterns occur very frequently in algebra. We sometimes call them special products, and their patterns should be memorized.

Special Products

Square of a Binomial

$$(p + q)^2 = p^2 + 2pq + q^2$$
$$(p - q)^2 = p^2 - 2pq + q^2$$

Cube of a Binomial

$$(p + q)^3 = p^3 + 3p^2q + 3pq^2 + q^3$$
$$(p - q)^3 = p^3 - 3p^2q + 3pq^2 - q^3$$

Difference of Two Squares

$$(p + q)(p - q) = p^2 - q^2$$

Demonstrations of the cube formulas are Problems 71 and 72 in the exercises. The binomial theorem includes a formula for $(p + q)^n$, for *all* natural numbers n. The binomial theorem can be found in Section 9.3.

EXAMPLE 4 Find each product.

(a) $(2x + 11)^2$ (b) $(2 - s)^3$ (c) $(6a - 5k)(6a + 5k)$

Solutions (a) $(2x + 11)^2 = (2x)^2 + 2 \cdot 2x \cdot 11 + 11^2 = 4x^2 + 44x + 121$

(b) $(2 - s)^3 = 2^3 - 3 \cdot 2^2 s + 3 \cdot 2s^2 - s^3 = 8 - 12s + 6s^2 - s^3$

(c) $(6a - 5k)(6a + 5k) = (6a)^2 - (5k)^2 = 36a^2 - 25k^2$ □

Division of Polynomials

There are several ways of writing a division problem. Each statement below means P divided by Q.

$$P \div Q \qquad \frac{P}{Q} \qquad P/Q \qquad Q\overline{)\,P}$$

P is called the **dividend,** and Q is called the **divisor.**

When the divisor is a monomial, the division can be performed by dividing the divisor into each term of the dividend and using the properties of exponents.

EXAMPLE 5 Perform the indicated divisions.

(a) $\dfrac{-18x^3y^5}{6x^2y^7}$ (b) $(4t^5 + 7t^3 - t) \div 2t^3$

Solutions (a) $\dfrac{-18x^3y^5}{6x^2y^7} = \dfrac{-3x}{y^2}$ Properties of exponents

(b) $(4t^5 + 7t^3 - t) \div 2t^3 = \dfrac{4t^5 + 7t^3 - t}{2t^3}$

$$= \frac{4t^5}{2t^3} + \frac{7t^3}{2t^3} - \frac{t}{2t^3}$$

$$= 2t^2 + \frac{7}{2} - \frac{1}{2t^2}$$ Properties of exponents □

When the divisor is not a monomial, the division may require long division. Long division of polynomials can be done very much like the division of real numbers. The steps are illustrated in Example 6.

EXAMPLE 6 Perform the indicated division: $\dfrac{2x^4 - 3x^3 + 11x - 7}{x^2 - x + 2}$.

Solution We use the long division algorithm. Both polynomials must be in standard form. Notice that we must insert $0x^2$ for the missing x^2 term (or leave sufficient space in the divisor) as x^2 terms will occur in the quotient.

$$x^2 - x + 2\overline{)\,2x^4 - 3x^3 + 0x^2 + 11x - 7}$$

What is $2x^4 \div x^2$? It is $2x^2$, so we write

$$\begin{array}{r}
2x^2 \\
x^2 - x + 2\overline{)\,2x^4 - 3x^3 + 0x^2 + 11x - 7}
\end{array}$$

Next we multiply $x^2 - x + 2$ by $2x^2$ and subtract the product.

$$\begin{array}{rl}
& 2x^2 \\
x^2 - x + 2 & \overline{)\,2x^4 - 3x^3 + 0x^2 + 11x - 7} \\
& \underline{\;\;2x^4 - 2x^3 + 4x^2} \\
& \qquad\; - x^3 - 4x^2 \qquad\qquad \text{Subtract!}
\end{array}$$

We bring down the $11x$ and repeat the last two steps.

$$\begin{array}{rl}
& 2x^2 - x \\
x^2 - x + 2 & \overline{)\,2x^4 - 3x^3 + 0x^2 + 11x - 7} \\
& \underline{\;\;2x^4 - 2x^3 + 4x^2} \\
& \qquad\; - x^3 - 4x^2 + 11x \\
& \qquad\; \underline{- x^3 + x^2 - 2x} \\
& \qquad\qquad\; - 5x^2 + 13x \qquad \text{Subtract!}
\end{array}$$

We bring down the -7 and repeat those two steps again.

$$\begin{array}{rl}
& 2x^2 - x - 5 \\
x^2 - x + 2 & \overline{)\,2x^4 - 3x^3 + 0x^2 + 11x - 7} \\
& \underline{\;\;2x^4 - 2x^3 + 4x^2} \\
& \qquad\; - x^3 - 4x^2 + 11x \\
& \qquad\; \underline{- x^3 + x^2 - 2x} \\
& \qquad\qquad\; - 5x^2 + 13x - 7 \\
& \qquad\qquad\; \underline{- 5x^2 + 5x - 10} \\
& \qquad\qquad\qquad\quad 8x + 3 \qquad \text{Subtract!}
\end{array}$$

Since the polynomial $8x + 3$ is of lower degree than the polynomial $x^2 - x + 2$, the algorithm stops. We have learned that

$$\frac{2x^4 - 3x^3 + 11x - 7}{x^2 - x + 2} = 2x^2 - x - 5 + \frac{8x + 3}{x^2 - x + 2} \qquad \square$$

Notice that the answer is written in the form

$$\frac{\text{Polynomial}}{\text{Divisor}} = \text{quotient} + \frac{\text{remainder}}{\text{divisor}}$$

Most errors made in long division occur at the subtraction step. It should be done carefully. Notice that the algorithm contains repetitive steps that end when the subtraction step yields a polynomial of lower degree than the divisor.

Order of Operations

Since polynomials represent real numbers, the order of operations from Section 1.1 remains in effect.

EXAMPLE 7 Perform the indicated operations.

(a) $(x + 3)^2 \div x - 2$ (b) $3y^2 - 2y + 5y(1 - y)^2$

Solutions (a) According to the order of operations we do exponentiations *first.*

$$\begin{aligned}(x + 3)^2 \div x - 2 &= (x^2 + 6x + 9) \div x - 2 \\ &= x + 6 + \frac{9}{x} - 2 && \text{Divide by } x. \\ &= x + 4 + \frac{9}{x} && \text{Collect like terms.}\end{aligned}$$

(b)
$$\begin{aligned}3y^2 - 2y + 5y(1 - y)^2 &= 3y^2 - 2y + 5y(1 - 2y + y^2) && \text{Exponentiation first} \\ &= 3y^2 - 2y + 5y - 10y^2 + 5y^3 && \text{Multiplication next} \\ &= 5y^3 - 7y^2 + 3y && \text{Addition and subtraction last}\end{aligned}$$

□

PROBLEM SET 1.3

Warm-ups

In Problems 1 through 4, evaluate each polynomial when x is −1 and y is 2. See Example 1.

1. $2x^3 - 3x^2 - 4x - 5$

2. $2y^3 - 3xy^2 - 4x^2y - 5x^3$

3. $-5x^5y^2 + 3x^3y^3$

4. $-3x^4y - 2xy^4 - 26$

In Problems 5 and 6, approximate the value of each polynomial to the nearest ten-thousandth when x is $-\sqrt{7}$. See Example 2.

5. $x^3 + 6x^2 - 13x - 1$

6. $-\frac{2}{121}x^6 + \frac{3}{11}x^5 + 40$

In Problems 7 through 12, perform the indicated operation. See Example 3.

7. $-2ab(5ab^2)$

8. $2x^3(1 - x^2)$

9. $(2x - 7y)(3x - y^2)$

10. $(3x + 2y)(2x - y)$

11. $(v^2 - 5w)(2v^3 + 3vw - w^3)$

12. $(a + b + c)(x + y + z)$

In Problems 13 through 16, find each product. See Example 4.

13. $(7x - 3y)^2$

14. $(4x + 9u)^2$

15. $(3 + 2g)^3$

16. $(2k - 3j)^3$

In Problems 17 and 18, perform the indicated operation. See Example 5.

17. $\dfrac{(3s^2 - 6s + 3)}{s}$

18. $\dfrac{3r^4t^5}{6rt^8}$

In Problems 19 and 20, perform the indicated divisions, writing answers in the form

$\dfrac{\textit{Polynomial}}{\textit{Divisor}} = \textit{quotient} + \dfrac{\textit{remainder}}{\textit{divisor}}$. *See Example 6.*

19. $\dfrac{2x^3 - x^2 - 5x + 2}{2x - 3}$

20. $\dfrac{6t^3 + t^2 + t - 5}{3t + 2}$

In Problems 21 through 24, perform the indicated operations. See Example 7.

21. $(2x^2 - x) \div x + 3$

22. $(z - 1)(z + 1)^2$

23. $2t + 7t^2 + t^2(3t - 2)^2$

24. $5x - 2x[3 + 2(4 - 3x)(1 + x)^2]$

Practice Exercises

In Problems 25 through 62, perform the indicated operation.

25. $(2x^2 - 3x + 4) + (9 - 2x - x^2)$

26. $(z^3 + 2z^2 - 5z) - (3z^3 - 3z - 11)$

27. $(2x + 3)^2$

28. $(5k - 6)^2$

29. $x^2 - 9 - (x - 3)^2$

30. $(4 + t)^2 - 16 + t^2$

31. $(2x^2 - 3x + 4) - (9 - 2x - x^2)$

32. $(z^3 + 2z^2 - 5z) + (3z^3 - 3z - 11)$

33. $(2y - 7)(y + 8)$

34. $(11x + 5)(x^2 - 2)$

35. $(6z - 13w)(6z + 13w)$

36. $(9x + y)(9x - y)$

37. $(3x - 2a)(9x^2 + 4a^2)$

38. $(s + t^2)(s^2 - 3t)$

39. $\dfrac{y^3 + 5y^2 - 10y + 18}{y + 7}$

40. $(3x^4 - x^2 - 3x + 1) \div (x - 1)$

41. $(2x + y)^3$

42. $(r - 4t)^3$

43. $(2x^2 + 15y)(y - 3x + 11x^2)$

44. $(1 - k + k^2)(k - 2k^2 - 22)$

45. $\dfrac{q^6 - q^5 - p^3q + p^3}{pq}$

46. $\dfrac{3x^4 - 18x^2 + 3x - 2}{-3x^4}$

47. $(10t^2 - 17)^2$

48. $(16 + 8s^3)^2$

49. $(16 + 25x^4)(-25x^4 + 16)$

50. $(-13d + 14)(14 + 13d)$

51. $(4x^2 + 5)(16x^4 - 20x^2 + 25)$

52. $(6k^2 - 7b^3)(36k^4 + 42b^3k^2 + 49b^6)$

53. $\dfrac{(x - y)^2}{x} + y$

54. $(2k - 9)^2(2k + 1)$

55. $(x^2 - y^2) - (x - y)^2$

56. $(3s + 2t) - (9s + 4t^2)$

57. $\dfrac{6s^3 + 4s^2 + s - 5}{3s^2 - s - 2}$

58. $\dfrac{3x^4 - 2x^3 + x^2 - 2x + 4}{3x^2 + x - 1}$

59. $(x^5 + x^4 - 4x + 1) \div (x^2 - 2)$

60. $(4x^5 - 6x^3 + 2x^2 + 1) \div (2x^2 + 1)$

61. $(2x^3 - x^2 + 5x + 2) \div (2x + 1)$

62. $\dfrac{2x^3 - 3x^2y - 2xy^2 + y^3}{2x + y}$

In Problems 63 through 66, approximate the value of each polynomial to the nearest thousandth when the given number is substituted for the variable.

63. $3x^3 + 2x^2 - x - 1$; x is -6.1010

64. $11 - t^2 + t^3 + 2t^6$; t is -1.0101

65. $z^9 - 2z^8 + \sqrt{2}z^7$; $z \approx 0.9876543$

66. $\sqrt{11}y^{11} - \pi y^{10} + \dfrac{5}{11}y$; $y \approx 2.6711403$

67. Find a polynomial that gives the perimeter of a rectangle if the width is given by $x^2 - 1$ and the length is given by $3(x - 1)^2$. What is the perimeter of such a rectangle if x is 5 units?

68. Find a polynomial that gives the area of a triangle if the base is $2k - 1$ units and the altitude is $3k + 1$ units. What is the area of such a triangle if k is 71 centimeters?

69. Jim Levert's Rib Shack has heavy weekend business but tends to slack off during midweek. Jim is trying to figure out whether he should close Tuesday through Thursday. His net profit in dollars for each day, before utilities, is given by the polynomial $59x^2 - 531x + 1277$, where x is 1 for Friday, 2 for Saturday, 3 for Sunday, etc. His utilities are $93 a day if he is open, and $14 a day if he is closed. Should Jim close Tuesday through Thursday? If he wants to close three consecutive days, is there a better choice?

70. The net profit of Madelyn Gould's sport fishing boat is heavily influenced by the weather. Her yearly net in dollars is given by the polynomial $1.12t^2 - 1134$, where t is the number of days she can fish. If she gets out 110 days next year, how much will she net? If there are 280 good fishing days in a particular year, how much will it cost Captain Gould to take a 1-week vacation during good weather?

71. Prove that for any real numbers p and q,

$$(p + q)^3 = p^3 + 3p^2q + 3pq^2 + q^3.$$

72. Prove that for any real numbers p and q,

$$(p - q)^3 = p^3 - 3p^2q + 3pq^2 - q^3.$$

73. Some 4000 years ago, the Babylonians were familiar with Pythagoras' great theorem, that in any right triangle the sum of the squares of the length of the legs equals the square of the length of the hypotenuse. In fact, they developed the following scheme for creating "Pythagorean triples," positive integers x, y, and z such that

$$x^2 + y^2 = z^2$$

Choose any two positive integers p and q, such that $p > q$. Let

$$x = p^2 - q^2$$

$$y = 2pq$$

$$z = p^2 + q^2$$

Prove these are always Pythagorean triples. Find four sets of Pythagorean triples.

In Problems 74 through 81, assume that m and n are natural numbers. In Problems 74 through 77, evaluate each polynomial when x is −1.

74. $x^n + 1$

75. $x^{2n} + 1$

76. $x^{2n+1} + 1$

77. $x^{2n-1} + 1$

In Problems 78 through 81, perform the indicated operation.

78. $(x^n + 1)(x^m + 1)$

79. $(z^n - 5)^2$

80. $(x^m - y^n)^3$

81. $\left(\dfrac{1}{x^{n+1}} + x^{n+1}\right)\left(\dfrac{1}{x^{n+1}} - x^{n+1}\right)$

Challenge Problems

82. Prove that $x^2 - xy + y^2 \geq 0$ for all real numbers x and y. (*Hint:* How is $(x - y)^2$ related to zero?) Don't forget that one or both of the variables may represent a negative number.

83. Suppose x and y are two numbers such that $x - y = 2$ and $x^2 + y^2 = 8$. Find $x^3 - y^3$.

IN YOUR OWN WORDS . . .

84. Explain how to multiply a binomial by a trinomial.

1.4 FACTORING

We learned to add, subtract, multiply, and divide polynomials in the previous section. Often it is useful to write a polynomial as a product. This process is called **factoring.** For example, we can factor $x^2 - x - 2$ by writing it as $(x - 2)(x + 1)$.

A polynomial P can always be written as $1 \cdot P$. If this is the only way to factor a polynomial, the polynomial is said to be **prime** or **irreducible.** When a polynomial is written as a product of prime polynomials, it is **factored completely** or it is written in its **prime factorization.**

When deciding if a polynomial is prime, it is important to designate the set of numbers in which factoring is to be done. For example, the polynomial $x^2 - 7$ is prime if we are using the set of integers as coefficients, but since $x^2 - 7 = (x - \sqrt{7})(x + \sqrt{7})$, it can be factored if we are using real numbers as coefficients. In this section, we will factor using *integer coefficients.*

Factoring with the Distributive Property

The distributive property provides a tool for factoring polynomials that have common factors. We often say we are **taking out a common factor.**

EXAMPLE 1 Factor each polynomial.

(a) $-7a^4b^3 + 14a^2b^2 - 7a^2b$ (b) $a(r - 1) - b(1 - r)$

Solutions (a) There is a common factor of $-7a^2b$ in each term. We factor it out with the distributive property.

$$-7a^4b^3 + 14a^2b^2 - 7a^2b = -7a^2b(a^2b^2 - 2b + 1)$$

The polynomial could also be factored as

$$-7a^4b^3 + 14a^2b^2 - 7a^2b = 7a^2b(-a^2b^2 + 2b - 1)$$

Often it is convenient to factor out the *negative* coefficient.

(b) Remember that $r - 1$ and $1 - r$ are opposites. That is, $1 - r = -(r - 1)$. We replace $(1 - r)$ with $-(r - 1)$.

$$\begin{aligned} a(r - 1) - b(1 - r) &= a(r - 1) - b[-(r - 1)] \\ &= a(r - 1) + b(r - 1) \\ &= (r - 1)(a + b) \end{aligned}$$

Factoring with Special Products

The special product formulas that we learned in the previous section are valuable as factoring formulas. Three other products are also useful for factoring. Consider the product.

$$(p + q)(p^2 - pq + q^2) = p^3 - p^2q + pq^2 + qp^2 - pq^2 + q^3$$
$$= p^3 + q^3$$

This is called the **sum of two cubes.** In a similar manner we obtain a formula called the **difference of two cubes.**

Factoring Formulas

Difference of Two Squares

$$p^2 - q^2 = (p - q)(p + q)$$

Sum or Difference of Two Cubes

$$p^3 + q^3 = (p + q)(p^2 - pq + q^2)$$
$$p^3 - q^3 = (p - q)(p^2 + pq + q^2)$$

Square of a Binomial

$$p^2 + 2pq + q^2 = (p + q)^2$$
$$p^2 - 2pq + q^2 = (p - q)^2$$

EXAMPLE 2 Factor each polynomial completely.

(a) $16x^2 - 4y^2$ (b) $16a^3 - 54b^3$ (c) $25t^2 + 20t + 4$
(d) $9g^2 + h^2$ (e) $25 - (x + k)^2$

Solutions (a) $16x^2 - 4y^2 = 4(4x^2 - y^2)$ Common factor
$= 4(2x - y)(2x + y)$

(b) $16a^3 - 54b^3$ looks like the difference of two cubes, but, neither 16 nor 54 is a perfect cube. However, there is a common factor of 2 in each term. It is a good idea to factor out common factors first.

$16a^3 - 54b^3 = 2(8a^3 - 27b^3)$ Common factor
$= 2(2a - 3b)(4a^2 + 6ab + 9b^2)$ Difference of two cubes

(c) $25t^2 + 20t + 4 = (5t + 2)^2$

(d) $9g^2 + h^2$ is the sum of squares and cannot be factored using integer coefficients.

(e) $25 - (x + k)^2$ is the difference of two squares if we treat $(x + k)^2$ as one term.

$$25 - (x + k)^2 = [5 - (x + k)][5 + (x + k)]$$
$$= (5 - x - k)(5 + x + k)$$

□

Factoring Trinomials

Some trinomials can be written as a product of two binomials. For example, the trinomial

$$10x^2 - 7x - 12$$

can be factored to the product of two binomials,

$$(2x - 3)(5x + 4)$$

EXAMPLE 3 Factor $6x^2 + 23x + 20$ if possible.

Solution If this trinomial factors with integer coefficients we have

Factors of 6

$$6x^2 + 23x + 20 = (\square\, x + \square)(\square\, x + \square)$$

Factors of 20

The possible factors of 6 are $1 \cdot 6$ and $2 \cdot 3$, in either order. The factors of 20 are $1 \cdot 20$, $2 \cdot 10$, and $4 \cdot 5$. It should take only one or two tries to find

$$6x^2 + 23x + 20 = (2x + 5)(3x + 4)$$ □

Practice makes factoring trinomials easier. The more practice, the less trial and error.

EXAMPLE 4 Factor each trinomial.

(a) $x^4 - x^2 - 6$ (b) $2x^4 + 5x^2 + 2$

Solutions (a) $x^4 - x^2 - 6 = (\square\, x^2 + \square)(\square\, x^2 - \square)$
$= (x^2 + 2)(x^2 - 3)$

(b) $2x^4 + 5x^2 + 2 = (2x^2 + 1)(x^2 + 2)$ □

Factoring by Grouping

Polynomials with more than three terms can sometimes be factored by **grouping.**

EXAMPLE 5 Factor each polynomial.

(a) $x^4 - x^3 - 2x + 2$ (b) $x^4 - x^2 - 2x - 1$

Solutions (a) $x^4 - x^3 - 2x + 2 = (x^4 - x^3) - (2x - 2)$ Notice the sign in $(2x - 2)$.

$$= x^3(x - 1) - 2(x - 1)$$

$$= (x - 1)(x^3 - 2) \quad x - 1 \text{ is a common factor.}$$

(b) Sometimes it is difficult to decide how to group the terms. In this example, we group three terms together.

$$x^4 - x^2 - 2x - 1 = x^4 - (x^2 + 2x + 1) \quad \text{Grouping}$$

$$= x^4 - (x + 1)^2$$

$$= [x^2 - (x + 1)][(x^2 + (x + 1)]$$

$$= (x^2 - x - 1)(x^2 + x + 1) \quad \text{Watch the signs!}$$ □

There is often more than one way to group. Consider part (a) of Example 5. We *could* have grouped the first and third terms and the second and fourth.

$$x^4 - x^3 - 2x + 2 = (x^4 - 2x) - (x^3 - 2)$$

$$= x(x^3 - 2) - (x^3 - 2)$$

$$= (x^3 - 2)(x - 1)$$

General Procedure for Factoring

Now that we have looked at several types of factoring, we summarize with a procedure for factoring a polynomial completely.

Procedure for Factoring a Polynomial Completely

1. Factor out the greatest common factor if there is one. This should be done first.
2. Count the number of terms.
3. If the polynomial is a binomial, check for difference of squares or cubes, or sum of cubes.
4. If the polynomial is a trinomial, look for two binomial factors.
5. If the polynomial has more than three terms, try grouping.
6. Make sure that each factor is prime.
7. Check to see if the product of the factors is the original polynomial.

EXAMPLE 6 Factor each polynomial completely.

(a) $x^4 + x^3 - 4x^2 - 4x$ (b) $x^4 - 625$

Solutions (a) $x^4 + x^3 - 4x^2 - 4x = x(x^3 + x^2 - 4x - 4)$ Common factor

$= x[x^2(x + 1) - 4(x + 1)]$ Grouping

$= x(x + 1)(x^2 - 4)$ Common factor

$= x(x + 1)(x - 2)(x + 2)$ Difference of two squares

(b) $x^4 - 625 = (x^2 - 25)(x^2 + 25)$ Difference of two squares

$= (x - 5)(x + 5)(x^2 + 25)$ Difference of two squares

□

PROBLEM SET 1.4

Warm-ups

In Problems 1 through 26, factor each polynomial completely. For Problems 1 through 4, see Example 1.

1. $3a^2bc - 6ab^2c^2 + 9abc$ **2.** $-5x^2 - 5x$ **3.** $y(z + t) - x(z + t)$ **4.** $m(a - b) + n(b - a)$

For Problems 5 through 10, see Example 2.

5. $25 - 4x^2$ **6.** $8 + z^3t^3$ **7.** $8y^3 - 1$ **8.** $9y^2 + 6y + 1$

9. $16x^2 + y^2$ **10.** $9m^2 - 16n^2$

For Problems 11 through 18, see Example 3.

11. $x^2 + 3x + 2$ **12.** $r^2 - 6r + 5$ **13.** $z^2 - 4z - 5$ **14.** $x^2 + x - 2$

15. $10x^2 + 21x + 9$ **16.** $3t^2 - 5t + 2$ **17.** $6x^2 - x - 1$ **18.** $4x^2 + 5x - 6$

For Problems 19 and 20, see Example 4.

19. $x^4 + x^2 - 42$ **20.** $2y^4 + 3y^2 - 2$

For Problems 21 through 24, see Example 5.

21. $x^3 + 4x^2 + x + 4$ **22.** $y^2 - x^2 - 4x - 4y$ **23.** $x^4 - x^3 + 4x - 4$ **24.** $2xt + 2t - x - 1$

For Problems 25 and 26, see Example 6.

25. $x^8 - 16$ **26.** $x^4 - x^3 + 4x^2 - 4x$

Practice Exercises

In Problems 27 through 74, determine if the polynomial is factored, factored completely, or not factored. If possible, factor each polynomial completely.

27. $-6t^2 + 6$ **28.** $-2x^2 - 3x$ **29.** $90x^2y^4z^2 - 54x^3y^3z^3$

30. $72s^2t^9 + 60s^3t^3$ **31.** $27 - t^3$ **32.** $8x^3 + 1$

33. $r^2(x - 1) + s^2(1 - x)$ **34.** $x^2(2 - 3y) + 3y - 2$ **35.** $3y^4 - 5y^2 + 2$

36. $t^4 - t^2 - 6$ **37.** $b(49 - 16a^2)$ **38.** $15(225 - 144x^2)$

39. $x^8 - 1$ **40.** $1 - w^8$ **41.** $x^3 + x^2 + 2x$

42. $z^3 + z^2 - 12z$ **43.** $24a^2x - 6x - 12a^2 + 3$ **44.** $2ax - 2ay + x - y$

45. $3 - x + 4x^2$ **46.** $6 - 7w + 2w^2$ **47.** $125w^3 - 1$

48. $27b^3 + 8$ **49.** $x^2 + x + 1$ **50.** $4x^2 + 9$

51. $(3t - 1)(t + 2)$

52. $(x + y)(2x - y)$

53. $9x^4 + 9x^2$

54. $8y^3 + 32y$

55. $x^2y^3 - y$

56. $t^3 - 25t^5x^2$

57. $2at - a + 2bt - b$

58. $4 - x^2 + 2xy - y^2$

59. $x^4 - x^2 + x^3y - xy$

60. $a^2 - x^2 - 2x - 1$

61. $a^3 - 4a^2 - 5a$

62. $3x^3 + 2x^2 - x$

63. $x^6 - x^3 - 2$

64. $y^6 - 6y^3 - 16$

65. $6x^4 - x^2 - 1$

66. $2x^4 - x^2 + 1$

67. $a^2(2 - b) + r^2(b - 2)$

68. $9(v - 4) + x^2(4 - v)$

69. $a^3 + 27b^3$

70. $64 - 125x^3$

71. $32c^3d^2 - 72c^2d^3$

72. $81h^2v^4 - 54h^3v^3$

73. $4x^2 - (a + b)^2$

74. $9 - (x - y)^2$

75. Show that if n is an integer, then $\dfrac{n^2 + n}{2}$ is also an integer.

In Problems 76 through 79, factor each polynomial. Assume that n is a natural number.

76. $x^{2n} - 1$

77. $x^{2n} - x^n - 2$

78. $x^{5n} + x^{4n}$

79. $x^{3n} - 1$

Challenge Problems

80. Show that if n is an integer, then $\dfrac{2n^3 + 3n^2 + n}{6}$ is also an integer. (*Hint:* Factor the numerator completely and then try a few numbers.)

81. Factor $x^6 - 1$ first as the difference of two squares and then factor it as the difference of two cubes. Which factorization is the prime factorization?

IN YOUR OWN WORDS . . .

82. Explain why the statement, "The sum of two squares is prime," is not necessarily true.

83. Discuss the factoring of polynomials to include reasons and methods.

84. Why do we emphasize looking for a common factor *first?*

1.5 RATIONAL EXPRESSIONS

Combining the ideas of fractions with those of polynomials, we can obtain a fraction whose numerator and denominator are polynomials. Such fractions are known as rational expressions. A **rational expression** is the quotient of two polynomials. Rational expressions are often called *algebraic fractions* or simply *fractions.* We use the terms *numerator* and *denominator* just as we would in arithmetic. Some examples of rational expressions are

$$\frac{x^2 + 2}{6} \qquad \frac{x^2 - 3x + 23}{x^3 + 16x^2 - 1} \qquad \frac{2}{11k} \qquad \frac{(x - 1)(x + 2)}{x}$$

We evaluate rational expressions just like we do polynomials. However, rational expressions are fractions, so we must be careful about denominators with value zero.

EXAMPLE 1 Evaluate $\dfrac{x + 2}{x - 1}$ when x is the given number.

(a) 3 (b) -2 (c) 0 (d) 1

Solutions (a) When x is 3, $\dfrac{x+2}{x-1} = \dfrac{3+2}{3-1} = \dfrac{5}{2}$

(b) When x is -2, $\dfrac{x+2}{x-1} = \dfrac{-2+2}{-2-1} = \dfrac{0}{-3} = 0$

Remember that zero divided by any positive or negative number is zero.

(c) When x is 0, $\dfrac{x+2}{x-1} = \dfrac{0+2}{0-1} = \dfrac{2}{-1} = -2$

(d) When x is 1, $\dfrac{x+2}{x-1} = \dfrac{1+2}{1-1} = \dfrac{3}{0}$

But, $\dfrac{3}{0}$ does not represent a real number because division by zero is undefined. Therefore, we *cannot* evaluate this rational expression when x is 1. We say that the rational expression $\dfrac{x+2}{x-1}$ is **undefined** when x is 1. □

Definition: Undefined Rational Expression

If P and Q are polynomials, the rational expression

$$\frac{P}{Q}$$

is *undefined* for all numbers that make Q have value zero.

The set of permissible numbers that can be assigned to the variable in an algebraic expression is called the **domain** of the variable.

EXAMPLE 2 Determine the numbers for which each rational expression is undefined.

(a) $\dfrac{x-4}{x^2-x-12}$ (b) $\dfrac{8}{x^4+1}$

Solutions (a) First, we factor the denominator.

$$\frac{x-4}{x^2-x-12} = \frac{x-4}{(x+3)(x-4)}$$

Now we can see that this denominator is zero when x is -3 or 4. Therefore, the rational expression is undefined when x is -3 or 4.

(b) $\dfrac{8}{x^4+1}$

Notice that x^4 is nonnegative for any real numbers. Therefore, $x^4 + 1$ is *never* less than 1. This rational expression is defined for *all* real numbers. □

There are many ways to write fractions representing the same number. For example, $\frac{2}{3}$ may be written as $\frac{10}{15}$, $\frac{-12}{-18}$, or $\frac{222}{333}$. Which form we use depends on what we are doing with the fraction. The same idea extends to rational expressions, as stated in the following principle.

Fundamental Principle of Rational Expressions

If A, B, and C are polynomials, then

$$\frac{AC}{BC} = \frac{A}{B}$$

where each rational expression is defined.

This principle states that we can divide the numerator and denominator of a rational expression by the same nonzero *factor.* Another way of looking at the Fundamental Principle of Rational Expressions is that the common factor C can be divided out of $\frac{A \cdot C}{B \cdot C}$ to give $\frac{A}{B}$. When we write a rational expression so that the numerator and denominator have no common factor, we say that the fraction is in **lowest terms.**

The Fundamental Principle applies when each rational expression is defined. This means that the common factor must not have value zero. So, when reducing rational expressions, we must note any such restrictions placed on the variables. Consider the fraction $\frac{x^2 - 3x}{2x^2 - 5x - 3}$. We factor the numerator and denominator so that we can reduce using the Fundamental Principle.

$$\frac{x^2 - 3x}{2x^2 - 5x - 3} = \frac{x(x-3)}{(2x+1)(x-3)}$$

$$= \frac{x}{2x+1} \qquad \text{Fundamental Principle}$$

Notice that this expression is undefined if $2x + 1 = 0$. However, that is evident from the reduced form. What is *not evident* from the reduced form is that x cannot be 3 either! The Fundamental Principle *does not* allow dividing by zero. We must show this restriction.

$$\frac{x^2 - 3x}{2x^2 - 5x - 3} = \frac{x}{2x+1}; \; x \neq 3$$

The operations of addition, subtraction, multiplication, and division are performed with rational expressions in the same manner as they are with rational numbers.

Definitions: Operations with Rational Expressions

If A, B, C, and D are polynomials and if each rational expression is defined,

Addition $\quad \dfrac{A}{C} + \dfrac{B}{C} = \dfrac{A + B}{C}$

Subtraction $\quad \dfrac{A}{C} - \dfrac{B}{C} = \dfrac{A - B}{C}$

Multiplication $\quad \dfrac{A}{B} \cdot \dfrac{C}{D} = \dfrac{AC}{BD}$

Division $\quad \dfrac{A}{B} \div \dfrac{C}{D} = \dfrac{A}{B} \cdot \dfrac{D}{C}$

$$= \frac{AD}{BC}$$

Adding and Subtracting

Addition and subtraction of rational expressions are performed using the same procedures that we use for rational numbers. If the denominators are the same, add or subtract the numerators and keep the common denominator. If the denominators are not the same, we usually rewrite each fraction with the **least common denominator** (LCD). We use the Fundamental Principle of Rational Expressions to do this.

Procedure to Find the Least Common Denominator

1. Factor each denominator completely, using exponents.
2. List all different prime factors from all denominators.
3. The LCD is the product of the factors in step 2, each raised to the highest power of that factor in any single denominator.

EXAMPLE 3 Perform the operation indicated and write the answers in lowest terms.

(a) $\dfrac{4}{ab^2c^4} + \dfrac{1}{a^2b^2c}$ (b) $\dfrac{3}{z^2 + 6z + 9} - \dfrac{5}{z^2 - 2z - 15}$

Solutions (a) The LCD is $a^2b^2c^4$.

$$\frac{4}{ab^2c^4} + \frac{1}{a^2b^2c} = \frac{4a}{a^2b^2c^4} + \frac{c^3}{a^2b^2c^4} = \frac{4a + c^3}{a^2b^2c^4}$$

(b) First we factor the denominators.

$$\begin{aligned}\frac{3}{z^2 + 6z + 9} - \frac{5}{z^2 - 2z - 15} &= \frac{3}{(z + 3)^2} - \frac{5}{(z + 3)(z - 5)} \\ &= \frac{3(z - 5)}{(z + 3)^2(z - 5)} - \frac{5(z + 3)}{(z + 3)^2(z - 5)} \\ &= \frac{3(z - 5) - 5(z + 3)}{(z + 3)^2(z - 5)} \\ &= \frac{3z - 15 - 5z - 15}{(z + 3)^2(z - 5)} \\ &= \frac{-2z - 30}{(z + 3)^2(z - 5)}\end{aligned}$$

□

Multiplying and Dividing

It is unnecessary to find a common denominator when multiplying or dividing rational expressions. In multiplication, the numerators are multiplied to form the new numerator, and the denominators are multiplied to obtain the new denominator. In division, the dividend is multiplied by the reciprocal of the divisor.

EXAMPLE 4 Perform the operation indicated and reduce the answers to lowest terms.

(a) $\frac{36x^2}{(-z)^2} \cdot \frac{z^2}{27x}$ (b) $\frac{s - t}{t^2} \cdot \frac{t}{t - s}$ (c) $\frac{(y - z)^4}{3k^2} \div (y - z)$

Solutions (a) $\frac{36x^2}{(-z)^2} \cdot \frac{z^2}{27x} = \frac{36x^2z^2}{(-z)^2 \cdot 27x}$ Definition of multiplication

$= \frac{36x^2z^2}{27xz^2}$

$= \frac{4x}{3};\ xz \neq 0$ $9xz^2$ is a common factor.

(b) $\frac{s - t}{t^2} \cdot \frac{t}{t - s} = \frac{(s - t)t}{t^2(t - s)}$ Definition of multiplication

$= \frac{(s - t)t}{-t^2(s - t)}$ Since $s - t$ and $t - s$ are opposites

$$= -\frac{1}{t};\ s \neq t$$ $t(s - t)$ is a common factor.

(c) $$\frac{(y-z)^4}{3k^2} \div (y-z) = \frac{(y-z)^4}{3k^2} \cdot \frac{1}{y-z}$$ Definition of division

$$= \frac{(y-z)^4}{3k^2(y-z)}$$ Definition of multiplication

$$= \frac{(y-z)^3}{3k^2};\ y \neq z$$ □

Complex Fractions

In algebra, quotients are seldom written in the form $P \div Q$ but rather as fractions like $\frac{P}{Q}$. If either P or Q in such a fraction is itself a fraction, then the expression is called a **complex fraction.** Complex fractions may be simplified by one of two general methods.

Two Procedures for Simplifying Complex Fractions

Division Method

1. Write the numerator and the denominator as single fractions.
2. Multiply the numerator by the reciprocal of the denominator.
3. Write the resulting fraction in lowest terms.

Multiplication Method

1. Find the LCD of all the fractions in the main fraction.
2. Multiply the numerator and the denominator by the LCD.
3. Write the resulting fraction in lowest terms.

A complex fraction may be simplified by either method. In general, it is usually better to use the division method when the complex fraction contains a single term in its numerator and one term in its denominator. If the numerator *or* denominator of a complex fraction contains more than one term, it is usually better to use the multiplication method.

EXAMPLE 5 Simplify each of the following complex fractions.

(a) $$\frac{\frac{1}{x+1}}{\frac{1}{x-1}}$$ (b) $$\frac{1 + \frac{1}{t+4}}{\frac{t+5}{t^2-16}}$$

Solutions (a) Note that in the complex fraction $\dfrac{\dfrac{1}{x+1}}{\dfrac{1}{x-1}}$, $x \neq 1$ and $x \neq -1$.

Since the numerator and the denominator each contain a single term, we will use the division method.

$$\frac{\dfrac{1}{x+1}}{\dfrac{1}{x-1}} = \frac{1}{x+1} \cdot \frac{x-1}{1} \qquad \text{Division}$$

$$= \frac{1(x-1)}{(x+1)1} \qquad \text{Multiplication}$$

$$= \frac{x-1}{x+1};\ x \neq 1$$

(b) In $\dfrac{1 + \dfrac{1}{t+4}}{\dfrac{t+5}{t^2-16}}$, $t \neq -4$ and $t \neq 4$.

The numerator has more than one term. We will use the multiplication method. The LCD is $(t + 4)(t - 4)$.

$$\frac{1 + \dfrac{1}{t+4}}{\dfrac{t+5}{t^2-16}} = \frac{\left(1 + \dfrac{1}{t+4}\right)(t+4)(t-4)}{\left(\dfrac{t+5}{(t+4)(t-4)}\right)(t+4)(t-4)}$$

$$= \frac{(t+4)(t-4) + (t-4)}{t+5}$$

$$= \frac{(t+5)(t-4)}{t+5}$$

$$= t - 4;\ t \neq \pm 4;\ t \neq -5$$

□

PROBLEM SET 1.5

Warm-ups

In Problems 1 through 4, determine the numbers for which each expression is undefined. See Examples 1 and 2.

1. $\dfrac{s+1}{s(s-1)}$ **2.** $\dfrac{y}{y^2+y-12}$ **3.** $\dfrac{x^2-16}{(x-4)^2}$ **4.** $\dfrac{(x-4)^2}{x^2-16}$

In Problems 5 through 20, perform the operation indicated and reduce the answers to lowest terms. For Problems 5 through 12, see Example 3.

5. $\frac{1}{x+1}+\frac{1-x}{x+1}$

6. $\frac{2-3y}{y^2+1}-\frac{1-2y}{y^2+1}$

7. $\frac{1}{r^3s^2}+\frac{3}{rs^4}-\frac{5}{r^2s^3}$

8. $\frac{2}{x+2}+\frac{3}{x-3}-\frac{1}{x^2-x-6}$

9. $\frac{x}{x-2}-\frac{x}{x+2}$

10. $\frac{10}{x-2}-\frac{x}{1-2x}$

11. $\frac{w}{w^2-4w+4}+\frac{2}{w^2-4}$

12. $\frac{2x}{x^2-9x+14}-\frac{x-1}{x^2-8x+7}$

For Problems 13 through 20, see Example 4.

13. $\frac{x-3}{x^2}\cdot\frac{x^3}{x-3}$

14. $\frac{(2k)^3g}{25}\div\frac{2kg^2}{85}$

15. $\frac{s-t}{s+t}\div(t-s)$

16. $\frac{1}{k^4-5k^2+6}\cdot(k^4-4)$

17. $\frac{x^3-8}{2x-8}\cdot\frac{-8}{x^2+2x+4}$

18. $\frac{50-2y^2}{y^2+2y}\div\frac{4y+20}{4-y^2}$

19. $\frac{2x^2-8x-42}{x^2+2x+1}\div\frac{4x^2+8x-12}{x+x^2}$

20. $(t-s)\div\frac{s-t}{t}$

In Problems 21 through 24, simplify each complex fraction. See Example 5.

21. $\dfrac{\frac{12}{5s+5t}}{\frac{18}{s+t}}$

22. $\dfrac{x^{-1}-1}{x^{-1}+1}$

23. $\dfrac{\frac{2}{q}+\frac{1}{p}}{\frac{2p^2-pq-q^2}{pq}}$

24. $\dfrac{2+\frac{1}{m-n}}{1-\frac{2}{m-n}}$

Practice Exercises

In Problems 25 through 50, perform the operation indicated. Write the answers in lowest terms.

25. $\frac{2}{x+5}+\frac{3}{x-5}$

26. $\frac{1}{p-3}+\frac{1}{p+3}$

27. $\frac{x^2-2x}{36K}\cdot\frac{9K^3+18K}{x^2-4}$

28. $\frac{121y^2}{16-j^2}\cdot\frac{16-4j}{44y^4-11y^2}$

29. $(x^2-x)\cdot\frac{y^2}{x^4-x^2}$

30. $\frac{x+1}{3x^3-9x^2}\cdot(12x-36)$

31. $\frac{p}{p-q}\div\frac{q}{q-p}$

32. $\frac{2x-5}{J}\div\frac{5-2x}{M}$

33. $\frac{x^2-y^2}{s^2-2st+t^2}\div\frac{sx+tx+sy+ty}{sx-tx-sy+ty}$

34. $\frac{u^2+4uv+4v^2}{j^2-4k^2}\div\frac{ju+2jv-2ku-4kv}{ju-jv+2ku-2kv}$

35. $\frac{x+2}{x^2-1}-\frac{1}{x^2+2x+1}$

36. $\frac{w-1}{4+4w+w^2}-\frac{1}{w^2-4}$

37. $\frac{27+8y^3}{x^3}\div(3+2y)$

38. $\frac{a^3-b^3}{a^2-b^2}\cdot(a+b)^2$

39. $\frac{2x^2+6x+4}{x^2-4x+3}\cdot\frac{x^2-x-6}{4x^2-4x-8}$

40. $\frac{y^2+5y+6}{3y^2-9y+6}\cdot\frac{9y^2+9y-18}{y^2-y-12}$

41. $1-\frac{7}{y+2}$

42. $\frac{2}{z+5}+1$

43. $\frac{-28}{z^2-2z-3}+\frac{7}{z-3}$

44. $\frac{28}{x^2+3x-10}+\frac{4}{x+5}$

45. $\frac{2}{q^2-1}-\frac{1}{q+1}$

46. $\frac{t+2}{(t-2)^2}-\frac{t+2}{t-2}$

47. $\dfrac{3}{5-k}+\dfrac{4}{k-5}$

48. $\dfrac{x}{x-1}-\dfrac{1}{1-x}$

49. $\dfrac{1}{x+1}+\dfrac{2}{x+2}-\dfrac{3}{x+3}$

50. $\dfrac{2}{2s-1}-\dfrac{1}{s+1}+\dfrac{1}{s-1}$

In Problems 51 through 60, simplify each complex fraction.

51. $\dfrac{\dfrac{x^2y}{x+5}}{\dfrac{x}{x+5}}$

52. $\dfrac{\dfrac{6z}{r+2}}{\dfrac{30z^3}{r+2}}$

53. $\dfrac{x^{-1}+2y^{-1}}{3x^{-1}-4y^{-1}}$

54. $\dfrac{\dfrac{1}{k}-\dfrac{1}{k-1}}{\dfrac{1}{k}+\dfrac{1}{k-1}}$

55. $\dfrac{\dfrac{3t^2+5t}{t^2-25}}{\dfrac{2}{t-5}+\dfrac{1}{t+5}}$

56. $\dfrac{\dfrac{1}{x-5}-\dfrac{1}{x+3}}{\dfrac{8x^2+8}{x^2-2x-15}}$

57. $\dfrac{\dfrac{1}{x+h}-\dfrac{1}{x}}{h}$

58. $\dfrac{\dfrac{2}{x+h+1}-\dfrac{2}{x+1}}{h}$

59. $\dfrac{\dfrac{x+h+3}{x+h-1}-\dfrac{x+3}{x-1}}{h}$

60. $\dfrac{\dfrac{2(x+h)}{x+h-7}-\dfrac{2x}{x-7}}{h}$

In Problems 61 and 62, n, p, and s are natural numbers. Perform the operations indicated.

61. $\dfrac{x^{2n}-x^n-6}{x^{2p}+2x^p-3}\cdot\dfrac{x^{2p}-1}{x^{2n}+4x^n+4}$

62. $\dfrac{y^{3s}-8}{y^{2s}+2y^s-3}\div\dfrac{y^{2s}-4}{y^{2s}+3y^s}$

63. Use the numbers in the tables to investigate the value of $\dfrac{1}{x^2}$ at positive and negative numbers very close to 0. Does $\dfrac{1}{x^2}$ have a largest value?

x	−0.5	−0.3	−0.1	−0.01	−0.001	−0.00001
$1/x^2$						

x	0.5	0.3	0.1	0.01	0.001	0.00001
$1/x^2$						

What is the value of $\dfrac{1}{x^2}$ if x is a very large positive number? What is its value if x is a very large negative number (that is, a negative number with a large absolute value)? Does $\dfrac{1}{x^2}$ have a smallest value?

64. Use the numbers in the tables to investigate the value of $\dfrac{1}{x-1}$ at numbers very close to 1. Does $\dfrac{1}{x-1}$ have a largest value? Does $\dfrac{1}{x-1}$ have a smallest value?

x	1.5	1.3	1.1	1.01	1.001	1.00001
$1/(x-1)$						

x	0.5	0.8	0.9	0.99	0.999	0.99999
$1/(x-1)$						

65. Use the numbers in the table to investigate the value of $\dfrac{x^2-4}{x-2}$ at numbers very close to 2.

x	2.5	2.3	2.1	2.01	2.001	2.00001
$(x^2-4)/(x-2)$						

x	1.5	1.8	1.9	1.99	1.999	1.99999
$(x^2-4)/(x-2)$						

Compare the values obtained for $\dfrac{x^2-4}{x-2}$ in the tables with the value of $x+2$. What is true about $\dfrac{x^2-4}{x-2}$ and $x+2$?

Challenge Problems

In Problems 66 through 70, perform the operations indicated and simplify each expression.

66. $1+\dfrac{1}{1+\dfrac{1}{1+1}}$

67. $1+\dfrac{1}{1+\dfrac{1}{1+1+1}}$

68. $1+\dfrac{1}{1+\dfrac{1}{1+1+\cdots+1}}$ (There are n ones in the denominator of the denominator.)

69. $x+\dfrac{x}{x+\dfrac{x}{x+\dfrac{1}{x}}}$

70. $\dfrac{1+\dfrac{1}{1-\dfrac{1}{x}}}{1-\dfrac{1}{1+\dfrac{1}{x}}}$

71. If $\dfrac{A}{B}=\dfrac{C}{D}$, prove $\dfrac{A\pm B}{B}=\dfrac{C\pm D}{D}$.

IN YOUR OWN WORDS . . .

72. Explain what we mean when we say that a rational expression is undefined.

73. Explain what the Fundamental Principle of Rational Expressions allows us to do.

74. Explain how an LCD is used in adding rational expressions and also how the LCD can be used in simplifying a complex fraction.

1.6 RADICALS AND RATIONAL EXPONENTS

There are two real numbers whose squares are 25. They are 5 and -5. Likewise, there are two real numbers whose squares are 11. They are real numbers; however, they are not integers or rational numbers. They are irrational numbers. We let $\sqrt{11}$ be the *positive* number whose square is 11, and $-\sqrt{11}$ be the *negative* number whose square is 11.

$$(\sqrt{11})^2 = 11 \qquad (-\sqrt{11})^2 = 11$$

The number $\sqrt{11}$ is called the **principal square root** of 11. Every nonnegative real number has a principal square root.

Is there a real number whose square is -9? No, there is no such real number. Expressions such as $\sqrt{-9}$ are undefined in the set of real numbers. (In the next section we will examine the set of complex numbers, where such expressions are defined.)

> **Square Roots of Negative Numbers**
>
> If $k > 0$, then $\sqrt{-k}$ is undefined in the set of real numbers.

There is a real number whose cube is 27. It is 3. Also, there is a real number whose cube is -27. It is -3. Is there a real number whose cube is 5? Yes, it is denoted by $\sqrt[3]{5}$. Furthermore, there is a real number whose cube is -5. It is denoted $\sqrt[3]{-5}$. In fact, $\sqrt[3]{-5} = -\sqrt[3]{5}$. We call $\sqrt[3]{5}$ the **cube root** of 5. Every real number has a cube root.

In general, there are roots for every natural number n, denoted by

$$\sqrt[n]{q}$$

where $(\sqrt[n]{q})^n = q$. We call $\sqrt[n]{q}$ a **radical.** $\sqrt{}$ is called a **radical sign.** The real number q is called the **radicand,** and the natural number n is called the **index.** If the index is omitted, the radical is a square root.

Definition: $\sqrt[n]{q}$

If n is an *even* natural number,

1. $\sqrt[n]{q}$ is $\begin{cases} \text{a real number when } q \geq 0. \\ \textit{not} \text{ a real number when } q < 0. \end{cases}$
2. $\sqrt[n]{q}$ is the *nonnegative* real number such that $(\sqrt[n]{q})^n = q$.

If n is an *odd* natural number,

1. $\sqrt[n]{q}$ is a real number for any real number q.
2. $\sqrt[n]{q}$ is the real number such that $(\sqrt[n]{q})^n = q$.

A Property of *Odd* Indexes

If n is an *odd* natural number, then

$$\sqrt[n]{-q} = -\sqrt[n]{q}$$

EXAMPLE 1 Find each root.

(a) $\sqrt{36}$ (b) $\sqrt[3]{-8}$ (c) $\sqrt{-16}$ (d) $\sqrt[5]{32}$ (e) $\sqrt[3]{\dfrac{125}{27}}$ (f) $-\sqrt{100}$

Solutions (a) Because 6 is positive and $6^2 = 36$,

$$\sqrt{36} = 6$$

(b) Since $(-2)^3 = -8$,

$$\sqrt[3]{-8} = -2$$

(c) $\sqrt{-16}$ is not a real number since -16 is negative.

(d) $\sqrt[5]{32} = 2$ because $2^5 = 32$.

(e) $\sqrt[3]{\dfrac{125}{27}} = \dfrac{5}{3}$ since $\left(\dfrac{5}{3}\right)^3 = \dfrac{125}{27}$.

(f) $-\sqrt{100} = -10$ because 10 is positive and $10^2 = 100$. □

We must be very careful when finding *even* roots if there are variables in the radicand. For example, what is $\sqrt{x^2}$? If we know that x is *not negative,* then $\sqrt{x^2} = x$. However, suppose x is -3. Then to say that $\sqrt{x^2} = x$ would be to say that $\sqrt{(-3)^2} = -3$, which is incorrect! Notice that if n is *even,* then $p^n = |p|^n$. Therefore $\sqrt[n]{p^n} = \sqrt[n]{|p|^n} = |p|$ because $|p|$ is always nonnegative. If n is *odd,* then $\sqrt[n]{p^n} = p$ directly. We have shown the following property.

Simplifying $\sqrt[n]{p^n}$

$$\sqrt[n]{p^n} = \begin{cases} |p| & \text{if } n \text{ is } even \\ p & \text{if } n \text{ is } odd \end{cases}$$

EXAMPLE 2 Find each root.

(a) $\sqrt[4]{16t^4}$ (b) $\sqrt{z^4}$ (c) $\sqrt[3]{-s^3}$ (d) $\sqrt[3]{\dfrac{x^3}{y^6}}$

Solutions (a) $\sqrt[4]{16t^4} = |2t|$ since $|2t|$ is nonnegative and $|2t|^4 = 16t^4$
$= 2|t|$

(b) $\sqrt{z^4} = |z^2| = z^2$ since z^2 is nonnegative.

(c) $\sqrt[3]{-s^3} = -s$ since $(-s)^3 = -s^3$.

(d) $\sqrt[3]{\dfrac{x^3}{y^6}} = \dfrac{x}{y^2}$ since $\left(\dfrac{x}{y^2}\right)^3 = \dfrac{x^3}{y^6}$. □

The following properties of radicals are useful.

Properties of Radicals

If $\sqrt[n]{p}$ and $\sqrt[n]{q}$ represent real numbers, then

$\sqrt[n]{p \cdot q} = \sqrt[n]{p} \cdot \sqrt[n]{q}$ Root of a product

$\sqrt[n]{\dfrac{p}{q}} = \dfrac{\sqrt[n]{p}}{\sqrt[n]{q}}$; $q \neq 0$ Root of a quotient

The proofs of these properties are found in the exercises. They are used to simplify roots of products and quotients.

We consider a radical to be in **simplified form** when the power of any factor in the radicand is less than the index of the radical.

EXAMPLE 3 Simplify each radical.

(a) $\sqrt{75}$ (b) $\sqrt[3]{-16x^3}$

Solutions (a) $\sqrt{75} = \sqrt{5^2 \cdot 3}$ — Factor.

$= \sqrt{5^2} \cdot \sqrt{3}$ — Root of a product

$= 5\sqrt{3}$

(b) $\sqrt[3]{-16x^3} = \sqrt[3]{(-2)^3 \cdot 2 \cdot x^3}$ — Factor.

$= \sqrt[3]{(-2)^3} \cdot \sqrt[3]{2} \cdot \sqrt[3]{x^3}$ — Root of a product

$= -2 \cdot \sqrt[3]{2} \cdot x$

$= -2x\sqrt[3]{2}$ □

Operations with Radicals

Arithmetic with radicals follows the rules for arithmetic with real numbers. We can combine *like terms* in addition and subtraction. Only radicals with the same index and the same radicand can be combined when collecting like terms.

EXAMPLE 4 Perform the operation indicated. Express each answer in simplified form.

(a) $\sqrt{5} - 6\sqrt{5}$ (b) $\sqrt{45x} - \sqrt{20x}$ (c) $\sqrt[3]{-16} + \sqrt[3]{54} - \sqrt[3]{2}$

Solutions (a) $\sqrt{5} - 6\sqrt{5} = -5\sqrt{5}$ — Combine like terms.

(b) $\sqrt{45x} - \sqrt{20x} = \sqrt{9 \cdot 5x} - \sqrt{4 \cdot 5x}$

$= 3\sqrt{5x} - 2\sqrt{5x}$ — Combine like terms.

$= \sqrt{5x}$

(c) $\sqrt[3]{-16} + \sqrt[3]{54} - \sqrt[3]{2} = \sqrt[3]{-8 \cdot 2} + \sqrt[3]{27 \cdot 2} - \sqrt[3]{2}$

$= -2\sqrt[3]{2} + 3\sqrt[3]{2} - \sqrt[3]{2}$ — Combine like terms.

$= 0$ □

The Root of a Product Property shows how to multiply radicals that have the same index.

EXAMPLE 5 Perform the operations indicated. Express the answers in simplified form.

(a) $\sqrt[3]{-2xy} \cdot \sqrt[3]{4x^2y^4}$

(b) $\sqrt{x} \cdot \sqrt{x}$

(c) $\sqrt{3}(\sqrt{7} - \sqrt{3})$

(d) $(\sqrt{2} + 3\sqrt{3})(\sqrt{2} - 3\sqrt{3})$

(e) $(\sqrt{3} - 3\sqrt{7})^2$

Solutions (a) $\sqrt[3]{-2xy} \cdot \sqrt[3]{4x^2y^4} = \sqrt[3]{(-2xy)(4x^2y^4)}$ — Root of a Product Property

$= \sqrt[3]{-8x^3y^5}$

$= -2xy\sqrt[3]{y^2}$

(b) $\sqrt{x} \cdot \sqrt{x} = (\sqrt{x})^2$

$= x$ — Definition of square root

(c) $\sqrt{3}(\sqrt{7} - \sqrt{3}) = \sqrt{3} \cdot \sqrt{7} - \sqrt{3}\sqrt{3}$ — Distributive Property

$= \sqrt{21} - 3$

(d) $(\sqrt{2} + 3\sqrt{3})(\sqrt{2} - 3\sqrt{3}) = (\sqrt{2})^2 - (3\sqrt{3})^2$ — Difference of two squares

$= 2 - 3^2(\sqrt{3})^2$

$= 2 - 9(3)$

$= 2 - 27 = -25$

(e) $(\sqrt{3} - 3\sqrt{7})^2 = (\sqrt{3})^2 - 2 \cdot \sqrt{3} \cdot 3\sqrt{7} + (3\sqrt{7})^2$ — Square a binomial.

$= 3 - 6\sqrt{21} + 63$

$= 66 - 6\sqrt{21}$ □

Expressions such as $\sqrt{2} + 3\sqrt{3}$ and $\sqrt{2} - 3\sqrt{3}$ that occur in part (d) of Example 6 are called **conjugates.** Notice how the product of conjugates became a rational number. This is well to remember, as it occurs often in mathematics.

EXAMPLE 6 Perform the divisions indicated and simplify.

(a) $\dfrac{\sqrt{24}}{\sqrt{3}}$ (b) $\dfrac{\sqrt[3]{-16}}{\sqrt[3]{2}}$

Solutions (a) $\dfrac{\sqrt{24}}{\sqrt{3}} = \sqrt{\dfrac{24}{3}}$ — Root of a Quotient Property

$= \sqrt{8} = 2\sqrt{2}$

(b) $\dfrac{\sqrt[3]{-16}}{\sqrt[3]{2}} = \sqrt[3]{\dfrac{-16}{2}}$ — Root of a Quotient Property

$= \sqrt[3]{-8} = -2$ □

Rationalizing Numerators and Denominators

Consider $\dfrac{\sqrt{7}}{\sqrt{3}}$. We can write $\dfrac{\sqrt{7}}{\sqrt{3}} = \sqrt{\dfrac{7}{3}}$. However, there are many times when neither of these forms is desirable. We can remove a radical from the denominator or numerator by a procedure called **rationalizing.** We will give examples of several techniques for rationalizing numerators or denominators.

The first technique for rationalizing applies if there is a square root factor involved.

EXAMPLE 7 Rationalize the denominator in each of the following expressions.

(a) $\dfrac{2\sqrt{2}}{5\sqrt{6}}$ (b) $\sqrt{\dfrac{7}{8}}$

Solutions (a) $\sqrt{6}$ is a factor in the denominator.

$$\frac{2\sqrt{2}}{5\sqrt{6}} = \frac{2\sqrt{2}\cdot\sqrt{6}}{5\sqrt{6}\cdot\sqrt{6}} = \frac{2\sqrt{12}}{5(6)} = \frac{2\sqrt{4\cdot 3}}{30}$$

$$= \frac{4\sqrt{3}}{30} = \frac{2\sqrt{3}}{15}$$

(b)

$$\sqrt{\frac{7}{8}} = \frac{\sqrt{7}}{\sqrt{8}} = \frac{\sqrt{7}}{2\sqrt{2}}$$

$$= \frac{\sqrt{7}\cdot\sqrt{2}}{2\sqrt{2}\cdot\sqrt{2}} = \frac{\sqrt{14}}{4}$$

□

If a radical of index higher than 2 is a factor of the denominator, we modify the procedure as shown in the next example.

EXAMPLE 8 Rationalize the denominator in $\dfrac{2}{\sqrt[3]{3}}$.

Solution We can change the denominator to $\sqrt[3]{3^3}$ if we multiply by $\sqrt[3]{3^2}$ or $\sqrt[3]{9}$.

$$\frac{2}{\sqrt[3]{3}} = \frac{2\cdot\sqrt[3]{9}}{\sqrt[3]{3}\cdot\sqrt[3]{9}}$$

$$= \frac{2\sqrt[3]{9}}{\sqrt[3]{27}} = \frac{2\sqrt[3]{9}}{3}$$

□

The third technique for rationalizing applies to a sum or difference of two terms containing square roots. We create the special product, *the difference of two squares,* using conjugates.

EXAMPLE 9 Rationalize the numerator or denominator as indicated in each of the following expressions.

(a) $\dfrac{2}{5-\sqrt{3}}$; the denominator (b) $\dfrac{\sqrt{3}+5\sqrt{2}}{\sqrt{3}-5\sqrt{2}}$; the numerator

Solutions (a) Multiply the numerator and denominator by the conjugate of the denominator, $5+\sqrt{3}$.

$$\frac{2}{5-\sqrt{3}}=\frac{2(5+\sqrt{3})}{(5-\sqrt{3})(5+\sqrt{3})}$$
$$=\frac{10+2\sqrt{3}}{5^2-(\sqrt{3})^2}=\frac{10+2\sqrt{3}}{25-3}$$
$$=\frac{10+2\sqrt{3}}{22}=\frac{5+\sqrt{3}}{11}$$

(b) Here, we are to rationalize the *numerator*. We multiply the numerator and denominator by the conjugate of the numerator, $\sqrt{3}-5\sqrt{2}$.

$$\frac{\sqrt{3}+5\sqrt{2}}{\sqrt{3}-5\sqrt{2}}=\frac{(\sqrt{3}+5\sqrt{2})(\sqrt{3}-5\sqrt{2})}{(\sqrt{3}-5\sqrt{2})(\sqrt{3}-5\sqrt{2})}$$
$$=\frac{3-50}{3-10\sqrt{6}+50}=\frac{-47}{53-10\sqrt{6}}$$

□

Rational Exponents

Radicals are used to define rational exponents in a manner consistent with the properties of exponents that we have already studied. Suppose that we wanted to define $2^{1/2}$. Using the existing properties we have $(2^{1/2})^2 = 2^1$ by the second property of exponents. What number squared is 2? It is $\sqrt{2}$ (or $-\sqrt{2}$). This leads to our definition of an exponent of the form $\frac{1}{n}$.

Definition: Exponent of the Form $\frac{1}{n}$

If n is a natural number,

$$x^{1/n}=\sqrt[n]{x};\ x \geq 0 \text{ if } n \text{ is even}$$

EXAMPLE 10 Simplify each expression.

(a) $9^{1/2}$ (b) $27^{1/3}$ (c) $-16^{1/4}$ (d) $(-32)^{1/5}$

Solutions (a) $9^{1/2}=\sqrt{9}=3$

(b) $27^{1/3}=\sqrt[3]{27}=3$

(c) $-16^{1/4}=-\sqrt[4]{16}$ The base is 16, *not* -16.
$=-2$

(d) $(-32)^{1/5} = \sqrt[5]{-32}$ The base is -32.
$= -2$ □

The properties of exponents allow us to include any rational number as an exponent. Notice how we can simplify numbers such as $81^{3/4}$. In fact, there are two approaches.

$$81^{3/4} = (81^{1/4})^3 \qquad 81^{3/4} = (81^3)^{1/4}$$
$$= (\sqrt[4]{81})^3 \qquad = \sqrt[4]{81^3}$$
$$= 3^3 \qquad = \sqrt[4]{531{,}441}$$
$$= 27 \qquad = 27$$

Notice that the first approach is easier in computations than the second approach.

Exponent Form $\Longleftrightarrow$ Radical Form

If $\sqrt[n]{x}$ represents a real number and m is an integer, then

$$x^{m/n} = (\sqrt[n]{x})^m \quad \text{or} \quad x^{m/n} = \sqrt[n]{x^m}$$

EXAMPLE 11 Simplify each expression.

(a) $4^{3/2}$ (b) $4^{-1/2}$ (c) $-8^{-2/3}$ (d) $(-27)^{-2/3}$

Solutions (a) $4^{3/2} = (\sqrt{4})^3 = 2^3 = 8$

(b) $4^{-1/2} = (4^{1/2})^{-1} = \dfrac{1}{4^{1/2}} = \dfrac{1}{\sqrt{4}} = \dfrac{1}{2}$

(c) $-8^{-2/3} = -(\sqrt[3]{8})^{-2}$ The base is 8.

$= -\dfrac{1}{(\sqrt[3]{8})^2} = -\dfrac{1}{2^2} = -\dfrac{1}{4}$

(d) $(-27)^{-2/3} = (\sqrt[3]{-27})^{-2}$ The base is -27.

$= \dfrac{1}{(\sqrt[3]{-27})^2} = \dfrac{1}{(-3)^2} = \dfrac{1}{9}$ □

EXAMPLE 12 Perform the operations indicated. Simplify answers. All variables represent nonnegative real numbers.

(a) $\dfrac{x^{1/2}}{x^{1/3}}$ (b) $(x^{-1/2} - 1)^2$ (c) $\left(\dfrac{x^{1/2}y}{z^{-1/2}}\right)^2$

Solutions (a) $\dfrac{x^{1/2}}{x^{1/3}} = x^{1/2-1/3}$ Quotient with the same base

$= x^{1/6}$

(b) $(x^{-1/2} - 1)^2 = (x^{-1/2})^2 - 2x^{-1/2} + 1$ $(p - q)^2 = p^2 - 2pq + q^2$

$= x^{-1} - 2x^{-1/2} + 1$

$= \dfrac{1}{x} - \dfrac{2}{x^{1/2}} + 1$

(c) $\left(\dfrac{x^{1/2}y}{z^{-1/2}}\right)^2 = \dfrac{(x^{1/2}y)^2}{(z^{-1/2})^2}$ Quotient to a power

$= \dfrac{xy^2}{z^{-1}} = xy^2z$ □

Remembering how we factored polynomials can help us factor expressions with rational exponents and radicals.

EXAMPLE 13 Factor each expression ($x > 0$).

(a) $x - 1$ as the difference of two squares
(b) $x^{2/3} - 2x^{1/3} - 8$ as a trinomial
(c) $x^2 - 7$ as the difference of two squares
(d) $y^3 + 7$ as the sum of two cubes

Solutions (a) We think of x as $(\sqrt{x})^2$.

$x - 1 = (\sqrt{x} + 1)(\sqrt{x} - 1)$ Difference of two squares

(b) Here we think of $x^{2/3}$ as $(x^{1/3})^2$ and factor as a trinomial.

$x^{2/3} - 2x^{1/3} - 8 = (x^{1/3} - 4)(x^{1/3} + 2)$

(c) $x^2 - 7 = (x - \sqrt{7})(x + \sqrt{7})$ Difference of two squares

(d) Since $7 = (\sqrt[3]{7})^3$ we have the sum of two cubes.

$y^3 + 7 = (y + \sqrt[3]{7})(y^2 - \sqrt[3]{7}y + (\sqrt[3]{7})^2)$
$= (y + \sqrt[3]{7})(y^2 - \sqrt[3]{7}y + \sqrt[3]{49})$ □

EXAMPLE 14 Approximate each of the following roots in three significant digits.

(a) $\sqrt{7300}$ (b) $\sqrt[3]{-0.011295}$

Solutions (a) $\sqrt{7300}$

Every scientific calculator has a square root key.

Press [7300] [$\sqrt{\ }$], and the display shows **85.44003745**.

Therefore, $\sqrt{7300} \approx 85.4$.

Some of the superscientific or graphing calculators require the sequence of keystrokes for roots to be the reverse of the above. That is,

[$\sqrt{\ }$] [7300] [ENTER] or [$\sqrt{\ }$] [7300] [EXE]

(b) $\sqrt[3]{-0.011295}$

Many calculators have a cube root key. With such a key, this problem can be done as in part (a).

If not, remember that $\sqrt[3]{A} = A^{1/3}$.

[.011295] [+/−] [y^x] [(] [1] [÷] [3] [)] [=]

The display shows **−.2243686002**

Thus, $\sqrt[3]{-0.011295} \approx -0.224$.

Some calculators will give an error message if a negative base is used with the [y^x] key. If that is the case, the sign can be easily settled without the calculator, and the root found with a positive base. □

PROBLEM SET 1.6

Warm-ups

In Problems 1 through 10, find each root. For Problems 1 through 5, see Example 1.

1. $\sqrt{49}$ **2.** $\sqrt[3]{-27}$ **3.** $\sqrt{-25}$ **4.** $-\sqrt[4]{81}$ **5.** $\sqrt[3]{\frac{8}{27}}$

For Problems 6 through 10, see Example 2.

6. $\sqrt{16t^2}$ **7.** $\sqrt[4]{\frac{t^4}{16}}$ **8.** $\sqrt{z^6}$ **9.** $\sqrt[3]{-8s^3}$ **10.** $\sqrt[3]{\frac{a^6}{b^3}}$

In Problems 11 through 17, simplify each radical. See Example 3.

11. $\sqrt{48}$ **12.** $\sqrt[3]{16x^3}$ **13.** $-2\sqrt[3]{-16x^6}$ **14.** $\sqrt{\frac{x^7}{y^4}}$

15. $\sqrt{(x-y)^2}$ **16.** $\sqrt{x^2-y^2}$ **17.** $\sqrt{\frac{x^2}{y^2}}$

In Problems 18 through 28, perform the operations indicated. For Problems 18 through 21, see Example 4.

18. $2\sqrt{8} - \sqrt{18}$ **19.** $\sqrt[3]{54} - \sqrt[3]{-16} + 2\sqrt[3]{2}$ **20.** $\sqrt{50x} - \sqrt{32x}$ **21.** $2\sqrt{5y} + \sqrt{20y}$

For Problems 22 through 26, see Example 5.

22. $\sqrt{2xy} \cdot \sqrt{6x}$ **23.** $(5\sqrt{3})^2$ **24.** $\sqrt{6}(\sqrt{5} + \sqrt{7})$

25. $(\sqrt{2} - 3\sqrt{3})(\sqrt{2} - \sqrt{3})$ **26.** $(\sqrt{6} - \sqrt{5})^2$

For Problems 27 and 28, see Example 6.

27. $\dfrac{\sqrt{48}}{\sqrt{3}}$

28. $\dfrac{\sqrt[3]{-32}}{\sqrt[3]{-2}}$

In Problems 29 through 35, rationalize as indicated. For Problems 29 through 31, see Example 7.

29. $\dfrac{\sqrt{5}}{\sqrt{3}}$; numerator

30. $\dfrac{3\sqrt{6}}{7\sqrt{2}}$; denominator

31. $\sqrt{\dfrac{5}{18}}$; denominator

For Problems 32 and 33, see Example 8.

32. $\dfrac{1}{\sqrt[3]{9}}$; denominator

33. $\dfrac{7}{\sqrt[4]{4}}$; denominator

For Problems 34 and 35, see Example 9.

34. $\dfrac{4}{7+\sqrt{2}}$; denominator

35. $\dfrac{\sqrt{3}-5\sqrt{2}}{2\sqrt{3}+\sqrt{2}}$; numerator

In Problems 36 through 45, simplify each expression. For Problems 36 through 39, see Example 10.

36. $25^{1/2}$

37. $8^{1/3}$

38. $(-27)^{1/3}$

39. $-16^{1/2}$

For Problems 40 through 45, see Example 11.

40. $8^{2/3}$

41. $-4^{3/2}$

42. $(-27)^{2/3}$

43. $16^{-1/2}$

44. $8^{-2/3}$

45. $(-8)^{-2/3}$

In Problems 46 through 49, perform the operation indicated. See Example 12.

46. $x^{1/4} \cdot x^{1/2}$

47. $x^{-1/2}(x^{1/3} + x^{1/2})$

48. $(x^{1/2} + 3)^2$

49. $\left(\dfrac{x^{1/3}y^{1/2}}{z^2}\right)^3$

In Problems 50 through 52, factor each expression. See Example 13.

50. $x^2 - 5$

51. $x^{2/5} - 3x^{1/5} - 10$

52. $11 - y^3$

In Problems 53 through 56, approximate each to three significant digits. See Example 14.

53. $\sqrt{3457}$

54. $\sqrt[3]{-0.98421}$

55. $\sqrt[5]{35}$

56. $\sqrt[6]{3590.4329}$

Practice Exercises

In Problems 57 through 80, simplify each expression if possible.

57. $\sqrt{-100}$

58. $\sqrt[4]{-81}$

59. $\sqrt[4]{32s^7t^8}$

60. $\sqrt[3]{-54x^7y^3}$

61. $-\sqrt{48x^2}$

62. $-\sqrt{125t^5}$

63. $\sqrt[3]{\dfrac{x^4}{8y^6}}$

64. $\sqrt[4]{\dfrac{16x^4}{y^8}}$

65. $\sqrt{x^2+4}$

66. $\sqrt[3]{8-y^3}$

67. $\sqrt[4]{(x-y)^4}$

68. $\sqrt{(x+2)^2}$

69. $\sqrt{4x^2}$

70. $\sqrt{135x^4y^5}$

71. $25^{1/2}$

72. $-25^{1/2}$

73. $-25^{-1/2}$

74. $25^{-1/2}$

75. $(-25)^{1/2}$

76. $(-25)^{-1/2}$

77. $125^{-2/3}$

78. $125^{2/3}$

79. $-125^{-2/3}$

80. $(-125)^{2/3}$

In Problems 81 through 112, perform the operations indicated and simplify the answers.

81. $\sqrt{50} - \sqrt{18} + \sqrt{200}$

82. $\sqrt[3]{-81} - \sqrt[3]{24} + 2 \cdot \sqrt[3]{192}$

83. $2\sqrt{20x} - 3\sqrt{45x}$

84. $11\sqrt{27xz} - 7\sqrt{12xz}$

85. $\sqrt[3]{x^7y} - 2x\sqrt[3]{-8x^4y} + x^2\sqrt[3]{27xy}$

86. $2t\sqrt[4]{16s^{12}t} + st\sqrt[4]{81s^8t} - s^2t\sqrt[4]{s^4t}$

87. $\sqrt{3xy} \cdot \sqrt{2xy}$

88. $\sqrt{5ry} \cdot \sqrt{10ry}$

89. $(\sqrt{5x^2})^2$

90. $(3\sqrt{3y^3})^2$

91. $\sqrt[4]{9x^2} \cdot \sqrt[4]{9x^6}$

92. $\sqrt[3]{-2a} \cdot \sqrt[3]{8a^2}$

93. $2\sqrt{3}(\sqrt{3} - 2)$

94. $4\sqrt{6}(\sqrt{2} - 2)$

95. $(7\sqrt{6} - 2\sqrt{3})(2\sqrt{6} + \sqrt{3})$

96. $(\sqrt{5} - 2\sqrt{3})(\sqrt{5} - \sqrt{3})$

97. $(\sqrt{2} + 1)(\sqrt{2} - 1)$

98. $(\sqrt{11} + \sqrt{7})(\sqrt{11} - \sqrt{7})$

99. $(2\sqrt{3} + \sqrt{5})^2$

100. $(5\sqrt{2} - 2\sqrt{5})^2$

101. $\dfrac{\sqrt{125}}{\sqrt{5}}$

102. $\dfrac{\sqrt[3]{-3}}{3\sqrt{81}}$

103. $x^{1/3} \cdot x^{1/2}$

104. $x^{1/4} \cdot x^{-1/2}$

105. $x^{-1/3}(x^{-1/3} - x^{1/3})$

106. $x^{2/3}(1 - x)$

107. $(x^{1/2} + x^{-1/2})^2$

108. $(x - x^{1/2})^2$

109. $\left(\dfrac{x^{-2}}{y^{1/2}z^{-1/3}}\right)^6$

110. $\left(\dfrac{a^2b^4}{c^{-4}}\right)^{1/2}$

111. $(x^2y^{1/2})^{-2}$

112. $(2s^{1/3}t)^{-3}$

In Problems 113 through 124, rationalize as indicated.

113. $\dfrac{1}{\sqrt{8}}$; denominator

114. $\dfrac{1}{\sqrt{18}}$; denominator

115. $\dfrac{5}{\sqrt[3]{3}}$; denominator

116. $\dfrac{4}{\sqrt[3]{2}}$; denominator

117. $\dfrac{\sqrt{7} - 1}{6}$; numerator

118. $\dfrac{\sqrt{5} - \sqrt{3}}{2}$; numerator

119. $\dfrac{3}{\sqrt{2} + \sqrt{3}}$; denominator

120. $\dfrac{5}{\sqrt{7} - \sqrt{2}}$; denominator

121. $\dfrac{\sqrt{2 + h} - \sqrt{2}}{h}$; numerator

122. $\dfrac{\sqrt{x + h} - \sqrt{x}}{h}$; numerator

123. $\dfrac{2 - \sqrt{x_1}}{4 - x_1}$; numerator

124. $\dfrac{\sqrt{7} - \sqrt{t_1}}{7 - t_1}$; numerator

In Problems 125 through 130, guess which expression is the larger one. Check with a calculator.

125. $2^{-1/2}$; $2^{-1/3}$

126. $(\sqrt{2})^{\sqrt{2}}$; $(\sqrt{3})^{\sqrt{3}}$

127. $\sqrt{\dfrac{1}{7}}$; $\dfrac{1}{7}$

128. $\sqrt{\dfrac{2}{3}}$; $\sqrt[3]{\dfrac{2}{3}}$

129. $\pi^{1/2}$; $\left(\dfrac{22}{7}\right)^{1/2}$

130. $125^{1/125}$; $225^{1/225}$

131. Approximate $\dfrac{5}{\sqrt{7} - \sqrt{2}}$ and approximate $\sqrt{7} + \sqrt{2}$. Just looking at a calculator, is it possible to decide if the expressions are in fact equal? Look at Problem 120.

132. For what numbers is $\sqrt{x}$ greater than x? (*Hint:* Look at Problem 127.)

133. In a certain region the number of people with an income exceeding x dollars can be calculated by the expression $\dfrac{14.7 \times 10^{10}}{x^{3/2}}$. How many people have incomes over \$25,000? (This is Pareto's Law of Distribution of Income.)

134. Approximate $\sqrt[6]{335}$ and approximate $\sqrt{\sqrt[3]{335}}$ and $\sqrt[3]{\sqrt{335}}$. What seems to be true about $\sqrt[m]{\sqrt[n]{x}}$? Prove it.

135. Use $\sqrt{x^2} = |x|$ to prove the Absolute Value Property $|s|^2 = s^2$.

136. Use $\sqrt{x^2} = |x|$ to prove the Absolute Value Property $|s - t| = |t - s|$.

The properties of exponents are useful in Problems 137 and 138.

137. Prove the root of a product formula; if $\sqrt[n]{p}$ and $\sqrt[n]{q}$ are real numbers, then

$$\sqrt[n]{pq} = \sqrt[n]{p} \cdot \sqrt[n]{q}$$

138. Prove the root of a quotient formula; if $\sqrt[n]{p}$ and $\sqrt[n]{q}$ are real numbers, then

$$\sqrt[n]{\frac{p}{q}} = \frac{\sqrt[n]{p}}{\sqrt[n]{q}}$$

Challenge Problems

139. Work Problems 123 and 124 by factoring.

140. The average or arithmetic mean of two numbers a and b is given by $\frac{a+b}{2}$. The geometric mean is given by $(ab)^{1/2}$. Prove that for positive numbers a and b,

$$\frac{a+b}{2} \geq (ab)^{1/2}$$

IN YOUR OWN WORDS . . .

141. Explain why $\frac{2}{\sqrt{2}}$ is $\sqrt{2}$ by discussing how to rationalize the denominator and then show that $\frac{2}{\sqrt{2}} = \sqrt{2}$ by using laws of exponents. Which method is best to use?

1.7 COMPLEX NUMBERS

We have learned that the radical $\sqrt{Q}$ is a representation of the nonnegative number whose square is Q. Since there are no real numbers whose squares are negative, $\sqrt{Q}$ was defined only for nonnegative radicands. However, in order to have solutions to equations like $x^2 + 1 = 0$, we need a definition for $\sqrt{-1}$. Clearly, as no real number has a square of -1, we must enlarge our number system to include a "number" whose square is -1 if there is to be any consistency in our use of the symbol $\sqrt{}$.

This new number whose square is -1 we call i.

Definition: *i*

i is a number such that

$$i^2 = -1$$

The new set of numbers, the real numbers and the new number i, form a new number system. If it is to have all the nice properties we have come to expect and appreciate, it must be *closed*. That is, when two numbers are added or multiplied, a number from the number system should be the result. To achieve closure in this system, we must add still more new numbers to the system. Closure is obtained if we add *every* number of the form $a + bi$ to the system, where a and b are real numbers. We call this new, enlarged number system the set of **complex numbers.**

Definition: The Set of Complex Numbers

The set of complex numbers is the set of all numbers that can be written as

$$a + bi$$

where a and b are real numbers and i has the property that

$$i^2 = -1$$

Some examples of complex numbers are

$$\pi + \sqrt{11}i \qquad 4i \qquad -8$$

Notice that $4i$ is a complex number because it can be written in the form $0 + 4i$. Also, as -8 may be written in the form $-8 + 0 \cdot i$, it is also a complex number. As a matter of fact, *all real numbers* are also complex numbers. Thus the set of real numbers is a subset of the set of complex numbers.

Now we have a sensible definition of $\sqrt{-1}$. $\sqrt{-1} = i$.

Definition: Negative Radicand

If k is any positive real number,

$$\sqrt{-k} = \sqrt{k}i$$

Notice that the i is *not* under the radical. To emphasize this, $\sqrt{k}i$ may be written as $i\sqrt{k}$. Thus, if k is a positive number, then the negative number $-k$ has two square roots, $\sqrt{k}i$ and $-\sqrt{k}i$. We call $\sqrt{k}i$ the **principle square root** of $-k$. Every negative real number has a principle square root in the set of complex numbers.

When working with negative radicands it is **important** to rewrite them with positive radicands **before doing any arithmetic!** The rules of arithmetic "work" only with nonnegative radicands. So, we must rewrite $\sqrt{-2}$ as $\sqrt{2}i$ before we use these rules. For example,

$\sqrt{-2} \cdot \sqrt{-3}$ **does not** equal $\sqrt{(-2)(-3)}$ which is $\sqrt{6}$

$\sqrt{-2} \cdot \sqrt{-3}$ equals $\sqrt{2}i \cdot \sqrt{3}i$ which is $\sqrt{6}i^2$ or $\sqrt{6}(-1)$ or $-\sqrt{6}$

Notice the importance of rewriting $\sqrt{-k}$ as $\sqrt{k}i$ **before** doing any arithmetic.

A complex number written in the form $a + bi$, where a and b are real numbers, is said to be in **standard form.** The real number a in the standard form is called the **real part** of the complex number, and the real number b is called the **imaginary part.**

EXAMPLE 1 Identify the real part and the imaginary part of each complex number.

(a) $11 + 6i$ (b) $7 - \sqrt{-3}$ (c) $\sqrt{-18}$ (d) $\sqrt{18}$

Solutions (a) $11 + 6i$

The real part is 11, and the imaginary part is 6.

(b) $7 - \sqrt{-3} = 7 - \sqrt{3}i$

$= 7 + (-\sqrt{3})i$

The real part is 7, and the imaginary part is $-\sqrt{3}$.

(c) $\sqrt{-18} = \sqrt{18}i$

$= 3\sqrt{2}i$

$= 0 + 3\sqrt{2}i$

The real part is 0, and the imaginary part is $3\sqrt{2}$.

(d) $\sqrt{18} = 3\sqrt{2} + 0 \cdot i$

The real part is $3\sqrt{2}$, and the imaginary part is 0. □

Powers of i

We know that $i^2 = -1$. What about the other powers of i? Using the definitions and properties of exponents we see

$$i^0 = 1$$
$$i^1 = i$$
$$i^2 = -1$$
$$i^3 = i^2 \cdot i = (-1)i = -i$$
$$i^4 = i^3 \cdot i = -i \cdot i = -i^2 = -(-1) = 1$$
$$i^5 = i^4 \cdot i = 1 \cdot i = i$$
$$i^6 = i^5 \cdot i = i \cdot i = -1$$
$$i^7 = i^6 \cdot i = -1 \cdot i = -i$$
$$i^8 = i^7 \cdot i = -i \cdot i = -i^2 = -(-1) = 1$$
$$\vdots$$

Notice that

$$i^0 = i^4 = i^8 = i^{12} = \cdots = 1$$
$$i^1 = i^5 = i^9 = i^{13} = \cdots = i$$
$$i^2 = i^6 = i^{10} = i^{14} = \cdots = -1$$
$$i^3 = i^7 = i^{11} = i^{15} = \cdots = -i$$

The powers of i repeat in groups of four. This pattern suggests that we can find any whole-number power of i by dividing the exponent by 4 and examining the remainder. Notice that 15 divided by 4 leaves a remainder of 3, and $i^{15} = i^3$. In the same manner, 14 divided by 4 leaves a remainder of 2, and $i^{14} = i^2$.

EXAMPLE 2 Evaluate each power of i.

(a) i^{19} (b) i^{77} (c) i^{464} (d) $(-i)^{26}$

Solutions (a) Four divides into 19 four times, with a remainder of 3, so

$$i^{19} = i^3 = -i$$

(b) Division of 77 by 4 gives a remainder of 1, so

$$i^{77} = i^1 = i$$

(c) Division of 464 by 4 gives a remainder of 0, so

$$i^{464} = i^0 = 1$$

(d) $(-i)^{26} = i^{26}$ since 26 is an *even* exponent. Now, if we divide 26 by 4, we have a remainder of 2.

$$= i^2 = -1$$

□

Operations with Complex Numbers

In order for the set of complex numbers to form a meaningful set of numbers, we need a few definitions.

Definitions: Complex Numbers

Let $a + bi$ and $c + di$ be complex numbers.

Equality

$$a + bi = c + di \text{ if and only if } a = c \text{ and } b = d$$

Addition

$$(a + bi) + (c + di) = (a + c) + (b + d)i$$

Subtraction

$$(a + bi) - (c + di) = (a - c) + (b - d)i$$

Multiplication

$$(a + bi) \cdot (c + di) = (ac - bd) + (ad + bc)i$$

(We will discuss division separately.)

Rather than memorize the definitions above, we usually do arithmetic with complex numbers the same way we do arithmetic with polynomials, remembering to replace powers of i by 1, -1, i, or $-i$.

EXAMPLE 3 Suppose $Z = 3 + 5i$ and $W = 2 - 3i$. Perform the operations indicated.

(a) $Z + W$ (b) $Z - W$ (c) ZW (d) $(Z + W)(Z - W)$
(e) $Z^2 - W^2$ (f) $(Z + W)^2$ (g) $Z^2 + 2ZW + W^2$

Solutions (a) $Z + W = (3 + 5i) + (2 - 3i)$
$= 5 + 2i$

(b) $Z - W = (3 - 5i) - (2 - 3i)$
$= (3 + 5i) + (-2 + 3i)$
$= 1 + 8i$

(c) $ZW = (3 + 5i)(2 - 3i)$
$= 6 - 9i + 10i - 15i^2$
$= 6 + i - 15(-1)$
$= 21 + i$

(d) $(Z + W)(Z - W) = (5 + 2i)(1 + 8i)$
$= 5 + 40i + 2i + 16i^2$
$= 5 + 42i + 16(-1)$
$= -11 + 42i$

(e) $Z^2 - W^2 = (3 + 5i)^2 - (2 - 3i)^2$
$= 9 + 30i + 25i^2 - (4 - 12i + 9i^2)$
$= 5 + 42i + 16(-1)$
$= -11 + 42i$

(f) $(Z + W)^2 = (5 + 2i)^2$
$= 25 + 20i + 4i^2$
$= 21 + 20i$

(g) $Z^2 + 2ZW + W^2 = (3 + 5i)^2 + 2(21 + i) + (2 - 3i)^2$
$= (9 + 30i + 25i^2) + (42 + 2i) + (4 - 12i + 9i^2)$
$= 55 + 20i + 34(-1)$
$= 21 + 20i$ □

Notice from the demonstration in parts (d), (e), (f), and (g) that the special products apply for complex numbers.

$$(Z + W)(Z - W) = Z^2 - W^2$$
$$(Z + W)^2 = Z^2 + 2ZW + W^2$$

The complex numbers $a + bi$ and $a - bi$ are called **complex conjugates.** Note that in standard form conjugates differ only in their middle sign. We denote the conjugate of the complex number Z by $\overline{Z}$. That is, if Z is $-11 - 4i$, then $\overline{Z}$ is $-11 + 4i$.

EXAMPLE 4 Suppose $Z = -11 - 4i$ and $W = p + qi$ (p and q real numbers). Perform the operations indicated.

(a) $\overline{Z}^2$ (b) $W\overline{W}$

Solutions (a) $\overline{Z}^2 = (-11 + 4i)^2$
$= 121 - 88i + 16i^2$
$= 105 - 88i$

(b) $W\overline{W} = (p + qi)(p - qi)$
$= p^2 - (qi)^2$
$= p^2 - q^2(-1)$
$= p^2 + q^2$ □

Notice from part (b) in the above example that the product of a complex number and its conjugate is always a real number and, in fact, *a nonnegative real number.*

The Product of Complex Conjugates

$$(a + bi)(a - bi) = a^2 + b^2$$

Division

We divide complex numbers by writing the quotient as a fraction; then we rationalize the denominator and write the resulting complex number in standard form.

EXAMPLE 5 Perform the operation indicated.

(a) $(2 - 3i) \div (4 + 5i)$ (b) i^{-63}

Solutions (a) $(2 - 3i) \div (4 + 5i) = \dfrac{2 - 3i}{4 + 5i}$

Now we rationalize the denominator by multiplying both the numerator and the denominator by the *conjugate of the denominator.*

$$= \frac{(2-3i)(4-5i)}{(4+5i)(4-5i)}$$

$$= \frac{8-10i-12i+15i^2}{4^2+5^2}$$ The product of complex conjugates

$$= \frac{-7-22i}{41}$$

$$= -\frac{7}{41} - \frac{22}{41}i$$

(b) $i^{-63} = \frac{1}{i^{63}}$ Definition of negative exponent

$= \frac{1}{i^3}$ 63 divided by 4 leaves a remainder of 3.

$= \frac{1}{-i}$ $i^3 = -i$

$= \frac{1 \cdot i}{-i \cdot i} = \frac{i}{1} = i$ Rationalize. The conjugate of $-i$ is i. □

PROBLEM SET 1.7

Warm-ups

In Problems 1 through 8, write each complex number in standard form and identify the real and imaginary parts. See Example 1.

1. $4 + \sqrt{-4}$ **2.** $\sqrt{-49}$ **3.** $\sqrt{81} - 8$ **4.** $-\sqrt{-72}$

5. $-\sqrt{-12} - \sqrt{12}$ **6.** $1 + (\sqrt{-2})^2$ **7.** $1 + \sqrt{(-2)^2}$ **8.** $(\sqrt{-2})^3$

In Problems 9 through 12, evaluate each power of i. See Example 2.

9. i^{34} **10.** i^{1271} **11.** $(-i)^{41}$ **12.** $(-i)^{40}$

In Problems 13 through 30, perform the operations indicated. For Problems 13 through 16, see Example 3.

13. $(-4 + 3i) + (3 + 5i)$ **14.** $(-4 + 3i) - (3 + 5i)$

15. $(-4 + 3i)(3 + 5i)$ **16.** $(-4 + 3i)^2$

In Problems 17 through 24, let $C = 1 - 7i$ and $D = -2 + 4i$. See Examples 3 and 4.

17. $D + \overline{D}$ **18.** $D - \overline{D}$ **19.** C^2 **20.** $\overline{C}C$

21. $\overline{C} - D$ **22.** $(C - \overline{D})^2$ **23.** $C^2 - \overline{D}^2$ **24.** $C\overline{C} + D\overline{D}$

For Problems 25 through 30, see Example 5.

25. $(5 - 2i) \div (5 + 2i)$ **26.** $\frac{3i}{2 - 5i}$

27. i^{-365} **28.** $(17 - 12i) \div i$

29. $\frac{1}{1 + i}$ **30.** $\frac{6 + 3i}{3}$

Practice Exercises

In Problems 31 through 74, perform the operation indicated.

31. $(12 - 8i) + (3 + 5i)$
32. $(4 + 9i) + (-5 + i)$
33. $(-1 + 3i)(2 - i)$
34. $(7 + i)(-2 + 2i)$
35. $(8 - 9i) - (-8 + 3i)$
36. $(11 - 6i) - (-4 + 6i)$
37. $\sqrt{-9}(1 + \sqrt{-5})$
38. $(2 - \sqrt{-7})\sqrt{-4}$
39. $(16 - 3i) \div (2 + 3i)$
40. $(5 + 12i) \div (1 - 2i)$
41. $(4 - 3i)i$
42. $2i(6 + 11i)$
43. $\dfrac{4 - 3i}{i}$
44. $\dfrac{6 + 11i}{2i}$
45. $(5 + 2i)^2$
46. $(2 - 3i)^2$
47. $(\sqrt{-2} + \sqrt{18})\sqrt{-2}$
48. $(\sqrt{12} - \sqrt{-3})\sqrt{-3}$
49. $(1 - \sqrt{-2})^2$
50. $(\sqrt{-3} + 2)^2$
51. $\sqrt[3]{-8} - \sqrt{-8}$
52. $\sqrt{-27} + \sqrt[3]{-27}$
53. $(2 + i)^2 \cdot i$
54. $(1 - 2i)^2 \cdot 2i$
55. $(1 - 2i)^{-2}$
56. $(2 + i)^{-2}$
57. $\sqrt{-18} - \sqrt{-8}$
58. $\sqrt{-12} + \sqrt{-27}$
59. $(1 + \sqrt{-12})(1 - \sqrt{-12})^{-1}$
60. $(2 + \sqrt{-8})^{-1}(2 - \sqrt{-8})$

In Problems 61 through 74, let $Z_1 = 3 + 2i$, $Z_2 = 2 - 3i$, and $Z_3 = 1 - i$.

61. Z_1Z_2
62. Z_1Z_3
63. $Z_1\overline{Z}_1$
64. $Z_2\overline{Z}_2$
65. $Z_1^2 - Z_2^2$
66. $Z_2^2 - Z_3^2$
67. $(Z_1 - Z_2)(Z_1 + Z_2)$
68. $(Z_2 + Z_3)(Z_2 - Z_3)$
69. $(Z_1 - Z_2)^2$
70. $(Z_2 - Z_3)^2$
71. $Z_1Z_3^{-1}$
72. $Z_3Z_2^{-1}$
73. Z_1^3
74. Z_3^4

75. Show that for all nonzero complex numbers Z, $Z\overline{Z}$ is a positive, real number.

76. Prove that the sum of the conjugates of two complex numbers is equal to the conjugate of the sum of the complex numbers.

77. Prove that the product of the conjugates of two complex numbers is equal to the conjugate of the product of the complex numbers.

Challenge Problems

78. Using complex numbers, factor $x^2 + 4$. [*Hint:* $(2i)^2 = -4$.]

79. Using complex numbers, factor $x^2 + \sqrt{7}$.

In Problems 80 through 83, write each power of i as -1, 1, $-i$, or i if possible. If not possible explain why. (k an integer)

80. i^{4k}
81. i^{4k+2}
82. i^{2k}
83. i^{2k+1}

In Problems 84 through 87, let $Z_1 = 1$, $Z_2 = -\dfrac{1}{2} + \dfrac{\sqrt{3}}{2}i$, and $Z_3 = -\dfrac{1}{2} - \dfrac{\sqrt{3}}{2}i$.

84. Find Z_1^3.
85. Find Z_2^3.
86. Find Z_3^3.

87. From the results of Problems 84, 85, and 86 discuss the complex cube roots of 1.

IN YOUR OWN WORDS . . .

88. What was the reason for introducing the new number i?

89. Describe the relationship between the sets of real numbers, rational numbers, irrational numbers, integers, natural numbers, and the set of complex numbers.

CHAPTER SUMMARY

GLOSSARY

Additive identity: The number zero is the additive identity.

Base in the expression x^n: The base is x.

Binomial: A sum of two monomials.

Coefficient in the expression bx^n: The coefficient is b.

Complex numbers: Numbers that can be written in the form $a + bi$, where a and b are real numbers and i has the property that $i^2 = -1$.

Conjugate of $a + bi$ is $a - bi$.

Element or member of a set: The objects in the set. The symbol $\in$ means "is an element of."

Empty or null set: A set with no elements, written as $\{\ \}$ or $\varnothing$.

To **evaluate** a polynomial: To find the value of the polynomial given number(s) for the variable(s).

Exponent in the expression x^n: The exponent is n.

To **factor** a polynomial: To write the polynomial as a product.

To **factor** a polynomial **completely:** To write the polynomial as a product of prime polynomials.

The **intersection** of sets A and B is the set of all elements that belong to both A and B and is written as $A \cap B$.

Monomial in one variable: An expression of the form bx^n, where n is a whole number and b is a real number. The **degree** is n.

Multiplicative identity: The number 1 is the multiplicative identity.

Opposite or **additive inverse of p:** The number that when added to p gives a sum of 0.

Polynomial: A sum of monomials.

Rational expression: The quotient of two polynomials.

Reciprocal or **multiplicative inverse of p:** The number that when multiplied by p gives a product of 1.

Set: A collection of objects.

Standard form of a polynomial: The monomial of highest degree is written first, and other monomials are written in decreasing order of degree.

Subset of a set A: A set with the property that each of its elements is in set A. B is a subset of A is written as $B \subseteq A$.

Trinomial: A sum of three monomials.

The **union** of set A and B is the set of all members that belong either to A or B and is written as $A \cup B$.

PROPERTIES OF ADDITION AND MULTIPLICATION OF REAL NUMBERS

CLOSURE	$p + q$ is a real number. pq is a real number.
COMMUTATIVE	$p + q = q + p$ $pq = qp$
ASSOCIATIVE	$p + (q + r) = (p + q) + r$ $p(qr) = (pq)r$
IDENTITY	$p + 0 = p = 0 + p$ $p \cdot 1 = p = 1 \cdot p$

INVERSE $p + (-p) = 0 = (-p) + p$

$$p \cdot \frac{1}{p} = 1 = \frac{1}{p} \cdot p;\ p \neq 0$$

DISTRIBUTIVE $p(q + r) = pq + pr$

ORDER OF OPERATIONS

If grouping symbols are present, perform operations inside them, starting with the innermost symbol of grouping in the following order.

1. Perform any exponentiations.
2. Perform all multiplications and divisions in order from left to right.
3. Perform all subtractions and additions from left to right.

If grouping symbols are not present, perform operations in the order given above.

ABSOLUTE VALUE

The absolute value of a real number p is its distance from zero.

$$|p| = \begin{cases} p & \text{if } p \geq 0 \\ -p & \text{if } p < 0 \end{cases}$$

PROPERTIES OF ABSOLUTE VALUE

1. $|p| \geq 0$
2. $|p \cdot q| = |p||q|$
3. $\left|\frac{p}{q}\right| = \frac{|p|}{|q|};\ q \neq 0$
4. $|p| = |-p|$
5. $|p - q| = |q - p|$
6. $|p|^2 = p^2$

EXPONENT DEFINITIONS

If n is a natural number,

$$x^n = x \cdot x \cdot x \cdots x \ (n \text{ factors of } x)$$

$$x^0 = 1;\ x \neq 0$$

$$x^{-n} = \frac{1}{x^n};\ x \neq 0$$

$$x^{1/n} = \sqrt[n]{x};\ x \geq 0 \text{ if } n \text{ is even.}$$

PROPERTIES OF EXPONENTS

1. $x^m \cdot x^n = x^{m+n}$ Product with same base
2. $(x^m)^n = x^{mn}$ Power of a power
3. $(xy)^n = x^n y^n$ Product to a power

4. $\left(\dfrac{x}{y}\right)^n = \dfrac{x^n}{y^n}$; $y \neq 0$ Quotient to a power

5. $\dfrac{x^m}{x^n} = x^{m-n}$; $x \neq 0$ Quotient with same base

FACTORING FORMULAS

Square of a binomal

$$(p + q)^2 = p^2 + 2pq + q^2$$

$$(p - q)^2 = p^2 - 2pq + q^2$$

Difference of two squares

$$(p + q)(p - q) = p^2 - q^2$$

Sum or difference of two cubes

$$(p + q)(p^2 - pq + q^2) = p^3 + q^3$$

$$(p - q)(p^2 + pq + q^2) = p^3 - q^3$$

***n*TH ROOTS OF REAL NUMBERS**

If n is an even natural number,

1. $\sqrt[n]{x}$ is a real number only if $x \geq 0$.
2. $\sqrt[n]{x}$ is the positive real number such that $(\sqrt[n]{x})^n = x$.

If n is an odd natural number,

1. $\sqrt[n]{x}$ is a real number for all real numbers x.
2. $\sqrt[n]{x}$ is the real number such that $(\sqrt[n]{x})^n = x$.
3. $\sqrt[n]{-x} = -\sqrt[n]{x}$

DEFINITION OF *i*

i is a number such that $i^2 = -1$.

SQUARE ROOTS OF NEGATIVE NUMBERS

If $k > 0$, then $\sqrt{-k} = \sqrt{k}i$

REVIEW PROBLEMS

In Problems 1 through 8, simplify each expression. Do not leave zero or negative exponents or absolute values in the answer.

1. $\dfrac{-2x^{-1}}{y^{-3}}$

2. $\left|\dfrac{-\pi^2}{4}\right|$

3. $-(-x^{-2}y)^0$

4. $|x^2 + 4|$

5. $\sqrt[3]{27x^3}$ **6.** $\sqrt{16x^4}$ **7.** $-8^{2/3}$ **8.** $(-27)^{-2/3}$

In Problems 9 through 41, perform the indicated operations.

9. $-5^2 - \dfrac{-14}{2} - (-3)^2$ **10.** $|12 - 15|^2 - \sqrt{3^2 + 4^2}$ **11.** $(-3x^4)^2$

12. $\dfrac{x^2 + 2x}{k^4 - 4} \cdot \dfrac{5k^3 - 10k}{x^2 - 4}$ **13.** $\dfrac{3y - 7}{y} \div (7 - 3y)$ **14.** $(3x^2 + 5) + (x^3 - x^2 - 7)$

15. $(2t - 5)^2$ **16.** $(7x + 3) - (16x - 1)$ **17.** $(x^2 - 3)(x + 2)$

18. $\dfrac{3}{x^2 + x - 2} - \dfrac{1}{x - 1}$ **19.** $\sqrt{54} - 2\sqrt{24}$ **20.** $\sqrt[3]{6x} \cdot \sqrt[3]{4x^5}$

21. $\dfrac{3 + 4i}{3 - i}$ **22.** $\dfrac{3t^4 - 25t^2 + 4t + 5}{t^2 - 3t}$ **23.** $(1 + i)(1 - i)$

24. $(2\sqrt{3} - \sqrt{2})(\sqrt{2} - \sqrt{3})$ **25.** $\sqrt{6}(2\sqrt{6})^2$ **26.** $\dfrac{\sqrt{12}}{\sqrt{3}}$

27. $x^{1/2}(3 - x^{-1/2})$ **28.** $\left(\dfrac{-t^{1/3}}{s^{-1/3}}\right)^3$ **29.** $(y^{1/4} + 1)^2$

30. $2\sqrt{-2}(5 - \sqrt{-8})$ **31.** $(2r + 1)^3$ **32.** $\dfrac{1}{2 - i}$

33. $-25^{1/2} + 4^{3/2}$ **34.** $\dfrac{8 + k^3}{1 - m} \div \dfrac{2 + k}{m - 1}$ **35.** $i^7 \cdot i^{51}$

36. $\sqrt{-27} \cdot \sqrt{-48}$ **37.** $125^{1/3} - (-8^{-2/3})$ **38.** $(2 + \sqrt{3}i) \cdot (3i)$

39. $\dfrac{4x^3 + 8x^2 - 7x + 11}{2x^2 - 3x + 1}$ **40.** $3i^{15}(6 - 7i)$ **41.** $\dfrac{4r^3 - 3r + 1}{4r}$

In Problems 42 through 45, evaluate each expression when x is -3.

42. $x^3 - x^2 + 2x$ **43.** $x^{-1} + x^{-2}$ **44.** $\dfrac{x^2 - 16}{x + 3}$ **45.** $\sqrt{-3x^3}$

In Problems 46 through 51, factor each polynomial completely.

46. $x^8 - 25$ **47.** $x^4 - 2x^2 + 1$ **48.** $-6x^5 + 3x$ **49.** $x^2 - 3x - 4$

50. $8a^3 + 125$ **51.** $2x^4 - 2x^3 + 8x^2 - 8x$

In Problems 52 through 55, simplify each complex fraction.

52. $\dfrac{x^{-1} - 1}{x^{-1} + 1}$ **53.** $\dfrac{x^{-1} - 2y^{-2}}{x^{-2} + y^{-1}}$ **54.** $\dfrac{\dfrac{1}{a} + \dfrac{1}{b}}{\dfrac{a^2 - ab - 2b^2}{ab}}$ **55.** $\dfrac{\dfrac{x}{x + 3} - \dfrac{2}{x - 3}}{\dfrac{x - 6}{x^2 - 9}}$

In Problems 56 through 63, rationalize the denominator or numerator as indicated.

56. $\dfrac{4}{\sqrt{2}}$; denominator **57.** $\dfrac{\sqrt{3}}{\sqrt{6}}$; numerator **58.** $\dfrac{5}{3 - \sqrt{5}}$; denominator

59. $\dfrac{1 + \sqrt{2}}{6}$; numerator **60.** $\dfrac{\sqrt{7}}{\sqrt{2} - 3}$; numerator **61.** $\dfrac{\sqrt{7}}{\sqrt{2} - 3}$; denominator

62. $\dfrac{\sqrt{3} - \sqrt{5}}{\sqrt{3} + \sqrt{5}}$; numerator **63.** $\dfrac{\sqrt{3} - \sqrt{5}}{\sqrt{3} + \sqrt{5}}$; denominator

In Problems 64 through 67, approximate each expression to six decimal places.

64. $(45.6 - 3\pi^2)^3$

65. $\sqrt[5]{134.5398}$

66. $\dfrac{-5.69 + \sqrt{14.2^2 - 4(2.33)(-14.6)}}{2(2.33)}$

67. $\dfrac{2.317 \times 10^6 - 14.854 \times 10^{-5}}{(4.5 \times 10^{-4})(11.47 \times 10^9)}$

LET'S NOT FORGET . . .

68. What form is more useful?

(a) $\sqrt{50}$ or $5\sqrt{2}$ if the problem asks for an approximation.

(b) $x^2 - x - 6$ (multiplied) or $(x - 3)(x + 2)$ (factored) if the problem is $\dfrac{1}{x^2 - x - 6} + \dfrac{1}{x + 2}$.

69. Watch the role of negative signs.

(a) $-4^{-2} = ?$

(b) $-8^{-2/3} = ?$

70. From memory.

(a) $(3x - 7)^2 = ?$

(b) $\sqrt{x^2} = ?$

(c) $27u^3 - s^3 = ?$

71. With a calculator approximate each expression to four decimals.

(a) 124.578^7

(b) $\sqrt[3]{-0.00325}$

(c) $\dfrac{2.34 \times 10^6 - 5.578 \times 10^{-5}}{7.2^3 + 12.1^{-4}}$

CHAPTER 2

Equations and Inequalities in One Variable

CONNECTIONS

For nearly 4000 years mathematicians have attempted to solve problems using equations and inequalities as models. The Babylonians made the earliest attempts using the rhetorical approach: writing equations in words instead of symbols. The earliest description of how to solve equations by working on both sides of the equation appeared in a book by the Arabian mathematician Al-Khowarizmi. In fact, we get our word ''algebra'' from the Arabic word *al-jabr,* which was prominent in the title of his book. European mathematicians learned of Al-Khowarizmi's methods in the twelfth century when his book was translated into Latin. Keen interest in finding ways to solve cubic and other higher-degree polynomial equations developed.

Historically, methods for solving equations and inequalities can be put in three categories: numerical, algebraic, and graphical. In solving an equation or inequality numerically, one often starts with a guess. Then better and better guesses lead to better approximations, and sometimes to exact solutions. The algebraic approach involves manipulating symbols with skills such as factoring, adding fractions, and simplifying complex fractions. The graphical approach uses graphs to find solutions. Like the numerical method, it may lead only to approximations. It is important to realize that all methods rely on an understanding of the mathematical concepts that underlie the procedures and algorithms used in any approach to solving equations and inequalities.

This chapter explores the mathematical concepts involved in solving several types of equations and inequalities. Algebraic methods for solving them are presented. Representing problem situations with mathematical models is also presented. Chapters 3 and 4 develop graphing skills and show the relationship between the algebraic methods in this chapter and graphing methods. We will study polynomial equations in more depth in Chapter 5.

2.1 USING LINEAR EQUATIONS

An **equation** is a statement that two numbers are equal. This statement may not be true. For example, if x is a number, then the statement

$$11 + x = 15$$

is an equation. This statement is true if x is 4, and false otherwise. The letter x is called a **variable,** and a number that, when substituted for the variable, makes the statement true is called a **solution** or **root.** Thus 4 is the only solution to the equation above. The collection of all solutions to an equation is called the **solution set.** To **solve** an equation means to find its solution set.

Two equations that have the same solution set are called **equivalent.** The three equations

$$2(x^2 - 1) = 3x$$

$$2x^2 - 3x - 2 = 0$$

$$(2x + 1)(x - 2) = 0$$

are equivalent because each has $\left\{-\frac{1}{2}, 2\right\}$ as its solution set.

An equation that can be written in the form $ax + b = 0$, where a and b are real numbers with $a \neq 0$, is called a **linear equation** in one variable. Two tools are essential in solving equations.

Two Tools Used to Solve Equations

Addition Property of Equality

If the same number is added to (or subtracted from) both sides of an equation, the resulting equation is equivalent to the first equation.

Multiplication Property of Equality

If both sides of an equation are multiplied (or divided) by the same nonzero number, the resulting equation is equivalent to the first equation.

Often in mathematics we use the phrase "if and only if" or a double arrow $\Leftrightarrow$. If a, b, and c are real numbers, the Addition Property of Equality can be stated as

$$a = b \quad \text{if and only if} \quad a + c = b + c$$

$$a = b \quad \Leftrightarrow \quad a + c = b + c$$

The meaning of these statements is a combination of two if-then statements:

"If $a = b$, then $a + c = b + c$" and "if $a + c = b + c$, then $a = b$."

(The second if-then statement is called the *converse* of the first statement, and vice versa).

To prove the Addition Property of Equality, we must either write the proof using if and only if or prove the two if-then statements. Let's prove the two if-then statements.

First let's prove: If $a = b$, then $a + c = b + c$.

$a + c = a + c$	Reflexive Property
$a + c = b + c$	Substitution ($a = b$)

To prove: If $a + c = b + c$, then $a = b$.

$a = a + (c - c)$	Additive identity
$a = (a + c) - c$	Associative Property
$a = (b + c) - c$	Substitution ($a + c = b + c$)
$a = b + (c - c)$	Associative Property
$a = b$	Additive identity

In a similar manner, we could prove the Multiplication Property of Equality.

EXAMPLE 1 Solve the equation, $6(y - 3) = 3y - 2\left(9 - \frac{1}{2}y\right)$

Solution $6(y - 3) = 3y - 2\left(9 - \frac{1}{2}y\right)$

$6y - 18 = 3y - 18 + y$	Distributive Property
$6y - 18 = 4y - 18$	Addition
$2y - 18 = -18$	Addition Property of Equality
$2y = 0$	Addition Property of Equality

$y = \frac{0}{2} = 0$ Multiplication Property of Equality

$\{0\}$ The solution set □

EXAMPLE 2 Solve each equation.

(a) $13 - 2y = 2(5 - y)$ (b) $3t - 21 = (t - 7)3$

Solutions (a) $13 - 2y = 2(5 - y)$

$13 - 2y = 10 - 2y$ Distributive Property

$13 = 10$ Addition Property

As there are *no values for y* that will make this statement true, the equation has *no solutions*. However, every equation has a solution set. In this case it must be a set with no elements.

The solution set is the empty set $\varnothing$.

Be sure to write the empty set as $\varnothing$ or $\{\ \}$, *never* $\{\varnothing\}$, as this set contains one element!

Equations with no solutions are often called **contradictions.**

(b) $3t - 21 = (t - 7)3$

$3t - 21 = 3t - 21$ Distributive Property

$-21 = -21$

Notice how this equation differs from the one in part (a). The statement $-21 = -21$ is true *no matter what number is assigned to t!* Therefore, every number is a solution of the given equation. Such equations are called **identities.** Since we are working in the set of real numbers, the solution set is the set of all real numbers.

The solution set is R. □

Solving Equations with a Calculator

To use a calculator to solve an equation, the idea is to do all the arithmetic with the calculator in one series of keystrokes.

EXAMPLE 3 Approximate (to the nearest thousandth) the solution to $1.25x - 13.65 = 5.61x - 87.3$.

Solution

$1.25x - 13.65 = 5.61x - 87.3$

$1.25x - 5.61x = -87.3 + 13.65$ Collect the x-terms on one side.

$(1.25 - 5.61)x = -87.3 + 13.65$ Factor out x.

$$x = \frac{-87.3 + 13.65}{1.25 - 5.61}$$

Now do the calculation on a calculator.

$$x \approx 16.892$$

The solution is approximately 16.892. □

Word Problems

Mathematics is most useful to us when we can apply it to the real problems of everyday life. Many such real-life problems lead to equations. We call an equation that represents a problem a **mathematical model** of the problem. Forming mathematical models of real problems is a skill we will develop. In a textbook, word problems are used to imitate real-life situations.

The first thing to do when confronted with a word problem is to read it to find out what *kind* of problem it is and to discover what must be found. Do not become discouraged if it seems too complicated! Next, assign a letter, such as x, to one of the things to be found. *Make this assignment statement in writing*. This first step may be the most important step in the solution of the problem. Several examples may be helpful.

EXAMPLE 4 Suppose Bill Ross and Linda Boyd each leave the Thompson-Washington airport at the same time in their private airplanes, traveling in opposite directions. If they are 1240 miles apart after 4 hours, and Bill's average rate is 40 miles per hour greater than Linda's, what are their average rates?

Solution This particular example is a distance-rate-time *(DRT)* problem or simply a distance problem. We are looking for Linda's average rate and Bill's average rate. We decide to assign the letter x to Linda's rate. *We make this initial assignment in writing.*

Let x be Linda's rate in miles per hour

Then Bill's average rate is $x + 40$ miles per hour

Notice that the problem can be divided into two distinct pieces: Bill's part and Linda's part. Although it may take a little thought to find the pieces, most distance problems can be divided into two parts. Now we make a simple sketch of the situation (Figure 2.1), label it, and find the distance, rate, and time *for each piece*.

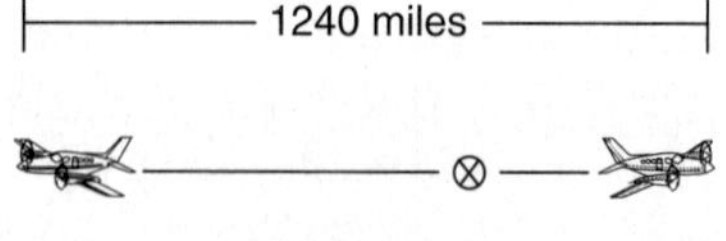

Fig. 2.1

Bill's Plane			***Linda's Plane***
	$D =$	$D =$	
	$R =$	$R =$	
	$T =$	$T =$	

Notice that one of the three *DRT* variables is given to us *without calculation* in the problem statement. It is T (time). Both Bill and Linda flew for 4 hours. A second *DRT* variable can be found in the assignment statement. It is R (rate). Linda's rate is

x miles per hour, and Bill's is $x + 40$ miles per hour. If we enter these in the table we have

Bill's Plane			**Linda's Plane**
	$D =$	$D =$	
	$R = x + 40$	$R = x$	
	$T = 4$	$T = 4$	

Once we have two of the three *DRT* variables established, we *do not* search the problem statement for the third but rather use the formula $D = RT$ or one of its alternate forms, $R = D/T$ or $T = D/R$. In this case we need D, so we use $D = RT$ to complete the table.

Bill's Plane			**Linda's Plane**
	$D = (x + 40)4$	$D = 4x$	
	$R = x + 40$	$R = x$	
	$T = 4$	$T = 4$	

The last piece of information and the problem statement should produce a mathematical model of the problem. Since Bill traveled $(x + 40)4$ miles and Linda traveled $4x$ miles and they went in opposite directions until they were 1240 miles apart, it must be true that the sum of their individual distances is 1240 miles.

$$(x + 40)4 + 4x = 1240$$

$$4x + 160 + 4x = 1240 \qquad \text{Distributive Property}$$

$$8x = 1240 - 160 \qquad \text{Addition Property}$$

$$x = 1080/8 \qquad \text{Multiplication Property}$$

$$x = 135 \qquad \text{Linda's rate}$$

$$x + 40 = 175 \qquad \text{Bill's rate}$$

Do not stop yet! We answer word problems with a written statement.

Linda's average rate was 135 miles per hour and Bill's was 175 miles per hour. □

Notice that the final statement answers the original question and makes sense. We check word problems by ensuring our answer does both.

EXAMPLE 5 Susana Chero, a silversmith in Monterrey, has 300 ounces of 15% silver alloy that she wants to upgrade to 20% silver. How many ounces of a 50% alloy should she add?

Solution This problem is a typical example of a mixture problem, another application that will lead to a linear equation. Our first step is always to make an assignment statement.

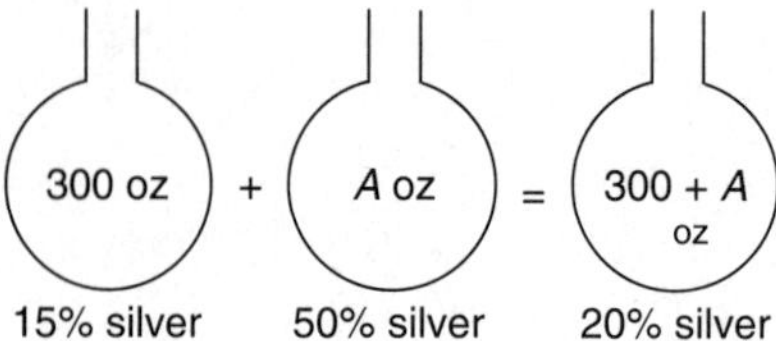

Fig. 2.2

Let A be the amount of the 50% alloy added (ounces). Mixture problems are usually clarified with a schematic diagram like the one in Figure 2.2. This diagram illustrates that she adds A ounces of 50% silver to 300 ounces of 15% silver to obtain $(300 + A)$ ounces of 20% silver. As there is exactly the same amount of silver on either side of the diagram, we can form the equation

$$300 \cdot \frac{15}{100} + A \cdot \frac{50}{100} = (300 + A)\frac{20}{100}$$

Notice that we changed percent to hundredths as we formed the mathematical model. It is perfectly acceptable to write hundredths in decimal form.

In either case, we clear fractions (decimals) by multiplying both sides by 100. In both cases we get

$$
\begin{aligned}
300(15) + A(50) &= (300 + A)(20) && \text{Multiplication Property}\\
4500 + 50A &= 6000 + 20A && \text{Distributive Property}\\
50A - 20A &= 6000 - 4500 && \text{Addition Property}\\
30A &= 1500\\
A &= \frac{1500}{30} = 50
\end{aligned}
$$

We check to see if this makes sense: 50 ounces added to the original 300 ounces of the 15% silver alloy makes a total of 350 ounces of the new alloy. The new alloy contains $50 \cdot 0.50 + 300 \cdot 0.15 = 25 + 45 = 70$ ounces of silver. Since $\frac{70}{350} = 0.20$ our answer makes sense.

She must add 50 ounces to the 50% alloy. □

Word problems, such as the ones in this section, provide important mathematical training. Modeling real-life situations into the language of mathematics, so its powerful tools can be applied, is the heart of engineering and the sciences. It is important to take a methodical approach to such problems. The following procedure has been found to be very effective for solving word problems.

A Procedure for Solving Word Problems

1. Read the problem and assign a variable, such as x, to represent one of the quantities to be found. (Write this assignment down.)
2. Express all other quantities to be found in terms of x (or the variable chosen).
3. Draw a figure or picture if possible. Label it.
4. Reread the problem and form an equation (the mathematical model).
5. Solve the equation and find the values of all the quantities to be found.

> **6.** Check the values in the *original problem statement*. They should answer the question and make sense.
>
> **7.** Write an answer to the original question.

EXAMPLE 6 Jim Hauther finally hit the Florida lottery for \$200,000! After exactly half went to pay taxes and he spent half of what remained, he invested some of what he had left in a money market account that pays $9\frac{1}{4}\%$ annually and the rest in a certificate of deposit (CD) that pays $7\frac{1}{2}\%$ annually. If Jim receives \$4100 a year in interest from his two investments, how much did he invest in each account?

Solution Investment or interest problems usually depend on the simple interest formula $I = PRT$, where I is the amount of interest, P is the principal, R is the interest rate, and T is the time. It is important that the units of time be the same for both R and T. The first step is to make an assignment statement.

Let x be the amount invested in the money market account (in dollars).

We also need to know how much he invested in the CD. Of his original windfall, half went to taxes, leaving him with \$100,000. But he spent half of that, leaving him a total of \$50,000 to invest. He invested x dollars in the money market account, so he has $(50{,}000 - x)$ dollars for the CD. So we write

Then $(50{,}000 - x)$ is invested in the CD (in dollars).

Now that we know how much he has invested and where, we use the simple interest formula.

$$I = PRT$$

His total yearly interest is \$4100, so I is 4100 and T is 1. However, as it comes from two investments, it is the *sum* of the individual interest amounts. P for the money market account is x, and R is 9.25%, while P for the CD is $(50{,}000 - x)$, and R is 7.5%. That is,

Total interest = interest from money market + interest from CD

Therefore we have the equation

$$4100 = x \cdot 0.0925 \cdot 1 + (50{,}000 - x) \cdot 0.075 \cdot 1$$

Multiply both sides by 10,000 to clear decimals.

$41{,}000{,}000 = 925x + (50{,}000 - x) \cdot 750$	Multiplication Property
$41{,}000{,}000 = 925x + 37{,}500{,}000 - 750x$	Distributive Property
$3{,}500{,}000 = 175x$	Addition Property
$\dfrac{3{,}500{,}000}{175} = x$	Multiplication Property

$$20{,}000 = x \qquad \text{Money market portion}$$

$$50{,}000 - x = 30{,}000 \qquad \text{CD portion}$$

Jim invested \$20,000 in the money market account and \$30,000 in the CD. □

The next example illustrates a type of word problem often called "work problems."

EXAMPLE 7 Homer can lay a certain section of pipe in 2 hours working alone, while George takes 3 hours to lay the section if he is working alone. How long does it take Homer and George to lay the same length of pipe if they work together?

Solution First we make an assignment statement.

Let T be the time it takes them to lay the section when working together (hours).

Homer takes 2 hours to lay the section of pipe, so he works at the rate of $\frac{1}{2}$ of a section an hour. By the same reasoning, George works at a rate of $\frac{1}{3}$ of a section per hour. Therefore, in T hours Homer lays $T \cdot \frac{1}{2}$ of the pipe and Goerge lays $T \cdot \frac{1}{3}$ of the pipe. But in T hours they lay 1 section (from the assignment statement). So we have the mathematical model

$$\frac{T}{2} + \frac{T}{3} = 1$$

We clear fractions by multiplying both sides by the LCD.

$$6\left(\frac{T}{2} + \frac{T}{3}\right) = 6 \cdot 1 \qquad \text{Multiplication Property}$$

$$3T + 2T = 6 \qquad \text{Distributive Property}$$

$$5T = 6$$

$$T = \frac{6}{5}$$

It will take them $\frac{6}{5}$ hours (or 1 hour and 12 minutes) to lay the section together.

□

EXAMPLE 8 A company makes portable computers. The company has fixed costs (rent, insurance, etc.) of \$45,000 per month and variable costs (labor, materials, etc.) of \$1200 per computer produced. The computers sell for \$1500 each.

(a) How many computers must be sold each month so that cost will equal revenue? This is called the *breakeven point*.
(b) How many computers must be sold before the company makes a profit? Profit is revenue minus cost.

Solutions (a) If x is the number of computers sold each month, then cost (C) and revenue (R) are related to the number of computers sold each month by the equations

$$C = 45{,}000 + 1200x$$
$$R = 1500x$$

If $C = R$, then we have

$$45{,}000 + 1200x = 1500x$$
$$45{,}000 = 300x$$
$$150 = x$$

The company must sell 150 computers to make cost equal revenue (to break even).

(b) Profit (P) is revenue (R) minus cost (C). So, we have a formula for profit.

$$P = R - C$$
$$P = 1500x - (45{,}000 + 1200x)$$
$$P = 300x + 45{,}000$$

To make a profit, P must be a nonnegative number. If $P = 0$, then $x = 150$, and this is the breakeven point. So, if x is greater than 150, the company will make a profit. The company must sell more than 150 computers to make a profit. □

Literal Equations

Manipulating formulas is another important skill in today's world. Formulas, often called *literal equations,* usually take the form of one variable set equal to an expression containing other variables. Often it is convenient to have the formula solved for a different variable. We do this by solving for the "new" key variable in the same way other equations are solved. However, since these equations are used as formulas, we do not usually write solution sets. Instead we leave the equation as a formula.

EXAMPLE 9 The formula for the volume of a right circular cylinder (like a tin can) is $V = \pi r^2 h$, where V is the volume, r is the radius, and h is the height of the cylinder. Solve for h.

Solution

$$V = \pi r^2 h$$

We wish to solve this formula for h, so we divide both sides by πr^2.

$$\frac{V}{\pi r^2} = \frac{\pi r^2 h}{\pi r^2} \qquad \text{Multiplication Property}$$

$$\frac{V}{\pi r^2} = h \qquad \text{Common factors}$$

It is customary to write the formula so the new key variable is on the left side of the equation.

$$h = \frac{V}{\pi r^2}$$

□

EXAMPLE 10 $I = \dfrac{nE}{R + nr}$ is a formula for the amount of current in a battery of n cells, where I is the current, R and r stand for resistance, and E is the electromotive force for one cell. Solve this equation for n.

Solution

$$I = \frac{nE}{R + nr}$$

$$I(R + nr) = nE \qquad \text{Clear fractions.}$$

$$IR + Inr = nE \qquad \text{Distributive Property}$$

$$IR = nE - Inr \qquad \text{Collect terms with } n \text{ on one side.}$$

$$IR = n(E - Ir) \qquad \text{Factor.}$$

$$\frac{IR}{E - Ir} = n \qquad \text{Divide both sides by } E - Ir.$$

$$n = \frac{IR}{E - Ir}$$

□

PROBLEM SET 2.1

Warm-ups

In Problems 1 through 6, solve each equation. See Examples 1, 2, and 3.

1. $2x + 3 = -x - 3$

2. $9(3y - 1) - 2(3 - 4y) = 10$

3. $3(s + 1) - 3 = 2(s - 1) + s + 2$

4. $\dfrac{2x}{5} + \dfrac{x}{2} = 9$

5. $1.24z - 1.06 = 2.05z - 5.7013$

6. $\dfrac{7x - 3}{7} - \dfrac{5x + 7}{5} = 1$

For Problems 7 through 9, see Example 4.

7. Michael Smith sets off for Idaho Falls, a distance of 164 miles, at a constant rate in his old truck. After 2 hours, he is forced to reduce his average speed by 16 miles per hour because of excessive front-end vibration. If the total trip takes 3 hours, what was his initial average speed?

8. A Delta Airlines 767 leaves Atlanta headed for Dallas, at an average speed of 600 miles per hour, at the same time that an American Airlines L-1011 leaves Dallas headed for Atlanta at an average speed of 480 miles per hour. If the flight takes the L-1011 24 minutes longer than the 767, how long did each plane fly?

9. Janet traveled from home to Bad Axe at an average rate of 50 miles per hour. She returned at an average rate of 55 miles per hour, and the trip was 15 minutes shorter. How far it it from Janet's home to Bad Axe?

For Problems 10 through 12, see Example 5.

10. How many liters of a 70% alcohol solution should be added to 100 liters of a 14% alcohol solution to obtain a 20% solution?

11. Dr. Hope, a pharmacist at Revco Drugs, has both 40% and 15% alcohol syrups available to fill a prescription for 1 fluid ounce of cough syrup that is 25% alcohol. How many fluid ounces of each alcohol syrup should Dr. Hope use?

12. Jamie has 20 gallons of a 40% H_2SO_4 (sulfuric acid) solution. How much should she drain and replace with pure H_2SO_4 to obtain 20 gallons of a 64% solution?

For Problems 13 through 15, see Example 6.

13. Miguel Ubico wishes to divide \$120,000 between a safe investment that pays $8\frac{1}{2}\%$ and a riskier one that pays $10\frac{3}{4}\%$. How should he divide his investment in order to have total yearly interest of \$12,000?

14. Justine invested part of \$80,000 in a CD paying 8.65% per annum and the rest in a money market account paying 9.85%. If the money market account earns \$5660 more a year than the CD, how much does Justine have invested in each account?

15. F. Lee Early invested \$200,000 in two real estate deals last year. He made a profit of 18% on one deal but lost 2% on the other. If Lee's net profit for 1 year was \$11,200, how much did he invest in each deal?

In Problems 16 through 18, see Example 7.

16. The cold water tap can fill a certain sink in 40 seconds if running alone. The hot water tap can fill the same sink in 1 minute if running alone. How long will it take the taps to fill the sink if they are both turned on?

17. Ray can wash a sink full of dishes in 45 minutes. If Joan helps, it takes only 20 minutes. How long would it take Joan to wash the dishes if she worked alone?

18. It takes Eldon 90 minutes to mow his lawn. If Jim can mow it in an hour, how long will it take them to mow it working together?

In Problems 19 and 20, see Example 8.

19. Virginia, a screen printer, has a contract to print some T-shirts for a mathematics conference. She has determined that she has fixed costs of \$500 and variable costs of \$2.50 per shirt.

(a) If she sells each shirt for \$7.50, how many shirts must she sell to break even?

(b) When will she make a profit?

(c) How many shirts must Virginia sell to make a profit of \$1000?

20. The Basket Bakery has fixed costs of \$840 per week. It costs \$1.10 to make a dozen giant chocolate chip cookies. If the chocolate chip cookies sell for \$3.50 a dozen,

(a) How many dozen cookies must be sold each week to break even?

(b) How many dozen cookies must be sold each week to make a profit of \$1000?

In Problems 21 through 24, solve the given formula for the indicated variable. See Examples 9 and 10.

21. $V = \frac{1}{3}\pi r^2 h$, for h (volume of a cone)

22. $A = \frac{1}{2}bh$, for b (area of a triangle)

23. $F = \frac{9}{5}C + 32$, for C (Fahrenheit-Celsius)

24. $A = P + Prt$, for P (interest)

Practice Exercises

In Problems 25 through 38, solve each equation.

25. $(2x - 1)7 = 5x + 2$

26. $6(2x - 3) = 10x - 4$

27. $2(3 - \sqrt{2}s) = 3(2 - \sqrt{2}s)$

28. $(5k - 11)2 - 3(7 - 3k) = 0$

29. $\dfrac{0.1k + 0.3}{5} = \dfrac{0.1k + 0.7}{4}$

30. $\dfrac{0.4 - 0.2x}{6} - \dfrac{0.3 - 1.2x}{7} = 0$

31. $\dfrac{1 - 3x}{2} - \dfrac{2 + 9x}{6} = 1$

32. $\dfrac{15r - 8}{5} + \dfrac{1 - 6r}{2} = -\dfrac{11}{10}$

33. $(x - 3)^2 - x^2 = 5$

34. $(z + 2)^2 = (4 - z)^2$

35. $w + \frac{2w - 2}{6} = \frac{4w - 1}{3}$

36. $2x - \frac{2x + 9}{8} = \frac{4 + 7x}{4}$

37. $\frac{1}{2}(3x - 4) = \frac{3 - 2x}{3} - \frac{1}{5}$

38. $\frac{2}{3}x + \frac{1}{2}(x + 1) - \frac{1}{4}(1 - x) = 1$

In Problems 39 through 46, solve the given formula for the indicated variable.

39. $F = G\frac{m_1 m_2}{R^2}$, for m_1 (gravitational force)

40. $\frac{P_1 V_1}{T_1} = \frac{P_2 V_2}{T_2}$, for T_2 (gas law)

41. $v = -32t + v_0$, for t (velocity in free fall)

42. $S = 2\pi r^2 + 2\pi rh$, for h (surface area of a cylinder)

43. $\frac{1}{R} = \frac{1}{R_1} + \frac{1}{R_2}$, for R_2 (resistance)

44. $R = \frac{R_1 R_2}{R_1 + R_2}$, for R_1 (resistance)

45. $S = \frac{a}{1 - r}$, for r (sum in a geometric series)

46. $S = \frac{n}{2}[2a + (n - 1)d]$, for d (sum in an arithmetic series)

47. The Video Store rents videos for \$2.50 each. The cost of processing one video rental is \$0.35. Rent and other fixed costs are \$2623.00 per month. If last month's profit was \$2367.15, how many videos were rented last month? How many videos must be rented to break even each month?

48. A taxi meter that is properly calibrated charges 25¢ for four short blocks (120 feet = 1 short block). This does not include \$1.50 drop-off fee, an extra 25¢ for every 75 seconds of waiting time, and 50¢ for night fare. Juan Reynoso estimates that it costs him 14¢ per mile to operate his cab. Last week when he picked up 56 fares, 7 of which were night fares, and has 12.5 minutes of waiting time, he collected \$2504.00. What was his profit last week?

49. In the auto industry, the efficiency of an object moving through the air is known as the *coefficient of drag* (CD). It can be calculated by the formula

$$\text{CD} = \frac{\text{drag force}}{\text{dynamic pressure} \times \text{frontal area}}.$$

Thus, the higher the CD, the more wind resistance. Ford's Taurus has a CD of 0.33. If the drag force is decreased by 10%, what will the CD for a Taurus be?

50. The formula

$$\frac{\lambda v}{c} = (2.426 \times 10^{-10})\sqrt{1 - \frac{v^2}{c^2}}$$

gives the wavelength in centimeters for particles traveling near the speed of light, where v is the velocity of the particle, c is the speed of light, and λ is the wavelength. If a particle is traveling at 0.75 of the speed of light, approximate the wavelength of the particle to four significant digits.

51. If P dollars is deposited in a savings account, the amount accumulated (A) in the account after t years earning simple interest at a rate of r per year is given by the formula $A = P(1 + rt)$. What must the simple interest rate be for \$750 to grow to \$1000 in 2 years?

52. Using the formula in Problem 51, how many years would it take an investment earning 10% simple interest to double? To triple?

53. If P dollars is invested at an annual interest rate of r compounded once a year, then the amount (A) accumulated after t years is given by the formula $A = P(1 + r)^t$. If an account has accumulated \$1250 in 2 years with 12.5% interest compounded annually, how much was deposited originally (to the nearest cent)?

54. The Rule of 72 is a method of approximating the number of years (t) required to double the amount invested (P) with interest compounded annually at a rate of r%. $t \approx 72/r$. Thus, an amount invested at 8% compounded annually will double in approximately $\frac{72}{8}$ or 9 years.

(a) Approximate the doubling time with the Rule of 72 for the rates given in the accompanying table.

(b) Use the formula in Problem 53 to approximate the amount (A) in an account if \$1000 is invested at each rate for the doubling time.

(c) How accurate is the Rule of 72?

r	RULE OF 72 DOUBLING TIME	A
18%		
12%		
8%		

55. Dr. Sylvia Krebs borrowed \$5000 to pay for her tuition at graduate school. She agreed to pay the bank in four equal annual installments (X). The bank charged 14.2% interest compounded annually. Solve the following equation to find out how much her payments were.

$$(1 + 0.142)^3X + (1 + 0.142)^2X + (1 + 0.142)X + X = 5000(1 + 0.142)^4$$

56. How much would Sylvia's payments in Problem 55 be if the interest rate had been 12.2%?

57. The radiator in Karen's old car holds 6 gallons. It is currently full and contains 8% ethylene glycol antifreeze. Karen wishes to winterize by increasing the mixture to 20% ethylene glycol. How many gallons should she drain and replace with a 40% solution to winterize her radiator?

58. An automobile radiator contains 8 quarts of coolant. If the coolant mixture is 30% antifreeze, how much of the mixture should be drained and replaced by pure antifreeze to obtain 8 quarts of a coolant mixture that is 50% antifreeze?

59. Teresa is investigating three preferred stocks. One stock yields 7.25%, another yields 6.95% and the third yields 8.55%. Can she divide $120,000 between *two* of the stocks and get an annual return of $10,000? If so, what are her options?

60. Kevin has $34,000 invested at 7%. He would like to divide $10,000 between a risky common stock paying 14% and a CD paying 6.6% so that his total yearly interest is $3595. How should he do this?

61. Valerie and Allison race to the Rib Shack. Although Val has a 6-minute head start, they arrive at the same time. If Allison runs for 18 minutes and her average speed is 2 miles per hour greater than Val's, what is the average speed of each? How far is it to the Rib Shack?

62. Two joggers are on the same path. If one of them whose average rate is 6 miles per hour has a 10-minute head start over the other whose average rate is 8 miles per hour, how long will it take the faster jogger to catch the slower jogger?

In Problems 63 through 65, use the chart below.

Fats and Oils	*Saturated Fatty Acid (%)*
Coconut oil	92
Butterfat	77
Lard	41
Chicken fat	34
Corn oil	12
Canola oil	9

63. Steve listened to his wife and has decided to lower the amount of saturated fatty acid in his diet. He plans to start with his cookies. His favorite cookies are store-bought chocolate chip made with coconut oil. His wife's homemade chocolate chip cookies are made with canola oil. If he eats a total of a dozen chocolate chip cookies, approximately how many of his wife's cookies must he eat so that his total amount of saturated fatty acid is 30%? (Assume that each cookie contains k ounces of oil.)

64. Steve eats 12 sandwiches per week, either chicken or grilled cheese. Approximately how many chicken sandwiches and cheese sandwiches can he eat to maintain a 45% saturated fatty acid content each week? (Assume that a sandwich contains k grams of fat and that cheese contains butterfat.)

65. Steve loves his mother-in-law's biscuits which are made with lard. His wife makes delicious muffins with corn oil. If he eats a total of 24 biscuits and muffins, approximately how many of each kind should he eat to make his saturated fatty acid content 28%. (Assume that each biscuit or muffin contains k ounces of oil.)

66. Hiromi Salah ran the marathon (26 miles) in Tokyo in 2 hours and 15 minutes. The first 7 miles of the race he averaged 10 miles per hour, on the middle part of the course he ran at an average of 9 miles per hour, and on the last leg he averaged 13.2 miles per hour. Approximately to the nearest tenth of a mile, how far did he run at 13.2 miles per hour.

67. In the first heat of the track-and-field events, the Spanish 400-meter relay team came in third. The time for the first leg was 0.1 second more than the time for the last leg. The time for the second leg was one full second more than the time for the first leg, and the time for the third leg was 0.8 second more than the time for the second leg. The average speed of each racer for each leg was 10.9, 9.8, 9.1, and 11 meters per second, respectively. Approximate the time recorded by the Spanish team to the nearest tenth of a second.

68. A large drain pipe can empty a full tank in 2 hours. A small drain pipe can empty the same full tank in 3 hours. The tank is empty and both drain pipes closed when a garden hose is turned on into the tank. It takes the hose 12 hours to completely fill the tank. If both drain pipes are opened the instant the tank is full, how long will it take the two drain pipes to empty the tank while the garden hose continues to run?

69. Vesta Masters can paint a room in 6 hours. Jack Glover can paint the same room in 4 hours, and Alice Glover can paint it in 8 hours. How long will it take all three of them to paint a room twice as big working together if Alice starts $\frac{1}{2}$ hour late and quits $\frac{1}{2}$ hour before they finish?

70. A new document shredder has been moved into the office on a trial basis. It shreds 60 pages a minute, while the old model shreds 40 pages a minute. How long will it take both shredders, working together, to shred a 2000-page report?

Challenge Problems

71. The cistern at the Laconia Friendship Community Center has three roof drains that empty into it and two outlet taps. If the cistern is full, the little tap can empty it in 24 hours alone, while the big tap can empty it in 12 hours working alone. In a severe rainstorm, with both outlet taps closed, either the right or the left roof drain, working alone, can fill the cistern in 8 hours, while it would take the center roof drain only 6 hours to fill the cistern working alone. A long, hard rainstorm hits the center at noon when the cistern is empty, both outlet taps are closed, and all the roof drains are working. At 1:00 P.M. both outlets are opened. At 2:00 P.M. the right roof drain becomes stopped up with leaves and fails to work. At what time will the cistern be full?

72. A column of tanks is moving across the desert at a steady speed of 36 miles per hour. A messenger travels from the front of the column to the rear of the column and then immediately returns to the front. If the messenger travels at a constant rate of 54 miles per hour and his round trip takes 18 minutes, how long is the column?

IN YOUR OWN WORDS . . .

73. Describe a linear equation.

74. What steps do we usually take in solving a linear equation?

2.2 LINEAR INEQUALITIES

Just as equations are statements about the equality of two numbers, **inequalities** are statements that one number is *larger* or *smaller* than another. These statements need not be true always. For example,

$$x > 5$$

is an inequality that is true if x is a number *larger* (or greater) than 5, and false otherwise, while

$$x < 5$$

is an inequality that is true if x is *smaller* (or less) than 5, and false otherwise. Often we wish to include equality along with inequality. For example, x **less than or equal to** 7 can be written

$$x \leq 7$$

while we write, x **greater than or equal to** 4 as

$$x \geq 4$$

The collection of numbers that make an inequality true is called the **solution set** of the inequality. To **solve** an inequality is to find its solution set. Often the solution set of an inequality cannot be written by listing its members. We use set builder notation. For example, to write the solution set for the inequality

$$x > 5$$

we need to write the set of all numbers greater than 5. In set builder notation,

$$\{x \mid x > 5\}$$

Two inequalities with the same solution set are called **equivalent.**

Solving Linear Inequalities

Linear inequalities can be solved in a manner very similar to the method used to solve linear equations.

> **Tools Used to Solve Linear Inequalities**
>
> ***Addition Property***
>
> If the same number is added to (or subtracted from) both sides of an inequality, the resulting inequality is equivalent to the first inequality.
>
> ***Multiplication by a Positive Number***
>
> If both sides of an inequality are multiplied (or divided) by the same *positive* number, the resulting inequality is equivalent to the first inequality.
>
> ***Multiplication by a Negative Number***
>
> If both sides of an inequality are multiplied (or divided) by the same *negative* number, and the inequality symbol is *reversed,* the resulting inequality is equivalent to the first inequality.

Just as for equations, we use the above tools to form equivalent inequalities until we have an inequality of the form $x < a$, $x > a$, $x \leq a$, or $x \geq a$. Then we write the solution set. Notice how this works in the following examples.

EXAMPLE 1 Solve $\dfrac{x+1}{3} - \dfrac{x+2}{2} \leq \dfrac{1}{6}$.

Solution

$$\frac{x+1}{3} - \frac{x+2}{2} \leq \frac{1}{6}$$

Multiply both sides by 6 to clear fractions.

$$6\left(\frac{x+1}{3} - \frac{x+2}{2}\right) \leq 6\left(\frac{1}{6}\right) \qquad \text{Multiplication by a positive number}$$

$$2(x+1) - 3(x+2) \leq 1$$

$$2x + 2 - 3x - 6 \leq 1 \qquad \text{Distributive Property}$$

$$-x - 4 \leq 1$$

$$-x \leq 5 \qquad \text{Addition Property}$$

Then multiply both sides by -1 to get x on the left. Note that -1 is a *negative* number, so the direction of the inequality must be *reversed.*

$$(-1)(-x) \geq (-1) \cdot 5 \qquad \text{Multiplication by a negative number}$$

$$x \geq -5$$

$$\{x \mid x \geq -5\} \qquad \text{The solution set}$$

□

EXAMPLE 2 Solve the following inequalities.

(a) $2(x + 6) > 3(x + 1) - x + 3$ (b) $\dfrac{3x - 2}{4} \geq \dfrac{6x + 3}{8} + 1$

Solutions (a) $2(x + 6) > 3(x + 1) - x + 3$

$2x + 12 > 3x + 3 - x + 3$ Distributive Property

$2x + 12 > 2x + 6$

$12 > 6$ Addition Property

As this statement is true for *all* values of x, the solution set is the set of all real numbers.

R The solution set

(b) $\dfrac{3x - 2}{4} \geq \dfrac{6x + 3}{8} + 1$

$2(3x - 2) \geq 6x + 3 + 8$ Multiplication Property

$6x - 4 \geq 6x + 11$ Distributive Property

$-4 \geq 11$ Addition Property

Since this statement is *never* true no matter what number x is, there are no solutions. The solution set is $\varnothing$. □

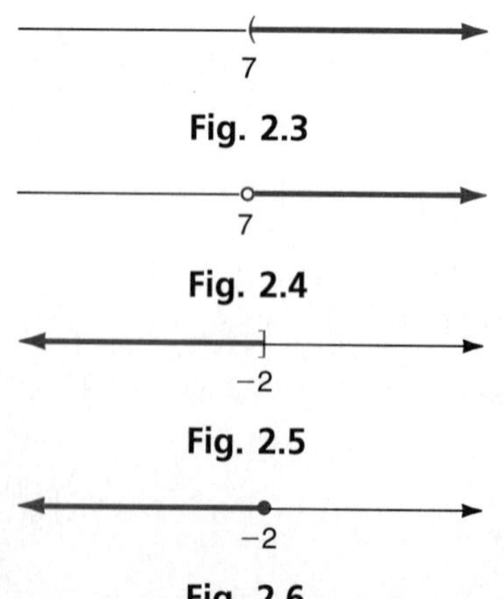

Fig. 2.3

Fig. 2.4

Fig. 2.5

Fig. 2.6

Graphing Inequalities

It is often convenient and instructive to draw a picture of the solution set of an inequality. This picture of the solution set is called the **graph** of the inequality.

To graph the set $\{x \mid x > 7\}$, we construct a copy of the number line and locate all numbers greater than 7 on it. They lie to the right of 7. We darken that portion of the number line and indicate with an arrowhead that it continues indefinitely. We indicate with a **parenthesis** that the number 7 *does not* belong to the graph in Figure 2.3. Other notations use a hollow circle to indicate that a number does not belong to the graph in Figure 2.4.

To graph the set $\{x \mid x \leq -2\}$, we draw a number line and darken that portion of the line to the left of -2. We use a **bracket** to indicate that the number -2 *belongs* to the graph in Figure 2.5. Other notations use a darkened circle to indicate that a number belongs to the graph (Figure 2.6).

EXAMPLE 3 Solve each inequality and graph each solution set.

(a) $2 - \frac{3}{4}x > x$ (b) $2(x + 3) \le 3(2x - 5) + 1$

Solutions (a) $2 - \frac{3}{4}x > x$

$8 - 3x > 4x$	Multiplication Property
$8 > 7x$	Addition Property
$\frac{8}{7} > x$	Multiplication Property

It is sometimes convenient to write this inequality the other way around in a solution set.

$\left\{x \mid x < \frac{8}{7}\right\}$	The solution set

Fig. 2.7

Now we graph the solution set, as instructed (Figure 2.7).

(b) $2(x + 3) \le 3(2x - 5) + 1$

$2x + 6 \le 6x - 15 + 1$	Distributive Property
$20 \le 4x$	Addition Property
$5 \le x$	Multiplication Property
$\{x \mid x \ge 5\}$	Solution set (Figure 2.8) □

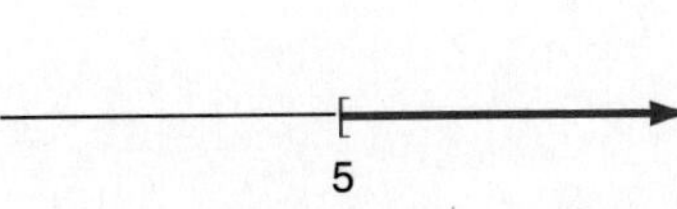

Fig. 2.8

Compound Inequalities

A **compound inequality** is a statement made by combining two inequalities with the word **and** or **or.** For example, the statement

$$x < -1 \quad \text{or} \quad x > 5$$

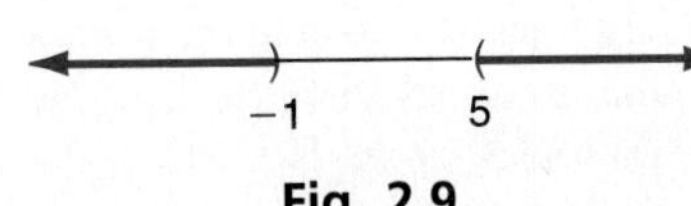

Fig. 2.9

is a compound inequality. This statement is true if x is less than -1 *or* if x is greater than 5. Its graph is shown in Figure 2.9. The use of the word "and" is illustrated with the compound inequality

$$x \le 6 \quad \text{and} \quad x > -3$$

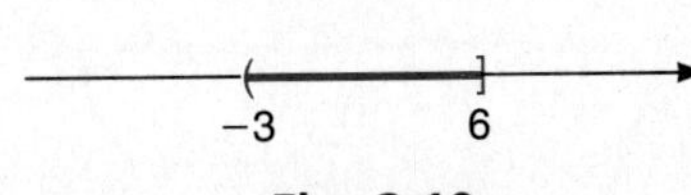

Fig. 2.10

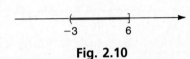

This statement will be true if x is less than or equal to 6 *and* if x is greater than -3. Its graph is shown in Figure 2.10. Notice from the graph that if x is in the solution set of the compound inequality, x must be *between* -3 and 6. Therefore, we usually write compound inequalities such as

$$x \le 6 \quad \text{and} \quad x > -3$$

in the **compact form**

$$-3 < x \le 6$$

The compact form will *never* be written using *both* $<$ and $>$ symbols. Such a statement as $2 < x > 5$ is nonsense. All parts of the compound inequality must make sense separately and together.

EXAMPLE 4 Solve each of the following compound inequalities and graph the solution set for each.

(a) $x + 3 < 0$ or $2x - 3 \geq 1$ (b) $1 \leq 2x + 5 < 15$

Solutions (a) $x + 3 < 0$ or $2x - 3 \geq 1$

We use the Addition Property on each of the inequalities separately and then we divide by the coefficient of x in the second inequality.

$x < -3$ or $2x \geq 4$	Addition Property
$x < -3$ or $x \geq 2$	Multiplication Property
$\{x \mid x < -3 \text{ or } x \geq 2\}$	Solution set (Figure 2.11)

Fig. 2.11

(b) $1 \leq 2x + 5 < 15$

Here we apply the properties to all three ''sides'' of this compound inequality at the same time. We subtract 5 and then divide by 2.

Fig. 2.12

$-4 \leq 2x < 10$	Addition Property
$-2 \leq x < 5$	Multiplication Property
$\{x \mid -2 \leq x < 5\}$	Solution set (Figure 2.12)

□

EXAMPLE 5 Write a set that represents the graph in Figure 2.13.

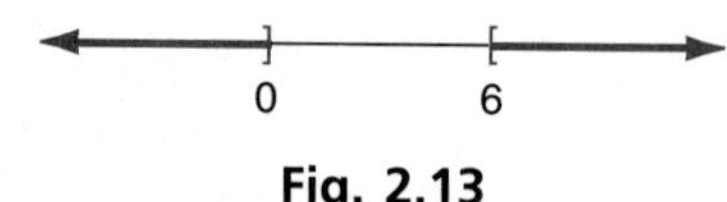

Fig. 2.13

Solution Notice that this is a combination of two graphs. A number would belong to this graph if it were less than or equal to 0, or greater than or equal to 6. So it is the inequalities $x \leq 0$ and $x \geq 6$ connected with the word ''or.''

$\{x \mid x \leq 0 \text{ or } x \geq 6\}$ Solution set □

EXAMPLE 6 Write a set that represents the graph in Figure 2.14.

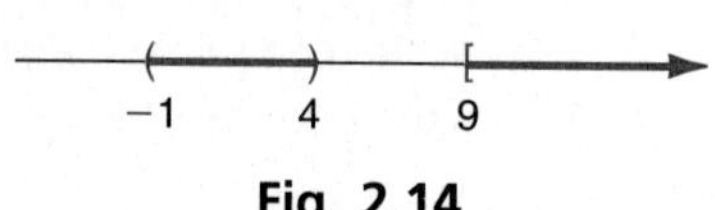

Fig. 2.14

Solution This is a combination of three graphs. However, the set represented can be described by looking at the picture.

$$\{x \mid -1 < x < 4 \quad \text{or} \quad x \geq 9\}$$

□

Interval Notation

Fig. 2.15

Fig. 2.16

Fig. 2.17

Another common way to write sets is with **interval notation.** Suppose p and q are two real numbers with p less than q. We write

$$p < q$$

and locate the numbers on the number line as shown in Figure 2.15. Consider the set of numbers *between* p and q. That is, the set $\{x \mid p < x < q\}$. The graph of this set is shown in Figure 2.16. We write this same set in interval notation as (p, q). It is called the **open interval** between p and q. If we *include* p and q, we have the set $\{x \mid p \le x \le q\}$ whose graph is shown in Figure 2.17. We write this set in interval notation as $[p, q]$. It is called the **closed interval** between p and q. We summarize in the box.

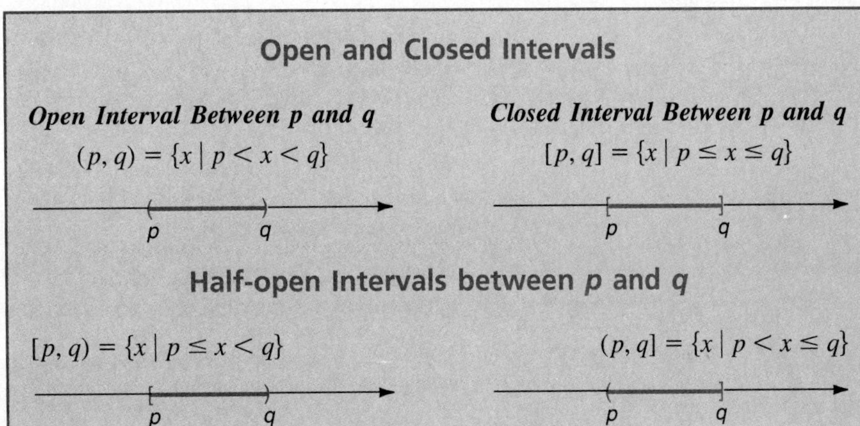

Open and Closed Intervals

Open Interval Between p and q

$$(p, q) = \{x \mid p < x < q\}$$

Closed Interval Between p and q

$$[p, q] = \{x \mid p \le x \le q\}$$

Half-open Intervals between *p* and *q*

$$[p, q) = \{x \mid p \le x < q\}$$

$$(p, q] = \{x \mid p < x \le q\}$$

Interval notation can be used to indicate sets that have no largest member or sets that have no smallest member.

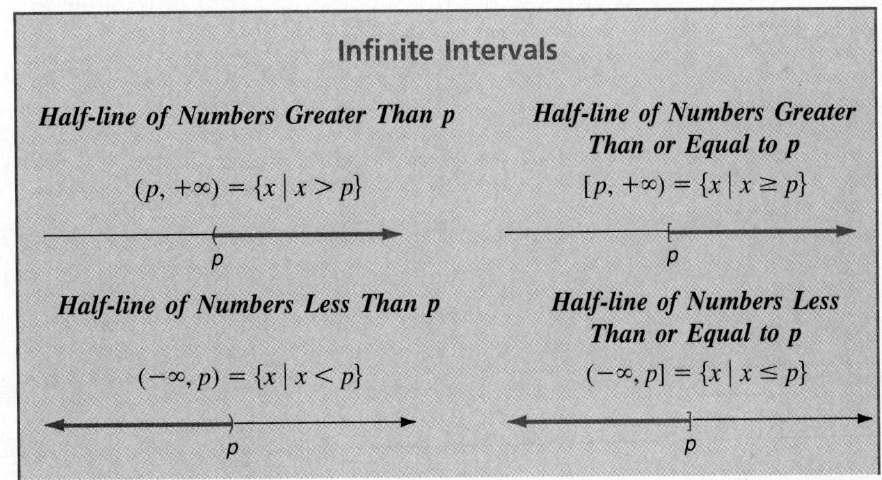

Infinite Intervals

Half-line of Numbers Greater Than p

$$(p, +\infty) = \{x \mid x > p\}$$

Half-line of Numbers Greater Than or Equal to p

$$[p, +\infty) = \{x \mid x \ge p\}$$

Half-line of Numbers Less Than p

$$(-\infty, p) = \{x \mid x < p\}$$

Half-line of Numbers Less Than or Equal to p

$$(-\infty, p] = \{x \mid x \le p\}$$

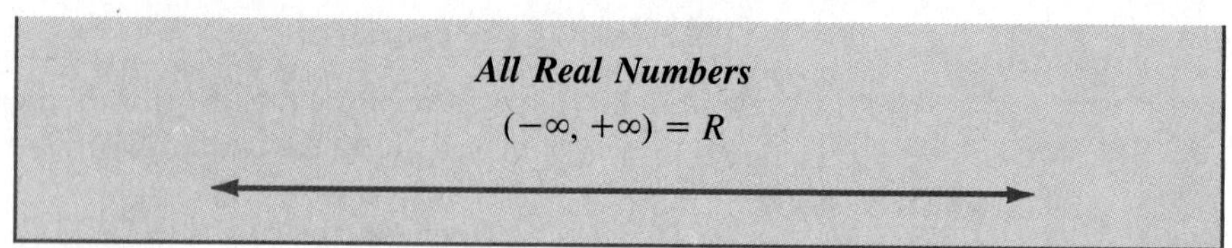

Now we have *three* ways to indicate the solution set of an inequality: set builder notation, interval notation, and the graph.

EXAMPLE 7 Solve each of the following inequalities, write each solution set in interval and set builder notation, and graph each solution set.

(a) $3(4 - x) \leq 17$ (b) $5x < 15$ or $x + 3 \geq 8$
(c) $-3 \leq 4x - 3 \leq 9$ (d) $2 < 3(x + 1) \leq 6$ and $x \geq 0$

Solutions (a)

$$\begin{aligned} 3(4 - x) &\leq 17 \\ 12 - 3x &\leq 17 && \text{Distributive Property} \\ -3x &\leq 5 && \text{Addition Property} \\ x &\geq -\frac{5}{3} && \text{Multiplication Property} \end{aligned}$$

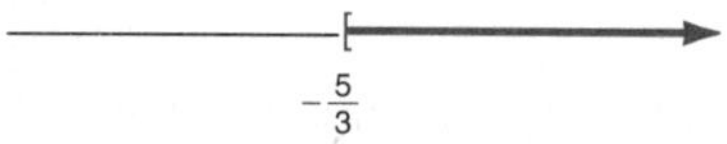

Now, following instructions, we write the solution set in both notations and graph it (Figure 2.18).

$-\frac{5}{3}$

Fig. 2.18

Set builder notation: $\left\{x \mid x \geq -\frac{5}{3}\right\}$ Interval notation: $\left[-\frac{5}{3}, +\infty\right)$

(b) $5x < 15$ or $x + 3 \geq 8$

This is a compound inequality. We divide the first inequality by 5 and subtract 3 in the second.

$x < 3$ or $x \geq 5$

Set builder notation: $\{x \mid x < 3 \text{ or } x \geq 5\}$
Interval notation: $(-\infty, 3) \cup [5, +\infty)$

3 5

Fig. 2.19

Notice that we used the set connector union for "or" in the interval notation. The graph is shown in Figure 2.19.

(c) $-3 \leq 4x - 3 \leq 9$

This is another compound inequality. We add 3 to all three sides and then divide by 4.

$0 \leq 4x \leq 12$
$0 \leq x \leq 3$

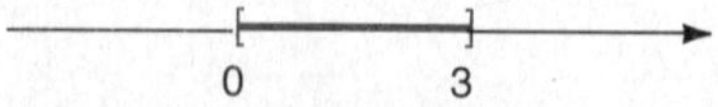

0 3

Set builder notation: $\{x \mid 0 \leq x \leq 3\}$ Interval notation: $[0, 3]$
The graph is shown in Figure 2.20.

Fig. 2.20

(d) $2 < 3(x + 1) \le 6$ and $x \ge 0$

$$2 < 3x + 3 \le 6 \quad \text{and} \quad x \ge 0$$
$$-1 < 3x \le 3 \quad \text{and} \quad x \ge 0$$
$$-\frac{1}{3} < x \le 1 \quad \text{and} \quad x \ge 0$$

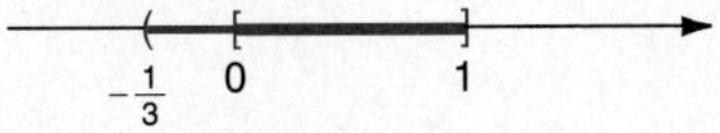

Fig. 2.21

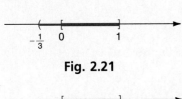

The solution set for the compound inequality is the intersection of the solution sets for $-\frac{1}{3} < x \le 1$ and $x \ge 0$. The graph in Figure 2.21 is helpful.
Set builder notation: $\{x \mid 0 \le x \le 1\}$ Interval notation: $[0, 1]$
The graph is shown in Figure 2.22. □

Fig. 2.22

Inequalities model real-life problems just as equations do.

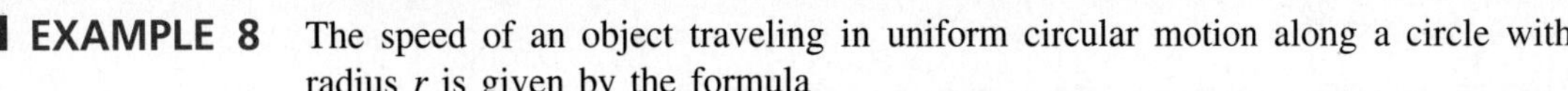

EXAMPLE 8 The speed of an object traveling in uniform circular motion along a circle with radius r is given by the formula

$$v = \frac{2\pi r}{T}$$

where v is speed and T is the time needed to make 1 revolution. A yo-yo is whirled in a circle making 2 revolutions each second. What must the radius of the circle be if the velocity is not to exceed 48 feet per second?

Solution Let r be the radius of the circle measured in feet.

T is $\frac{1}{2}$ second since 2 revolutions are made in 1 second.

$$v = \frac{2\pi r}{T}$$

If the velocity is not to exceed 48 feet per second, then

$$v \le 48$$
$$\frac{2\pi r}{0.5} \le 48$$
$$4\pi r \le 48$$
$$r \le \frac{48}{4\pi} \approx 3.82$$

The radius must not exceed $\frac{12}{\pi}$ feet which is approximately 3.82 feet. □

EXAMPLE 9 One 8-ounce cup of regular brewed percolated coffee contains 100 milligrams to 150 milligrams of caffeine. Miss Delia Hood drinks $3\frac{1}{2}$ cups of coffee and eats a small chocolate bar containing 15 milligrams of caffeine before noon every day. If her cup holds 6 ounces, how much caffeine does she consume every morning?

Solution Let c be the number of milligrams of caffeine in one 8-ounce cup of coffee. Then,

$$100 < c < 150$$

Since c is the number of milligrams of caffeine in 8 ounces, her cup has $\frac{6}{8}c$ or $0.75c$ milligrams of caffeine. The amount of caffeine in the $3\frac{1}{2}$ cups of coffee that she drinks is $(0.75) \cdot (3.5)c$. So, her total consumption of caffeine is $(0.75)(3.5)c + 15$. Thus,

$$(0.75)(3.5)(100) + 15 < (0.75)(3.5)c + 15 < (0.75)(3.5)150 + 15$$

$$277.5 < (0.75)(3.5)c + 15 < 408.75$$

Miss Delia consumes between $277\frac{1}{2}$ and $408\frac{3}{4}$ milligrams of caffeine every morning. □

PROBLEM SET 2.2

Warm-ups

In Problems 1 through 6, solve each inequality.
For Problems 1 through 4, see Example 1.

1. $2x - 4 > 5(x - 2)$

2. $\frac{1}{2}(x - 3) \le \frac{1}{3}(x - 5)$

3. $\frac{x - 2}{3} + \frac{x - 1}{4} \ge \frac{1}{4}$

4. $2(2v + 1) < 3(1 + v) + v$

For Problems 5 and 6, see Example 2.

5. $\frac{1}{4}z - \frac{2}{5}(z - 1) \le \frac{8 - 3z}{20}$

6. $2(t + 3) - 3(2t - 5) < 1 - 4t$

In Problems 7 and 8, solve each inequality and graph the solution set. See Example 3.

7. $\frac{2 - x}{5} - \frac{1}{2} < x$

8. $3(1 - 2x) \ge 5(x - 2) + 1$

In Problems 9 and 10, solve each inequality and graph the solution set for each. See Example 4.

9. $-1 < 3x - 10 \le 5$

10. $2(2s + 5) \le 2$ or $3 - 2s < 4$

In Problems 11 and 12, write sets that represent each graph. See Examples 5 and 6.

11.

−1 2

12. 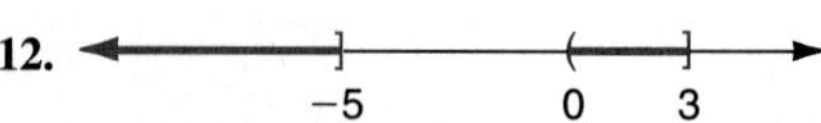

In Problems 13 through 16, solve each inequality, write the solution sets using both interval and set builder notation, and graph each solution set. See Example 7.

13. $(5 - 2w)2 \ge 7$

14. $0 < 5x - 3 \le 4$

15. $3x - 1 \le 8$ or $5 - x < 4$

16. $2 - y < 0$ or $0 \le 3y + 6 \le 6$

In Problems 17 through 19, use inequalities to work each problem. See Examples 8 and 9.

17. No engine operating between absolute temperatures T_1 and T_2 can be more efficient than $1 - \frac{T_2}{T_1}$. That is,

$$E \leq 1 - \frac{T_2}{T_1}$$

If a steam engine is operating between 560°C and 85°C, what is its maximum efficiency expressed as a percent? (Absolute temperature is measured in kelvins and given by the formula $K = C + 273$.)

18. Eddie wants to know what grades on the third and fourth tests will result in an A (90 or above) in history. Eddie's first two grades are 91 and 84. Find three possibilities for Eddie to make an A.

19. A 12-ounce can of cola contains between 35 and 65 milligrams of caffeine, while one Excedrin tablet contains 65 milligrams of caffeine. Dr. T. J. Jackson drank five glasses (8 ounces per glass) of cola and took two Excedrin tablets while grading the final exam for a calculus class. How much caffeine did Dr. Jackson consume?

Practice Exercises

In Problems 20 through 27, write sets in interval notation that represent each graph.

20. 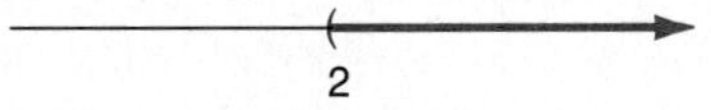 2

21. 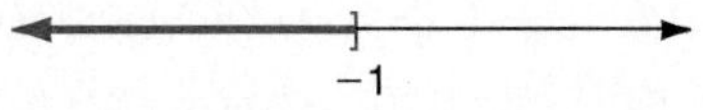 -1

22. 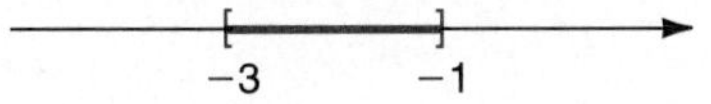 -3 -1

23. 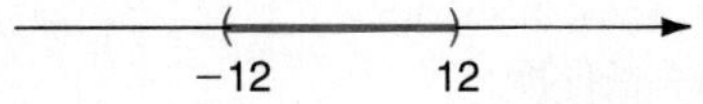 -12 12

24. 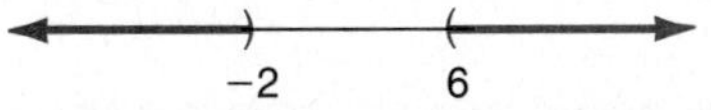 -2 6

25. 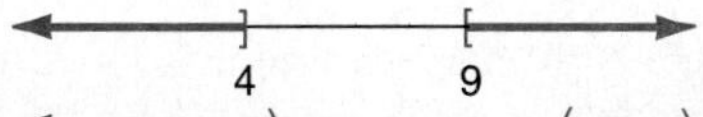 4 9

26. 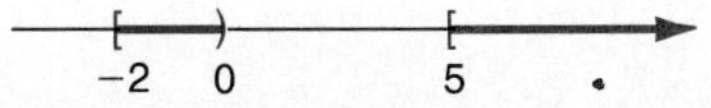 -2 0 5

27. 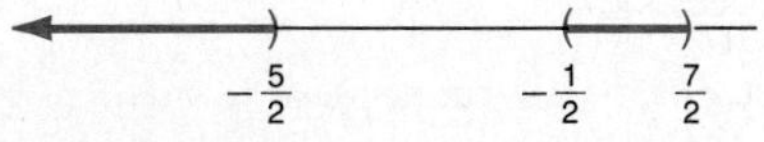 $-\frac{5}{2}$ $-\frac{1}{2}$ $\frac{7}{2}$

In Problems 28 through 41, solve each inequality, write the solution sets using both interval and set builder notation, and graph each solution set.

28. $x - 6 < 5x + 4$

29. $4t - 11 > t + 4$

30. $\frac{1}{2}t - \frac{1}{3} \leq \frac{1}{3}t + \frac{1}{6}$

31. $\frac{x}{3} + \frac{1}{2} \geq \frac{3x}{4} - \frac{1}{6}$

32. $\frac{3s + 2}{-3} \geq \frac{5s - 1}{4}$

33. $\frac{x - 5}{2} \leq \frac{2x + 11}{-5}$

34. $5.1(x + 4.2) - (x - 6) > 2[2.1x - (x + 3)]$

35. $-3[5p - (2p - 1.1)] < 0.2(p - 1) - 3(0.1 - p)$

36. $4 < 2 - x \leq 10$

37. $-2 \leq 3 - 2y < 11$

38. $\frac{1}{2}y + 2 \leq \frac{1}{3}y + \frac{3}{2}$ or $3(y - 1) > 3$

39. $4(2x + 3) < 5x + 3$ or $3 - \frac{x}{4} \leq \frac{1}{2} - \frac{x}{5}$

40. $1 < 2x - 1 < 5$ and $4x - 6 \geq 0$

41. $3x - 7 \leq 5$ and $\frac{3}{2}x + 4 > 4$

In Problems 42 through 55, solve each inequality. (Write solution sets in interval notation.)

42. $0.5(x + 0.3) < 0.2x - 0.1$

43. $0.3t + 2.1 > 7(t - 0.4)$

44. $12t + 17 \geq (3t - 11)4$

45. $6x - 13 \leq (x - 7)6$

46. $\frac{1}{3}(w + 2) > \frac{1}{4}(w - 2)$

47. $\frac{1}{5}(3 - x) < (6 - x)\frac{1}{4}$

48. $-\frac{1}{2}(x - 6) \leq \frac{2}{3}(x + 6)$

49. $\frac{3}{2}(2g + 3) \geq -\frac{1}{3}(2g + 3)$

50. $3[5 - (z - 2)] - 2z \geq 6z - 7$

51. $2[7 - (x - 1)] - 2x \leq -4 - 3x$

52. $6 - 7x - 3(3x - 4) > -8(2x - 7)$

53. $3(7x + 11) < 4(3x - 7) - 3(5 - 3x)$

54. $-2(v + 3) \leq -3(2 + v) + v$

55. $4 - 5(3 - x) - x \geq 2(x - 3) - (10 - 2x)$

Problems 56 through 60 are word problems involving inequalities. Answer each with a sentence.

56. A rectangular picture frame is to be made from a special framing stock. What is the largest size picture that can be framed if the width is to be two-thirds the length and there is only a 74-inch piece of the special stock and an additional 1 inch of framing is required for each corner?

57. Christopher poured 0.15 kilograms of tea at 100°C into his 0.5-kilogram mug. If he likes to drink his tea between 74°C and 86°C, what should the original temperature of the mug be?

To calculate the temperature of the tea in the mug, the heat gained by the mug must be equal to the heat lost by the tea. If we assume that no heat is transferred to or from the outside, the relationship is expressed by the formula

$$0.2(T - t_m)m_m = (t_t - T)m_t$$

with variables representing:

T: final temperature of the mug with tea in it (in degrees Celsius)

t_m: temperature of the mug (in degrees Celsius)

t_t: temperature of the tea (in degrees Celsius)

m_t: mass of the tea (in kilograms)

m_m: mass of the mug (in kilograms)

58. What will the speed (in meters per second) of the tip of the minute hand on a clock be if the length of the minute hand is between 0.5 meter and 1.5 meters? (Use the formula in Example 8.)

59. Marion Green has 2 yards of old lace that she wishes to use for the border of a hand-made rectangular scarf. She plans to sew one row of lace around the outside and one row of lace 1 inch from the outside. If the length of the scarf is to be twice the width, what dimensions are possible for such a scarf?

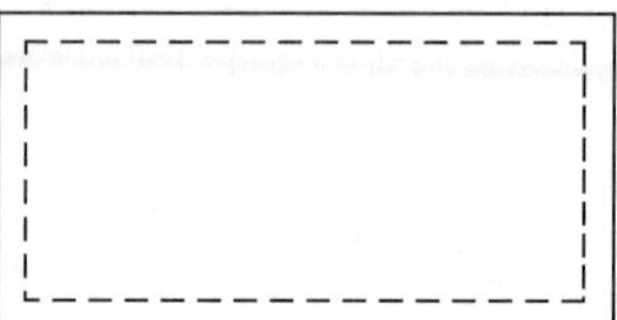

60. An engine operates between 1250°K and 500°K. If its efficiency is 35% what percent of its maximum efficiency is this? Use the formula in Problem 17.

61. Prove that the sum of two positive numbers is a positive number.

62. Prove that the sum of two negative numbers is a negative number.

63. If $c > 0$ and $a > b$, prove that $ac > bc$.

64. If $c < 0$ and $a < b$, prove that $ac > bc$.

65. If $a < b$ and $c < d$, prove or disprove with an example that $a + c < b + d$.

66. If x and y are positive numbers with $x < y$, prove that $\frac{1}{y} < \frac{1}{x}$. Is the result true if $0 > x > y$?

67. If x and y are positive numbers with $x < y$, prove that $x^2 < y^2$. Is the result true if $0 > x > y$?

Challenge Problems

In Problems 68 through 75, solve each inequality; a, b, and c are constants, with $b > 0$ and $c < 0$.

68. $bx > c$

69. $cx > b$

70. $-bx > a$

71. $-cx > a$

72. $c^2x + x \leq ab$

73. $-a^2x \leq bc + x$

74. $2x < c$ or $5x > b$

75. $3c \leq bx + a < a^2$

IN YOUR OWN WORDS . . .

76. Compare solving linear equations with solving linear inequalities.

2.3 ABSOLUTE VALUE EQUATIONS AND INEQUALITIES

The absolute value of a number is its distance from zero on the number line. The definition is

$$|q| = \begin{cases} q & \text{if } q \geq 0 \\ -q & \text{if } q < 0 \end{cases}$$

Thus we see that the absolute value of 6 is 6, because 6 is greater than zero. So,

$$|6| = 6$$

However, the absolute value of -6 is also 6, because -6 is *negative,* and so

$$|-6| = -(-6) = 6$$

Absolute Value Equations

Absolute value equations can be solved by using the definition of absolute value. Since the absolute value of a number is its distance from 0, there is also a geometric approach. We will consider three types of absolute value equations:

1. $|X| = k$, where k is a real number constant
2. $|X| = |V|$, where X and V are variable expressions
3. $|X| = V$, where X and V are variable expressions

Let's begin with the first type. We consider two cases: first, $k \geq 0$, and second, $k < 0$. Let's solve $|x| = 6$. That is, we are to find the set of numbers that make the statement true. If x is 6, it is certainly true, and if x is -6, it is also true. It is clearly false if x is any other real number. Thus the solution set is $\{-6, 6\}$.

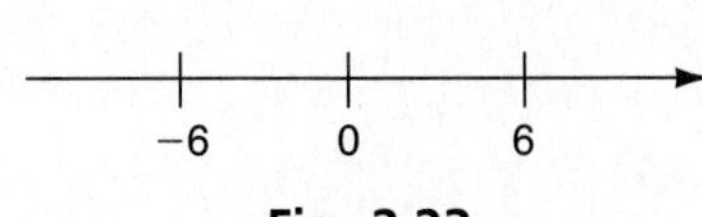

Fig. 2.23

The geometric approach relies on the fact that $|x|$ represents the distance of numbers x from 0. The solutions to $|x| = 6$ are numbers that are 6 units from 0. There are two such numbers, -6 and 6 (Figure 2.23).

Next, let's solve the equation $|x| = -6$. Here we are to find the set of numbers whose *absolute value* is -6. Since absolute value is a distance, all absolute values are nonnegative. There are no such numbers whose absolute value is -6. The solution set is the empty set $\varnothing$.

Absolute Value Equations

Type I: $|X| = k$, where k is a real number constant.

1. If $k \geq 0$, then $|X| = k$ is equivalent to the two equations

$$X = k \quad \text{or} \quad X = -k$$

2. If $k < 0$, the solution set for the equation $|X| = k$ is the empty set.

EXAMPLE 1 Solve each equation.

(a) $|x| = \dfrac{1}{2}\pi$ (b) $|2x - 5| - 4 = 0$ (c) $|3y + 12| = 0$

Solutions (a) $|x| = \frac{1}{2}\pi$

As π is positive, $\frac{1}{2}\pi$ is positive, and so the given equation is equivalent to the two equations

$$x = \frac{1}{2}\pi \qquad \text{or} \qquad x = -\frac{1}{2}\pi$$

and the solution set is

$$\left\{-\frac{1}{2}\pi, \frac{1}{2}\pi\right\}$$

(b) $|2x - 5| - 4 = 0$

The X inside the absolute value need not be a simple variable. It stands for the entire expression inside the absolute value bars. However, we must write the equation in the form $|X| = k$ *before we decide what to do.*

$$|2x - 5| = 4 \qquad \text{Addition Property of Equality}$$

This now fits the pattern. The one absolute value equation is equivalent to a pair of equations.

$$2x - 5 = 4 \qquad \text{or} \qquad 2x - 5 = -4$$

We solve both these equations.

$$2x = 9 \qquad \text{or} \qquad 2x = 1 \qquad \text{Addition Property of Equality}$$

$$x = \frac{9}{2} \qquad \text{or} \qquad x = \frac{1}{2} \qquad \text{Multiplication Property of Equality}$$

$$\left\{\frac{1}{2}, \frac{9}{2}\right\} \qquad \text{Solution set}$$

(c) $|3y + 12| = 0$

Since 0 is *not negative,* the equation fits the pattern when $k = 0$. However, as 0 and -0 represent the same number, we just get one equation.

$$3y + 12 = 0$$

$$3y = -12$$

$$y = -4$$

$$\{-4\} \qquad \text{Solution set}$$

□

EXAMPLE 2 Solve $|2x - 5| + 4 = 0$.

Solution

$$|2x - 5| + 4 = 0$$

Again we must first put the equation in the form $|X| = k$.

$$|2x - 5| = -4$$

Now this equation fits the pattern, $|X| = k$, where $k < 0$, and we have an empty solution set.

$$\varnothing$$

□

Suppose we are dealing with the situation $|X| = |V|$, where both X and V are variable expressions. As $|V|$ is never negative, $|X| = |V|$ is equivalent to the pair of equations

$$X = |V| \qquad \text{or} \qquad X = -|V|$$

But, since $|V|$ is either V or $-V$, these simplify to $X = V$ or $X = -V$.

Absolute Value Equations

Type II: $|X| = |V|$, where X and V are variable expressions.
$|X| = |V|$ is equivalent to

$$X = V \qquad \text{or} \qquad X = -V$$

EXAMPLE 3 Solve the equation, $|8 - x| - |x + 1| = 0$

Solution

$$|8 - x| - |x + 1| = 0$$

We must write this in the form $|U| = |V|$ before writing a pair of equivalent equations.

$$|8 - x| = |x + 1|$$

This is now in the Type II form, $|X| = |V|$.

$$8 - x = x + 1 \qquad \text{or} \qquad 8 - x = -(x + 1)$$

$$7 = 2x \qquad \text{or} \qquad 8 - x = -x - 1$$

$$\frac{7}{2} = x \qquad \text{or} \qquad 8 = -1$$

There are no solutions from the second equation, so the solution set is

$$\left\{\frac{7}{2}\right\}$$

□

An absolute value equation like

$$|x - 2| = 2x + 1$$

is a trickier proposition. Notice that if $x < -\frac{1}{2}$, the right-hand side of this equation is *negative,* so such a value *cannot* be a solution, no matter how it was found! The best approach is to solve the pair of equations and throw out any candidate solutions that make the left-hand side negative. In this case we solve

$$\begin{aligned} x - 2 &= 2x + 1 &\quad \text{or} \quad x - 2 &= -(2x + 1) \\ -3 &= x &\quad \text{or} \quad x - 2 &= -2x - 1 \\ -3 &= x &\quad \text{or} \quad 3x &= 1 \\ -3 &= x &\quad \text{or} \quad x &= \frac{1}{3} \end{aligned}$$

We have *candidate solutions* of -3 and $\frac{1}{3}$. Each candidate must be checked in the original equation to make sure it does not make the right side negative. The candidate -3 gives the right side a value of -5; thus it is *not* a solution of the original equation. The candidate $\frac{1}{3}$, however, gives the right side the value $\frac{5}{3}$ and is a solution. Thus the solution set is $\left\{\frac{1}{3}\right\}$.

Absolute Value Equations

Type III: $|X| = V$, where X and V are variable expressions
The solutions to the equation $|X| = V$ are *included in* the solutions to

$$X = V \quad \text{or} \quad X = -V$$

Remember that any candidate solutions that fail to check in the original equation are not solutions of $|X| = V$.

EXAMPLE 4 Solve the equation, $|2x + 1| = x - 1$

Solution First we solve the pair of equations

$$\begin{aligned} 2x + 1 &= x - 1 &\quad \text{or} \quad 2x + 1 &= -(x - 1) \\ x &= -2 &\quad \text{or} \quad 2x + 1 &= -x + 1 \\ & & 3x &= 0 \\ & & x &= 0 \end{aligned}$$

Then we check the candidates in the original equation. First, -2.
Left side: $|2x + 1| = |2(-2) + 1| = |-3| = 3$

Right side: $x - 1 = -2 - 1 = -3$
Thus -2 does not check.
Now we check 0.
Left side: $|2x + 1| = |2 \cdot 0 + 1| = 1$
Right side: $x - 1 = 0 - 1 = -1$
Likewise 0 does not check. Since there are no other candidates, the original equation has no solutions.

$\varnothing$ Solution set □

Boundary Numbers

The multiplication and addition tools we used to solve linear inequalities are not sufficient to solve most *nonlinear* inequalities. Now we will examine absolute value inequalities, which are one important type of nonlinear inequality, and develop a technique that extends to all inequalities.

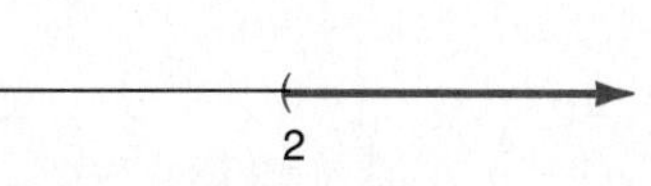

Fig. 2.24

First, let's look at a simple linear inequality. If we examine the graph in Figure 2.24 of the solution set for the inequality $3x - 1 > 5$, we see that *every number* to the right of 2 *belongs* to the solution set and *no numbers* to the left of 2 belong to the solution set. The number 2 plays an important role. It divides the number line into two regions, one of which belongs to the solution set *in its entirety* and the other of which contains *no* solutions of the inequality. We call 2 a **boundary number** for the inequality $3x - 1 > 5$. Boundary numbers divide the number line into distinct regions, each of which *belongs to the solution set in its entirety,* or contains *no members of the solution set.*

How can we obtain the boundary numbers for a given inequality? Notice that if we solve the *equation* obtained by replacing the inequality symbol with an equals symbol, we obtain the boundary number. We call the equation thus formed the **associated equation.**

$$3x - 1 > 5 \quad \text{The given inequality}$$
$$3x - 1 = 5 \quad \text{The associated equation}$$
$$3x = 6$$
$$x = 2$$

> **Boundary Numbers**
>
> Solutions to the equation formed by replacing the inequality symbol of an inequality with an equals symbol are *boundary numbers* for that inequality.

It may seem that this method is a lot of fuss over a simple inequality. As a matter of fact, it is. However, this technique works equally well for inequalities that cannot be solved with linear techniques. The reason it works relies on the fact that in any interval determined by the boundary numbers, the inequality will be either true or false throughout the entire interval. The proof depends on concepts that we have not yet discussed but will be fully developed in calculus.

The Method of Boundary Numbers for Solving Inequalities

1. Find the boundary numbers.
2. Locate the boundary numbers on a number line.
3. Determine which of the regions formed by the boundary numbers make the inequality true by testing one number from each region.
4. Shade only the regions that test true.
5. Check the boundary numbers themselves.
6. Write the solution set.

Absolute Value Inequalities

Let's solve an inequality using the techniques of boundary numbers.

EXAMPLE 5 Solve $|5 - 2x| \leq 1$.

Solution Find the boundary numbers.

$$|5 - 2x| = 1$$

$$5 - 2x = 1 \quad \text{or} \quad 5 - 2x = -1$$

$$-2x = -4 \quad \text{or} \quad -2x = -6$$

$$x = 2 \quad \text{or} \quad x = 3$$

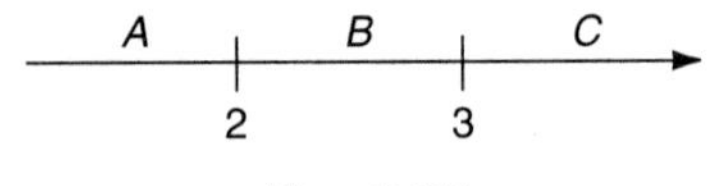

Fig. 2.25

Locate the boundary numbers on a number line (Figure 2.25). The next step is to test regions A, B, and C. A table such as the one that follows can be made, or the regions can be shaded as they are tested.

REGION	TEST NUMBER	ORIGINAL INEQUALITY $\lvert 5 - 2x \rvert \leq 1$	TRUE OR FALSE	IN SOLUTION SET?
A	0	$\lvert 5 - 2 \cdot 0 \rvert \leq 1$	F	No
B	$\frac{5}{2}$	$\left\lvert 5 - 2 \cdot \frac{5}{2} \right\rvert \leq 1$	T	Yes
C	6	$\lvert 5 - 2 \cdot 6 \rvert \leq 1$	F	No

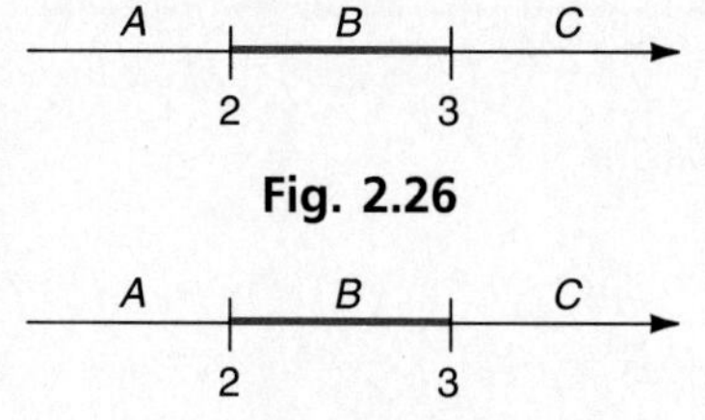

Fig. 2.26

Fig. 2.27

Shade the regions that test true (Figure 2.26). Check the boundary numbers themselves. As the original inequality $|5 - 2x| \leq 1$ includes equality, both boundary numbers belong to the solution set (Figure 2.27). Now, we can write the solution set by inspection of its graph.

$$[2, 3] \qquad \text{Solution set}$$ □

The table used in the previous examples may seem a bit formal. It is often convenient to test the regions in order from left to right (or from right to left) and shade, or not shade, each region.

EXAMPLE 6 Solve $\left|\dfrac{x-3}{2}\right| \leq |x + 5|$.

Solution Find the boundary numbers.

$$\left|\frac{x-3}{2}\right| = |x + 5|$$

$$\frac{x-3}{2} = x + 5 \quad \text{or} \quad \frac{x-3}{2} = -(x + 5)$$

$$x - 3 = 2x + 10 \quad \text{or} \quad x - 3 = -2x - 10$$

$$-13 = x \quad \text{or} \quad 3x = -7$$

$$x = -\frac{7}{3}$$

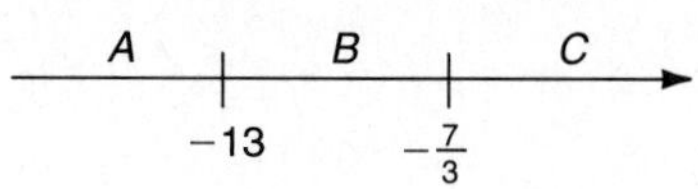

Fig. 2.28

The boundary numbers are -13 and $-\dfrac{7}{3}$. We locate them on the number line in Figure 2.28.

REGION	TEST NUMBER	ORIGINAL INEQUALITY $\left\|\frac{x-3}{2}\right\| \leq \|x+5\|$	TRUE OR FALSE	IN SOLUTION SET?
A	-15	$\left\|\frac{-18}{2}\right\| \leq \|-10\|$	T	Yes
B	-5	$\left\|\frac{-8}{2}\right\| \leq \|-5 + 5\|$	F	No
C	0	$\left\|\frac{-3}{2}\right\| \leq \|5\|$	T	Yes

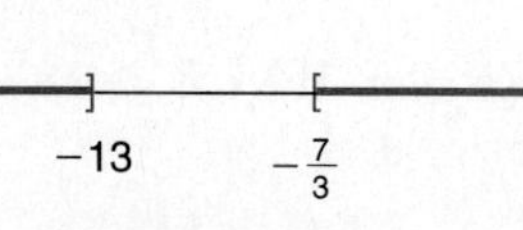

Fig. 2.29

Equality is permitted in the original inequality, so both boundary numbers belong to the solution set (Figure 2.29).

We write the solution set.

$$(-\infty, -13] \cup \left[-\frac{7}{3}, +\infty\right)$$ □

EXAMPLE 7 Solve $|4 - 3x| > 4x$.

Solution Find the boundary numbers.

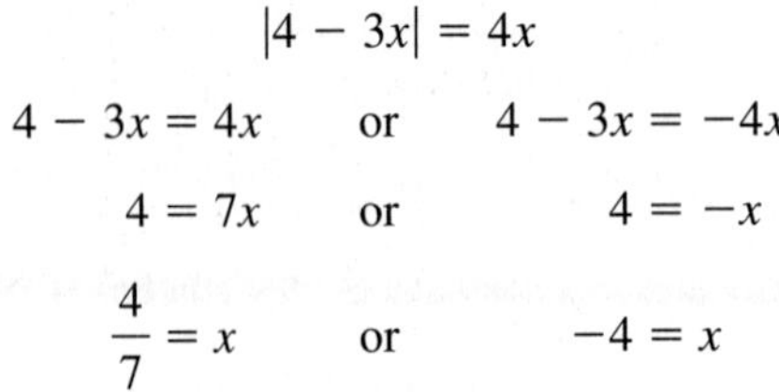

$$|4 - 3x| = 4x$$

$$4 - 3x = 4x \quad \text{or} \quad 4 - 3x = -4x$$

$$4 = 7x \quad \text{or} \quad 4 = -x$$

$$\frac{4}{7} = x \quad \text{or} \quad -4 = x$$

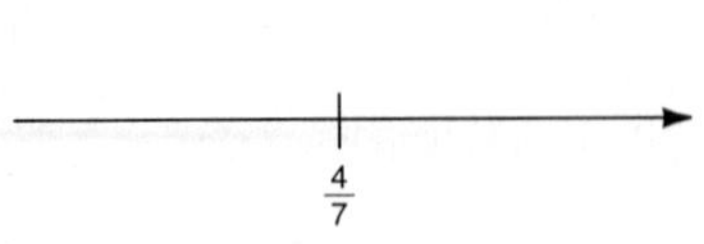

Fig. 2.30

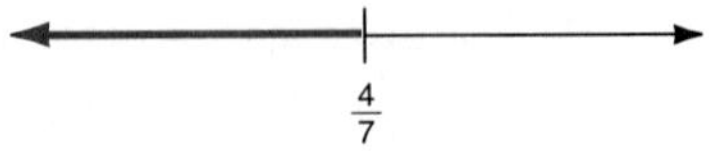

Remember, we *must check* in this type of absolute value equation. Note that $\frac{4}{7}$ checks in $|4 - 3x| = 4x$, while -4 does not. So -4 *is not* a solution to the associated equation; thus, -4 *is not* a boundary number. We locate the only boundary number on a number line (Figure 2.30). Since 0 is in the left region we test with it. The inequality $4 > 0$ is true, so we shade the left region. We test the right region with 4. The inequality $8 > 16$ is false, so we do not shade the right region (Figure 2.31). However, equality is not permitted in the original inequality $|4 - 3x| > 4x$, so the boundary number *does not* belong to the solution set (Figure 2.32).

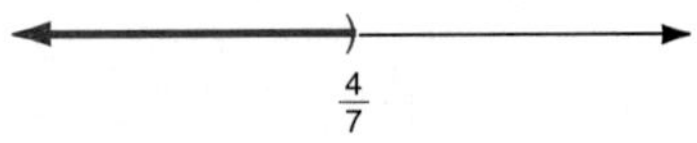

Fig. 2.31

Fig. 2.32

$$\left(-\infty, \frac{4}{7}\right)$$

□

EXAMPLE 8 Solve each inequality.

(a) $|2x + 1| < -5$ (b) $|2x + 1| > -5$

Solutions (a) $|2x + 1| < -5$

We find the boundary numbers by solving the equation

$$|2x + 1| = -5$$

However, this equation *has no solutions!* Therefore, there are no boundary numbers. So, when we locate the boundary numbers on a number line, we get one region, *the entire line* (Figure 2.33). We can test with *any* number. If we choose 0, we see that the inequality $1 < -5$ is false, so 0 is not in the solution set. Thus, no numbers are in the solution set. The solution set is the empty set.

Fig. 2.33

$$\varnothing \quad \text{Solution set}$$

We could have written the solution set by inspection. There is no number that will make an absolute value less than to a negative number.

(b) $|2x + 1| > -5$

Notice that the associated equation for this inequality is exactly the same as the associated equation for the inequality in part (a). Again we have no boundary numbers and thus a single region (Figure 2.34). However, when we test with 0

Fig. 2.34

Fig. 2.35

this time, we get the inequality $1 > -5$, which is *true*. Therefore 0 belongs to the solution set, which means *all numbers* belong. The solution set is the set of all real numbers (Figure 2.35).

$$(-\infty, +\infty) \qquad \text{Solution set}$$ □

Absolute Value Inequalities and Equivalent Compound Inequalities

Absolute value inequalities of the forms

$$|x| < q \qquad |x| > q$$
$$|x| \leq q \qquad |x| \geq q$$

with $q > 0$, arise frequently in mathematics. Often the solution set is not required, but rather an equivalent compound inequality without the absolute value symbol. A geometric interpretation of these absolute value inequalities shows the equivalent compound inequalities. Since $|x|$ represents the distance between x and 0, each of these inequalities can be described in terms of distance from 0.

Solutions to $|x| < q$ consist of all numbers whose distance from 0 is *less than* q units. This is the set of numbers between $-q$ and q. Likewise, solutions to $|x| \leq q$ are numbers between $-q$ and q *and including* $-q$ and q.

ABSOLUTE VALUE INEQUALITY	GRAPH OF SOLUTION SET	COMPOUND INEQUALITY
$\|x\| < q$	$-q$ 0 q	$-q < x < q$
$\|x\| \leq q$	$-q$ 0 q	$-q \leq x \leq q$

Solutions to $|x| > q$ consist of all numbers whose distance from 0 is *greater than* q, and solutions to $|x| \geq q$ are all numbers whose distance from 0 is *greater than or equal to* q.

ABSOLUTE VALUE INEQUALITY	GRAPH OF SOLUTION SET	COMPOUND INEQUALITY
$\|x\| > q$	$-q$ 0 q	$x < -q$ or $x > q$
$\|x\| \geq q$	$-q$ 0 q	$x \leq -q$ or $x \geq q$

Absolute Value Inequalities and Equivalent Compound Inequalities

If $q > 0$ and X is a variable expression, then

$$|X| < q \qquad \text{is equivalent to} \qquad -q < X < q$$
$$|X| > q \qquad \text{is equivalent to} \qquad X < -q \text{ or } X > q$$

The results hold if the $<$ and $>$ are replaced by $\leq$ and $\geq$.

EXAMPLE 9 Write each absolute value inequality as an equivalent compound inequality and simplify.

(a) $|3x - 7| < 5$ (b) $|1 - 2x| \geq 6$

Solutions (a) $|3x - 7| < 5$

This is of the form $|X| < q$, where $q > 0$ and X is an expression. Thus, we can write

$$-5 < 3x - 7 < 5$$

This can be simplified by adding 7 to all three sides and then dividing by 3.

$$2 < 3x < 12$$

$$\frac{2}{3} < x < 4$$

(b) $|1 - 2x| \geq 6$

This is of the form $|X| \geq q$, where $q > 0$ and X is an expression. Therefore,

$$1 - 2x \leq -6 \quad \text{or} \quad 1 - 2x \geq 6$$

$$-2x \leq -7 \quad \text{or} \quad -2x \geq 5$$

$$x \geq \frac{7}{2} \quad \text{or} \quad x \leq -\frac{5}{2}$$

□

An application of absolute value is the distance between two numbers on the number line. Recall that if a and b are two real numbers, the distance between them is $|a - b|$. So, $|x - 7|$ represents the distance between x and 7.

EXAMPLE 10 Graph each set described and then write an absolute value equation or inequality whose solution set is the given numbers.

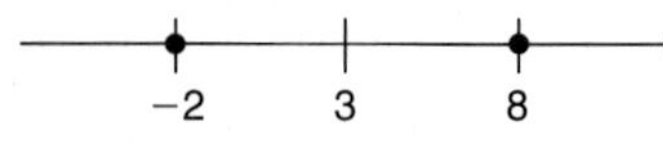

Fig. 2.36

(a) The set of numbers 5 units from 3.

(b) The set of numbers within 5 units of 3.

(c) The set of numbers greater than or equal to 5 units from 3.

Solutions (a) There are two numbers that are 5 units from 3. They are -2 and 8 (Figure 2.36). If x is a number whose distance from 3 is 5, then the distance between x and 3 is 5 units. So, we have $|x - 3| = 5$.

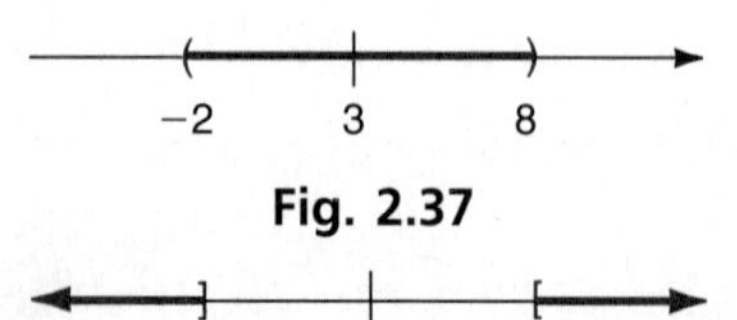
Fig. 2.37

Fig. 2.38

(b) The numbers that are within 5 units of 3 are the numbers between -2 and 8 (Figure 2.37). If x is a number whose distance from 3 is within 5 units, then $|x - 3| < 5$.

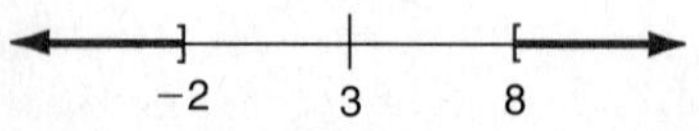

(c) The numbers greater than or equal to 5 units from 3 are to the left of -2 and to the right of 8 and include -2 and 8 (Figure 2.38). If x is such a number, we have $|x - 3| \geq 5$.

□

GRAPHING CALCULATOR BOX

Absolute Value; Checking Equations and Testing Inequalities

A calculator that has a function memory can be used to check equations and inequalities. Suppose we are solving the inequality $\left|\frac{2}{5}x - \frac{1}{5}\right| < \left|\frac{1}{15}x + \frac{1}{15}\right|$. We find the boundary numbers to be $\frac{2}{7}$ and $\frac{4}{5}$. We choose -1, 0.5, and 2 as test numbers in each of the regions. Let's test these numbers in the original inequality. First, let's investigate absolute value on the calculator.

Absolute Value

TI-81

Press [Y=] and select Y_3. Press [ABS].

CASIO fx-7700G

Absolute value is in the [MATH] menu (above [GRAPH]). Press [F3] followed by [F1]. With ABS on the screen, enter the left side:

[(] [2] [X,θ,T] [÷] [5] [−] [1] [÷] [5] [)]

as Y_3 or f_3. In a similar manner, enter the right side as Y_4 or f_4.

Testing

TI-81

We have stored the left side in the [Y=] menu as Y_3, and the right side in Y_4. Next, we use the *test menu* to test the original inequality when x is -1. Store -1 in X.

Press [Y-VARS] [3] to call Y_3. Next, we press [TEST] (above [MATH] on the left side) to access the test menu. Notice the choices we have. We are testing for *less than,* so we press [5]. This returns us to the computation screen with $Y_3 <$ showing on the action line. Now we call Y_4 with [Y-VARS] [4]. $Y_3 < Y_4$ is on the screen. Press [ENTER] to complete the test.

The test menu gives 1 for *true* and 0 for *false.* The result is 0; so -1 *is not* a solution. Store 0.5 in X and repeat the procedure. The result is 1. Thus, 0.5 is a solution. Testing 2 gives 0, which says that 2 is not a solution. The solution set is the interval $\left(\frac{2}{7}, \frac{4}{5}\right)$.

CASIO fx-7700G

Each side is entered into function memory, [F MEM]. Store each test number in X and recall each side to evaluate each side separately. Compare the results visually.

Equations can be checked in the same manner on both calculators. On the TI-81, choose [1] (=) on the test menu.

PROBLEM SET 2.3

Warm-ups

In Problems 1 through 10, solve each equation. For Problems 1 through 4, see Example 1.

1. $|x + 6| = 3$

2. $|2 - 3x| - 5 = 0$

3. $2 + |2 - x| = 2$

4. $\left|\dfrac{2x + 1}{4}\right| = 1$

For Problems 5 and 6, see Example 2.

5. $-3 = |6 + 5x|$

6. $7 + \left|\dfrac{x - 3}{2} - 1\right| = 0$

For Problems 7 and 8, see Example 3.

7. $|2 - 3x| = |x + 8|$

8. $|x - 3| - |2x + 3| = 0$

For Problems 9 and 10, see Example 4.

9. $|2x + 1| + x = 0$

10. $|3x - 1| - x = 0$

In Problems 11 through 24, solve each inequality. For Problems 11 through 14, see Example 5.

11. $|3w + 2| > 4$

12. $|2x - 1| \leq 7$

13. $|4 - 3x| \geq 2$

14. $|5 - s| < 1$

In Problems 15 through 18, see Example 6.

15. $|3y - 2| \leq |y + 8|$

16. $\left|\dfrac{1}{2}x - 3\right| > |11x + 7|$

17. $|-4x - 5| < |-5x|$

18. $\left|\dfrac{7 - 3k}{3}\right| \geq \left|\dfrac{k + 1}{5}\right|$

For Problems 19 through 22, see Example 7.

19. $|2x + 3| > 3x - 10$

20. $\left|\dfrac{y + 5}{6}\right| < y$

21. $\left|\dfrac{2}{3}t + \dfrac{1}{2}\right| \leq \dfrac{1}{2}t + \dfrac{2}{3}$

22. $2x - 3 \geq |2x + 3|$

For Problems 23 and 24, see Example 8.

23. $|3r - 4| < -\dfrac{1}{2}$

24. $|3r - 4| > -\dfrac{1}{2}$

In Problems 25 and 26, write each absolute value inequality as an equivalent compound inequality and simplify. See Example 9.

25. $|2x + 5| \leq 7$

26. $3 < |3x - 1|$

In Problems 27 and 28, write an absolute value equation or inequality whose solution set is the given set. See Example 10.

27. The set of numbers 17 units from 0

28. The set of numbers within 0.5 units of -4

Practice Exercises

In Problems 29 through 48, solve each equation.

29. $|2(3 - 4x) - 1| = 4$

30. $|3 - 5(1 - t)| = 11$

31. $|12t + 15| + 6 = 1$

32. $8 - |4x + 3| = 22$

33. $8 - |2x - 7| = 8$

34. $\left|\dfrac{2w - 1}{17}\right| - 12 = -12$

35. $\left|\dfrac{1}{2}z + 8\right| = \dfrac{1}{3}$

36. $\left|4 - \dfrac{x}{5}\right| = \dfrac{4}{7}$

37. $\left|\dfrac{2}{7}s - \dfrac{1}{3}\right| = \dfrac{1}{7}$

38. $\dfrac{5}{13} = \left|\dfrac{2x}{13} - \dfrac{3}{2}\right|$

39. $|2x + 1| = |x - 2|$

40. $|2 - 3x| = |1 - 2x|$

41. $\left|\dfrac{x}{3}\right| = |x + 2|$

42. $\left|\dfrac{s + 2}{2}\right| = \left|\dfrac{s}{9}\right|$

43. $|2x + 1| = x - 2$

44. $3 - 2z = |5z + 11|$

45. $|x - 3| = x + 2$ **46.** $|3 - 2y| = 2y + 4$ **47.** $|3x - 2| = |2 - 3x|$ **48.** $|4C - 1| = |1 - 4C|$

In Problems 49 through 72, solve each inequality.

49. $|x - 6| < 0.001$ **50.** $|x - 1| < 0.005$ **51.** $|3 - 2x| < 4$ **52.** $|5 - 3w| > 11$

53. $\left|\frac{1}{2}t + 2\right| - 1 \le 1$ **54.** $4 + \left|3 - \frac{x}{3}\right| \ge 9$ **55.** $\left|x - \frac{11}{3}\right| + \frac{7}{3} > 1$ **56.** $12 < \left|\frac{6}{7} - 2t\right| + \frac{3}{7}$

57. $|2x + 3| < 0$ **58.** $|2x + 3| > 0$ **59.** $|2x + 3| \le 0$ **60.** $|2x + 3| \ge 0$

61. $|3r + 7| < |r - 1|$ **62.** $|2x + 5| > |2 - 5x|$ **63.** $|t + 3| \le 2t - 1$ **64.** $3x + 2 \ge |5 - 4x|$

65. $7x + |5x + 4| > 5$ **66.** $|6t - 1| + 1 < t - 1$ **67.** $|4t + 11| > \left|\frac{2}{3}t + 1\right|$ **68.** $\left|1 - \frac{3}{5}x\right| < |3 - x|$

69. $6 - |3s - 2| > 6$ **70.** $2 - |3 - 5x| < 2$ **71.** $\left|\frac{y + 8}{12}\right| < \frac{y - 1}{10}$ **72.** $\frac{2x - 7}{8} > \left|\frac{2x + 5}{6}\right|$

73. The operating range for a certain new computer is given by the inequality $|90 - T| < 40$, where T is the room temperature in degrees Fahrenheit. What is the operating range for the computer?

74. The number of vibrations per second made by sound waves is called the *frequency*. One vibration per second is called a *hertz*. The frequency f of sound that can be heard by human ears satisfies the inequality

$$|f - 11{,}000| < 9000$$

What is the range of frequencies that can be heard by human ears?

75. Drug Z must be stored at temperatures T given by the formula $|T - 9| < 6$. Drug Y must be stored at temperatures T given by the formula $|T - 15| < 5$. If T is measured in degrees Celsius, at what range of temperatures can both drugs be stored?

76. The acceptable level of concentration C in milligrams per cubic centimeter of a drug is given by the formula

$$|C - 0.02| < 1$$

What is the acceptable concentration range?

77. A client of the fitness center has a predicted body density of 1.0573 with an error of 0.0062. The actual body density is then described with the inequality $|d - 1.0573| < 0.0062$. Brozek's formula for calculating body fat is

$$\text{BF} = \frac{4.570}{d} - 4.142$$

where d is body density. Use this formula to find the range of the percent of the client's body fat.

78. Find an absolute value inequality whose solution set is $(-3, 9)$.

79. If $|x - 5| < 0.001$, prove that $|2x - 10| < 0.002$.

80. If $|x - 2| < 0.001$, prove that $|4 - 2x| < 0.002$.

81. If $0 < |x - a| < \delta$, prove that $|mx - ma| < |m|\delta$.

Challenge Problems

In Problems 82 through 85, solve each equation. k is a constant real number.

82. $|2x + 1| = k;\ k > 0$ **83.** $|2x + k| = 1$ **84.** $\left|\frac{2x}{3} - 1\right| = -k;\ k < 0$ **85.** $\left|\frac{7x - 3}{k}\right| = 1$

In Problems 86 through 95, solve each inequality. Assume m, p, and q are real numbers with p > 0 and q < 0.

86. $|x + m| > p$ **87.** $|x + m| < p$ **88.** $|x + m| > q$ **89.** $|x + m| < q$

90. $|x - p| \le 0$ **91.** $|x - q| \le 0$ **92.** $|x - m| \ge 0$ **93.** $|x - m| > 0$

94. $|x + p| \le p$ **95.** $|x + q| \le q$

IN YOUR OWN WORDS . . .

96. Discuss the relationship between absolute value and distance.

97. Suppose q is a real number. Discuss the absolute value of $-q$.

98. If $\delta > 0$ and a is a real number, solve $|x - a| < \delta$ and write the solution set in interval notation. Using a number line, explain the relationship among δ, a, $a - \delta$, $a + \delta$, and x.

2.4 QUADRATIC EQUATIONS AND INEQUALITIES

An equation that can be written in the form $ax^2 + bx + c = 0$, where a, b, and c are real numbers with $a \neq 0$, is called a **quadratic equation.** When written in this manner, it is said to be in **standard form.** The two tools used to solve linear equations, adding the same number to both sides and multiplying both sides by the same nonzero number, are not enough to solve quadratic and higher-degree equations. However, the set of complex numbers has the following important property which will provide the other tool necessary to solve equations of degree higher than 1.

> **Property of Zero Products**
>
> The statement $pq = 0$ is *true* if either $p = 0$ or $q = 0$, and *false* if neither p nor q is 0.

We will use the Property of Zero Products to find the solutions to quadratic equations. The solutions may be a nonreal, complex number. The equation must be in standard form and factored before we can use this property.

EXAMPLE 1 Solve each equation.

(a) $2x^2 + 3x = 0$ (b) $3x^2 - 12 = 0$ (c) $(x + 3)(x + 1) = -1$

Solutions (a) $2x^2 + 3x = 0$

$x(2x + 3) = 0$ Factor

$x = 0$ or $2x + 3 = 0$ Property of Zero Products

$2x = -3$

$x = -\dfrac{3}{2}$

$\left\{-\dfrac{3}{2}, 0\right\}$ Solution set

(b) $3x^2 - 12 = 0$

$3(x^2 - 4) = 0$ Common factor

$3(x - 2)(x + 2) = 0$ Difference of two squares

$x - 2 = 0$ or $x + 2 = 0$ Property of Zero Products

$x = 2$ or $x = -2$

$\{-2, 2\}$ Solution set

A common abbreviation for $\{-2, 2\}$ is $\{\pm 2\}$. Also, we often write

$$x = \pm k \text{ which means } x = k \text{ or } x = -k.$$

(c) $(x + 3)(x + 1) = -1$
Here it is very tempting to set each factor on the left equal to -1 and solve. *That would be wrong!* We know that if $a \cdot b = 0$, then $a = 0$ or $b = 0$, but zero is the only number with that property. First, we must write the equation in standard form.

$$x^2 + 4x + 3 = -1$$

$$x^2 + 4x + 4 = 0 \qquad \text{Standard form}$$

Now, with zero on the right side, we factor the left.

$$(x + 2)(x + 2) = 0$$

$$\{-2\} \qquad \text{Solution set}$$

We obtained only one number in the solution set because both factors were the same. We call such solutions **repeated roots** or **roots of multiplicity 2.** □

Often mathematical models of real-life problems involve quadratic equations.

EXAMPLE 2 Billy Jones is designing his home so that the length of his family room is 3 feet less than twice its width. If he can carpet the room with 28 square yards of carpet, what will its dimensions be?

Solution We *always* start a word problem with a written assignment statement.

Let W be the width of the family room in feet.

From the wording of the problem we get another useful statement.

Then $2W - 3$ is its length.

Since the area of a rectangle is length times width we know that the area of the family room is $(2W - 3)W$, so we can carpet it with $2W^2 - 3W$ square *feet* of carpet. If we change 28 square yards to $28 \cdot 9$ or 252 square feet, we get the equation

$$2W^2 - 3W = 252$$

$$2W^2 - 3W - 252 = 0 \qquad \text{Standard form}$$

$$(2W + 21)(W - 12) = 0 \qquad \text{Factor}$$

$$W = -\frac{21}{2} \quad \text{or} \quad W = 12 \qquad \text{Solutions to the model}$$

However, $-\frac{21}{2}$ does not make sense in the problem. So we conclude that the width is 12 feet. The dimensions of the family room are 12 feet by 21 feet. □

Square Root Property

We can solve equations of the form $x^2 = A$ by factoring.

$$x^2 = A$$

$$x^2 - A = 0$$

$$x^2 - (\sqrt{A})^2 = 0 \qquad \text{Substitution}$$

$$(x - \sqrt{A})(x + \sqrt{A}) = 0 \qquad \text{Difference in two squares}$$

$$x = \pm\sqrt{A} \qquad \text{Property of Zero Products}$$

$$\{\pm\sqrt{A}\} \qquad \text{Solution set}$$

We have shown the following useful property.

Theorem: Square Root Property

$$X^2 = A$$

is equivalent to the pair of equations

$$X = \pm\sqrt{A}$$

EXAMPLE 3 Solve each equation using the Square Root Property.

(a) $x^2 = 125$ (b) $2x^2 + 16 = 0$ (c) $(x - 4)^2 = 24$ (d) $t^2 + 6t + 9 = 2$

Solutions

(a) $x^2 = 125$

$x = \pm\sqrt{125}$ Square Root Property

$x = \pm 5\sqrt{5}$

$\{\pm 5\sqrt{5}\}$ Solution set

(b) $2x^2 + 16 = 0$

$2x^2 = -16$

$x^2 = -8$ Divide by 2

$x = \pm\sqrt{-8}$ Square Root Property

$x = \pm\sqrt{8}i$ Negative radicand

$x = \pm 2\sqrt{2}i$

$\{\pm 2\sqrt{2}i\}$ Solution set

(c) $(x - 4)^2 = 24$

Note that the X in the Square Root Property need not be a single variable. It can be an expression, such as $x - 4$.

$x - 4 = \pm\sqrt{24}$ Square Root Property

$x = 4 \pm 2\sqrt{6}$

$\{4 \pm 2\sqrt{6}\}$ Solution set

(d) $t^2 + 6t + 9 = 2$

$(t + 3)^2 = 2$ Factor.

$t + 3 = \pm\sqrt{2}$ Square Root Property

$t = -3 \pm \sqrt{2}$

$\{-3 \pm \sqrt{2}\}$ Solution set □

Completing the Square

Notice the quadratic equation in part (d) of Example 3, $t^2 + 6t + 9 = 2$. We solved it *without* first writing it in standard form. However, in standard form, $t^2 + 6t + 7 = 0$, the left side cannot be factored. This suggests a general method for solving quadratic equations. We use the addition property to format one side of the equation as a perfect trinomial square, factor, and use the square root property. This is called **completing the square.**

To complete the square, we start with $x^2 + kx$ and make it into a perfect trinomial square.

Completing the Square

To make $x^2 + kx$ a perfect square, add $\left(\frac{k}{2}\right)^2$.

EXAMPLE 4 Solve $x^2 - 8x + 4 = 0$ by completing the square.

Solution First, we get the constant term out of the way.

$$x^2 - 8x = -4$$

Next, we divide the coefficient of x by 2, square the result, and *add it to both sides.* In this case the coefficient of x is -8. Half of it squared is $\left(\frac{-8}{2}\right)^2 = (-4)^2 = 16$. We add this to both sides of the equation.

$$x^2 - 8x + 16 = -4 + 16$$

Notice that the left side is a perfect trinomial square.

$$(x - 4)^2 = 12$$

$x - 4 = \pm\sqrt{12}$ Square Root Property

$$x = 4 \pm 2\sqrt{3}$$

$$\{4 \pm 2\sqrt{3}\} \qquad \text{Solution set}$$ □

EXAMPLE 5 Solve the equation $3y^2 + 6y + 11 = 0$ by completing the square.

Solution The algorithm we are using to complete the square requires the coefficient of the squared term to be 1. We must divide by 3.

$$y^2 + 2y + \frac{11}{3} = 0$$

$$y^2 + 2y = -\frac{11}{3}$$

$$y^2 + 2y + \left(\frac{2}{2}\right)^2 = -\frac{11}{3} + \left(\frac{2}{2}\right)^2 \qquad \text{Add } \left(\frac{2}{2}\right)^2 \text{ to both sides.}$$

$$y^2 + 2y + 1 = -\frac{11}{3} + 1$$

$$(y + 1)^2 = -\frac{8}{3}$$

$$y + 1 = \pm\sqrt{-\frac{8}{3}} = \pm\sqrt{\frac{8}{3}}i \qquad \text{Square Root Property}$$

$$y = -1 \pm \frac{2\sqrt{6}}{3}i$$

$$\left\{-1 \pm \frac{2\sqrt{6}}{3}i\right\} \qquad \text{Solution set}$$ □

The Quadratic Formula

The method of completing the square can be used to solve *any* quadratic equation, but it is somewhat cumbersome. However, if we solve the general quadratic equation by that method, we get a formula that can be used for solving *any* quadratic equation.

$$ax^2 + bx + c = 0;\ a \neq 0 \qquad \text{Standard form}$$

$$x^2 + \frac{b}{a}x + \frac{c}{a} = 0 \qquad \text{Divide by } a.$$

$$x^2 + \frac{b}{a}x = -\frac{c}{a} \qquad \text{Subtract } \frac{c}{a}.$$

$$x^2 + \frac{b}{a}x + \left(\frac{b}{2a}\right)^2 = -\frac{c}{a} + \left(\frac{b}{2a}\right)^2 \qquad \text{Complete the square.}$$

$$\left(x + \frac{b}{2a}\right)^2 = -\frac{c}{a} + \frac{b^2}{4a^2} \qquad \text{Factor.}$$

$$\left(x + \frac{b}{2a}\right)^2 = -\frac{4ac}{4a^2} + \frac{b^2}{4a^2}$$

$$\left(x + \frac{b}{2a}\right)^2 = \frac{b^2 - 4ac}{4a^2}$$

$$x + \frac{b}{2a} = \pm\sqrt{\frac{b^2 - 4ac}{4a^2}} \qquad \text{Square Root Property}$$

$$x = -\frac{b}{2a} \pm \frac{\sqrt{b^2 - 4ac}}{\sqrt{4a^2}}$$

$$x = -\frac{b}{2a} \pm \frac{\sqrt{b^2 - 4ac}}{2|a|}$$

The absolute value appears in the denominator because $\sqrt{a^2} = |a|$. However, as $|a|$ is either a or $-a$ and $\pm$ is in front of this term, we can eliminate the absolute value symbol.

$$x = -\frac{b}{2a} \pm \frac{\sqrt{b^2 - 4ac}}{2a}$$

The two fractions on the right have the same denominator. If we add them, we obtain the quadratic formula.

The Quadratic Formula

The quadratic equation $ax^2 + bx + c = 0$, $a \neq 0$, is equivalent to

$$x = \frac{-b \pm \sqrt{b^2 - 4ac}}{2a}$$

EXAMPLE 6 Use the quadratic formula to solve $2x^2 + 3x - 7 = 0$.

Solution Comparing the given equation with the general quadratic equation $ax^2 + bx + c = 0$, we see that a is 2, b is 3, and c is -7. Therefore, $2x^2 + 3x - 7 = 0$ is equivalent to

$$x = \frac{-3 \pm \sqrt{3^2 - 4(2)(-7)}}{2 \cdot 2} \qquad \text{Quadratic Formula}$$

$$x = \frac{-3 \pm \sqrt{9 + 56}}{4} = \frac{-3 \pm \sqrt{65}}{4} \qquad \text{Simplify}$$

$$\left\{\frac{-3 \pm \sqrt{65}}{4}\right\} \qquad \text{Solution set}$$

EXAMPLE 7 Solve each equation.

(a) $3z^2 - 12z + 12 = 0$ (b) $2t^2 = 5t - 4$

Solutions (a) $3z^2 - 12z + 12 = 0$

We notice a common factor of 3, which we divide out. (The formula would work anyway, but if we divide out the 3s, we will have smaller numbers in the calculations.)

$$z^2 - 4z + 4 = 0 \qquad \text{Divide both sides by 3.}$$

Now, $a = 1$, $b = -4$, and $c = 4$.

$$z = \frac{-(-4) \pm \sqrt{16 - 4(1)(4)}}{2 \cdot 1} \qquad \text{Quadratic formula}$$

$$= \frac{4 \pm \sqrt{16 - 16}}{2}$$

$$z = \frac{4 \pm 0}{2} = 2$$

$$\{2\} \qquad \text{Solution set}$$

(The number 2 is a root of multiplicity 2.)

(b) $2t^2 = 5t - 4$

Before we use the quadratic formula, we must write the quadratic in standard form.

$$2t^2 - 5t + 4 = 0$$

$$t = \frac{-(-5) \pm \sqrt{25 - 4(2)(4)}}{2 \cdot 2} \qquad \text{Quadratic formula}$$

$$= \frac{5 \pm \sqrt{25 - 32}}{4}$$

$$= \frac{5 \pm \sqrt{-7}}{4} = \frac{5 \pm \sqrt{7}i}{4} \qquad \text{Negative radicand}$$

$$t = \frac{5}{4} \pm \frac{\sqrt{7}}{4}i \qquad \text{Complex number in standard form}$$

$$\left\{\frac{5}{4} \pm \frac{\sqrt{7}}{4}i\right\} \qquad \text{Solution set}$$

□

A hand-held calculator is an invaluable tool for approximating solutions to quadratic equations. The next example not only shows a calculator approach to the quadratic formula but also illustrates two skills that are important in calculator work. They are *chaining the entire expression* and *doing simple arithmetic in your head.* By "chaining the entire expression," we mean evaluating the whole expression with a sequence of keystrokes *without writing down any intermediate results.* It is very important in long calculator computations to let the calculator do the work. It can store the intermediate results and recall them for later computation.

EXAMPLE 8 Approximate the solutions to the quadratic equation $2x^2 + \sqrt{3}x - \pi = 0$ to four significant digits.

Solution

$$x = \frac{-\sqrt{3} \pm \sqrt{(\sqrt{3})^2 - 4(2)(-\pi)}}{2 \cdot 2} \qquad \text{Substitution}$$

We calculate each solution from the right-hand expression in one chain of calculations without writing down any intermediate steps. As we work, we will do some of the simplest arithmetic in our heads. A particularly good strategy in such evaluations is to calculate the complicated square root first, put it into memory, and continue to approximate the first solution. Then a lot of the work is already done for the second solution.

First, we compute $\sqrt{3 + 8\pi}$

[3] [+] [8] [×] [π] [=] [√]

We store it and complete the calculation.

[STO] [−] [3] [√] [=] (This completes the numerator, taking the + part of the ±.)

We complete the calculation: [÷] [4] [=].

The display shows .8929949628, an approximation of the first solution. The second solution, which takes the − of the ±, is easily done using the [RCL] key.

[3] [√] [+/−] [−] [RCL] [=] [÷] [4] [=]

We now read the second solution, −1.759020367, on the display. So the solutions to four significant digits are −1.759 and 0.8930. □

EXAMPLE 9 The height y (in feet) of a freely falling body can be approximated with the formula

$$y = y_0 + v_0 t - 16t^2$$

where y_0 is the initial height in feet, v_0 is the initial velocity in feet per second, and t is the elapsed time in seconds. A helicopter hovering at 500 feet drops relief supplies to the ground. How long will it take the supplies to hit the ground?

Solution First make a written assignment statement.

Let t be the time (in seconds) it takes for the supplies to hit the ground.

The initial height and initial velocity occur when $t = 0$. So,

$y_0 = 500$

$v_0 = 0$ (Since the supplies are dropped and the helicopter is not moving)

Since t is the number of seconds it takes for the supplies to hit the ground, the height is 0. So,

$$y = 0$$

So the formula gives

$$y = y_0 + v_0 t - 16t^2$$

$$0 = 500 - 16t^2 \qquad \text{Substitution}$$

$$-500 = -16t^2$$

$$\frac{-500}{-16} = t^2$$

$$\pm 5.59 \approx t \qquad \text{Square Root Property}$$

The supplies will hit the ground after approximately 6 seconds. □

The Discriminant

In the quadratic formula, notice the important role played by the expression under the radical symbol, $b^2 - 4ac$. This expression, called the **discriminant,** gives the following information about the nature of the solutions of the quadratic equation.

The Discriminant

$b^2 - 4ac$	Nature of Solutions
Positive	Two distinct real solutions
Zero	One real solution of multiplicity 2
Negative	Two complex solutions

EXAMPLE 10 Use the discriminant to determine the nature of the solutions.

(a) $x^2 + x + 1 = 0$ (b) $4x^2 - x = 3$ (c) $4y^2 - 12y + 9 = 0$

Solutions (a) $x^2 + x + 1 = 0$

$b^2 - 4ac = 1^2 - 4(1)(1) = 1 - 4 = -3$

The discriminant is negative. The quadratic has two complex roots.

(b) $4x^2 - x = 3$

$4x^2 - x - 3 = 0$ Standard form

$b^2 - 4ac = (-1)^2 - 4(4)(-3) = 1 + 48 = 49$

The discriminant is positive. The quadratic has two distinct real solutions.

(c) $4y^2 - 12y + 9 = 0$

$b^2 - 4ac = (-12)^2 - 4(4)(9) = 144 - 144 = 0$

The discriminant is zero. The quadratic has one real solution of multiplicity 2. □

Quadratic Inequalities

If we replace the equality symbol in a quadratic equation with an inequality symbol, we will have a quadratic inequality. Since we know how to solve quadratic equations, we can solve quadratic inequalities using the method of boundary numbers.

EXAMPLE 11 Solve each inequality.

(a) $x^2 - 4x + 3 < 0$ (b) $-7x^2 + 5x + 2 \leq 0$

Solutions (a) $x^2 - 4x + 3 < 0$

First, we find the boundary numbers.

$x^2 - 4x + 3 = 0$ Associated equation

$(x - 1)(x - 3) = 0$

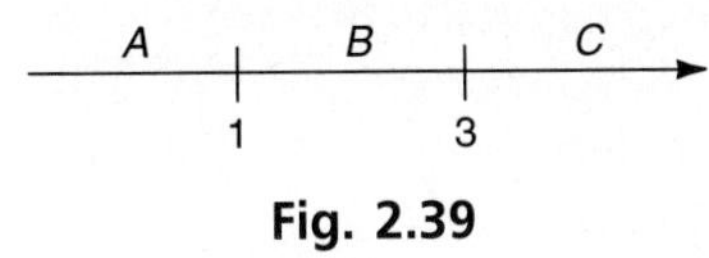

Fig. 2.39

The boundary numbers are 1 and 3.

Next, we locate the boundary numbers on a number line (Figure 2.39). We test each region.

REGION	TEST NUMBER	ORIGINAL INEQUALITY $x^2 - 4x + 3 < 0$	TRUE OR FALSE	IN SOLUTION SET?
A	0	$0^2 - 4 \cdot 0 + 3 < 0$	F	No
B	2	$2^2 - 4 \cdot 2 + 3 < 0$	T	Yes
C	5	$5^2 - 4 \cdot 5 + 3 < 0$	F	No

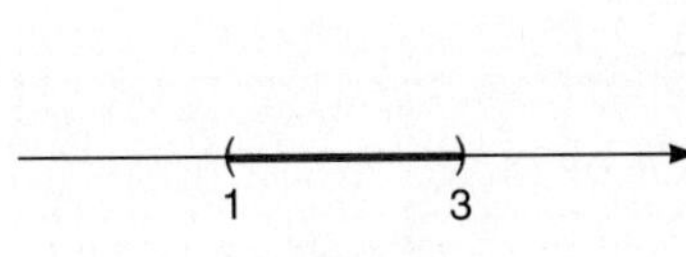

Fig. 2.40

Since both boundary numbers came from solving the associated equation and equality is *not* included, they are not in the solution set (Figure 2.40). We write the solution set from the figure. In interval notation,

$(1, 3)$ Solution set

(b) $-7x^2 + 5x + 2 \le 0$

Find the boundary numbers.

$-7x^2 + 5x + 2 = 0$	The associated equation
$7x^2 - 5x - 2 = 0$	Multiply by -1.
$(7x + 2)(x - 1) = 0$	Factor.

Fig. 2.41

The boundary numbers are $-\dfrac{2}{7}$ and 1 (Figure 2.41).

REGION	TEST NUMBER	ORIGINAL INEQUALITY $-7x^2 + 5x + 2 \le 0$	TRUE OR FALSE	IN SOLUTION SET?
A	-1	$-7(-1)^2 - 5 + 2 \le 0$	T	Yes
B	0	$-7 \cdot 0^2 - 0 + 2 \le 0$	F	No
C	2	$-7 \cdot 2^2 - 5 \cdot 2 + 2 \le 0$	T	Yes

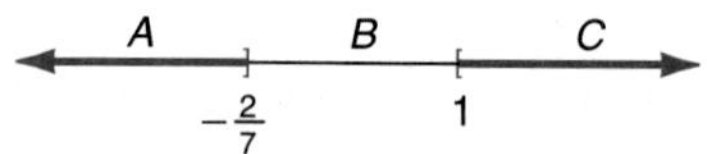

Fig. 2.42

Both boundary numbers are in the solution set shown in Figure 2.42.

$$\left(-\infty, -\frac{2}{7}\right] \cup [1, +\infty) \qquad \text{Solution set}$$

□

EXAMPLE 12 Solve each inequality.

(a) $x^2 - 4x + 2 > 0$ (b) $x^2 - 2x + 2 > 0$

Solutions (a) $x^2 - 4x + 2 > 0$

Find the boundary numbers.

$x^2 - 4x + 2 = 0$	Associated equation
$x = \dfrac{4 \pm \sqrt{(-4)^2 - 4(1)(2)}}{2 \cdot 1}$	Quadratic formula
$x = \dfrac{4 \pm \sqrt{16 - 8}}{2} = 2 \pm \sqrt{2}$	Solutions of the associated equation

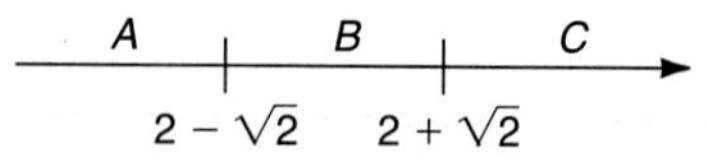

Fig. 2.43

The boundary numbers are $2 - \sqrt{2}$ and $2 + \sqrt{2}$. We locate the boundary numbers on a number line (Figure 2.43). Since 2 is greater than $\sqrt{2}$, 0 is in region A and we use it for a test number. A number that we are certain is in region B is 2. For region C, we pick a number we are sure is in that region.

REGION	TEST NUMBER	ORIGINAL INEQUALITY $x^2 - 4x + 2 > 0$	TRUE OR FALSE	IN SOLUTION SET?
A	0	$0^2 - 4 \cdot 0 + 2 > 0$	T	Yes
B	2	$2^2 - 4 \cdot 2 + 2 > 0$	F	No
C	10	$10^2 - 4 \cdot 10 + 2 > 0$	T	Yes

$2-\sqrt{2}$ $2+\sqrt{2}$

Fig. 2.44

Since both boundary numbers come from solving the associated equation and equality is not included in the original inequality, neither boundary number is included in the solution set shown in Figure 2.44.

$$(-\infty, 2-\sqrt{2}) \cup (2+\sqrt{2}, +\infty) \qquad \text{Solution set}$$

(b) $x^2 - 2x + 2 > 0$

Find the boundary numbers.

$x^2 - 2x + 2 = 0$ Associated equation

$$x = \frac{2 \pm \sqrt{(-2)^2 - 4(1)(2)}}{2 \cdot 1} \qquad \text{Quadratic formula}$$

$$x = \frac{2 \pm \sqrt{4-8}}{2} = 1 \pm i \qquad \text{Solutions of the associated equation}$$

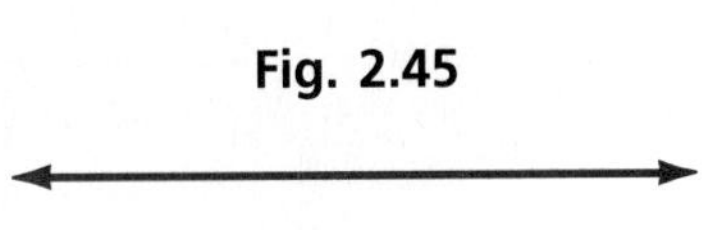

Fig. 2.45

Fig. 2.46

Since neither $1 - i$ nor $1 + i$ is represented on the number line (it has only *real numbers*), there are no boundary numbers. We have only one region—the entire line (Figure 2.45). We test with 0.

Since $0^2 - 2 \cdot 0 + 2 > 0$ is true, the entire line is in the solution set (Figure 2.46).

$$(-\infty, +\infty) \qquad \text{Solution set}$$

□

EXAMPLE 13 The Farmer's Market estimates that at a price of p dollars per basket of peaches, the weekly cost C and revenue R (in hundreds of dollars) are given by the formulas

$$C = 45 - 2p$$

$$R = 21p - 2p^2$$

At what prices will the Farmer's Market have a loss?

Solution The Farmer's Market will have a loss if revenue is less than cost. That is, when

$$R < C$$

$21p - 2p^2 < 45 - 2p$ Substitute for R and C.

This is a quadratic inequality. We solve by the method of boundary numbers.

$21p - 2p^2 = 45 - 2p$ The associated equation

$0 = 2p^2 - 23p + 45$ Standard form for the quadratic equation

$0 = (2p - 5)(p - 9)$ Factor.

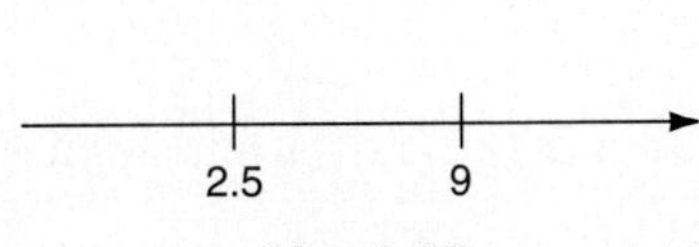

Fig. 2.47

The boundary numbers are 2.5 and 9 (Figure 2.47). Testing each region in the inequality gives a solution set of $(-\infty, 2.5) \cup (9, +\infty)$. Since price cannot be negative or zero, we do not use the solutions in the interval $(-\infty, 0]$. The Farmer's Market will have a loss if the price per basket of peaches is between 0 and \$2.50 or if the price per basket is greater than \$9.00.

□

GRAPHING CALCULATOR BOX

Approximating Solutions to Quadratic Equations with the Quadratic Formula

Approximating the solutions of quadratic equations with the quadratic formula is a good example of formula evaluation. In all cases, the solutions to the equation $ax^2 + bx + c = 0$ are given by the formula $x = \dfrac{-b \pm \sqrt{b^2 - 4ac}}{2a}$. Some solutions may be complex numbers.

Because the calculator will not evaluate the square root of a negative number, we will calculate the discriminant first. Put the discriminant, $b^2 - 4ac$, in function memory: Y_1 or f_1.

TI-81

Press [Y =] to enter the function list and arrow the cursor to Y_1.

[ALPHA] [B] [x^2] [−] [4] [ALPHA] [A] [ALPHA] [C]

CASIO fx-7700G

[ALPHA] [B] [x^2] [−] [4] [ALPHA] [A] [ALPHA] [C]

Store this in function memory f_1. [F MEM] [F1] [1] [AC]

Enter the quadratic formula as Y_2 or f_2. Notice how we use the discriminant stored in Y_1 or f_1.

TI-81

We press [Y =] to enter the function list and arrow the cursor to Y_2. Enter [(] [(−)] [ALPHA] [B] [+] [√] [Y-VARS] [1] [)] [÷] [(] [2] [ALPHA] [A] [)]

CASIO fx-7700G

We enter the formula. Press

[(] [(−)] [ALPHA] [B] [+] [√] [(] [F MEM] [F2] [1] [)] [)] [÷] [(] [2] [ALPHA] [A] [)]

We store this in function memory f_2. [F MEM] [F1] [2] [AC]

Notice that we used the + case of the ± in the formula. Now we need only to put the numbers a, b, and c into the memory locations A, B, and C and then recall the formula and evaluate it if the solutions are real numbers.

Positive Discriminant

Let's approximate solutions to the quadratic equation $2x^2 + 3x - 4 = 0$. Store 2 in A, 3 in B, and −4 in C.

TI-81

[Y-VARS] [2] [ENTER] and read the approximate root, 0.8507810594.

For the second root, go back to the function memory (Y=), edit the + to a − in Y_2, and again use Y-VARS to approximate the other root: −2.350781059.

CASIO fx-7700G

F MEM F2 2 EXE yields the first approximation, 0.8507810594.
Use the left arrow and edit the + to a − and press EXE to approximate the second root, −2.350781059.

Negative Discriminant

Now let's repeat the procedure for the equation $x^2 + 2x + 2 = 0$. Store 1, 2, and 2 in A, B, and C, respectively, and evaluate the discriminant. We get −4. The *negative* discriminant signals complex roots. The quadratic formula can be written as $x = \dfrac{-b \pm \sqrt{D}}{2a}$, where D is the discriminant. The solutions are $-1 \pm i$. (Paper and pencil may be necessary for calculations.)

PROBLEM SET 2.4

Warm-ups

In Problems 1 through 17, solve each equation.
In Problems 1 through 6, solve by factoring. See Example 1.

1. $3x^2 - 2x = 0$
2. $2x^2 + 5x - 3 = 0$
3. $9s^2 - 24s + 16 = 0$
4. $x^2 = x$
5. $3z^2 = 27$
6. $(3t + 5)(2t + 1) = -1$

For Problem 7, see Example 2.

7. The width of a rectangle is $\frac{2}{3}$ inch more than one-third the length. What are the dimensions of the rectangle if the area is 65 square inches.

8. An air search for a lost private plane is being conducted over a rectangular section of Holly Springs National Forest. The perimeter of the search zone is 160 miles, and its area is 1500 square miles. What are the dimensions of the search zone?

In Problems 9 through 11, use the Square Root Property to solve. See Example 3.

9. $x^2 = 27$
10. $w^2 + 24 = 0$
11. $(t + 8)^2 = 12$

In Problems 12 through 14, solve by completing the square. See Examples 4 and 5.

12. $x^2 + 2x - 4 = 0$
13. $x^2 + 1 = 6x$
14. $2t^2 + 6t + 8 = 0$

In Problems 15 through 17, use the quadratic formula. See Examples 6 and 7.

15. $x^2 + x + 1 = 0$
16. $3w^2 + 2 = w$
17. $2s = s^2 + 6$

In Problems 18 and 19, approximate the roots of each quadratic equation to five significant digits. See Example 8.

18. $3x^2 - \sqrt{11}x = 4\pi$
19. $2\sqrt{3}y^2 + 9\pi y + 1 = 0$

For Problems 20 and 21, see Example 9.

20. How many seconds would it take a stone dropped from a cliff 64 feet high to reach the foot of the cliff?

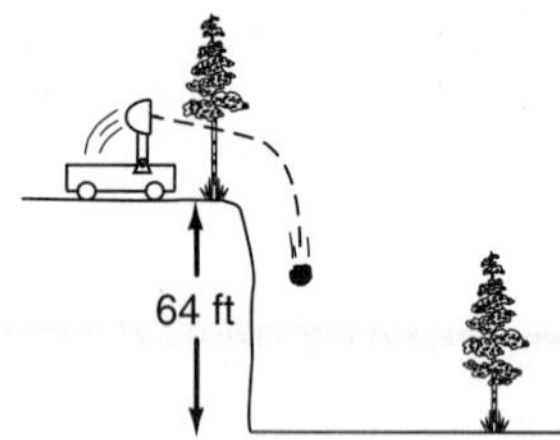

21. An object is shot vertically upward with an initial velocity of 96 feet per second from the top of a 256-foot building. How many seconds will it take the object to strike the ground?

In Problems 22 through 24, use the discriminant to determine the nature of the solutions. See Example 10.

22. $x^2 + 2x - 1 = 0$ **23.** $t^2 - 5t + 7 = 0$ **24.** $25y^2 + 70y + 49 = 0$

In Problems 25 through 30, solve each inequality. See Examples 11 and 12.

25. $x^2 > 25$ **26.** $12s^2 + 5s - 2 \le 0$ **27.** $-6y^2 + 6 \ge 5y$

28. $x^2 + 6x + 4 \ge 0$ **29.** $x^2 + 14 > 6x$ **30.** $3x \le 4x^2$

For Problems 31 and 32, see Example 13.

31. The length of a pendulum L is given by the formula

$$L = \frac{gT^2}{4\pi^2}$$

where T is the time required to make one complete swing (period) in seconds and $g \approx 32$ feet per sec^2. What will the period be to the nearest thousandth of a second if the length of the pendulum is not to exceed 6 inches.

32. The population of a small town is predicted by the equation $P = t^2 - 80t + 2000$, where t is the number of years after 1985. When will the population be greater than 4000?

Practice Exercises

In Problems 33 through 56, solve each equation by any method.

33. $x^2 = 27$ **34.** $3t^2 = 375$ **35.** $16y^2 = 3y$ **36.** $x^2 + \frac{11}{2}x = 0$

37. $x^2 - 6x + 9 = 0$ **38.** $w^2 + \frac{2}{5}w + \frac{1}{25} = 0$ **39.** $4y^2 - 7y - 2 = 0$ **40.** $3x^2 + 5x + 2 = 0$

41. $5t^2 + \frac{7}{4}t - \frac{3}{2} = 0$ **42.** $\frac{4}{3}x^2 - \frac{1}{3}x = \frac{1}{2}$ **43.** $x^2 - 6x + 1 = 0$ **44.** $r^2 + 2r = 4$

45. $t^2 - 4t = 4$ **46.** $8x + 16 = x^2$ **47.** $(x + 2)(x - 5) = 30$ **48.** $2(x - 3)(2x + 3) = 22$

49. $z^2 + z + 3 = 0$ **50.** $z^2 + z - 1 = 0$ **51.** $6x^2 + 13x - 28 = 0$ **52.** $12x^2 - 29x + 15 = 0$

53. $3t^2 - 19t = 14$ **54.** $6y^2 + 15y = 9$ **55.** $\left(x - \frac{1}{2}\right)^2 = -\frac{1}{4}$ **56.** $\left(x + \frac{1}{2}\right)^2 = 2$

In Problems 57 through 68, solve each inequality.

57. $(x - 3)^2 > 16$ **58.** $(2x - 5)^2 - 36 < 0$ **59.** $2x^2 - 5x - 12 \le 0$ **60.** $5x^2 - 4x \le 33$

61. $\frac{9}{5}x + 2 > \frac{9}{5}x^2$ **62.** $2 - \frac{1}{3}x < 5x^2$ **63.** $x^2 + 2x \ge 6$ **64.** $2x^2 - 3x < 4$

65. $10x \ge 28 + x^2$ **66.** $2x^2 - 3x + 4 > 0$ **67.** $3 - 7x \le 6x^2$ **68.** $12x^2 - 25x + 12 < 0$

69. Two years ago, Dixie Glover invested \$45,000 in a savings account at Trust Company. The account now has \$53,464.50 in it. If the interest has been compounded annually, approximate the rate of interest. [If P dollars is invested at an annual rate of r compounded once a year, after t years it will yield A dollars calculated by $A = P(1 + r)^t$.]

70. The area of a circle is given by the formula $A = \pi r^2$. By how many centimeters should a circle of radius 6 centimeters be increased to make the area 48π square centimeters.

71. A rectangular garden is to be made with a uniform border completely around the outside. The garden and border completely fill a 12 feet by 16 feet rectangle. If the area taken up by the border is to be no more than one-half the total area of garden and border, what can the width of the border be? (Figure 2.48)

72. A rectangular brass sheet is to be made into a box with no top by cutting a square piece from each corner and turning up the sides. If the perimeter of the brass sheet is 50 inches, the squares cut from the corners are 2 inches on a side, and the volume of the resulting box is to be less than 132 cubic inches, what are the dimensions of the brass sheet? (Figure 2.49)

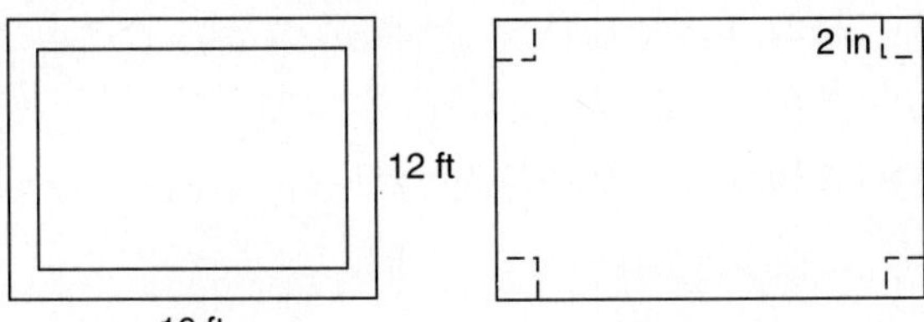

Fig. 2.48 Fig. 2.49

Challenge Problems

80. Solve $3 < x^2 - 2x < 15$.

81. The ALTA AA-2 tennis team wishes to have a water jug designed so that it will hold enough water for a long three-set match. The water jug presently being used is shown in Figure 2.50. The team feels that they need one-third more water. If the height remains the same, by approximately how many inches would the radius have to be increased? If the manufacturer can measure only in a half, quarter, eighth, or sixteenth of an inch, what should the new radius be?

IN YOUR OWN WORDS . . .

82. Given a quadratic equation to solve, explain how to decide what method to use to solve it.

73. After extensive research, the market research department for the Condor Company concluded that the revenue (R) that could be generated by manufacturing a new product could be calculated by the formula $R = 5000p - 25p^2$, where p is the price (in dollars) per unit sold. They also concluded that the cost (in dollars) of manufacturing the new product would be $C = 400{,}000 - 1500p$.

(a) At what price would the Condor Company break even? ($R = C$.)

(b) Profit is revenue minus cost. ($P = R - C$.) At what prices would the Condor Company make a profit?

(c) Guess what the maximum profit would be.

74. Find the dimensions of a rectangle whose length is 3 meters less than twice its width and whose area is 65 square meters.

75. The weight W of a person d units from the center of the earth can be calculated by the formula

$$W = W_e\left(\frac{r}{d}\right)^2$$

where W_e is the person's weight on the earth and r is the radius of the earth. Approximately how many miles above the earth would a 150-pound-person weigh 100 pounds? (Use $r \approx 3963$ miles.)

76. Use the formula in Problem 75 to approximate how many miles above the earth a person would weigh one-half his or her weight on earth.

77. If the discriminant of a quadratic equation is a perfect square, prove that the equation has two rational roots.

78. If a quadratic equation has two rational roots, $\frac{s}{t}$ and $\frac{u}{v}$, write the equation in standard form with no fractions.

79. If r_1 and r_2 are roots of a quadratic equation, prove that $r_1 + r_2 = -\frac{b}{a}$ and that $r_1 r_2 = \frac{c}{a}$.

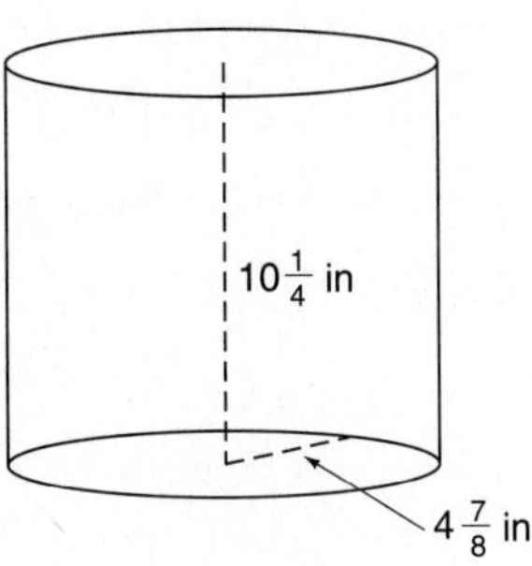

Fig. 2.50

2.5 FRACTIONAL EQUATIONS AND INEQUALITIES

An equation that contains a variable in a denominator is called a **fractional equation.** In this section we will examine fractional equations and fractional inequalities.

While studying linear equations, we solved equations that contained fractions. The procedure we used was to clear fractions by multiplying both sides of the equation by the least common denominator of all the fractions in the equation. If any denominator contains a variable, there is a possibility of multiplication by zero. (Remember, we can multiply both sides of an equation by any *nonzero* number.)

The best strategy is to clear fractions and solve the resulting equation for possible solutions. Then check to see if any of the possible solutions would make any denominator of the original equation have a value of zero. If so, that number cannot be included in the solution set.

EXAMPLE 1 Solve each equation.

(a) $\dfrac{4}{x+2} - 1 = 0$ (b) $\dfrac{1}{x+2} + \dfrac{3}{x-7} = \dfrac{-9}{(x+2)(x-7)}$

Solutions (a) $\dfrac{4}{x+2} - 1 = 0$

First we multiply both sides by $x + 2$ to clear fractions.

$$(x+2)\frac{4}{x+2} - 1(x+2) = (x+2)\cdot 0$$
$$4 - x - 2 = 0$$
$$x = 2$$

Since 2 does not make a denominator have value 0, we write the solution set

$\{2\}$ Solution set

(b) $\dfrac{1}{x+2} + \dfrac{3}{x-7} = \dfrac{-9}{(x+2)(x-7)}$

Multiply both sides by $(x + 2)(x - 7)$.

$$(x+2)(x-7)\left(\frac{1}{x+2} + \frac{3}{x-7}\right) = (x+2)(x-7)\frac{-9}{(x+2)(x-7)}$$
$$(x-7)\cdot 1 + (x+2)\cdot 3 = -9$$
$$x - 7 + 3x + 6 = -9$$
$$4x - 1 = -9$$
$$4x = -8$$
$$x = -2$$

Since -2 makes the denominator in $\frac{1}{x+2}$ have value zero, -2 is not a solution. There are no other possible solutions. So, the solution set is the empty set.

$$\varnothing \qquad \text{Solution set}$$ □

Sometimes a fractional equation leads to a quadratic equation.

EXAMPLE 2 Solve $\dfrac{t+1}{t-2} = \dfrac{8}{t+1} + \dfrac{9}{t^2 - t - 2}$.

Solution

$$\frac{t+1}{t-2} = \frac{8}{t+1} + \frac{9}{(t-2)(t+1)} \qquad \text{Factor.}$$

The LCD is $(t + 1)(t - 2)$.

$$(t+1)(t-2)\left(\frac{t+1}{t-2}\right) = (t+1)(t-2)\left(\frac{8}{t+1} + \frac{9}{(t-2)(t+1)}\right)$$

$$(t+1)^2 = 8(t-2) + 9 \qquad \text{Common factors}$$

$$t^2 + 2t + 1 = 8t - 16 + 9$$

$$t^2 - 6t + 8 = 0 \qquad \text{Quadratic equation in standard form}$$

$$(t-2)(t-4) = 0$$

$$t = 2 \quad \text{or} \quad t = 4 \qquad \text{Property of Zero Products}$$

In this example, notice that replacing t with 2 in the original equation makes two of the denominators have a value of zero, while 4 does not make any denominator have a value of zero. Thus 4 is in the solution set, and 2 is not.

$$\{4\} \qquad \text{Solution set}$$ □

Procedure to Solve Fractional Equations

1. Multiply both sides of the equation by the LCD to clear fractions.
2. Find possible solutions by solving the resulting equation.
3. Discard possible solutions that make any denominator in the original equation have a value of zero.
4. Write the solution set.

Recalling the definition of negative exponents, we see that equations containing them are often just equations containing fractions.

EXAMPLE 3 Solve $3z^{-2} - z^{-1} - 4 = 0$.

Solution Since $z^{-2} = \frac{1}{z^2}$, we clear fractions in this equation by multiplying both sides by z^2.

$$z^2(3z^{-2} - z^{-1} - 4) = z^2 \cdot 0$$

$$3 - z - 4z^2 = 0 \qquad \text{Distributive Property}$$

$$(3 - 4z)(1 + z) = 0 \qquad \text{Factor.}$$

$$z = \frac{3}{4} \quad \text{or} \quad z = -1 \qquad \text{Property of Zero Products}$$

Neither of the two candidate solutions makes a denominator in the original equation have value zero, so they are both in the solution set.

$$\left\{-1, \frac{3}{4}\right\} \qquad \text{Solution set}$$

□

EXAMPLE 4 Carol Todd leaves Norwich University in her car on an important business trip. She drives 200 miles to Laconia, New Hampshire. On the return trip she slows her average speed by 10 miles per hour in order to enjoy the scenery. If the return trip takes one more hour than the trip to Laconia, what was her average speed in both directions?

Solution Clearly this is a distance-rate-time *(DRT)* problem. We make an assignment statement.

Let x be her average rate from Norwich to Laconia (mph).

Then $x - 10$ is her average rate on the return trip (mph).

Notice that there are two distinct parts, as in most *DRT* problems.

Norwich to Laconia	*Return Trip*
$D = 200$	$D = 200$
$R = x$	$R = x - 10$
$T =$	$T =$

Now we use the formula $T = D/R$ to find the time on each leg.

Norwich to Laconia	*Return Trip*
$D = 200$	$D = 200$
$R = x$	$R = x - 10$
$T = \frac{200}{x}$	$T = \frac{200}{x - 10}$

Since the problem states that the return trip took *1 hour longer,* we have the equation

$$\frac{200}{x-10} - \frac{200}{x} = 1$$

$$200x - 200(x - 10) = x(x - 10) \qquad \text{Multiply by } x(x - 10).$$

$$200x - 200x + 2000 = x^2 - 10x$$

$$0 = x^2 - 10x - 2000 \qquad \text{Quadratic equation in standard form}$$

$$0 = (x + 40)(x - 50) \qquad \text{Factor.}$$

$$x = -40 \quad \text{or} \quad x = 50$$

The numbers -40 and 50 are solutions to the *mathematical model.* However, only 50 is a solution to the *problem.*

Her average speed from Norwich to Laconia was 50 miles per hour, and her average speed on the return trip was 40 miles per hour. □

EXAMPLE 5 At a Parke Davis experimental lab there is a chemical reagent vat with two input pipes; one is marked 411, and the other is marked 12. If both pipes are opened, the vat will fill in 3 hours. How long does it take each pipe working alone to fill the vat if pipe 411 takes 2 hours longer than pipe 12? Approximate the answer to the nearest minute.

Solution The first thing we always do is make an assignment statement.

Let x be the time (in hours) that it takes pipe 12 to fill the tank working alone.

Then it takes $x + 2$ hours for pipe 411 to fill the tank working alone.

Therefore, in 1 hour,

1. Pipe 12 fills $\frac{1}{x}$ of the vat.

2. Pipe 411 fills $\frac{1}{x+2}$ of the vat.

3. Both pipes, working together, fill $\frac{1}{3}$ of the vat.

So,

$$\frac{1}{x} + \frac{1}{x+2} = \frac{1}{3}$$

$$3(x + 2) + 3x = x(x + 2) \qquad \text{LCD of } 3x(x + 2)$$

$$3x + 6 + 3x = x^2 + 2x$$

$$0 = x^2 - 4x - 6 \qquad \text{Standard form}$$

$$x = \frac{4 \pm \sqrt{16 - 4 \cdot 1(-6)}}{2} \qquad \text{Quadratic formula}$$

$$x = \frac{4 \pm \sqrt{16 + 24}}{2}$$

$$x = \frac{4 \pm \sqrt{40}}{2} = \frac{4 \pm 2\sqrt{10}}{2}$$

$$x = 2 \pm \sqrt{10}$$

$$x \approx -1.162 \qquad \text{or} \qquad x \approx 5.162$$

The negative solution of the mathematical model does not fit the problem. So we have the single solution 5.162.

It takes pipe 12 approximately 5 hours and 10 minutes and pipe 411 approximately 7 hours and 10 minutes to fill the vat when working alone. □

Fractional Inequalities

Just as there are fractional equations, there are fractional inequalities. As in other nonlinear inequalities, the method of boundary numbers reduces the problem of solving an inequality to the problem of solving an equation.

A fractional inequality has boundary numbers at all solutions of the associated equation just like other inequalities. In addition, a fractional inequality has a boundary number where any denominator has a value of zero. That is, at numbers not in the domain of any rational expression in the inequality. Furthermore, boundary numbers where a denominator has a value of zero are *never* in the solution set of the inequality. We call these boundary numbers **free boundary numbers.** A free boundary number is never in a solution set.

Free Boundary Numbers

Numbers not in the domain of a rational expression in an inequality are free boundary numbers for that inequality.

EXAMPLE 6 Solve $\dfrac{x+5}{x-5} < 2$.

Solution Find the boundary numbers.

$$\frac{x+5}{x-5} = 2 \qquad \text{The associated equation}$$

Since this is a fractional inequality, we get boundary numbers wherever the denominator is 0, as well as at the solutions of the associated equation. We have a free boundary number at 5. Now we solve the associated equation.

$$x + 5 = 2(x - 5)$$

$$x + 5 = 2x - 10$$

$$15 = x$$

Fig. 2.51

The boundary numbers are 5 and 15 (Figure 2.51).

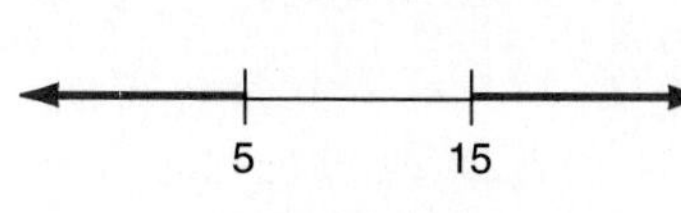

Fig. 2.52

Next, we test the regions (Figure 2.52).

Left region, test 0:	$\dfrac{0 + 5}{0 - 5} < 2$	True
Middle region, test 6:	$\dfrac{6 + 5}{6 - 5} < 2$	False
Right region, test 20:	$\dfrac{20 + 5}{20 - 5} < 2$	True

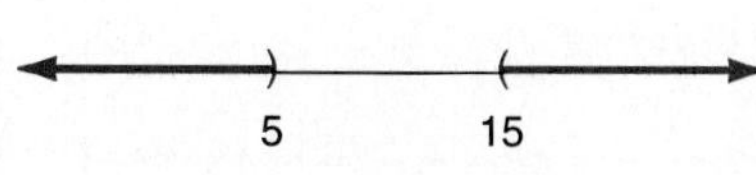

Fig. 2.53

The boundary number 15 does not belong to the solution set because equality is not included in the original inequality and it came from solving the associated equation. The boundary number 5 does not belong because it is a free boundary number, and free boundary numbers *never* belong to the solution set (Figure 2.53).

$(-\infty, 5) \cup (15, +\infty)$ Solution set □

 **EXAMPLE 7** Solve $\dfrac{2x + 10}{x + 1} \leq x + 1$.

Solution Find the boundary numbers. -1 is a free boundary number.

$$\frac{2x + 10}{x + 1} = x + 1 \qquad \text{Associated equation}$$

$$2x + 10 = (x + 1)(x + 1)$$

$$2x + 10 = x^2 + 2x + 1$$

$$0 = x^2 - 9$$

$$0 = (x - 3)(x + 3) \qquad \text{Factor.}$$

$$x = 3 \quad \text{or} \quad x = -3 \qquad \text{Property of Zero Products}$$

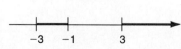

Fig. 2.54

Fig. 2.55

The boundary numbers are -3, -1, and 3 (Figure 2.54).
We test the regions (Figure 2.55).

Region A, test -5:	$\dfrac{0}{-4} \le -5 + 1$	False
Region B, test -2:	$\dfrac{-4 + 10}{-2 + 1} \le -2 + 1$	
	$\dfrac{6}{-1} \le -1$	True
Region C, test 0:	$\dfrac{0 + 10}{0 + 1} \le 0 + 1$	False
Region D, test 5:	$\dfrac{10 + 10}{5 + 1} \le 5 + 1$	
	$\dfrac{20}{6} \le 6$	True

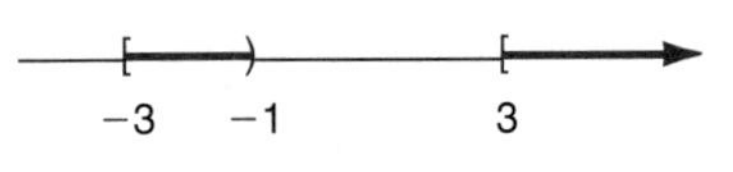

Fig. 2.56

The boundary numbers -3 and 3 *are* in the solution set because they came from solving the associated equation and the original problem allows equality, while the boundary number -1 *is not* in the solution set because it is a free boundary number (Figure 2.56).

$[-3, -1) \cup [3, +\infty)$ Solution set □

EXAMPLE 8 Two wires are connected in parallel. The resistance in the first wire is 0.03 ohm, and the resistance in the second wire is R_2. The formula

$$R = \frac{0.03\,R_2}{R_2 + 0.03}$$

where R is the total resistance, expresses the relationship between the resistances. If total resistance must be less than 0.01 ohm, what must the resistance in the second wire be?

Solution Because $R < 0.01$, we form the inequality

$$\frac{0.03\,R_2}{R_2 + 0.03} < 0.01$$

First we find the boundary numbers.
This fractional inequality has a free boundary number, -0.03.
We solve the associated equation for other boundary numbers.

$$\frac{0.03\,R_2}{R_2 + 0.03} = 0.01$$

$$0.03R_2 = 0.01(R_2 + 0.03) \quad \text{Clear fractions.}$$

$$0.03R_2 = 0.01R_2 + 0.0003$$

$$0.02R_2 = 0.0003$$

$$R_2 = 0.015$$

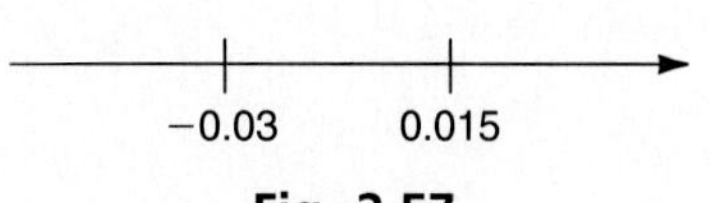

Fig. 2.57

The boundary numbers are -0.03 and 0.015 (Figure 2.57). Using -1 as a test number in the left region, we have $\frac{-0.03}{-1 + 0.03} < 0.01$ which is false. Using 0 as a test number in the middle region, we have $\frac{0}{0 + 0.03} < 0.01$ which is true. Using 1 as a test number from the right region, we have $\frac{0.03}{1 + 0.03} < 0.01$ which is false. The boundary numbers are not included in the solution set. The solution set for the mathematical model (the inequality) is the middle region (Figure 2.58). To answer the question, we must use only the positive numbers in this region. The resistance in the second wire must be between 0 and 0.015 ohm. □

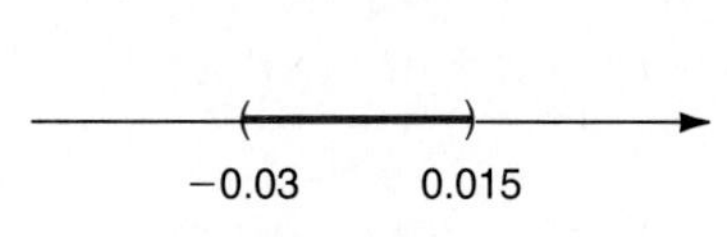

Fig. 2.58

PROBLEM SET 2.5

Warm-ups

In Problems 1 through 6, solve each equation.
For Problems 1 through 4, see Examples 1 and 2.

1. $\frac{3}{x - 1} = \frac{5}{x - 1} - 1$

2. $\frac{1}{x} + \frac{x + 3}{7} = \frac{5}{7x}$

3. $\frac{y + 1}{y + 2} - \frac{y + 3}{y + 4} = \frac{y - 6}{y^2 + 6y + 8}$

4. $\frac{7}{2t^2 - 5t - 3} = \frac{7}{t - 3} - \frac{2}{2t + 1}$

For Problems 5 and 6, see Example 3.

5. $2x^{-2} + 9x^{-1} = 18$

6. $4v^{-2} + 15v^{-1} - 4 = 0$

For Problems 7 and 8, see Example 4.

7. Larry can row 30 miles down the Columbia River and back in 8 hours. If the rate of the current is 5 miles per hour, find the rate Larry would row in still water.

8. John McDowell entered a 26-mile bike race. He was forced to walk the last 2 miles of the race because of mechanical difficulties. If his riding speed was 12 miles per hour faster than his walking speed and his time was 2 hours, find his walking speed.

For Problems 9 and 10, see Example 5.

9. There is a hot water tap and a cold water tap for a large storage tank. The hot tap can fill the tank in 5 hours. The two taps together can fill the tank in 2 hours less than the cold tap alone. How long would it take the cold water tap alone to fill the tank?

10. Shopping together, Jorge and Marie Alejos can complete their grocery shopping in 2 hours. Shopping alone, it takes Jorge 3 hours longer than it takes Marie shopping alone. How long does it take each to do the grocery shopping alone?

In Problems 11 through 14, solve each inequality. See Examples 6 and 7.

11. $\frac{3}{2x - 5} < 1$

12. $\frac{1}{1 - x} \geq x$

13. $\frac{2}{t - 3} \leq \frac{1}{t + 2}$

14. $\frac{1}{y - 2} \geq \frac{1}{y + 2}$

For Problems 15 and 16, see Example 8.

15. The concentration C in milligrams per cubic centimeter t minutes after a drug is administered is given by the formula

$$C = \frac{0.16}{(t+1)^2}$$

When will the concentration level be less than 0.04?

16. The number of units demanded, x, and the price per unit in dollars, p, are related by the formula

$$p = \frac{150}{4+x}$$

How many units are demanded if the price per unit is not to exceed $3?

Practice Exercises

In Problems 17 through 48, solve each equation or inequality.

17. $\frac{25}{x+2} - \frac{9}{x-2} = 16$

18. $\frac{x+1}{x-2} - \frac{8}{x+1} = \frac{9}{x^2-x-2}$

19. $\frac{3}{x} + \frac{x}{x+4} = \frac{10}{x^2+4x}$

20. $\frac{1}{x} + \frac{5}{3x} = \frac{x+2}{3}$

21. $\frac{1}{w-5} + \frac{w}{w-2} = \frac{3}{w^2-7w+10}$

22. $\frac{2}{t+7} + \frac{16}{t^2+6t-7} + \frac{t}{t-1} = 0$

23. $\frac{x+4}{x} \leq 0$

24. $\frac{2t-5}{t} \geq 0$

25. $\frac{7}{3x+6} + 3 = \frac{8}{3x-3}$

26. $\frac{-2}{x+1} - \frac{5}{3-x} = 1$

27. $\frac{-3}{y+4} + \frac{y}{y-3} = \frac{21}{y^2+y-12}$

28. $\frac{5}{z^2+3z-4} + \frac{1}{z+4} = \frac{z}{z-1}$

29. $\frac{18}{x^2+x-6} + \frac{x-1}{x^2+5x+6} = \frac{12}{x^2-4}$

30. $\frac{t}{t^2-t-2} - \frac{1}{2t^2+t-1} = \frac{1}{2t^2-5t+2}$

31. $x^{-2} + 5x^{-1} - 6 = 0$

32. $3y^{-2} + y^{-1} = 4$

33. $\frac{x-5}{x} > 2$

34. $5 > \frac{3r+1}{r-1}$

35. $4z^{-2} + 27 = 21z^{-1}$

36. $3 - 2x^{-1} = 5x^{-2}$

37. $\frac{2}{5x^2-15x-50} + \frac{x+1}{x^2-x-6} = \frac{-8}{5x^2-40x+75}$

38. $\frac{k}{k^2+3k+2} + \frac{1}{k^2+4k+3} = \frac{2}{k^2+5k+6}$

39. $x^{-2} - 4x^{-1} + 4 > 0$

40. $x^{-2} - 10x^{-1} + 25 < 0$

41. $x^{-2} + 6x^{-1} + 9 \leq 0$

42. $x^{-2} + \frac{1}{8}x^{-1} + \frac{1}{16} \geq 0$

43. $\frac{x+9}{x+1} \leq x$

44. $w \leq \frac{2w+25}{w+2}$

45. $\frac{7}{3+x} - \frac{1}{3-x} \leq \frac{-6}{9-x^2}$

46. $1 \geq \frac{3}{x}$

47. $\frac{2t}{t+2} \geq \frac{t}{t-2}$

48. $\frac{2x+1}{x-1} \geq \frac{x-1}{x+1}$

49. An Amtrak train leaves St. Louis at 8:00 A.M. bound for Kansas City, a distance of 250 miles. After traveling 130 miles it is forced to reduce its average speed by 5 miles per hour because of mechanical difficulties. If the train arrives in Kansas City at noon, what was its original average speed?

50. Teresa entered a 17-mile race. She ran for 15 miles until a charley horse forced her to walk the rest of the race. Her running speed was 3 miles per hour faster than her walking speed. If it took Teresa no more than 2 hours to complete the race, what could her walking speed be?

51. Water from the garden hose and the fill pipe can fill a pool in 4 hours. If only the garden hose is used, it takes 6 hours longer than if only the fill pipe is used. Find the time it takes the fill pipe to fill the pool working alone.
52. Billy and Donna row their racing shell 5 miles upriver to the ugly stump and return to the dock where they started. If the current is 1 mile per hour and it took twice as long to travel upstream as it did to travel downstream, what was their rate in still water?
53. Kitty and Charles drive their boat upstream from Texas Point 20 miles to the highway bridge and then return downstream to Texas Point. If the boat averages 8 miles per hour in still water and the trip upstream took twice as long as the trip downstream, how fast is the current between Texas Point and the highway bridge?
54. Boyle's Law gives the relationship between pressure and volume assuming constant temperature.

$$V_2 = \frac{V_1P_1}{P_2}$$

As an air bubble with a volume of 4.2 cubic centimeters moves from the bottom of a pond to the top of the pond, the pressure decreases by 1.3 atmospheres. If the volume of the bubble at the top is 8.6 cubic centimeters, approximate to the nearest tenth of an atmosphere the pressure at the bottom and at the top of the pond.
55. A company produces red and blue gem clips. The relationship between the number of each color of gem clip produced is

$$N_r = \frac{740}{N_b + 15} + 20$$

where N_r is the number of thousands of red and N_b is the number of thousands of blue produced. If not more than 40,000 red gem clips are manufactured, how many blue ones will be manufactured?
56. Solve the formula in Problem 55 for N_b.
57. Assume that $b \neq 0$. Prove that $\frac{a}{b} = 0$ if and only if $a = 0$.

Challenge Problems

In Problems 58 and 59, solve each equation.

58. $a^2x^{-2} - 2ax^{-1} + 1 \geq 0$

59. $a^2x^{-2} - 2ax^{-1} - 1 = 0$

Solve the following inequality.

60. $\frac{2x}{x-3} \leq \frac{x-1}{x+1}$

IN YOUR OWN WORDS . . .

61. Explain how to solve a fractional inequality.

2.6 OTHER TYPES OF EQUATIONS AND INEQUALITIES

In this chapter we have studied linear equations, absolute value equations, quadratic equations, fractional equations, and the inequalities associated with each of these types of equations. Of course there are many types of equations we have not studied. Any two number representations with an equals symbol between them is an equation. In this section we will look at a few commonly occurring equation types that do not fit into the above categories.

The first type we will consider is polynomial equations in one variable of degree higher than 2. A polynomial equation is an equation that can be written as a polynomial equal to zero. For example, a quadratic equation is a second-degree polynomial equation. Like quadratic equations, solving higher-degree equations depends on factoring and using the Property of Zero Products. That property can be extended to any finite number of factors.

Extended Property of Zero Products

The statement $pqr = 0$ is *true* if $p = 0$ or $q = 0$ or $r = 0$, and *false* if p nor q nor r is 0.

EXAMPLE 1 Solve $x^3 - x^2 = 30x$.

Solution This is a polynomial equation. As we did with quadratic equations, we write the equation first as a polynomial in standard form equal to zero. This is called **standard form** for a polynomial equation. *The zero is crucial* because we want to use the Property of Zero Products.

$$x^3 - x^2 - 30x = 0 \qquad \text{Standard form}$$

Next, we try to factor so we can use the Property of Zero Products.

$x(x^2 - x - 30) = 0$ — Common factor

$x(x - 6)(x + 5) = 0$ — Factor completely.

$x = 0 \quad \text{or} \quad x = 6 \quad \text{or} \quad x = -5$ — Extended Property of Zero Products

$\{-5, 0, 6\}$ — Solution set □

EXAMPLE 2 Solve the equation, $2x^4 + 2x = 3x^3 + 3$

Solution

$$2x^4 + 2x = 3x^3 + 3$$

$$2x^4 - 3x^3 + 2x - 3 = 0 \qquad \text{Standard form}$$

This is a four-term polynomial without a common factor. We try to factor by grouping.

$$x^3(2x - 3) + (2x - 3) = 0$$

$(2x - 3)(x^3 + 1) = 0$ — Common factor

$(2x - 3)(x + 1)(x^2 - x + 1) = 0$ — Factor completely.

$x = \dfrac{3}{2} \quad \text{or} \quad x = -1 \quad \text{or} \quad x^2 - x + 1 = 0$ — Property of Zero Products

$x = \dfrac{1 \pm \sqrt{1 - 4}}{2}$ — Quadratic formula

$$x = \frac{1 \pm \sqrt{3}i}{2}$$

$$\left\{-1, \frac{3}{2}, \frac{1}{2} \pm \frac{\sqrt{3}}{2}i\right\} \qquad \text{Solution set} \qquad \square$$

The method of boundary numbers gives us an approach to any inequality when we can solve the associated equation.

EXAMPLE 3 Solve each inequality.

(a) $x^4 - 5x^3 - 4x^2 + 20x \leq 0$ (b) $x^3 - x^2 + 4x > 4$

Solutions (a) $x^4 - 5x^3 - 4x^2 + 20x \leq 0$

First, we find the boundary numbers.

$$\begin{aligned} x^4 - 5x^3 - 4x^2 + 20x &= 0 && \text{The associated equation} \\ x(x^3 - 5x^2 - 4x + 20) &= 0 && \text{Common factor} \\ x[x^2(x - 5) - 4(x - 5)] &= 0 && \text{Grouping} \\ x(x - 5)(x^2 - 4) &= 0 && \\ x(x - 5)(x - 2)(x + 2) &= 0 && \text{Factor completely.} \\ x &= -2, 0, 2, 5 && \text{The boundary numbers} \end{aligned}$$

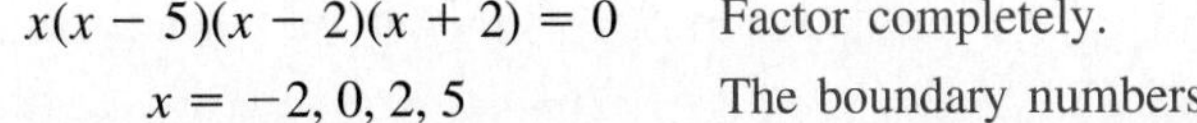

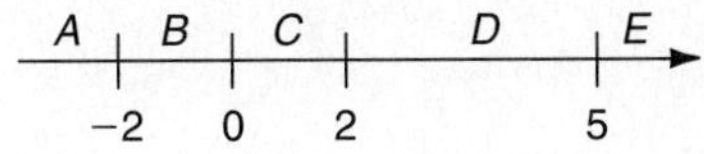

Fig. 2.59

Next, we locate the boundary numbers on a number line (Figure 2.59). Now we check each region in the original inequality.

Region A, test -3: $(-3)^4 - 5(-3)^3 - 4(-3)^2 + 20(-3) \leq 0$
$81 + 135 - 36 - 60 \leq 0$
$120 \leq 0$ False

Region B, test -1: $(-1)^4 - 5(-1)^3 - 4(-1)^2 + 20(-1) \leq 0$
$1 + 5 - 4 - 20 \leq 0$
$-18 \leq 0$ True

Region C, test 1: $1^4 - 5 \cdot 1^3 - 4 \cdot 1^2 + 20 \cdot 1 \leq 0$
$1 - 5 - 4 + 20 \leq 0$
$12 \leq 0$ False

Region D, test 3: $3^4 - 5 \cdot 3^3 - 4 \cdot 3^2 + 20 \cdot 3 \leq 0$
$81 - 135 - 36 + 60 \leq 0$
$-30 \leq 0$ True

Region E, test 6: $6^4 - 5 \cdot 6^3 - 4 \cdot 6^2 + 20 \cdot 6 \leq 0$
$1296 - 1080 - 144 + 120 \leq 0$
$192 \leq 0$ False

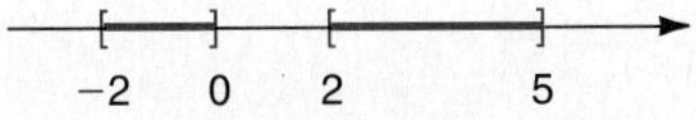

−2 0 2 5

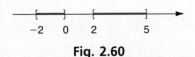

Fig. 2.60

As the original problem allowed equality and all the boundary numbers came from solving the associated equation, they all belong to the solution set (Figure 2.60). Now we can write the solution set.

$$[-2, 0] \cup [2, 5] \qquad \text{Solution set}$$

(b) $x^3 - x^2 + 4x > 4$

Find the boundary numbers.

$x^3 - x^2 + 4x = 4$	Associated equation
$x^3 - x^2 + 4x - 4 = 0$	Standard form
$x^2(x - 1) + 4(x - 1) = 0$	Grouping
$(x - 1)(x^2 + 4) = 0$	Factored
$x = 1, \pm 2i$	Solutions of the associated equation

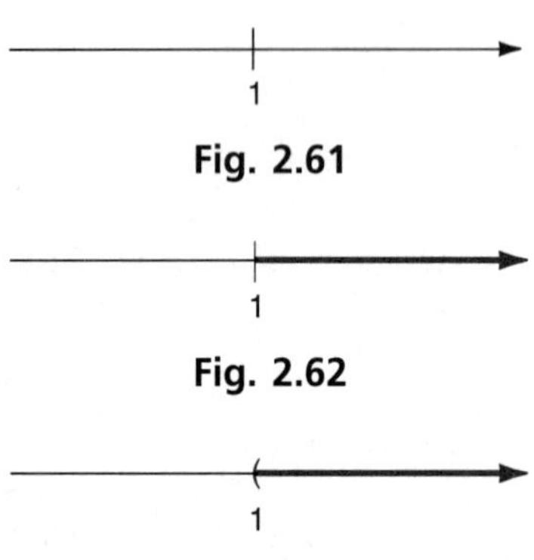

Fig. 2.61
Fig. 2.62
Fig. 2.63

Although $2i$ and $-2i$ are perfectly good solutions of the associated equation, they have no representation on the number line and are *not* boundary numbers. Our only boundary number is 1 (Figure 2.61).

Left region, test 0:	$0 > 4$	False
Right region, test 2:	$8 - 4 + 8 > 4$	True (Figure 2.62)

The boundary number 1 *does not* belong because the original inequality does not allow equality (Figure 2.63).

$(1, +\infty)$ Solution set □

We must be very careful when solving such simple looking equations as

$$x^3 = 8$$

If we invent a "cube root property" and write

$$x = \sqrt[3]{8}$$

$$x = 2$$

we will have lost two solutions! The Square Root Property came from *factoring,* and we must factor here.

$$x^3 - 8 = 0$$

$$(x - 2)(x^2 + 2x + 4) = 0$$

$$x = 2 \quad \text{or} \quad x^2 + 2x + 4 = 0$$

$$x = \frac{-2 \pm \sqrt{4 - 16}}{2}$$

$$x = \frac{-2 \pm \sqrt{-12}}{2}$$

$$x = \frac{-2 \pm \sqrt{12}i}{2} = -1 \pm \sqrt{3}i$$

$$\{2, -1 \pm \sqrt{3}i\}$$ □

Other common varieties of nonlinear equations are those containing square roots or nth roots. The general strategy is to raise both sides to a suitable power in order to obtain a solvable equation then *check all candidate solutions.* The check is not for reassurance. It is a necessary part of the procedure because squaring both sides *does not* produce an equivalent equation. Consider the equation $x = 2$. Its solution set is $\{2\}$. But, if we square both sides, we get $x^2 = 4$, whose solution set is $\{-2, 2\}$. The equations are not equivalent. However, all solutions to the original equation are contained in the solution set of the squared equation.

EXAMPLE 4 Solve each equation.

(a) $\sqrt{x+3} - 1 = x$ (b) $\sqrt{2-7t} - \sqrt{t+3} = 3$

Solutions (a) $\sqrt{x+3} - 1 = x$

This is an equation containing square roots. Our strategy is to isolate a radical *before* squaring both sides.

$\sqrt{x+3} = x + 1$	Isolate a radical.
$(\sqrt{x+3})^2 = (x+1)^2$	Square both sides.
$x + 3 = x^2 + 2x + 1$	
$0 = x^2 + x - 2$	Standard form
$0 = (x+2)(x-1)$	Factor.
$x = -2, 1$	Candidate solutions

We must check both candidates.
Check -2.
Left side: $\sqrt{-2+3} - 1 = \sqrt{1} - 1 = 0$
Right side: -2
When x is -2, the left side and the right side of the equation represent *different* numbers; -2 is not a solution.
Check 1.
Left side: $\sqrt{1+3} - 1 = \sqrt{4} - 1 = 1$
Right side: 1
Therefore 1 is a solution.

$\{1\}$	Solution set

(b) $\sqrt{2-7t} - \sqrt{t+3} = 3$

We isolate a radical and square both sides. Either radical will do. Let's isolate $\sqrt{2-7t}$.

$\sqrt{2-7t} = \sqrt{t+3} + 3$	Isolate a radical.
$(\sqrt{2-7t})^2 = (\sqrt{t+3} + 3)^2$	Square both sides.
$2 - 7t = t + 3 + 6\sqrt{t+3} + 9$	$(a+b)^2 = a^2 + 2ab + b^2$
$-10 - 8t = 6\sqrt{t+3}$	
$-5 - 4t = 3\sqrt{t+3}$	Divide by 2.

The equation still contains a square root. It is isolated, so we square again.

$$(-5 - 4t)^2 = (3\sqrt{t + 3})^2$$
$$25 + 40t + 16t^2 = 9(t + 3)$$
$$25 + 40t + 16t^2 = 9t + 27$$
$$16t^2 + 31t - 2 = 0$$
$$(t + 2)(16t - 1) = 0$$

We have possible roots of -2 and $\frac{1}{16}$. We must check them both.

Check -2.
Left side: $\sqrt{2 - 7(-2)} - \sqrt{-2 + 3} = \sqrt{16} - \sqrt{1} = 3$
Right side: 3
-2 checks.

Check $\frac{1}{16}$.

Left side: $\sqrt{2 - 7 \cdot \frac{1}{16}} - \sqrt{\frac{1}{16} + 3} = \sqrt{\frac{32 - 7}{16}} - \sqrt{\frac{1 + 48}{16}}$

$$= \sqrt{\frac{25}{16}} - \sqrt{\frac{49}{16}} = \frac{5}{4} - \frac{7}{4} = -\frac{1}{2}$$

Right side: 3

$\frac{1}{16}$ does not check.

$\{-2\}$ Solution set □

Equations with Fractional Exponents

Equations containing fractional exponents may require raising both sides to a power, but *factoring* should be investigated first.

EXAMPLE 5 Solve the equation, $x^{2/3} + 3x^{1/3} = 4$

Solution $x^{2/3} + 3x^{1/3} = 4$

As $\frac{1}{3}$ powers are cube roots, it is tempting to cube both sides. However, let's try to factor first.

$$x^{2/3} + 3x^{1/3} - 4 = 0$$

$(x^{1/3} - 1)(x^{1/3} + 4) = 0$ Factor.

$x^{1/3} = 1$ or $x^{1/3} = -4$ Property of Zero Products

$x = 1$ or $x = -64$ Cube both sides.

Both of these candidates check.

$\{-64, 1\}$ Solution set □

PROBLEM SET 2.6

Warm-ups

In Problems 1 through 10, solve each equation or inequality.
For Problems 1 through 4, see Examples 1 and 2.

1. $x^5 - 16x = 0$

2. $x^4 - 3x^2 - 4 = 0$

3. $3z^4 - z^3 - 24z + 8 = 0$

4. $y^4 + 16y^2 = 10y^3$

For Problems 5 and 6, see Example 3.

5. $x^3 > 4x$

6. $A^3 + 7A^2 \leq 4A + 28$

For Problems 7 and 8, see Example 4.

7. $\sqrt{4r + 1} + 5 = r$

8. $\sqrt{x + 5} - 3 = \sqrt{2x - 8}$

For Problems 9 and 10, see Example 5.

9. $w^{2/3} = w^{1/3}$

10. $x^{2/3} + x^{1/3} = 12$

Practice Exercises

In Problems 11 through 48, solve each equation.

11. $x^4 - 81 = 0$

12. $64 = s^4$

13. $2x^4 - 16x = 0$

14. $5x^4 = 625x$

15. $t^4 - 5t^2 + 6 = 0$

16. $2x^4 - 7x^2 + 3 = 0$

17. $x^3 - x^2 - 4x + 4 = 0$

18. $9s + 18 = s^3 + 2s^2$

19. $x - 1 = \sqrt{2x^2 - 3x - 1}$

20. $\sqrt{8y^2 - 5y + 7} = 3y - 1$

21. $27r^3 = 8$

22. $-125 = 64x^3$

23. $x^{2/3} = 4$

24. $v^{4/3} - 81 = 0$

25. $t^4 - t^3 = 6t^2$

26. $x^2(6x^2 + 7x) = 5x^2$

27. $\sqrt{x + 3} = x - 3$

28. $\sqrt{2w} = w - 4$

29. $\sqrt{x} + 3 = x + 1$

30. $3 + \sqrt{x - 4} = x - 7$

31. $x^3 + x = x^2 + 1$

32. $4t^3 + 3 = 12t^2 + t$

33. $2y^4 + 13y^2 + 15 = 0$

34. $12 + 17x^2 + 6x^4 = 0$

35. $3x^4 - 2x^3 + 16 = 24x$

36. $3 - 5t = 3t^3 - 5t^4$

37. $\sqrt{3 - 2z} - \sqrt{4 + z} = 2$

38. $\sqrt{2 - 7x} = \sqrt{x + 3} + 3$

39. $\sqrt{x} - \sqrt{x - 9} = 1$

40. $\sqrt{x} + \sqrt{x + 4} = 3$

41. $w^{2/3} - w^{1/3} - 2 = 0$

42. $z^{2/5} - 2z^{1/5} = 3$

43. $2x - x^{3/2} = 0$

44. $t = t^{3/8}$

45. $x^{-4} - 4x^{-2} + 4 = 0$

46. $x^{-4} - 10x^{-2} + 25 = 0$

47. $\sqrt{3 - x} + \sqrt{x + 1} = \sqrt{2x + 6}$

48. $\sqrt{x} + \sqrt{x + 4} = \sqrt{2x + 4}$

In Problems 49 through 54, solve each inequality.

49. $x^4 + 4x^3 + 3x^2 > 0$

50. $(x - 1)^2(x^2 - 5x) \leq 0$

51. $x^3 + x^2 + x + 1 \geq 0$

52. $6x^3 + 15x < 4x^2 + 10$

53. $x^4 < x^2 + 12$

54. $2x^4 - 6 \geq x^2$

55. Is $\sqrt{2}$ a solution (or an approximate solution to at least five significant digits) to the following equation?

$$\sqrt{3}x^3 + 1.73205x^2 - 3.58912x - \sqrt{10} = 0$$

56. Is $\frac{\pi}{4}$ a solution (or an approximate solution to at least five significant digits) to the following equation?

$$\sqrt{2}x^4 + 3.316625x^2 - \sqrt{7}x - 0.505100 = 0$$

In Problems 57 through 60, approximate each solution to five significant digits.

57. $x^3 = 7$

58. $3r^3 + 17 = 0$

59. $2t^4 - 13t^2 + 15 = 0$

60. $12x^4 + 6 = 17x^2$

61. The time required for a boat to cross a river and return to the same point is given by the formula

$$t = \frac{2d}{\sqrt{v^2 - s^2}}$$

where d is the width of the river, v is the speed of the boat, and s is the speed of the current. If the river's speed is 3 miles per hour, the width of the river is 0.5 miles, and the round trip takes 15 minutes, what is the speed of the boat?

62. The base of a rectangular box is a square whose sides are 2 centimeters less than the height of the box. If the volume of the box is 16 cubic centimeters, find the dimensions of the box.

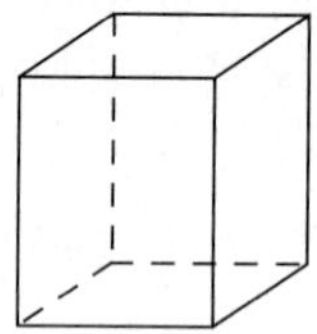

Challenge Problems

63. Anne Tidmore bought a cone of yogurt at TCBY. Her niece wanted a taste. They agree to split the yogurt by letting Anne eat the top semicircle part and the niece eat the part in the cone. If the yogurt is packed firmly into the cone as shown in the figure at right, who gets the most yogurt? (The volume of a cone is $\frac{1}{3}\pi r^2 h$, and the volume of a sphere is $\frac{4}{3}\pi r^3$.)

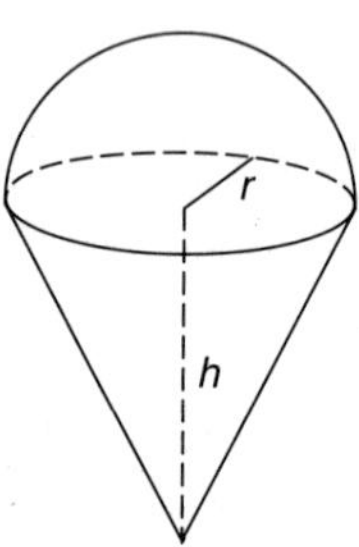

IN YOUR OWN WORDS . . .

64. Explain the purpose for checking equations containing square roots.

65. Explain how factoring can be used in solving equations.

CHAPTER SUMMARY

GLOSSARY

Boundary numbers: Numbers that border the graph of the solution set of an inequality.

Compound inequality: A statement made by combining two inequalities with the word *and* or *or.*

Contradiction: An equation that is always false.

Discriminant: The value of $b^2 - 4ac$ from the quadratic expression $ax^2 + bx + c$.

Equation: A statement that two numbers are equal.

Equivalent equations: Equations that have the same solution set.

Equivalent inequalities: Inequalities that have the same solution set.

Fractional equation: An equation that contains a variable in a denominator.

Graph of an inequality: A picture of the solution set on a number line.

Identity: An equation that is always true.

Inequality: A statement that one number is greater or less than another number.

Linear equation in one variable: An equation that can be written in the form $Ax + B = 0$, where $A \neq 0$.

Mathematical model: A mathematical structure that represents a situation from the physical world.

Quadratic equation in one variable: An equation that can be written in the form $ax^2 + bx + c = 0$, where $a \neq 0$.

Radical equation: An equation containing a variable under a radical.

Solution or **root:** A number that makes an equation or inequality a true statement when it replaces the variable.

Solution set: The set of all solutions of an equation or inequality.

Solve: Find the solution set.

Standard form of a polynomial equation: An equation written as a polynomial in standard form equal to 0.

Variable: A letter representing a number.

USING THE PROPERTIES OF EQUALITY

1. Add or subtract the same number from both sides.
2. Multiply or divide both sides by the same nonzero number.

USING THE PROPERTIES OF INEQUALITY

1. Add or subtract the same number from both sides.
2. Multiply or divide both sides by the same *positive* number.
3. Multiply or divide both sides by the same *negative* number and *reverse the direction of the inequality symbol.*

A PROCEDURE FOR SOLVING WORD PROBLEMS

1. Assign a variable to represent a key quantity in the problem (usually one of quantities to be found).
2. Express all quantities to be found in terms of the variable.
3. Draw a picture or figure if possible. Label it.
4. Reread the problem and form an equation (the mathematical model).
5. Solve the equation and find the values of the quantities to be found.
6. Check the values in the original problem statement. They should answer the question and make sense.
7. Write an answer to the original question.

ABSOLUTE VALUE EQUATIONS

1. If $q \geq 0$, then $|X| = q$ is equivalent to $X = q$ or $X = -q$.
2. If $q < 0$, then there are no solutions to the equation $|X| = q$.
3. $|X| = |V|$ is equivalent to $X = V$ or $X = -V$.
4. The solutions to the equation $|X| = V$ are included in the solutions to the pair of equations $X = V$ or $X = -V$. *Each candidate must be checked.*

THE METHOD OF BOUNDARY NUMBERS FOR SOLVING INEQUALITIES

1. Find the boundary numbers.
2. Locate the boundary numbers on a number line.
3. Determine which of the regions formed by the boundary numbers make the inequality true by testing a number from each region.
4. Shade only the regions that test true.
5. Check the boundary numbers themselves.
6. Write the solution set.

PROPERTY OF ZERO PRODUCTS

The statement $pq = 0$ is *true* if either $p = 0$ or $q = 0$, and *false* if neither p nor q is 0.

SQUARE ROOT PROPERTY

The equation $X^2 = Q$ is equivalent to the pair of equations $X = \pm\sqrt{Q}$.

QUADRATIC FORMULA

The quadratic equation $ax^2 + bx + c = 0$, $a \neq 0$, is equivalent to
$$x = \frac{-b \pm \sqrt{b^2 - 4ac}}{2a}$$

A PROCEDURE FOR SOLVING FRACTIONAL EQUATIONS

1. Clear fractions by multiplying both sides by the LCD.
2. Solve the resulting equation to find candidate solutions.
3. Discard candidates that make any denominator in the original equation have value 0.
4. Write the solution set.

A PROCEDURE FOR SOLVING RADICAL EQUATIONS

1. Isolate one radical (containing a variable).
2. Raise both sides to an appropriate power.
3. Solve the resulting equation. (This may require repeating steps 1 and 2.)
4. Check all possible solutions in the original equation.
5. Write the solution set.

REVIEW PROBLEMS

In Problems 1 through 22, solve each equation.

1. $3(2x - 5) = 2x + 5(1 - x)$

2. $\dfrac{3w + 4}{5} = \dfrac{w + 2}{2} - 3$

3. $|2 - 3y| - 4 = 10$

4. $|2x| = |x - 5|$

5. $7x^2 - 3x = 0$

6. $4t^2 - 9 = 0$

7. $\dfrac{1}{x} - \dfrac{1}{x - 3} = \dfrac{x + 4}{x^2 - 3x}$

8. $\dfrac{3}{x + 1} - \dfrac{2}{x - 1} + 1 = 0$

9. $z^4 - 64 = 0$

10. $2x^4 - x^2 - 21 = 0$

11. $\sqrt{x - 1} + \sqrt{x + 4} = 5$

12. $12w^2 - w = 6$

13. $6 + |3 - 2x| = 4$

14. $x^2 - 3x - 1 = 0$

15. $2k^5 = 162k$

16. $7[2(3 - t) - 3(2 - t)] = 5(1 - t)$

17. $\dfrac{x + 3}{x - 1} = x + 3$

18. $|2s - 1| + s = 1$

19. $x^6 + 4 = 4x^4 + x^2$

20. $\dfrac{5}{x + 2} - \dfrac{3}{x - 5} = \dfrac{x^2 - 8x - 6}{x^2 - 3x - 10}$

21. $|3s - 7| = \left|\dfrac{6 - 5s}{2}\right|$

22. $\sqrt{3x + 4} - \sqrt{2x - 5} = 2$

In Problems 23 through 30, solve each inequality.

23. $\dfrac{1}{2}(2x - 7) < \dfrac{2}{3}(2 + x)$

24. $5(z - 2) \geq \dfrac{7z - 3}{2}$

25. $|6 - z| + 3 > 5$

26. $\frac{1}{x} \le \frac{1}{x-2}$

27. $6x^2 - 5x - 6 < 0$

28. $t^3 + t^4 > 6t^2$

29. $|t^2 - 1| \ge |t^2 - t|$

30. $x^2 + 1 < 4x$

In Problems 31 through 40, solve each equation or inequality.

31. $2[3 - 5(2t - 1)] + 7t = 4 - 3t$

32. $12 - 5x > 3[x - 2(4 - x)]$

33. $\frac{x}{x^2 - 5x - 14} \ge \frac{2}{x - 7}$

34. $\sqrt{y - 1} - \sqrt{2y + 5} + 2 = 0$

35. $r^2 + 3r + 5 = 0$

36. $x^4 + 8x = 0$

37. $x^3 > 4x$

38. $|q - 11| \le 2q - 1$

39. $(t^2 - 10)^2 = 36$

40. $x^2 + 2x = 2$

41. Solve for R_1: $\frac{1}{R_T} = \frac{1}{R_1} + \frac{1}{R_2} + \frac{1}{R_3}$.

42. Solve for S: $\frac{1}{2}L = \frac{SR - 1}{S + 1}$.

In Problems 43 through 48, approximate the solutions to four significant digits.

43. $3x^2 + 2x - 4 = 0$

44. $2w^2 - w - 5 = 0$

45. $3z^2 + 3\sqrt{7}z - \frac{\pi}{2} = 0$

46. $\sqrt{6}x^2 - \pi x = \sqrt{5}$

47. $z^2 - z + 4 = 0$

48. $z^2 + 4z + 10 = 0$

In Problems 49 through 54, find the discriminant for each equation and determine the nature of the solutions.

49. $7x^2 + 11x - 5 = 0$

50. $-6t^2 + 17t - 13 = 0$

51. $3z^2 - 8z + 6 = 0$

52. $-5x^2 + \sqrt{17}x - 1 = 0$

53. $\frac{11}{17}x^2 - \frac{4}{9}x + 1 = 0$

54. $\pi k^2 + \frac{13}{11}k + \frac{1}{13} = 0$

55. How many liters of a 15% brine solution should be added to 20 liters of a 30% brine solution to obtain a 25% brine solution?

56. A. J. invested $24,000 in an oil painting and a cloisonné vase. She made a profit of 20% on the painting but lost 5% on the vase. If A. J.'s net profit was 13.75%, how much did she pay for each object?

57. The Benoit Asphalt Paving Company can pave a certain section of road in 2 days, while a local construction company would take 3 days to pave the same section of road. How long will it take them working together?

58. Jon and Tamra run on the same track. Jon runs at the rate of $\frac{1}{6}$ mile per minute, and Tamra runs at the rate of $\frac{1}{8}$ mile per minute. Tamra starts first, and in 3 minutes Jon overtakes her. How long after Tamra started did Jon start?

59. Vu Cao works 20 miles from the Civic Center, and Ly Tan works 32 miles from the Civic Center. They plan to meet at the Civic Center after work to view a new art exhibit. If they leave work at the same time and travel at the same average rate, Vu Cao will arrive $\frac{1}{2}$ hour before Ly Tan. How long will it take Ly Tan to drive from work to the Civic Center?

60. A wood lot of pulpwood trees is in the form of a right triangle with a hypotenuse of approximately $677\frac{1}{2}$ feet. The longest of the two legs is 48.9 feet longer than the shortest side. To the nearest half-foot, what are the lengths of the two sides?

61. A vat has two identical input valves and a drain. If the drain is closed and the two valves are open, the tank will fill in 40 minutes less time than the drain can empty a full vat. If the drain and both input valves are open, it takes $\frac{1}{2}$ hour to fill the vat. If the drain is closed, how long will it take one of the two input valves to fill the vat working alone?

62. A piece of wire $23\frac{3}{4}$ inches long is cut from a piece of copper wire and bent into a square. The remaining piece of wire is also bent into a square. If the sum of the areas of the two squares is approximately 984.45 square inches, what was the original length of the copper wire to the nearest hundredth of an inch?

63. The tolerance level for a new antibiotic is given by the formula

$$|t - 0.004| < 1$$

What is the range of the tolerance level?

64. If $|r| < 1$, then what value can $2r + 7$ have?

65. If the Fahrenheit reading is greater than 100°, what will the Celsius reading be? ($F = \frac{9}{5}C + 32$.)

66. The top of a cone of height 10 centimeters and radius 5 centimeters is cut off as shown in Figure 2.64. If the volume of the part remaining is $196\pi/3$, find x.

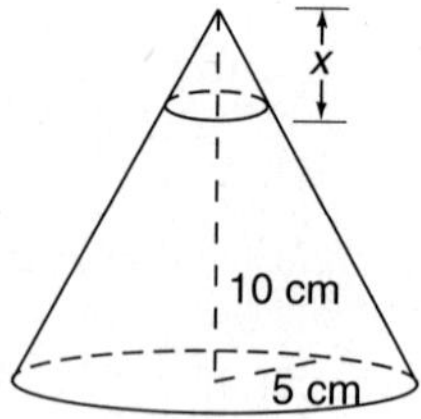

Fig. 2.64

LET'S NOT FORGET . . .

67. Which form is more useful?

(a) $\dfrac{-2 \pm \sqrt{12}}{4}$ or $\dfrac{-1 \pm \sqrt{3}}{2}$ if approximating solutions to a quadratic equation

(b) $(x + 2)(x - 1) = 0$ or $x(x + 1) = 2$ if solving the equation

68. Watch the role of negative signs.

(a) Solve $\dfrac{2}{x - 3} - \dfrac{1}{x + 3} = 0.$

(b) $-7^2 = ?$

69. From memory.

(a) State the quadratic formula.

(b) $(\sqrt{x + 2} + 5)^2 = ?$

(c) $\sqrt{x^2y^2} = ?$

(d) $8x^3 + 1 = ?$

70. Approximate with a calculator to 5 significant digits.

(a) The solutions to $5x^2 - 4x - \sqrt{2} = 0$

(b) $\dfrac{\sqrt{53} - 0.006^3}{25(2.04) + \sqrt[3]{3}}$

CHAPTER 3

Graphs and Functions

CONNECTIONS

Many of the modern ideas of mathematics have their roots in the seventeenth century. The development of geometry had been proceeding slowly, and the new thoughts of algebra were more and more abstract. Two Frenchmen of entirely different mold and character, Rene Descartes (1596–1650) and Pierre de Fermat (1601–1665), collaborated in the development of analytic geometry, which united the fields of algebra and geometry and laid the groundwork for the great discoveries that followed.

The idea that created analytic geometry was the rectangular coordinate system. Graphs of some equations in two variables on a Cartesian coordinate system are geometric figures such as lines, circles, and parabolas. The ability to visualize mathematical equations as graphs laid the groundwork for great discoveries. One such discovery by one of the founding fathers of calculus, Gottfried Leibniz, was the concept of function which we introduce in Section 3.5.

Today, mathematicians continue to rely on pictures to solve problems. Graphing calculators and computers draw graphs easily and accurately. The impact of this technology on mathematics will probably rival the development of analytic geometry.

In this chapter we see that the equations and inequalities in one variable that we solved in Chapter 2 are connected to the graphs of equations in two variables. We begin by graphing lines, circles, and parabolas on a Cartesian coordinate system. Finally, in Section 3.6 we connect the two great ideas, coordinate graphs and functions.

3.1 THE CARTESIAN COORDINATE SYSTEM

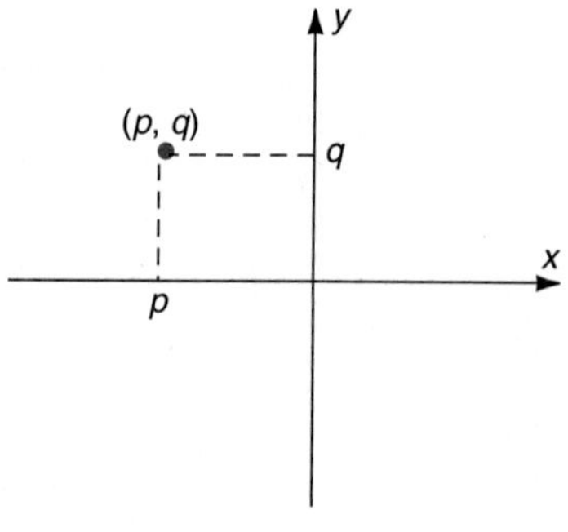

Fig. 3.1

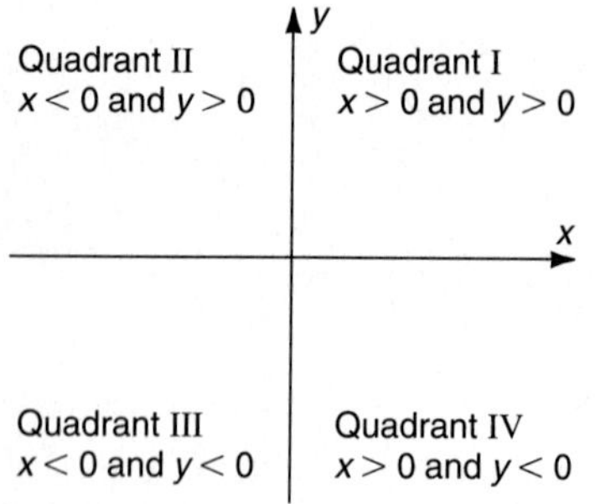

Fig. 3.2

If two copies of the number line, one horizontal and one vertical, are placed so that they intersect at the zero point of each line, a pair of axes is formed. The horizontal number line is called the ***x*-axis,** and the vertical number line the ***y*-axis.** The point where the lines intersect is called the **origin.** We call this a **rectangular coordinate system** or a **Cartesian coordinate system** (named for the French mathematician René Descartes).

Suppose p and q are two real numbers and written in the form (p, q). This is called an **ordered pair.** For our purposes, the order of the numbers is important. We associate a point on the coordinate system with this ordered pair by finding the first number of the ordered pair p on the x-axis, and the second number q on the y-axis. If we draw lines through these points perpendicular to the axes of the coordinate system, the lines will intersect in a single point. This unique point corresponds to the ordered pair (p, q) and is called its **graph.** To plot a point means to graph its ordered pair (see Figure 3.1).

The first number of an ordered pair is called the **abscissa** or **first coordinate** or ***x*-coordinate,** and the second number, the **ordinate** or **second coordinate** or ***y*-coordinate.**

The four regions made by the axes are called **quadrants.** The graph of every ordered pair is either in one of the four quadrants or on one of the axes (see Figure 3.2).

Distance

The distance between two numbers on the number line is given by the absolute value of their difference. That is, the distance between the numbers p and q on the number line is $|p - q|$. What about the distance between two points plotted on our coordinate system? Remembering the Pythagorean Theorem, this distance can easily be found. The Pythagorean Theorem states that the square of the length of the hypotenuse of any right triangle equals the sum of the squares of the lengths of the other two sides (Figure 3.3).

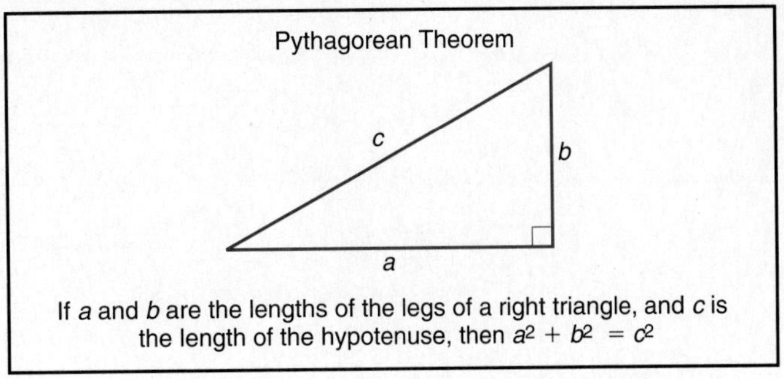

Pythagorean Theorem

c b a

If a and b are the lengths of the legs of a right triangle, and c is the length of the hypotenuse, then $a^2 + b^2 = c^2$

Fig. 3.3

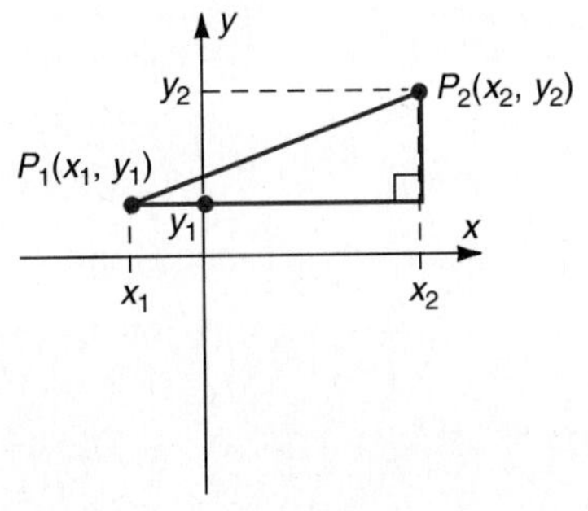

Fig. 3.4

Suppose two points P_1 and P_2 have coordinates (x_1, y_1) and (x_2, y_2). What is the distance between P_1 and P_2?

Notice in Figure 3.4 that the distance from P_1 to P_2 is the length of the hypotenuse of a right triangle. Also note that the length of one side is the same as the distance between x_1 and x_2 on the x-axis, which is $|x_2 - x_1|$, and the length of the other side is the same as the distance between y_1 and y_2 on the y-axis, which is $|y_2 - y_1|$.

So, if we let d be the distance between P_1 and P_2, by the Pythagorean Theorem,

$$d^2 = |x_2 - x_1|^2 + |y_2 - y_1|^2$$

Now, because $|q|^2 = q^2$ for any real number q, and because distance is not negative,

$$d = \sqrt{(x_2 - x_1)^2 + (y_2 - y_1)^2}$$

The Distance Formula

The distance between the points (x_1, y_1) and (x_2, y_2) is given by

$$d = \sqrt{(x_2 - x_1)^2 + (y_2 - y_1)^2}$$

Since $(x_2 - x_1)^2 = (x_1 - x_2)^2$ and $(y_1 - y_2)^2 = (y_2 - y_1)^2$ it does not matter which point we call P_1 or P_2. The distance formula says that the distance between two points is the square root of the sum of the difference between the x's squared and the difference between the y's squared.

EXAMPLE 1 Find the distance between the points $(4, -7)$ and $(-1, 3)$ (Figure 3.5).

Solution We substitute into the distance formula.

$$d = \sqrt{(x_1 - x_2)^2 + (y_1 - y_2)^2}$$

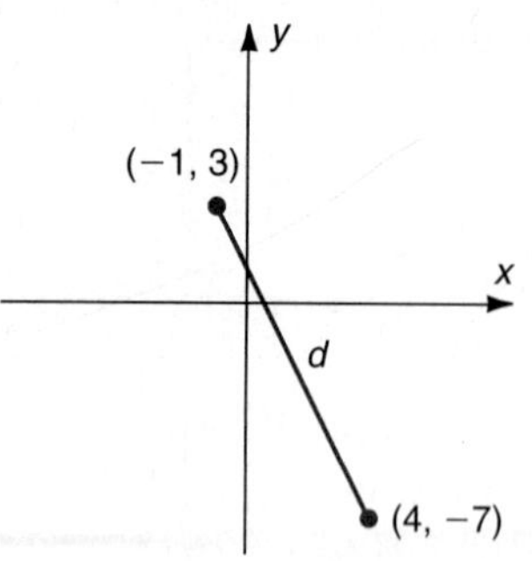

Fig. 3.5

$$= \sqrt{(4-(-1))^2 + (-7-3)^2}$$
$$= \sqrt{5^2 + (-10)^2}$$
$$= \sqrt{25 + 100}$$
$$= \sqrt{125}$$
$$= 5\sqrt{5}$$

The distance is $5\sqrt{5}$ units. □

EXAMPLE 2 Approximate (to three decimal places) the distance between the points $\left(\frac{1}{2}, -\frac{1}{6}\right)$ and $\left(\frac{13}{8}, \frac{5}{12}\right)$.

Solution Since the problem asks for an approximation, we will set up the distance formula and then use a calculator.

$$d = \sqrt{(x_1 - x_2)^2 + (y_1 - y_2)^2}$$
$$= \sqrt{\left(\frac{1}{2} - \frac{13}{8}\right)^2 + \left(-\frac{1}{6} - \frac{5}{12}\right)^2}$$

Enter these keystrokes, being careful not to write down intermediate results.

[( ] [1] [÷] [2] [−] [13] [÷] [8] [ )] [x^2] [+]
[( ] [1] [+/−] [÷] [6] [−] [5] [÷] [12] [ )] [x^2] [=] [$\sqrt{\ }$]

and read on the display 1.267242194. The distance is approximately 1.267 units, or $d \approx 1.267$. □

Another useful formula gives the coordinates of the point halfway between two points.

The Midpoint Formula

The midpoint between (x_1, y_1) and (x_2, y_2) is

$$\left(\frac{x_1 + x_2}{2}, \frac{y_1 + y_2}{2}\right)$$

Proof: Consider the three points (x_1, y_1), (x_2, y_2), and $\left(\frac{x_1 + x_2}{2}, \frac{y_1 + y_2}{2}\right)$.

First we will show that the distances d_1 and d_2 in Figure 3.6 are equal. Then, to show that M lies on the line containing P_1 and P_2, we show that $d_1 + d_2 = d$.

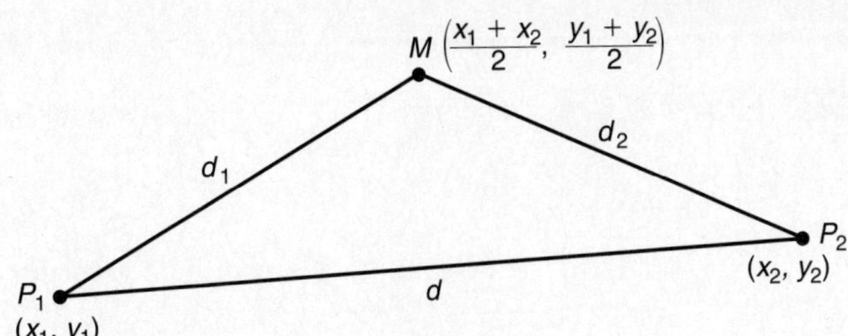

Fig. 3.6

$$d_1 = \sqrt{\left(x_1 - \frac{x_1 + x_2}{2}\right)^2 + \left(y_1 - \frac{y_1 + y_2}{2}\right)^2} = \sqrt{\left(\frac{x_1 - x_2}{2}\right)^2 + \left(\frac{y_1 - y_2}{2}\right)^2}$$

$$d_2 = \sqrt{\left(\frac{x_1 + x_2}{2} - x_2\right)^2 + \left(\frac{y_1 + y_2}{2} - y_2\right)^2} = \sqrt{\left(\frac{x_1 - x_2}{2}\right)^2 + \left(\frac{y_1 - y_2}{2}\right)^2}$$

Therefore, $d_1 = d_2$. Now we examine $d_1 + d_2$.

$$d_1 + d_2 = 2d_1 = 2\sqrt{\left(\frac{x_1 - x_2}{2}\right)^2 + \left(\frac{y_1 - y_2}{2}\right)^2}$$

$$= \sqrt{4\left(\frac{x_1 - x_2}{2}\right)^2 + 4\left(\frac{y_1 - y_2}{2}\right)^2} \quad \text{Root of a product}$$

$$= \sqrt{4 \cdot \frac{(x_1 - x_2)^2}{4} + 4 \cdot \frac{(y_1 - y_2)^2}{4}} \quad \text{Square of a quotient}$$

$$= \sqrt{(x_1 - x_2)^2 + (y_1 - y_2)^2}$$

But this is d, the distance between (x_1, y_1) and (x_2, y_2). Therefore, $\left(\frac{x_1 + x_2}{2}, \frac{y_1 + y_2}{2}\right)$ must lie on the line containing (x_1, y_1) and (x_2, y_2) and, since $d_1 = d_2$, it must be the midpoint.

EXAMPLE 3 Find the coordinates of the point halfway between points $(-4, 7)$ and $(8, -12)$ (Figure 3.7).

Solution The x-coordinate is found by substituting into $\frac{x_1 + x_2}{2}$.

$$\frac{-4 + 8}{2} = \frac{4}{2} = 2 \quad \text{The } x\text{-coordinate is 2.}$$

Likewise, the y-coordinate is found by substituting into $\frac{y_1 + y_2}{2}$.

$$\frac{7 + (-12)}{2} = \frac{-5}{2} \quad \text{The } y\text{-coordinate is } -\frac{5}{2}.$$

The point halfway between $(-4, 7)$ and $(8, -12)$ is $\left(2, -\frac{5}{2}\right)$. □

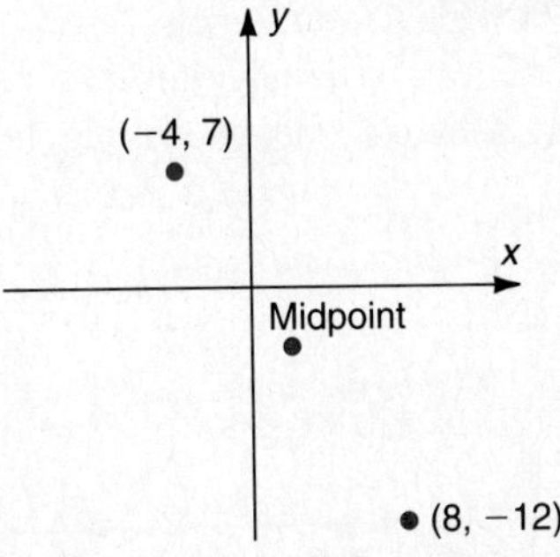

Fig. 3.7

EXAMPLE 4 The point $\left(-\frac{1}{2}, 2\right)$ is halfway between the point (3, 5) and another point. Find the other point (Figure 3.8).

Solution Let (x, y) be the coordinates of the other point. Then,

$$-\frac{1}{2} = \frac{x+3}{2} \quad \text{and} \quad 2 = \frac{y+5}{2}$$

Solving each of these equations,

$$-1 = x + 3 \qquad 4 = y + 5$$
$$-4 = x \qquad -1 = y$$

The other point is $(-4, -1)$. □

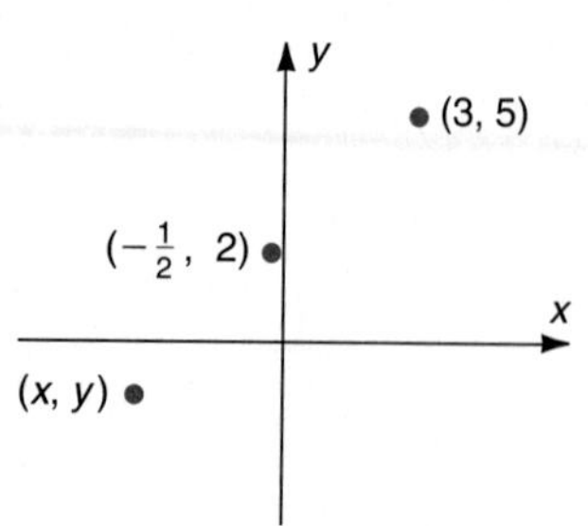

Fig. 3.8

Graph of an Equation in Two Variables

In Chapter 2 we learned how to find the solution set for various kinds of equations in *one variable.* In many applications, the mathematical model is often an equation in *two variables.* The Cartesian coordinate system allows us to draw pictures of solution sets for equations in two variables. This picture of the solution set is called the **graph** of the equation.

> **Definition: Graph of an Equation in Two Variables**
>
> The graph of an equation in two variables is the graph of the solution set of the equation.

Let's consider finding the solution set for an equation in *two variables,* such as $x^2 + y^2 = 9$. A solution to the equation consists of a number for x and a number for y that make the equation a true statement. For example, if x is 0 and y is 3, the equation is true. We can write this solution as an ordered pair with the number x listed first and y listed second. So, (0, 3) is a solution. Notice that there are other solutions. For instance, $(0, -3)$, $(3, 0)$, and $(-3, 0)$ are also solutions.

How many solutions to $x^2 + y^2 = 9$ are there? If we let x be a real number, such as $\sqrt{5}$, then we can find a corresponding number for y, if there is one, by solving the equation

$$(\sqrt{5})^2 + y^2 = 9$$
$$5 + y^2 = 9$$
$$y^2 = 4$$
$$y = \pm 2 \qquad \text{Square Root Property}$$

This gives two more solutions, $(\sqrt{5}, 2)$ and $(\sqrt{5}, -2)$.

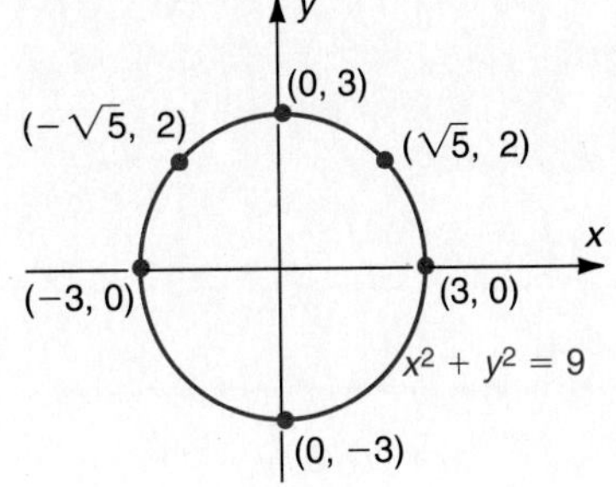

Fig. 3.9

It turns out that there is an infinite number of solutions. If we plot the solutions we have found on a Cartesian coordinate system, it appears that the points lie on a circle centered at the origin with radius 3. In fact every solution to the equation $x^2 + y^2 = 9$ will be on this circle, and every point on the circle represents a solution to the equation $x^2 + y^2 = 9$. This circle in Figure 3.9 is the graph of the equation.

Plotting many points is a useful technique for graphing solution sets of equations. Often this is our only approach. As we study more algebra, we will learn to recognize the graphs of various kinds of equations, and then we can sketch a graph by plotting a few special points. Without knowing what a graph looks like, it is impossible to know how to connect a set of points.

EXAMPLE 5

Plot several points on the graph of each equation and sketch a guess for each graph.

(a) $y - x^2 = 0$ (b) $y = |x| - 1$

Solutions

(a) Tables are often used to record the x- and y-coordinates of points. Calculations will be easier if we solve the equation for y and then choose numbers for x and find corresponding numbers for y. Figure 3.10 shows the ordered pairs from the table plotted on the graph. To see that the graph is actually a smooth curve, we could plot several more points with x-coordinates between -1 and 1.

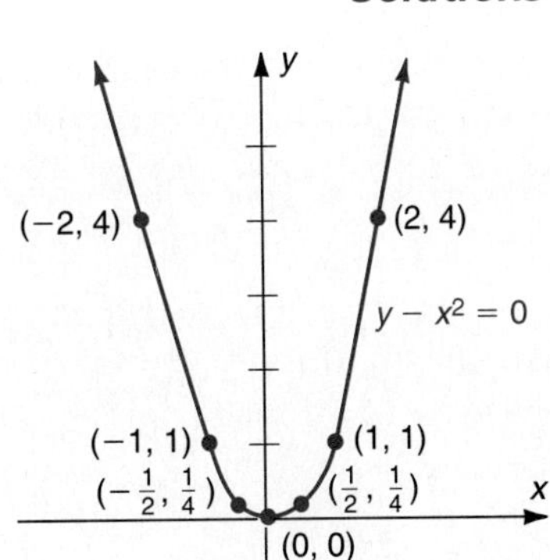

Fig. 3.10

$y = x^2$							
x	-2	-1	$-\frac{1}{2}$	0	$\frac{1}{2}$	1	2
y	4	1	$\frac{1}{4}$	0	$\frac{1}{4}$	1	4

We connected the points in a smooth curve. The arrowheads indicate that the graph continues. The graph in Figure 3.10 is called a *parabola.*

(b) The points plotted in Figure 3.11 suggest that the graph might be a parabola. However, by plotting many points we can see that this graph is a V which has a sharp point at the vertex. □

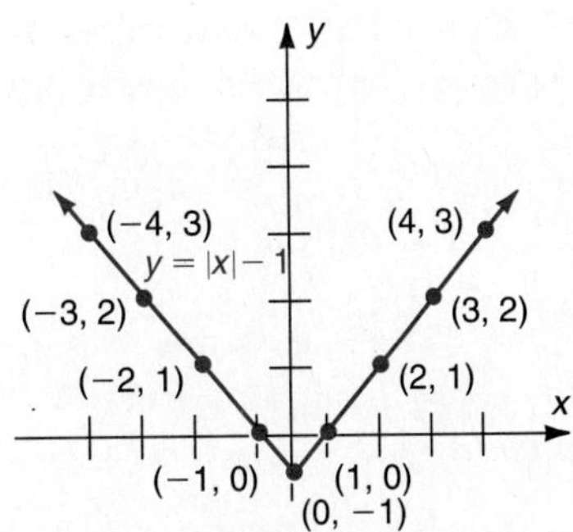

Fig. 3.11

Notice that if a point is located on either of the coordinate axes, one of its coordinates is zero.

> **Definitions: Intercepts**
>
> An **x-intercept** is the coordinate on the x-axis where the graph crosses the x-axis.
>
> A **y-intercept** is the coordinate on the y-axis where the graph crosses the y-axis.

Often the intercepts are the easiest points to find on a graph.

> **To Find the Intercepts of the Graph of an Equation**
>
> **1.** To find the x-intercept(s), replace y with 0 and solve for x.
>
> **2.** To find the y-intercept(s), replace x with 0 and solve for y.

EXAMPLE 6 Find the intercepts for the graph of each equation.

(a) $2x^2 + y^2 = 8$ (b) $y = x^2 - 2x + 3$

Solutions (a) If we let x be 0, then we have

$$0 + y^2 = 8$$

$$y = \pm\sqrt{8} = \pm 2\sqrt{2}$$

The y-intercepts are $2\sqrt{2}$ and $-2\sqrt{2}$. [The coordinates are $(0, 2\sqrt{2})$ and $(0, -2\sqrt{2})$.]

If we let y be 0, then

$$2x^2 = 8$$

$$x^2 = 4$$

$$x = \pm 2$$

The x-intercepts are 2 and -2. [The coordinates are $(2, 0)$ and $(-2, 0)$.]

(b) $y = x^2 - 2x + 3$

If we let y be 0, we have

$0 = x^2 - 2x + 3$

This cannot be factored. The discriminant is $(-2)^2 - 4(1)(3)$ or -8. This means that the solutions to the equation $0 = x^2 - 2x + 3$ are not real numbers but are complex numbers. Thus, there are no real solutions, which means that the graph has no x-intercepts.

If we let x be 0, we have $y = 3$. The y-intercept is 3. □

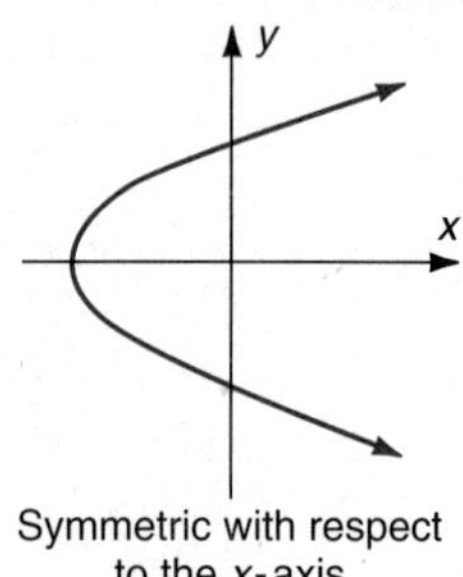

Symmetric with respect to the x-axis
(a)

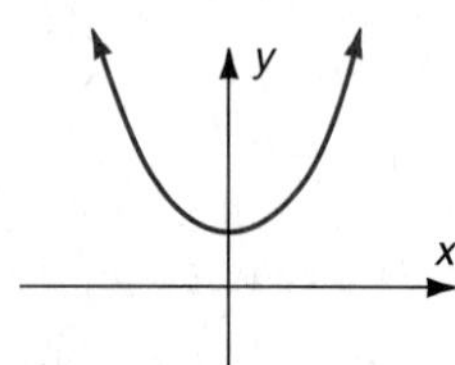

Symmetric with respect to the y-axis
(b)

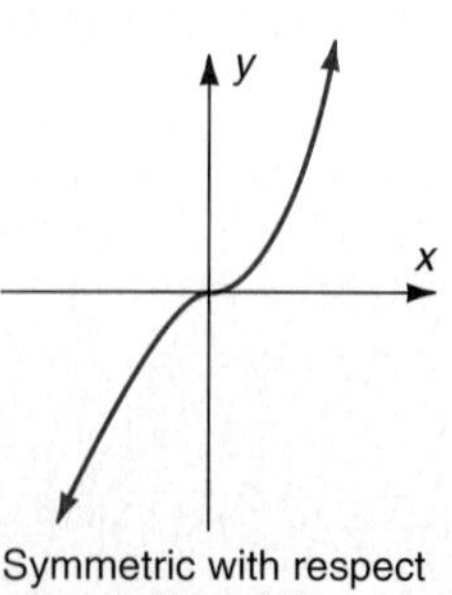

Symmetric with respect to the origin
(c)

Fig. 3.12

Symmetry

The behavior of a graph is sometimes described by using the idea of **symmetry.** Each of the graphs in Figure 3.12 has a type of symmetry.

Symmetry with respect to the x-axis: Considering the x-axis as a dividing line, each half of the graph is a mirror image of the other half through the x-axis.

Symmetry with respect to the y-axis: Considering the y-axis as a dividing line, each half of the graph is a mirror image of the other half through the y-axis.

Symmetry with respect to the origin: Each half of the graph is a mirror image of the other half through the origin.

Symmetry can be useful in sketching a graph.

Definition: Symmetry

1. A graph is **symmetric with respect to the x-axis** if whenever (x, y) is on the graph, then $(x, -y)$ is also on the graph.
2. A graph is **symmetric with respect to the y-axis** if whenever (x, y) is on the graph, then $(-x, y)$ is also on the graph.
3. A graph is **symmetric with respect to the origin** if whenever (x, y) is on the graph, then $(-x, -y)$ is also on the graph.

EXAMPLE 7 Suppose a graph contains the points $(-2, 3)$, $(1, 5)$, $(-4, -3)$, and $(5, -7)$.

(a) If the graph is symmetric with respect to the x-axis, name four more points on the graph.

(b) If the graph is symmetric with respect to the y-axis, name four more points on the graph.

(c) If the graph is symmetric with respect to the origin, name four more points on the graph.

Solutions (a) Symmetry with respect to the x-axis gives $(-2, -3)$, $(1, -5)$, $(-4, 3)$ and $(5, 7)$. That is, if (x, y) is on the graph, then $(x, -y)$ is on the graph.

(b) Symmetry with respect to the y-axis gives $(2, 3)$, $(-1, 5)$, $(4, -3)$, and $(-5, -7)$. That is, if (x, y) is on the graph, then $(-x, y)$ is on the graph.

(c) Symmetry with respect to the origin gives $(2, -3)$, $(-1, -5)$, $(4, 3)$, and $(-5, 7)$. That is, if (x, y) is on the graph, then $(-x, -y)$ is on the graph.

Plotting the points in each case might be helpful. □

The following list tells how to decide whether a graph is symmetric with respect to the axes or the origin.

Testing for Symmetry with Respect to Axes or Origin

1. The graph of an equation is symmetric with respect to the x-axis if replacing y with $-y$ in the equation results in an equivalent equation.

2. The graph of an equation is symmetric with respect to the y-axis if replacing x with $-x$ in the equation results in an equivalent equation.
3. The graph of an equation is symmetric with respect to the origin if replacing x with $-x$ and y with $-y$ in the equation results in an equivalent equation.

EXAMPLE 8 Use intercepts and symmetry to sketch the graph of each equation.

(a) $x = y^2 - 1$ (b) $y = x^3$

Solutions (a) Replacing x with $-x$ in the equation $x = y^2 - 1$ gives

$$-x = y^2 - 1$$

which is not equivalent to the original equation. So, the graph is not symmetric with respect to the y-axis. Replacing y with $-y$ in $x = y^2 - 1$ gives

$$x = (-y)^2 - 1$$
$$x = y^2 - 1$$

which is the original equation. The graph is symmetric with respect to the x-axis. Replacing x with $-x$ and y with $-y$ gives

$$-x = (-y)^2 - 1$$
$$-x = y^2 - 1$$

which is not equivalent to the original equation. The graph is not symmetric with respect to the origin.

To find the y-intercepts, replace x with 0. To find the x-intercepts, replace y with 0.

$$0 = y^2 - 1 \qquad\qquad x = 0^2 - 1$$
$$1 = y^2 \qquad\qquad x = -1$$
$$\pm 1 = y$$

The x-intercept is -1.

The y-intercepts are 1 and -1.

Some more points on the graph can be found by choosing a number for y and finding the corresponding number for x. They are listed in the following table.

x	3	8
y	2	3

Symmetry with the x-axis gives other points listed in the following table.

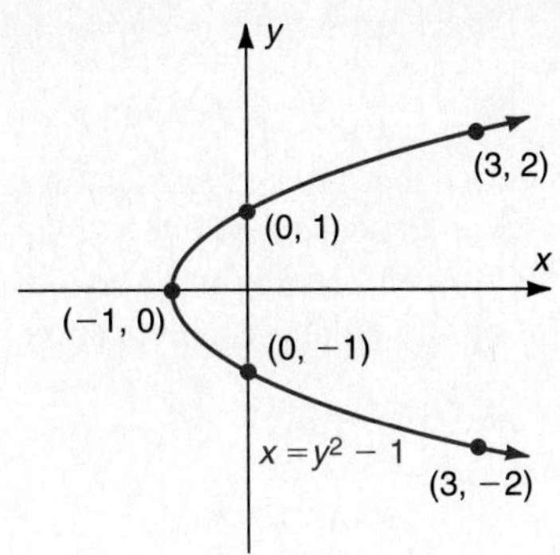

Fig. 3.13

x	3	8
y	-2	-3

Using the symmetry, intercepts, and a few points, we are able to sketch a graph (Figure 3.13).

(b) We test $y = x^3$ for symmetry.
Replacing y with $-y$ gives

$$-y = x^3$$
$$y = -x^3$$

which is not equivalent to the original equation. So, the graph is not symmetric with respect to the x-axis.
Replacing x with $-x$ gives

$$y = (-x)^3$$
$$y = -x^3$$

Thus, the graph is not symmetric with respect to the y-axis.
Substituting $-x$ for x and $-y$ for y, we have

$$-y = (-x)^3$$
$$-y = -x^3$$
$$y = x^3$$

which is the original equation. This graph is symmetric with respect to the origin. Finding the intercepts tells us that the graph goes through the origin. The following table lists some points on the graph.

x	0	$\frac{1}{2}$	1	2
y	0	$\frac{1}{8}$	1	8

Symmetry with the origin gives more points which are listed in the following table.

x	$-\frac{1}{2}$	-1	-2
y	$-\frac{1}{8}$	-1	-8

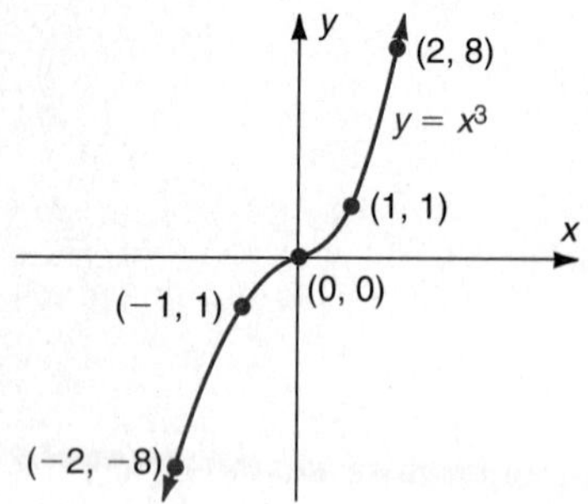

Fig. 3.14

When we plot these points and use symmetry, it allows us to draw the graph in Figure 3.14. □

It is important to understand the relationship between a graph and an equation. The graph of an equation is a picture of the solution set of the equation. Every point on the graph of an equation corresponds to a solution of the equation. Likewise, every solution of an equation corresponds to a point on the graph of the equation.

EXAMPLE 9 Decide if the given ordered pair is on the graph of the given equation.

(a) $3x + 5y = 9$; $(-5, 3)$
(b) $x^2 - y^2 = 1$; $(\sqrt{2}, 1)$

Solutions (a) If x is -5 and y is 3, then $3x + 5y$ becomes $3(-5) + 5(3)$ which is 0. Since $0 \neq 9$, the statement $3(-5) + 5(3) = 9$ is not true, and $(-5, 3)$ is not a solution to $3x + 5y = 9$ and is not on the graph of $3x + 5y = 9$.

(b) Substituting $\sqrt{2}$ for x and 1 for y into $x^2 - y^2$ gives $(\sqrt{2})^2 - 1^2$ which is 1. So $(\sqrt{2}, 1)$ is a solution to $x^2 - y^2 = 1$ and is therefore on the graph of $x^2 - y^2 = 1$. □

GRAPHING CALCULATOR BOX

Setting Windows

The display on a graphing calculator is called a *window* or *viewing rectangle.* Because the display shows only a portion of a coordinate system, our first consideration is to set an appropriate window. Imagine the display as a rectangle with a set of axes at its center. The size of the rectangle is fixed by setting minimum and maximum x- and y-coordinates.

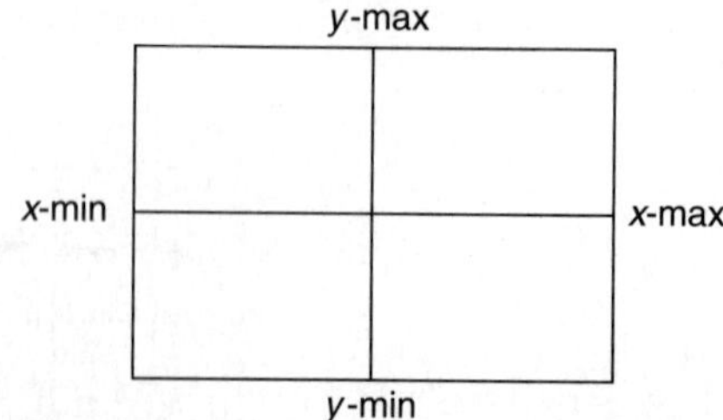

The Default Window Setting

Graphing calculators have a standard window which is often good for starting a graph. It is called the **default setting.** On the TI-81, it is −10 to 10 on both axes. On the CASIO fx-7700G, they are −4.7 to 4.7 on the x-axis and −3.1 to 3.1 on the y-axis.

TI-81

Press [ZOOM] followed by [6]. A coordinate system will appear with the default setting. Return to the computation screen with [CLEAR].

CASIO fx-7700G

Press [RANGE] for the range menu. Notice the word INIT on the status line. Press [F1] for the initial or default setting. Notice the range settings on the menu. Press [RANGE] [RANGE] to return to the computation screen.

Setting the Window

The [RANGE] key allows us to change the size of the viewing rectangle. Let's set the viewing rectangle at [−2, 2] for both x and y.

TI-81

The [RANGE] key is located on the first row of keys. Use the editing arrows and number keys to set Xmin to −2 and Xmax to 2. Be careful to use the [(−)] key to enter a negative number. Set Ymin and Ymax in a similar manner. To quit the range screen, press [QUIT].

CASIO fx-7700G

The [RANGE] key is located on the third row of keys. Set Xmin and press [EXE]. Set Xmax, Ymin, and Ymax the same way. To exit the range screen, press [RANGE] as many times as is necessary. Notice that there are two range screens, one for x- and y-coordinates and one for polar coordinates.

PROBLEM SET 3.1

Warm-ups

In Problems 1 through 6, find the distance between the two given points. See Example 1.

1. (8, 6), (5, 2)
2. (1, 2), (−5, 5)
3. (−7, 3), (1, −3)
4. (−1, −2), (2, 2)
5. (−1, −2), (−3, −4)
6. (−4, 1), (5, 0)

In Problems 7 through 9, approximate (to four decimal places) the distance between the two given points. See Example 2.

7. $\left(\frac{2}{7}, 2\right), \left(\frac{17}{3}, -\frac{3}{7}\right)$
8. $\left(\frac{\sqrt{3}}{2}, \frac{1}{2}\right), \left(\frac{1}{2}, \frac{\sqrt{3}}{2}\right)$
9. $(\pi, -2\pi), \left(-\frac{\pi}{2}, \pi\right)$

In Problems 10 through 12, find the midpoint between the two given points. See Example 3.

10. $(-5, 1), (3, 1)$

11. $\left(\frac{1}{5}, -\frac{1}{2}\right), \left(-\frac{2}{5}, -1\right)$

12. $(-2\sqrt{7}, 10), (6\sqrt{7}, 10)$

In Problems 13 and 14, find the indicated point. See Example 4.

13. The point $(-3, 2)$ is halfway between the point $(-8, -4)$ and another point. Find the other point.

14. The point $\left(\frac{7}{3}, \frac{1}{3}\right)$ is halfway between the point $\left(-\frac{7}{3}, \frac{11}{3}\right)$ and another point. Find the other point.

In Problems 15 through 18, plot 10 points on the graph of each equation and sketch a guess for each graph. See Example 5.

15. $3x + y = 3$

16. $y = \sqrt{x}$

17. $x^2 + 2y^2 = 4$

18. $y = 2^x$

In Problems 19 through 22, find the intercepts of the graph of each equation. See Example 6.

19. $x^2 - y^2 = 4$

20. $x^2 + y^2 = 4$

21. $y = |x| + 2$

22. $y = \sqrt[3]{x}$

In Problems 23 through 26, use symmetry and intercepts to sketch the graph of each equation. See Examples 6, 7, and 8.

23. $x^2 - y^2 = 1$

24. $x^2 + y^2 = 1$

25. $y = 1 - |x|$

26. $y = x^2 - 2$

In Problems 27 through 30, determine if the given point is on the graph of the given equation. See Example 9.

27. $y = x^3 - 3x^2 + 4$; $(0, 4)$

28. $y = \dfrac{1}{x}$; $(0, 1)$

29. $y = \dfrac{x+3}{x^2 - x + 1}$; $(1, 4)$

30. $x^2 + y^2 - 2x + 4y - 4 = 0$; $(1, -2)$

Practice Exercises

In Problems 31 through 36, find the distance between the two points.

31. $(7, 0), (0, 7)$

32. $(0, 6), (6, 0)$

33. $\left(\frac{1}{2}, \frac{2}{3}\right), \left(-1, -\frac{1}{6}\right)$

34. $\left(\frac{1}{4}, -\frac{3}{8}\right), \left(1, -\frac{1}{4}\right)$

35. $(2\sqrt{3}, -\sqrt{2}), (\sqrt{3}, 2\sqrt{2})$

36. $(5\sqrt{2}, \sqrt{3}), (-\sqrt{2}, -\sqrt{3})$

In Problems 37 through 42, classify the triangle formed by the given points as a right triangle, an isosceles triangle, or an equilateral triangle.

37. $(1, 2), (1, 6), (4, 2)$

38. $(-2, 6), (5, 4), (-2, 2)$

39. $(6, 0), (2, -2), (2, 2)$

40. $(0, 4), (-3, 6), (-2, 1)$

41. $(0, -2), (3, -4), (7, 2)$

42. $(-3, 1), (3, 9), (7, 6)$

43. What is the radius of a circle centered at the origin and passing through the point $\left(\frac{1}{2}, \frac{\sqrt{3}}{2}\right)$?

44. Find the radius of a circle passing through the point $(-2, 4)$ with center at $(0, 2)$.

45. Find all points with a y-coordinate of 4 that are 5 units from the point $(1, 1)$.

46. Find all points with an x-coordinate of 5 that are $\sqrt{34}$ units from the point $(2, -3)$.

In Problems 47 through 50, find all numbers t such that the distance between the two given points is 10 units.

47. $(0, 4), (t, -2)$

48. $(3, 2), (t, 2)$

49. $(-2, t), (t, 12)$

50. $(-3, 12), (5, -2t)$

In Problems 51 and 52, approximate t to four significant digits so that the distance between the two given points is 7 units.

51. $(t, -2)$, $(5, 1)$

52. $(3, t)$, $(t, -5)$

53. Suppose $(-1, 5)$ represents the midpoint between $(2, -1)$ and P. Find the coordinates of P.

54. Suppose $(14, -5)$ represents the midpoint between $(52, 131)$ and P. Find the coordinates of P.

In Problems 55 through 70, sketch the graph of each equation. Label all intercepts and look for symmetry.

55. $y = 2x + 3$

56. $3x - y = 6$

57. $y = x^2 + 5$

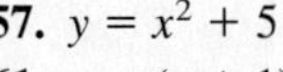

58. $y = x^2 - 4$

59. $x = y^2 - 4$

60. $x = y^2 + 1$

61. $y = (x + 1)^2$

62. $y = (x - 2)^2$

63. $y = \sqrt{x + 2}$

64. $y = \sqrt{x - 2}$

65. $x = y^2 - y - 2$

66. $x = y^2 + 3y - 10$

67. $y = |x| + 2$

68. $y = |x| - 3$

69. $y = x^3 - x$

70. $y = x(x + 2)(x - 3)$

71. A ball is thrown straight up from the ground. The graph in Figure 3.15 shows the relationship between the height (h) in feet of the ball and time (t) in seconds that has elapsed since the ball was thrown. (The graph does not represent the path of the ball.)

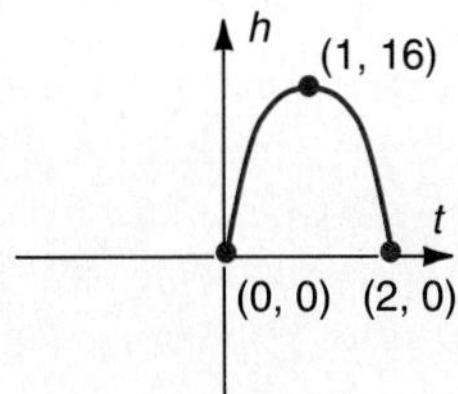

Fig. 3.15

(a) What do the t-intercepts mean?
(b) How high up does the ball go?
(c) Describe the symmetry of the graph. What does this mean?

72. Steve Carlson dropped his sunglasses from a window 30 meters above the ground. The graph in Figure 3.16 shows the relationship between the time (t) in seconds the sunglasses have been falling and their distance (s) in meters from the ground.

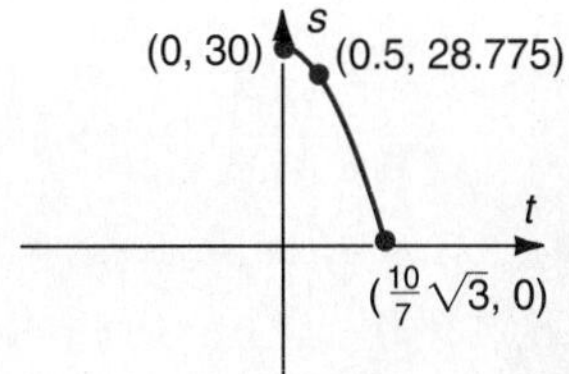

Fig. 3.16

(a) Interpret the intercepts.
(b) How far have the sunglasses fallen after 0.5 seconds?

In Problems 73 and 74, suppose $(x_1 + p, y_1 + q)$ represents the midpoint between (x_1, y_1) and (x_2, y_2).

73. Find the coordinates x_2 in terms of x_1 and p, and y_2 in terms of y_1 and q.

74. Find p and q in terms of x_1, x_2, y_1, and y_2.

Challenge Problems

75. Sketch the graph of $y = |x(x + 2)(x - 3)|$.

76. Sketch the graph of $y = \sqrt{4 - x^2}$.

77. Sketch the graph of each equation. Look for symmetry and find intercepts.
(a) $|x| = |y|$ (b) $x^{2/3} + y^{2/3} = 1$

IN YOUR OWN WORDS . . .

78. Without using the formula itself, explain the distance formula.

79. Explain the relationship between the graph of an equation and the equation.

3.2 LINES

In Section 3.1 we saw how the Cartesian coordinate system allows us to graph the solution set of an equation in two variables. In this section we continue this idea by focusing on **linear equations in two variables.**

> **Definition: Linear Equation in Two Variables**
>
> A linear equation in two variables is an equation that can be written in the form
>
> $$Ax + By = C$$
>
> where A, B, and C are constants, with A and B not both zero. This is called the **standard form.**

Graphing Lines

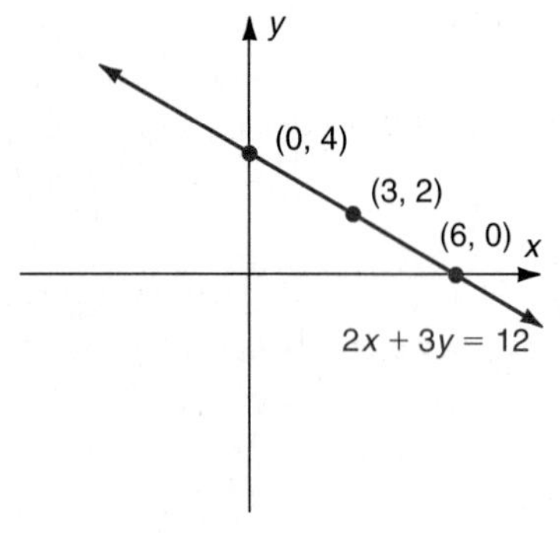

Fig. 3.17

Let's consider graphing a linear equation in two variables, such as $2x + 3y = 12$. Remember that the graph is a picture of the solution set for the equation. The easiest points to find are the intercepts. If x is 0, then y is 4. Thus, (0, 4) is on the graph. If y is 0, then x is 6. So, (6, 0) is on the graph. If we let x be any real number, then we can find a number y that will make $2x + 3y = 12$ a true statement. For example, if x is 3, then y is 2. Hence, (3, 2) is another point on the graph. The graph contains an infinite number of ordered pairs. If we plot (0, 4), (6, 0), and (3, 2), it appears that these points lie on the same line. They do. (See Figure 3.17.) Furthermore, every point on this line represents an ordered pair in the solution set of the equation $2x + 3y = 12$, and every solution to $2x + 3y = 12$ will be on this line.

> **The Graph of an Equation of the Form**
>
> $$Ax + By = C$$
>
> is a line.
>
> A, B, and C are constants, with A and B not both zero.

Since two points determine a line, we need to find two solutions of such an equation in order to draw its graph. Often the two easiest points to find are the intercepts. A third point can be found as a check.

EXAMPLE 1 Write each equation in standard form and graph each equation.

(a) $2x + 3y = 3$ (b) $x = 2y$ (c) $x = 1$ (d) $y = -2$

Solutions (a) $2x + 3y = 3$ is in standard form, with $A = 2$, $B = 3$, and $C = 3$. To find the y-intercept, replace x with 0 and solve for y. This gives us the point (0, 1). To find the x-intercept, replace y with 0 and solve for x. This gives us the point $\left(\frac{3}{2}, 0\right)$ (Figure 3.18).

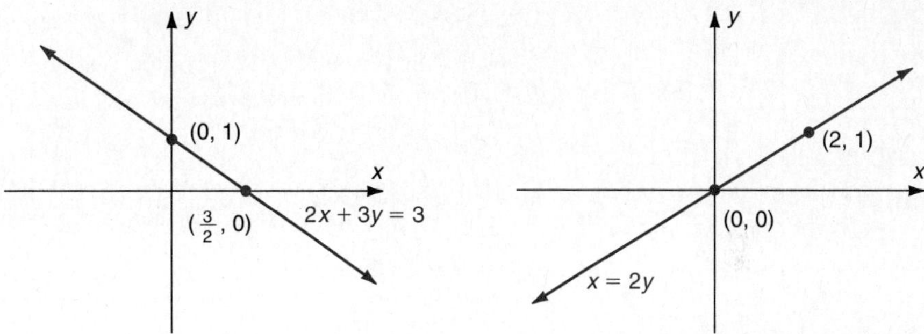

Fig. 3.18 **Fig. 3.19**

(b) The equation $x = 2y$ is not in standard form. We write it as $x - 2y = 0$ and see that $A = 1$, $B = -2$, and $C = 0$. Finding the intercepts gives only one point, (0, 0)! That means that this line goes through the origin. We find any other point on the line. For example, if y is 1, then x is 2. So, (2, 1) is on the line (Figure 3.19).

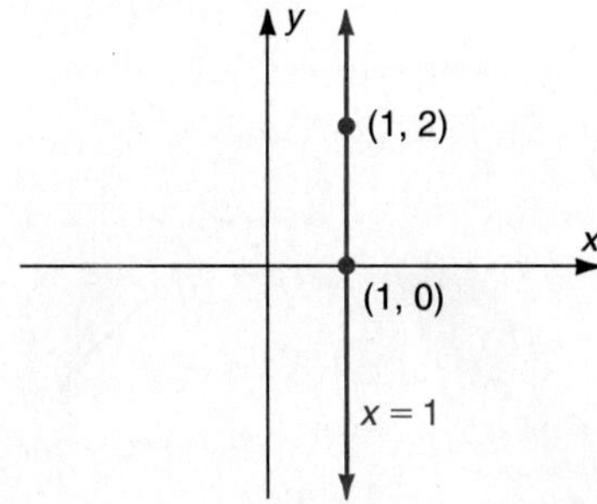

Fig. 3.20

(c) Standard form for $x = 1$ is $x + 0y = 1$, with $A = 1$, $B = 0$, and $C = 1$. If we let y be 0, then x is 1. So, (1, 0) is a point on the line. If we let x be 0, then we find that $0 = 1$. Thus, there is no y-intercept. What kind of line does not have a y-intercept? This line must be parallel to the y-axis. Such a line is called a *vertical line.* Any point on this line has coordinates of the form $(1, k)$, where k is any real number (Figure 3.20).

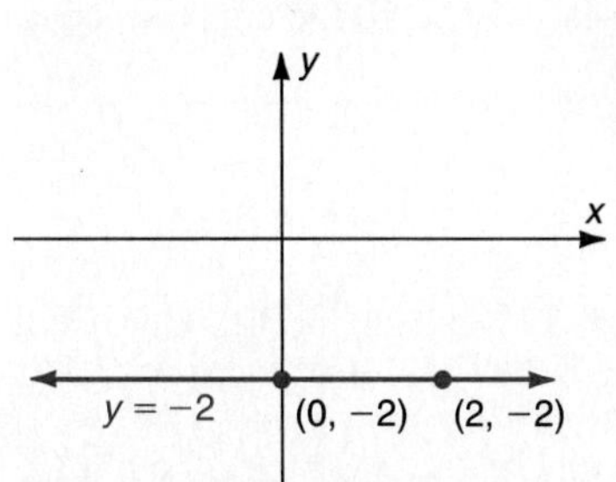

Fig. 3.21

(d) Standard form for $y = -2$ is $0x + y = -2$, with $A = 0$, $B = 1$, and $C = -2$. As in the last example, points on this line will be of the form $(k, -2)$, where k is any real number. The graph will be a line parallel to the x-axis. Such a line is called a *horizontal line* (Figure 3.21). □

Linear equations in two variables are used by scientists, engineers, mathematicians, and businesspeople to express relationships between two variables. Usually letters that are more appropriate to the situation are used rather than x and y.

EXAMPLE 2 Jason fires a rocket vertically upward with an initial velocity of 98 meters per second. The relationship between the velocity v in meters per second after t seconds of flight is given by $v + 9.8t = 98$.

(a) Graph this equation using the horizontal axis for time.

(b) What is the velocity after 2 seconds?

(c) When will the velocity be 0?

(d) Approximate to the nearest second when the velocity will be 10 meters per second.

Solutions (a) Let the horizontal axis be time and the vertical axis be velocity. The t-intercept is 10, and the v-intercept is 98 (Figure 3.22).

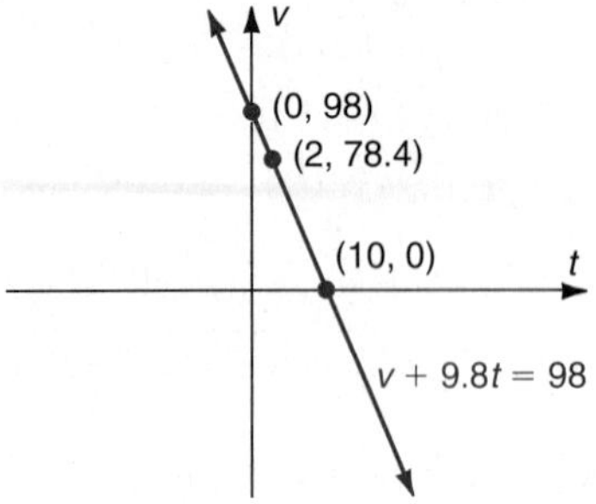

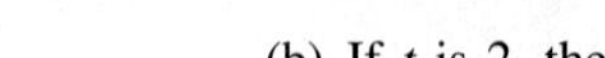

(b) If t is 2, then

$$\begin{aligned} v + 9.8(2) &= 98 \\ v &= 98 - 9.8(2) \\ v &= 78.4 \end{aligned}$$

The velocity after 2 seconds is 78.4 meters per second.
On the graph (Figure 3.22) this is represented by the point (2, 78.4).

(c) If v is 0, then

$$\begin{aligned} 0 + 9.8t &= 98 \\ t &= 10 \end{aligned}$$

Fig. 3.22

After 10 seconds of flight, the velocity will be 0 meters per second.
This corresponds to the point (10, 0) on the graph. The question could have been answered directly from Figure 3.22.

(d) If the velocity is 10 meters per second, then

$$\begin{aligned} 10 + 9.8t &= 98 \\ 9.8t &= 88 \\ t &= \frac{88}{9.8} \end{aligned}$$

Using a calculator, we see that $t \approx 9$. The velocity will be 10 meters per second after about 9 seconds of flight. □

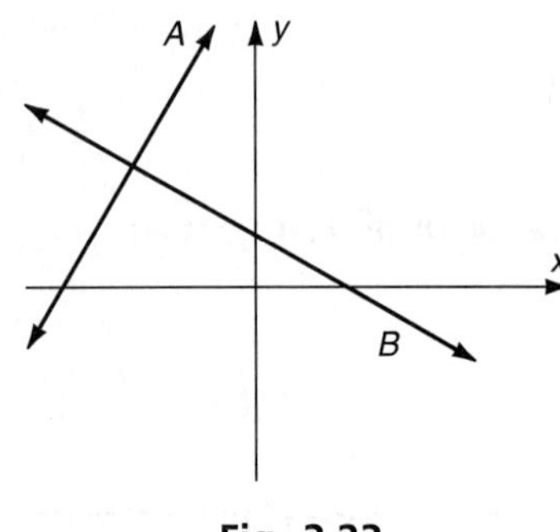

Fig. 3.23

Slope

Look at the two lines in Figure 3.23. Notice how line A rises from left to right, while line B slopes downward. (Always look at graphs as if moving along them from left to right.)

Line A is **increasing,** and line B is **decreasing.** It turns out that perhaps the most important property of a line is the rate at which it is increasing or decreasing. This property is called the **slope** of the line. To measure the slope, consider any two points on the line, say P_1 whose coordinates are (x_1, y_1) and P_2 whose coordinates are (x_2, y_2) (Figure 3.24).

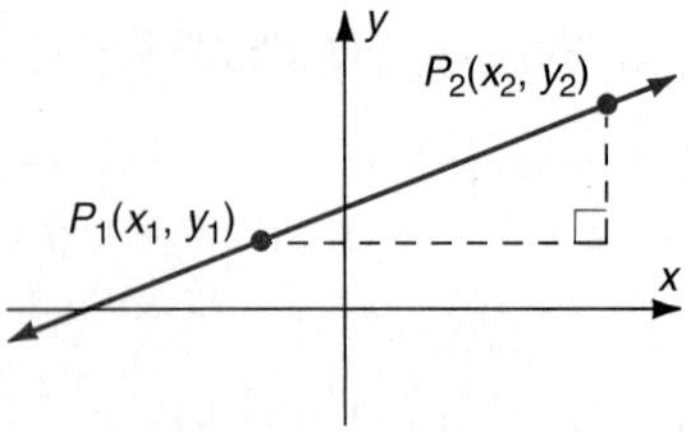

From P_1 to P_2, the vertical change or *rise* is given by the difference between the y-coordinates, while the horizontal change or *run* is given by the difference between the x-coordinates. That is,

Fig. 3.24

$$\text{Rise} = y_2 - y_1 \qquad \text{and} \qquad \text{Run} = x_2 - x_1$$

We define the slope as slope = rise/run. We use the letter m for the slope. Therefore, the slope of a line containing the points (x_1, y_1) and (x_2, y_2) is given by the following formula.

Definition: The Slope of a Line

The slope of the line through the points (x_1, y_1) and (x_2, y_2) is given by the formula

$$m = \frac{y_2 - y_1}{x_2 - x_1};\ x_1 \neq x_2$$

It is important to notice that the slope of a line is independent of how the two points are chosen. Suppose any other two points are chosen, say P_3 with coordinates (x_3, y_3) and P_4 with coordinates (x_4, y_4).

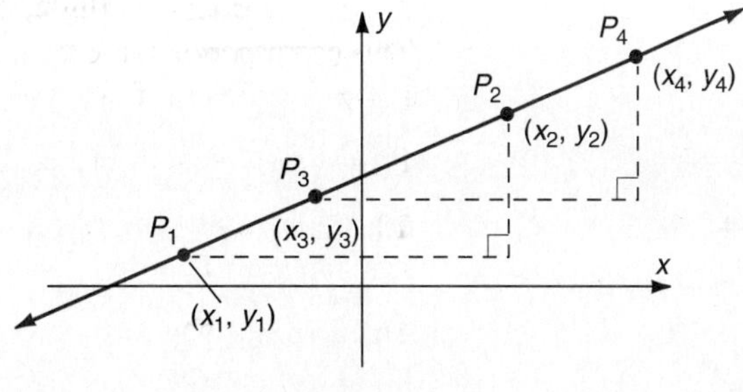

Fig. 3.25

Since the two triangles in Figure 3.25 are similar, the ratios of their corresponding sides are equal. Therefore

$$\frac{y_4 - y_3}{x_4 - x_3} = \frac{y_2 - y_1}{x_2 - x_1} = m$$

EXAMPLE 3 Find the slope of the line containing the points $(3, -5)$ and $(-4, 1)$.

Solution The slope formula requires that we know the coordinates of two points on the line. We have two points, so we substitute into the slope formula.

$$m = \frac{y_2 - y_1}{x_2 - x_1}$$

$$= \frac{-5 - 1}{3 - (-4)} = \frac{-6}{7} = -\frac{6}{7}$$

The slope of the line containing $(3, -5)$ and $(-4, 1)$ is $-\frac{6}{7}$. □

EXAMPLE 4 Use the slope formula to find the slope of the graph of each equation.

(a) $2x + 3y = 6$ (b) $x = 2$ (c) $y = 1$

Solutions (a) The graph of $2x + 3y = 6$ is a line. Two points on that line are needed to find the slope. If the intercepts are found, we have two points. Letting y be 0 gives 3 as the x-intercept, and letting x be 0 gives 2 as the y-intercept. Using these points, (3, 0) and (0, 2), the slope is given by

$$m = \frac{2-0}{0-3} = -\frac{2}{3}$$

The slope is $-\frac{2}{3}$.

(b) $x = 2$ is the equation of a vertical line. Two points on this line are (2, 0) and (2, 1).

$$m = \frac{1-0}{2-2} = \frac{1}{0} \text{ which is undefined}$$

Since the x-coordinates of all the points on a vertical line are the same, the run is zero and we are unable to calculate the slope. Thus *a vertical line does not have a defined slope.* The graph of the equation $x = 2$ has no slope. Sometimes we say that a vertical line has undefined slope.

(c) The graph of $y = 1$ is a horizontal line. Two points on the graph are (0, 1) and (1, 1).

$$m = \frac{1-1}{1-0} = \frac{0}{1} = 0$$

As the y-coordinates of all the points on a horizontal line are the same, the rise is zero and so the slope is 0. □

Vertical and Horizontal Lines

The graph of the equation $x = p$ is a *vertical* line with x-intercept of p. It has no defined slope.

The graph of the equation $y = q$ is a *horizontal* line with y-intercept of q. It has slope 0.

Equations of Lines

Let's look at another form of the linear equation in two variables. Suppose a line has slope m and y-intercept b. The equation of this line can be obtained from the slope formula. Since the y-intercept is b, the point $(0, b)$ is on the line. If (x, y) is any other point on the line, then the slope formula gives the following.

$$m = \frac{y - b}{x - 0} \qquad \text{Definition of slope}$$

$$mx = y - b \qquad \text{Clear fractions.}$$

$$mx + b = y \qquad \text{or} \qquad y = mx + b$$

We call $y = mx + b$ the slope-intercept form.

Slope-Intercept Form

An equation of a line with slope m and y-intercept of b is given by

$$y = mx + b$$

EXAMPLE 5 Rewrite $5x - 6y = 18$ in slope-intercept form and determine the slope and y-intercept.

Solution Solve $5x - 6y = 18$ for y.

$$-6y = -5x + 18$$

$$y = \frac{5}{6}x - 3$$

This is the slope-intercept form, where the slope is $\frac{5}{6}$ and the y-intercept is -3. □

It is convenient to use the slope-intercept form when drawing the graph of a linear equation. The y-intercept gives one point immediately, and then we use the fact that the slope is the ratio of the rise to the run to get a second point.

EXAMPLE 6 Sketch the graph of each linear equation.

(a) $y = \frac{1}{2}x - 1$ (b) $y = -\frac{2}{3}x$

Solutions (a) Since the equation $y = \frac{1}{2}x - 1$ is in slope-intercept form, read the y-intercept directly from the equation. Since b is -1, this graph crosses the y axis at -1. The slope is $\frac{1}{2}$. So, working from the y-intercept of -1, we rise 1 while we run 2 (from left to right). This gives a second point on the graph. These are enough to let us complete the sketch shown in Figure 3.26.

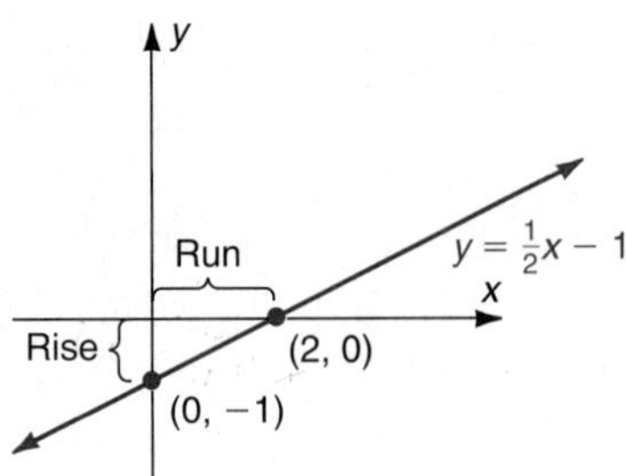

Fig. 3.26

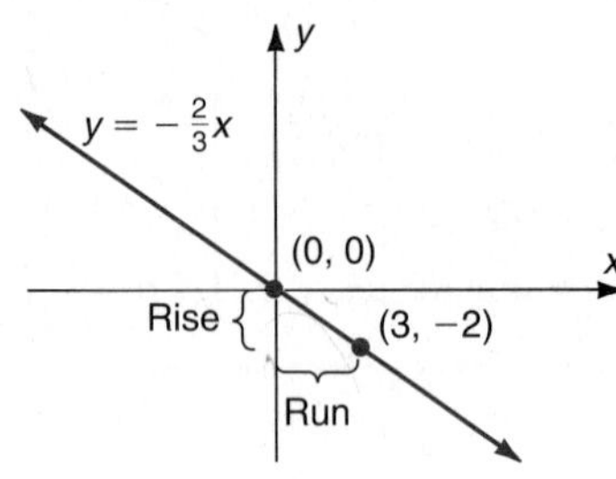

Fig. 3.27

(b) Reading from the equation $y = -\dfrac{2}{3}x$, we see that the slope is $-\dfrac{2}{3}$ and the y-intercept is 0. Think of $-\dfrac{2}{3}$ as $\dfrac{-2}{3}$. A rise of -2 means to go down 2 units. So, from the origin we rise -2 while we run 3 (Figure 3.27). □

Writing an Equation of a Line

Often in solving problems, a line is given and an equation having that line as its graph must be found. Two pieces of information are needed to find an equation of a line:

1. The slope of the line.
2. The coordinates of any point on the line.

Substituting this information into the slope formula and simplifying will produce the desired equation.

Suppose that a line has slope m and contains the point (x_1, y_1). In other words, we know a point and the slope of a line. Now if (x, y) is *any other point* on this line, then we get from the slope formula

$$m = \frac{y - y_1}{x - x_1}$$

Multiplying both sides of this equation by $x - x_1$ produces the **point-slope** form

$$m(x - x_1) = y - y_1$$

which is usually written

$$y - y_1 = m(x - x_1)$$

The Point-Slope Form

An equation of the line with slope m and containing the point (x_1, y_1) is given by

$$y - y_1 = m(x - x_1)$$

EXAMPLE 7 Draw the graph of and find an equation of a line with slope 2 that contains the point $(1, -2)$.

Solution We plot the point $(1, -2)$ and from this point rise 2 while we run 1 (since $2 = \dfrac{2}{1}$) as shown in Figure 3.28.

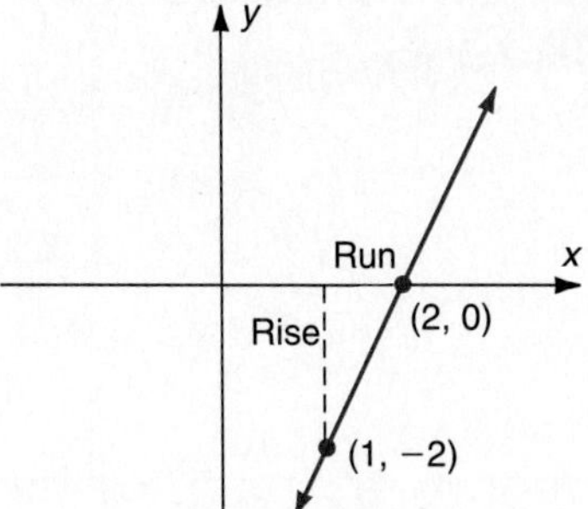

Fig. 3.28

To find an equation, we use the point-slope form.

$$y - y_1 = m(x - x_1)$$

$$y - (-2) = 2(x - 1) \qquad (x_1, y_1) = (1, -2)$$

$$y + 2 = 2x - 2$$

$$4 = 2x - y$$

$$2x - y = 4 \qquad \text{Standard form}$$

□

To Write an Equation of a Line That Has Slope

1. Find m, the slope of the line.
2. Find (x_1, y_1), the coordinates of any point on the line.
3. Substitute into the point-slope formula $y - y_1 = m(x - x_1)$.
4. Write the equation in standard or slope-intercept form, if desired.

EXAMPLE 8 Find an equation for the line containing the points $(-1, 3)$ and $(2, -1)$.

Solution First find the slope of the line containing the two points.

$$m = \frac{3 - (-1)}{-1 - 2} = \frac{4}{-3} = -\frac{4}{3}$$

Next, pick either of the two given points to be (x_1, y_1) and substitute into the point-slope form of the linear equation.

$$y - y_1 = m(x - x_1)$$

$$y - (-1) = -\frac{4}{3}(x - 2) \qquad \text{Using } (2, -1) \text{ as } (x_1, y_1)$$

$$3(y + 1) = -4(x - 2) \qquad \text{Multiply both sides by 3.}$$

$$3y + 3 = -4x + 8$$

$$4x + 3y = 5 \qquad \text{Standard form}$$

The graph of $4x + 3y = 5$ contains the points $(-1, 3)$ and $(2, -1)$. □

A special case occurs if the slope and y-intercept are known. It is convenient to substitute directly into the slope-intercept form.

EXAMPLE 9 Find an equation of the line with slope -1 and y-intercept of 5.

Solution Using the slope intercept form $y = mx + b$ gives

$$y = -x + 5. \qquad \square$$

Parallel and Perpendicular Lines

> **Theorem: Parallel and Perpendicular Lines**
>
> Suppose L_1 is a line with slope m_1, L_2 is a line with slope m_2 and neither line is vertical.
>
> **(a)** L_1 and L_2 are parallel lines if and only if their slopes are the same; that is, $m_1 = m_2$.
>
> **(b)** L_1 and L_2 are perpendicular to each other if and only if their slopes are negative reciprocals of each other; that is, $m_1 m_2 = -1$ or $m_1 = -1/m_2$.

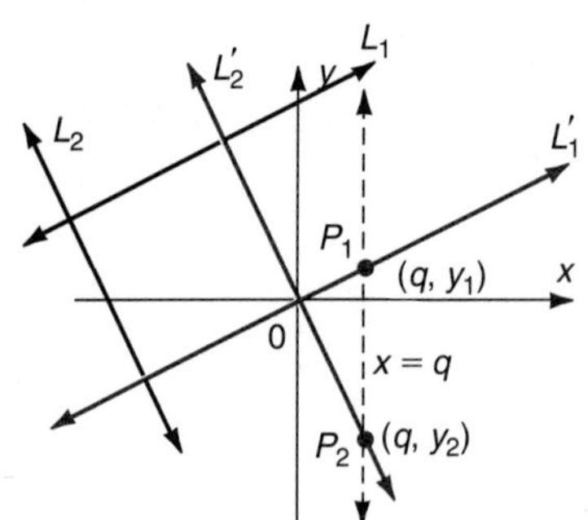

Fig. 3.29

We will prove the second part of this theorem and leave part (a) for the exercises.

Proof: Let L_1' and L_2' be lines containing the origin parallel to L_1 and L_2, respectively. By part (a), the slope of L_1' is m_1 and the slope of L_2' is m_2. L_1 is perpendicular to L_2 if and only if L_1' is perpendicular to L_2'. Let q be any real number different from 0. Since L_1' and L_2' both have defined slope, neither is vertical. Thus they both intersect the line $x = q$ as shown in Figure 3.29.

Consider the point P_1 with coordinates (q, y_1). Since L_1' has slope m_1,

$$m_1 = \frac{y_1 - 0}{q - 0}$$

$$y_1 = qm_1$$

In the same manner,

$$y_2 = qm_2$$

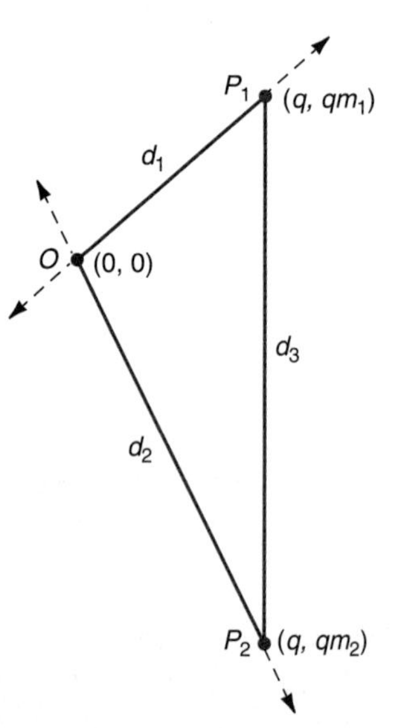

Fig. 3.30

We examine the triangle OP_1P_2 in Figure 3.30. L_1' is perpendicular to L_2' if and only if $\angle P_1OP_2$ is a right angle. By the Pythagorean Theorem, $\angle P_1OP_2$ is a right angle if and only if $d_1^2 + d_2^2 = d_3^2$.

By the distance formula,

$$d_1^2 = q^2 + q^2m_1^2$$

$$d_2^2 = q^2 + q^2m_2^2$$

$$d_3^2 = (q - q)^2 + (qm_1 - qm_2)^2$$

$$= 0 + q^2m_1^2 - 2q^2m_1m_2 + q^2m_2^2$$

Therefore

$$d_1^2 + d_2^2 = q^2m_1^2 + q^2m_2^2 + 2q^2$$

and

$$d_3^2 = q^2m_1^2 + q^2m_2^2 - 2q^2m_1m_2$$

So,

$$d_1^2 + d_2^2 = d_3^2 \quad \text{if and only} \quad \text{if } 2q^2 = -2q^2m_1m_2$$

$$2 = -2m_1m_2 \qquad \text{Since } q \neq 0$$

$$m_1m_2 = -1$$

□

Therefore L_1 is perpendicular to L_2 if and only if $m_1m_2 = -1$.

EXAMPLE 10 Find an equation for the line through the point $(4, -7)$ that is parallel to the graph of $x + 2y = 6$.

Solution First, find the slope of the graph of the given equation by writing it in slope-intercept form.

$$x + 2y = 6$$

$$2y = -x + 6$$

$$y = -\frac{1}{2}x + 3.$$

Note that the given line has slope $-\frac{1}{2}$. The line we are looking for is parallel to the given line, so its slope is the same. Since $(4, -7)$ is on our line, we have enough information to find an equation.

$$y - y_1 = m(x - x_1)$$

$$y - (-7) = -\frac{1}{2}(x - 4)$$

$$-2(y + 7) = x - 4 \qquad \text{Multiply both sides by } -2 \text{ to clear fractions,}$$

$$-2y - 14 = x - 4$$

$$-10 = x + 2y \quad \text{or} \quad x + 2y = -10 \qquad \text{Standard form}$$

□

EXAMPLE 11 Find an equation of the line through the origin perpendicular to the graph of $2x + 3y = 7$.

Solution First, find the slope. Rewrite the given equation in slope-intercept form.

$$2x + 3y = 7$$

$$3y = -2x + 7$$

$$y = -\frac{2}{3}x + \frac{7}{3}$$

The slope of the graph of the given equation is $-\frac{2}{3}$. Therefore, the slope of a line perpendicular to this line is $\frac{3}{2}$. Thus we are looking for a line through the origin with slope $\frac{3}{2}$. The coordinates of the origin are (0, 0). So, we have

$$y - y_1 = m(x - x_1)$$

$$y - 0 = \frac{3}{2}(x - 0)$$

$$y = \frac{3}{2}x$$

$$3x - 2y = 0 \qquad \text{Standard form}$$ □

Since a vertical line has no defined slope, it must be treated in another way. It is usually best to treat horizontal and vertical lines as exceptions when writing their equations. They are easy to identify (slope of zero or slope undefined), and it is simple to write their equations.

EXAMPLE 12 Find an equation of the line satisfying the stated conditions.

(a) Parallel to the x-axis and containing $(-3, 4)$
(b) Containing $(3, 5)$ and $(3, 8)$

Solutions (a) Since the line is parallel to the x-axis, it is a horizontal line. The equation of a horizontal line is of the form $y = q$. Thus, an equation of the line is $y = 4$.

(b) The two points determine a vertical line. The equation of a vertical line has the form $x = p$, where p is the x-coordinate of any point on the line. Thus, the equation is $x = 3$. □

Forms of the Linear Equation in Two Variables

Standard Form

$$Ax + By = C \quad (A \text{ and } B \text{ not both zero})$$

Slope-Intercept Form

$$y = mx + b$$

Point-Slope Form

$$y - y_1 = m(x - x_1)$$

GRAPHING CALCULATOR BOX

Graphing Lines

Let's graph $y = -2x + 1$ on the default window. First, set the range for the default window.

TI-81

Use [MODE] to set the calculator in Norm, Function, Connected, Rect.

Press [Y =] [(−)] [2] [X | T] [+] [1] [GRAPH]

CASIO fx-7700G

Before graphing with the Casio, make sure that the graphing type (G-type) is set as REC/CON as shown on the screen when the calculator is first turned on. [MODE] [+] will do this. Also, it is a good idea to clear the graphing screen ([F5] [EXE]). To enter the equation, press [GRAPH] [−] [2] [X,θ,T] [+] [1] [EXE]

The editing arrows [←], [→], [↑], and [↓] will allow us to scroll across the graph. Press [F2] [F5] to return to original graph.

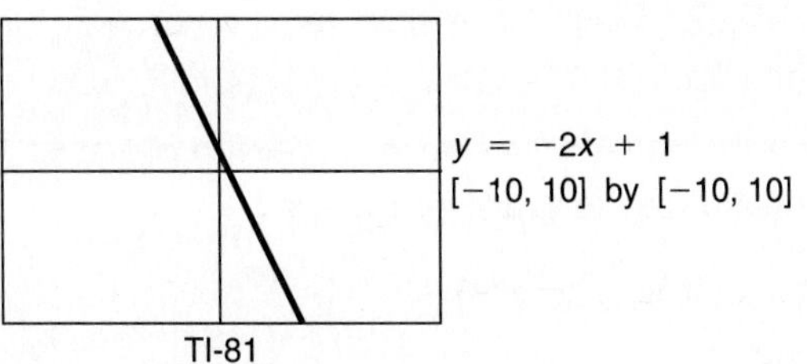

TI-81

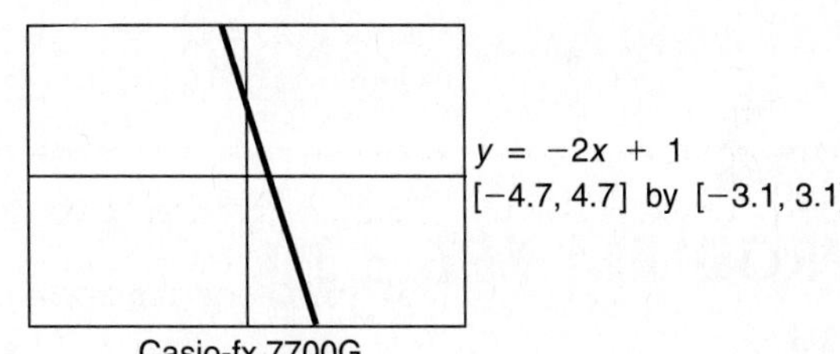

Casio-fx 7700G

Graphing Another Line

Let's graph $x - 3y = -15$ and show both intercepts. With the Casio it will be necessary to clear the screen of previous graphs. [F5] [EXE] will clear the screen when the graph is displayed. We must write the equation in slope-intercept form before entering it.

$y = \frac{1}{3}x + 5$. (A convenient way to enter $\frac{1}{3}x + 5$ is [X | T] [÷] [3] [+] [5].)

We graph and see that the default window does not show what we want. The y-intercept is 5, and the x-intercept is -15. So, let's set the window at $[-20, 5]$ for x and $[-1, 10]$ for y. Press [RANGE] and enter the settings. Regraph and note the difference.

Two Graphs on the Same Screen

Let's display two graphs on the same window. Let $y_1 = -\frac{1}{2}x + 3$ and $y_2 = \frac{1}{4}x + 1$.

TI-81

Press [Y =] and enter the equation for Y_1. Notice that the equal sign on the display for Y_1 is dark. The dark equal sign indicates that the equation *will be graphed* when GRAPH is pressed. Arrow down one space and enter Y_2. Again the equals sign is darkened. Place the cursor on the equal sign in Y_1 and press ENTER. Notice that the equal sign is no longer darkened. Pressing ENTER again toggles between off and on. With *both* equals signs dark press GRAPH, and the graph of both equations should appear.

CASIO fx-7700G

Clear the screen of previous graphs. Using the F MEM menu, enter the first equation as f_1 and the second as f_2. To graph f_1, press GRAPH F2 1 EXE, and to graph f_2, GRAPH F2 2 EXE. (To remove the menu from the screen, press PRE.)

We now have a screen showing two graphs. However, the window needs adjusting. Experiment with various settings. Notice that the settings of $[-8, 8]$ for x and $[-6, 6]$ for y give a viewing rectangle that contains the interesting features of the graphs.

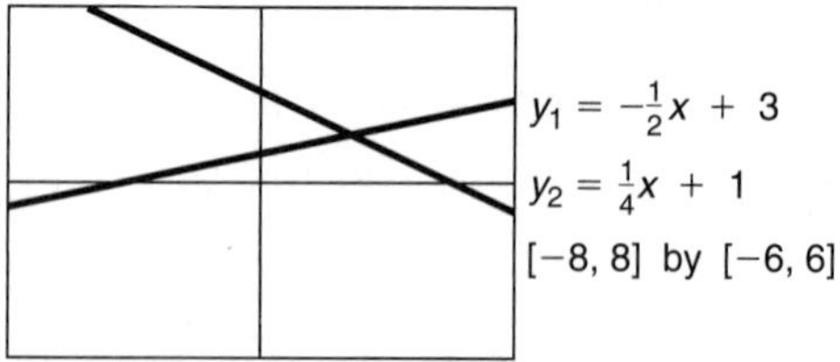

$y_1 = -\frac{1}{2}x + 3$

$y_2 = \frac{1}{4}x + 1$

$[-8, 8]$ by $[-6, 6]$

PROBLEM SET 3.2

Warm-ups

In Problems 1 through 4, write each equation in standard form and graph each equation. See Example 1.

1. $2x = 4 - y$

2. $y = 2x$

3. $x + 3 = 0$

4. $y - 2 = 0$

For Problems 5 and 6, see Example 2.

5. Mary fires a rocket vertically upward with an initial velocity of 49 meters per second. The relationship between the velocity v in meters per second after t seconds of flight is given by $v + 9.8t = 49$.
(a) Graph this equation using the horizontal axis for time.
(b) What is the velocity after 2 seconds?
(c) When will the velocity be 0?
(d) Approximate to the nearest second when the velocity will be 10 meters per second.
(e) As time increases, what happens to the velocity?

6. Manuel jogs at an average rate of 6 kilometers per hour. The relationship of his distance d in kilometers to his running time t in hours is given by $d - 6t = 0$.
(a) Graph this equation using the horizontal axis as time.
(b) How far can he run in 2 hours?
(c) How long will it take him to run 24 kilometers?
(d) As time increases, what happens to the distance?

In Problems 7 through 10, find the slope of the line containing the two given points. See Example 3.

7. $(2, 0), (5, -1)$

8. $(-2, -1), (-1, 2)$

9. $(5, 0), (0, 5)$

10. $\left(\frac{1}{2}, \frac{1}{3}\right), \left(1, -\frac{1}{3}\right)$

In Problems 11 through 14, find the slope of the graph of each equation. Use the slope formula. See Example 4.

11. $2x - y = 2$

12. $3x + y = 0$

13. $x + 4 = 0$

14. $y = 5$

In Problems 15 through 18, write each equation in slope-intercept form and find the slope and y-intercept of the graph of each equation. Sketch the graph of each using the slope and y-intercept. See Examples 5 and 6.

15. $3x + y = 1$

16. $2x - 3y = 6$

17. $2x - y = 0$

18. $3x - 2y = 0$

In Problems 19 through 30, find an equation for a line satisfying the given conditions. Write each equation in standard form.

For Problems 19 through 22, see Example 7.

19. The line contains $(1, 2)$ and has slope 3.

20. The line contains $(-1, 1)$ and has slope -2.

21. The line with slope -1 and x-intercept -3.

22. The line with x-intercept -2 and y-intercept 3.

For Problems 23 and 24, see Example 8.

23. The line contains $(2, 3)$ and $(4, 5)$.

24. The line contains $(-1, 4)$ and $(2, 3)$.

For Problems 25 and 26, see Example 9.

25. The line with y-intercept 2 and slope $\frac{1}{2}$.

26. The line passing through the origin with slope $\frac{1}{2}$.

For Problems 27 and 28, see Examples 10 and 11.

27. The line through $(1, -5)$ parallel to the graph of $y = 6x - 5$.

28. The line perpendicular to the graph of $3x + y = -17$ containing the point $(4, -11)$.

For Problems 29 and 30, see Example 12.

29. The line with undefined slope through the point $(2, 8)$.

30. The line perpendicular to the y-axis with y-intercept of 0.

Practice Exercises

In Problems 31 through 44, find the slope of each line defined by the given conditions.

31. Contains $(-5, -3)$ and $(1, -7)$

32. Contains $(3, -6)$ and $(0, 9)$

33. With equation $3x - 2y = 8$

34. With equation $x - 2y = 6$

35. With equation $y = \frac{1}{4}x$

36. With equation $y = \frac{3}{7}x$

37. With equation $x + 2y = 0$

38. With equation $3x - y = 0$

39. Parallel to the x-axis

40. Parallel to the y-axis

41. With x-intercept 2 and y-intercept -1

42. With x-intercept 0 and y-intercept 5

43. Perpendicular to the graph of $y = \frac{5}{7}x - 3$

44. Perpendicular to the graph of $y = -\frac{3}{2}x + 7$

In Problems 45 through 56, graph each equation.

45. $x + 2y = 6$

46. $4x + 2y = 2$

47. $3x + 5y = 0$

48. $6x - 2y = 0$

49. $y = -\frac{2}{3}x$

50. $y = -\frac{4}{3}x$

51. $y = \frac{3}{2}x + 1$

52. $y = \frac{1}{2}x - 3$

53. $x = \frac{3}{2}$

54. $x = -\frac{1}{2}$

55. $3y + 3 = 0$

56. $5y = 10$

In Problems 57 through 72, write an equation of the line satisfying the given conditions. Write the equation in standard form.

57. Slope $-\frac{1}{2}$ passing through the origin

58. Slope $\frac{2}{3}$ and passing through $(3, -1)$

59. Contains $(2, 7)$ and $(-1, 7)$

60. Contains $(5, 3)$ and $(5, -4)$

61. No slope and contains $(2, 4)$

62. Slope 0 and contains $(-9, 11)$

63. Parallel to the graph of $3x - y = 4$ through $(1, 1)$

64. Perpendicular to the graph of $x + 5y = 0$ through $(1, 1)$

65. Slope 4 and y-intercept -3

66. Slope -5 and y-intercept 7

67. Horizontal through $\left(\frac{1}{7}, -\frac{1}{7}\right)$

68. Vertical through $\left(-\frac{2}{3}, \frac{2}{3}\right)$

69. Perpendicular to x-axis and contains $(7, 3)$

70. Parallel to x-axis and contains $(-4, 8)$

71. x-intercept 3 and y-intercept 3

72. x-intercept 7 and y-intercept 7

73. The velocity v in feet per second of a stone thrown vertically upward from the ground with an initial velocity of 128 feet/second after t seconds is given by $v + 32t = 128$.
(a) Graph this equation and explain the meaning of the intercepts. Use the horizontal axis for time.
(b) What is the slope of the graph?
(c) What does this mean?

74. The relationship between temperature in Celsius and Fahrenheit degrees is given by $F = \frac{9}{5}C + 32$.
(a) Graph this equation. Use the horizontal axis for Celsius.
(b) Change 32°F to degrees Celsius.
(c) Change 0°F to degrees Celsius.

75. The relationship between simple interest earned at 10% for 1 year on an investment of P dollars is given by $I = 0.10P$.
(a) Graph this equation using the horizontal axis as the P-axis.
(b) What does the 10% rate mean in terms of the graph of the equation?
(c) Interpret the meaning of the part of the graph that lies to the left of the I-axis.

76. A company makes portable computers. The company has fixed costs of \$45,000 per month and variable costs of \$1200 per computer produced.
(a) If q represents the number of computers produced in 1 month and C represents the total costs for the company per month, write an equation that shows the relationship between C and q. Assume this relationship is linear.
(b) Graph this equation using the horizontal axis as the q-axis.
(c) What is the meaning of the C-intercept?

77. A company makes copper bracelets. Data show that 10 bracelets are sold when the price is \$75 and 30 bracelets are sold when the price is \$50. The relationship between price and quantity demanded is called the *demand curve*. The demand curve for the copper bracelets is the line shown in the figure.

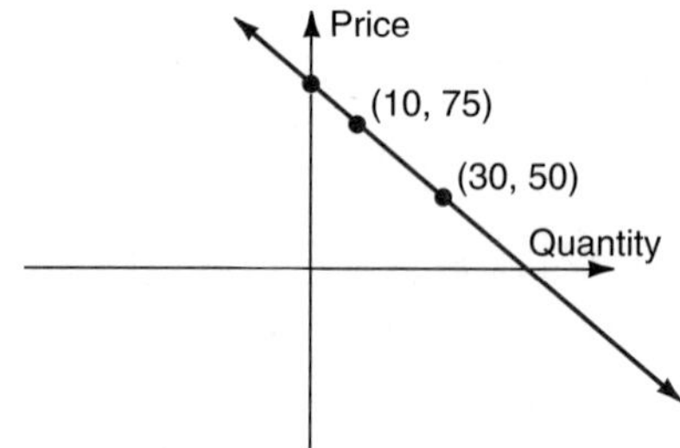

(a) Do the portions of the line in quadrants II and IV have any meaning for the company?
(b) Write an equation for the demand curve.
(c) Explain what the intercepts mean.
(d) If the price of a bracelet is \$60, predict how many bracelets will be sold.

78. When the price is \$40, a company can supply 30 radios. When the price is \$65, the company can supply 75 radios. The relationship between price and number of items supplied to the market is called the *supply curve*. The graph of this supply curve is the line shown in the figure.

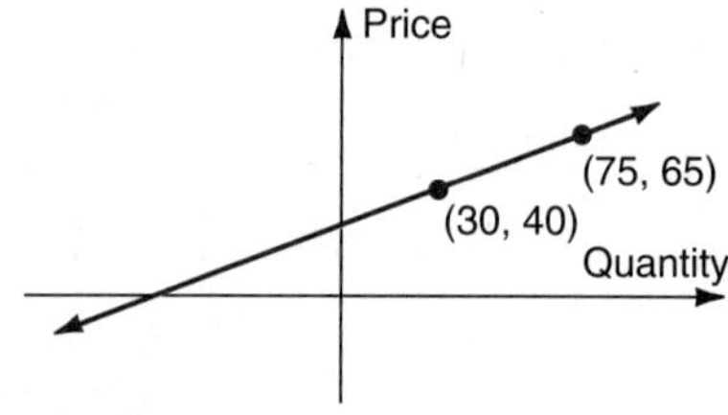

(a) Do the portions of the line in quadrants II and III have any meaning for the company?

(b) Write an equation for the supply curve.
(c) Explain the meaning of the intercepts.
(d) Predict how many radios can be supplied when the price is \$100.

79. How is the graph of each equation listed below related to the graph $2x - 3y = 7$?
(a) $2x - 3y = 15$
(b) $2x - 3y = -7$
(c) $2x - 3y = 20$
(d) $2x - 3y = 0$

80. Draw a line with slope 2. Write an equation for a line with slope 2.

81. Consider the triangle whose vertices are $(0, 2)$, $(6, 0)$, and $(8, 4)$. Find the equations of the three lines connecting the midpoints of the sides.

82. Prove: Two distinct lines are parallel if and only if both have the same slope or neither has a defined slope.

83. Write the equation $Ax + By = C$ in slope-intercept form. What assumption must be made before we can solve for y?

Challenge Problems

84. Prove: If L is a line with defined slope and x-intercept p, and q is any real number, $q \neq p$, then the point $[q, M(q - p)]$ is on L if and only if M is the slope of L.

85. If a line has x-intercept a and y-intercept b, show that an equation of the line is $\frac{x}{a} + \frac{y}{b} = 1$. This is called the **intercept form** of the equation of a line. Write $4x + 2y = 8$ in intercept form.

86. Prove that the diagonals of a rectangle bisect each other. (*Hint:* Draw a rectangle with one vertex at the origin of a coordinate system and label the other vertices.)

For Graphing Calculators

In Problems 87 through 90, find an appropriate window that includes both intercepts of each line and their point of intersection.

87. $x + 2y = 3$; $x - 2y = 1$

88. $y = x$; $2x - y = 5$

89. $25.2x + 15.8y = -22.9$; $y = 0.33x + 1.5$

90. $60x - 25y = 100$; $240x - 100y = 1000$

91. Find an appropriate window that shows both intercepts of the line through the point $(-2, 4)$ and perpendicular to the graph of $y = -\frac{1}{3}x + 5$.

In Problems 92 and 93, discuss the effect of each window setting on the graph of $y = 2x + 5$.

92. $[-10, 10]$ by $[-10, 10]$

93. $[-5, 0]$ by $[-10, 10]$

IN YOUR OWN WORDS . . .

94. Is the graph of $2x + y = 5$ increasing or decreasing? Explain the relationship between the slope of a line and increasing or decreasing lines.

95. Explain the relationship between the graph $y = 2x + 3$ and the solution to the equation $2x + 3 = 0$.

96. Compare the mathematical concept of slope with the meaning of the road signs shown in Figure 3.31.

Diamond-shaped signs warn of upcoming long, steep grades.

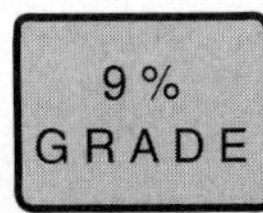

Square informational signs are posted along the grade itself.

Fig. 3.31

3.3 CIRCLES

A **circle** is defined as the set of all points that are the same distance from a given point. The distance is called the **radius,** and the given point the **center** of the circle.

Let (x, y) be any point on a circle with center (h, k) and radius r as shown in Figure 3.32.

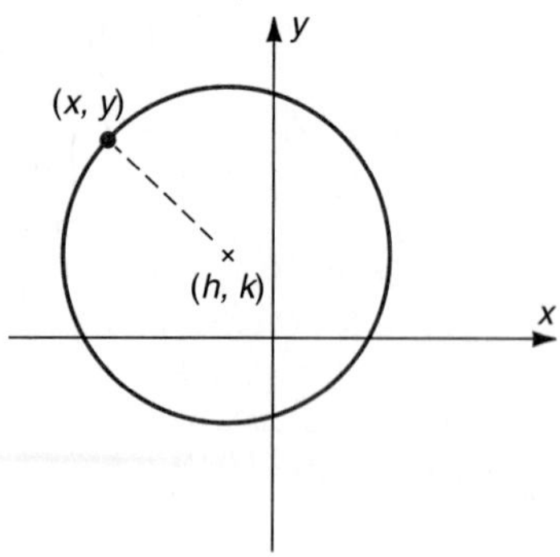

Fig. 3.32

Since (x, y) must be r units from the center of the circle, the distance formula gives

$$r = \sqrt{(x-h)^2 + (y-k)^2}$$

$$r^2 = (x-h)^2 + (y-k)^2 \qquad \text{Square both sides.}$$

This is an equation whose graph is a circle of radius r with center at the point (h, k). Since r is a distance, it must be nonnegative. If r is 0, then we have a circle of 0 radius which is called a *point circle* or a *degenerate circle.*

The Circle

The graph of

$$(x-h)^2 + (y-k)^2 = r^2$$

is a circle of radius $r (r \geq 0)$ with center at the point (h, k). If the circle has its center at the origin, the equation becomes $x^2 + y^2 = r^2$.

Graphing Circles

EXAMPLE 1 Graph $x^2 + y^2 = 1$.

Solution This is the form $x^2 + y^2 = r^2$, where r^2 is 1. Since r is a distance and must be positive, r must be 1. So this is a circle of radius 1 with center at the origin. This circle is called the *unit circle* and is very important in trigonometry (Figure 3.33).

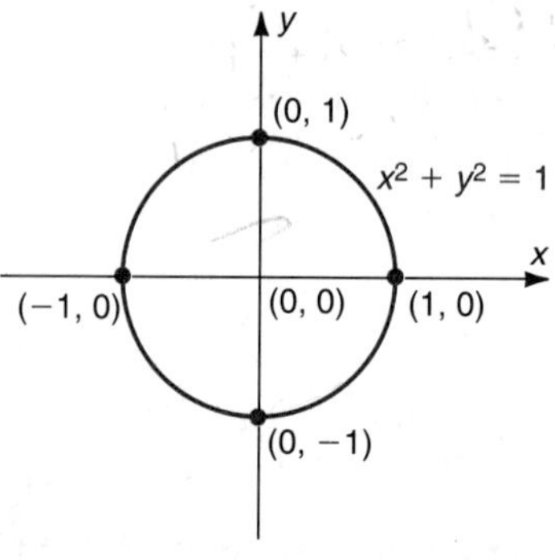

Fig. 3.33

□

EXAMPLE 2 Graph the equation $(x-2)^2 + (y-3)^2 = 1$.

Solution We compare the equation with the general equation of a circle.

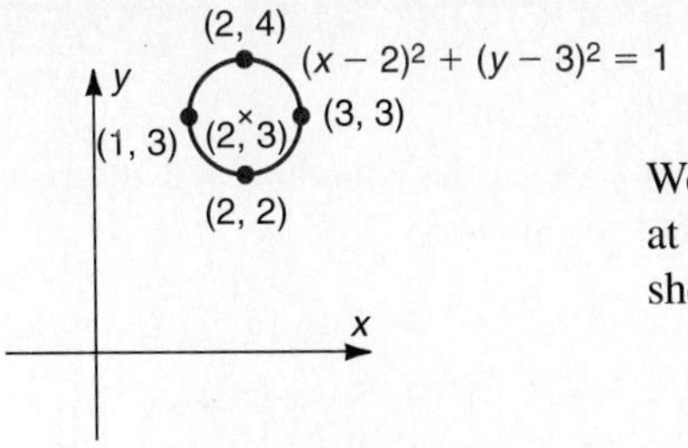

$$(x - h)^2 + (y - k)^2 = r^2$$

$$(x - 2)^2 + (y - 3)^2 = 1^2$$

We see that h is 2, k is 3, and r is 1. So, the graph in Figure 3.34 is a circle centered at (2, 3) with radius 1. It is as if a circle of radius 1 has been drawn on a transparent sheet and moved until its center coincides with the point (2, 3). □

(2, 4)
$(x - 2)^2 + (y - 3)^2 = 1$
(1, 3) (2, 3) (3, 3)
(2, 2)
y
x

Fig. 3.34

EXAMPLE 3 Graph the equation $(x + 1)^2 + y^2 = 3$.

Solution Notice that we can rewrite the given equation so that it matches the general form.

$$(x - h)^2 + (y - k)^2 = r^2$$

$$[x - (-1)]^2 + (y - 0)^2 = (\sqrt{3})^2$$

where h is -1, k is 0, and $r = \sqrt{3}$. Since $\sqrt{3} \approx 1.73$, we sketch the graph in Figure 3.35.

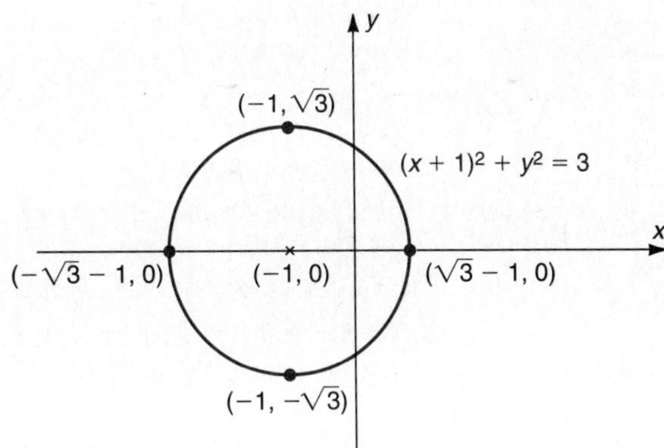

Fig. 3.35 □

Notice how we can rewrite the equation of a circle. Suppose we have the equation of the circle with center at $(-2, 1)$ and radius 2.

$$(x + 2)^2 + (y - 1)^2 = 4$$

$$x^2 + 4x + 4 + y^2 - 2y + 1 = 4 \quad \text{Squaring}$$

$$x^2 + y^2 + 4x - 2y + 1 = 0 \quad \text{Combining like terms}$$

If the equation is *given* in this form, we must be able to repeat these steps *in reverse order* in order to determine the center and radius. This process requires completing the square before graphing.

EXAMPLE 4 Sketch the graph of $x^2 + y^2 - 2x = 3$.

Solution We note that there is both an x^2-term and an x-term in the equation. We combine them to form the $(x - h)^2$ term by completing the square.

$$x^2 + y^2 - 2x = 3$$

First we group the x-terms together.

$$(x^2 - 2x) + y^2 = 3$$

To complete the square, we need to add 1 to $x^2 - 2x$. We simply cannot add it on because that would change the equation. However, we can add 0 without harming the equation. So, we *add and subtract 1*.

$$(x^2 - 2x + 1 - 1) + y^2 = 3$$

Now, group the terms in the parentheses to form a perfect square.

$$(x^2 - 2x + 1) - 1 + y^2 = 3$$

Add 1 to the both sides of the equation.

$$(x^2 - 2x + 1) + y^2 = 3 + 1$$

Now, factor $(x^2 - 2x + 1)$,

$$(x - 1)^2 + y^2 = 4$$

or to put it in the form of a circle,

$$(x - h)^2 + (y - k)^2 = r^2$$

$$(x - 1)^2 + (y - 0)^2 = 2^2$$

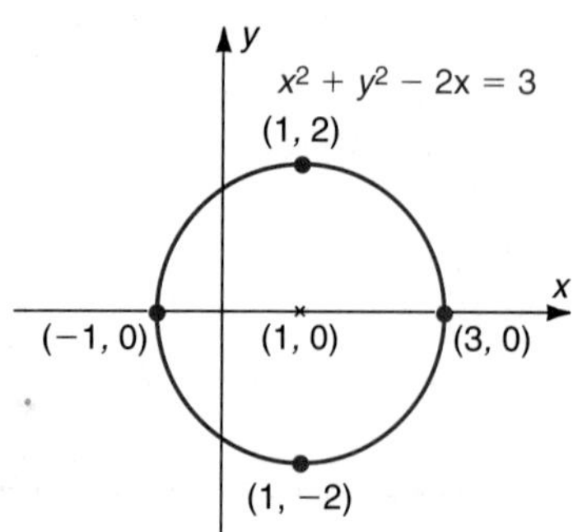

Fig. 3.36

We see that h is 1 and k is 0. So, the center of the circle is (1, 0) and the radius is 2 as shown in Figure 3.36. □

Often we must complete the square of both x and y.

EXAMPLE 5 Graph the equation $2x^2 + 2y^2 + 8x - 12y + 24 = 0$.

Solution $2x^2 + 2y^2 + 8x - 12y + 24 = 0$

Divide both sides of the equation by 2.

$x^2 + y^2 + 4x - 6y + 12 = 0$

Subtract 12 from both sides of the equation and group the x-terms together and the y-terms together.

$(x^2 + 4x) + (y^2 - 6y) = -12$

Now we complete the square twice. We must add and subtract 4 to complete the square in x, and add and subtract 9 to complete the square in y.

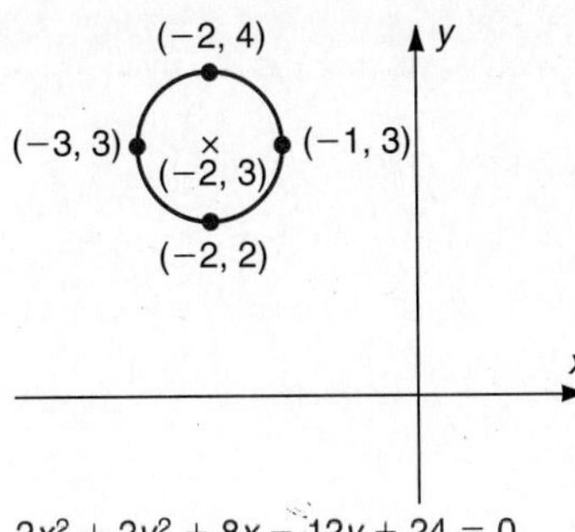

Fig. 3.37

$$(x^2 + 4x + 4 - 4) + (y^2 - 6y + 9 - 9) = -12$$
$$(x^2 + 4x + 4) - 4 + (y^2 - 6y + 9) - 9 = -12$$
$$(x^2 + 4x + 4) + (y^2 - 6y + 9) = -12 + 4 + 9$$
$$(x + 2)^2 + (y - 3)^2 = 1$$

To determine the numbers h and k, we must be careful to write the equation in the form

$$(x - h)^2 + (y - k)^2 = r^2$$
$$[x - (-2)]^2 + (y - 3)^2 = 1^2$$

So we see that the center is $(-2, 3)$ and the radius is 1 as shown in Figure 3.37. □

EXAMPLE 6 Find the center and radius of the graph of each equation.

(a) $(x + 3)^2 + (y - 4)^2 = 0$ (b) $x^2 + (y + 3)^2 = -7$

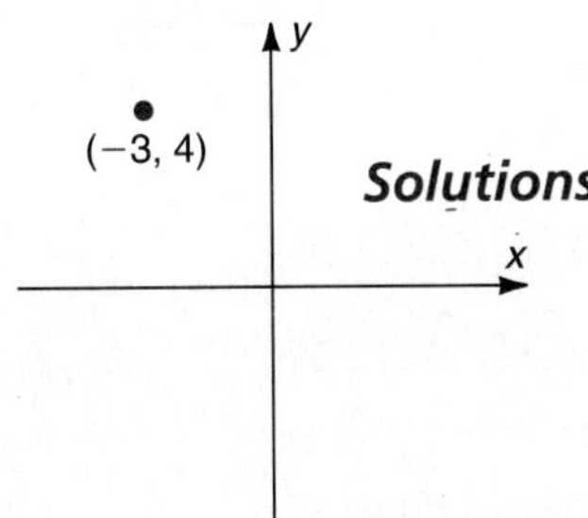

Fig. 3.38

Solutions (a) This is a point circle. The graph is the point $(-3, 4)$ as shown in Figure 3.38.

(b) Since the left side of this equation is never negative, there are are no points in the plane that satisfy the equation. The graph is the empty set. □

The graphs in Example 6 are called **degenerate** circles.

EXAMPLE 7 Find an equation of a circle with center $(-3, 6)$ and radius 4.

Solution Knowing the center and radius gives an equation for the circle. We simply substitute into the equation $(x - h)^2 + (y - k)^2 = r^2$.

$$[x - (-3)]^2 + (y - 6)^2 = 4^2$$

An equation of the desired circle is $(x + 3)^2 + (y - 6)^2 = 16$. □

EXAMPLE 8 Find the intercepts of the graph of $(x - 1)^2 + (y + 2)^2 = 5$.

Solution Substitute 0 for x to find the y-intercepts.

$$(0 - 1)^2 + (y + 2)^2 = 5$$
$$1 + (y + 2)^2 = 5$$
$$(y + 2)^2 = 4$$
$$y + 2 = \pm 2 \qquad \text{Square Root Property}$$

$$y = -2 \pm 2$$

$$y = 0 \quad \text{or} \quad y = -4$$

The y-intercepts are 0 and -4.
Substitute 0 for y to find the x-intercepts.

$$(x - 1)^2 + 4 = 5$$

$$(x - 1)^2 = 1$$

$$x - 1 = \pm 1 \qquad \text{Square Root Property}$$

$$x = 0 \quad \text{or} \quad x = 2$$

The x-intercepts are 0 and 2. □

Semicircles

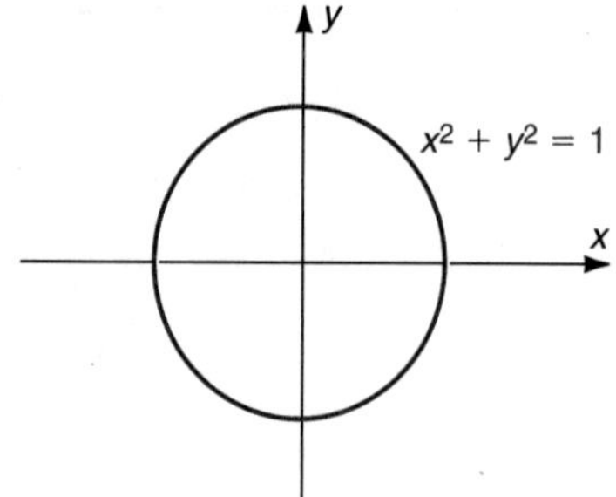

Fig. 3.39

What is the equation of a semicircle? Consider the semicircle that is the top half of the unit circle $x^2 + y^2 = 1$ shown in Figure 3.39.

Let's solve the equation $x^2 + y^2 = 1$ for y.

$$x^2 + y^2 = 1$$

$$y^2 = 1 - x^2$$

$$y = \pm\sqrt{1 - x^2} \qquad \text{Square Root Property}$$

$$y = -\sqrt{1 - x^2} \quad \text{or} \quad y = \sqrt{1 - x^2}$$

Since the top half of the circle has positive y-coordinates and the bottom half has negative y-coordinates, the equation of the top semicircle is $y = \sqrt{1 - x^2}$ and the equation of the bottom semicircle is $y = -\sqrt{1 - x^2}$ as shown in Figure 3.40.

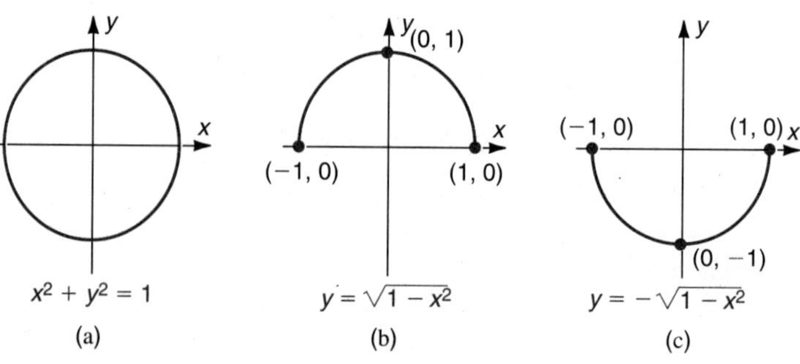

Fig. 3.40

Likewise if we solve the equation $x^2 + y^2 = 1$ for x, we can find equations for the left and right halves of the unit circle (Figure 3.41).

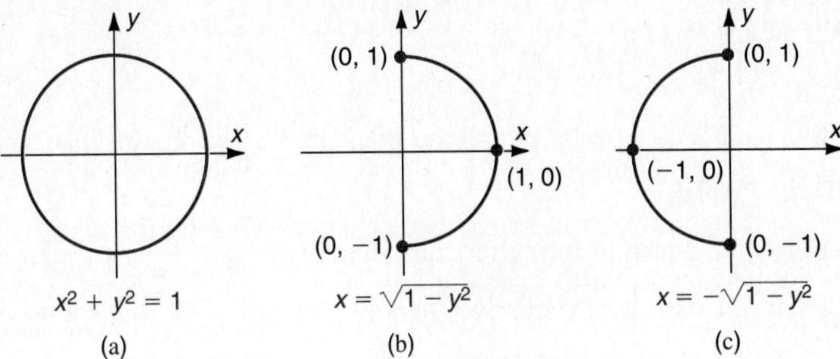

Fig. 3.41

$$x^2 + y^2 = 1$$

$$x^2 = 1 - y^2$$

$$x = \pm\sqrt{1 - y^2} \quad \text{Square Root Property}$$

EXAMPLE 9 Graph the equation $y = \sqrt{4 - x^2}$.

Solution This is the top half of the circle $x^2 + y^2 = 4$ as shown in Figure 3.42.

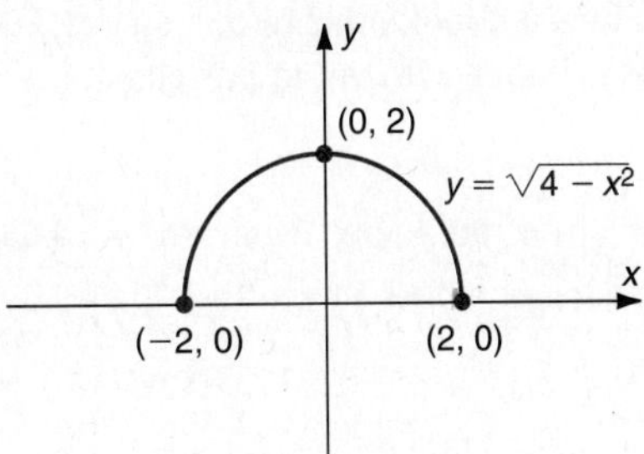

Fig. 3.42

□

GRAPHING CALCULATOR BOX

Graphing Circles

Let's graph the circle $x^2 + y^2 = 4$. To enter the equation, we must first solve the equation for y.

$$x^2 + y^2 = 4$$

$$y^2 = 4 - x^2$$

$$y = \pm\sqrt{4 - x^2} \quad \text{Square Root Property}$$

$$y = \sqrt{4 - x^2} \quad \text{or} \quad y = -\sqrt{4 - x^2}$$

Our strategy is to graph both semicircles at the same time. Let's set the range at [−4, 4] by [−4, 4].

TI-81

Enter one equation in Y_1 and the other in Y_2. Make sure that both equal signs are dark. Press [GRAPH].

CASIO fx-7700G

Enter one equation in the function memory [F MEM] as f_1 and the other as f_2.

Press [GRAPH] [F2] [1] [EXE] to draw one half of the circle and [GRAPH] [F2] [2] [EXE] for the other half.

CONNECTING TWO EQUATIONS AS A MULTISTATEMENT

Clear the screen ([F5] [EXE]) and repeat the process, replacing the first [EXE] with [↵] (found over the [EXE] key).

[GRAPH] [F MEM] [F2] [1] [↵] [GRAPH] [F MEM] [F2] [2] [EXE]

Notice that [↵] moves the cursor to the next line on the display. This connects the two graph commands and is called a *multistatement.*

It is a good idea to clear the graph screen after drawing a graph. Use [CLEAR] on the TI-81 and [F5] [EXE] on the Casio-fx 7700G.

The displayed graph does not look much like a circle! The stretching in the x-direction is caused by the nonsquare shape of the viewing rectangle. It can be corrected.

TI-81

Select "square" from the zoom menu. Press [ZOOM] [5]. Notice how the shape has been corrected.

Look at the range values to see how they have been changed by "square."

CASIO fx-7700G

Change the viewing window to [−6, 6] by [−4, 4] and regraph. A fairly regular graph can be made by setting the window so that its length is $\frac{3}{2}$ its width.

PROBLEM SET 3.3

Warm-ups

In Problems 1 through 18, sketch the graph of each equation.
For Problems 1 and 2, see Example 1.

1. $x^2 + y^2 = 4$

2. $x^2 + y^2 = 7$

For Problems 3 and 4, see Example 2.

3. $(x + 1)^2 + (y - 2)^2 = 4$

4. $(x - 1)^2 + (y + 1)^2 = 9$

For Problems 5 and 6, see Example 3.

5. $x^2 + (y - 1)^2 = 5$

6. $(x - 2)^2 + y^2 = 8$

For Problems 7 through 12, see Example 4.

7. $x^2 + y^2 - 2x = 8$

8. $x^2 + y^2 - 4y = 12$

9. $x^2 + y^2 + 8y = 4$

10. $x^2 + y^2 + 6x = -1$

11. $x^2 + y^2 - 10x + 16 = 0$

12. $x^2 + y^2 - 12y + 32 = 0$

For Problems 13 through 16, see Example 5.

13. $x^2 + y^2 + 2x - 4y = 4$

14. $x^2 + y^2 + 6x - 8y = 0$

15. $2x^2 + 2y^2 - 2x - 4 = 0$

16. $3x^2 + 3y^2 - 9y - 12 = 0$

For Problems 17 and 18, see Example 6.

17. $x^2 + y^2 + 4y + 5 = 0$

18. $x^2 + y^2 - 2x = -1$

In Problems 19 and 20, find an equation of the circle satisfying the given conditions. See Example 7.

19. Center $(7, 5)$; radius $\sqrt{5}$

20. Center $(-4, 0)$; radius 8

In Problems 21 and 22, find the intercepts. See Example 8.

21. $(x - 3)^2 + (y - 1)^2 = 5$

22. $(x + 1)^2 + y^2 = 10$

In Problems 23 and 24, sketch the graph of each equation. See Example 9.

23. $x = \sqrt{9 - y^2}$

24. $y = \sqrt{16 - x^2}$

Practice Exercises

In Problems 25 through 52, find the center, radius, and intercepts of each circle.

25. $x^2 + y^2 = 9$

26. $x^2 + y^2 = 6$

27. $x^2 + (y - 2)^2 = 16$

28. $(x - 1)^2 + y^2 = 12$

29. $(x + 3)^2 + (y + 2)^2 = 49$

30. $(x + 1/2)^2 + (y - 3/2)^2 = 9/4$

31. $x^2 + y^2 - 4x = -5$

32. $x^2 + y^2 - 2y = -15$

33. $x^2 + y^2 - 8y = 0$

34. $x^2 + y^2 - 6x = 16$

35. $2x^2 + 2y^2 + 16x = -14$

36. $3x^2 + 3y^2 + 12y = 0$

37. $4x^2 + 4y^2 + 8y = 28$

38. $5x^2 + 5y^2 + 30x = 15$

39. $-2x^2 - 2y^2 - 24x + 26 = 0$

40. $-3x^2 - 3y^2 + 30y + 33 = 0$

41. $x^2 + y^2 - 4x - 2y = 11$

42. $x^2 + y^2 - 8x - 6y = 0$

43. $x^2 + y^2 - 4x + 4y + 9 = 0$

44. $x^2 + y^2 + 12x - 2y + 16 = 0$

45. $x^2 + y^2 + 6x - 6y + 18 = 0$

46. $x^2 + y^2 - 10x + 2y + 26 = 0$

47. $2(x + 2)^2 + 2(y + 3)^2 = 50$

48. $4(x + 3/2)^2 + 4(y - 1/2)^2 = 9$

49. $x^2 + y^2 + 4x = 0$

50. $x^2 + y^2 + 2y = 0$

51. $x^2 + y^2 + 12x + 4y = 0$

52. $x^2 + y^2 + 8x + 10y = 0$

In Problems 53 through 60, find an equation of the circle satisfying the given conditions.

53. Center $(-5, -7)$; radius 0

54. Center $(-3, 0)$; radius 2

55. Center $(4, -7)$; tangent to y-axis

56. Center $(-8, -3)$; tangent to x-axis

57. End points of a diameter are $(-4, -3)$ and $(4, 1)$.

58. End points of a diameter are $(-2, 8)$ and $(4, -2)$.

59. Center at the origin and passes through $(5, -1)$

60. Center at $(1, -2)$ and passes through $(2, -5)$

In Problems 61 through 68, sketch the graph of each semicircle.

61. $y = -\sqrt{4 - x^2}$

62. $y = -\sqrt{9 - x^2}$

63. $x = \sqrt{25 - y^2}$

64. $x = \sqrt{36 - y^2}$

65. $y = \sqrt{4 - x^2}$

66. $y = \sqrt{16 - x^2}$

67. $x = -\sqrt{4 - y^2}$

68. $x = -\sqrt{9 - y^2}$

69. Graph the equation $x^2 + y^2 = 25$ and then, on the same set of coordinate axes, graph the equation $y = \frac{4}{3}x$. Guess the coordinates of the point(s) of intersection. Check to see if this guess is in the solution set of *both* equations.

70. Graph the equation $x^2 + y^2 = 4y + 5$ and then, on the same set of coordinate axes, graph the equation $x - y = -5$. Guess the coordinates of the point(s) of intersection. Check to see if this guess is in the solution set of *both* equations.

Challenge Problems

71. A triangle is inscribed in the circle with equation $x^2 + y^2 = r^2 (r > 0)$ so that two vertices of the triangle are end points of a diameter of the circle and the third vertex of the triangle is on the circle. Prove that the triangle is a right triangle.

72. Find an equation of the line tangent to the graph of $x^2 + y^2 = 1$ at the point $\left(\frac{1}{2}, \frac{\sqrt{3}}{2}\right)$. $\left(\textit{Hint:}\text{ The radius passing through } \left(\frac{1}{2}, \frac{\sqrt{3}}{2}\right) \text{ is perpendicular to the tangent line.}\right)$

IN YOUR OWN WORDS . . .

73. Explain how to find the center and radius of a circle by completing the square.

3.4 PARABOLAS

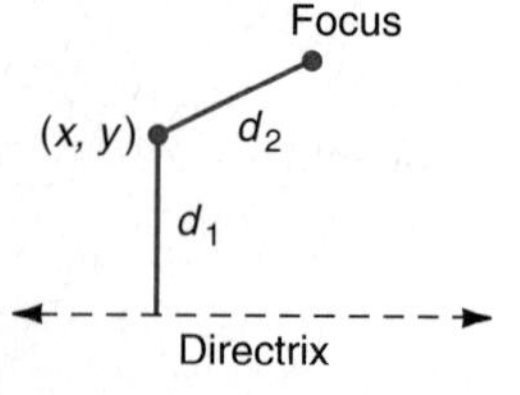

Fig. 3.43

A **parabola** is the set of all points in a plane that are equidistant from a fixed point and a fixed line (Figure 3.43). The fixed point is called the **focus,** and the fixed line is called the **directrix.** The parabola is the set of points (x, y) such that $d_1 = d_2$. Sketching the points results in a curve like the one in Figure 3.44.

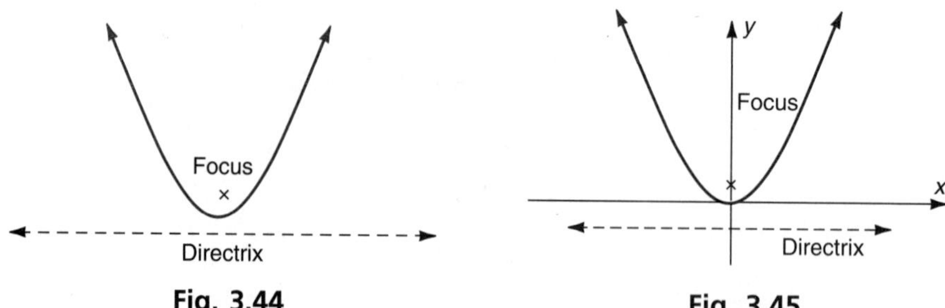

Fig. 3.44 **Fig. 3.45**

We place a set of axes so that the origin is halfway between the focus and the directrix at the **vertex** of the parabola (Figure 3.45). The equation of this graph can be written in the form $y = ax^2$.

Parabolas are common in everyday life. The cross section of the reflector in a flashlight is in the shape of a parabola. The cables of the Verrazano Narrows Bridge hang in parabolic form, and a parabola was utilized in the design of the 200-inch telescope on Mount Palomar.

Graphing Parabolas

We begin by graphing the equation $y = x^2$ (Figure 3.46). The parabola is drawn as a smooth curve. There are no sharp places. We make an extra effort to draw parabolas as smooth curves. The lowest point on the parabola in Figure 3.46, (0, 0), is called the **vertex.** A line through the vertex about which the parabola is symmetric is called the **axis of symmetry** (Figure 3.47). Notice that this parabola is symmetric about the y-axis.

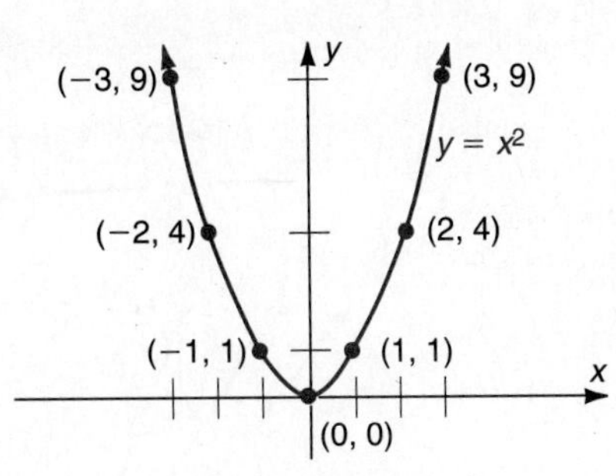

Fig. 3.46

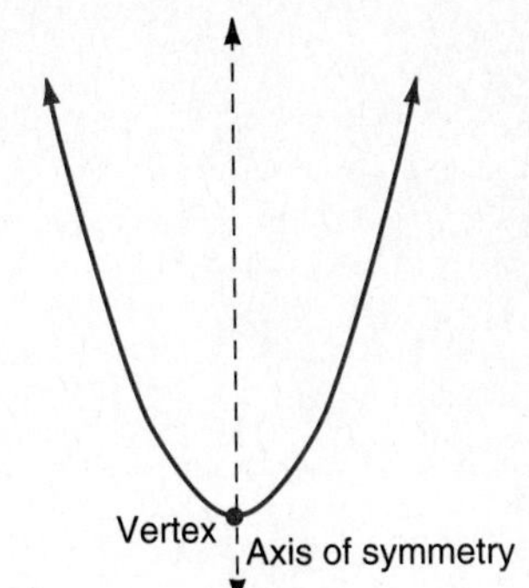

Fig. 3.47

EXAMPLE 1 Graph the equation $y = x^2 + 1$ and identify the vertex and axis of symmetry.

Solution Adding 1 to the right side of the equation $y = x^2$ will simply add 1 to each y-value. The shape should still be a parabola. Again we make a table of selected solutions and sketch the graph (Figure 3.48).

x	y
−2	5
−1	2
0	1
1	2
2	5

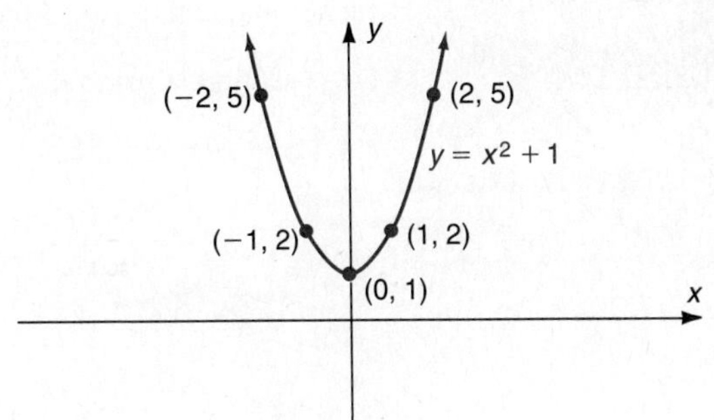

Fig. 3.48

By inspection, we see that the vertex is the point (0, 1) and the axis of symmetry is the y-axis. Compare this graph with the graph of $y = x^2$ that we drew earlier in the section. Notice that the size and shape of the two curves are just the same. The graph of $y = x^2 + 1$ is exactly the graph of $y = x^2$ *shifted up* 1 unit. Therefore it is also a parabola. □

Thus, the graph of the equation $y = x^2 + k$ is a parabola with exactly the same shape as the graph of $y = x^2$ with its vertex at $(0, k)$.

EXAMPLE 2 Graph $y = 2x^2$, $y = x^2$, and $y = \frac{1}{2}x^2$ on the same set of axes.

Solution In Figure 3.49 you can see that, as before, we graph the three curves. Compare these three parabolas. Notice the effect that the coefficient of x^2 has on the shape of the graph. Considering the graph of $y = x^2$ as a basic parabola shape, the coefficient 2 in $y = 2x^2$ makes a thinner parabola, and in $y = \frac{1}{2}x^2$ the coefficient $\frac{1}{2}$ makes a wider one.

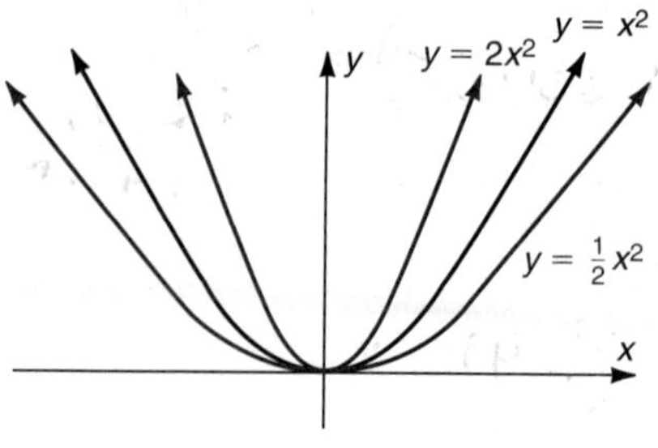

Fig. 3.49

□

EXAMPLE 3 The primary mirror of a reflecting telescope is ground and polished to a shape that is a parabola in cross section. John Tomlins has purchased a mirror blank to grind for the primary mirror in his new telescope. The blank is a circular disk of glass 10 inches in diameter and 1 inch thick as in Figure 3.50. He intends to grind it so that its cross section is parabolic, $\frac{3}{8}$ inch thick at the center, and 1 inch thick at the edge as shown in Figure 3.51.

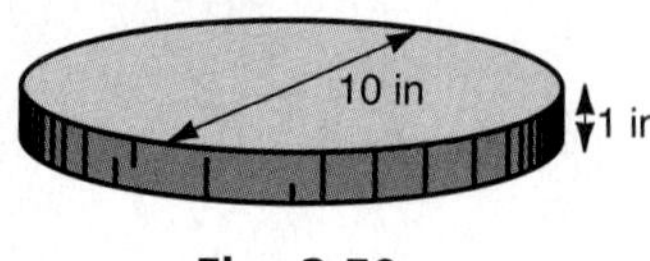

Fig. 3.50

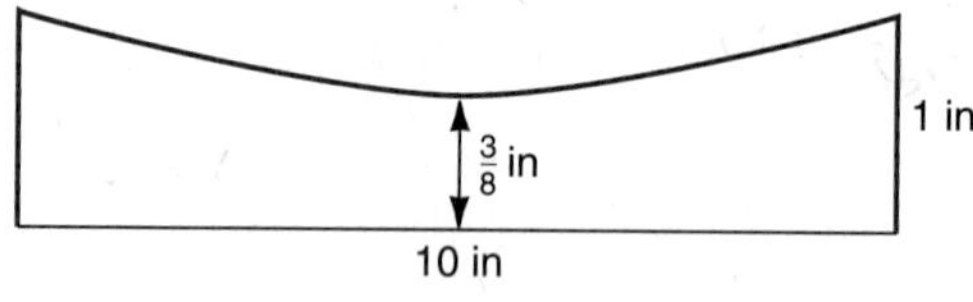

Fig. 3.51

How thick should the mirror be 1 inch from the center? 2.5 inches from the center? 4 inches from the center?

Solutions To answer these questions, we need the equation of the cross-section parabola. Place a coordinate system as shown in Figure 3.52.

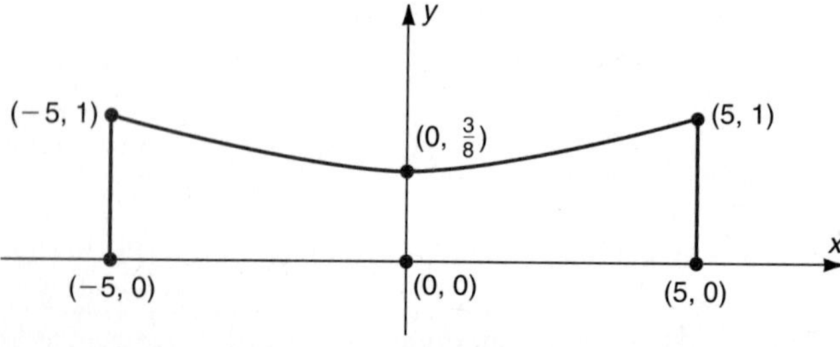

Fig. 3.52

Since the axis of symmetry of the parabola is the y-axis, the equation of the parabola is of the form $y = ax^2 + k$, where y is the thickness of the disk and x is the distance from the center. When x is 0, y is $\frac{3}{8}$.

$$\frac{3}{8} = a \cdot 0 + k$$

Therefore, k is $\frac{3}{8}$. Now when x is 5, y is 1, so

$$1 = a \cdot 5^2 + \frac{3}{8}$$

$$\frac{5}{8} = 25a$$

$$a = \frac{1}{40}$$

Thus, the equation of the parabola is $y = \frac{1}{40}x^2 + \frac{3}{8}$.

When x is 1,

$$y = \frac{1}{40} + \frac{3}{8} = \frac{16}{40} = \frac{2}{5}$$

When x is 2.5,

$$y = \frac{1}{40}(2.5)^2 + \frac{3}{8} = \frac{17}{32}$$

When x is 4,

$$y = \frac{1}{40} \cdot 4^2 + \frac{3}{8} = \frac{16}{40} + \frac{3}{8} = \frac{31}{40}$$

The thickness 1 inch from the center should be $\frac{2}{5}$ inch. The thickness 2.5 inches from the center should be $\frac{17}{32}$ inch. The thickness 4 inches from the center should be $\frac{31}{40}$ inch. □

EXAMPLE 4 Graph $y = -x^2$.

Solution Plotting a few points gives the graph in Figure 3.53. Notice that this is just the graph of $y = x^2$ turned upside down. □

We can now make some observations about graphing parabolas. The graph of $y = ax^2 + k$ is a parabola with vertex at the point $(0, k)$. It opens downward if a is negative, and it opens upward if a is positive. It is thinner than the graph of $y = x^2$ if $|a| > 1$, and wider if $|a| < 1$.

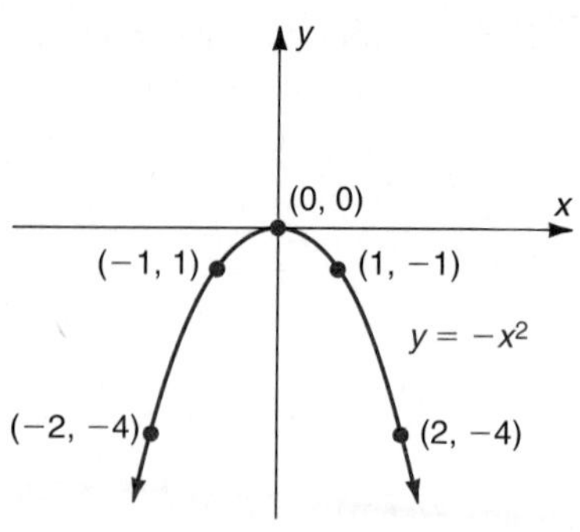

Fig. 3.53

EXAMPLE 5 Graph $y = (x - 1)^2$.

Solution Plotting several points gives the graph in Figure 3.54. The vertex of this parabola is (1, 0). In fact, it is the graph of $y = x^2$ shifted 1 unit to the right. Notice that the axis of symmetry is the line $x = 1$. □

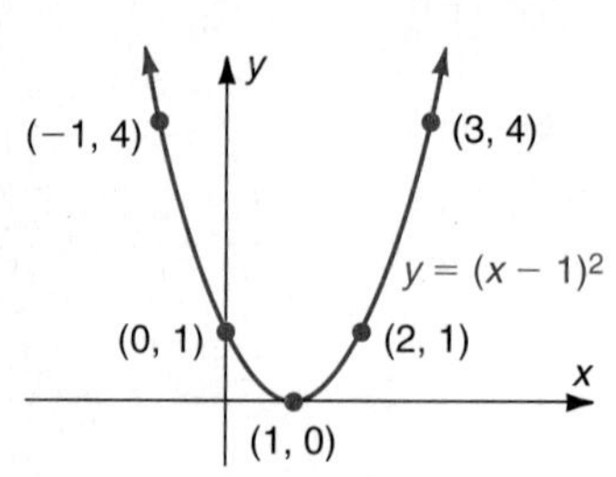

Fig. 3.54

EXAMPLE 6 Graph $y = -2(x - 1)^2 + 2$.

Solution Plot several points to see the parabola in Figure 3.55. Its vertex is (1, 2). It is thinner than the graph of $y = x^2$, and it opens downward. □

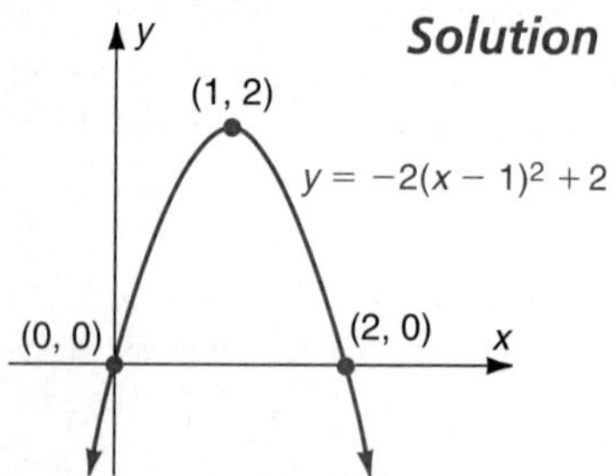

Fig. 3.55

The graph of $y = a(x - h)^2 + k$ is a parabola with vertex at the point (h, k). Its axis of symmetry is the line $x = h$. The parabola opens upward if a is positive, and downward if a is negative.

EXAMPLE 7 Find the vertex and axis of symmetry for each parabola:

(a) $y = (x + 1)^2 + 2$ (b) $y = x^2 - 1$ (c) $y = (x + 3)^2$

Solutions (a) To match the form of $y = a(x - h)^2 + k$ we see that $y = (x + 1)^2 + 2$ must be written as $y = [x - (-1)]^2 + 2$. So h is -1, and k is 2. Thus the vertex is $(-1, 2)$. The axis of symmetry is the line $x = -1$.

(b) $y = x^2 - 1$
$y = (x - 0)^2 - 1$
So the vertex is $(0, -1)$, and the axis of symmetry is the line $x = 0$.

(c) $y = (x + 3)^2$
$y = [x - (-3)]^2 + 0$
Thus the vertex is $(-3, 0)$, and the axis of symmetry is the line $x = -3$. □

As in the case of the circle, it is often necessary to complete the square in order to find the vertex and graph a parabola.

EXAMPLE 8 Complete the square and find the vertex for the graph of $y = x^2 - 8x$.

Solution $y = x^2 - 8x$

We add 16 for a perfect square. So, just as we did with the circle, we can add zero without harming the equation. So we add and subtract 16.

$$y = x^2 - 8x + 16 - 16$$

$$y = (x - 4)^2 - 16$$

The vertex is at $(4, -16)$. □

EXAMPLE 9 Graph $y = -2x^2 + 4x - 3$.

Solution To graph the equation of a parabola we write it in the form $y = a(x - h)^2 + k$. In this form we can identify the vertex and the axis of symmetry. In order to write the equation in the desired form we must complete the square.

$$y = -2x^2 + 4x - 3$$

The coefficient of x^2 must be 1 for the completing-the-square procedure to work. We factor -2 out of the first two terms.

$$y = -2(x^2 - 2x) - 3$$

Now we complete the square.

$$y = -2(x^2 - 2x + 1 - 1) - 3$$

$$y = -2(x^2 - 2x + 1) + 2 - 3$$

$$= -2(x - 1)^2 - 1$$

The vertex is $(1, -1)$, and the parabola opens downward as shown in Figure 3.56. The axis of symmetry is $x = 1$. □

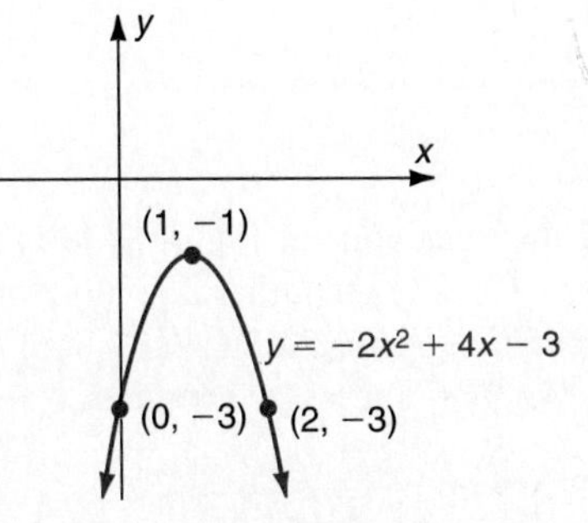

Fig. 3.56

EXAMPLE 10 A howitzer fires a shell whose trajectory is given by the equation

$$h = 1.924x - 0.633x^2$$

where h is the height in kilometers and x is the lateral distance from the gun, also in kilometers. Find the maximum height the shell obtains during flight. How far from the howitzer does the shell impact on the ground? Assume level ground and give approximations to the nearest meter.

Solution We see from the equation that the path of the shell is a parabola. We find the vertex and axis of symmetry and then sketch the graph.

$$h = -0.633x^2 + 1.924x$$

$$= -0.633\left(x^2 - \frac{1.924}{0.633}x\right) \qquad \text{Factor out } -0.633.$$

Then we complete the square.

$$= -0.633\left[x^2 - \frac{1.924}{0.633}x + \left(\frac{1.924}{2(0.633)}\right)^2 - \left(\frac{1.924}{2(0.633)}\right)^2\right]$$

$$= -0.633\left(x - \frac{0.962}{0.633}\right)^2 + \frac{(0.962)^2}{0.633}$$

$$h \approx -0.633(x - 1.520)^2 + 1.462$$

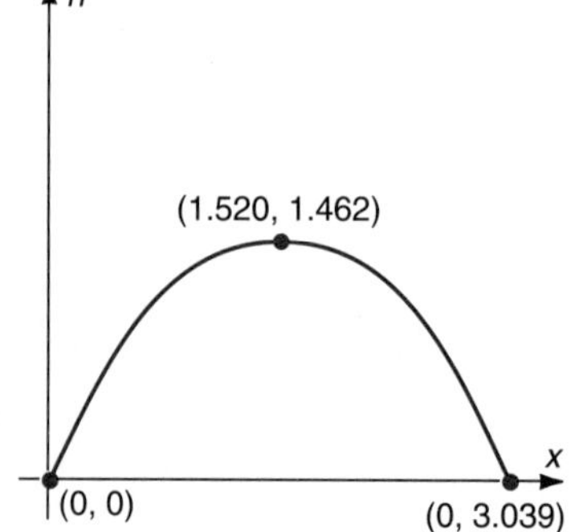

Fig. 3.57

Therefore the vertex is approximately the point (1.520, 1.462), and the axis of symmetry is approximately the line $x = 1.520$. So we see from Figure 3.57 that the highest point reached by the shell is at the vertex of the parabola and is approximately 1.462 kilometers or 1462 meters to the nearest meter. Since the x-intercepts of the graph occur when h is 0, we have

$$0 = -0.633x^2 + 1.924x$$

$$= (-0.633x + 1.924)x$$

$$x = 0 \qquad \text{or} \qquad x = \frac{1.924}{0.633} \approx 3.039$$

The shell impacts the ground approximately 3039 meters from the howitzer. □

EXAMPLE 11 Graph the equation $x = y^2$.

Solution Plot a few points. Pick a number for y first and then calculate x. Figure 3.58 is a parabola of the same general shape as the graph of $y = x^2$ except that it is lying on its side, opening to the right. In fact, if we interchange the roles of x and y (and h and k) in the equation $y = a(x - h)^2 + k$, we get similar figures, except they open left-right instead of up-down. □

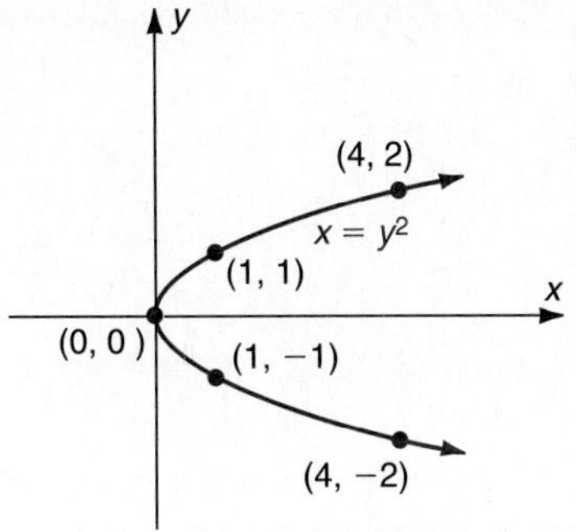

Fig. 3.58

> **Parabolas**
>
> The graph of $y = a(x - h)^2 + k$
> is a parabola with vertex at the point (h, k).
> It opens upward if a is positive, and downward if a is negative.
> Its axis of symmetry is the line $x = h$.
>
> The graph of $x = a(y - k)^2 + h$
> is a parabola with vertex at the point (h, k).
> It opens to the right if a is positive, and to the left if a is negative.
> Its axis of symmetry is the line $y = k$.

Notice that the graphs of the two quadratic forms

$$y = ax^2 + bx + c;\ a \neq 0$$

$$x = dy^2 + ey + f;\ d \neq 0$$

are both parabolas. We can complete the square to find the vertex and axis of symmetry.

The coordinates of the points on the graph *1 unit* either side of the axis of symmetry are usually easy to find. This information is usually sufficient to make a sketch of the graph of the equation.

EXAMPLE 12 Graph the equation $x + 2y^2 = -5 + 8y$.

Solution Solve the equation for x. (Note the y^2 term.)

$$x = -2y^2 + 8y - 5$$

Before completing the square it is necessary to factor -2 from the first two terms.

$$x = -2(y^2 - 4y) - 5$$

Now we can complete the square inside the parentheses.

$$x = -2(y^2 - 4y + 4 - 4) - 5$$

$$x = -2(y^2 - 4y + 4) + 8 - 5$$

$$x = -2(y - 2)^2 + 3.$$

This is a parabola with vertex at (3, 2), opening to the left and with the line $y = 2$ as its axis of symmetry. We find the coordinates of the points *1 unit on either side of the axis of symmetry* by substituting 1 and 3 for y in the equation $x = -2(y - 2)^2 + 3$.

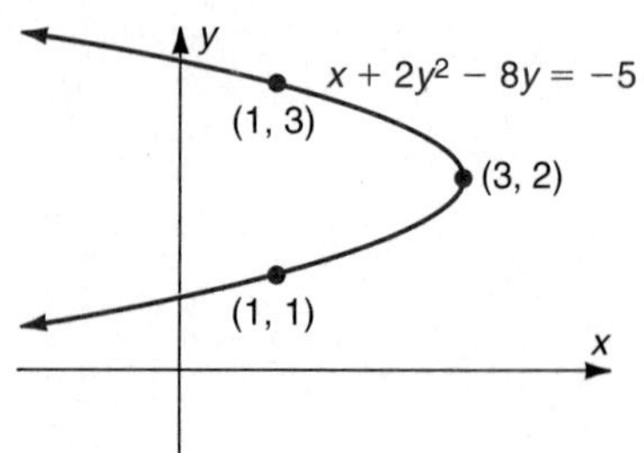

Fig. 3.59

$$x = -2(1 - 2)^2 + 3 = -2 \cdot 1 + 3 = 1$$

Therefore, (1, 1) is on the graph

$$x = -2(3 - 2)^2 + 3 = -2 \cdot 1 + 3 = 1$$

and (1, 3) is also on the graph (Figure 3.59). □

The x- and y-intercepts are another aid in sketching a graph. If they exist, they can be found in a straightforward manner.

EXAMPLE 13 Graph each equation and label intercepts and vertices.

(a) $y = (x + 1)^2 - 2$ (b) $x = y^2 - 2$

Solutions (a) The graph of $y = (x + 1)^2 - 2$ is a parabola with vertex at $(-1, -2)$. It opens upward.

To find the y-intercepts, we let x be 0.

$$y = 1^2 - 2$$

$$y = -1$$

The y-intercept is -1.

To find the x-intercepts we let y be 0.

$$0 = (x + 1)^2 - 2$$

$$2 = (x + 1)^2$$

$$\pm\sqrt{2} = x + 1 \quad \text{Square Root Property}$$

$$-1 \pm \sqrt{2} = x$$

The x-intercepts are $-1 \pm \sqrt{2}$.

$$-1 + \sqrt{2} \approx 0.414 \quad \text{and} \quad -1 - \sqrt{2} \approx -2.414$$

The graph is sketched in Figure 3.60.

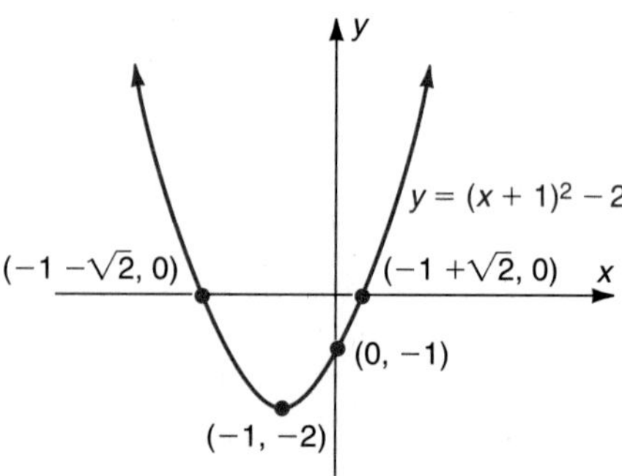

Fig. 3.60

(b) The graph of $x = y^2 - 2$ is a parabola with vertex at $(-2, 0)$. It opens to the right.

Let x be 0 to find the y-intercepts.

$$0 = y^2 - 2$$

Let y be 0 to find the x-intercepts.

$$x = -2$$

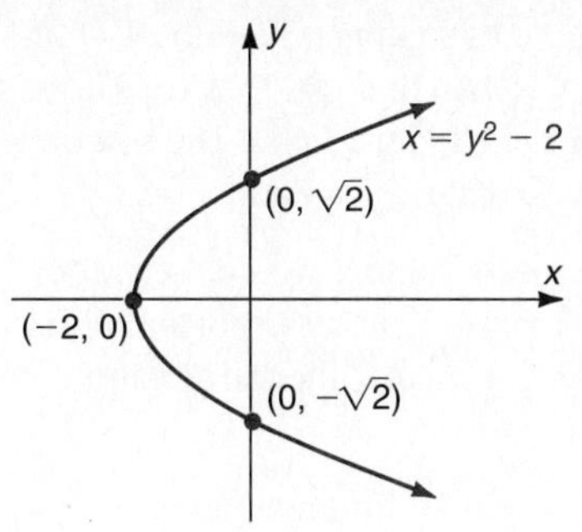
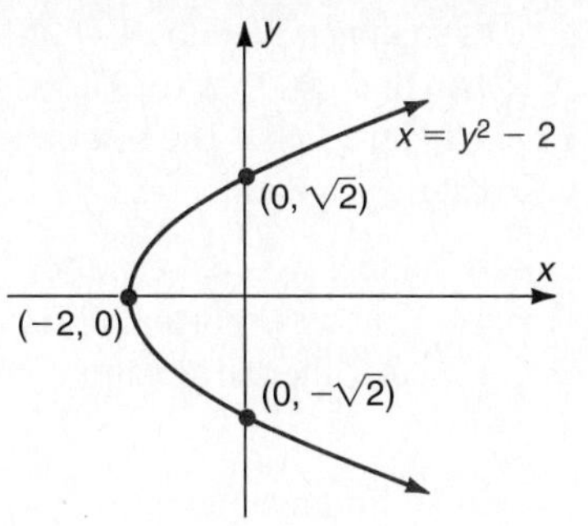

Fig. 3.61

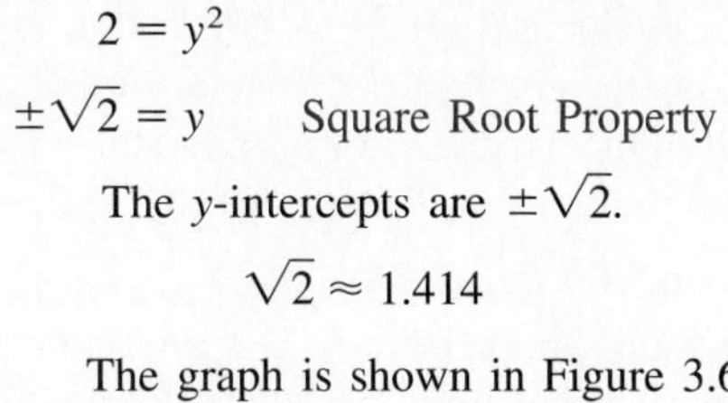

$$2 = y^2$$ The x-intercept is -2.

$$\pm\sqrt{2} = y \qquad \text{Square Root Property}$$

The y-intercepts are $\pm\sqrt{2}$.

$$\sqrt{2} \approx 1.414$$

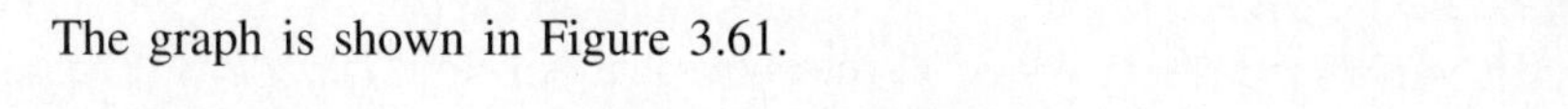

The graph is shown in Figure 3.61. □

Quadratic Inequalities and Graphs of Parabolas

Understanding the behavior of a graph helps us to see the relationship between the graph of an equation in two variables and solutions to inequalities in one variable which we studied in Chapter 2. Consider the graph of $y = x^2 - x - 6$ shown in Figure 3.62.

Fig. 3.62

The y-coordinates of points on the parabola are either positive (above the x-axis), negative (below the x-axis), or zero (on the x-axis). Notice that the x-intercepts, -2 and 3, determine where the y-coordinates change from positive to negative. Remember that the y-coordinates are 0 at the x-intercepts. The y-coordinates are negative when x is between -2 and 3. That is, when x is in the interval $(-2, 3)$. The y-coordinates are positive when x is less than -2 or when x is greater than 3. That is when x is in the interval $(-\infty, -2) \cup (3, +\infty)$.

In Chapter 2, we solved quadratic inequalities in one variable. We can now see the relationship between the graph of a parabola with a vertical axis of symmetry and the boundary numbers for quadratic inequalities in one variable. Notice that there are *no breaks* in the graph of a parabola. This property is called **continuity** and is important. However, a careful definition must wait until calculus. Continuity is the property that makes the method of boundary numbers work. Let's solve the inequality $x^2 - x - 6 \geq 0$ using the technique of boundary numbers.

First, we find the boundary numbers by solving the associated equation.

$$x^2 - x - 6 = 0$$

$$(x - 3)(x + 2) = 0$$

$$x = 3 \quad \text{or} \quad x = -2$$

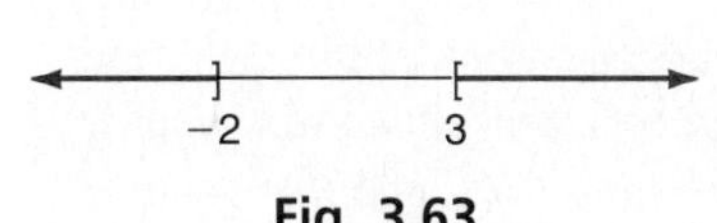

Fig. 3.63

The boundary numbers are -2 and 3. Locate the boundary numbers on a number line and test each region (Figure 3.63). Since equality is allowed in $x^2 - x - 6 \geq 0$, the boundary numbers are included. We see that the solution set is $(-\infty, -2] \cup [3, +\infty)$.

Notice that the boundary numbers are the x-intercepts of the graph. Since $y = x^2 - x - 6$ on the parabola, solving the inequality $x^2 - x - 6 \geq 0$ is the same problem as finding where $y \geq 0$ on the parabola.

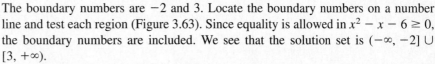

EXAMPLE 14 Use the graph of $y = x^2 - x - 6$ shown in Figure 3.64 to solve each inequality.

(a) $x^2 - x - 6 \geq 0$ (b) $x^2 - x - 6 < 0$ (c) $x^2 - x - 6 \leq 6$

Solutions

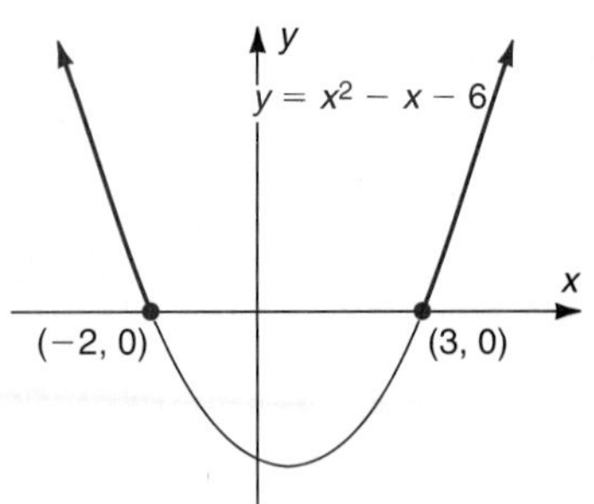

Fig. 3.64

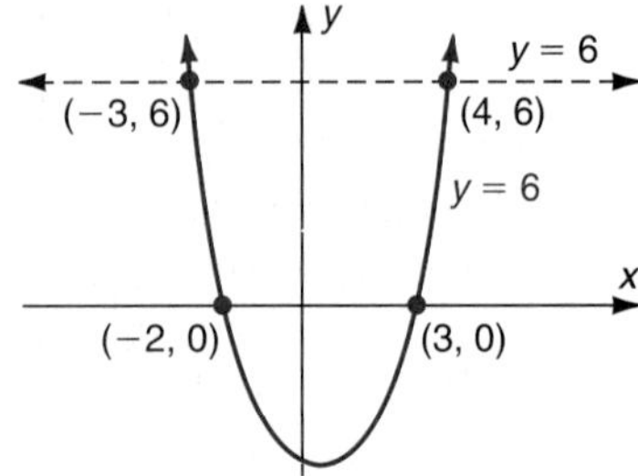

Fig. 3.65

(a) Solving the inequality $x^2 - x - 6 \geq 0$ is the same as finding where the y-coordinates on the graph of $y = x^2 - x - 6$ are greater than or equal to zero. This is the part of the parabola that is on or above the x-axis (Figure 3.64). The solution set for $x^2 - x - 6 \geq 0$ is $(-\infty, -2] \cup [3, +\infty)$ in interval notation.

(b) Solving the inequality $x^2 - x - 6 < 0$ is the same as finding where the y-coordinates on the graph of $y = x^2 - x - 6$ are less than 0. This is the portion of the parabola that is below the x-axis. The solution set in the interval notation is $(-2, 3)$.

(c) Solving the inequality $x^2 - x - 6 \leq 6$ is the same as finding points on the parabola having y-coordinates less than or equal to 6 (Figure 3.65).

We must first find where the y-coordinates are *equal to* 6. In other words, we must solve the equation $x^2 - x - 6 = 6$.

$$x^2 - x - 12 = 0 \qquad \text{Standard form}$$

$$(x - 4)(x + 3) = 0 \qquad \text{Factor.}$$

$$x = 4 \qquad \text{or} \qquad x = -3$$

So, the solution set for $x^2 - x - 6 \leq 6$ is the interval $[-3, 4]$. □

GRAPHING CALCULATOR BOX

Using Trace and Graphing Parabolas

***USING* TRACE**

Not only can a graphing calculator draw graphs, but it can also approximate coordinates of points on a graph. The TRACE key allows us to move a point along a graph and shows the current coordinates of the point as it moves.

Let's approximate the x-intercepts and vertex of the graph of the equation $y = x^2 - x - 1$. The graph of this equation is a parabola that opens upward. The intercepts and vertex can be found algebraically, but let's see what the calculator can find. We must choose a window that will contain the vertex and intercepts. Since the y-intercept is -1 and the parabola opens up, let's set $[-5, 5]$ for both x and y. GRAPH the function. Let's call the negative x-intercept x_1 and the positive one x_2.

Press TRACE and notice that the blinking cursor appears. Also, the coordinates of the cursor point are written at the bottom of the screen. Use the ← or → editing key to move the cursor along the parabola. Since the y-coordinate of the x-intercepts is 0, we find the y-coordinates closest to 0.

TI-81

$x_1 \approx -.5789474 \qquad y_1 \approx -0.0858726$

$x_2 \approx 1.6315789 \qquad y_1 \approx 0.03047091$

Move the cursor to the vertex. To find the lowest point, compare the y-coordinates as the cursor moves along the graph. Find the smallest y-coordinate. The smallest y-coordinate is approximately -1.249307. So, the approximate coordinates of the vertex are $(0.47368421, -1.249307)$.

CASIO fx-7700G

$x_1 \approx -0.638297 \qquad y_1 \approx 0.045722$

$x_2 \approx 1.5957446 \qquad y_1 \approx -0.049343$

The smallest y-coordinate is approximately -1.248981. So, the approximate coordinates of the vertex are $(0.532, -1.249)$.

The x-intercepts are solutions to the equation $x^2 - x - 1 = 0$. With the quadratic formula, we find them to be approximately -0.6180 and 1.6180.

The TRACE feature depends on the dimensions of the window for its accuracy. Set the range at $[-2, 5]$ for both x and y and redraw the graph. Use TRACE again to approximate the vertex. Now we find the coordinates of the vertex are approximately $(0.50526316, -1.249972)$. Since by completing the square we know the vertex is $(0.5, -1.25)$, this window gives a better approximation.

Graphing Horizontal Parabolas

An equation to be graphed must first be solved for y before entering it into the calculator. Parabolas with a vertical axis of symmetry, such as $y = x^2 - x - 1$, can be entered with no algebraic manipulation. However, to graph a parabola with a horizontal axis of symmetry, we graph the top half and the bottom half of the parabola on the same screen, just as we did for circles.

Let's graph $x = y^2 + 2y$. To solve the equation for y, we must first complete the square and use the Square Root Property to write the equation as $y = -1 \pm \sqrt{x + 1}$. Enter $y = -1 + \sqrt{x + 1}$ as one equation and $y = -1 - \sqrt{x + 1}$ as another. Then GRAPH (use the multistatement feature on the Casio).

Moving From One Graph to Another in TRACE

The up and down arrows allow us to move the cursor from one graph to another. On the **CASIO fx-7700G** the graphs must be graphed using the multistatement command.

PROBLEM SET 3.4

Warm-ups

In Problems 1 through 4, identify the vertex and the axis of symmetry. Graph each equation. See Examples 1 and 2.

1. $y = x^2 + 3$ **2.** $y = x^2 - 3$ **3.** $y = 4x^2$ **4.** $y = \frac{1}{2}x^2 + 2$

For Problems 5 and 6, see Example 3.

5. The reflector in a large flashlight has a parabolic cross section. It is 8 centimeters in diameter and 4 centimeters deep. (See Figure 3.66 on p. 200.)

(a) Find the distance D_1.
(b) Find the distance D_2.

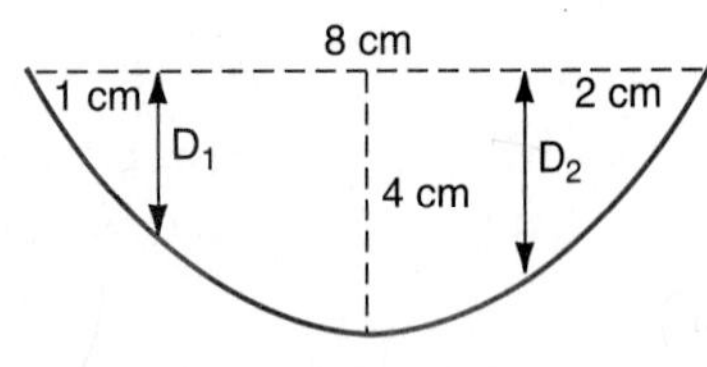

Fig. 3.66

6. If air resistance is neglected, the path of a projectile is very closely approximated by a parabola. Suppose a football is kicked from the 13-yard line in a Packers-Steelers game and it reaches its maximum height of 33 yards as it crosses the 40-yard line. If the kicker's foot met the ball 1 yard above the ground, approximately how high is the ball when it crosses midfield? (Assume the ball travels parallel to the sidelines. See Figure 3.67.)

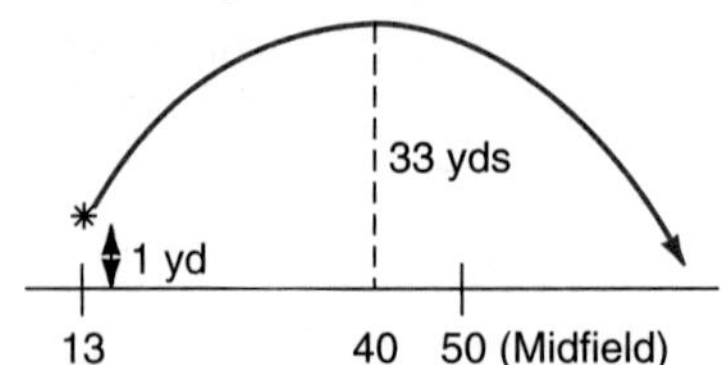

Fig. 3.67

In Problems 7 through 22, sketch the graph of each equation. For Problems 7 through 14, see Examples 4 through 7.

7. $y = (x - 3)^2$

8. $y = (x - 2)^2$

9. $y = (x + 2)^2 + 3$

10. $y = (x + 4)^2 - 2$

11. $y = 2(x + 2)^2 - 3$

12. $y = \frac{1}{2}(x - 5)^2 + 1$

13. $y = -(x - 4)^2 - 3$

14. $y = -(x + 2)^2 - 1$

For Problems 15 through 18, see Example 8.

15. $y = x^2 + 4x$

16. $y = x^2 - 2x$

17. $y = x^2 - 4x + 2$

18. $x^2 + x = y$

For Problems 19 through 22, see Example 9.

19. $y = -3x^2 - 12x - 8$

20. $-2x^2 + 4x = y$

21. $y = -x^2 + 2x + 2$

22. $y = -x^2 - 4x$

For Problems 23 and 24, see Example 10.

23. The daily profit made by Bill's Bagelry is given by $P = \frac{2}{3}x - \frac{1}{1800}x^2$, where x is the number of bagels Bill cooks per day and P is in dollars. How many bagels per day should Bill make to maximize his profit?

24. The total cost C of producing x pocket calculators per day is $C = \frac{1}{2}x^2 - 100x + 5150$. How many calculators should be made to minimize costs?

In Problems 25 through 28, sketch the graph. See Examples 11 and 12.

25. $x = y^2 - 4y + 2$

26. $y^2 + 2y = x$

27. $x = -2y^2 + 4y$

28. $x = 3y^2 + 18y + 29$

In Problems 29 and 30, find the intercepts. See Example 13.

29. $y = x^2 - 2x - 3$

30. $x = y^2 + 4y - 12$

In Problems 31 and 32, use the graph of $y = x^2 - x - 2$ to solve the inequality. See Example 14.

31. $x^2 - x - 2 \le 0$

32. $x^2 - x - 2 < 4$

Practice Exercises

In Problems 33 through 38, match the graph with the correct equation.

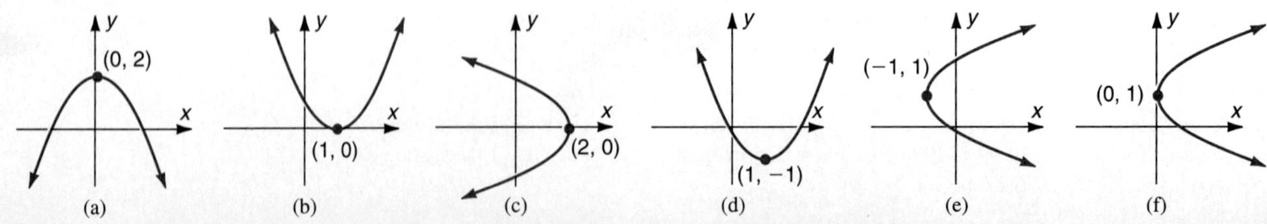

33. $y = (x - 1)^2$
34. $x = (y - 1)^2$
35. $x = (y - 1)^2 - 1$
36. $y = (x - 1)^2 - 1$
37. $y = 2 - x^2$
38. $x = 2 - y^2$

In Problems 39 through 62, the graph of each equation is a parabola. Find the vertex of each and sketch the graph. Label all intercepts.

39. $x = (y - 2)^2$
40. $x = -y^2 + 1$
41. $y = (x - 1)^2 + 2$
42. $y = -(x - 2)^2 - 3$
43. $x = -(y + 2)^2 + 1$
44. $x = 2(y + 1)^2 - 2$
45. $y = \frac{1}{2}\left(x + \frac{2}{3}\right)^2 + \frac{1}{3}$
46. $y = \frac{2}{3}\left(x + \frac{1}{2}\right)^2 - \frac{2}{3}$
47. $y = -(x + 3)^2 + 3$
48. $y = 2(x + 2)^2 - 1$
49. $x = \frac{1}{3}y^2 + 2$
50. $x = \frac{1}{2}y^2 - 3$
51. $x = -2y^2 - 8y - 8$
52. $y = -2x^2 - 2x - 1$
53. $x^2 + 3y = 3$
54. $y^2 - 2x - 4 = 0$
55. $y^2 + x = 0$
56. $x^2 - y = 0$
57. $x^2 + 2 = y$
58. $x = -y^2 + 1$
59. $3y^2 - 3 = x$
60. $y = -2x^2 - 2$
61. $x^2 - 2y + 4 = 0$
62. $y^2 + 3x = 3$

In Problems 63 through 68, graph both equations on one set of coordinate axes. Guess the points of intersection. Show that they are the points of intersection by substituting in both equations.

63. $y = x^2 - 12x + 38;\ x - y = 2$
64. $y = -2x^2 + 16x - 22;\ 2x + y = 14$
65. $x = y^2 + 3;\ x^2 + y^2 + 12x = 65$
66. $y + x^2 = 4;\ x^2 + y^2 - 10y + 20 = 0$
67. $y = 2x^2 + 1;\ 3x^2 = y + 3$
68. $y = x^2 - 4x + 5;\ 3x^2 + y = 5$

69. Approximate (to three decimal places) the x-intercepts for $y = 4x^2 - 3x - 5$.

70. Approximate (to three decimal places) the y-intercepts for $x = 5y^2 + y - 7$.

71. Find an equation of all parabolas containing the origin whose vertex is at the point (2, 1).

72. Find an equation of all parabolas containing the point (3, 2) whose vertex is at the origin.

73. Find an equation of a parabola containing the points (2, 6) and (6, 6) whose vertex is at (4, −2).

74. Find an equation of a parabola containing the points (−5, −1) and (−5, 3) whose vertex is at (1, 1).

75. A pair of binoculars is dropped straight down from a window 20 feet above the ground. The relationship between distance above the ground (s, measured in feet) and time (t, measured in seconds) that the binoculars have been falling is given by the equation $s = 20 - 16t^2$.
(a) Graph the parabola and interpret the intercepts. Use the horizontal axis as the t-axis.
(b) Where are the binoculars after falling 0.5 second?
(c) How long will it take the binoculars to hit the ground?
(d) Approximate to the nearest hundredth of a second when the binoculars will be 5 feet below the window.

76. A rocket is fired straight upward from the top of a building 100 meters high. The relationship between distance above the ground (s, measured in meters) and time (t, measured in seconds) since the rocket was launched is given by the equation $s = 100 + 180t - t^2$.
(a) Graph this equation using the horizontal axis for time.
(b) How high will the rocket go?
(c) Approximate the t-intercepts to five decimal places. Interpret what they mean.

77. The Steel Company produces grade X steel and grade Y steel using the same production process. The relationship between the number of tons of each grade of steel that can be produced is given by the equation $X = 16 - Y^2 - 6Y$.
(a) Graph this equation.
(b) What is the largest amount of grade X that can be produced?
(c) What is the largest amount of grade Y that can be produced?

In Problems 78 through 83, use graphing to solve each inequality.

78. $1 - x^2 \leq 0$
79. $1 - x^2 \geq 0$
80. $1 - x^2 > -1$
81. $x^2 - 2x - 2 > 0$
82. $x^2 - 2x - 2 \leq 0$
83. $x^2 - 2x - 2 \geq 1$

84. This problem investigates the relationship between the discriminant of $ax^2 + bx + c = 0$, where $a \neq 0$, and the graph of $y = ax^2 + bx + c$. For each equation use the discriminant to determine the nature of the solutions.

(a) $x^2 - 2x + 2 = 0$

(b) $x^2 + 2x = 0$

(c) $x^2 + 2x + 1 = 0$

Graph each parabola.

(a) $y = x^2 - 2x + 2$

(b) $y = x^2 + 2x$

(c) $y = x^2 + 2x + 1$

Make a generalization about the graph of $y = ax^2 + bx + c$ and the discriminant of the equation $ax^2 + bx + c = 0$.

Challenge Problems

85. The slope of the tangent line to the parabola $y = ax^2 + bx + c$ at the point (p,q) is given by $m = 2ap + b$. Find an equation of the line perpendicular to the tangent line (called the **normal** line) of the graph of the equation $y = 2x^2 - 3x - 5$ at the point $(3, 4)$.

86. Find an equation of the normal line to the parabola $y = ax^2 + bx + c$ at the point (p, q). See Problem 85.

For Graphing Calculators

Use the TRACE *key to approximate as indicated. Use algebraic methods to determine the accuracy of the approximation.*

87. $y = 2x^2 - 5x + 1$; coordinates of the vertex of the graph

88. $y = 5 + 7x - x^2$; largest y-coordinate on the graph

89. $x = \left(y + \dfrac{\sqrt{5}}{3}\right)^2 - 1.0089$; y-intercepts

90. $x = -y^2 + 2\sqrt{2}y - 2 + \pi$; y-intercepts

91. $\pi x^2 + 2.4x - \pi^2 = 0$; solutions

92. $\sqrt{11}x^2 + 3x < \sqrt{7}$; solutions

IN YOUR OWN WORDS . . .

93. Describe the procedure used to rewrite an equation of the form $y = Ax^2 + Bx + C$ into the form $y = p(x - h)^2 + k$.

94. Explain the relationship between the graph of $y = x^2 - x - 2$ and solutions to the equation $x^2 - x - 2 = 0$ and solutions to the inequality $x^2 - x - 2 < 0$.

95. Solve the inequality $x^2 + 2x < 3$ by graphing the parabola $y = x^2 + 2x$ and the line $y = 3$. Then solve the same inequality by graphing $y = x^2 + 2x - 3$. Compare the two approaches.

3.5 FUNDAMENTAL CONCEPTS OF FUNCTIONS

In everyday life, objects are often paired together. A person has a name. A postal zip code is assigned to a geographic region. A student is assigned a grade in mathematics. A hockey player has a number on his uniform. Often, particularly in science, business, and engineering, one number is paired with another. For example, the *amount* of a radioactive element present determines the *rate* of decay, and the *price* of a manufacturer's product determines the *amount* of yearly profit. This pairing of one number with another is basic to the idea of *function* and is one of the great unifying concepts of mathematics. It lies at the heart of calculus.

A function involves two sets and a rule of assignment.

Definition: Function

A **function** is a rule that assigns to each member in one set (the **domain**) exactly one member from another set (the **range**).

Consider assigning group numbers to students in a large class by assigning to each student the last digit of his or her social security (SS) number. This is a function. The *domain* is the set of SS numbers of students in the class. The *range* is the set of integers, 0 through 9 (providing the class is large enough). The *rule of assignment* is "Drop the first eight digits from the SS number."

EXAMPLE 1 Consider the assignments made by Figures 3.68, 3.69, and 3.70. Is each a function?

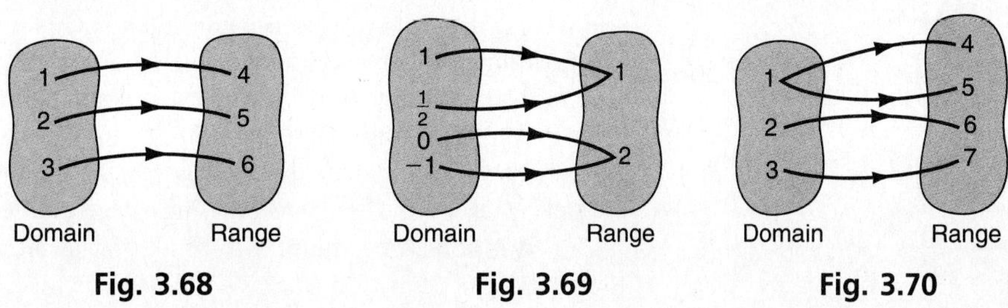

Fig. 3.68 **Fig. 3.69** **Fig. 3.70**

Solutions (a) Figure 3.68 shows a function. The rule assigns to each number in the domain exactly one number in the range.

(b) Figure 3.69 also shows a function. Its domain is $\left\{-1, 0, \frac{1}{2}, 1\right\}$, and its range is $\{1, 2\}$. It does not matter that both 1 and $\frac{1}{2}$ are paired with 1 or that both -1 and 0 are paired with 2. To each number in the domain exactly one number in the range is assigned.

(c) Figure 3.70 does *not* show a function. Both 4 and 5 are paired with the domain element 1. This violates the definition of a function which states that to each number in the domain *exactly one* number in the range is assigned. □

There are several ways of giving the rule that defines a function. Using pictures, as in Example 1, is one way. Sometimes the assignments are listed as ordered pairs, the first number being from the domain and the second from the range. Tables are also used. The most common method of assignment is with a formula.

EXAMPLE 2 The domain of a function is $\{x \mid -2 \leq x \leq 2$ and x is an integer$\}$ and the rule is "To each number in the domain assign its square."

(a) Write the assignments as ordered pairs.

(b) Find the range.

Solutions (a) The assignments listed as ordered pairs are $(-2, 4), (-1, 1), (0, 0), (1, 1), (2, 4)$.

(b) The range is $\{0, 1, 4\}$.

Notice that to each number in the domain exactly one number in the range is assigned. □

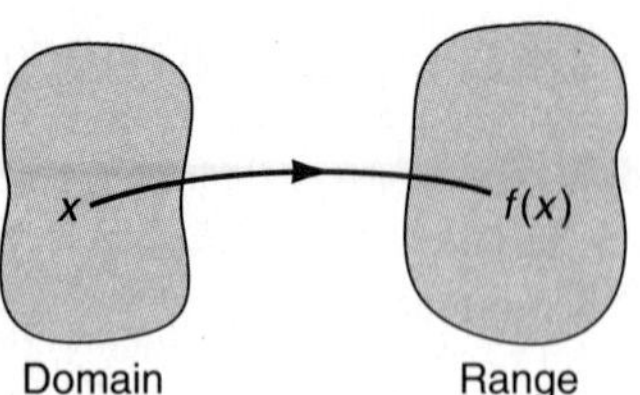

Fig. 3.71

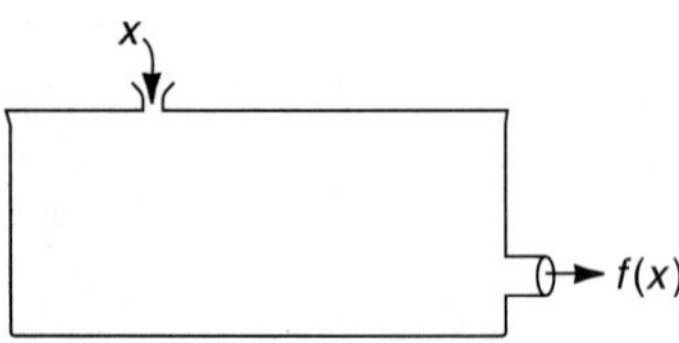

Fig. 3.72

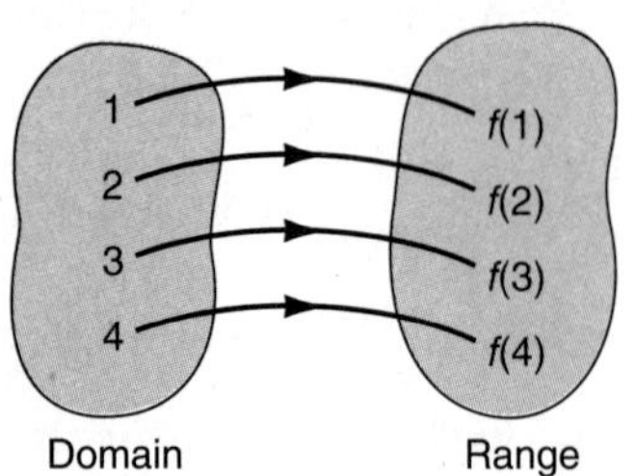

Fig. 3.73

Using x to represent an element in the domain to which the element y in the range is assigned, we can think of the rule as defining a set of ordered pairs (x, y). In this case, we can give the rule with an equation such as $y = x^2$. We refer to x as the **independent variable,** and y as the **dependent variable.** The number y depends on the number x. The domain consists of all possible real numbers that can substitute for the independent variable, while the range consists of all possible values of the dependent variable.

In mathematical applications of functions the rule is often given using functional notation. We use letters such as f, g, and h to name functions. If f is a function and x is a number in its domain, then $f(x)$ is a symbol used to indicate the corresponding number in the range. We read $f(x)$ as "f of x" or "f evaluated at x." This could be represented by writing the ordered pair $(x, f(x))$ or as shown in Figure 3.71. *Note:* In using functional notation, remember that $f(x)$ may look like f times x, but it *does not* mean that. It means that x is an element of the domain of the function f and "f of x" is the corresponding range element.

It is sometimes helpful to think of x as an input number and $f(x)$ as an output number. A "function machine" illustrates this idea in Figure 3.72. A number for x goes into the function machine, and a function value comes out.

Suppose the rule for a function f with domain $\{1, 2, 3, 4\}$ is "Assign to each number 3 less than its square." This rule could be written in functional notation as $f(x) = x^2 - 3$. We often use x as the independent variable as a matter of convenience or habit. However, we can use any letter. $f(x) = x^2 - 3$, $f(s) = s^2 - 3$, and $f(t) = t^2 - 3$ all define the *same function.* The letter x acts as a placeholder. $f(\ ) = (\ )^2 - 3$ is a good way of thinking about this function.

If x is a number in the domain, then it is paired with $f(x)$ in the range. So, we have the pairings shown in Figure 3.73.

Evaluating Functions

When we find the value of $f(x)$, we **evaluate** the function at x. We often call $f(x)$ the function value at x. Remember that $f(1)$ is a name for the number in the range paired with 1 from the domain. To calculate the value of $f(1)$, use the rule $f(x) = x^2 - 3$ and replace x with 1. Thus,

$$f(1) = 1^2 - 3 = 1 - 3 = -2$$

$$f(2) = 2^2 - 3 = 4 - 3 = 1$$

$$f(3) = 3^2 - 3 = 9 - 3 = 6$$

$$f(4) = 4^2 - 3 = 16 - 3 = 13$$

EXAMPLE 3 If $h(x) = x^2 - 2x$ with domain of all real numbers, find

(a) $h(0)$ (b) $h(-2)$ (c) $h(2\sqrt{3})$ (d) $h\left(\frac{\pi}{2}\right)$

Solutions (a) $h(0) = 0^2 - 2(0) = 0 - 0 = 0$ Replace x with 0 in $h(x)$.

(b) $h(-2) = (-2)^2 - 2(-2) = 4 + 4 = 8$ Replace x with -2 in $h(x)$.

(c) $h(2\sqrt{3}) = (2\sqrt{3})^2 - (2)2\sqrt{3} = 12 - 4\sqrt{3}$ Replace x with $2\sqrt{3}$.

(d) $h\left(\frac{\pi}{2}\right) = \left(\frac{\pi}{2}\right)^2 - (2)\frac{\pi}{2} = \frac{\pi^2}{4} - \pi$ Replace x with $\frac{\pi}{2}$ in $h(x)$.

□

EXAMPLE 4 If $g(x) = x^2 + 3$ with domain of all real numbers, find

(a) $g(a)$ (b) $g(a) + g(b)$ (c) $g(x + h)$ (d) $\frac{g(x + h) - g(x)}{h}$

Solutions (a) $g(a) = a^2 + 3$ Replace x with a in $g(x)$.

(b) $g(a) + g(b) = a^2 + 3 + b^2 + 3 = a^2 + b^2 + 6$

(c) $g(x + h) = (x + h)^2 + 3 = x^2 + 2xh + h^2 + 3$ Replace x with $x + h$ in $g(x)$.

(d)
$$\frac{g(x + h) - g(x)}{h} = \frac{(x + h)^2 + 3 - (x^2 + 3)}{h} = \frac{x^2 + 2xh + h^2 + 3 - x^2 - 3}{h} = \frac{2xh + h^2}{h} = \frac{h(2x + h)}{h} = 2x + h;\ h \neq 0$$

□

Finding the Domain

It is common practice to state the rule for a function with a formula but not specify the domain. When this happens, we assume that the domain is the largest subset of real numbers for which the rule has meaning and produces real numbers.

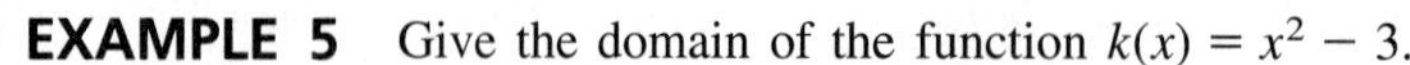

EXAMPLE 5 Give the domain of the function $k(x) = x^2 - 3$.

Solution $k(x)$ is a real number for any number x. So the domain of the function k is R. □

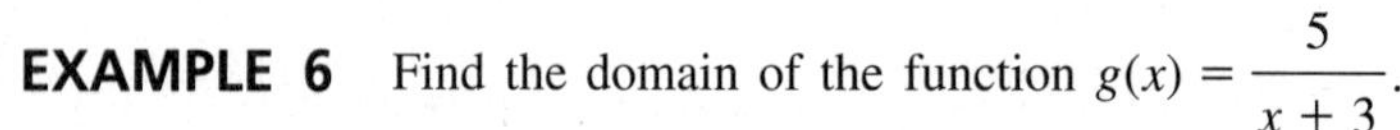

EXAMPLE 6 Find the domain of the function $g(x) = \dfrac{5}{x + 3}$.

Solution We can see that $g(x)$ is a real number for any number x except when x is -3. $g(-3)$ is $\dfrac{5}{-3 + 3}$ or $\dfrac{5}{0}$ which is undefined. So $g(-3)$ is not a real number. Thus the domain of g is $(-\infty, -3) \cup (-3, +\infty)$. □

EXAMPLE 7 Find the domain of each function.

(a) $g(x) = \sqrt{x}$ (b) $h(x) = \sqrt{x^2 - x - 2}$

Solutions (a) If $g(x) = \sqrt{x}$, then $g(x)$ is a real number for all nonnegative numbers x, that is, for $x \geq 0$. (If x is -5, then $g(-5)$ is $\sqrt{-5}$ which is not a real number.) So the domain of g is $[0, +\infty)$.

(b) If $h(x) = \sqrt{x^2 - x - 2}$, then $h(x)$ will be a real number if $x^2 - x - 2 \geq 0$. We solve the inequality by the technique of boundary numbers. The associated equation is

$$x^2 - x - 2 = 0$$

$$(x - 2)(x + 1) = 0$$

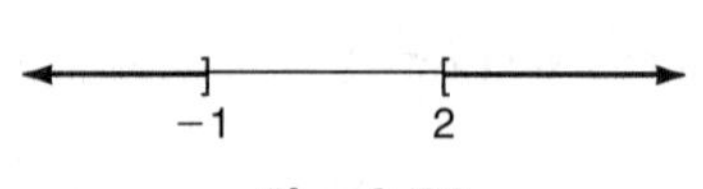

Fig. 3.74

The boundary numbers are -1 and 2 as shown in Figure 3.74. Testing each of the regions gives a solution set for $x^2 - x - 2 \geq 0$ of $(-\infty, -1] \cup [2, +\infty)$. So the domain of h is $(-\infty, -1] \cup [2, +\infty)$. □

A function can be evaluated only at numbers in its domain. If $f(x) = \sqrt{x}$, then $f(-2)$ is undefined since -2 is not in the domain of the function.

Writing a Function

Functional notation often allows us to express the relationship between two quantities in a meaningful way.

EXAMPLE 8 Express the area and the perimeter of the rectangle in Figure 3.75 as a function of its width w.

Solution To write the area as a function of the width, we will use w as the independent variable and A as the name of the function.

$A(w) = w(w + 7)$ The area of a rectangle is length times width.

To write the perimeter as a function of the width, we use w as the independent variable and P as the name of the function.

$P(w) = 2w + 2(w + 7)$ The perimeter is the sum of the lengths of the sides.

$P(w) = 4w + 14$ □

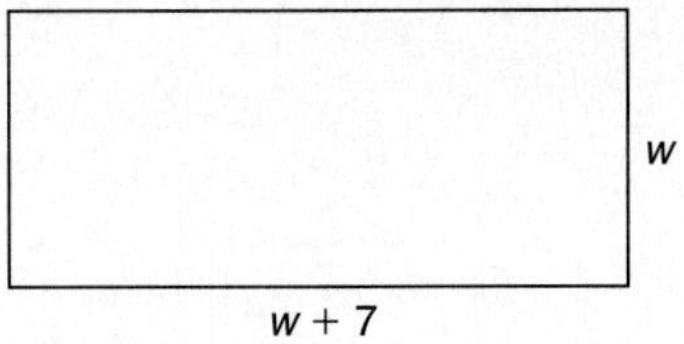

w

$w + 7$

Fig. 3.75

Often an equation in two variables determines one variable as a function of the other. For example, the formula for the circumference of a circle is $C = 2\pi r$. The variable C is given as a function of r. We could write $C(r) = 2\pi r$. The equation $C = 2\pi r$ can be solved for r.

$$C = 2\pi r$$

$$r = \frac{C}{2\pi}$$

We have written r *as a function* of C. We could write $r(C) = \dfrac{C}{2\pi}$.

EXAMPLE 9 Write each variable as a function of the other *if possible.*

(a) $2S + 3T = 1$ (b) $2v^2 - \frac{1}{3}s = 1$

Solutions (a) $2S + 3T = 1$

We solve the equation for S.

$2S = 1 - 3T$

$S = \dfrac{1 - 3T}{2}$ or, in function notation, $S(T) = \dfrac{1 - 3T}{2}$. We have found S as a function of T. Also, we can solve the original equation for T.

$T = \dfrac{1 - 2S}{3}$ and $T(S) = \dfrac{1 - 2S}{3}$

We have written T as a function of S.

(b) $2v^2 - \frac{1}{3}s = 1$

We can find s as a function of v easily.

$6v^2 - s = 3$

$6v^2 - 3 = s$ and $s(v) = 6v^2 - 3$

However, finding v as a function of s is another matter. We solve for v.

$$v^2 = \frac{s+3}{6}$$

$$v = \pm\sqrt{\frac{s+3}{6}} \qquad \text{Square Root Property}$$

Notice that this is shorthand for *two* equations. It means

$$v = \sqrt{\frac{s+3}{6}} \quad \text{or} \quad v = -\sqrt{\frac{s+3}{6}}$$

Since v will have *two* values for each $s > -3$, v *cannot* be a function of s! □

PROBLEM SET 3.5

Warm-ups

In Problems 1 through 4, determine which assignments represent functions. Write the assignments as ordered pairs. See Examples 1 and 2.

1. 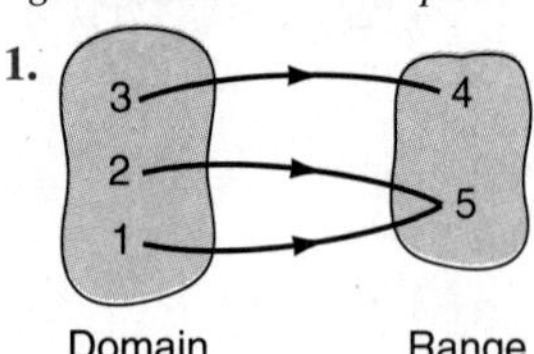

2. 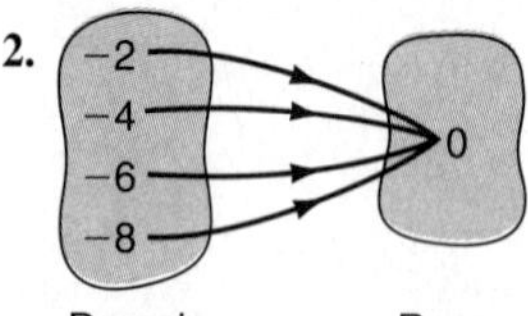

3. 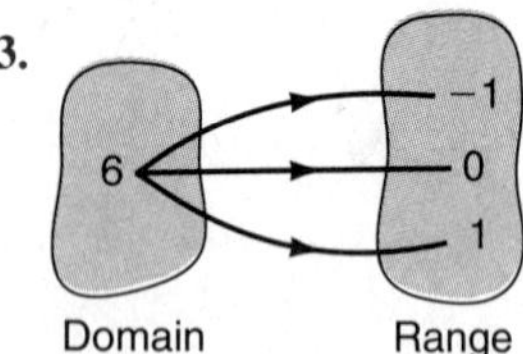

4. 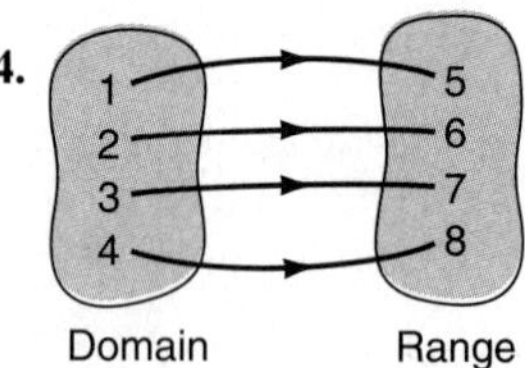

In Problems 5 through 24, if $f(x) = 2x^2 + 3x$ and $g(x) = |x - 1|$, find the indicated function values. See Examples 3 and 4.

5. $f(2)$ **6.** $f(-3)$ **7.** $g(0)$ **8.** $g\left(\frac{1}{2}\right)$

9. $f(\sqrt{2})$ **10.** $g(\pi)$ **11.** $f(t)$ **12.** $g(s)$

13. $f(-1)g(-1)$ **14.** $f(4) - f(2)$ **15.** $\dfrac{g(0)}{f(1)}$ **16.** $[g(-3)]^2$

17. $g(f(0))$ **18.** $f(g(5))$ **19.** $g(a + 2)$ **20.** $f(x - h)$

21. $f(x + h)$ **22.** $g(x + h)$ **23.** $\dfrac{f(x+h) - f(x)}{h}$ **24.** $\dfrac{f(2) - f(a)}{2 - a}$

In Problems 25 through 33, find the natural domain for each function. See Examples 5, 6, and 7.

25. $f(x) = 2x + 8$ **26.** $h(x) = x^2 - 3x + 4$ **27.** $g(x) = \dfrac{7}{x + 4}$

28. $F(x) = \dfrac{x + 2}{(x + 3)(x - 9)}$ **29.** $G(x) = \sqrt{x + 3}$ **30.** $H(x) = \sqrt{x^2 + x - 12}$

31. $k(x) = \sqrt[3]{x}$ **32.** $s(x) = \sqrt[3]{x^2 - 1}$ **33.** $R(x) = |x + 6|$

In Problems 34 and 35, see Example 8.

34. Express the area and circumference of a circle as a function of the radius.

35. Express the perimeter and area of a square as a function of the side.

In Problems 36 through 39, write each variable as a function of the other if possible. See Example 9.

36. $5k - 7j = 35$ **37.** $3x + 2y^2 - 7 = 0$ **38.** $x^2 - 6y + 2 = 0$ **39.** $x^2 + y^2 = 4$

Practice Exercises

In Problems 40 through 46, find the range of the function.

40. Domain is $\{-1, 0, 1\}$; $f(x) = x^3$.

41. Domain is $\{1, 4, 7, 11, 14\}$; $f(x) = x - 7$.

42. Domain is $\{x \mid x \text{ is an odd natural number}\}$; $f(x) = x + 1$.

43. Domain is $\{x \mid x \text{ is an even natural number}\}$; $f(x) = 2x - 1$.

44. Domain is $\{x \mid x \text{ is an integer}\}$; $f(x) = 2x$.

45. Domain is $\{x \mid x \text{ is a real number}\}$; $f(x) = 11$.

46. Domain is $\{x \mid x < 0\}$; $f(x) = -x$.

In Problems 47 through 62, find the domain of the given function.

47. $g(x) = 7x - 11$ **48.** $f(x) = \dfrac{1}{3}x$ **49.** $h(x) = x^4 - 4$ **50.** $g(x) = 3x^3 - 1$

51. $f(x) = \dfrac{1}{x + 8}$ **52.** $h(x) = \dfrac{5}{x - 2}$ **53.** $g(x) = \dfrac{3}{(x - 2)(x + 3)}$ **54.** $h(x) = \dfrac{x}{(x - 2)(x + 1)}$

55. $f(x) = |x - 2|$ **56.** $g(x) = |x + 1|$ **57.** $d(x) = \sqrt{3x + 1}$ **58.** $p(x) = \sqrt{5 - x}$

59. $h(x) = \sqrt{x^2 - 9}$ **60.** $f(x) = \sqrt{x^2 - 2x - 3}$ **61.** $g(x) = \sqrt[3]{x + 1}$ **62.** $f(x) = \sqrt[3]{1 - x^2}$

In Problems 63 through 70, evaluate each expression if $g(x) = \dfrac{1}{x + 1}$.

63. $g(0)$ **64.** $g(1)$ **65.** $g(-1)$ **66.** $2g(5)$

67. $3g(7)$ **68.** $g\left(\dfrac{1}{2}\right)$ **69.** $g(1)g(2)$ **70.** $g\left(-\dfrac{3}{2}\right)$

In Problems 71 through 78, evaluate each expression if $f(x) = x^2 - x + 3$ *and* $g(x) = x + 5$.

71. $f(3) + g(3)$ **72.** $f(3) - g(3)$ **73.** $f(3)g(3)$ **74.** $\dfrac{f(3)}{g(3)}$

75. $f(g(3))$ **76.** $g(f(3))$ **77.** $g(x^2)$ **78.** $[g(x)]^2$

In Problems 79 through 86, find $\dfrac{g(t + h) - g(t)}{h}$ *for each function when* $h \neq 0$.

79. $g(x) = 2x + 5$ **80.** $g(x) = 4 - x$ **81.** $g(x) = x^2 + 2x$ **82.** $g(x) = x^2 - 5$

83. $g(t) = \dfrac{1}{t}$ **84.** $g(t) = \dfrac{1 + t}{1 - t}$ **85.** $g(z) = z^3$ **86.** $g(y) = 1 - y^3$

87. A manufacturer produces scarves at a cost of \$22 per scarf. Fixed costs are \$500 per day. Express cost as a function of number of scarves sold. What would it cost to make 25 scarves?

88. The Red and White Taxi Company charges a flat rate of \$1.50 and an additional charge of \$0.25 for each mile. Express the total fare as a function of length of the ride. What is the fare for a 10-mile ride?

89. Tina Bigelow is fencing a rectangular pen for her horses. She plans to use the barn as one side. If she has 100 feet of fencing, express the area of the pen as a function of its width. Find the area of the pen if the width is (a) 10 feet, (b) 20 feet, (c) 30 feet. Guess what width would give a maximum area.

90. An open candy box is to be made from an 8 by 10 inch rectangular piece of cardboard by cutting out equal squares with side of x in from each corner and folding up the sides. Express the volume of the box as a function of x. Approximate (to the nearest hundredth of a cubic inch) the volume of the box if the length of a side of the square is $\sqrt{2}$ inches. Calculate the volume if x is 4.5 inches.

91. Two cars leave the Detroit Airport at 10:00 A.M. traveling in opposite directions. One car is traveling at a constant rate of 45 miles per hour, and the other at a constant rate of 50 miles per hour. Express the distance between them as a function of time. How far apart are the cars at 3:00 P.M. that same afternoon?

92. The top of a 20-foot ladder is sliding down a vertical wall at the rate of $\frac{1}{2}$ feet per second as the base moves away from the wall. Express the distance of the foot of the ladder from the base of the wall as a function of time. Approximate (to the nearest tenth of a foot) the distance of the ladder from the base of the wall after 5 seconds. Assume the ladder starts flat against the wall.

93. Two ships leave San Francisco at the same time, one headed west at 25 miles per hour and the other south at 15 miles per hour. Express the distance between the ships as a function of time. How far apart are the ships after 10 hours?

94. The lighthouse at Tybee Beach is $2\frac{1}{2}$ miles north of the marina. A sailboat at a buoy 10 miles due east of the marina is headed toward the marina at 6 miles per hour. Express the distance between the sailboat and the lighthouse as a function of time. How far from the lighthouse is the sailboat 30 minutes after leaving the buoy?

In Problems 95 and 96, express the volume of the cone shown as a function of its radius.
In Problems 97 and 98, express the volume of the cylinder shown as a function of its height.

95.

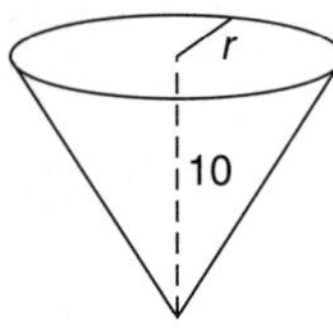

96.

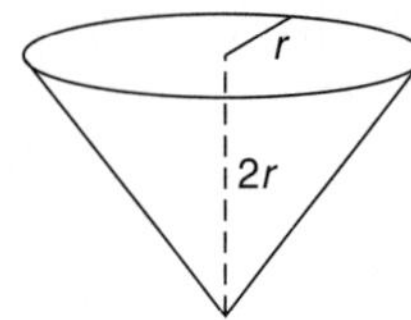

97.

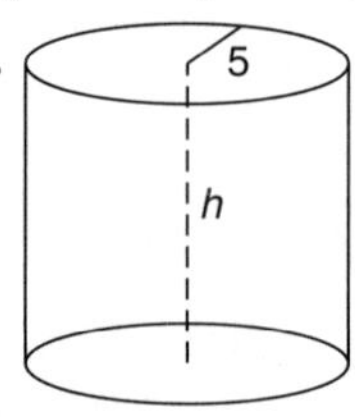

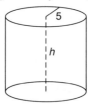

98.

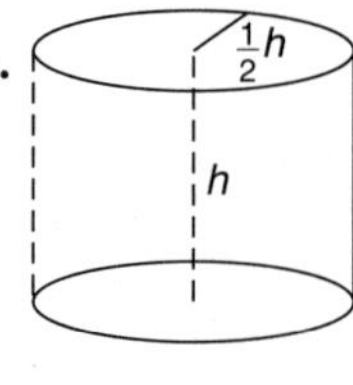

Challenge Problems

99. Find the domain of $f(x) = \dfrac{x + 4}{\sqrt{x^2 - 4}}$.

100. Find the domain of $g(x) = \sqrt{\dfrac{x + 1}{x - 2}}$.

101. Find the domain of $H(x) = \dfrac{\sqrt{x + 1}}{\sqrt{x - 2}}$.

IN YOUR OWN WORDS . . .

102. What is a function?

103. What is the domain of a function?

104. Explain the relationship between x and $f(x)$.

3.6 GRAPHS OF FUNCTIONS

To **graph** a function f means to graph the equation $y = f(x)$ for every x in the domain of the function. So, to graph $f(x) = x^2 + 1$, we graph the equation $y = x^2 + 1$. This graph is shown in Figure 3.76. It is a parabola opening upward with vertex at (0, 1). [Sometimes the vertical axis is labeled the $f(x)$ axis when it is known that only the function f will be displayed.]

In Figure 3.77 notice that elements in the domain are found on the *horizontal* axis, while elements in the range are found on the *vertical* axis.

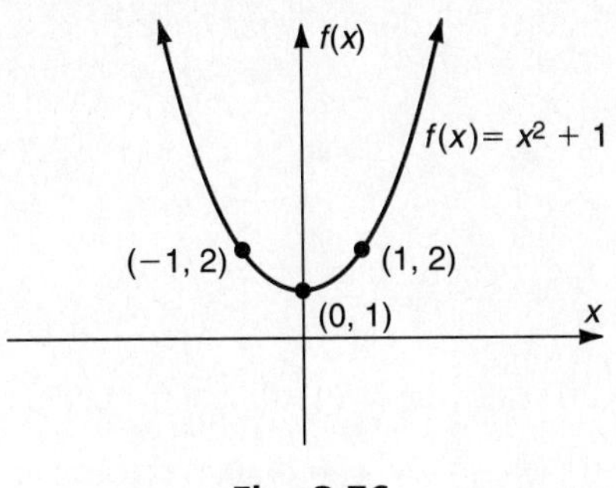

Fig. 3.76

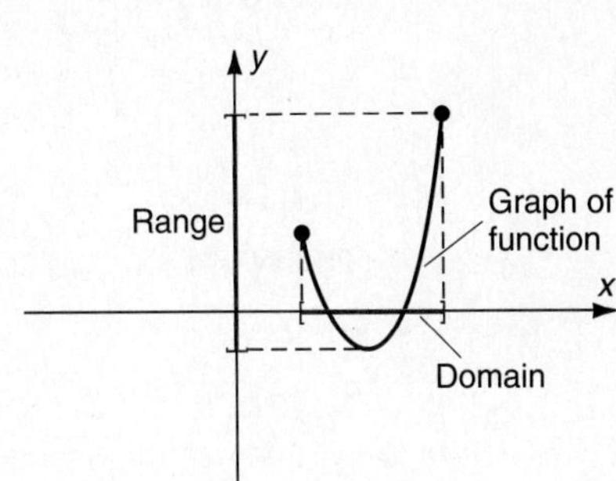

Fig. 3.77

EXAMPLE 1 Give the domain and range for each function graphed in Figure 3.78.

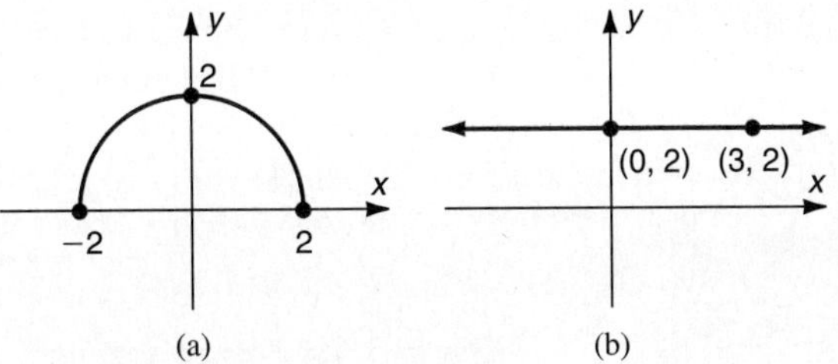

Fig. 3.78

Solutions (a) The points on the graph of f have x-coordinates between -2 and 2 and also include -2 and 2. So the domain of f is $[-2, 2]$. The y-coordinates of points on the graph of f are between 0 and 2, including 0 and 2. So the range of f is $[0, 2]$.

(b) The points on the graph of g use every real number as an x-coordinate. So the domain of g is R. The only y-coordinate for points on the graph of g is 2. So the range of g is $\{2\}$. □

EXAMPLE 2 Use the graph of f (Figure 3.79) to answer each question.

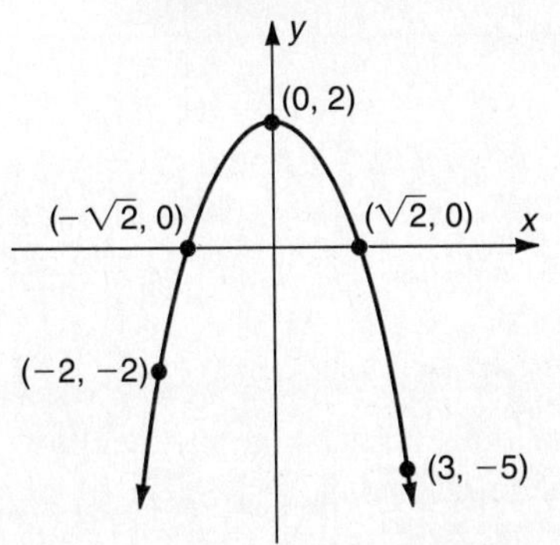

Fig. 3.79

(a) Find $f(3)$.

(b) Find $f(-2)$.

(c) Find $f(0)$.

(d) For what numbers is $f(x) = 0$?

(e) For what numbers is $f(x) > 0$?

(f) Is $f(1) < f(4)$?

Solutions (a) $f(3)$ is the y-coordinate when x is 3.
$f(3) = -5$

(b) $f(-2)$ is the y-coordinate when x is -2.
$f(-2) = -2$

(c) $f(0)$ is the y-coordinate when x is 0.
$f(0) = 2$

(d) $f(x) = 0$ means the y-coordinate is 0. This is true when x is $\pm\sqrt{2}$.

(e) $f(x) > 0$ means the y-coordinates are greater than 0. This is true when x is in the interval $(-\sqrt{2}, \sqrt{2})$.

(f) $f(1)$ is positive and $f(4)$ is negative. So, $f(1)$ is not less than $f(4)$. □

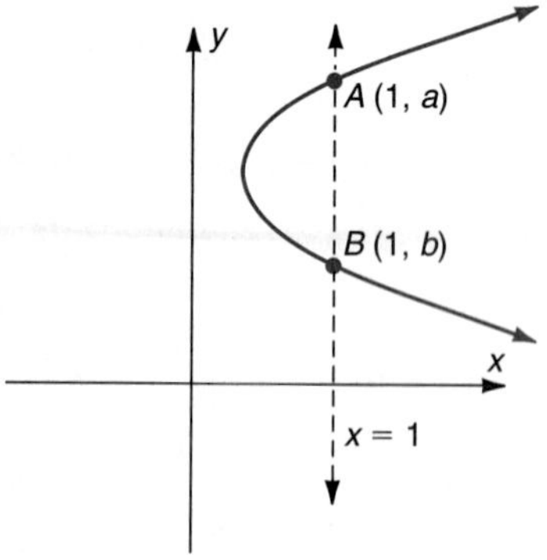

Fig. 3.80

How can we look at a graph and determine whether or not it is the graph of a function? When will one number in the domain be paired with more than one number? Look at the graph in Figure 3.80. Draw the line $x = 1$.
The line crosses the graph at points A and B. The x-coordinate of A and B is 1. However, A and B have different y-coordinates, say a and b. The number 1 in the domain is paired with *two* numbers, a and b. So the graph in Figure 3.80 cannot be the graph of a function.

Vertical Line Test

If every vertical line crosses a graph at no more than one point, then the graph is the graph of a function.

EXAMPLE 3 Is each graph in Figure 3.81 the graph of a function?

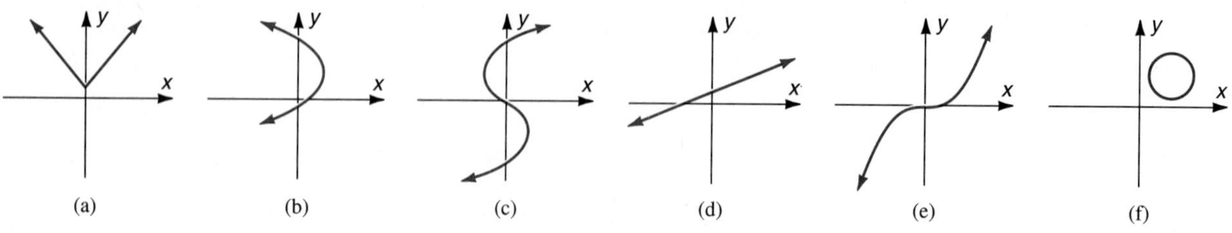

Fig. 3.81

Solutions The graphs (a), (d), and (e) are graphs of functions because they pass the vertical line test, but (b), (c), and (f) are not graphs of functions because the vertical lines shown in Figure 3.82 cross each graph at more than one point.

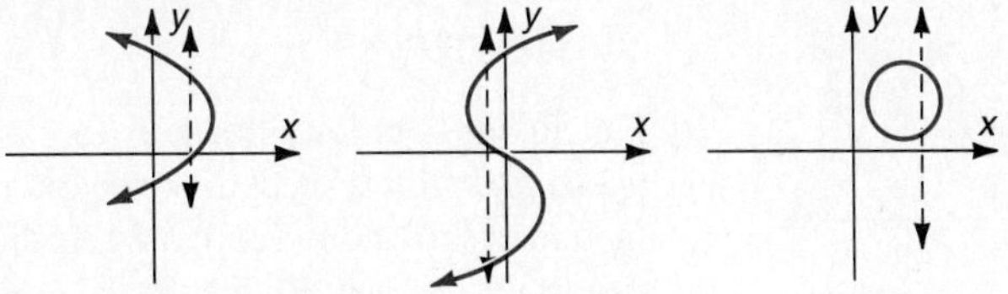

(b) (c) (f)

Fig. 3.82

Increasing and Decreasing Functions

The graph of a function helps us describe the behavior of a function. We usually look at graphs from left to right.

> **Definitions: Increasing, Decreasing, and Constant Functions**
>
> Let x_1 and x_2 be any two numbers from an interval in the domain of a function f.
>
> If $x_1 < x_2$ implies $f(x_1) < f(x_2)$, then f is **increasing** on the interval.
> If $x_1 < x_2$ implies $f(x_1) > f(x_2)$, then f is **decreasing** on the interval.
> If $f(x_1) = f(x_2)$, then f is **constant** on the interval.

Generally speaking, a function is increasing when it is rising from left to right, and decreasing when it is falling from left to right.

EXAMPLE 4 The graph of f is given in Figure 3.83. Determine the intervals over which f is increasing, decreasing, or constant.

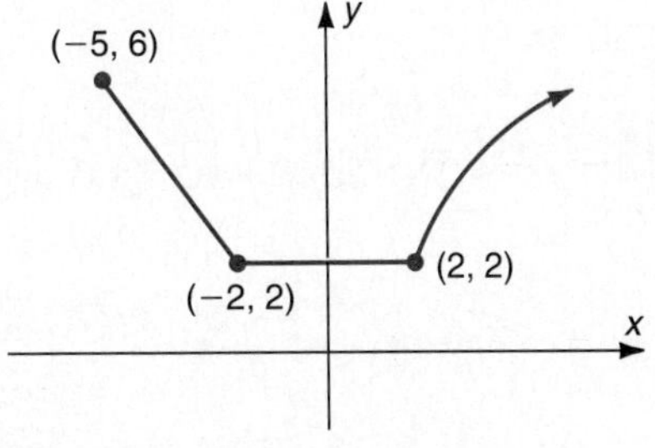

Fig. 3.83

Solution The function is decreasing over $[-5, -2]$, constant over $[-2, 2]$, and increasing over $[2, +\infty)$.

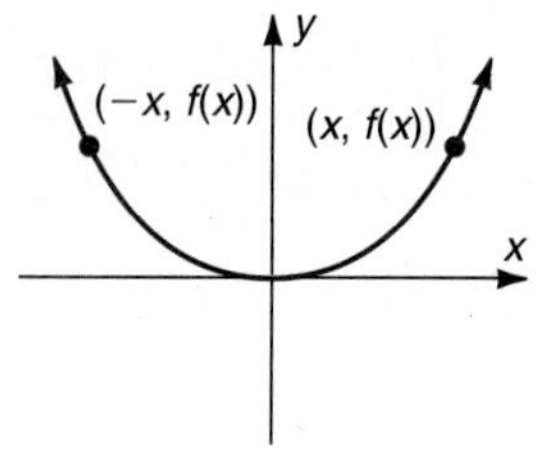

Symmetric with respect to the y-axis

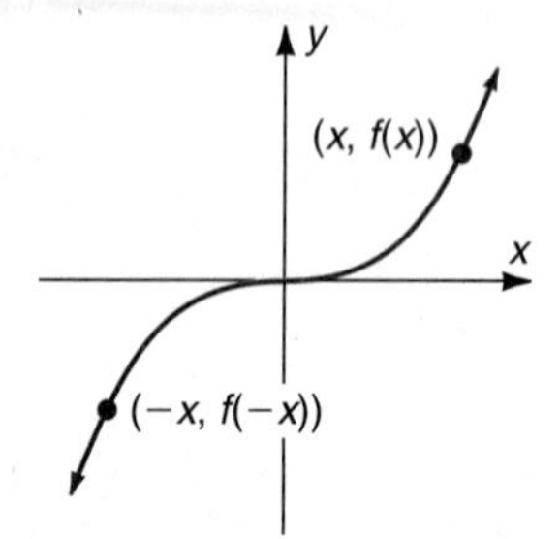

Symmetric with respect to the origin

Fig. 3.84

Symmetry

Functional notation allows us to express the idea of symmetry with respect to the y-axis and with respect to the origin in a convenient way. Consider the graphs of the functions shown in Figure 3.84. If the graph is symmetric with respect to the y-axis, then $f(x) = f(-x)$. Remember that in functional notation $f(x)$ is the y-coordinate for x. So, $f(-x)$ is the y-coordinate for $-x$. Similarly, if a graph is symmetric with respect to the origin, then $f(-x) = -f(x)$.

With functions, we do not have symmetry with respect to the x-axis. The graphs of *equations* are sometimes symmetric with the x-axis, but the graphs of *functions* never are. See Problem 70, in the exercises.

Definitions: Even and Odd Functions

A function f is **even** if $f(-x) = f(x)$ for every x in the domain of f.
The graph of an even function is symmetric with respect to the y-axis.

A function f is **odd** if $f(-x) = -f(x)$ for every x in the domain of f.
The graph of an odd function is symmetric with respect to the origin.

It is not necessary that a function be even or odd. In fact, most functions are neither even nor odd.

EXAMPLE 5 Determine whether each of these functions is even, odd, or neither odd nor even.

(a) $f(x) = x^2$ (b) $g(x) = 2x^3 - 3x$ (c) $h(x) = x^2 + 3x - 5$

Solutions (a)
$$\begin{aligned} f(x) &= x^2 \\ f(-x) &= (-x)^2 \\ &= x^2 \\ &= f(x) \end{aligned}$$
So, $f(-x) = f(x)$ and f is even.

(b)
$$\begin{aligned} g(x) &= 2x^3 - 3x \\ g(-x) &= 2(-x)^3 - 3(-x) \\ &= -2x^3 + 3x \\ &= -g(x) \end{aligned}$$
So, $g(-x) = -g(x)$ and g is odd.

(c)
$$\begin{aligned} h(x) &= x^2 + 3x - 5 \\ h(-x) &= (-x)^2 + 3(-x) - 5 \\ &= x^2 - 3x - 5 \end{aligned}$$
So, $h(-x) \neq h(x)$. Since $-h(x) = -x^2 - 3x + 5$, $h(-x) \neq -h(x)$.
So h is neither odd nor even. □

Notice in Example 5 that if a function has a polynomial rule and *each term* of the polynomial has *odd* degree, the function is odd, and if *each term* has *even* degree, the function is even. If a polynomial function has mixed degrees, even and odd, the function is *neither* even nor odd. This is a convenient property to remember, however, it applies only to functions that have a polynomial rule.

GRAPHING CALCULATOR BOX

Using Zoom

The zoom feature is convenient for changing the range of the viewing window to allow us to "zoom in on" an interesting part of a graph. Let's approximate the points of intersection of the graph of $y = \sqrt{4 - x^2}$ and the graph of $y = x^2$.

First graph both functions on the same screen with a window of $[-4, 4]$ by $[-4, 4]$. We will use the box feature of the zoom feature to approach each point of intersection.

TI-81

Press [ZOOM] and a menu appears. Select 1: Box by pressing [1] or [ENTER]. Notice that this also returns us to the graph. Next we make a box with the editing arrows. Press the right arrow key a few times and then the up arrow once or twice and notice a + moving like a cursor. Position the + approximately at the point (1, 1). Press [ENTER]. One corner of a box is fixed. We now fix the corner diagonally across. Arrow up to about (1, 2) and then right to about (2, 2). We see a little box. Press [ENTER], and the full screen becomes that box!

CASIO fx-7700G

Press [F2] which brings up the zoom menu. Press [F1] to select the box feature. The box feature on the Casio works exactly like the one on the TI-81.

The [TRACE] feature can be used to estimate the points of intersection. Or, to increase accuracy, we can use box again, and again, and again until [TRACE] gets us the desired accuracy. It shouldn't take too many tries to approximate the rightmost point of intersection as (1.2496, 1.5616), which is correct to five significant digits.

PROBLEM SET 3.6

Warm-ups

In Problems 1 through 4, use the graph of f to answer each question. See Examples 1 and 2.

(a) What is the domain and range of f?
(b) Find $f(1)$.
(c) For what numbers is $f(x) = 0$?
(d) For what numbers is $f(x) < 0$?
(e) Is $f(1) < f(4)$?

1.

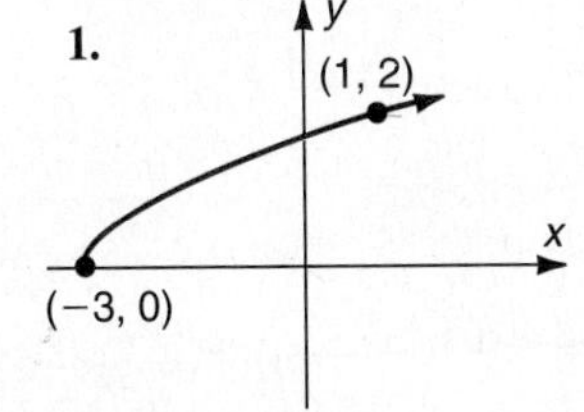

2.

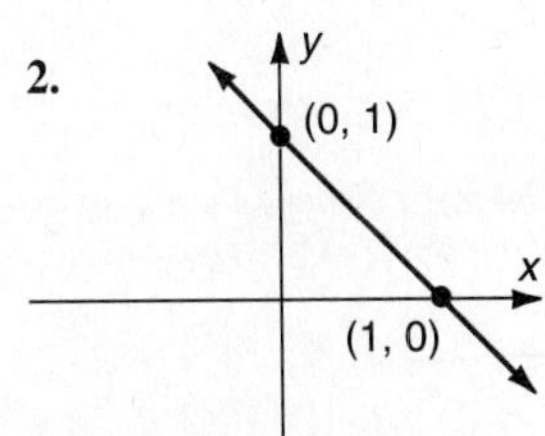

3.

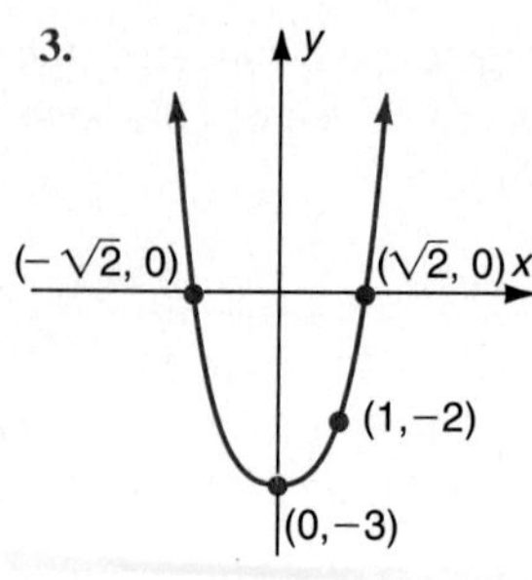

4.

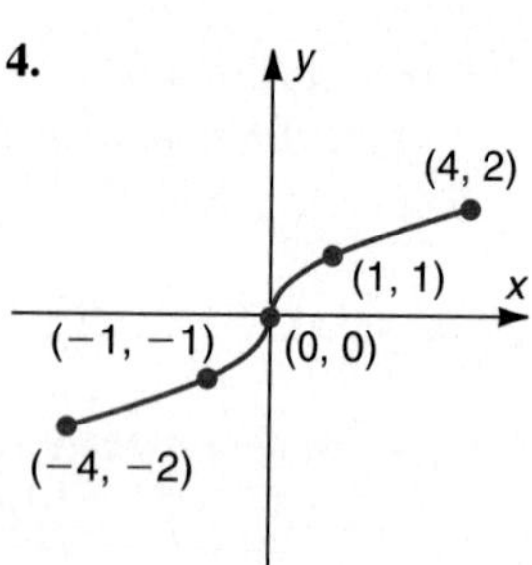

In Problems 5 through 8, does each graph represent the graph of a function? See Example 3.

5.

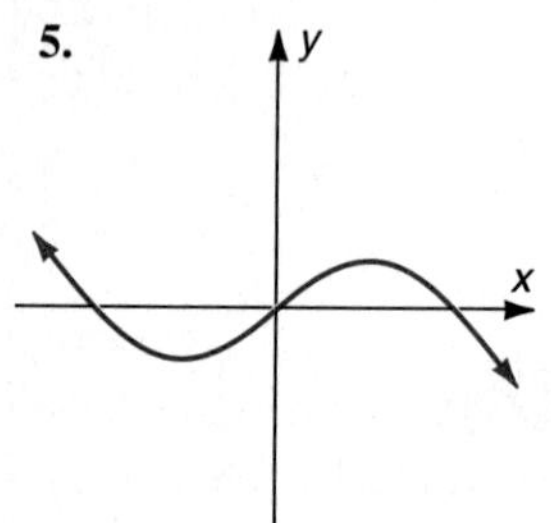

6.

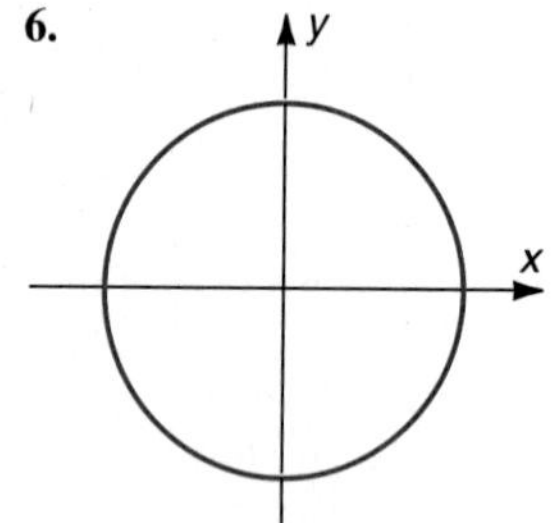

7.

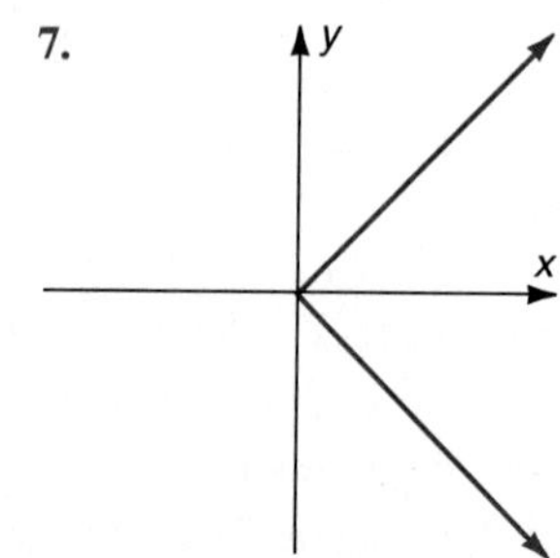

8.

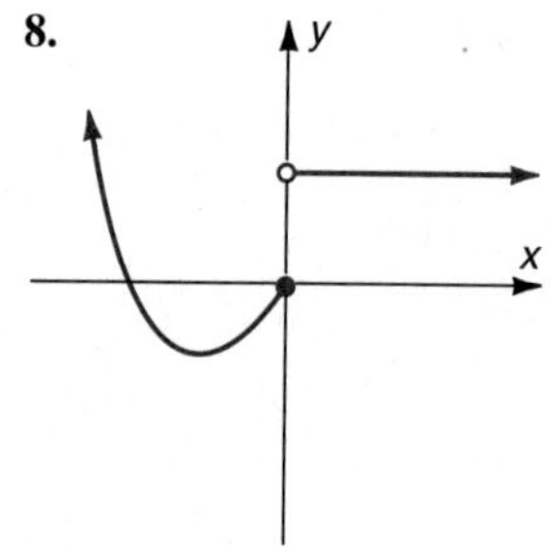

In Problems 9 and 10, give the intervals over which the functions shown are increasing, decreasing, or constant. See Example 4.

9.

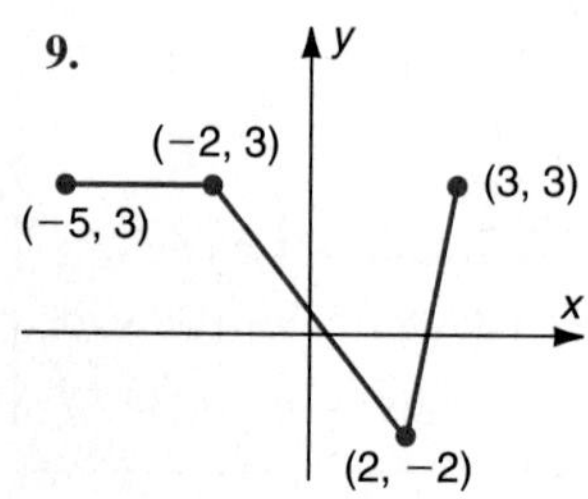

10.

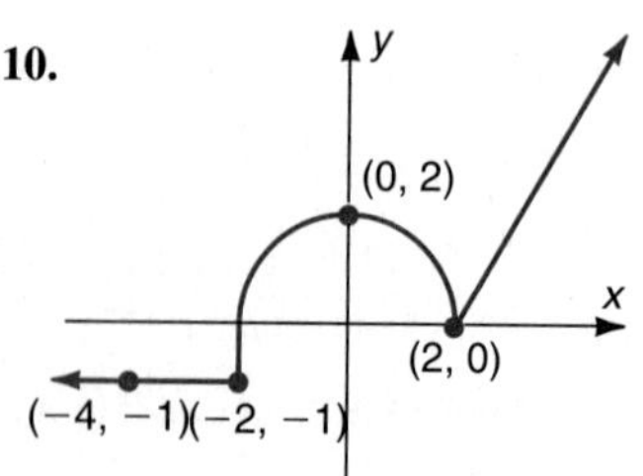

In Problems 11 through 14, determine whether each function is even, odd, or neither even nor odd. See Example 5.

11. $f(x) = |x|$ **12.** $f(x) = 3x^4 - 4x + 5$ **13.** $f(x) = x^3 + 3x$ **14.** $f(x) = \sqrt{1 - x^2}$

Practice Exercises

In Problems 15 through 18, find the domain and range of each function from its graph.

15.

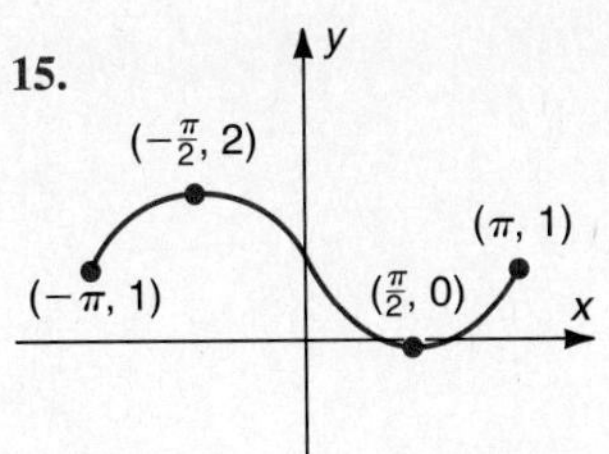

16.

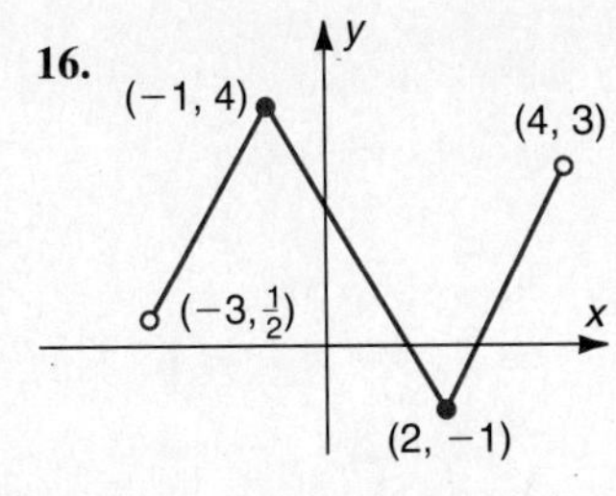

17.

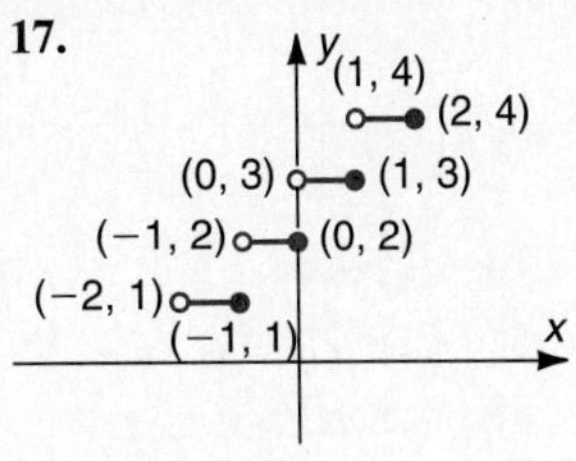

18.

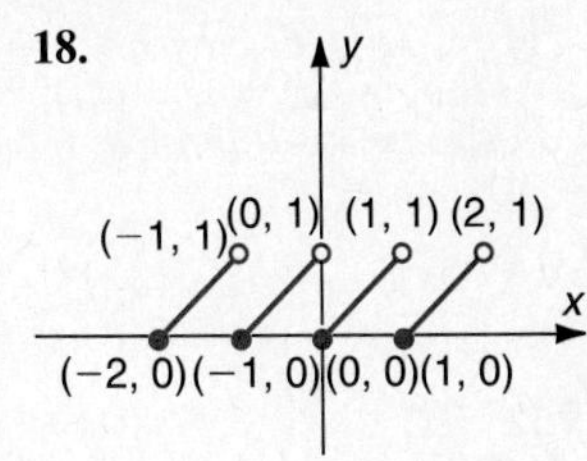

In Problems 19 through 28, use the graph of f shown in Figure 3.85 to answer each question.

19. Give the intervals over which the graph is increasing.

20. Give the intervals over which the graph is decreasing.

21. Find where $f(x) = 0$.

22. Find where $f(x) > 0$.

23. Is f even, odd, or neither even nor odd?

24. Give the domain and range for f.

25. Is $f(1) < f(-5)$?

26. Find $|f(-1)|$.

27. Find $|f(0) - f(-5)|$.

28. Find $\dfrac{f(2) - f(1)}{2 - 1}$

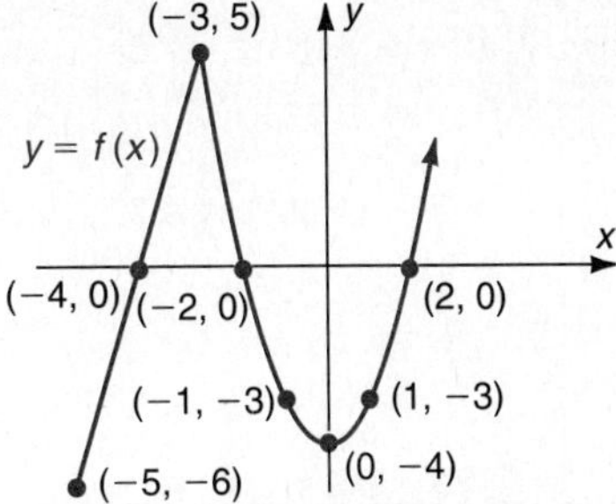

Fig. 3.85

In Problems 29 through 38, use the graph of f shown in Figure 3.86 to answer each question.

29. Give the intervals over which the graph is increasing.

30. Give the intervals over which the graph is decreasing.

31. Find where $f(x) = x$.

32. Find where $f(x) > 0$.

33. Is f even, odd, or neither even nor odd?

34. Give the domain and range for f.

35. Is $f(1) < f(-5)$?

36. Find $|f(-1)|$.

37. Find $|f(-1) - f(-5)|$.

38. Find $\dfrac{f(5) - f(2)}{5 - 2}$.

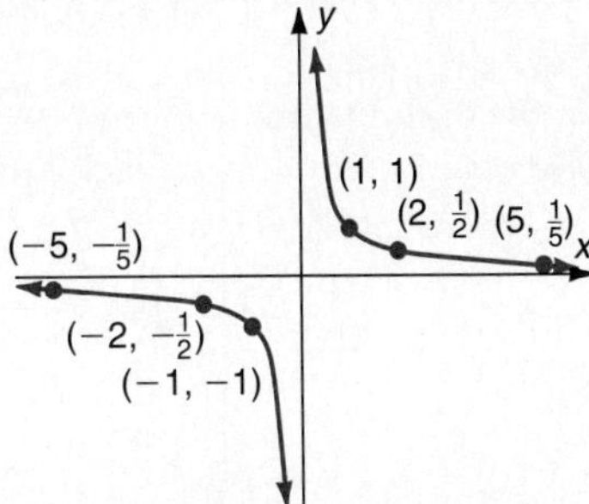

Fig. 3.86

In Problems 39 through 48, use the graph of g shown in Figure 3.87 to answer each question.

39. Find $g(p)$.

40. Find $g(q)$.

41. Is p positive or negative?

42. Is q positive or negative?

43. Is $g(p)$ positive or negative?

44. Is $g(q)$ positive or negative?

45. Give the intervals over which g is decreasing.

46. Give the intervals over which g is increasing.

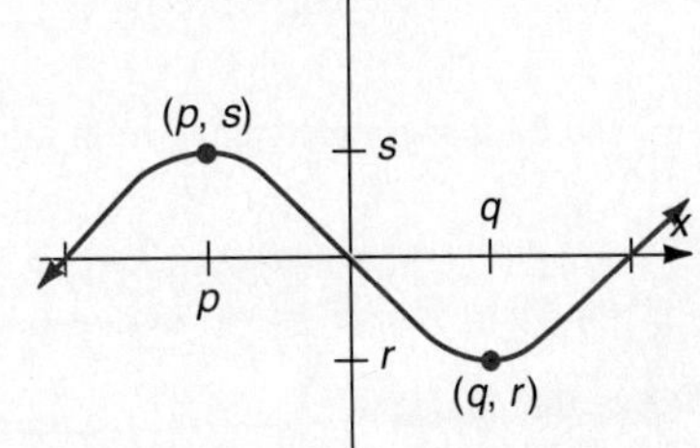

Fig. 3.87

47. Is $g(0) < g(p)$?

48. Is $g(p) - g(q) > 0$?

In Problems 49 through 60, determine whether each function is even, odd, or neither even nor odd.

49. $f(x) = x^2 + 2$

50. $f(x) = 4 - x^6$

51. $f(x) = |x| + 2$

52. $f(x) = |x| - 5$

53. $f(x) = |x + 3|$

54. $f(x) = |5 - x|$

55. $f(x) = \sqrt{x^2 + 4}$

56. $f(x) = \sqrt{7 + 3x^4}$

57. $f(x) = \sqrt[3]{x}$

58. $f(x) = \sqrt[3]{5x}$

59. $f(x) = x^{2/3}$

60. $f(x) = x^{3/5}$

61. The City Radio Company believes that dollars spent on advertising will result in more sales. The relationship between number of radios sold, $N(x)$, and dollars spent on advertising, x, is shown by the graph in Figure 3.88. When are sales increasing? When are sales decreasing?

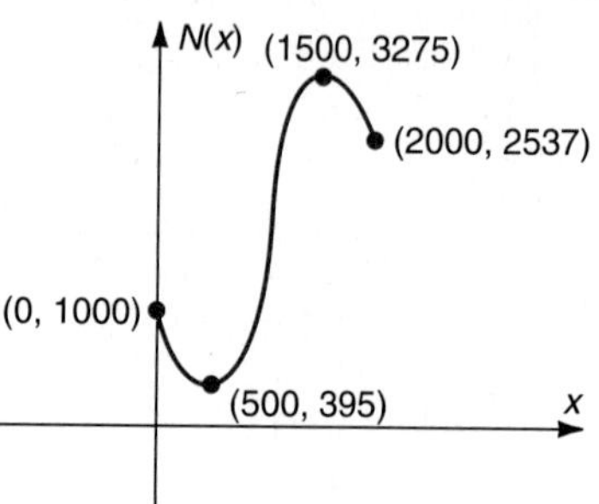

Fig. 3.88

62. A wound was treated with a medicine to reduce the bacteria present. The number of bacteria, $N(t)$, remaining t hours after the medicine was applied is shown by the graph in Figure 3.89. When is the number of bacteria increasing? When is the number of bacteria decreasing?

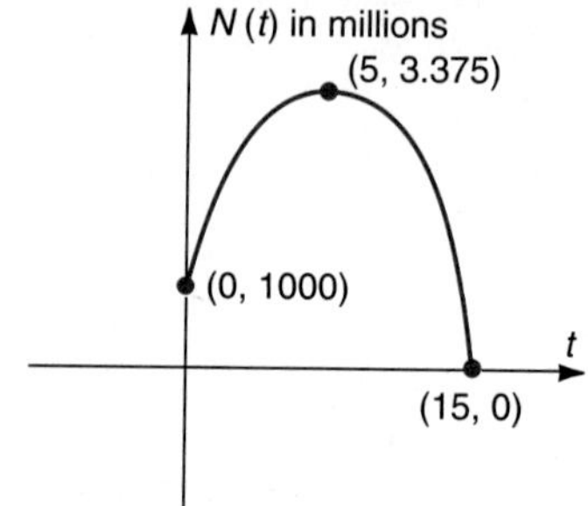

Fig. 3.89

In Problems 63 and 64, consider the graph of the function $G(x) = 8x - x^2$ shown in Figure 3.90.

63. Estimate the area of the shaded region in Figure 3.90 by adding the areas of the *lower rectangles* as indicated in Figure 3.91.

64. Estimate the area of the shaded region in Figure 3.90 by adding the areas of the *upper rectangles* as indicated in Figure 3.92.

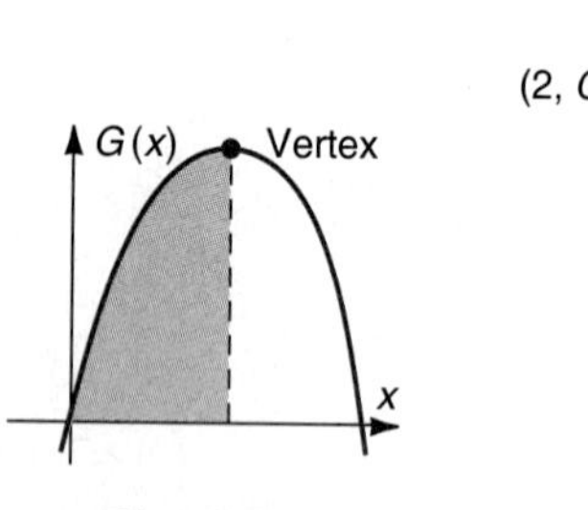

Fig. 3.90

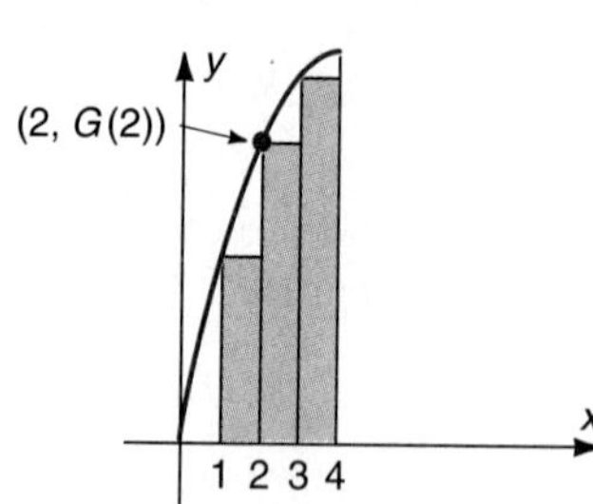

Fig. 3.91

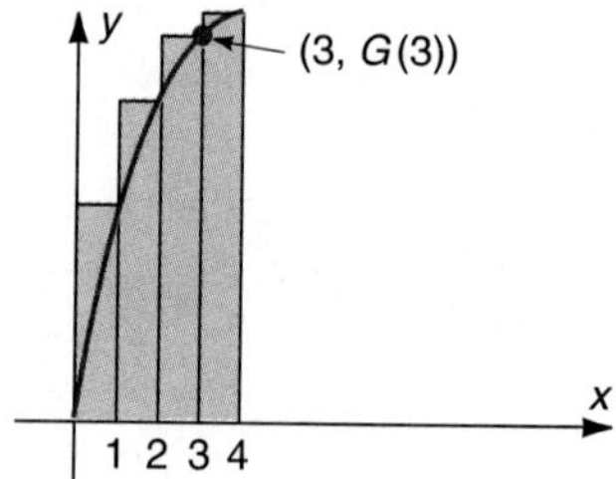

Fig. 3.92

65. Rework Problems 63 and 64 using a rectangle width of $\frac{1}{2}$ unit instead of 1 unit. Draw a conclusion.

66. Suppose f is an *odd* function. If $g(x) = \dfrac{1}{f(x)}$, is g odd, even, or neither? Prove the conjecture.

67. Suppose f is an *odd* function and g is an *even* function. If $h(x) = \dfrac{f(x)}{g(x)}$ is h odd, even, or neither? Prove the conjecture.

Challenge Problems

68. Suppose f is an *even* function. If $g(x) = f(f(x))$, is g odd, even, or neither? Prove the conjecture.

69. Suppose f is an *odd* function. If $g(x) = f(f(x))$, is g odd, even, or neither? Prove the conjecture.

70. Prove that the graph of a function cannot be symmetric with the x-axis.

71. Consider the graph of the function $f(x) = ax^2 + bx + c$; $a \neq 0$. Prove that the vertex of this parabola has coordinates $\left[-\frac{b}{2a}, f\left(-\frac{b}{2a}\right)\right]$

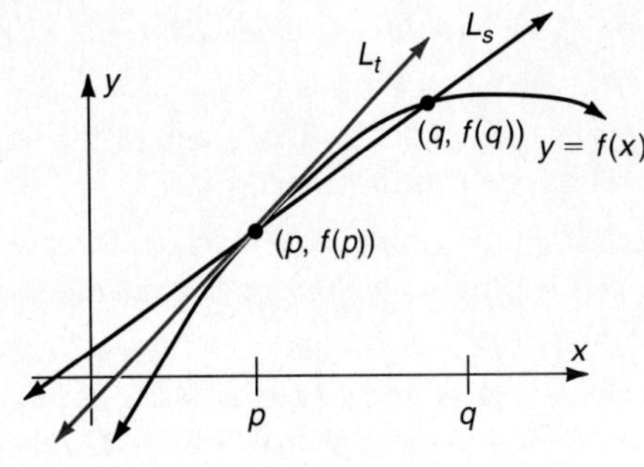

Fig. 3.93

72. The line L_s superimposed on the graph of the function f in Figure 3.93 is called a **secant line.** The line L_t is called the **tangent line to f at $(p, f(p))$.**

(a) What is the relationship of the expression $\dfrac{f(q) - f(p)}{q - p}$ to the line L_s?

(b) If the number p remains fixed but the number q becomes closer and closer to p, what is the relationship of the line L_s to the line L_t?

(c) If q gets "close" to p as described in part (b), what is the relationship between $\dfrac{f(q) - f(p)}{q - p}$ and the tangent line L_t?

For Graphing Calculators

For all points of intersection correct to five significant digits of each pair of graphs.

73. $y = \sqrt{x + 1}$; $y = x^3$

74. $y = -\sqrt{4 - x^2}$; $y = \sqrt{11}x$

75. $y = x^2 - x + 1$; $y = \dfrac{0.42331x + 1.99863}{1.98557}$

IN YOUR OWN WORDS . . .

76. Discuss the concept of even and odd functions.

77. Compare the domain and the range of a function.

78. If the domain and the rule of a function are specified, explain why it is unnecessary to specify the range.

CHAPTER SUMMARY

GLOSSARY

Cartesian coordinate system: A pair of perpendicular number lines intersecting at the zero of both lines. The point of intersection is called the **origin.**

Dependent variable: A letter representing elements from the range of a function, such as y in the equation $y = f(x)$.

Domain of a function: The specified input set of the function or, if not specified, the largest subset of real numbers for which the rule has meaning.

Function: A rule that assigns to each member in one set (the *domain*) exactly one member from another set (the *range*).

Graph of an equation in two variables: The graph of the solution set of the equation.

Graph of a function f: The graph of the equation $y = f(x)$ for every x in the domain of f.

Independent variable: A letter representing an arbitrary element from the domain of a function, such as x in the equation $y = f(x)$.

Linear equation in two variables: An equation that can be written in the form $Ax + By = C$, where A and B are not both 0.

Ordered pair: A pair of numbers written in the form (x, y). The first number is called the **x-coordinate,** and the second number is called the **y-coordinate.**

Parallel lines: Lines that have the same slope and do not intersect.

Perpendicular lines: Lines that intersect at right angles and have slopes whose product is -1.

Range of a function: The set of all function values.

x-intercept: The x-coordinate of a point at which a graph crosses the x-axis.

y-intercept: The y-coordinate of a point at which a graph crosses the y-axis.

DISTANCE FORMULA

The distance between two points (x_1, y_1) and (x_2, y_2) is given by the formula $d = \sqrt{(x_2 - x_1)^2 + (y_2 - y_1)^2}$.

MIDPOINT FORMULA

The point halfway between the points (x_1, y_1) and (x_2, y_2) is $\left(\dfrac{x_1 + x_2}{2}, \dfrac{y_1 + y_2}{2}\right)$.

SLOPE FORMULA

The slope of the line through the points (x_1, y_1) and (x_2, y_2) is given by the formula $m = \dfrac{y_2 - y_1}{x_2 - x_1}$, $x_2 \neq x_1$.

TO FIND INTERCEPTS OF A GRAPH

1. To find the x-intercept(s), replace y with 0 and solve for x.
2. To find the y-intercept(s), replace x with 0 and solve for y.

SYMMETRY OF THE GRAPH OF AN EQUATION

1. The graph of an equation is symmetric with respect to the x-axis if replacing y with $-y$ in the equation results in an equivalent equation.
2. The graph of an equation is symmetric with respect to the y-axis if replacing x with $-x$ in the equation results in an equivalent equation.
3. The graph of an equation is symmetric with respect to the origin if replacing x with $-x$ and y with $-y$ in the equation results in an equivalent equation.

VERTICAL AND HORIZONTAL LINES

The graph of the equation $x = p$ is a *vertical* line with x-intercept of p. It has no defined slope.

The graph of the equation $y = q$ is a *horizontal* line with y-intercept of q. It has slope 0.

TO WRITE AN EQUATION OF A LINE THAT HAS SLOPE

1. Find m, the slope of the line.
2. Find (x_1, y_1), the coordinates of any point on the line.
3. Substitute into the point-slope formula $y - y_1 = m(x - x_1)$.
4. Write the equation in standard or slope-intercept form, whichever is preferred.

FORMS OF THE LINEAR EQUATION IN TWO VARIABLES

STANDARD FORM $Ax + By = C$ (A and B not both zero)

SLOPE-INTERCEPT FORM $y = mx + b$

POINT-SLOPE FORM $y - y_1 = m(x - x_1)$

THE CIRCLE

The graph of $(x - h)^2 + (y - k)^2 = r^2$ is a circle of radius r ($r \geq 0$) with its center at the point (h, k). If the circle has its center at the origin, the equation becomes $x^2 + y^2 = r^2$.

THE PARABOLAS

The graph of $y = a(x - h)^2 + k$ is a parabola with vertex at the point (h, k).
It opens upward if a is positive, and downward if a is negative.
Its axis of symmetry is the line $x = h$.
The graph of $x = a(y - k)^2 + h$ is a parabola with vertex at the point (h, k).
It opens to the right if a is positive, and to the left if a is negative.
Its axis of symmetry is the line $y = k$.

VERTICAL LINE TEST

If every vertical line crosses a graph at no more than one point, then the graph is the graph of a function.

INCREASING, DECREASING, CONSTANT FUNCTIONS

Let x_1 and x_2 be any two numbers from an interval in the domain of a function f.

If $x_1 < x_2$ implies $f(x_1) < f(x_2)$, then f is **increasing** on the interval.
If $x_1 < x_2$ implies $f(x_1) > f(x_2)$, then f is **decreasing** on the interval.
If $f(x_1) = f(x_2)$, then f is **constant** on the interval.

EVEN AND ODD FUNCTIONS

A function f is *even* if $f(x) = f(-x)$ for every x in the domain of f.
The graph of f is symmetric with respect to the y-axis.
A function f is *odd* if $f(-x) = -f(x)$ for every x in the domain of f.
The graph of f is symmetric with respect to the origin.

REVIEW PROBLEMS

In Problems 1 through 12, find the distance between each pair of points. Also, find the midpoint of the segment joining each pair of points.

1. $(-4, 7), (3, 5)$
2. $(6, 0), (2, 2)$
3. $(9, -5), (0, 5)$
4. $(1, -4), (-3, 6)$
5. $(-3, -7), (8, -5)$
6. $(0, 0), (-2, -2)$
7. $(6, -1), (0, 4)$
8. $(5, -3), (-3, 2)$
9. $(-1, -8), (8, -1)$
10. $(4, 0), (-1, -2)$
11. $(-9, -1), (4, 0)$
12. $(7, -1), (4, -2)$

In Problems 13 through 22, find the slope of the line described.

13. Through $(1, -2)$ and $(4, 0)$
14. Through $(0, 0)$ and $(-2, 3)$
15. x-intercept -3; y-intercept 4
16. The graph of $3x + y = 5$

17. Parallel to the graph of $x + 2y = 6$

18. Perpendicular to the graph of $x = 6$

19. Through $(5, 7)$ and $(5, -3)$

20. Through $(-3, 4)$ and $(0, 4)$

21. Perpendicular to the graph of $y + 2 = 0$

22. Parallel to the graph of $y = \frac{2}{5}x$

In Problems 23 through 32, write an equation of the line satisfying the given conditions.

23. Slope $\frac{1}{3}$; through $(-2, 5)$

24. Containing $(4, 0)$ and $(3, -7)$

25. x-intercept 4; y-intercept -1

26. Slope -5; y-intercept $\frac{2}{5}$

27. Parallel to the graph of $3x + 2y = 0$; containing $(-2, -3)$

28. Through the origin; perpendicular to the graph of $y = 2x$

29. Through $(2, 4)$; perpendicular to the x-axis

30. Through $(8, -7)$; parallel to the x-axis

31. Perpendicular to the graph of $y = \frac{3}{4}x - 7$; containing $(5, 0)$

32. Parallel to the graph of $x = 5$; containing $(-3, -3)$

33. A United 767 is beginning its takeoff run at Indianapolis International Airport. When rolling at 30 knots, full power is applied. For the next 40 seconds, the speed of the plane in knots is given by the equation $S = 30 + 9t$, where t is time in seconds after application of full throttle.

(a) Graph this equation using the horizontal axis as time.
(b) How fast is the plane traveling after 5 seconds? 30 seconds?
(c) How long after application of full throttle will the plane reach its takeoff speed of 120 knots?

In Problems 34 through 36, match each graph in Figure 3.94 with the correct equation.

34. $x^2 + y^2 + 4x - 2y = 0$

35. $2x^2 + 4x = y - 8$

36. $2x + 5y = 10$

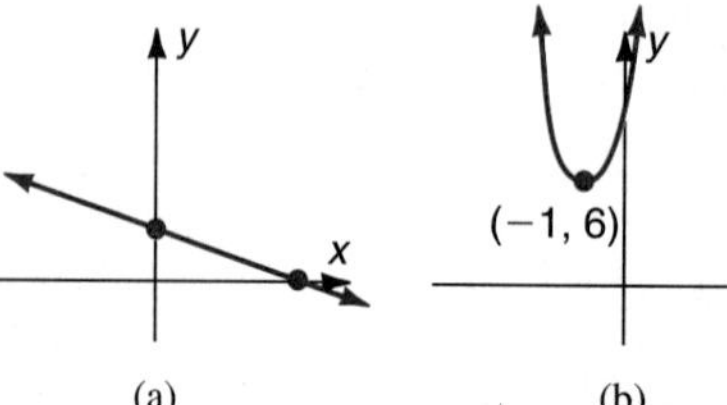

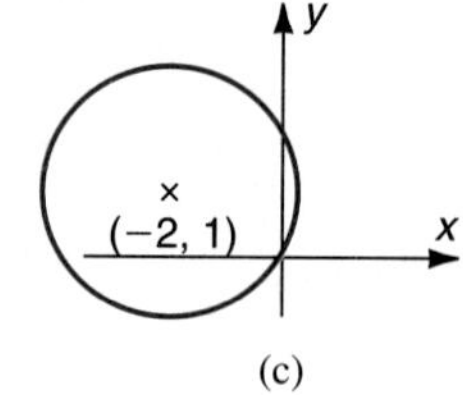

Fig. 3.94

In Problems 37 through 39, analyze each graph to answer the following questions:

(a) What is the domain and range of each function?
(b) Find $f(1)$.
(c) For what numbers is $f(x) > 0$?
(d) Is $f(1) > f(3)$?

37.

38.

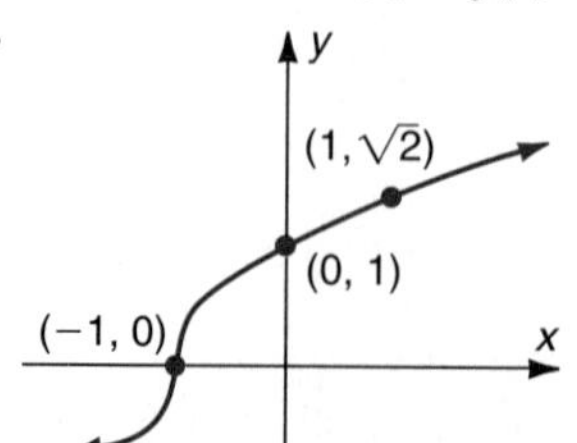

39.

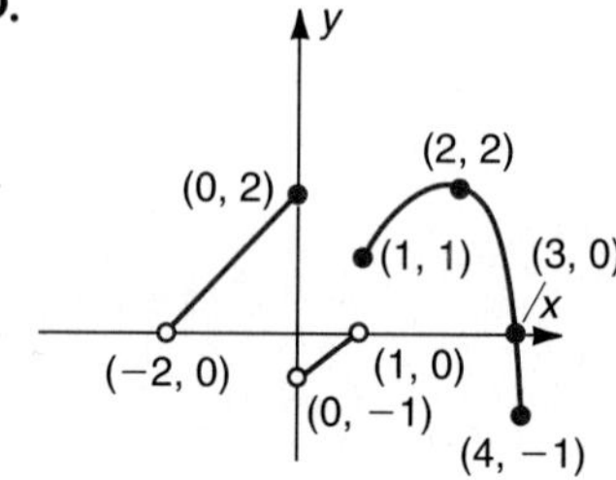

In Problems 40 through 55, sketch the graph of each equation. Label all intercepts and vertices.

40. $3x + y = 3$
41. $x^2 + (y - 1)^2 = 4$
42. $y = (x + 2)^2$
43. $x = y^2 + 3$
44. $3x + 2 = 0$
45. $x^2 - 2x + y^2 = 0$
46. $y = 2x^2 + 8x + 9$
47. $x = 2y^2 + 2y + 3$
48. $y = 4$
49. $x^2 + 2x + y^2 + 6y = -6$
50. $x = 3 - y^2$
51. $y = 4 - x^2$
52. $x^2 + 4x + y^2 + 4y + 12 = 0$
53. $y = \frac{1}{3}x$
54. $y = 2x - x^2$
55. $x = 3 - 4y - y^2$

In Problems 56 through 65, find the domain of the given function.

56. $g(x) = x - 1$
57. $f(x) = \frac{1}{3}x^4 + 3x^3 + 2x - 7$
58. $f(x) = \frac{4}{x + 3}$
59. $h(x) = \frac{5}{(x - 2)(x + 2)}$
60. $f(x) = |2 - x|$
61. $g(x) = |x^2 + 1|$
62. $d(x) = \sqrt{3x - 1}$
63. $p(x) = \sqrt{4 - x}$
64. $g(x) = \sqrt[3]{x^2 + 1}$
65. $f(x) = \sqrt[3]{8 - x^2}$

In Problems 66 through 73, if $g(x) = \frac{1}{x - 1}$, find:

66. $g(0)$
67. $g(-1)$
68. $g(x^2)$
69. $[g(x)]^2$
70. $g(x + h)$
71. $g\left(\frac{1}{2}\right)$
72. $\frac{g(4) - g(2)}{4 - 2}$
73. $\frac{g(x + h) - g(x)}{h}$; $h \neq 0$

In Problems 74 through 83, use the graph of f shown in Figure 3.95 to answer each question.

74. Give the intervals over which the graph is increasing.
75. Give the intervals over which the graph is decreasing.
76. Find where $f(x) = 0$.
77. Find where $f(x) > 0$.
78. Is f even, odd, or neither even nor odd?
79. Give the domain and range for f.
80. Is $f(-1) < f(3)$?
81. Find $|f(0) - f(3)|$.
82. Find $f(f(-1))$.
83. Find $\frac{f(2) - f(3)}{2 - 3}$.

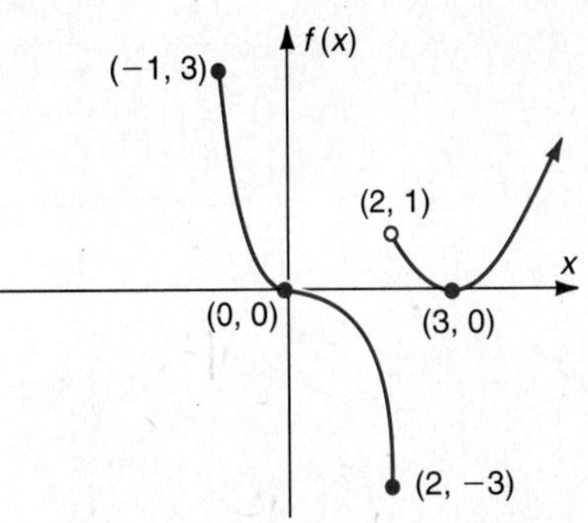

Fig. 3.95

In Problems 84 through 89, determine whether each function is even, odd, or neither even nor odd.

84. $f(x) = x^2 - 1$
85. $f(x) = 8x - 2x^7$
86. $k(x) = |x| + 4$
87. $g(x) = 3x^3 + 5$
88. $F(x) = \sqrt{x^2 - 4}$
89. $w(x) = \sqrt[3]{8x}$

90. The height (in feet) of a ball thrown upward from a building 50 feet high with an initial velocity of 64 feet per second is given by $s(t) = -16t^2 + 64t + 50$, where t is time in seconds. Find how high the ball will go.

91. A company that makes flags has determined that profit is given by $P(x) = -\frac{1}{10}x^2 + 8x + 700$, where x is the number of flags sold. What is the maximum profit? How many flags must be sold to make the maximum profit? Where is profit increasing? Decreasing?

92. The Metro Pool and Chemical Company treats a pool chemically for algae. If the algae content is given by $A(t) = 30t^2 - 900t + 9500$, where t is the number of days after the pool has been treated, how many days after a treatment will the algae content be a minimum? When is the algae content increasing? Decreasing?

In Problems 93 through 100, find $\frac{f(x+h) - f(x)}{h}$ *and simplify for* $h \neq 0$.

93. $f(x) = 7x + 3$

94. $f(x) = x^2$

95. $f(x) = x - x^2$

96. $f(x) = \frac{2}{x+1}$

97. $f(x) = \frac{1-x}{1+x}$

98. $f(x) = \sqrt{x}$ (Rationalize numerator to simplify.)

99. $f(x) = \frac{1}{\sqrt{x}}$

100. $f(x) = x - x^3$

LET'S NOT FORGET . . .

101. Which form is more useful?

$$f(x) = x^2 - x - 6 = (x-3)(x+2) = \left(x - \frac{1}{2}\right)^2 - \frac{25}{4}$$

(a) When solving the equation $f(x) = 0$?
(b) When graphing the function f?

102. $Ax + By = C$; $y = mx + b$; $y - y_1 = m(x - x_1)$
(a) When finding the slope of a line
(b) When graphing a linear equation
(c) When finding an equation of a line

103. Watch the role of negative signs.
Graph the equations $y = x^2 - 2x - 2$ and $y = 2 + 2x - x^2$ on the same coordinate system.

104. From memory.
(a) Find the distance between $(-2, 8)$ and $(10, 3)$.
(b) Find the midpoint between $(3, -7)$ and $(-1, -15)$.
(c) Find the slope of a line containing the points $(7, -3)$ and $(2, 5)$.

105. With a calculator.
Approximate to four decimal places the y-intercepts of $x = 6y^2 + 3y - 2$.

CHAPTER 4

Working with Functions

CONNECTIONS

In the last two sections of Chapter 3 we defined a function, explained the notation, and graphed functions utilizing our knowledge of the graph of an equation. An understanding of functions is essential for further study in mathematics. Those studying in preparation for calculus should realize that calculus is a *study* of functions. Graphing calculators are sometimes called *function machines* because they work with functions. In this chapter, we pursue the idea of function with emphasis on graphing the functions commonly used.

In Chapter 5 we continue the study of functions with emphasis on *polynomial equations.* In Chapter 6, we introduce the *exponential* function and its inverse, the *logarithmic* function.

An in depth look at concepts involved in graphing functions is important. Today, computers and graphing calculators can draw graphs of very complicated functions, freeing us from the drudgery of pencil and paper calculations. However, an understanding of the functions being graphed is necessary for precise use of such technology.

4.1 GRAPHING SOME BASIC FUNCTIONS

In the previous chapter we sketched the graphs of certain functions by the somewhat tedious method of plotting points and using symmetry. In this chapter we continue our study by looking at some particular functions that are so important in mathematics that we should recognize each graph by looking at the rule for the function.

Linear Functions

Definition: Linear Functions

A function of the form $f(x) = mx + b$, where m and b are real numbers, is called a **linear function.** The graph of a linear function is a line.

EXAMPLE 1 Graph each function.

(a) $g(x) = 3x - 1$ (b) $f(x) = x$ (c) $h(x) = 2$

Solutions (a) Recall that the graph of the *function* g is the graph of the *equation* $y = g(x)$. Therefore, we graph the equation $y = 3x - 1$. We recognize that this is the equation of a line in slope-intercept form with slope 3 and y-intercept -1 (Figure 4.1).

(b) Graph $y = x$. The graph is a line through the origin with slope 1 as shown in Figure 4.2. The linear function $f(x) = x$ is called the **identity function** because $f(x)$ always equals x.

(c) Graph the equation $y = 2$. It is the horizontal line shown in Figure 4.3. This is an example of a function of the form $f(x) = k$, where k is constant. Such a function is called a **constant function,** and its graph is a horizontal line.

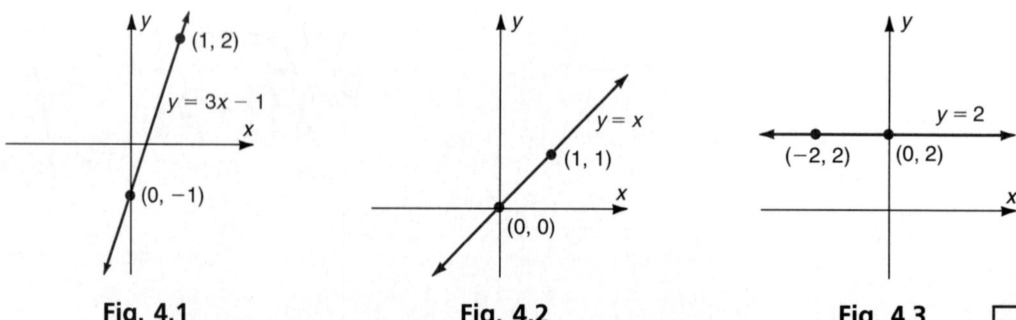

Fig. 4.1 Fig. 4.2 Fig. 4.3 □

EXAMPLE 2 Charles' Law says that when gas is heated at a constant pressure, the relationship between the temperature (in kelvin) of the gas and the volume of the gas is linear. The following information was recorded in a lab.

Temperature (kelvin)	*Volume (cubic centimeters)*
273	100
323	118.3

(a) Express the volume of the gas (V) as a function of temperature (T).

(b) Graph this function.

(c) As the temperature increases, what happens to the volume?

Solutions (a) Since the relationship is given to be *linear,* we can use the two points (273, 100) and (323, 118.3) and write an equation of the line. Notice that we are using temperature as the independent variable and volume as the dependent variable. Using the slope formula, we have

$$m = \frac{118.3 - 100}{323 - 273} = \frac{18.3}{50}$$

Now, the point-slope formula gives

$$V - 100 = \frac{18.3}{50}(T - 273)$$

$$V = 0.366T + 0.082$$

Using functional notation this becomes $V(T) = 0.366T + 0.082$.

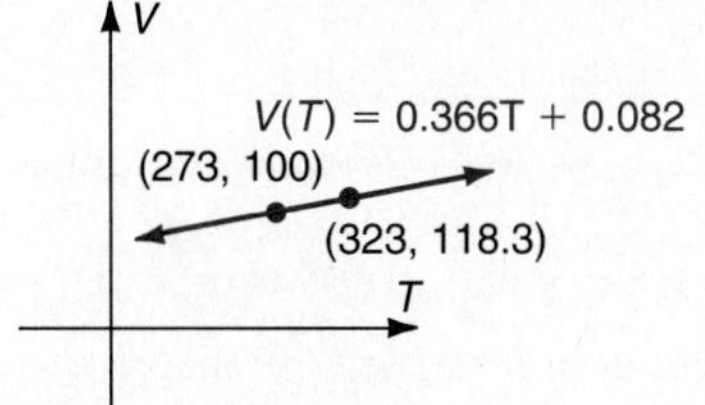

Fig. 4.4

(b) To graph the function, we plot the points (273, 100) and (323,118.3) as in Figure 4.4.

(c) Since the line is increasing, volume increases as temperature increases. □

Quadratic Functions

Definition: Quadratic Functions

A function of the form $f(x) = ax^2 + bx + c$, where a, b, and c are real numbers and $a \neq 0$, is called a **quadratic function.** The graph of a quadratic function is a parabola.

EXAMPLE 3 Graph $f(x) = x^2 - 2x + 3$.

Solution The graph of $y = x^2 - 2x + 3$ is a parabola. We write the equation in the form $y = a(x - h)^2 + k$ in order to find the vertex. To do this, complete the square.

$$y = x^2 - 2x + 3$$

First, group the x-terms together.

$$y = (x^2 - 2x) + 3$$

Then complete the square by taking one-half the coefficient of x and squaring it. Add and subtract this number.

$$y = (x^2 - 2x + 1 - 1) + 3$$

$$y = (x - 1)^2 + 2$$

The vertex is $(1, 2)$, and the parabola opens upward. The graph is shown in Figure 4.5. □

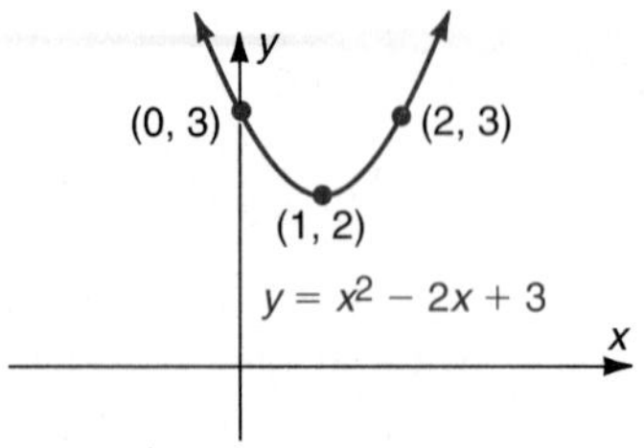

Fig. 4.5

The vertex (h, k) of these parabolas will be either the highest point or the lowest point on the parabola. The function $f(x) = ax^2 + bx + c$ has a **minimum** or **maximum** at h. The **minimum** or **maximum value** is $f(h) = k$.

EXAMPLE 4 Jim Byrne has 300 feet of chain link fencing. He wants to make the largest possible rectangular pen for his dogs. What should the dimensions be?

Solution Let w be the width and l be the length of the pen (both in feet). See the diagram in Figure 4.6. Since the perimeter must be 300, we have

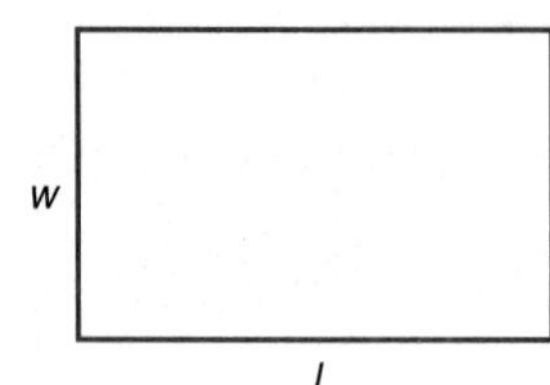

Fig. 4.6

$$2l + 2w = 300$$

$$l + w = 150$$

$$l = 150 - w$$

The largest pen would be the rectangle with the largest area. So, we write the area A as a function of the width.

$$A = lw$$

$$A(w) = (150 - w)w$$

$$= 150w - w^2 = -w^2 + 150w$$

This is a quadratic function. Its graph is a parabola opening *downward* as shown in Figure 4.7.

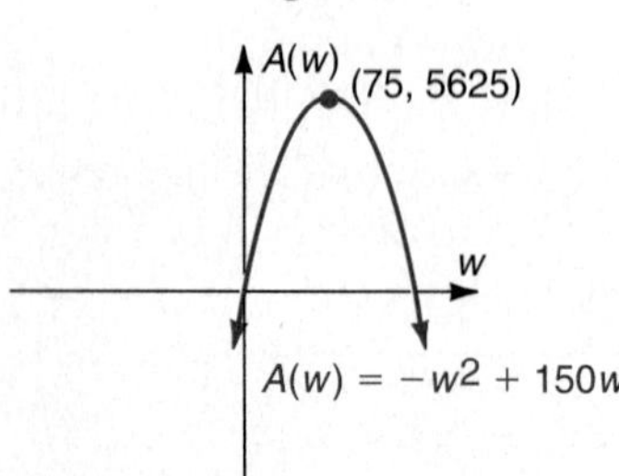

Fig. 4.7

The largest area will occur where the function A has its largest value. This is at the vertex of the parabola. We complete the square to find the vertex.

$$\begin{aligned} A(w) &= -w^2 + 150w \\ &= -(w^2 - 150w) \\ &= -(w^2 - 150w + 75^2 - 75^2) \\ &= -(w^2 - 150w + 75^2) + 75^2 \\ &= -(w - 75)^2 + 5625 \end{aligned}$$

The vertex is (75, 5625). Thus, the width of the pen should be 75 feet. Next, we find the length of the pen.

$$\begin{aligned} l &= 150 - w \\ &= 150 - 75 = 75 \end{aligned}$$

The largest pen would be a square with sides of 75 feet. □

If we are just looking for the maximum or minimum value of a quadratic function, it may be convenient to find the vertex with a formula. Consider the general quadratic function

$$f(x) = ax^2 + bx + c;\ a \neq 0$$

We complete the square.

$$f(x) = a\left(x^2 + \frac{b}{a}x\right) + c \qquad \text{Factor } a \text{ out of the } x\text{-terms.}$$

$$= a\left(x^2 + \frac{b}{a}x + \frac{b^2}{4a^2} - \frac{b^2}{4a^2}\right) + c$$

$$= a\left(x^2 + \frac{b}{a}x + \frac{b^2}{4a^2}\right) - \frac{b^2}{4a} + c = a\left(x + \frac{b}{2a}\right)^2 - \frac{b^2}{4a} + c$$

Now we see that the x-coordinate of the vertex is $-\frac{b}{2a}$. Therefore, the y-coordinate is $f\left(-\frac{b}{2a}\right)$ and the vertex of the parabola is at the point $\left(-\frac{b}{2a}, f\left(-\frac{b}{2a}\right)\right)$.

The Vertex of the Graph of a Quadratic Function

The graph of the function $f(x) = ax^2 + bx + c$, where $a \neq 0$, is a parabola with vertex at

$$\left(-\frac{b}{2a}, f\left(-\frac{b}{2a}\right)\right)$$

EXAMPLE 5 Find the minimum value of the function $g(t) = 3t^2 + 4t + 2$.

Solution Since the graph of $y = 3t^2 + 4t + 2$ is a parabola opening upward, the minimum value of the function will be the y-coordinate of the vertex. Because $-\frac{b}{2a} = -\frac{4}{2 \cdot 3} = -\frac{2}{3}$ the vertex is at $\left(-\frac{2}{3}, g\left(-\frac{2}{3}\right)\right)$.

$$g\left(-\frac{2}{3}\right) = 3\left(-\frac{2}{3}\right)^2 + 4\left(-\frac{2}{3}\right) + 2$$

$$= \frac{4}{3} - \frac{8}{3} + 2 = \frac{2}{3}$$

Therefore, the minimum value of the function g is $\frac{2}{3}$. □

Absolute Value Functions

Consider the graph of an absolute value function $f(x) = |x|$. It behaves much like the graph of a quadratic function.

EXAMPLE 6 Graph $f(x) = |x|$.

Solution Plotting several points shows us (Figure 4.8) that the graph is in the shape of a V. □

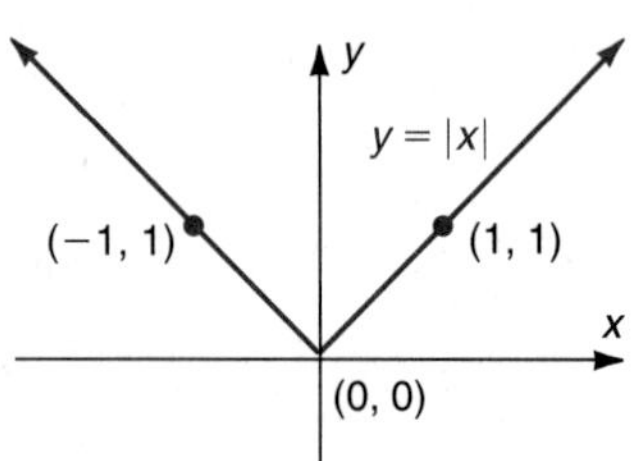

Fig. 4.8

Absolute value graphs can be shifted up and down or right and left just as we did with parabolas in Section 3.4.

> **Definition: Absolute Value Functions**
>
> A function in the form $f(x) = a|x - h| + k$ is called an **absolute value function.** Its graph is a V with vertex at the point (h, k). The V opens upward if a is positive, and downward if a is negative.

EXAMPLE 7 Graph $g(x) = |x + 2| - 1$.

Solution The vertex is $(-2, -1)$, and the graph opens upward. The graph is shown in Figure 4.9.

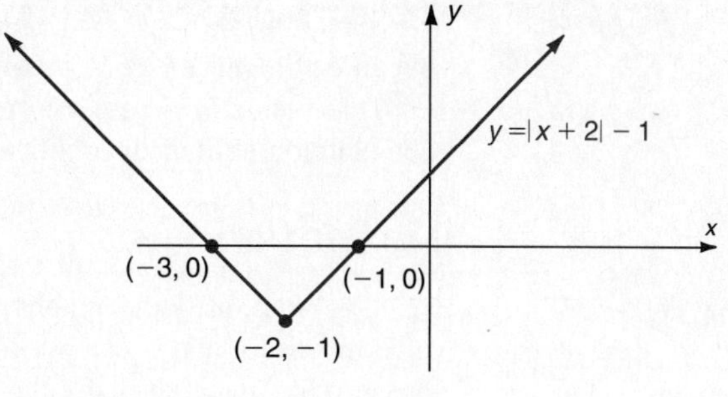

Fig. 4.9

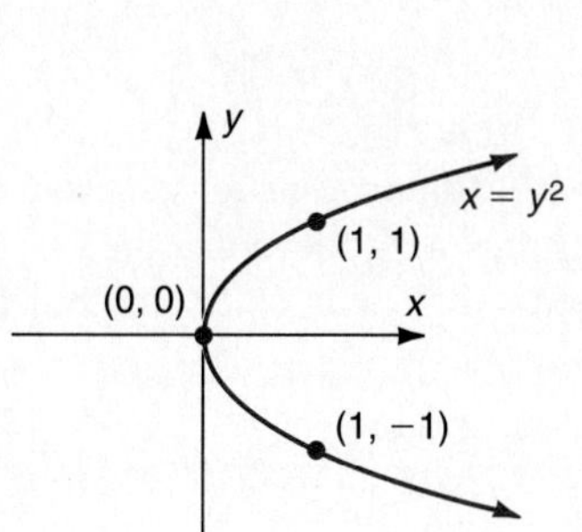

Fig. 4.10

Square Root Functions

Parabolas with a horizontal axis of symmetry are not graphs of functions. Consider the graph of $x = y^2$ shown in Figure 4.10.

Clearly, by the vertical line test, the graph of $x = y^2$ is not the graph of a function. However, it is interesting to notice that the top half and the bottom half are each the graph of a function (Figure 4.11).

How can we find a rule for these functions? Let's solve $x = y^2$ for y.

$$y^2 = x$$

$$y = \pm\sqrt{x} \qquad \text{Square Root Property}$$

Now we see *two* functions, one using the positive square root and the other using the negative square root. $f(x) = \sqrt{x}$ and $g(x) = -\sqrt{x}$ are the rules for the two functions (Figure 4.12).

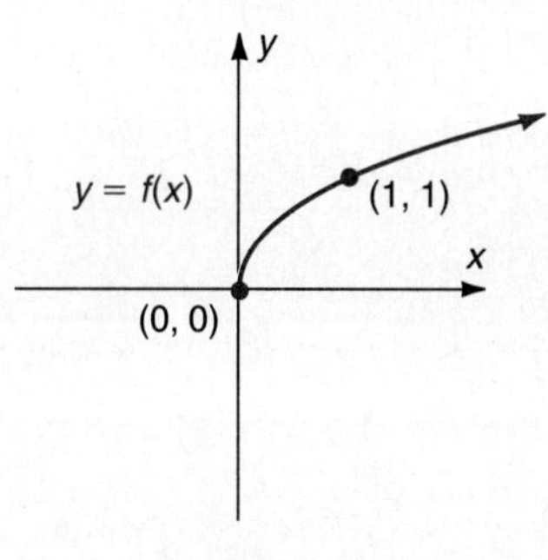

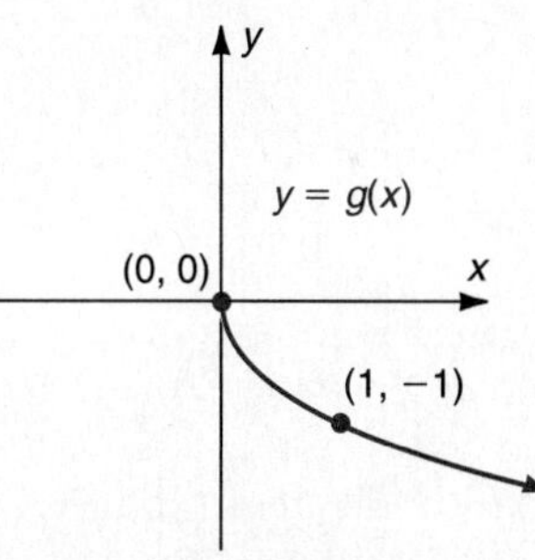

Fig. 4.11

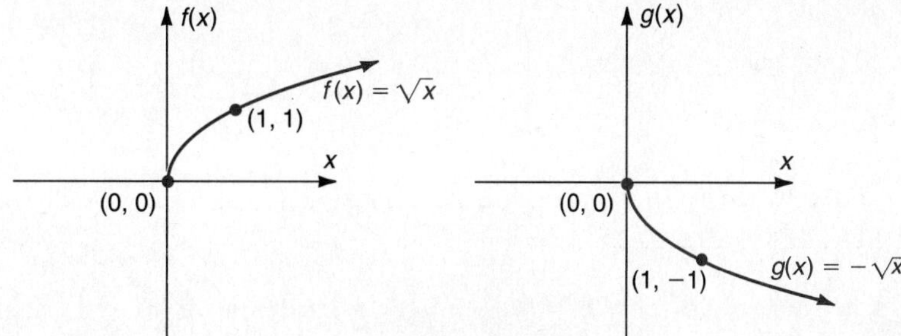

Fig. 4.12

These functions are called **square root** functions. Their graphs are sometimes called *half-parabolas.*

EXAMPLE 8 Sketch the graph of each function. Give the domain and range for each.

(a) $h(x) = \sqrt{x - 3}$ (b) $g(x) = -\sqrt{x} + 1$

Solutions The domain for h is $[3, +\infty)$, and the range is $[0, +\infty)$. The graph of h is the square root function shifted right 3 units. The domain for g is $[0, +\infty)$, and the range is $(-\infty, 1]$. Notice in Figure 4.13 that the graph of g is the bottom half of the square root function shifted up 1 unit. ☐

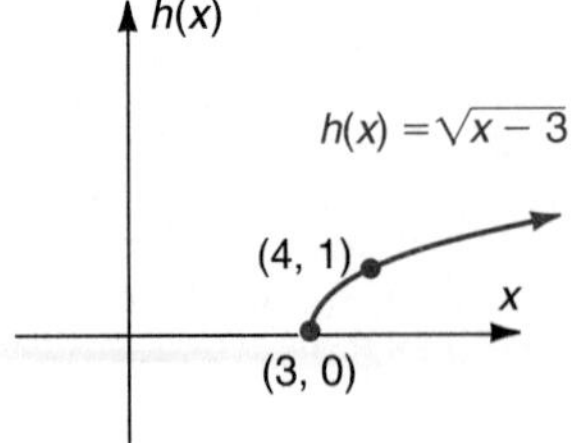

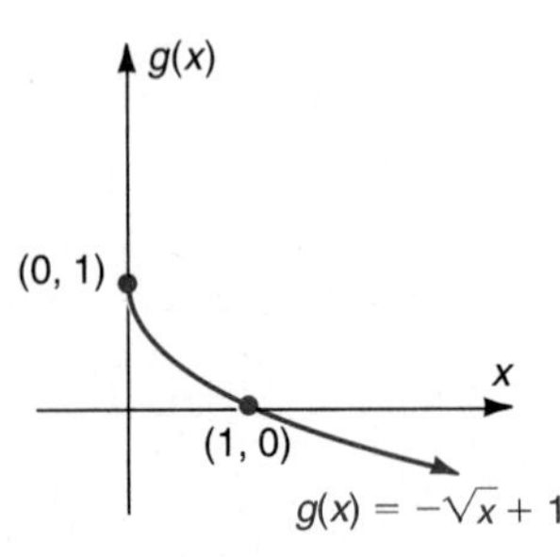

Fig. 4.13

Translations

We have examined the graphs of linear functions, quadratic functions, absolute value functions, and square root functions. In many cases we noticed that the graph was a basic shape shifted either vertically or horizontally. Such shifts are called **translations.**

> **Translations**
>
> If f is a function and $c > 0$, then
>
> The graph of $y = f(x) + c$ is the graph of $y = f(x)$ shifted *up* c units.
> The graph of $y = f(x) - c$ is the graph of $y = f(x)$ shifted *down* c units.
> The graph of $y = f(x + c)$ is the graph of $y = f(x)$ shifted c units to the *left.*
> The graph of $y = f(x - c)$ is the graph of $y = f(x)$ shifted c units to the *right.*

EXAMPLE 9 The graph of f is shown in Figure 4.14. Sketch the graph of

(a) $y = f(x) + 2$ (b) $y = f(x) - 1$
(c) $y = f(x + 3)$ (d) $y = f(x - 1) + 2$

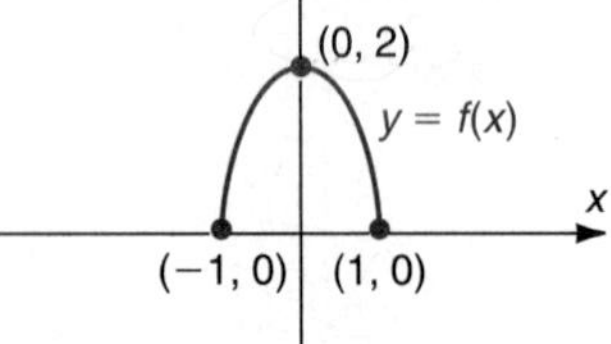

Fig. 4.14

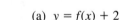

Solutions (a) $y = f(x) + 2$ (Figure 4.15a) (b) $y = f(x) - 1$ (Figure 4.15b)

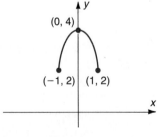

Fig. 4.15a

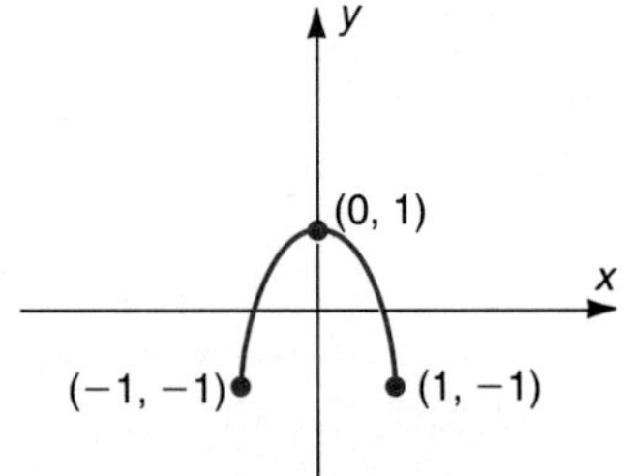

Fig. 4.15b

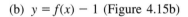

(c) $y = f(x + 3)$ (Figure 4.15c) (d) $y = f(x - 1) + 2$ (Figure 4.15d)

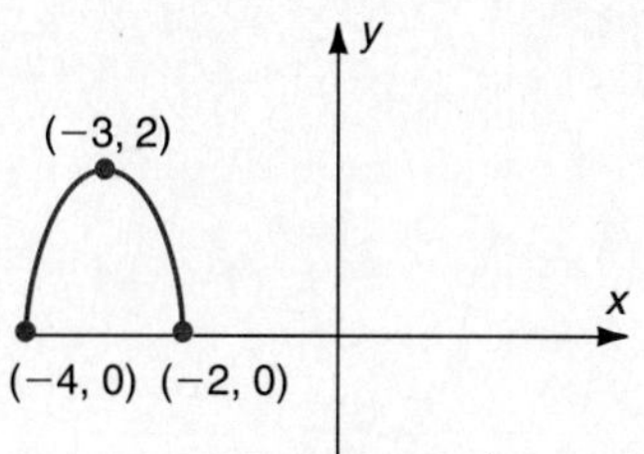

Fig. 4.15c

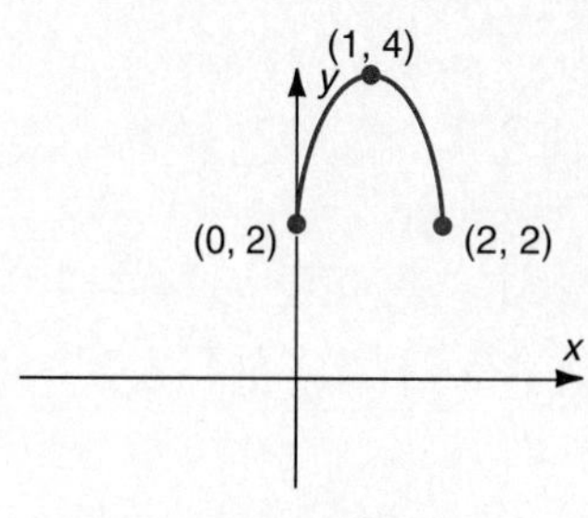

Fig. 4.15d

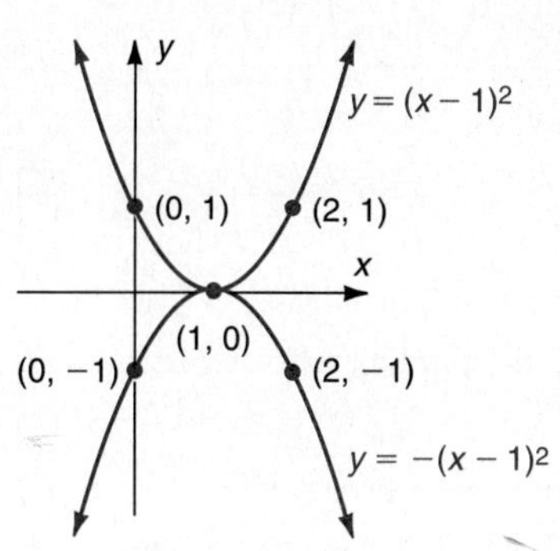

Fig. 4.16

Reflections

Consider the functions $f(x) = (x - 1)^2$ and $g(x) = -(x - 1)^2$. If we graph them on the same set of coordinate axes, we see that each is a *reflection* of the other in the x-axis, as shown in Figure 4.16. Likewise, if we examine the graphs of the functions $F(x) = \sqrt{x}$ and $G(x) = \sqrt{-x}$ shown in Figure 4.17, we see that they are *reflections* of each other in the y-axis.

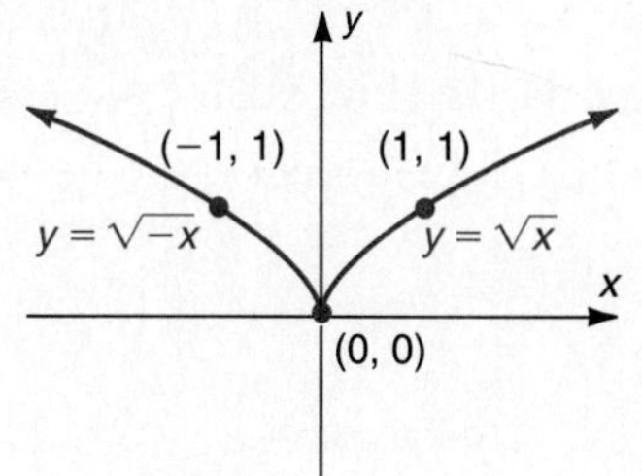

Fig. 4.17

Reflections

Suppose f is a function.

The graph of $y = -f(x)$ is the reflection of the graph of $y = f(x)$ in the x-axis.
The graph of $y = f(-x)$ is the reflection of the graph of $y = f(x)$ in the y-axis, provided $f(-x)$ exists.

EXAMPLE 10 The graph of f is shown in Figure 4.18. Sketch the graph of

(a) $y = -f(x)$

(b) $y = f(-x)$

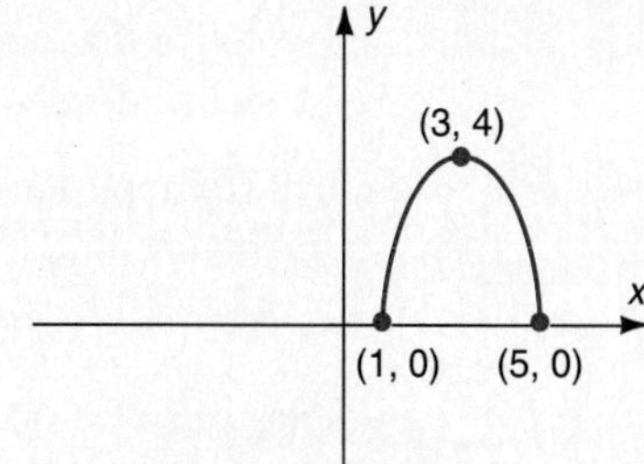

Fig. 4.18

Solutions (a) $y = -f(x)$ (Figure 4.19a)

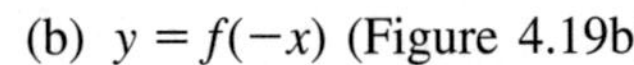

(b) $y = f(-x)$ (Figure 4.19b)

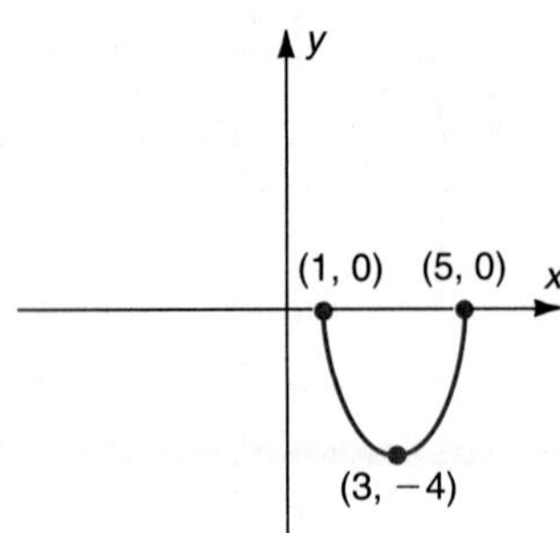

Fig. 4.19a

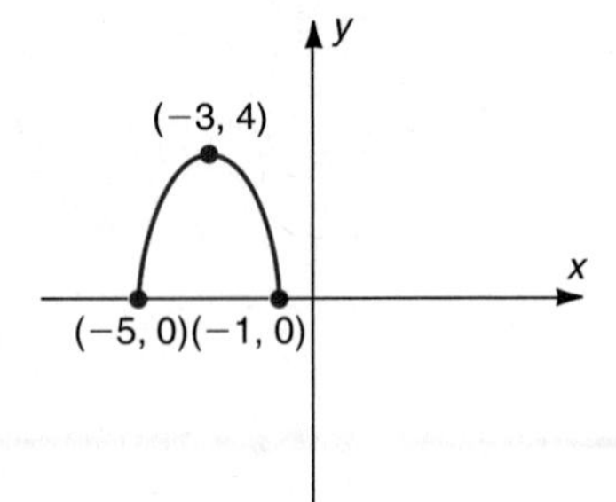

Fig. 4.19b □

Restricted Domains

When functions are used to model real situations, the application often dictates a restriction on the domain of the function.

EXAMPLE 11 Sketch the graph of each function.

(a) $f(x) = 4 - x^2;\ x < 1$ (b) $g(x) = 3x;\ x \geq -2$

Solutions (a) First we graph $y = 4 - x^2$. This graph is a parabola. Then to graph f, we use only those points on the parabola for which $x < 1$ (Figure 4.20).

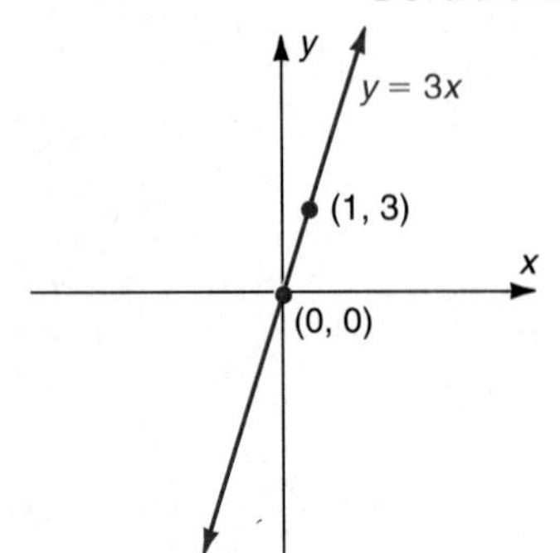

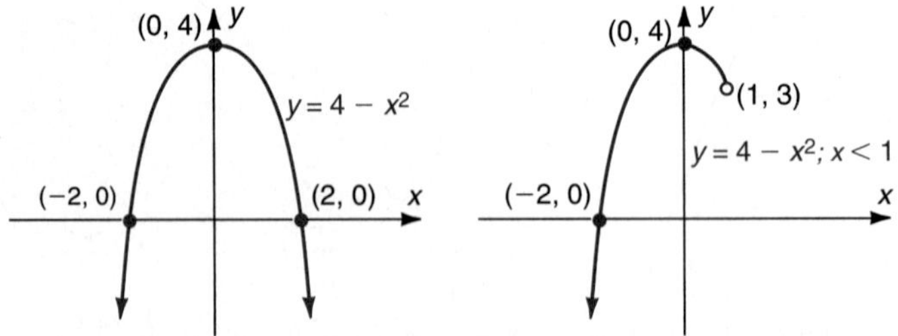

Fig. 4.20

Notice that we made a hollow circle at the point (1, 3). This is because 1 is not included in $x < 1$.

(b) We graph the line $y = 3x$ and then graph g which is the line where $x \geq -2$ (Figure 4.21). Notice that we made a solid dot at $(-2, -6)$ because -2 is included in $x \geq -2$. □

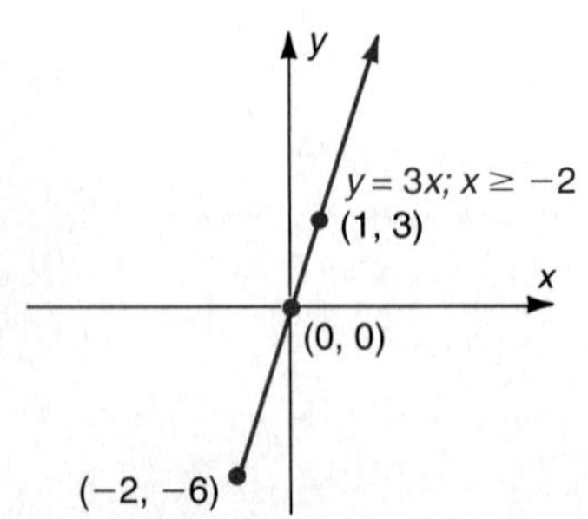

Fig. 4.21

Piecewise Defined Functions

Sometimes the rule for a function requires that the domain be split up into several pieces. This type of function is called a **piecewise defined** function.

EXAMPLE 12 Suppose $f(x) = \begin{cases} 1 & \text{if } x \leq -1 \\ x^2 + 1 & \text{if } x > -1 \end{cases}$

Find

(a) $f(3)$ (b) $f(-2)$ (c) $f(-1)$

Solutions (a) To find $f(3)$, we must first decide which part of the rule to use. The rule is broken into two parts depending on whether x is less than or equal to -1 or greater than -1. Since $3 > -1$, we use the bottom part; that is, $x^2 + 1$. So, $f(3) = 3^2 + 1 = 10$.

(b) To find $f(-2)$, we notice that $-2 < -1$. So, we use the top part of the rule. Thus $f(-2) = 1$.

(c) Notice that x equal to -1 is included in the top part of the rule. So, $f(-1) = 1$. □

EXAMPLE 13 Sketch the graph of $f(x) = \begin{cases} 1 & \text{if } x \leq -1 \\ x^2 + 1 & \text{if } x > -1 \end{cases}$

Solution The rule divides the domain into two parts: $x \leq -1$ and $x > -1$. Notice that every real number is included in one of the two parts. Thus, the domain is R. To sketch the graph, we graph each part of the rule using only the specified x-numbers. For $x \leq -1$, sketch $y = 1$. This is a horizontal line crossing the y-axis at $(0, 1)$ as shown in Figure 4.22a. Then we erase that portion of the graph *to the right of* $(-1, 1)$, as shown in Figure 4.22b, because the rule $f(x) = 1$ does not apply for that portion of the domain. Notice that we made a solid dot at $(-1, 1)$ because the top part of the rule *includes* -1.

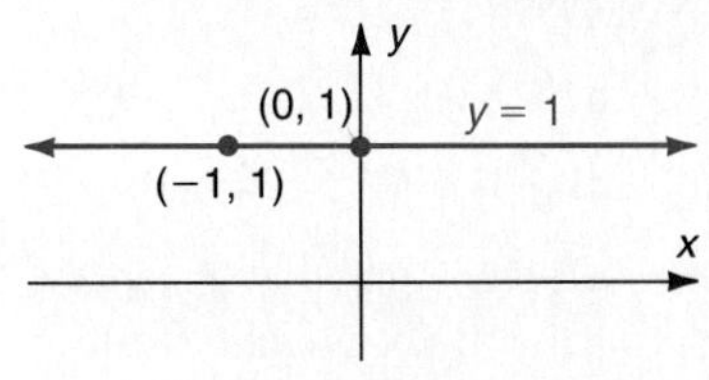

Fig. 4.22a

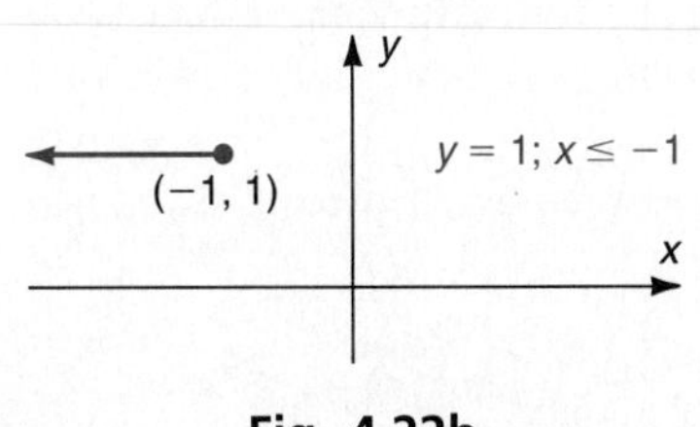

Fig. 4.22b

For $x > -1$, we sketch $y = x^2 + 1$ as shown in Figure 4.22c. This is a parabola. Next we erase that portion of the graph *to the left of* $x = -1$, as shown in Figure 4.22d, since this part of the rule does not apply there.

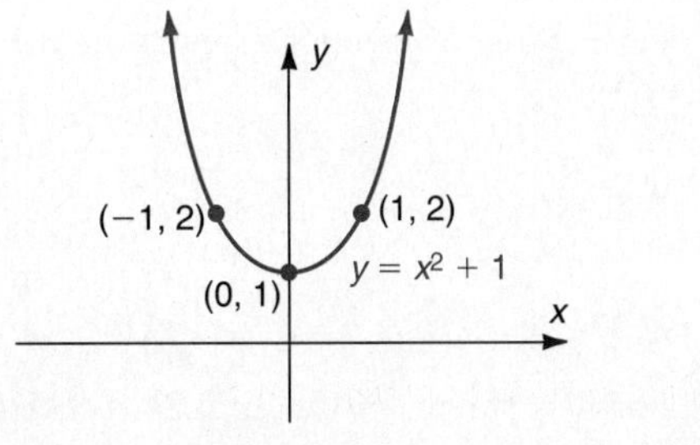

Fig. 4.22c

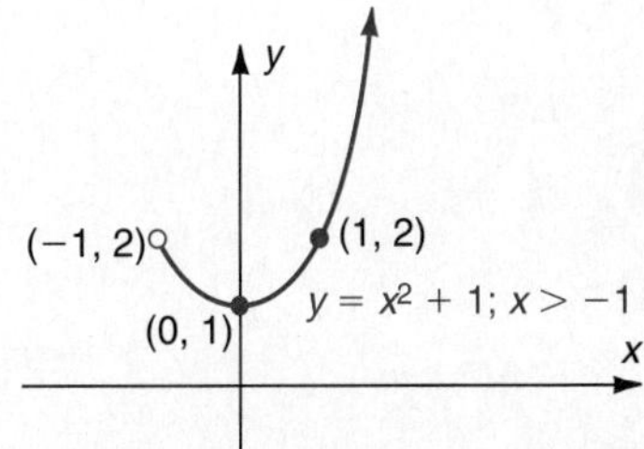

Fig. 4.22d

The point $(-1, 2)$ is a hollow circle because -1 is *not included* in this part.

Finally we put together the two parts shown in Figures 4.22b and 4.22d. This is the graph of f which is shown in Figure 4.23. Looking at the graph we see that the range is $[1, +\infty)$.

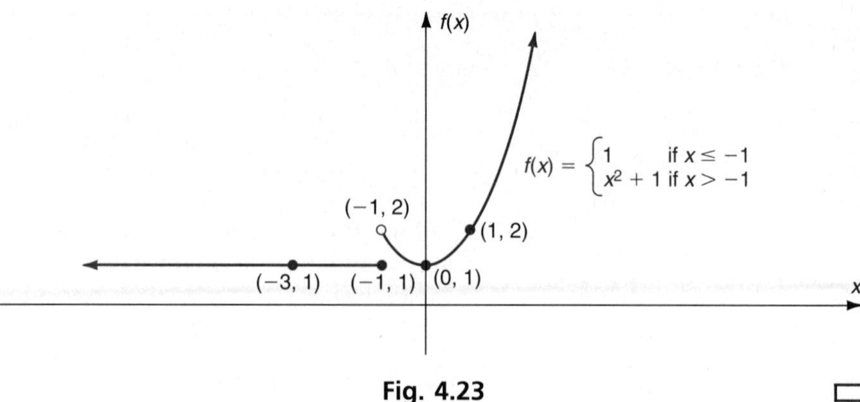

Fig. 4.23

□

EXAMPLE 14 A table of the recommended daily dietary energy intake for infants and children is given below.

Body Weight (pounds)	*Calories per Pound*
<13	49
13 to 29	44
29 to 44	41
>44	32

Using body weight as the independent variable, write a function that gives the recommended daily calorie intake. Graph this function for body weights from 5 pounds to 65 pounds.

Solution From the table, the recommended daily calorie intake for a 10-pound infant is 49 calories per pound or $49 \cdot 10$ calories. Thus, if C is the recommended daily calorie intake, then $C(10) = 490$. If we let x be the body weight, $C(x) = 49x$ for body weights less than 13 pounds. Continuing in this manner we form a piecewise defined function.

$$C(x) = \begin{cases} 49x & \text{if } x < 13 \\ 44x & \text{if } 13 \le x < 29 \\ 41x & \text{if } 29 \le x \le 44 \\ 32x & \text{if } x > 44 \end{cases}$$

The graph is shown in Figure 4.24. (As is often the case when lines are fitted to empirical data, the graph probably does not accurately represent the situation at the ends of each line.)

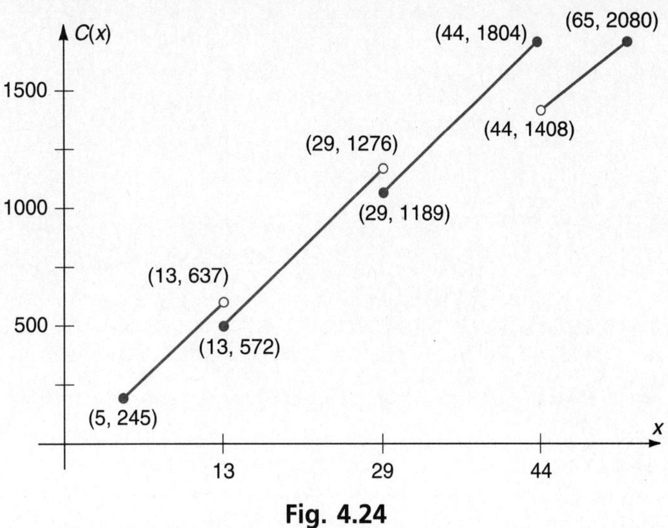

Fig. 4.24

□

EXAMPLE 15 A portion of the 1992 tax table gives the following information for single taxpayers.

TAXABLE INCOME IS		YOUR TAX IS
at least	but less than	
10,000	10,050	1,504
10,050	10,100	1,511
10,100	10,150	1,519
10,150	10,200	1,526
10,200	10,250	1,534
10,250	10,300	1,541
10,300	10,350	1,549
10,350	10,400	1,556

Graph this function using taxable income as the independent variable and tax due as the dependent variable.

Solution The domain consists of all possible number substitutions for the independent variable (Figure 4.25). So, the domain is [10,000, 10,400]. The right end point of each piece of the graph is a hollow circle because of the part of the rule that says ''but less than.''

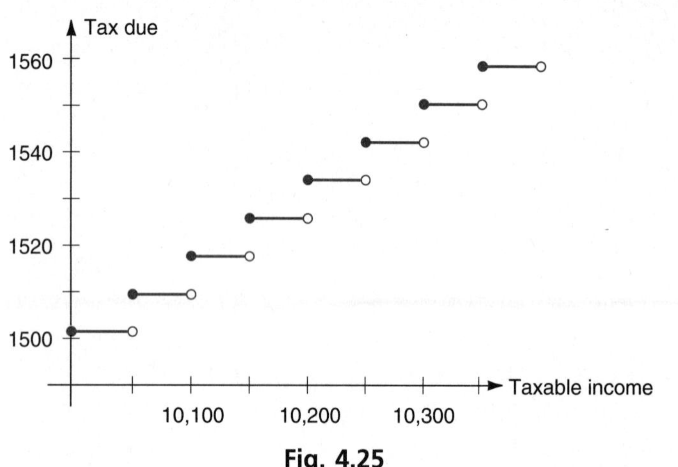

Fig. 4.25

□

Step Functions

The piecewise defined function in Example 15, Figure 4.25, is called a step function. The **greatest integer function** is also a step function. The symbol $[\![x]\!]$ means the greatest integer less than or equal to x. For example,

$[\![2]\!] = 2$	$[\![2.1]\!] = 2$	$[\![2.5]\!] = 2$	$[\![2.9]\!] = 2$	$[\![3]\!] = 3$
$[\![-3]\!] = -3$	$[\![-3.1]\!] = -4$	$[\![-3.5]\!] = -4$	$[\![-3.9]\!] = -4$	$[\![-4]\!] = -4$

If n is an integer, then $[\![n]\!] = n$.

EXAMPLE 16 Graph $f(x) = [\![x]\!]$.

Solution If x is in the interval $[-2, -1)$, then $[\![x]\!] = -2$.

If x is in the interval $[-1, 0)$, then $[\![x]\!] = -1$.

If x is in the interval $[0, 1)$, then $[\![x]\!] = 0$.

If x is in the interval $[1, 2)$, then $[\![x]\!] = 1$.

If x is in the interval $[2, 3)$, then $[\![x]\!] = 2$.

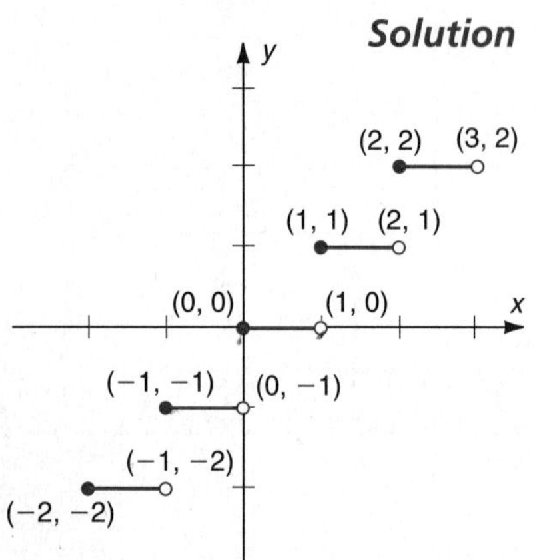

Fig. 4.26

The pattern continues as x moves through all real numbers. The graph is in steps along the intervals between integers, as seen in Figure 4.26. Notice that on each step the left end point is solid, while the right end point is hollow. The domain of the greatest integer function is $(-\infty, +\infty)$, and the range is the set of integers.

□

GRAPHING CALCULATOR BOX

Graphing Piecewise Defined Functions

Let's graph the function $y = \begin{cases} x^2 + 2x & \text{if } x \le 1 \\ 2x - 4 & \text{if } x > 1 \end{cases}$.

We sketch piecewise defined functions on a graphing calculator by sketching each *piece* as a separate graph and limit each domain. As a starting point, set the window at $[-10, 10]$ for both x and y.

TI-81

The TI has a *connected mode* and a *dot mode.* In connected mode the calculator connects the points it plots with a line, while in dot mode it does not. For graphing piecewise defined functions it should be in *dot mode.* Press [MODE] for the mode menu and find "connected dot" 5 lines down. Arrow down with the editing arrows and right so that "dot" is shaded. Press [ENTER] and [CLEAR]. We are now in dot mode. We press [Y =] and enter the function and its domain restrictions in the format $(x^2 + 2x)(x \le 1) + (2x - 4)(x > 1)$.

[(] [X|T] [X²] [+] [2] [X|T] [)] [(] [X|T] [TEST] [6] [1] [)]
[+] [(] [2] [X|T] [−] [4] [)] [(] [X|T] [TEST] [3] [1] [)]

Now [GRAPH].

CASIO fx-7700G

We need three new keys: the comma, a second function above the store arrow [→], and the left and right brackets which are [ALPHA] characters above the period [·] and the [EXP] keys. We enter the pieces of the function in the format

$x^2 + 2x,[-20, 1]$

$2x - 4,[1, 20]$

Note the commas and the brackets. The expression inside the brackets is the *domain of the graph shown on the calculator.* The window will show only that portion of the indicated domain that is inside the [RANGE] setting. The −20 and 20 are arbitrary.

[GRAPH] [X,θ,T] [x²] [+] [2] [X,θ,T] [,] [ALPHA] [[] [−20] [,] [1] [ALPHA] []] [↵]
[GRAPH] [2] [X,θ,T] [−] [4] [,] [ALPHA] [[] [1] [,] [20] [ALPHA] []] [EXE]

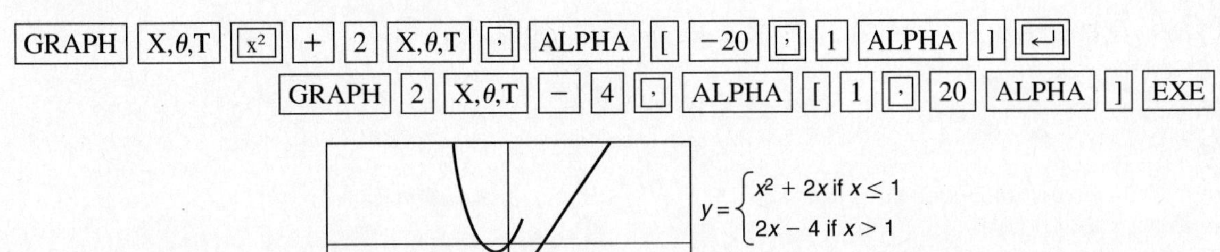

PROBLEM SET 4.1

Warm-ups

In Problems 1 through 4, sketch the graph of each linear function. See Example 1.

1. $f(x) = 3x + 1$
2. $g(x) = -5x$
3. $h(x) = \frac{2}{3}x + 1$
4. $f(x) = -\frac{1}{2}$

For Problems 5 and 6, see Example 2.

5. Dr. Joan Rupp has determined that the amount (in grams) of lactic acid removed from the blood is a linear function of the amount (in liters) of oxygen consumed during slow recovery. Use the following data to express the amount of lactic acid (A) as a function of the amount of oxygen (x) consumed.

Lactic Acid (grams)	*Oxygen (liters)*
40	5
64	7

6. Heart rate (in beats per minute) is a linear function of the number of liters of oxygen consumed per minute. Express this relationship using the following data from one subject.

Heart Rate (beats per minute)	*Oxygen (liters per minute)*
142	2.0
125	1.6

In Problems 7 through 14, sketch the graph of each quadratic function. See Example 3.

7. $h(x) = 2x^2 - 1$
8. $f(x) = 1 - x^2$
9. $g(x) = (x + 2)^2$
10. $f(x) = -(x - 2)^2 + 3$
11. $g(x) = x^2 + 2x + 3$
12. $g(x) = x^2 + 4x - 1$
13. $h(x) = -2x^2 + 2x$
14. $f(x) = 3x^2 - 6x$

For Problems 15 and 16, see Example 4.

15. The profit on a certain Amazon River basin tour is given by $P(x) = -(x - 120)^2 + 8500$, where x represents the number of tour tickets sold. How many tickets must be sold to obtain a maximum profit? What will the maximum profit be?

16. The sum of two numbers is 20. What is the largest possible value for the product of the two numbers? What are the numbers when the product is largest?

In Problems 17 and 18, find the required value of each function. See Example 5.

17. $f(x) = 4 - 6x - 3x^2$. Find the maximum value.
18. $g(t) = 5t^2 + 12t - 1$. Find the minimum value.

In Problems 19 through 26, sketch the graph of each absolute value function. See Examples 6 and 7.

19. $g(x) = |x + 4|$
20. $h(x) = |x - 3|$
21. $f(x) = |x - 1| - 2$
22. $g(x) = |x + 2| - 1$
23. $h(x) = |x| + 2$
24. $h(x) = |x| - 3$
25. $f(x) = 2 - |x + 3|$
26. $f(x) = 1 - |x|$

In Problems 27 through 32, sketch the graph of each square root function. See Example 8.

27. $f(x) = \sqrt{x - 2}$
28. $f(x) = \sqrt{x + 1}$
29. $f(x) = \sqrt{x} + 3$
30. $f(x) = \sqrt{x} - 1$
31. $f(x) = \sqrt{x - 1} + 2$
32. $f(x) = \sqrt{x + 2} - 1$

In Problems 33 through 36, use the graph shown in Figure 4.27 to graph each function. See Example 9.

33. $y = f(x) + 1$
34. $y = f(x + 2)$
35. $y = f(x + 1)$
36. $y = f(x + 2) - 1$

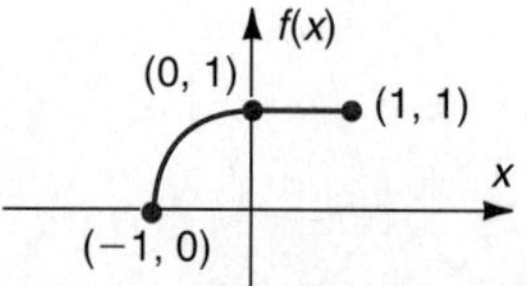

Fig. 4.27

In Problems 37 and 38, use the graph shown in Figure 4.28 to graph each function. See Example 10.

37. $y = -f(x)$

38. $y = f(-x)$

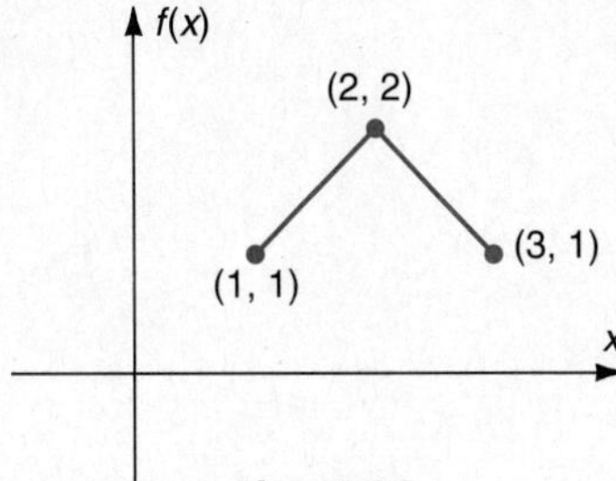

Fig. 4.28

In Problems 39 and 40, sketch the graph of each function. See Example 11.

39. $f(x) = x^2 + 2;\ x \geq -1$

40. $g(x) = 1 - |x|;\ x < 2$

In Problems 41 and 42, if $f(x) = \begin{cases} 2x & \text{if } x \leq 2 \\ 1 - x^2 & \text{if } x > 2 \end{cases}$, *find the indicated function values. See Example 12.*

41. $f(0)$

42. $f(5)$

In Problems 43 through 46, sketch the graph of each function. See Example 13.

43. $f(x) = \begin{cases} 2x^2 & \text{if } x < 0 \\ x + 2 & \text{if } x \geq 0 \end{cases}$

44. $g(x) = \begin{cases} -2 & \text{if } x \leq -3 \\ x + 2 & \text{if } x > -3 \end{cases}$

45. $f(x) = \begin{cases} x + 5 & \text{if } x < -1 \\ (x - 1)^2 & \text{if } x \geq -1 \end{cases}$

46. $h(x) = \begin{cases} |x| & \text{if } x \leq 2 \\ 3x - 4 & \text{if } x > 2 \end{cases}$

For Problems 47 and 48, see Example 14.

47. When the New York Stock Exchange established the commission structure of member firms, the commission for 100 shares was given (in part) by the following table.

Commission (dollars)	***Price (x dollars per share)***
$6.40 + 2x$	$x \leq 8$
$12 + 1.3x$	$8 < x \leq 25$
$22 + 0.9x$	$25 < x \leq 40$

Write a function C that gives the commission as a function of price per share for stocks selling at \$40 or less a share. Graph C.

48. Today, the member firms of the Stock Exchange can set their own commission structure. Suppose a discount broker set the following commissions for 50 shares.

Commission (dollars)	***Price (x dollars per share)***
$10 + x$	$x \leq 10$
$12 + 1.5x$	$10 < x \leq 30$
$25 + 1.2x$	$30 < x \leq 50$

Write a function C that gives the commission as a function of price per share for stocks selling at \$50 or less a share. Graph C.

In Problems 49 and 50, sketch the graph of each step function. See Example 15 and 16.

49. $f(x) = [\![2x]\!]$

50. $s(x) = [\![x - 1]\!]$

Practice Exercises

In Problems 51 through 80, sketch the graph of each function. Label the vertex of parabolas and absolute value functions.

51. $f(x) = 3x$

52. $g(x) = 2x + 3$

53. $f(x) = |x + 2|$

54. $g(x) = 3|x|$

55. $h(x) = \frac{1}{3}x + 2$

56. $f(x) = 1$

57. $f(x) = 4x^2$

58. $g(x) = x^2 + 3$

59. $h(x) = 3 - x^2$

60. $f(x) = 2 - x^2$

61. $f(x) = x^2 - 4;\ x < 1$

62. $g(x) = \frac{1}{2}x^2 + 3;\ x \geq 2$

63. $g(x) = (x - 3)^2$

64. $h(x) = (x + 1)^2$

65. $h(x) = \sqrt{x + 5}$

66. $f(x) = \sqrt{x - 4}$

67. $f(x) = -(x - 1)^2 + 4$

68. $f(x) = -2(x + 3)^2 + 1$

69. $g(x) = x^2 - 2x + 4$

70. $h(x) = x^2 + 6x + 1$

71. $f(x) = -x^2 - 2x - 3$

72. $f(x) = -2x^2 + 8x - 7$

73. $h(x) = |2x| + 1$

74. $h(x) = -|2x - 1| + 1$

75. $f(x) = [\![x - 2]\!]$

76. $f(x) = [\![x]\!] + 1$

77. $f(x) = \begin{cases} x - 1 & \text{if } x < 2 \\ x^2 & \text{if } x \geq 2 \end{cases}$

78. $g(x) = \begin{cases} x^2 - 1 & \text{if } x \leq -2 \\ -2 & \text{if } x > -2 \end{cases}$

79. $h(x) = \begin{cases} |x - 1| & \text{if } x < 2 \\ |x - 3| & \text{if } x \geq 2 \end{cases}$

80. $f(x) = \begin{cases} \sqrt{x - 1} & \text{if } x < 2 \\ 2x - 3 & \text{if } x \geq 2 \end{cases}$

In each Problem 81 through 86, sketch graphs of f and the functions

$$g(x) = f(x) + 2$$

$$h(x) = f(x + 2)$$

$$k(x) = 2f(x)$$

$$F(x) = -f(x)$$

$$G(x) = f(-x)$$

81. $f(x) = 3x$

82. $f(x) = 2x + 3$

83. $f(x) = x^2 + 2x$

84. $f(x) = x^2 - 4x$

85. $f(x) = 2\sqrt{x}$

86. $f(x) = |x - 1|$

87. The height (in feet) of a ball thrown upward from a building 150 feet high with an initial velocity of 64 feet per second is given by $s(t) = -16t^2 + 64t + 150$, where t is time in seconds. Find how high the ball will go.

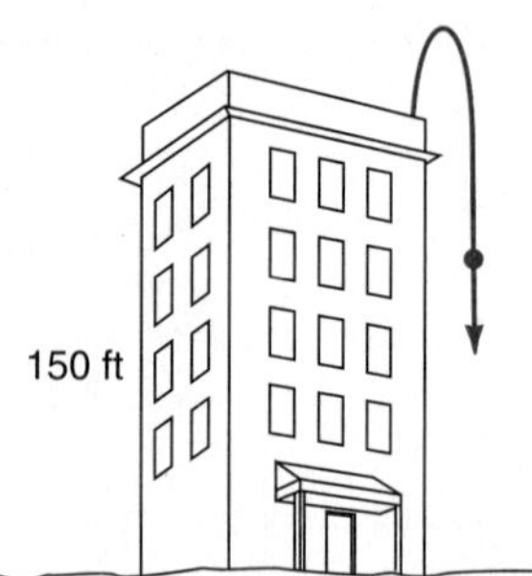

88. A company that makes beepers has determined that monthly profit (in thousands of dollars) is given by $P(x) = -x^2 + 20x - 36$, where x is the number of thousands of beepers sold. What is the maximum profit? How many beepers must be sold to make the maximum profit? For what production levels is profit increasing? Decreasing?

89. The Metro Pool and Chemical Company treats pools chemically for algae. If the algae content is given by $A(t) = 20t^2 - 800t + 8500$, where t is the number of days after the pool has been treated, how many days after a treatment will the algae content be a minimum? When is the algae content increasing? Decreasing?

90. The cost in thousands of dollars of making umbrellas is given by $C(x) = x^2 - 8x + 20$, where x is the number (in thousands) of umbrellas manufactured. How many umbrellas should be made to minimize the cost? What is the minimum cost? Where is cost increasing? Decreasing?

91. The 1991 Tax Rate Schedule X for single taxpayer filing status gives the formation in Table 4-1. Graph this function using taxable income as the independent variable. Label all points where the rule changes.

IF TAXABLE INCOME IS		TAX DUE IS	OF THE AMOUNT OVER
Over	But not over		
\$0 20,350 49,300	\$20,350 49,300	15% 3,052.50 + 28% 11,158.50 + 31%	\$0 20,350 49,300

Table 4-1

In Problems 92 and 93, simplify the ***difference quotient*** $D(x) = \dfrac{f(x+h) - f(x)}{h}$; $h \neq 0$, *for each function.*

92. The linear function $f(x) = mx + b$

93. The quadratic function $f(x) = ax^2 + bx + c$

Challenge Problems

94. If f is a linear function, show that $f\left(\dfrac{p+q}{2}\right) = \dfrac{f(p) + f(q)}{2}$.

95. If f is the quadratic function $f(x) = ax^2 + bx + c$, show that $f(p + q) = f(p) + f(q)$ if and only if $pq = \dfrac{c}{2a}$.

96. Graph $f(x) = \dfrac{|x|}{x}$.

97. Graph $f(x) = |x^2 - 4|$.

98. Write $f(x) = |x|$ as a piecewise defined function. Graph f using the piecewise definition.

In Problems 99 and 100, simplify the difference quotient. (See direction line for Problems 92 and 93.)

99. The absolute value function $f(x) = p|x - a| + b$. Write $D(x)$ as a piecewise defined function.

100. The square root function $f(x) = \sqrt{x}$. (*Hint:* Rationalize the *numerator*.)

IN YOUR OWN WORDS . . .

101. Give an example of a linear function and a quadratic function. Use the words "increasing" and "decreasing" to describe the graph of each function.

102. Consider the following four rules.

Rule 1: $y = \begin{cases} x - 1 & \text{if } x \leq 1 \\ x^2 & \text{if } x \geq 3 \end{cases}$ Rule 2: $y = \begin{cases} x - 1 & \text{if } x \leq 3 \\ x^2 & \text{if } x \geq 1 \end{cases}$

Rule 3: $y = \begin{cases} 2x & \text{if } x \leq 3 \\ x^2 & \text{if } x \geq 3 \end{cases}$ Rule 4: $y = \begin{cases} 2x & \text{if } x \leq 2 \\ x^2 & \text{if } x \geq 2 \end{cases}$

Explain why, or why not, each is the rule for a function.

4.2 GRAPHING POLYNOMIAL FUNCTIONS

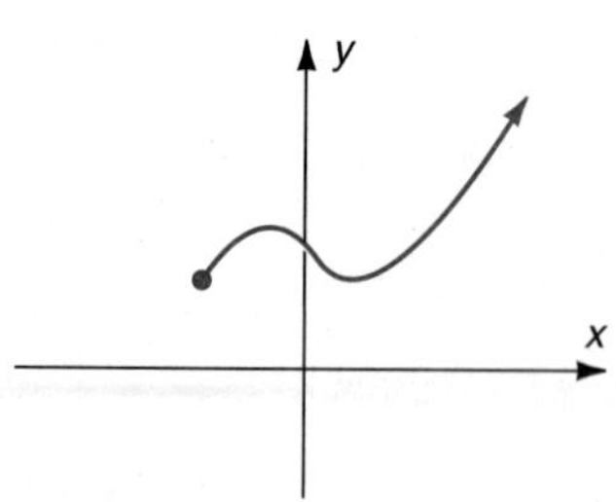

Domain is not all reals

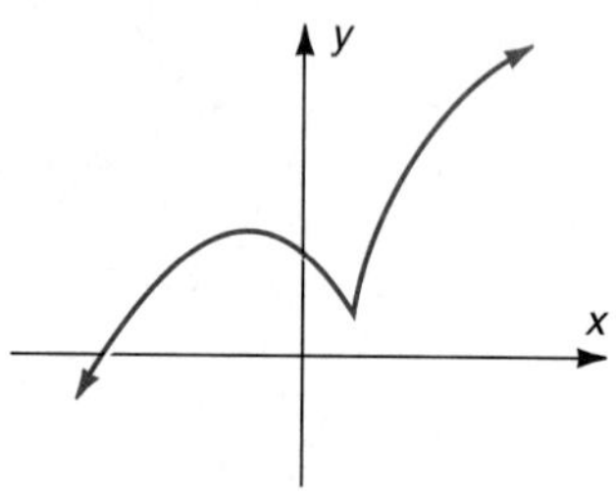

Sharp point in the graph

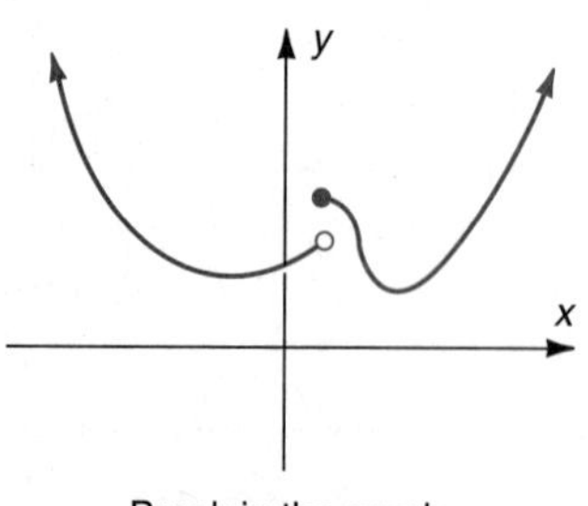

Break in the graph (Discontinuous)

Fig. 4.29

In Section 4.1, we studied linear and quadratic functions. The expressions that define such functions are, in fact, special cases of polynomials. For this reason, linear and quadratic functions belong to a larger family of functions called polynomial functions.

> **Definition: Polynomial Functions**
>
> A function of the form
>
> $$f(x) = a_n x^n + a_{n-1}x^{n-1} + \cdots + a_1 x + a_0, \text{ where } a_n \neq 0,$$
>
> each coefficient is a real number, and n is a whole number, is called a polynomial function of degree n in standard form.

The coefficient a_n is called the **leading coefficient.**

The graphs of zero- and first-degree polynomial functions are lines, and the graphs of second-degree polynomial functions are parabolas. The graphs of polynomial functions of degree higher than 2 have many different shapes. However, the graphs of polynomial functions share some properties.

(a) The domain is the set of all real numbers.

(b) The graphs are smooth curves with no sharp points.

(c) The graphs have no breaks. (In general, functions whose graphs have no breaks are called **continuous,** while those *with* breaks are called **discontinuous.**)

Each of the graphs in Figure 4.29 is *not* the graph of a polynomial function.

Power Functions

The simplest type of polynomial function is of the form $f(x) = x^n$. These functions are sometimes called **power functions.**

EXAMPLE 1 Sketch the graph of each function.

(a) $f(x) = x$ (b) $g(x) = x^2$ (c) $h(x) = x^3$

(d) $r(x) = x^4$ (e) $s(x) = x^5$ (f) $t(x) = x^6$

Solutions

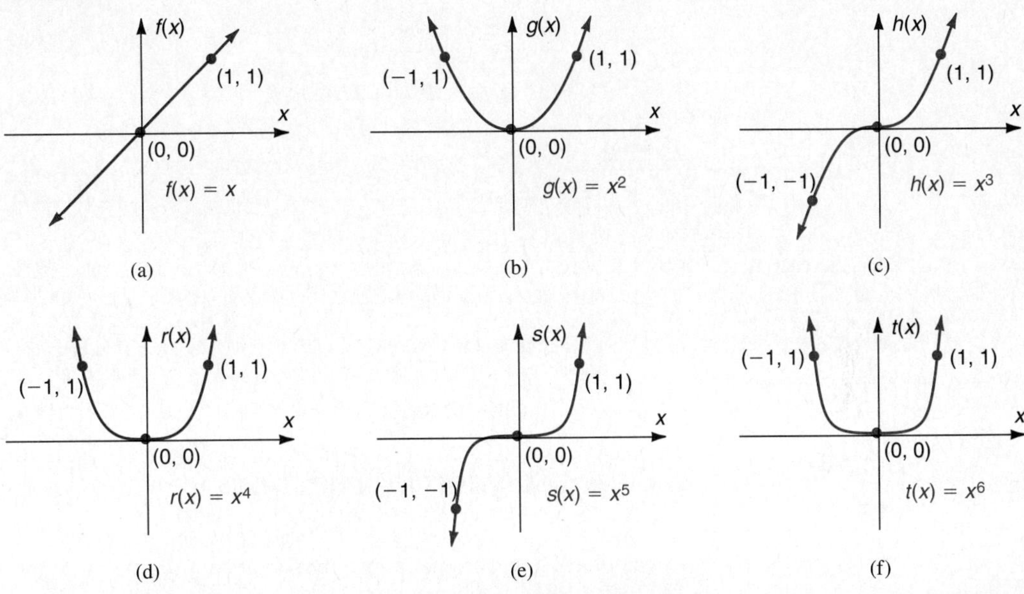

Fig. 4.30

The graphs of $r(x) = x^4$ and $t(x) = x^6$ may look like parabolas. However, they are not parabolas. A parabola is the graph of a second-degree polynomial function only. Notice that the graphs of *odd-degree* polynomials have the same general shape, and that the graphs of *even-degree* polynomials have the same general shape. Also note that the graphs of power functions get "flatter" around 0 as the power increases.

EXAMPLE 2 Sketch the graph of each function.

(a) $f(x) = x^3 + 1$ (b) $g(x) = (x + 1)^3$
(c) $h(x) = 1 - x^4$ (d) $r(x) = (x + 2)^6 - 1$

Solutions (a) $f(x) = x^3 + 1$ (Figure 4.31a) (b) $g(x) = (x + 1)^3$ (Figure 4.31b)

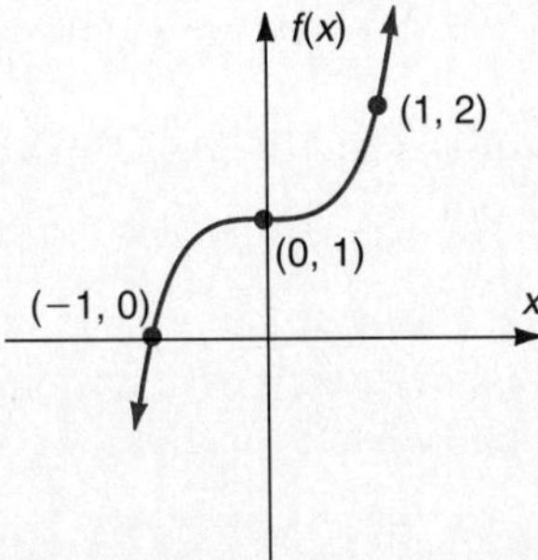

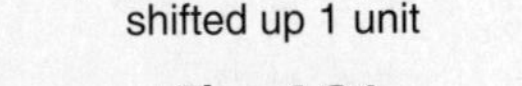
This graph is the graph of $y = x^3$ shifted up 1 unit

Fig. 4.31a

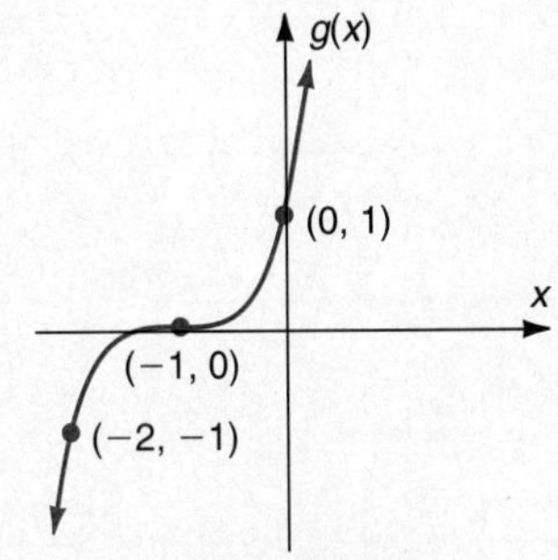

This graph is the graph of $y = x^3$ shifted left 1 unit

Fig. 4.31b

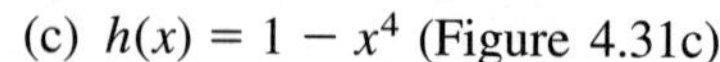

(c) $h(x) = 1 - x^4$ (Figure 4.31c)

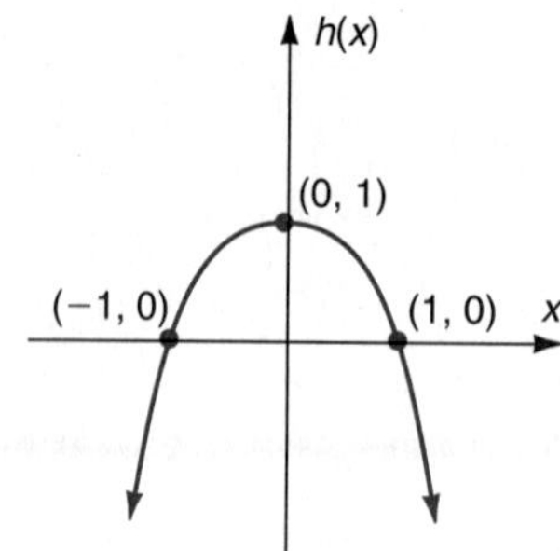

This graph is the graph of $y = x^4$ opening downward shifted up 1 unit

Fig. 4.31c

(d) $r(x) = (x + 2)^6 - 1$ (Figure 4.31d)

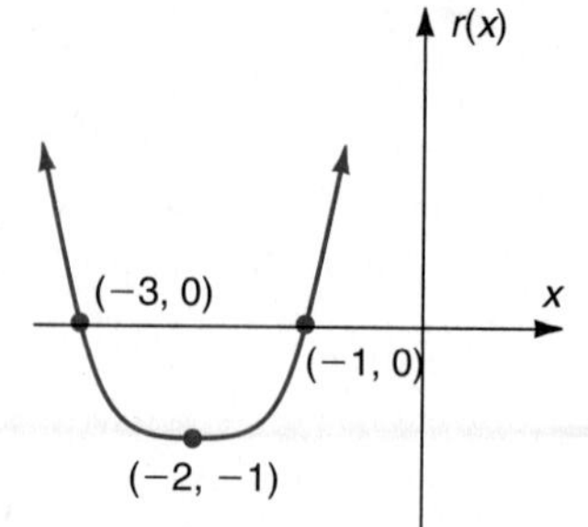

This graph is the graph of $y = x^6$ shifted left 2 units and down 1 unit

Fig. 4.31d □

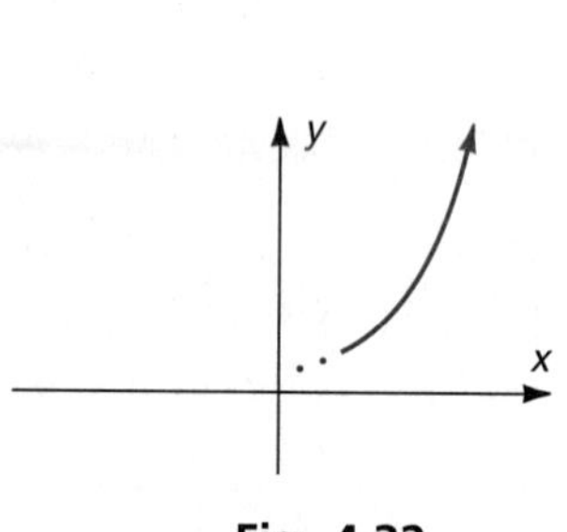

Fig. 4.32

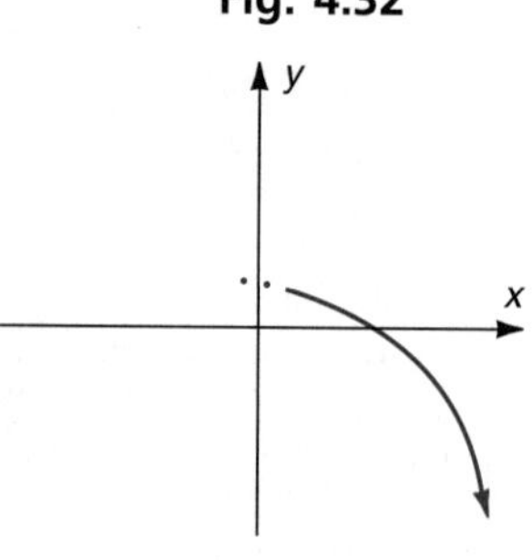

Fig. 4.33

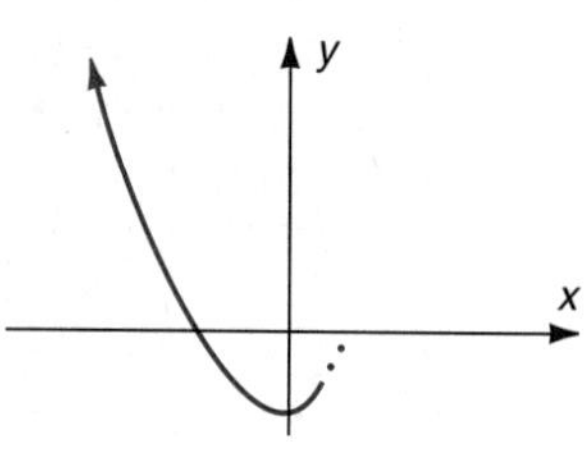

Fig. 4.34

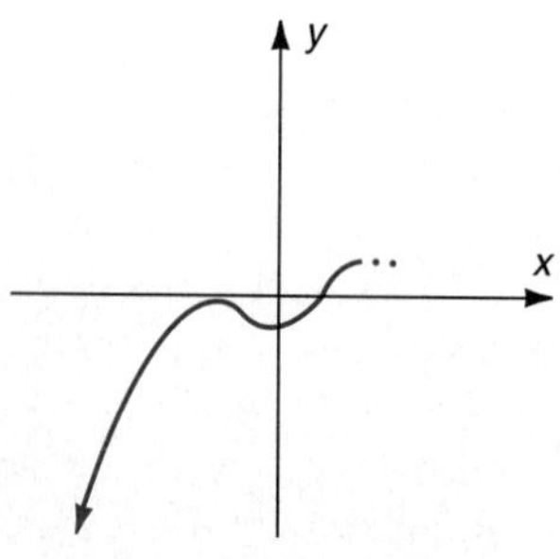

Fig. 4.35

Other Polynomial Functions

The details of the graphs of more complicated polynomial functions require some calculus. However, much information about the general shape of polynomial graphs can be found quite easily.

Noticė a feature shared by all the polynomial graphs we have examined so far. As the x-coordinates go to the right from the origin, the graph reaches a point where the y-coordinate rise steeply, never to return, or fall in the same manner. We say "$f(x)$ increases without bound as x increases without bound" and, in mathematical shorthand, write

$$f(x) \to +\infty \qquad \text{as} \qquad x \to +\infty$$

to describe the behavior at the *right* end of the curve in Figure 4.32, or, in the downward case, "$f(x)$ decreases without bound as x increases without bound" and write

$$f(x) \to -\infty \qquad \text{as} \qquad x \to +\infty$$

to describe the behavior at the *right* end of the curve in Figure 4.33. Furthermore, the graph has the same property to the left. That is, $f(x)$ increases or decreases without bound as x decreases without bound. We write

$$f(x) \to +\infty \qquad \text{as} \qquad x \to -\infty$$

to describe the behavior at the *left* end of the curve in Figure 4.34, or

$$f(x) \to -\infty \qquad \text{as} \qquad x \to -\infty$$

to describe the behavior at the *left* end of the curve in Figure 4.35.

Do *all* polynomial graphs have this property? To answer this question we examine the general polynomial function

$$f(x) = a_n x^n + a_{n-1}x^{n-1} + a_{n-2}x^{n-2} + \cdots + a_1 x + a_0$$

and factor out x^n.

$$= x^n\left(a_n + a_{n-1}\frac{x^{n-1}}{x^n} + a_{n-2}\frac{x^{n-2}}{x^n} + \cdots + a_1\frac{x}{x^n} + a_0\frac{1}{x^n}\right)$$

$$= x^n\left(a_n + a_{n-1}\frac{1}{x} + a_{n-2}\frac{1}{x^2} + \cdots + a_1\frac{1}{x^{n-1}} + a_0\frac{1}{x^n}\right)$$

Notice what happens to the expression inside the parentheses as x increases or decreases without bound. Every term *except* a_n gets closer and closer to zero. That is, as x gets very large, $f(x) \approx a_n x^n$. By the same reasoning, as x decreases without bound, $f(x) \approx a_n x^n$. That answers our question. The graph of every polynomial function acts like the power function $f(x) = a_n x^n$ when it gets far enough from the origin. This behavior depends on whether the leading coefficient a_n is positive or negative and whether n is odd or even. We illustrate these ideas in Figure 4.36.

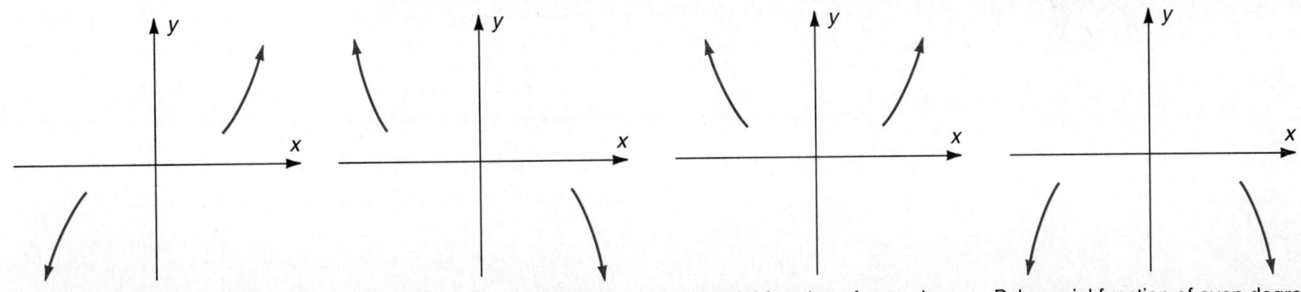

Fig. 4.36

Without calculus, we cannot determine the high and low points on the graphs. We can make a rough sketch of certain special polynomial functions by factoring to find the x-intercepts and then determine the general behavior of the graph by analyzing where the y-coordinates are positive or negative.

EXAMPLE 3 Sketch the graph of $f(x) = (x + 1)(x - 2)(x + 3)$.

Solution Before we graph f, notice that the polynomial is factored. If we multiply the factors, we see that f is a third-degree polynomial with a leading coefficient of 1. We graph

$$y = (x + 1)(x - 2)(x + 3)$$

by first finding the x-intercepts. The x-intercepts occur where y is 0.

$$0 = (x + 1)(x - 2)(x + 3)$$

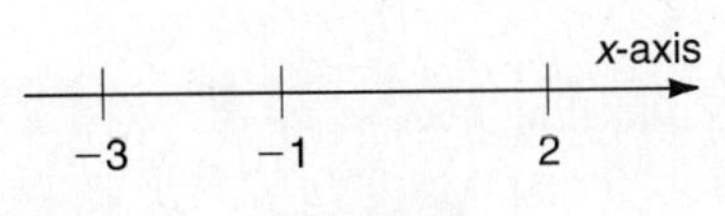

Fig. 4.37

The x-intercepts are -1, 2, and -3 by the Property of Zero Products. The graph must contain these points. The x-intercepts divide the x-axis into four intervals (Figure 4.37). Using interval notation, the intervals are $(-\infty, -3)$, $(-3, -1)$, $(-1, 2)$, and $(2, +\infty)$. Since polynomial functions are continuous (they have no breaks) the x-intercepts are the only places where the y-coordinates can change from positive to negative or from negative to positive. In each of the intervals determined by the x-intercepts, the y-coordinates will be either always positive or always negative. This means that we can choose one number in each interval and determine whether the y-coordinates are positive or negative. This analysis of the y-coordinates, together with the behavior at the ends of the graph, will give us a rough sketch of the graph.

INTERVAL ON x-AXIS	TEST NUMBER x	y-COORDINATE $f(x)$	GRAPH ABOVE OR BELOW x-AXIS
$(-\infty, -3)$	-4	-18	Below
$(-3, -1)$	-2	4	Above
$(-1, 2)$	0	-6	Below
$(2, +\infty)$	3	24	Above

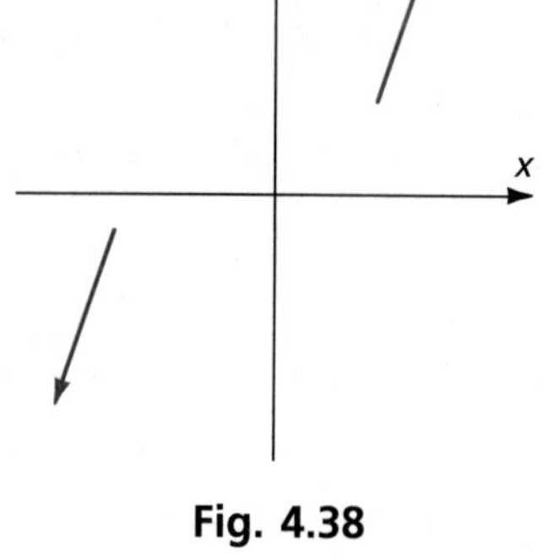

Fig. 4.38

This information tells us that the graph is below the x-axis to the left of -3 and between -1 and 2. The graph will be above the x-axis between -3 and -1 and to the right of 2.

Since f is an *odd-degree* polynomial function with a *positive* leading coefficient, its behavior at the ends will be of the form illustrated in Figure 4.38. To make an accurate sketch we could plot as many points as necessary. However, a rough sketch of the graph is shown in Figure 4.39.

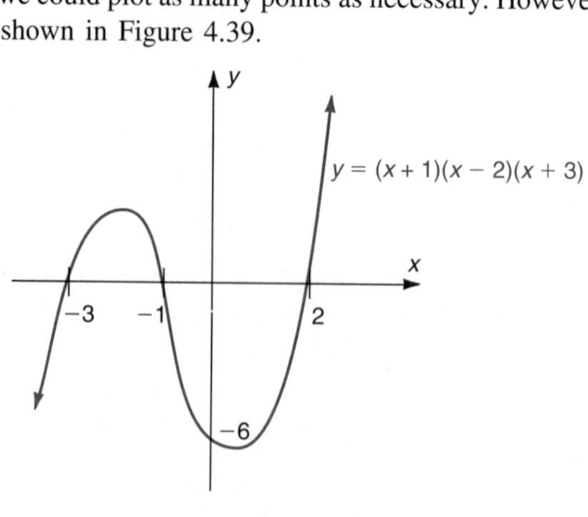

Fig. 4.39

□

EXAMPLE 4 Sketch the graph of $f(x) = (1 - x)(x + 2)(x + 4)$.

Solution Notice that this is a third-degree polynomial with a leading coefficient of -1. The x-intercepts are 1, -2, and -4. These points divide the x-axis into four intervals. See Figure 4.40. The intervals are $(-\infty, -4)$, $(-4, -2)$, $(-2, 1)$, and $(1, +\infty)$. We make a table describing the graph.

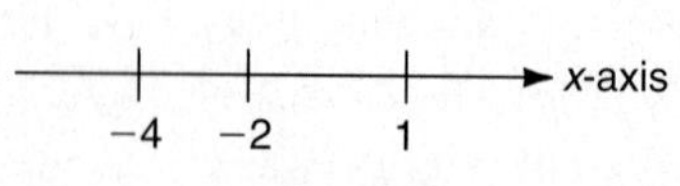

Fig. 4.40

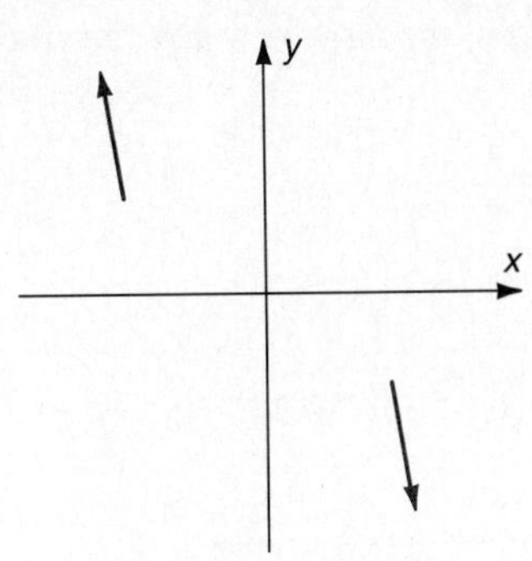

Fig. 4.41

INTERVAL ON x-AXIS	TEST NUMBER x	y-COORDINATE f(x)	GRAPH ABOVE OR BELOW x-AXIS
$(-\infty, -4)$	-5	18	Above
$(-4, -2)$	-3	-4	Below
$(-2, 1)$	0	8	Above
$(1, +\infty)$	2	-24	Below

Since f is an *odd-degree* polynomial function with a *negative* leading coefficient, its behavior at the ends will be of the form illustrated in Figure 4.41. The graph is shown in Figure 4.42.

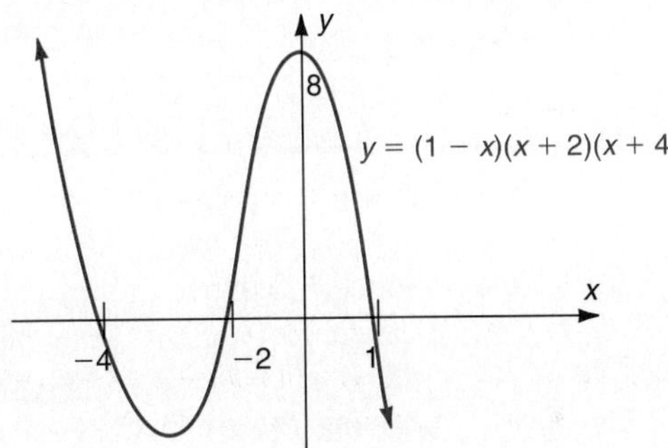

Fig. 4.42

□

EXAMPLE 5 Sketch the graph of $f(x) = (x - 2)(x - 1)(x + 1)(x + 2)$.

Solution If we multiply the factors, we see that f is a fourth-degree polynomial with a leading coefficient of 1. The x-intercepts are -2, -1, 1, and 2. The x-intercepts divide the x-axis into five intervals (Figure 4.43). The intervals are $(-\infty, -2)$, $(-2, -1)$, $(-1, 1)$, $(1, 2)$, and $(2, +\infty)$.

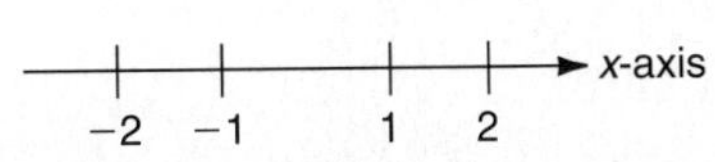

Fig. 4.43

INTERVAL ON x-AXIS	TEST NUMBER x	y-COORDINATE f(x)	GRAPH ABOVE OR BELOW x-AXIS
$(-\infty, -2)$	-3	40	Above
$(-2, -1)$	$-\frac{3}{2}$	$-\frac{35}{16}$	Below
$(-1, 1)$	0	4	Above
$(1, 2)$	$\frac{3}{2}$	$-\frac{35}{16}$	Below
$(2, +\infty)$	3	40	Above

Since f is an *even-degree* polynomial function with a *positive* leading coefficient, its behavior at the ends will be of the form illustrated in Figure 4.44. A rough sketch of the graph is shown in Figure 4.45.

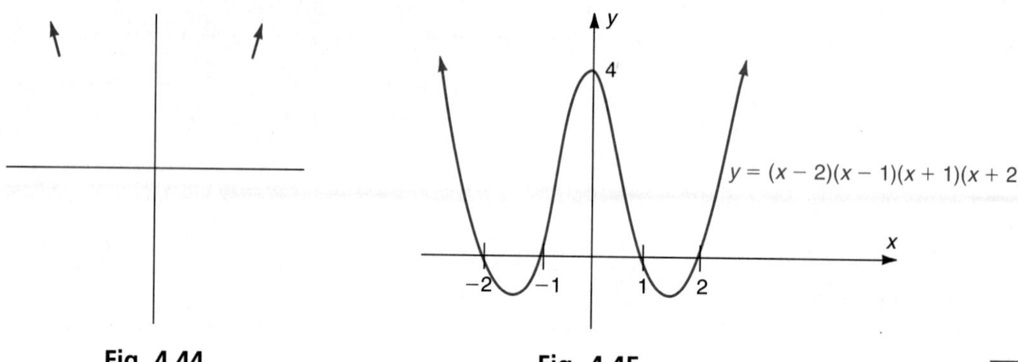

Fig. 4.44

Fig. 4.45

EXAMPLE 6 Sketch the graph of $f(x) = x(1 - x)(x + 2)(x + 4)$.

Solution Notice that this is a fourth-degree polynomial with a leading coefficient of -1. The x-intercepts are 0, 1, -2, and -4. These points divide the x-axis into the five intervals, $(-\infty, -4)$, $(-4, -2)$, $(-2, 0)$, $(0, 1)$, and $(1, +\infty)$. We make a table describing the graph.

INTERVAL ON x-AXIS	TEST NUMBER x	y-COORDINATE $f(x)$	GRAPH ABOVE OR BELOW x-AXIS
$(-\infty, -4)$	-5	-90	Below
$(-4, -2)$	-3	12	Above
$(-2, 0)$	-1	-6	Below
$(0, 1)$	$\frac{1}{2}$	$\frac{45}{16}$	Above
$(1, +\infty)$	2	-48	Below

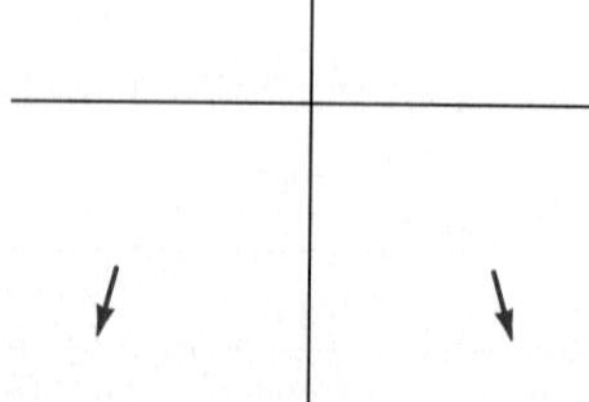

Fig. 4.46

Since f is an *even-degree* polynomial function with a *negative* leading coefficient, its behavior at the ends will be of the form illustrated in Figure 4.46. A rough sketch of the graph is shown in Figure 4.47.

$y = x(1 - x)(x + 2)(x + 4)$

Fig. 4.47

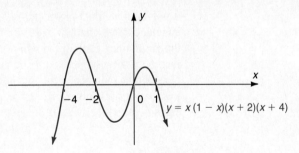

We have examined four cases of polynomial functions. In each case we observed that

1. The graph is smooth and has no breaks. (The function is continuous.)
2. The domain is R.
3. The combination of odd or even degree of the polynomial and positive or negative sense of the leading coefficient determines the behavior at the ends.
4. A rough sketch of the function could be made by finding the x-intercepts and determining whether the value of the function is positive or negative between each pair.

EXAMPLE 7 Sketch the graph of $g(x) = 5x^2 - 2x^3$.

Solution Notice that this is a third-degree polynomial with a negative leading coefficient. To find the x-intercepts, we solve the equation

$$0 = 5x^2 - 2x^3$$

$$0 = x^2(5 - 2x) \qquad \text{Factor.}$$

The x-intercepts are 0 and $\frac{5}{2}$. We use -1 as a test number in the interval $(-\infty, 0)$ and find that $g(-1) = 7$. So, the graph is above the x-axis throughout the interval $(-\infty, 0)$. We use 1 as a test number in the interval $\left(0, \frac{5}{2}\right)$ and note $g(1) = 3$. Thus, the graph is above the x-axis over the interval $\left(0, \frac{5}{2}\right)$. We use 3 as a test number in the interval $\left(\frac{5}{2}, +\infty\right)$ and note $g(3) = -9$. Thus, the graph is below the x-axis right of $\frac{5}{2}$.

From the information displayed in Figure 4.48 and knowing this is the graph of

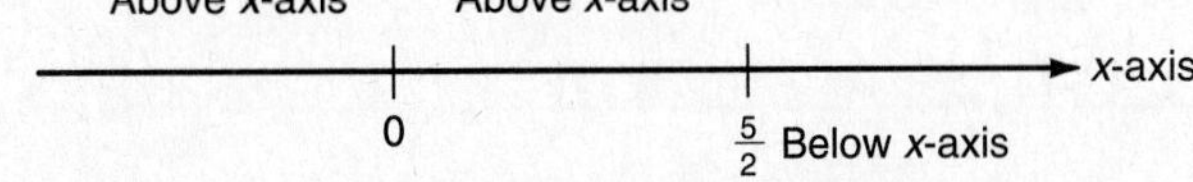

Fig. 4.48

a polynomial function of odd degree with a negative leading coefficient, we can conclude some characteristics of the graph which we display in Figure 4.49. Since the graph must contain the point (0, 0) but is positive on either side of (0, 0), it must just touch the x-axis at that point and *not cross it.* The graph is shown in Figure 4.50. Notice that the graph just touches the x-axis at the origin. This is a result of the x-intercept of 0 coming from a squared factor, x^2. The y-coordinates did not change signs at the x-intercept of 0. However, the signs of the y-coordinates changed at the other intercept, $\frac{5}{2}$.

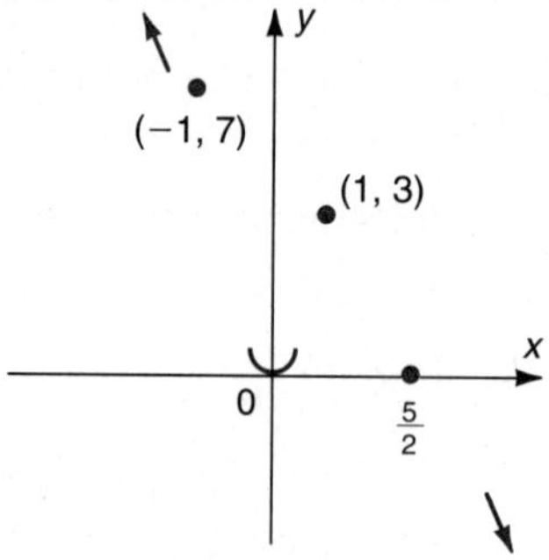

Fig. 4.49

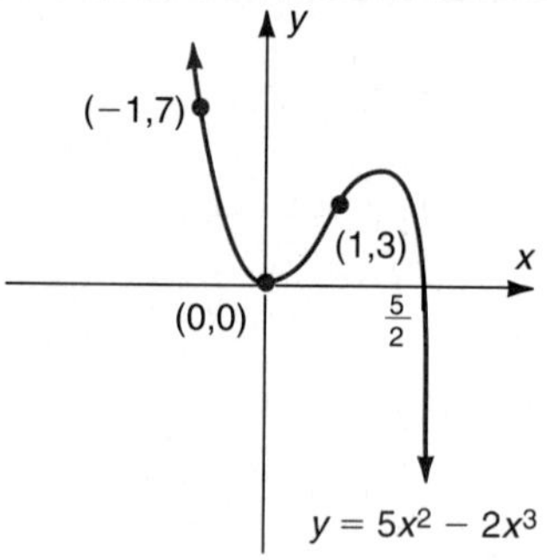

Fig. 4.50 □

EXAMPLE 8 Sketch the graph of $g(x) = \frac{1}{8}(x + 2)^2(x - 2)^2$.

Solution This is a fourth-degree polynomial function with a leading coefficient of $\frac{1}{8}$. The x-intercepts are ± 2. The graph as shown in Figure 4.51 will be above the x-axis. It will just touch the x-axis at both intercepts. (This can be verified by using test points in the intervals.)

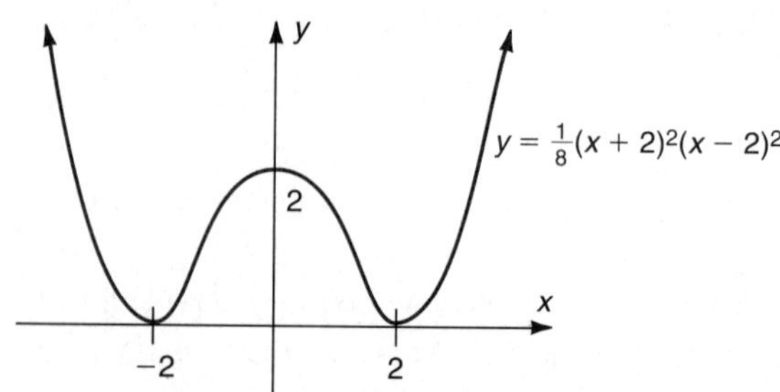

Fig. 4.51

Notice that this is an even-degree polynomial. The domain is R, while the range is $[0, +\infty)$. □

Sketching Polynomial Functions

1. Find the x-intercepts.
2. Determine the end behavior of the graph from the degree of the polynomial and whether the leading coefficient is positive or negative.
3. If an x-intercept was obtained from a linear factor to an *odd* power, the graph will *cross* the x-axis at that intercept.
4. If an x-intercept was obtained from a linear factor to an *even* power, the graph will *just touch* the x-axis at that intercept and not cross.

EXAMPLE 9 Sketch the graph of $h(x) = x^3(x + 1)^4(3x - 2)$.

Solution The leading coefficient is 3 (positive), and the degree of the polynomial is 8 (even). Thus, both "ends" of the graph will extend upward. The x-intercepts are 0, -1, and $\frac{2}{3}$. Since the x-intercept 0 came from a linear factor to an odd power, x^3, the graph will cross the x-axis at 0. At -1, the graph will just touch the x-axis because $(x + 1)^4$ is a linear factor to an even power. The graph will also cross the x-axis at $\frac{2}{3}$ since $(3x - 2)$ is a linear factor to an odd power. The graph is shown in Figure 4.52. □

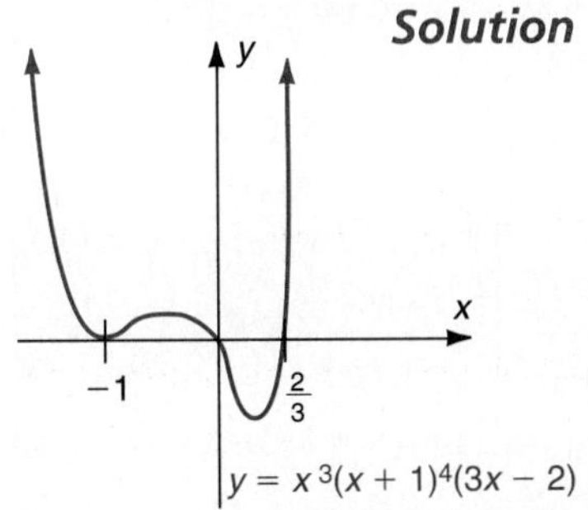

Fig. 4.52

GRAPHING CALCULATOR BOX

Graphing Polynomial Functions; Maxima and Minima

The graphing calculator is a fine tool for graphing polynomial functions. Start with the default window and graph the polynomial function $y = x^5 - 3x^4 - x^3 - x^2 + 5x - 2$. Remember, the domain of a polynomial function is $(-\infty, +\infty)$ and there are no breaks or sharp points in the graph. The screen is just a window looking at part of the graph. Ideally, the window should at least show all the high and low points as well as all intercepts of the graph. Change the range to $[-4, 4]$ for x and $[-40, 10]$ for y and regraph. Now we appear to have the shape of the graph. (We will see how to make sure in Chapter 5.)

Now let's check out some of the details. What is happening between 0 and 1 on the x-axis? Use the box feature of ZOOM to find out. After a box or two we see a screen showing two x-intercepts between 0 and 1. We can now sketch the graph of the function.

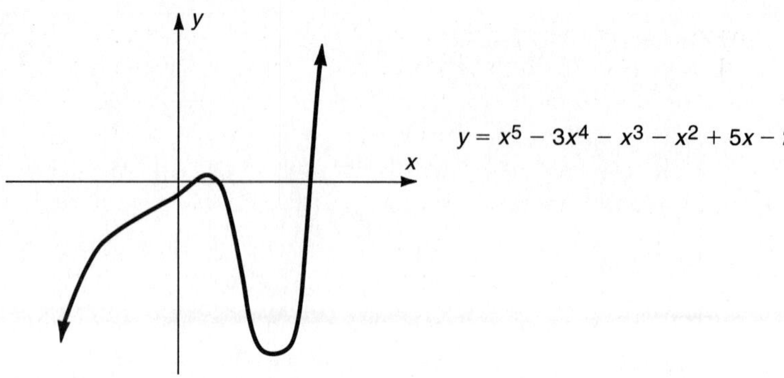

Relative Maxima and Minima

Set the range to [0, 1] for x and [−0.1, 0.3] for y and graph. Use TRACE to approximate the high point of this function between 0 and 1. Such a high point is called a **relative maximum** or **local maximum.**

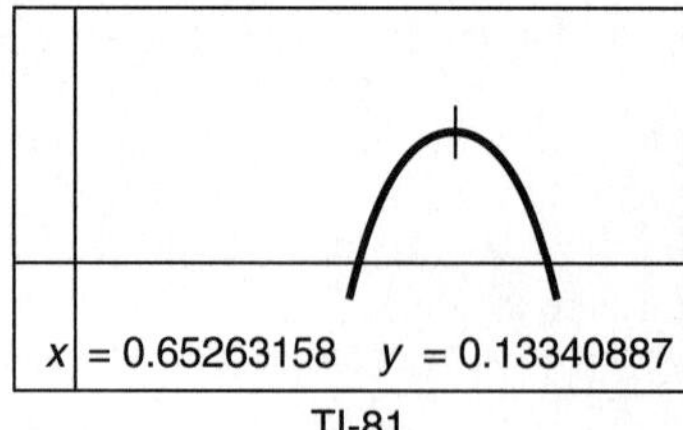

TI-81

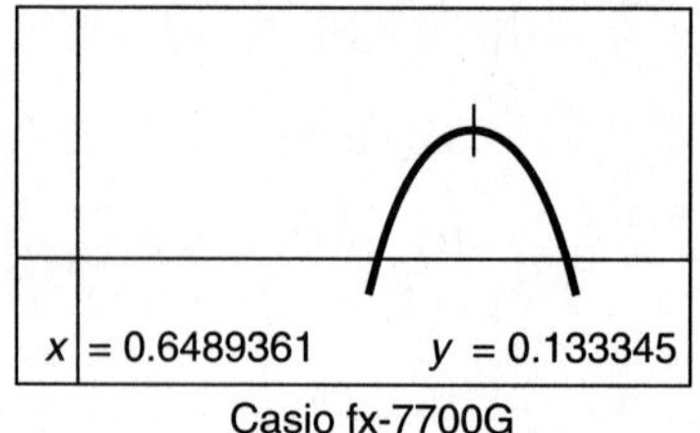

Casio fx-7700G

We can improve the accuracy by zooming and tracing until the digits in successive windows begin to agree. With a window of [0.6, 0.7] by [0.1, 0.14] we have,

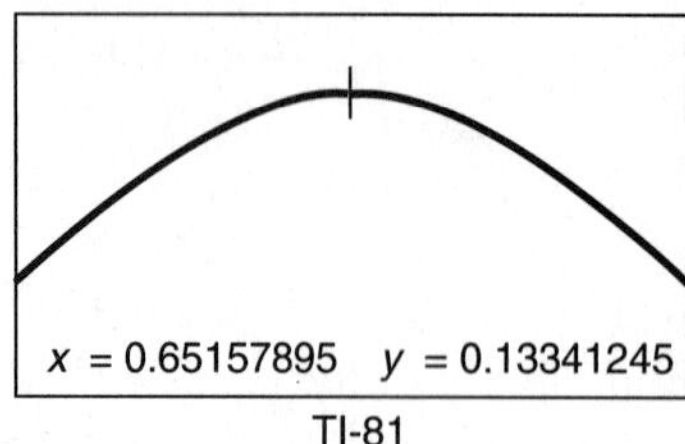

TI-81

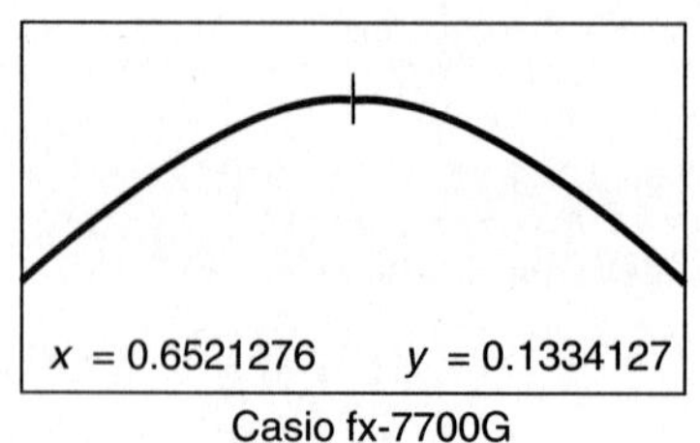

Casio fx-7700G

Zoom and trace again. With a window of [0.6, 0.675] by [0.1, 0.14] we have,

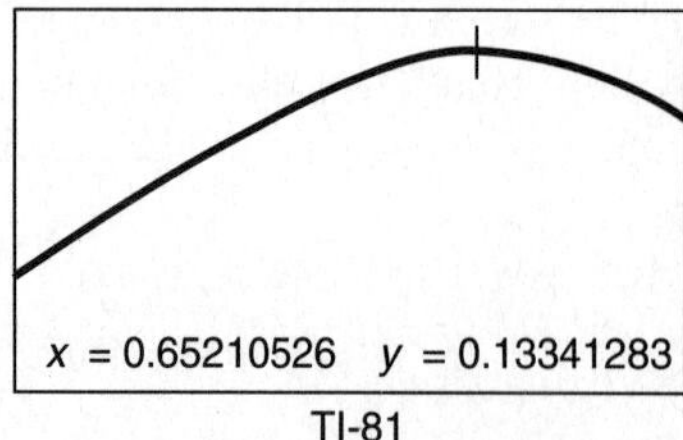

TI-81

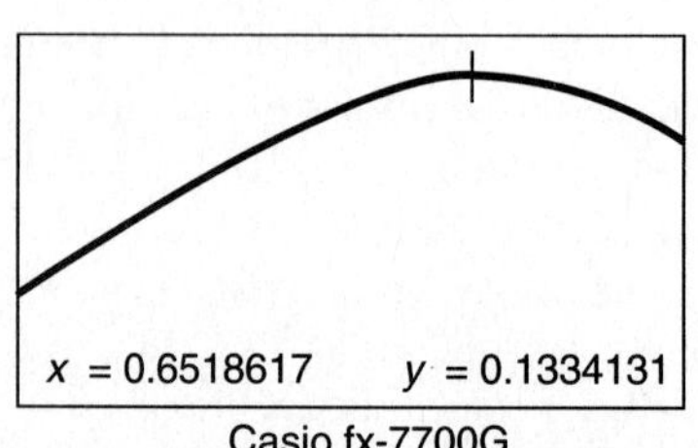

Casio fx-7700G

Now notice in the last two windows that the y-coordinate is the same to the ten-thousandths digit and the x-coordinate will round the same in the thousandths digit. The coordinates of the relative maximum to the nearest thousandth are: (0.652, 0.133). Since zoom boxes will produce different window dimensions than the dimensions shown above, slight differences will occur in the coordinates shown. However, the result will be the same.

Set the range back to $[-4, 4]$ and $[-40, 10]$ and regraph. Notice the low point between $x = 2$ and $x = 3$. Such a low point is called a **relative minimum** or **local minimum.** Approximate this relative minimum to the nearest thousandth. (Answer: (2.631, −31.660))

PROBLEM SET 4.2

Warm-ups

In Problems 1 through 16, sketch the graph of each function.

For Problems 1 through 4, see Examples 1 and 2.

1. $f(x) = x^3 + 2$ **2.** $f(x) = 1 - x^5$ **3.** $f(x) = (x - 3)^4 + 1$ **4.** $f(x) = -(x + 1)^6 - 2$

For Problems 5 and 6, see Example 3.

5. $f(x) = x(x + 3)(x - 2)$ **6.** $f(x) = (2x + 3)(x - 1)(x - 3)$

For Problems 7 and 8, see Example 4.

7. $f(x) = (2 - x)(x + 3)(x - 5)$ **8.** $f(x) = x(4 - 3x)(x + 2)$

For Problems 9 and 10, see Example 5 and 6.

9. $f(x) = x(x + 2)(x - 2)(x + 1)$ **10.** $f(x) = (3 - x)(x + 5)(x - 1)(x + 2)$

For Problems 11 through 16, see Examples 7 and 8.

11. $f(x) = (x - 3)^2(2x + 1)$ **12.** $f(x) = (x - 5)^2(3 - 4x)$ **13.** $f(x) = (x + 3)^2(x - 1)^2$

14. $f(x) = (4 + x)^2(4 - x)^2$ **15.** $f(x) = x(x + 3)^2(x - 2)^3$ **16.** $f(x) = (x - 1)^4(5x + 3)^2(2x - 7)$

Practice Exercises

In Problems 17 through 28, sketch the graph of each function. Indicate where $f(x) < 0$.

17. $f(x) = x(x + 1)(x - 1)$ **18.** $f(x) = x(x - 3)(x + 1)$ **19.** $f(x) = x^2(2x + 3)^2$

20. $f(x) = x(5x - 2)^2$ **21.** $f(x) = (2 - 3x)(x + 1)(x - 3)$ **22.** $f(x) = (5 - 3x)(2x + 5)^2$

23. $f(x) = x^5 - 9x^3$ **24.** $f(x) = x^3 - x^2 - 12x$ **25.** $f(x) = 2x^4(3x - 2)^2(x + 2)^4$

26. $f(x) = -2x^4(7x + 12)^2(4x - 9)^3$ **27.** $f(x) = (1 - 2x)(2 - x)(3 + x)^2$ **28.** $f(x) = (2 - x)^2(5 - x)^3$

In Problems 29 through 32, the graph of a polynomial function is given. Write a possible rule for each function.

29.

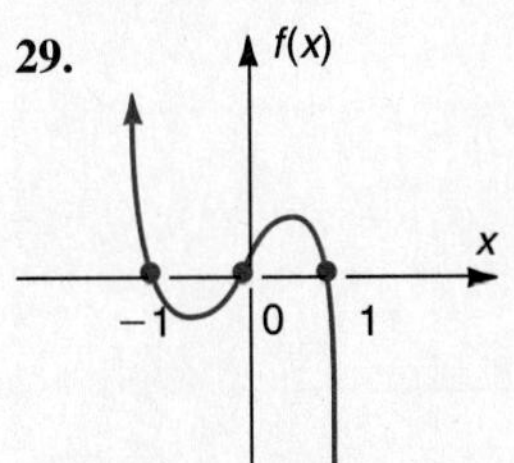

30.

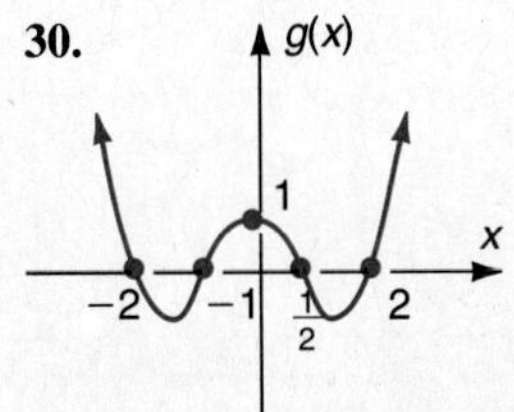

31.

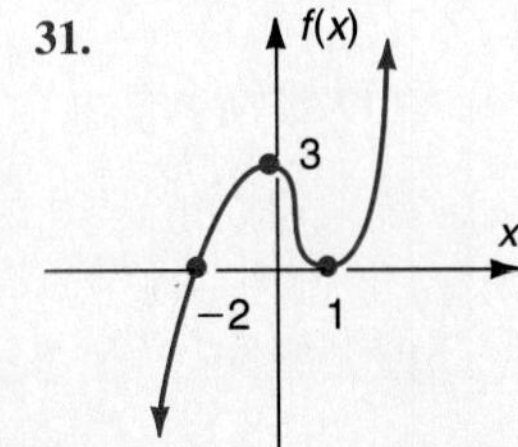

32.

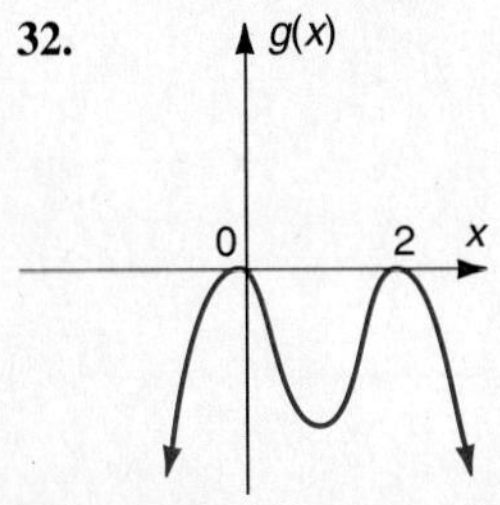

In Problems 33 through 36, the x-intercepts for a polynomial function are given. Write a polynomial function that has the given x-intercepts.

33. 2, 3, 4

34. $-2, 0, 2$

35. $-1, -\frac{1}{2}, \frac{1}{2}$

36. $-3, -2, -1$

In Problems 37 through 40, sketch each graph given that $f(x) = (x - 1)(x + 3)(x + 1)$.

37. $y = f(x) + 3$

38. $y = -f(x)$

39. $y = f(-x)$

40. $y = f(x - 1)$

41. Make up a rule for an odd polynomial function.

42. Make up a rule for an even polynomial function.

Challenge Problems

In Problems 43 through 45, sketch the graph of each function.

43. $f(x) = (x^2 + 1)(x - 3)$

44. $f(x) = (x^2 + 4)(x - 2)^2$

45. $f(x) = |x(x - 1)(x + 1)|$

46. A box is made by cutting the same size square from each corner of a rectangular piece of cardboard and then folding up the sides. If the cardboard is 15 centimeters by 20 centimeters, write a function that expresses the volume of the box in terms of the side of the square. Guess the size of the square that should be cut out to produce the maximum volume. Approximate the maximum volume.

For Graphing Calculators

In Problems 47 through 50, graph each function and approximate the coordinates of every relative maximum and relative minimum to the nearest thousandth.

47. $f(x) = 4x^3 - 11x^2 + 6x + 1$

48. $g(x) = 2 + x + \frac{2}{5}x^2 - \frac{1}{3}x^3$

49. $F(x) = x^4 - 3x^3 - 2x^2 + 2x + 1$

50. $G(x) = x^5 - 3x^3 + 2x + 1$

51. Jose Ortega predicts the total number of limes produced by an acre of his trees using the function

$$N(x) = -0.17x^3 + 6.8x^2 + 532x$$

where x is the number of trees and $N(x)$ is the number of limes produced. What is the maximum number of limes produced? How many trees per acre should Jose plant to produce the maximum number of limes? (Hint: Finding some function values might be helpful in setting the window.)

IN YOUR OWN WORDS . . .

52. Describe the similarities in the graphs of odd-degree polynomial functions.

53. Describe the similarities in the graphs of even-degree polynomial functions.

54. What is the relationship between an even-degree polynomial function and an even function?

55. What is the relationship between an odd-degree polynomial function and an odd function?

4.3 RATIONAL FUNCTIONS

We continue the study of functions by looking at the quotient of two polynomials.

> **Definition: Rational Functions**
>
> A function of the form $f(x) = \dfrac{g(x)}{h(x)}$, where $g(x)$ and $h(x)$ are polynomials, is called a rational function.

Unlike polynomial functions, rational functions may have breaks in their graphs. Breaks in the graph are created at numbers that are not in the domain of the rational function. These breaks occur at numbers for which the value of the polynomial in the denominator is 0.

EXAMPLE 1 Find the domain of each rational function.

(a) $f(x) = \dfrac{3}{x + 2}$ (b) $g(x) = \dfrac{x + 2}{x}$ (c) $h(x) = \dfrac{x - 3}{x^2 - 9}$ (d) $r(x) = \dfrac{1}{x^2 + 1}$

Solutions (a) Since the denominator will be 0 if x is -2, the domain of $f(x) = \dfrac{3}{x + 2}$ is the set of all real numbers except -2. In interval notation, $(-\infty, -2) \cup (-2, +\infty)$.

(b) The domain of $g(x) = \dfrac{x + 2}{x}$ consists of all real numbers except 0. Written in interval notation, the domain is $(-\infty, 0) \cup (0, +\infty)$.

(c) $h(x) = \dfrac{x - 3}{x^2 - 9} = \dfrac{x - 3}{(x - 3)(x + 3)}$

Although there is a common factor in the rational expression, the domain of the function includes all real numbers except ± 3. In interval notation the set is $(-\infty, -3) \cup (-3, 3) \cup (3, +\infty)$.

(d) No real numbers make the denominator of $r(x) = \dfrac{1}{x^2 + 1}$ have a value of 0. The domain is R. □

Vertical Asymptotes

The exclusion of numbers from the domain of a rational function creates a feature in the graph of the function that we have not seen before. Consider the rational function $f(x) = \dfrac{1}{x - 1}$. The domain does not include 1. However, every real number except 1 is in the domain. This means that all numbers very close to 1 are in the domain. The interesting feature of the graph occurs when x gets *very close* to 1.

There are two ways to get close to 1. We can approach it from the left side or from the right side as in Figure 4.53.

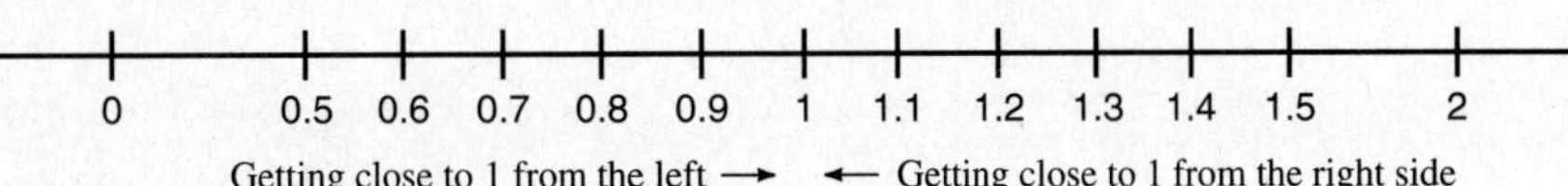

Fig. 4.53

Figure 4.54 shows a closer view.

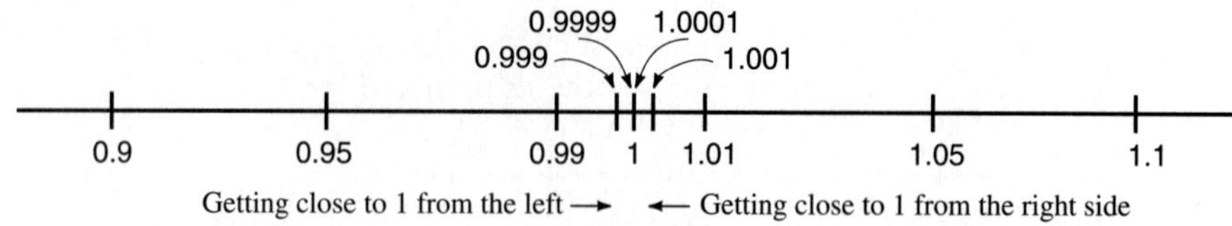

Fig. 4.54

We calculate the function values for x close to 1. This can be done quickly with a calculator using the reciprocal key [1/x]. For the function $f(x) = \dfrac{1}{x-1}$, for example, to calculate $f(1.05)$, we evaluate $\dfrac{1}{1.05-1}$ by entering [1.05] [−] [1] [=] [1/x] and read 20 on the display.

$$f(x) = \frac{1}{x-1}$$

$f(x)$ values for x getting close to 1 from the right

x	1.5	1.4	1.3	1.2	1.1	1.05	1.01	1.001	1.0001
$f(x)$	2.0	2.5	3.3	5.0	10.0	20.0	100.0	1000.0	10,000.0

We can see from the table that as x gets closer to 1 from the right side, $f(x)$ is getting very large. That is, $f(x) \to +\infty$ as $x \to 1$ from the right.

$$f(x) = \frac{1}{x-1}$$

$f(x)$ values for x getting close to 1 from the left

x	0.5	0.6	0.7	0.8	0.9	0.95	0.99	0.999	0.9999
$f(x)$	−2.0	−2.5	−3.3	−5.0	−10.0	−20.0	−100.0	−1000.0	−10,000.0

We see from this table that $f(x) \to -\infty$ as $x \to 1$ from the left.

To describe the behavior of x and $f(x)$ in this situation, we use some special notation that helps us write down the conclusions we made from the tables. For q any real number,

$x \to q$ means that x is approaching q from both sides.

$x \to q^+$ means that x is approaching q from the right side.

$x \to q^-$ means that x is approaching q from the left side.

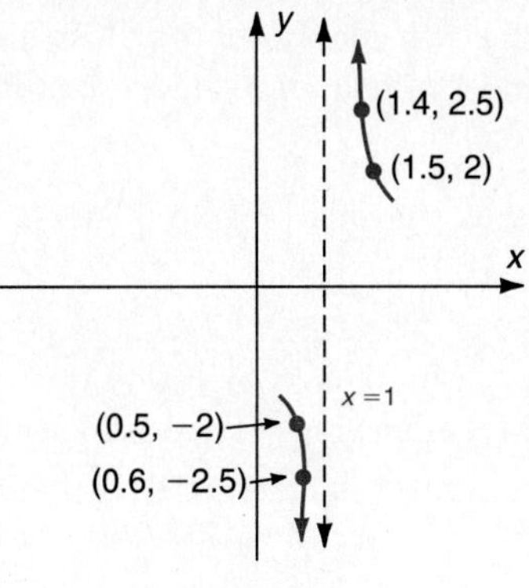

Fig. 4.55

With this notation, we can describe the behavior of f close to 1.

$$\text{If } x \to 1^-, \text{ then } f(x) \to -\infty.$$

This means that as x approaches 1 from the left, $f(x)$ *decreases* without bound.

$$\text{If } x \to 1^+, \text{ then } f(x) \to +\infty.$$

This means that as x approaches 1 from the right, $f(x)$ *increases* without bound.

If we look at the graph of f close to 1, we see that it is approaching the vertical line $x = 1$. However, the graph will never touch or cross the line $x = 1$, because 1 is not in the domain of f. We call the line $x = 1$ a **vertical asymptote.** We draw it as a dashed line to indicate that it is not part of the graph, as shown in Figure 4.55.

Horizontal Asymptotes

Thus far, we have looked at what happens on the graph of f when x is close to 1. We must look at other numbers in the domain to complete the graph. First we choose some numbers less than 1.

$$f(x) = \frac{1}{x - 1}$$

$f(x)$ values for x numbers less than 1

x	0	−1	−2	−3	−5	−10	−100	−1000	−10,000
$f(x)$	−1	−0.5	−0.3	−0.25	−0.17	−0.09	−0.0099	−0.00099	−0.0000999

This table suggests that as x decreases without bound, $f(x)$ gets closer and closer to 0 but stays negative. We write

$$\text{If } x \to -\infty, \text{ then } f(x) \to 0.$$

$$f(x) = \frac{1}{x - 1}$$

$f(x)$ values for x numbers greater than 1

x	2	3	4	5	10	100	1000	10,000	100,000
$f(x)$	1	0.5	0.33	0.25	0.11	0.01	0.001	0.0001	0.00001

This table suggests that as x increases without bound, $f(x)$ is positive but gets closer and closer to 0. We write

$$\text{If } x \to +\infty, \text{ then } f(x) \to 0.$$

These tables suggest that $f(x)$ approaches 0 as x increases without bound and

$f(x)$ approaches 0 as x decreases without bound. The graph gets closer and closer to the x-axis but never actually touches or crosses it. The equation of the x-axis is $y = 0$. We say that $y = 0$ is a **horizontal asymptote.**

We can now draw the graph of $f(x) = \dfrac{1}{x-1}$, as shown in Figure 4.56.

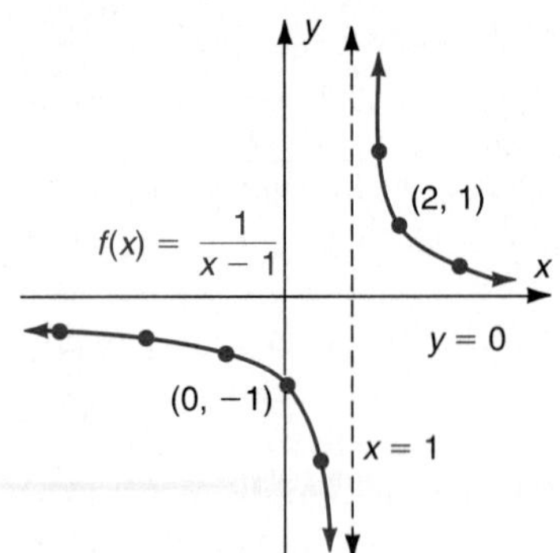

Fig. 4.56

Finding Asymptotes

The line $x = a$ is a vertical asymptote of the graph of the function f if at least one of the following statements is true.

1. If $x \to a^+$, then $f(x) \to +\infty$.
2. If $x \to a^+$, then $f(x) \to -\infty$.
3. If $x \to a^-$, then $f(x) \to +\infty$.
4. If $x \to a^-$, then $f(x) \to -\infty$.

The line $x = a$ is a vertical asymptote in each graph shown in Figure 4.57.

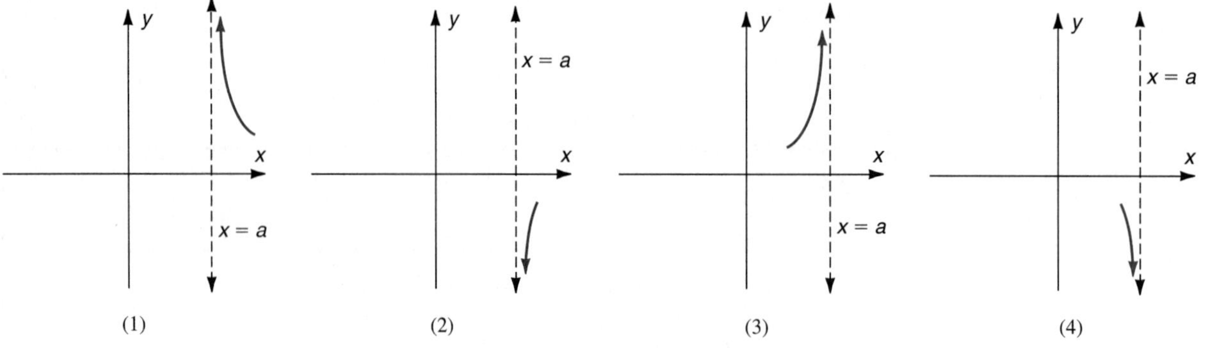

Fig. 4.57

The line $y = b$ is a horizontal asymptote of the graph of f if at least one of the following statements is true.

1. If $x \to +\infty$, then $f(x) \to b$.
2. If $x \to -\infty$, then $f(x) \to b$.

The line $y = b$ is a horizontal asymptote of each graph in Figure 4.58. Notice that it is possible for a graph to intersect a horizontal asymptote.

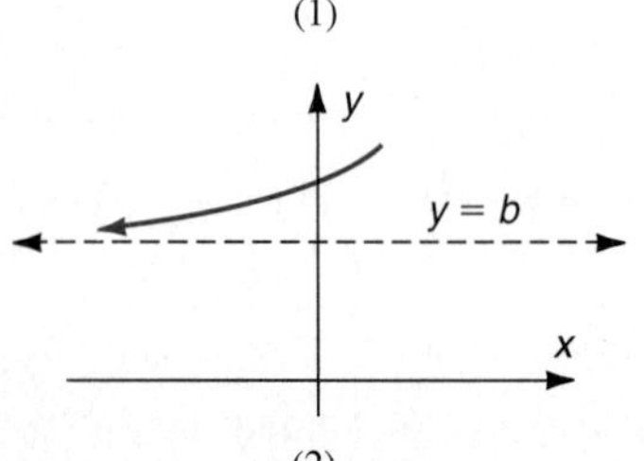

Fig. 4.58

Vertical Asymptotes of Rational Functions

Suppose $f(x) = \dfrac{g(x)}{h(x)}$ is a rational function in reduced form. If $h(a) = 0$, then $x = a$ will be a vertical asymptote.

The degree of the polynomials in the numerator and the denominator of a rational function determine the horizontal asymptotes. Consider the general rational function

$$f(x) = \frac{a_m x^m + a_{m-1}x^{m-1} + \cdots + a_1 x + a_0}{b_n x^n + b_{n-1}x^{n-1} + \cdots + b_1 x + b_0};\ a_m \neq 0,\ b_n \neq 0.$$

where both leading coefficients a_m and b_n are not zero.

There are three possibilities,

1. $m > n$. (The degree of the numerator is *greater than* the degree of the denominator.)
2. $m = n$. (The degree of the numerator is *equal to* the degree of the denominator.)
3. $m < n$. (The degree of the numerator is *less than* the degree of the denominator.)

We will look for horizontal asymptotes in examples of the three cases and leave proofs to the exercises.

Case 1: $m > n$ (The degree of the numerator is *greater than* the degree of the denominator.)

$$f(x) = \frac{2x^3 - 5x^2 + x + 14}{5x^2 + 9x - 1}$$

Divide the numerator and denominator by the highest power of x in the rational expression.

$$= \frac{2 - \frac{5}{x} + \frac{1}{x^2} + \frac{14}{x^3}}{\frac{5}{x} + \frac{9}{x^2} - \frac{1}{x^3}}$$

Notice that *every term* in the rational expression approaches zero as $x \to +\infty$ *except* the leading term in the numerator (which is the leading coefficient). Since the numerator approaches 2 and the denominator approaches 0 as $x \to +\infty$, there is no horizontal asymptote in the positive direction. However, since the same effect occurs as $x \to -\infty$, the graph of the function has no horizontal asymptotes.

Case 2: $m = n$ (The degree of the numerator is *equal to* the degree of the denominator.)

$$f(x) = \frac{2x^3 - 5x^2 + x + 14}{5x^3 + 9x - 1}$$

Divide the numerator and denominator by the highest power of x in the rational expression.

$$= \frac{2 - \frac{5}{x} + \frac{1}{x^2} + \frac{14}{x^3}}{5 + \frac{9}{x^2} - \frac{1}{x^3}}$$

Notice that in this case all the terms in the rational expression approach zero as $x \to +\infty$ or $x \to -\infty$, *except* the leading terms in *both* the numerator and the denominator. Thus, $f(x) \to \frac{2}{5}$ as $x \to +\infty$ or as $x \to -\infty$. The graph has a horizontal asymptote $y = \frac{2}{5}$.

Case 3: $m < n$ (The degree of the numerator is *less than* the degree of the denominator.)

$$f(x) = \frac{2x^3 - 5x^2 + x + 14}{5x^4 + 9x - 1}$$

Divide the numerator and denominator by the highest power of x in the rational expression.

$$= \frac{\frac{2}{x} - \frac{5}{x^2} + \frac{1}{x^3} + \frac{14}{x^4}}{5 + \frac{9}{x^3} - \frac{1}{x^4}}$$

In this case, all the terms in the rational expression approach zero as $x \to +\infty$ or $x \to -\infty$, *except* the leading term in the denominator. Thus, the numerator goes to zero but the denominator goes to 5. Therefore, $f(x) \to 0$, and the graph has $y = 0$ (the x-axis) for a horizontal asymptote.

Finding Horizontal Asymptotes of Rational Functions

Suppose $f(x) = \dfrac{a_m x^m + a_{m-1}x^{m-1} + \cdots + a_1 x + a_0}{b_n x^n + b_{n-1}x^{n-1} + \cdots + b_1 x + b_0}$, where $a_m \neq 0$ and $b_n \neq 0$, is a rational function in reduced form.

1. If $m > n$, then there is no horizontal asymptote.
2. If $m = n$, then $y = \dfrac{a_m}{b_n}$ is a horizontal asymptote of the graph of f.
3. If $m < n$, then $y = 0$ is a horizontal asymptote of the graph of f.

EXAMPLE 2 Find the vertical and horizontal asymptotes for each function.

(a) $f(x) = \dfrac{1}{x}$ (b) $g(x) = \dfrac{2x}{5x + 3}$ (c) $h(x) = \dfrac{-4x^2}{(x + 5)(x - 2)}$

(d) $t(x) = \dfrac{1}{x^2 + 1}$ (e) $s(x) = \dfrac{5x^4}{x + 8}$

Solutions (a) In $f(x) = \dfrac{1}{x}$, the degree of the numerator is 0 and the degree of the denominator is 1. So, $y = 0$ is a horizontal asymptote and $x = 0$ is a vertical asymptote.

(b) The degree of both the numerator and the denominator of $g(x) = \dfrac{2x}{5x + 3}$ is 1. Thus, $y = \dfrac{2}{5}$ is a horizontal asymptote, while $x = -\dfrac{3}{5}$ is a vertical asymptote. Notice that the horizontal asymptote is simply the ratio of the leading coefficients in the numerator and denominator.

(c) The degree of both the numerator and the denominator is 2 for $h(x) = \dfrac{-4x^2}{(x + 5)(x - 2)}$. So, $y = -4$ is the horizontal asymptote and $x = -5$ and $x = 2$ are vertical asymptotes.

(d) In $t(x) = \dfrac{1}{x^2 + 1}$, the degree of the numerator is 0, while the degree of the denominator is 2. Thus, $y = 0$ is the horizontal asymptote. Since the denominator will never have a value of 0, there is no vertical asymptote.

(e) Since the degree of the numerator is larger than the degree of the denominator in $s(x) = \dfrac{5x^4}{x + 8}$, there is no horizontal asymptote. The vertical asymptote is $x = -8$. □

Procedure for Sketching Rational Functions

1. Find any vertical asymptotes.
2. Find any horizontal asymptotes. Determine if the graph intersects any horizontal asymptotes.
3. Find x- and y-intercepts.
4. Plot several points, selecting one from each interval as determined by the x-intercepts and the vertical asymptotes.
5. Sketch the graph as a smooth curve over the domain of the function.

In Example 9 we illustrate the case where a rational function is not in reduced form.

EXAMPLE 3 Sketch the graph of $f(x) = \dfrac{-3}{x+2}$.

Solution *Vertical asymptote:* $x = -2$

Horizontal asymptote: $y = 0$

x-intercepts: Let y be 0.

$$0 = \frac{-3}{x+2}$$

$$0 = -3 \qquad \text{Clear fractions.}$$

There are no solutions. Hence, there are no x-intercepts. The graph does not touch the x-axis. Hence the graph does not cross the horizontal asymptotes.

y-intercept: Let x be 0. $\quad y = \dfrac{-3}{0+2} = \dfrac{-3}{2}$

The y-intercept is $-\dfrac{3}{2}$.

Plot selected points:

x	-1	0	-3	-4
$f(x)$	-3	$-\frac{3}{2}$	3	$\frac{3}{2}$

Sketch the graph. See Figure 4.59.

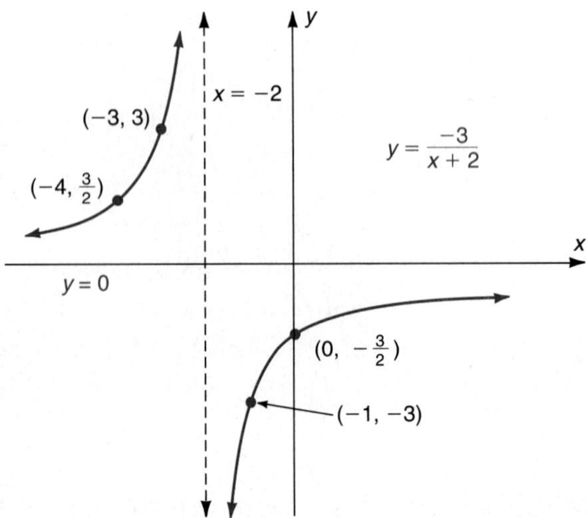

Fig. 4.59

□

EXAMPLE 4 Sketch the graph of $g(x) = \dfrac{2x + 3}{x - 1}$.

Solution *Vertical asymptote:* $x = 1$

Horizontal asymptote: $y = 2$

Does the graph intersect the horizontal asymptote? Let y be 2.

$$2 = \frac{2x + 3}{x - 1}$$

$$2(x - 1) = 2x + 3 \qquad \text{Clear fractions.}$$

$$2x - 2 = 2x + 3$$

$$-2 = 3$$

There are no solutions. The graph does not intersect the horizontal asymptote.

x-intercepts: Let y be 0.

$$0 = \frac{2x + 3}{x - 1}$$

$$0 = 2x + 3 \qquad \text{Clear fractions.}$$

$$-3 = 2x$$

$$-\frac{3}{2} = x$$

The x-intercept is $-\dfrac{3}{2}$.

y-intercept: Let x be 0.

$$y = \frac{2(0) + 3}{0 - 1} = \frac{3}{-1} = -3$$

The y-intercept is -3.

Plot selected points:

x	-2	-1	2	3
$f(x)$	$\frac{1}{3}$	$-\frac{1}{2}$	7	$\frac{9}{2}$

Sketch the graph. See Figure 4.60.

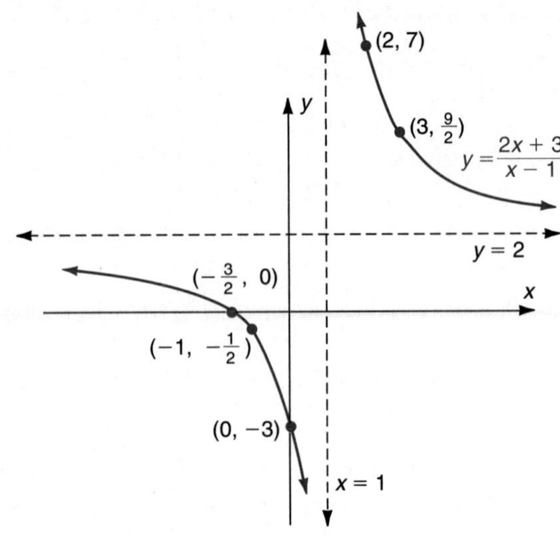

Fig. 4.60

□

EXAMPLE 5 Sketch the graph of $h(x) = \dfrac{3x^2 - 4x}{6 - 2x^2}$.

Solution *Vertical asymptotes:* $x = \pm\sqrt{3}$

Horizontal asymptote: $y = -\dfrac{3}{2}$

Does the graph intersect the horizontal asymptote? Let y be $-\dfrac{3}{2}$.

$$-\frac{3}{2} = \frac{3x^2 - 4x}{6 - 2x^2}$$

$$6x^2 - 8x = -18 + 6x^2 \qquad \text{Clear fractions.}$$

$$x = \frac{9}{4}$$

The graph crosses $y = -\dfrac{3}{2}$ at $\left(\dfrac{9}{4}, -\dfrac{3}{2}\right)$.

x-intercepts: Let y be 0. $\qquad 0 = \dfrac{3x^2 - 4x}{6 - 2x^2}$

$$0 = 3x^2 - 4x$$

$$x = 0 \qquad \text{or} \qquad x = \frac{4}{3}$$

The x-intercepts are 0 and $\frac{4}{3}$.

y-intercept: Let x be 0.

$$y = \frac{0}{6} = 0$$

The y-intercept is 0.

Plot selected points:

x	-1	1	-3
$h(x)$	$\frac{7}{4}$	$-\frac{1}{4}$	$-\frac{13}{4}$

Sketch the graph. See Figure 4.61.

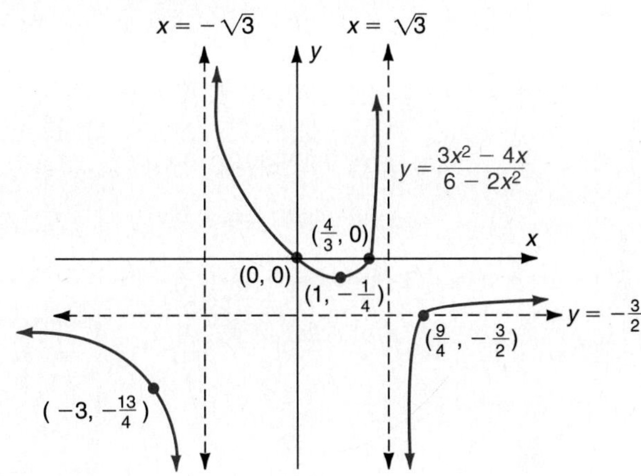

Fig. 4.61

□

EXAMPLE 6 Sketch the graph of $g(x) = \frac{1}{x^2 + 1}$.

Solution *Vertical asymptotes:* There are no vertical asymptotes.
Horizontal asymptote: $y = 0$

x-intercepts: Let y be 0. $\quad 0 = \frac{1}{x^2 + 1}$

$0 = 1 \qquad$ Clear fractions.

There are no solutions. The graph does not intersect the x-axis.

y-intercept: Let x be 0. $\quad y = \frac{1}{0^2 + 1} = \frac{1}{1} = 1$

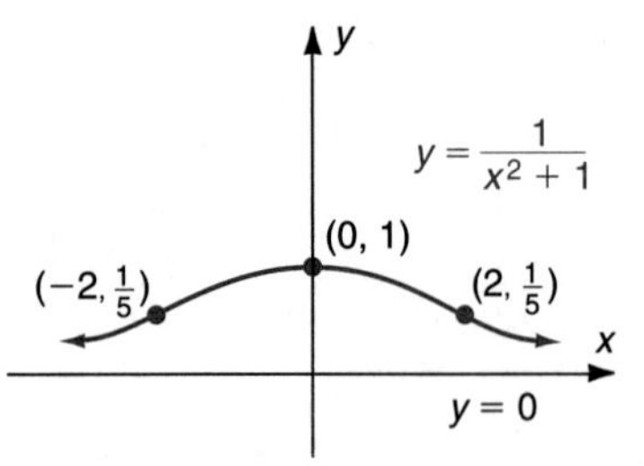

Fig. 4.62

The y-intercept is 1.

Plot selected points:

x	-2	2
$f(x)$	$\frac{1}{5}$	$\frac{1}{5}$

Sketch the graph. See Figure 4.62. □

EXAMPLE 7 Sketch the graph of $h(x) = \dfrac{x}{(x-1)(x+2)}$.

Solution *Vertical asymptotes:* $x = 1$, $x = -2$

Horizontal asymptote: $y = 0$

x-intercepts: Let y be 0.

$$0 = \frac{x}{(x-1)(x+2)}$$

Clear fractions.

$$0 = x$$

The x-intercept is 0. The graph intersects the horizontal asymptote when x is 0. It goes through the origin.

y-intercept: Let x be 0.

$$y = \frac{0}{(0-1)(0+2)} = 0$$

The y-intercept is 0.

Plot selected points:

x	-3	-1	2	3
$f(x)$	$-\frac{3}{4}$	$\frac{1}{2}$	$\frac{1}{2}$	$\frac{3}{10}$

Sketch the graph. See Figure 4.63.

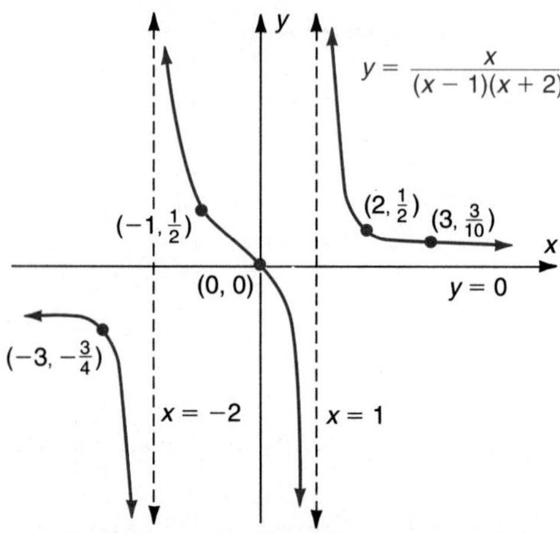

Fig. 4.63

□

Slant Asymptotes

If the degree of the numerator is *exactly 1* more than the degree of the denominator, the rational function will have a **slant asymptote.**

EXAMPLE 8 Sketch the graph of $h(x) = \dfrac{x^2 - 4}{x - 1}$.

Solution *Vertical asymptotes:* $x = 1$

Horizontal asymptote: There are no horizontal asymptotes.

Slant asymptote: To find the slant asymptote, use long division to rewrite the function.

$$h(x) = \frac{x^2 - 4}{x - 1} = x + 1 + \frac{-3}{x - 1}$$

Notice that $\dfrac{-3}{x - 1} \to 0$ as $x \to +\infty$, so $h(x) \to x + 1$ as $x \to +\infty$. The same thing happens as $x \to -\infty$. The slant asymptote is $y = x + 1$.

x-intercepts: Let y be 0.

$$0 = \frac{x^2 - 4}{x - 1}$$
$$0 = x^2 - 4$$
$$4 = x^2$$
$$\pm 2 = x$$

The x-intercepts are ± 2.

y-intercept: Let x be 0. $y = \dfrac{0^2 - 4}{0 - 1} = 4$

The y-intercept is 4.

Plot selected points:

x	-3	$-\frac{1}{2}$	$\frac{3}{2}$	3
$f(x)$	$-\frac{5}{4}$	$\frac{5}{2}$	$-\frac{7}{2}$	$\frac{5}{2}$

Sketch the graph. See Figure 4.64.

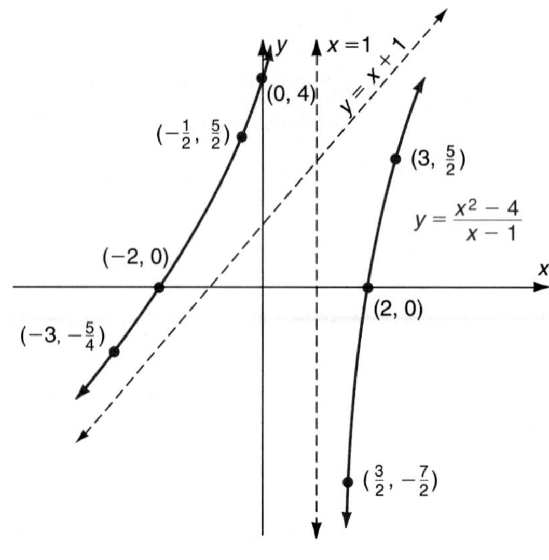

Fig. 4.64

□

Reduced Form

It is important to notice what happens if the rational function is not in reduced form.

EXAMPLE 9 Sketch the graph of $f(x) = \dfrac{x^2 - 4}{x - 2}$.

Solution

$$f(x) = \frac{x^2 - 4}{x - 2} = \frac{(x - 2)(x + 2)}{x - 2} = x + 2;\ x \neq 2$$

The graph of f is the same as the graph of $y = x + 2$ except $x \neq 2$. So, the graph of $f(x) = \dfrac{x^2 - 4}{x - 2}$ is a line with a hole in it as in Figure 4.65.

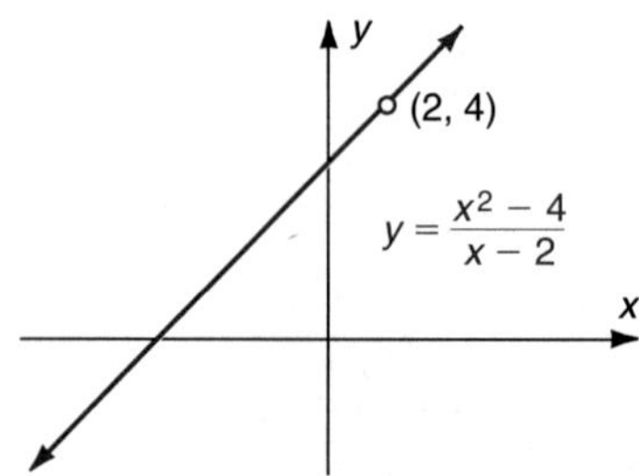

Fig. 4.65

□

GRAPHING CALCULATOR BOX

Graphing Rational Functions

Let's graph the rational function $y = \dfrac{1}{x-1}$ on the default window.

TI-81

Looking at the displayed graph, it is difficult to tell what is happening around 1 on the x-axis. Go to RANGE and change the window to [0, 2] by [−40, 40] and regraph. Now we have a complete graph of this function. Notice that the *asymptote* $x = 1$ appears to be displayed.

CASIO fx-7700G

The default window shows the behavior of this function quite clearly. Change the window to [0, 2] by [−40, 40] and regraph.

If a double line on the y-axis at large y ranges such as [−40, 40] appears, the y-scale is too small. (The Xscl and Yscl in the range menu control the spacing of the tick marks on the axes.) To correct this, enter the range menu and change Yscl to 10. Graph the rational function, $y = \dfrac{2x^3 + 3x^2 - 7x + 5}{x^3 - 5x^2 + 2x - 11}$ and describe its behavior.

TI-81

In the default window we seem to have a complete graph. There appears to be one x-intercept around −4, a y-intercept near −1, and an asymptote close to $x = 5$. We can improve the view of the graph left of the asymptote by changing the y interval to [−3, 3].

CASIO fx-7700G

The default window is too close to tell what's happening. Change to [−6, 10] by [−10, 10] and regraph. This view shows that there is an asymptote around $x = 5$ and seems to be a complete graph. However, to see what is happening to the left of the asymptote, change the y interval to [−3, 3].

Starting with this window, we can use zoom and trace to find the intercepts to any desired degree of accuracy. In the same manner, we can approximate the local maximum and minimum that appear near the origin. Finding the asymptote is another matter. The easiest and best method is to graph the denominator and find where it is 0 (crosses the x-axis).

Answers:	x-intercept	−2.965
(to three decimal places)	y-intercept	−0.455
	Relative minimum	(−0.768, −0.705)
	Relative maximum	(0.740, −0.192)
	Vertical asymptote	$x \approx 5.037$
	Horizontal asymptote	$y = 2$

The graph of the function:

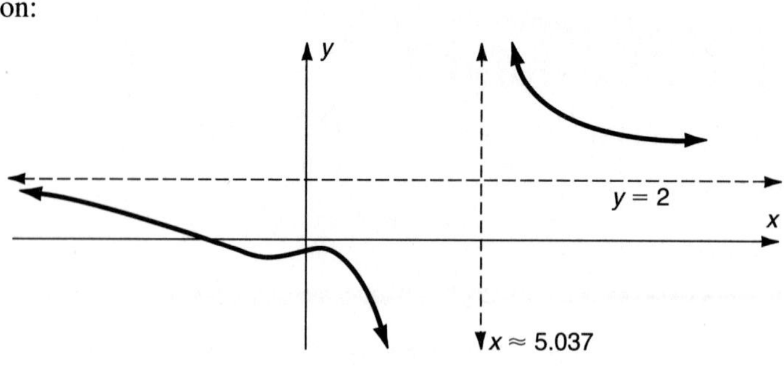

PROBLEM SET 4.3

Warm-ups

In Problems 1 through 4, find the domain of each function. See Example 1.

1. $f(x) = \dfrac{3}{x+3}$
2. $g(x) = \dfrac{x-2}{x}$
3. $h(x) = \dfrac{x+2}{x^2-4}$
4. $r(x) = \dfrac{1}{x^2+4}$

In Problems 5 through 8, find the horizontal and vertical asymptotes of each function. See Example 2.

5. $f(t) = \dfrac{1}{t^2}$
6. $g(x) = \dfrac{2x+3}{7x-4}$
7. $h(x) = \dfrac{-5x^2}{(x+3)(x-1)}$
8. $s(x) = \dfrac{3x^2}{x^2+1}$

In Problems 9 through 12, sketch the graph of each function. See Example 3.

9. $h(x) = \dfrac{2}{x+3}$
10. $s(x) = \dfrac{1}{x+1}$
11. $h(x) = \dfrac{-2}{2x-1}$
12. $g(t) = \dfrac{-1}{2t+3}$

In Problems 13 through 16, sketch the graph of each function. See Example 4.

13. $f(x) = \dfrac{x+1}{x-3}$
14. $h(x) = \dfrac{x-3}{x+2}$
15. $g(t) = \dfrac{3t+2}{2t}$
16. $s(x) = \dfrac{2x-1}{x+2}$

In Problems 17 through 20, sketch the graph of each function. See Examples 5 and 6.

17. $s(x) = \dfrac{3x^2-9x}{6-2x^2}$
18. $h(t) = \dfrac{2t^2-t}{t^2-2}$
19. $g(x) = \dfrac{1}{x^2+4}$
20. $h(x) = \dfrac{-2}{x^2+3}$

In Problems 21 and 22, sketch the graph of each function. See Example 7.

21. $g(x) = \dfrac{2x+1}{(x-2)(x+1)}$
22. $f(x) = \dfrac{x+2}{x(x-3)}$

In Problems 23 and 24, sketch the graph of each function. See Example 8.

23. $h(x) = \dfrac{x^2-x+3}{x-1}$
24. $g(x) = \dfrac{2x^2+3x+4}{x+2}$

In Problems 25 and 26, sketch the graph of each function. See Example 9.

25. $f(t) = \dfrac{t^2 - 9}{t + 3}$

26. $s(x) = \dfrac{x^2 - x - 12}{x - 4}$

Practice Exercises

In Problems 27 through 62, sketch the graph of each function.

27. $g(x) = \dfrac{1}{x^2}$

28. $f(x) = \dfrac{1}{(x - 2)^2}$

29. $s(x) = \dfrac{-1}{x^2}$

30. $f(x) = \dfrac{-1}{(x - 2)^2}$

31. $f(x) = \dfrac{4}{x + 4}$

32. $g(x) = \dfrac{3}{x - 2}$

33. $f(x) = \dfrac{-1}{x + 4}$

34. $s(t) = \dfrac{-3}{t - 2}$

35. $f(x) = \dfrac{x}{2x - 1}$

36. $f(x) = \dfrac{2x}{x + 2}$

37. $g(x) = \dfrac{x + 3}{2x - 3}$

38. $f(x) = \dfrac{2x - 1}{3x + 2}$

39. $h(t) = \dfrac{t}{2 - t}$

40. $g(x) = \dfrac{3 - 2x}{x}$

41. $s(x) = \dfrac{x}{(2x - 1)(x + 1)}$

42. $f(x) = \dfrac{2x}{(3x + 2)(x - 1)}$

43. $f(x) = \dfrac{2 - x}{x^2 - 1}$

44. $f(x) = \dfrac{2x + 1}{x^2 - 9}$

45. $g(x) = \dfrac{x + 4}{x^2}$

46. $s(t) = \dfrac{1 - 2t}{3t^2}$

47. $f(x) = \dfrac{x^2}{4x^2 - 1}$

48. $f(x) = \dfrac{(x + 2)(x - 3)}{(x + 1)(2x - 1)}$

49. $f(x) = \dfrac{x^2 + x}{(x - 1)^2}$

50. $f(x) = \dfrac{(x + 2)^2}{(x + 3)^2}$

51. $f(x) = \dfrac{-x}{x^2 + 1}$

52. $g(x) = \dfrac{(2x + 1)(x - 3)}{x^2 + 1}$

53. $s(x) = \dfrac{x^2 + x + 3}{x - 1}$

54. $h(x) = \dfrac{2x^2 - x - 8}{x + 2}$

55. $f(t) = \dfrac{2t^2 + t - 1}{2t - 1}$

56. $g(x) = \dfrac{x^2 - 9}{x + 3}$

57. $f(x) = \dfrac{x^2 - 2x - 2}{x + 1}$

58. $g(x) = \dfrac{2x^2 + 3x}{x + 2}$

59. $p(t) = \dfrac{6t^2 + t - 2}{3t + 2}$

60. $q(z) = \dfrac{3z^2 + 11z - 4}{z + 4}$

61. $g(x) = \dfrac{x^3 - x^2}{x - 1}$

62. $s(t) = \dfrac{t^3 + 1}{t + 1}$

In Problem 63, graph the function f using the techniques of this section. In Problems 64 through 68, sketch each graph by reflecting or translating the graph of f found in Problem 63.

63. $f(x) = \dfrac{1}{x}$

64. $y = -\dfrac{1}{x}$

65. $y = \dfrac{1}{x} + 1$

66. $y = \dfrac{1}{x} - 2$

67. $y = \dfrac{1}{x + 1}$

68. $y = \dfrac{1}{x - 2}$

In Problems 69 through 72, sketch each graph by reflecting or translating the graph of $g(x) = 1/x^2$ found in Problem 27.

69. $y = -\dfrac{1}{x^2}$

70. $y = \dfrac{1}{x^2} + 2$

71. $y = \dfrac{1}{x^2} - 1$

72. $y = \dfrac{2}{x^2}$

73. Give a rule for a rational function with a vertical asymptote of $x = 2$ and a horizontal asymptote of $y = 3$.

74. Give a rule for a rational function with the y-axis as a horizontal asymptote and the x-axis as a vertical asymptote.

75. Give a rule for a rational function with vertical asymptotes of $x = \pm 2$ and a horizontal asymptote of $y = 3$.

76. Give a rule for a rational function with no vertical asymptote and a horizontal asymptote of $y = 3$.

77. Boyle's Law states that pressure times volume is constant. That is, $PV = C$ or $P = C/V$. What would a sketch of the graph of this equation look like? If V increases, what happens to P? If P increases, what happens to V?

78. In manufacturing, unit cost is given by $U = T/n$, where T is the total cost and n is the number of units produced. If the total cost is made up of constant fixed costs F and variable costs of the form nV, we have $T = F + nV$ and we see that U is a function of n,

$$U(n) = \frac{F + nV}{n}$$

Sketch the graph of U taking F to be 20 and V to be 2. [For convenience in graphing, assume the domain of U is $[1, +\infty)$.]

Challenge Problems

79. Sketch the graph of $f(x) = \begin{cases} \dfrac{1}{x-1} & \text{if } x \neq 1 \\ 5 & \text{if } x = 1 \end{cases}$

80. Sketch the graph of $g(x) = \begin{cases} \dfrac{1}{x^2-1} & \text{if } |x| \neq 1 \\ 0 & \text{if } |x| = 1 \end{cases}$

81. In Problem 78, we *assumed* the domain of the function was $[1, +\infty)$ for ease in graphing. Read the problem again and determine the *actual* domain of U and draw the graph on the interval $[1, 5]$.

For Graphing Calculators

Graph each rational function and describe its behavior.

82. $y = \dfrac{x^2 + 2x + 2}{x^2 + x - 1}$

83. $y = \dfrac{3x^3 - 8x^2 - 5x - 22}{4x^3 + 16x^2 - 5x - 7}$

84. $y = \dfrac{x^4 - 3}{x^5 - 5}$

85. $y = \dfrac{x^4 - 3x^2 + x}{x^3 + 2x + 7}$

IN YOUR OWN WORDS . . .

86. Explain how to find the asymptotes of a rational function.

87. Explain how to find the x-intercepts and y-intercept of a rational function (if there are any).

4.4 ALGEBRA OF FUNCTIONS

If f and g are functions, we can make new functions by adding, subtracting, multiplying, and dividing the rules for f and g.

Definitions: Operations with Functions

$(f + g)(x) = f(x) + g(x)$	Sum function
$(f - g)(x) = f(x) - g(x)$	Difference function
$(fg)(x) = f(x) \cdot g(x)$	Product function
$\left(\dfrac{f}{g}\right)(x) = \dfrac{f(x)}{g(x)};\ g(x) \neq 0$	Quotient function

In general, the domain of these new functions consists of the intersection of the domains of f and g. However, the domain of the quotient function omits domain numbers for which g has value 0.

EXAMPLE 1 If $f(x) = x^2 - 1$ and $g(x) = x + 3$, find

(a) $(f + g)(x)$ (b) $(f - g)(x)$ (c) $(fg)(x)$ (d) $\left(\frac{f}{g}\right)(x)$

Solutions (a) $(f + g)(x) = f(x) + g(x)$

$$= (x^2 - 1) + (x + 3) = x^2 + x + 2$$

(b) $(f - g)(x) = f(x) - g(x)$

$$= (x^2 - 1) - (x + 3) = x^2 - x - 4$$

(c) $(fg)(x) = f(x) \cdot g(x)$

$$= (x^2 - 1)(x + 3) = x^3 + 3x^2 - x - 3$$

(d) $\left(\frac{f}{g}\right)(x) = \frac{f(x)}{g(x)}$

$$= \frac{x^2 - 1}{x + 3}$$ □

EXAMPLE 2 If $g(x) = x$ and $h(x) = x^3$, find

(a) $(g + h)(2)$ (b) $\left(\frac{h}{g}\right)(4)$

Solutions (a) $(g + h)(2) = g(2) + h(2)$

$$= 2 + 2^3 = 10$$

(b) $\left(\frac{h}{g}\right)(4) = \frac{h(4)}{g(4)}$

$$= \frac{4^3}{4} = 16$$ □

EXAMPLE 3 If $f(x) = x + 3$ and $g(x) = \sqrt{x^2 - 1}$, find a rule for $\frac{f}{g}$ and give its domain.

Solution $\left(\frac{f}{g}\right)(x) = \frac{f(x)}{g(x)}$

$$= \frac{x + 3}{\sqrt{x^2 - 1}}$$

The domain of $\frac{f}{g}$ is the intersection of the domains of f and g, *except* those numbers for which g has value 0. The domain of f is all real numbers. The domain of g is $(-\infty, -1] \cup [1, +\infty)$. The intersection of these domains is $(-\infty, -1] \cup [1, +\infty)$. However, we must exclude ± 1 from the domain of $\frac{f}{g}$ because ± 1 each make $\sqrt{x^2 - 1} = 0$. So, the domain of $\frac{f}{g}$ is $(-\infty, -1) \cup (1, +\infty)$. □

Composition of Functions

The next example illustrates another way of combining two functions.

EXAMPLE 4 If $f(x) = 5 - x$ and $g(x) = x^2$, find $f(g(x))$.

Solution

$f(g(x)) = f(x^2)$	Substitution
$= 5 - x^2$	Definition of f

□

The idea illustrated in Example 4 arises so often and is so useful that we call $f(g(x))$ the **composite function** of f and g, or ***f* composed with *g*.** The notation commonly used for the composite function is $(f \circ g)(x) = f(g(x))$. Some care must be taken when considering the domain of a composite function.

> **Definition: Composition of Functions**
>
> If f and g are functions, then
>
> $$(f \circ g)(x) = f(g(x))$$
>
> for all x in the domain of g such that $g(x)$ is in the domain of f.

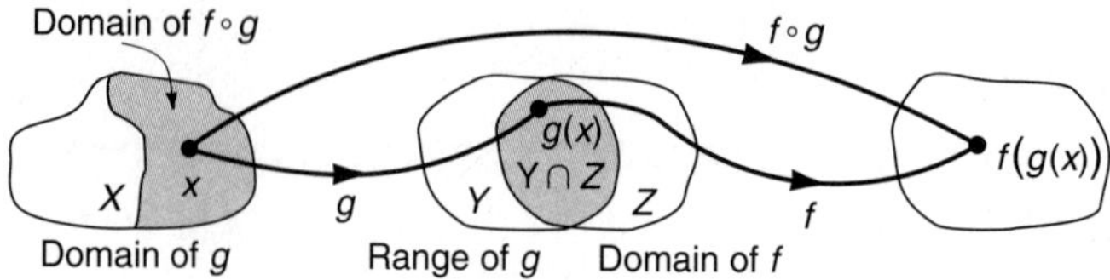

Fig. 4.66

Suppose f and g are functions with domains and ranges as shown in Figure 4.66. g has domain X and range Y, and f has domain Z. Notice that the *domain* of $f \circ g$ is that subset of the domain of g whose range is $Y \cap Z$, the intersection of the range of g and the domain of f. Then $f(g(x))$ is an element in the range of f. The composite function $f \circ g$ takes any *appropriate* element from X directly into the range of f.

EXAMPLE 5 If $f(x) = 2x - 1$ and $g(x) = 4x + 3$, find $(f \circ g)(x)$.

Solution

$(f \circ g)(x) = f(g(x))$	Definition of $f \circ g$
$= f(4x + 3)$	Substitution
$= 2(4x + 3) - 1$	Definition of f
$= 8x + 6 - 1$	
$= 8x + 5$	

Another way to evaluate $(f \circ g)(x)$ in this example is

$$\begin{aligned}(f \circ g)(x) &= f(g(x)) \\ &= 2g(x) - 1 \\ &= 2(4x + 3) - 1 \\ &= 8x + 6 - 1 \\ &= 8x + 5\end{aligned}$$

□

EXAMPLE 6 If $f(x) = 3x^2$ and $g(x) = 4x + 2$, find $(f \circ g)(x)$ and $(g \circ f)(x)$.

Solution

$$\begin{aligned}(f \circ g)(x) &= f(g(x)) \\ &= f(4x + 2) \\ &= 3(4x + 2)^2\end{aligned}$$

$$\begin{aligned}(g \circ f)(x) &= g(f(x)) \\ &= g(3x^2) \\ &= 4(3x^2) + 2 \\ &= 12x^2 + 2\end{aligned}$$

Notice that $(f \circ g)(x) \neq (g \circ f)(x)$. □

EXAMPLE 7 If $f(x) = x^2 + 3x - 1$ and $g(x) = 2x + 1$, find $(f \circ g)(3)$ and $(g \circ f)(3)$.

Solution

$(f \circ g)(3) = f(g(3))$	Definition of $(f \circ g)$
$= f(7)$	$g(3) = 2(3) + 1 = 7$

$$= 7^2 + 3(7) - 1$$

$$= 49 + 21 - 1 = 69$$

$(g \circ f)(3) = g(f(3))$ Definition of $(g \circ f)$

$= g(17)$ $f(3) = 9 + 9 - 1$

$= 34 + 1 = 35$ □

Notice in Example 7 that $(f \circ g)(3) = 69$, while $(g \circ f)(3) = 35$. In general, $(f \circ g)(x) \neq (g \circ f)(x)$. The next example is a case in which $(f \circ g)(x) = (g \circ f)(x)$. This situation is of importance. There is something very interesting going on that we will investigate further in Section 4.5.

EXAMPLE 8 If $f(x) = x^3 + 1$ and $g(x) = \sqrt[3]{x - 1}$, find

(a) $(f \circ g)(x)$ (b) $(g \circ f)(x)$ (c) $f(3)$ (d) $g(28)$ (e) $(f \circ g)(28)$ (f) $(g \circ f)(3)$

Solutions (a) $(f \circ g)(x) = f(g(x)) = f(\sqrt[3]{x - 1}) = (\sqrt[3]{x - 1})^3 + 1 = x - 1 + 1 = x$

(b) $(g \circ f)(x) = g(f(x)) = g(x^3 + 1) = \sqrt[3]{x^3 + 1 - 1} = \sqrt[3]{x^3} = x$

(c) $f(3) = 3^3 + 1 = 28$

(d) $g(28) = \sqrt[3]{28 - 1} = \sqrt[3]{27} = 3$

(e) $(f \circ g)(28) = f(g(28)) = f(3) = 28$

(f) $(g \circ f)(3) = g(f(3)) = g(28) = 3$ □

Notice in part (e) of Example 8 that $(f \circ g)(28) = 28$. Whatever the function g *did* to 28, the function f *undid.* In part (f) we see the same thing happening, except the functions occur the other way around. Parts (a) and (b) show that this effect will occur for all numbers in the appropriate domains.

The domain of $g \circ f$ consists of all elements in the domain of f for which $f(x)$ is in the domain of g. Therefore, the domain of $g \circ f$ is that part of the domain of f whose range is in the domain of g.

EXAMPLE 9 If $f(x) = \sqrt{x}$ and $g(x) = x^2$, find $(g \circ f)(x)$ and $(f \circ g)(x)$ and give the domain for each.

Solution The domain of f is $[0, +\infty)$. The range of f is $[0, +\infty)$. The domain of g is $\mathcal{R}$. The range of g is $[0, +\infty)$.

$$(g \circ f)(x) = g(f(x)) = g(\sqrt{x}) = (\sqrt{x})^2 = x;\ x \geq 0$$

The domain of $g \circ f$ consists of that subset of the domain of f for which $f(x)$ is in the domain of g. Since the domain of g is *all real numbers,* every $f(x)$ is in the domain of g. Therefore, the domain of $g \circ f$ is the domain of f, $[0, +\infty)$. This relationship is illustrated in Figure 4.67.

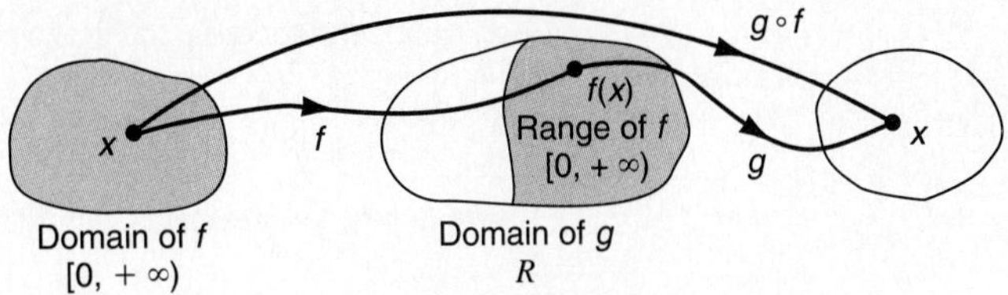

Fig. 4.67

$$(f \circ g)(x) = f(g(x)) = f(x^2) = \sqrt{x^2} = |x|$$

The domain of $f \circ g$ is R because the domain of g is R and the range of g is $[0, +\infty)$, *all of which* is in the domain of f, as shown in Figure 4.68.

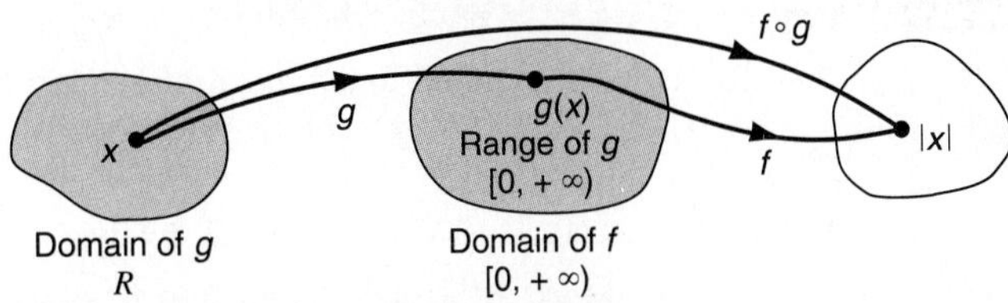

Fig. 4.68 □

There are times when we have a rule for a function h, and we would like to write h as the composition of two functions f and g. In other words, we would like to **decompose** a composite function. In such cases, we often refer to inner and outer functions. In the composite $(f \circ g)(x) = f(g(x))$, g is the inner function and f is the outer function.

EXAMPLE 10 Decompose each composite function.

(a) $h(x) = \sqrt{x + 3}$ (b) $h(x) = (3x - 5)^2$ (c) $h(x) = \dfrac{1}{7 - x^2}$

Solutions (a) $h(x) = \sqrt{x + 3}$

We are looking for two functions f and g so that $h(x) = f(g(x))$.
The outer function is f. It can be thought of as "the big picture." In this example the big picture is taking a square root. The inner function has the rule $x + 3$.
Let $f(x) = \sqrt{x}$ and $g(x) = x + 3$.
Then $f(g(x)) = f(x + 3) = \sqrt{x + 3}$, and we have $h(x) = (f \circ g)(x)$.

(b) $h(x) = (3x - 5)^2$

The obvious decomposition is squaring for the outer function, and the inner function is $3x - 5$.
Let $f(x) = x^2$ and $g(x) = 3x - 5$. Then, $f(g(x)) = f(3x - 5) = (3x - 5)^2$.
Thus, $h(x) = (f \circ g)(x)$.

(c) $h(x) = \dfrac{1}{7 - x^2}$

The outer function is the fraction, and the inner function is $7 - x^2$.

Let $f(x) = \dfrac{1}{x}$ and $g(x) = 7 - x^2$. Then $f(g(x)) = f(7 - x^2) = \dfrac{1}{7 - x^2}$.

So $h(x) = (f \circ g)(x)$. □

Decomposition is usually *not unique*. For example, another possibility for part (b) of Example 10 is $f(x) = (x - 5)^2$ and $g(x) = 3x$. Then, $h(x) = (f \circ g)(x)$. In part (c), another decomposition is $f(x) = \dfrac{1}{7 - x}$ and $g(x) = x^2$.

Partial Fractions

The rational functions, discussed in Section 4.3, are examples of quotient functions of the form $\dfrac{f}{g}$, where f and g are polynomials. Sometimes it is desirable to decompose such a quotient function into a sum of simpler fractions. To see how this is done, let's look at the sum of two simple rational expressions.

$$\frac{2}{x - 1} + \frac{1}{x + 1} = \frac{2(x + 1) + (x - 1)}{(x - 1)(x + 1)} = \frac{3x + 1}{(x - 1)(x + 1)}$$

Thus, we see that $\dfrac{3x + 1}{(x - 1)(x + 1)}$ is the sum of the two fractions $\dfrac{2}{x - 1}$ and $\dfrac{1}{x + 1}$. Now, if we consider this process in reverse, writing $\dfrac{3x + 1}{(x - 1)(x + 1)}$ as the sum of the two simpler fractions is called a **partial fraction decomposition** of the given quotient function. Example 11 illustrates a systematic procedure.

EXAMPLE 11 Find a partial fraction decomposition of the rational expression $\dfrac{x + 4}{(x - 2)(x + 1)}$.

Solution We can see the denominators of the simpler fractions are $x - 2$ and $x + 1$, so we let the numerators be the undetermined constants A and B. Then we add the fractions together and try to find A and B.

$$\frac{x + 4}{(x - 2)(x + 1)} = \frac{A}{x - 2} + \frac{B}{x + 1} = \frac{A(x + 1) + B(x - 2)}{(x - 2)(x + 1)}$$

Since these fractions are equal and have the same denominators, their numerators must be equal. That is,

$$x + 4 = A(x + 1) + B(x - 2)$$

and we want this to hold for *all* numbers x. So we substitute a number for x that makes the factor $x - 2$ equal zero. Let x be 2.

$$\text{If } x = 2 \qquad \text{then} \qquad 2 + 4 = A(2 + 1) + B \cdot 0$$
$$6 = 3A$$
$$A = 2$$

We find B in a similar manner. Suppose x is -1. Now $x + 1$ is zero.

$$\text{If } x = -1 \qquad \text{then} \qquad -1 + 4 = A \cdot 0 + B(-1 - 2)$$
$$3 = -3B$$
$$B = -1$$

Therefore,

$$\frac{x + 4}{(x - 2)(x + 1)} = \frac{2}{x - 2} + \frac{-1}{x + 1}$$

We can check this decomposition by adding the fractions on the right side together. □

The procedure illustrated in Example 11 will work as long as the degree of the numerator is *less than* the degree of the denominator and the factors of the denominator are linear and not repeated. If the degree of the numerator is equal to or greater than the degree of the denominator, the degree of the numerator can be reduced with long division. A *repeated* linear factor in the denominator requires a slight modification in the procedure.

EXAMPLE 12 Use partial fractions to decompose $\dfrac{1}{x^3 - 2x^2 + x}$.

Solution We factor the denominator completely.

$$\frac{1}{x^3 - 2x^2 + x} = \frac{1}{x(x - 1)^2}$$

In the case of repeated linear factors, we must have a fraction for each power of the factor, so we write

$$\frac{1}{x(x - 1)^2} = \frac{A}{x} + \frac{B}{x - 1} + \frac{C}{(x - 1)^2}$$

$$1 = A(x - 1)^2 + Bx(x - 1) + Cx \qquad \text{Equate numerators.}$$

$$\text{If } x = 0 \text{ then} \qquad 1 = A(-1)^2 + B \cdot 0 + C \cdot 0$$
$$A = 1$$

$$\text{If } x = 1 \text{ then} \qquad 1 = A \cdot 0 + B \cdot 0 + C(1)$$
$$C = 1$$

Now there is no substitution that will produce the value of B quite as easily as that of A and C, but any convenient number for x and the values of A and C will work.

If $x = 2$ then

$$1 = A(1) + B(2)(1) + C(2)$$

$$1 = (1)(1) + 2B + (1)(2)$$

$$B = -1$$

Therefore,

$$\frac{1}{x^3 - 2x^2 + x} = \frac{1}{x} + \frac{-1}{x - 1} + \frac{1}{(x - 1)^2}$$ □

This introduction to partial fraction decomposition is by no means complete. The principal use of partial fractions will be found in calculus where detailed techniques will be developed as needed.

PROBLEM SET 4.4

Warm-ups

In Problems 1 through 4, find $(f + g)(x)$, $(f - g)(x)$, $(fg)(x)$, and $\left(\frac{f}{g}\right)(x)$. See Example 1.

1. $f(x) = 2x$; $g(x) = 3x + 5$

2. $f(x) = x^2 - 7$; $g(x) = 2x + 3$

3. $f(x) = 3 - x^2$; $g(x) = \sqrt{x + 1}$

4. $f(x) = x^2 - x + 3$; $g(x) = x^2 + x$

In Problems 5 through 8, $f(x) = x^2 + 1$ and $g(x) = 1 - x^2$, find the value of each. See Example 2.

5. $(g - f)(0)$

6. $(fg)(-1)$

7. $\left(\frac{f}{g}\right)(2)$

8. $(f + g)(-2)$

In Problems 9 through 12, let $f(x) = 2(x - 3)$ and $g(x) = \sqrt{x + 2}$ and find the indicated domain. See Example 3.

9. Domain of $f + g$

10. Domain of $f - g$

11. Domain of $\frac{g}{f}$

12. Domain of fg

In Problems 13 through 16, find $(f \circ g)(x)$ and $(g \circ f)(x)$. See Examples 4, 5, 6, and 8.

13. $f(x) = 5x + 8$; $g(x) = 7x - 5$

14. $f(x) = 6x + 9$; $g(x) = \frac{2}{x}$

15. $f(x) = \frac{1}{x^2 + 2}$; $g(x) = x + 3$

16. $f(x) = x^3$; $g(x) = x^2 + 1$

In Problems 17 through 20, $f(x) = x - 1$ and $g(x) = 1 - x^2$. Find the value of each. See Example 7.

17. $(f \circ g)(-1)$

18. $(g \circ f)(0)$

19. $(f \circ f)(1)$

20. $(f \circ g)(1)$

In Problems 21 and 22, find the domain of $f \circ g$ and $g \circ f$. See Example 9.

21. $f(x) = \frac{1}{x + 2}$; $g(x) = x^2 + 4$

22. $f(x) = \sqrt{x + 2}$; $g(x) = x^2$

In Problems 23 and 24, decompose the given composite function into two functions. See Example 10.

23. $h(x) = \sqrt{3x + 2}$

24. $h(x) = \dfrac{2}{4x - 7}$

In Problems 25 through 30, find a partial fraction decomposition of each function. For Problems 25 through 28, see Example 11.

25. $f(x) = \dfrac{x - 5}{(x - 1)(x + 1)}$

26. $g(x) = \dfrac{8x - 5}{(x - 4)(2x + 1)}$

27. $h(x) = \dfrac{3x^2 - 6x - 1}{x^2 - 2x - 3}$

28. $F(x) = \dfrac{3x^3 + x^2 - 22x + 21}{x^2 + x - 6}$

For Problems 29 and 30, see Example 12.

29. $G(x) = \dfrac{x^2 + 7x + 6}{x^2(x + 2)}$

30. $H(x) = \dfrac{5x^2 + 9x + 1}{x^3 + 2x^2 + x}$

Practice Exercises

In Problems 31 through 42, use the graphs of f and g in Figure 4.69 to answer each question.

31. $f(-3) + g(-3)$

32. $f(0) - g(0)$

33. $f(-1) \cdot g(-1)$

34. $\dfrac{f(1)}{g(1)}$

35. $(f \circ g)(1)$

36. $(g \circ f)(0)$

37. $(f \circ g)(-1)$

38. $(g \circ f)(-1)$

39. $(g \circ g)(-3)$

40. $(g \circ g)(-1)$

41. Where is $f(x) > g(x)$?

42. Is $g(2) > f(2)$?

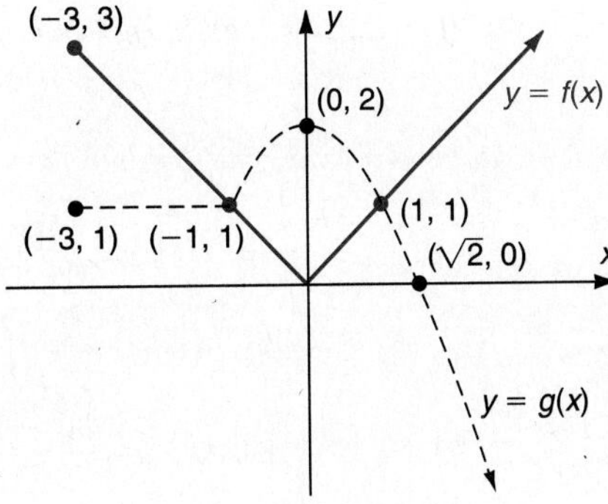

Fig. 4.69

In Problems 43 through 52, let $f(x) = \sqrt{2x + 5}$ and $g(x) = 4 - x^2$. Find the value of each.

43. $(g - f)(0)$

44. $(f + g)(-2)$

45. $(fg)(10)$

46. $(gf)(-2)$

47. $\left(\dfrac{f}{g}\right)(2)$

48. $\left(\dfrac{g}{f}\right)(1)$

49. $(f \circ g)(-1)$

50. $(g \circ f)(0)$

51. $(f \circ f)(1)$

52. $(g \circ g)(1)$

In Problems 53 through 58, find $(f + g)(x)$, $(f - g)(x)$, $(fg)(x)$ and $\left(\dfrac{f}{g}\right)(x)$.

53. $f(x) = 6x - 1$; $g(x) = \dfrac{1}{2}x$

54. $f(x) = x + 3$; $g(x) = \dfrac{2}{3}x - 3$

55. $f(x) = x^2 - x + 1$; $g(x) = x - 5$

56. $f(x) = 2x^2 + x + 7$; $g(x) = 2 - x$

57. $f(x) = \sqrt{x + 7}$; $g(x) = \sqrt{x + 4}$

58. $f(x) = \sqrt{3x}$; $g(x) = \sqrt{x + 8}$

In Problems 59 through 66, find $(f \circ g)(x)$ and $(g \circ f)(x)$ and give the domain for each.

59. $f(x) = x + 3;\ g(x) = x - 3$

60. $f(x) = x^3;\ g(x) = \sqrt[3]{x}$

61. $f(x) = \sqrt{x + 2};\ g(x) = x^2$

62. $f(x) = x^2;\ g(x) = \sqrt{x + 1}$

63. $f(x) = \dfrac{2}{x};\ g(x) = x^2 + 1$

64. $f(x) = 1 - 3x;\ g(x) = \dfrac{1}{3x^2}$

65. $f(x) = \dfrac{1}{x + 2};\ g(x) = \dfrac{2}{x}$

66. $f(x) = \dfrac{1}{x};\ g(x) = \dfrac{x}{x - 1}$

In Problems 67 through 72, decompose the composite function into two functions.

67. $h(x) = (x + 2)^3$

68. $h(x) = \sqrt[3]{x - 4}$

69. $h(x) = \dfrac{3}{x^2 - 5}$

70. $h(x) = \dfrac{1}{x^4 + 1}$

71. $h(x) = \sqrt{x^2 - 1}$

72. $h(x) = \sqrt{4 - x}$

In Problems 73 through 76, use the graphs of f and g in Figure 4.70 to find each function value.

73. $(f + g)(a)$

74. $\left(\dfrac{g}{f}\right)(a)$

75. $(f \circ g)(a)$

76. $(g \circ f)(a)$

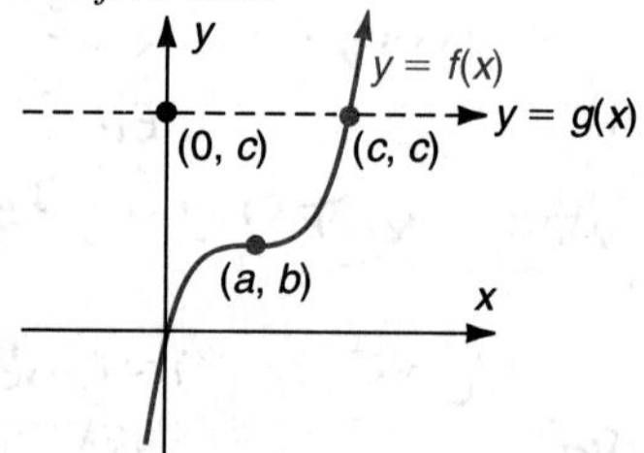

Fig. 4.70

In Problems 77 through 84, decompose each function into partial fractions.

77. $f(x) = \dfrac{3 - 5x}{(x + 1)(x - 3)}$

78. $g(x) = \dfrac{5x - 5}{(x + 3)(x - 2)}$

79. $f(x) = \dfrac{5x - 1}{2x^2 - 5x - 3}$

80. $g(t) = \dfrac{t - 8}{t^2 - t - 12}$

81. $f(t) = \dfrac{3t^2 - 7t + 2}{(t - 1)^2(t + 1)}$

82. $g(x) = \dfrac{x^2 + 5x - 9}{(2x - 1)(x + 2)^2}$

83. $F(x) = \dfrac{6x^2 - 3x - 2}{3x^2 - x - 2}$

84. $G(x) = \dfrac{2x^3 - 5x^2 - 5x + 15}{2x^2 - x - 6}$

Challenge Problems

85. Let $f(x) = \sqrt{1 - x^2}$ and $g(x) = \sqrt{x^2 - 1}$.
(a) Find the domain and range of $f \circ g$.
(b) Find the domain and range of $g \circ f$.

86. Let $f(x) = \sqrt{1 - x}$ and $g(x) = \sqrt{x - 1}$.
(a) Find the domain and range of $f \circ g$.
(b) Find the domain and range of $g \circ f$.

87. Let $f(x) = \dfrac{1}{x^2 - 1}$ and $g(x) = \sqrt{\dfrac{1}{x} - 1}$.
(a) Find the domain and range of $f \circ g$.
(b) Find the domain and range of $g \circ f$.

In Problems 88 and 89, decompose each rational expression with partial fractions.

88. $\dfrac{x^3 - 3x^2 + 8x - 4}{x^4 - 4x^3 + 4x^2}$

89. $\dfrac{-x^3 + 3x^2 - 15x + 1}{(x^2 - 1)^2}$

IN YOUR OWN WORDS . . .

90. Explain why 1 is not in the domain of $\dfrac{f}{g}$ if $f(x) = x - 1$ and $g(x) = 1 - x^2$.

91. Explain the composition of two functions.

4.5 INVERSE FUNCTIONS

Let's consider the functions shown in Figure 4.71.

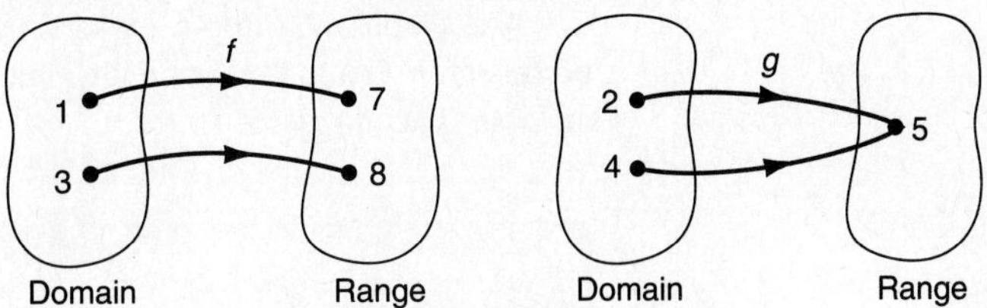

Fig. 4.71

Notice that each number in the domain of f is assigned to a different number in the range of f. However, this is not true for each number in the domain of g. We say that f is **one-to-one** and g *is not* one-to-one.

> **One-to-One Function**
>
> A function f is said to be one-to-one if every number in the range is paired with exactly one number in the domain; that is, for p and q in the domain of f,
>
> $$f(p) = f(q) \text{ implies } p = q$$

EXAMPLE 1 Figure 4.72 describes a function f. Is f one-to-one?

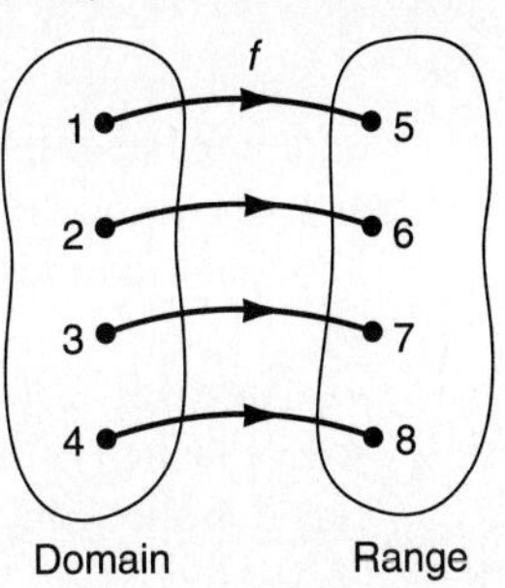

Fig. 4.72

Solution Notice that each number in the range is paired with one number in the domain. So f is one-to-one. □

EXAMPLE 2 A function f is graphed in Figure 4.73. Is f one-to-one?

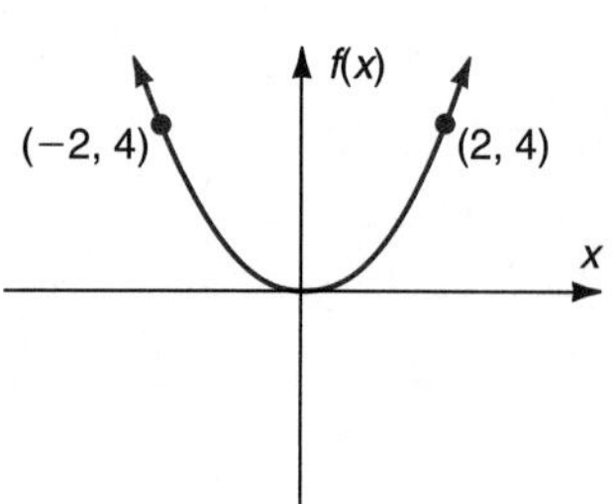

Fig. 4.73

Solution Notice that 4 is a number in the range. It is paired with -2 and 2. So f is not one-to-one. □

The graph in Example 2 was shown *not* to be the graph of a one-to-one function because $f(-2)$ and $f(2)$ have the same value. That is, they fall on the *same horizontal line*. This suggests a test.

> **Horizontal Line Test**
>
> If every horizontal line intersects the graph of a function at no more than one point, then the function is one-to-one.

EXAMPLE 3 Which of the graphs in Figure 4.74 represent one-to-one functions?

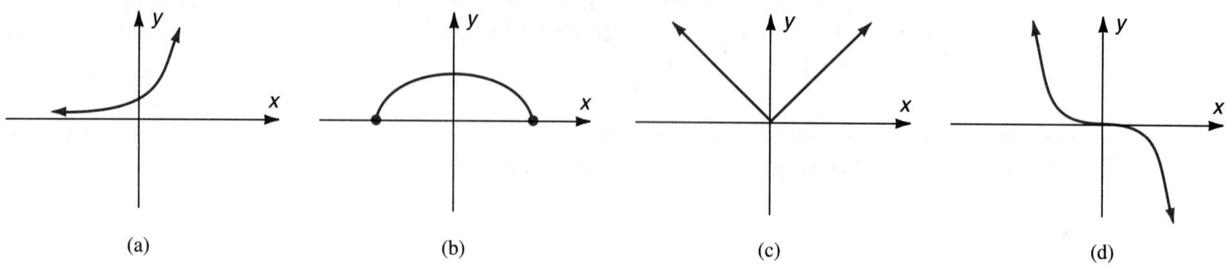

Fig. 4.74

Solutions All four graphs pass the vertical line test and thus are the graphs of functions. In addition, the graphs in parts (a) and (d) pass the horizontal line test and thus represent one-to-one functions. However, the graphs in parts (b) and (c) *do not* pass the horizontal line test and *are not* one-to-one functions. □

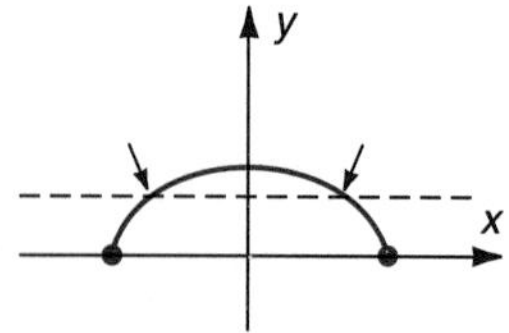

Fig. 4.74b

In many applications of functions, it is desirable to interchange the x- and y-coordinates of points on the graph of a function f—in other words to make a function that can "reverse" what f does. Such a function is named f^{-1} (read "f inverse"). *Note:* f^{-1} does not mean $\frac{1}{f}$.

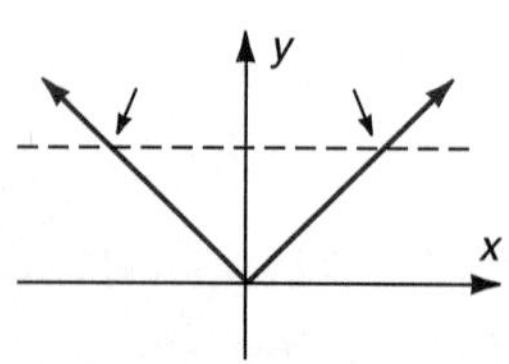

Fig. 4.74c

Consider a function f with domain $\{1, 2, 3, 4\}$ and range $\{5, 6, 7, 8\}$ with assignments as shown in Figure 4.75. f^{-1} is the function with domain $\{5, 6, 7, 8\}$ and range $\{1, 2, 3, 4\}$ and reverses the assignments made by f. This reversal can be done because f is one-to-one.

Notice that the domain of f^{-1} is the range of f, and that the range of f^{-1} is the domain of f. In other words, the domain and range are interchanged. We can see that f defines ordered pairs of (1, 5), (2, 6), (3, 7), and (4, 8), while f^{-1} defines (5, 1),

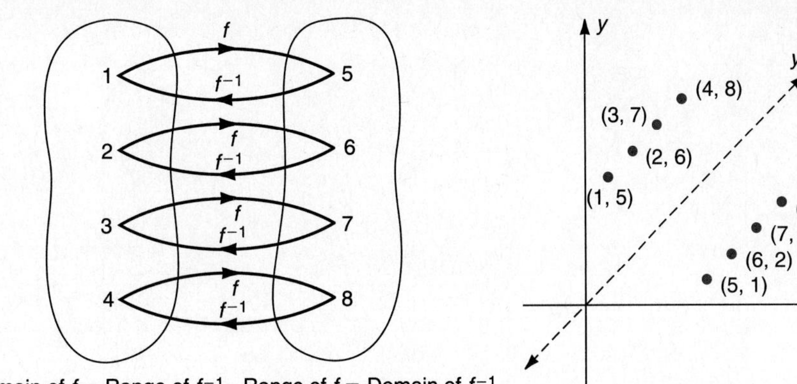

Fig. 4.75 **Fig. 4.76**

(6, 2), (7, 3), and (8, 4). Notice that in each case, the x- and y-coordinates are interchanged.

If we plot these points, we can observe in Figure 4.76 that the graphs of f and f^{-1} are symmetric with respect to the line $y = x$. That is, each is the mirror image of the other about the line $y = x$.

EXAMPLE 4 The graph of g is shown in Figure 4.77.

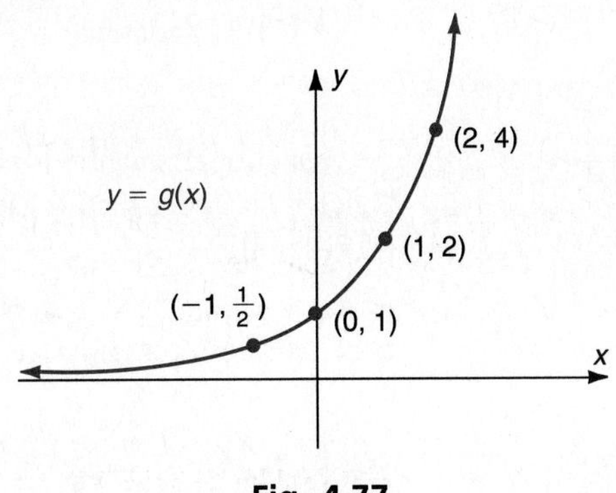

Fig. 4.77

(a) Is g a function?

(b) Is g one-to-one?

(c) Sketch the graph of g^{-1}.

(d) Find $g(0)$, $g(-1)$, $g^{-1}(1)$, $g^{-1}(2)$, and $g^{-1}(0)$.

Solutions (a) The graph of g passes the vertical line test. g is a function.

(b) The graph of g passes the horizontal line test. g is one-to-one.

(c) To graph g^{-1} we reflect the graph of g about the line $y = x$. Interchange the x- and y-coordinates.

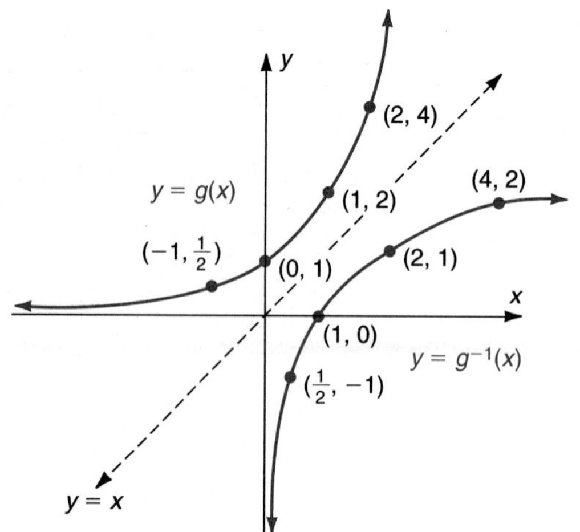

Fig. 4.78

(d) $g(0) = 1$	$g(0)$ is the y-coordinate of 0 on the graph of g.
$g(-1) = \dfrac{1}{2}$	$g(-1)$ is the y-coordinate of -1 on the graph of g.
$g^{-1}(1) = 0$	$g^{-1}(1)$ is the y-coordinate of 1 on the graph of g^{-1}.
$g^{-1}(2) = 1$	$g^{-1}(2)$ is the y-coordinate of 2 on the graph of g^{-1}.
$g^{-1}(0)$ is undefined	0 is not in the domain of g^{-1} because it is not in the range of g. □

Functional notation is very useful in giving names to the numbers in the range. It helps us understand what is happening with a function and its inverse (Figure 4.79).

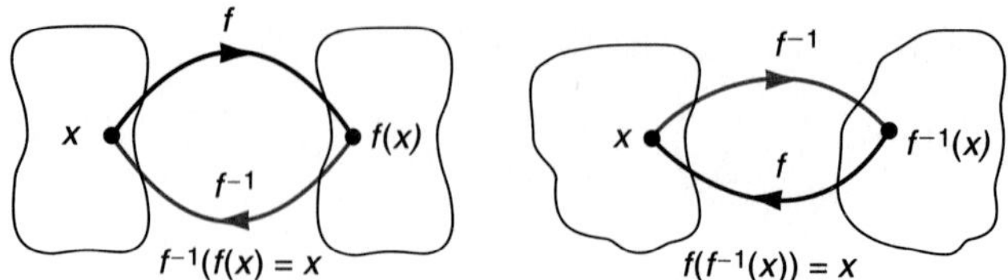

Fig. 4.79

Definition of f^{-1}

If f is a one-to-one function, then f^{-1} is the function such that

$$f(f^{-1}(x)) = x \text{ for all } x \text{ in the domain of } f^{-1} \text{ and}$$

$$f^{-1}(f(x)) = x \text{ for all } x \text{ in the domain of } f.$$

f^{-1} is called the inverse of f.

Not all functions have inverses. Functions as simple as $f(x) = x^2$ and $g(x) = |x|$ *do not* have inverses. Notice that f must be one-to-one in order for f^{-1} to exist. To verify that a function g is the inverse of a function f, we must show that $f(g(x)) = x$ for all x in the domain of g, and that $g(f(x)) = x$ for all x in the domain of f. In Example 8 in Section 4.4 we verified that the function $g(x) = \sqrt[3]{x - 1}$ is the inverse of $f(x) = x^3 + 1$.

EXAMPLE 5 Verify that $f(x) = \dfrac{3}{x - 1}$ and $g(x) = \dfrac{x + 3}{x}$ are inverses of each other.

Solution First, we must show that $f(g(x)) = x$ for all x in the domain of g. The domain of g is the set of all real numbers except 0, so we let x be any nonzero real number.

$$f(g(x)) = f\left(\frac{x + 3}{x}\right)$$

$$= \frac{3}{\dfrac{x + 3}{x} - 1} \qquad \text{Replace } x \text{ with } \frac{x + 3}{x} \text{ in } f(x).$$

$$= \frac{3x}{x + 3 - x} \qquad \text{Multiply the numerator and the denominator by } x.$$

$$= \frac{3x}{3} = x$$

Next, we must show that $g(f(x)) = x$ for all x in the domain of f. The domain of f contains all real numbers except 1. Let x be any number in the domain of f.

$$g(f(x)) = g\left(\frac{3}{x - 1}\right)$$

$$= \frac{\dfrac{3}{x - 1} + 3}{\dfrac{3}{x - 1}} \qquad \text{Replace } x \text{ with } \frac{3}{x - 1} \text{ in } g(x).$$

$$= \frac{3 + 3(x - 1)}{3} \qquad \text{Multiply the numerator and the denominator by } x - 1.$$

$$= \frac{3 + 3x - 3}{3} = \frac{3x}{3} = x$$

□

There is an algorithm that finds the inverse of simple one-to-one functions.

To Find a Rule for f^{-1}

1. Replace $f(x)$ with y.
2. Interchange x and y in $y = f(x)$.
3. Solve the equation for y, if possible.
4. Replace y with $f^{-1}(x)$.

EXAMPLE 6 If $f(x) = 5x - 2$, find $f^{-1}(x)$.

Solution The domain and range of f is R. So, the domain and range of f^{-1} is also R.

$$f(x) = 5x - 2$$

$$y = 5x - 2 \qquad \text{Replace } f(x) \text{ with } y.$$

$$x = 5y - 2 \qquad \text{Interchange } x \text{ and } y.$$

$$x + 2 = 5y \qquad \text{Solve for } y.$$

$$\frac{x + 2}{5} = y$$

$$f^{-1}(x) = \frac{x + 2}{5} \qquad \text{Replace } y \text{ with } f^{-1}(x).$$

□

EXAMPLE 7 If $f(x) = \sqrt{x + 2}$, find $f^{-1}(x)$.

Solution The graph of f will help us understand the domain and range of f. It will also aid in graphing f^{-1}. f is the square root function shifted 2 units left. We can sketch the graph of f^{-1} by reflecting the graph of f about the line $y = x$. The sketch is in Figure 4.80.

Since the domain and range of f will be reversed for f^{-1}, we will list each as we work.

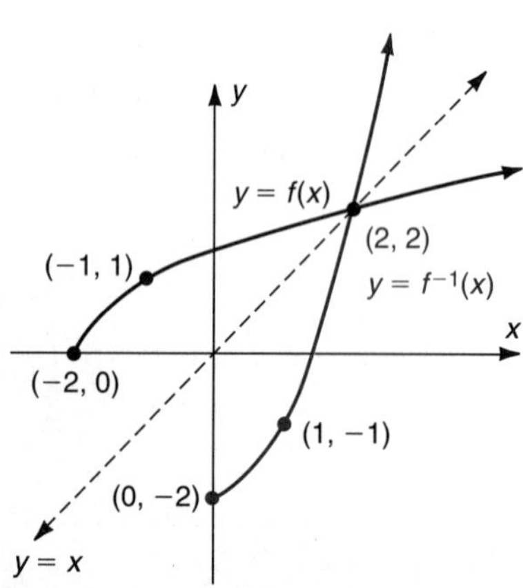

Fig. 4.80

$$y = \sqrt{x + 2};\ x \geq -2,\ y \geq 0 \qquad \text{Replace } f(x) \text{ with } y.$$

$$x = \sqrt{y + 2};\ y \geq -2,\ x \geq 0 \qquad \text{Interchange } x \text{ and } y.$$

$$x^2 = y + 2;\ y \geq -2,\ x \geq 0 \qquad \text{Square both sides.}$$

$$x^2 - 2 = y;\ y \geq -2,\ x \geq 0 \qquad \text{Solve for } y.$$

$$f^{-1}(x) = x^2 - 2;\ y \geq -2,\ x \geq 0 \qquad \text{Replace } y \text{ with } f^{-1}(x).$$

The restriction on x is part of the rule for f^{-1}. So, it must be listed. However, it is not necessary to list the range.

$$f^{-1}(x) = x^2 - 2;\ x \geq 0$$

□

GRAPHING CALCULATOR BOX

Graphing Inverse Functions

A graphing calculator that has the capability to graph parametric equations can graph many inverse functions. Both the TI-81 and the CASIO fx-7700G have this capability. Suppose f is a one-to-one function whose rule is explicitly stated in terms of some independent variable such as x. We can state the function in equivalent form with the parametric equations, $\begin{cases} x = t \\ y = f(t) \end{cases}$. The graph of these parametric equations is the same as the graph of $y = f(x)$. Therefore, we can graph f^{-1} by graphing the parametric equations, $\begin{cases} x = f(t) \\ y = t \end{cases}$.

Let's graph the function $f(x) = \sqrt{x + 2}$ and its inverse. First, we enter the function in parametric form: $\begin{cases} x = t \\ y = \sqrt{t + 2} \end{cases}$.

TI-81

Press [MODE], arrow down to the fourth line, select "Param" and [ENTER]. Press [Y=] and notice how the function memory has changed. Instead of functions Y_1 through Y_4, there are now three pairs; X_{1T}, Y_{1T} through X_{3T}, Y_{3T}. These are for graphing parametric equations. The cursor should be at X_{1T}. Press [X|T] and notice that T has appeared. Arrow down to Y_{1T} and enter the function $\sqrt{T + 2}$ in the usual manner. Before we graph, press [RANGE] and notice the new entries, T_{min}, T_{max}, and T_{step}. Set T_{min} to -2 and T_{max} to 10 and then set the window to $[-3, 6]$ for x and $[-3, 3]$ for y. Press [GRAPH] and you should see the graph of $f(x) = \sqrt{x + 2}$. To graph the inverse, return to [Y=] and enter $\sqrt{T + 2}$ in X_{2T} and T in Y_{2T} then regraph.

CASIO fx-7700G

Press [MODE] [SHIFT] [x] to select "PARAM" the parametric mode. Then call up the function memory [F MEM] and enter T as f_1 and $\sqrt{T + 2}$ as f_2. Press [RANGE] to set the window at $[-3, 6]$ for x and $[-3, 3]$ for y. Press [RANGE] once again and we see the parametric range values. Set "min:" to -2 and "max:" to 10 and "ptch" to 0.1. Press [RANGE] once more to return to the computation screen. In parametric mode, when we press [GRAPH] we see,

Graph(X, Y)=(

The calculator expects to receive parametric equations for X and Y separated by a comma and terminated with a closing parenthesis. We enter, [F2] [1] [,] [F2] [2] [)] [EXE] and see the graph. To graph the inverse, press [GRAPH] and repeat the same keystrokes interchanging the [1] and [2].

PROBLEM SET 4.5

Warm-ups

In Problems 1 through 4, determine whether each function is one-to-one. See Examples 1 and 2.

1.

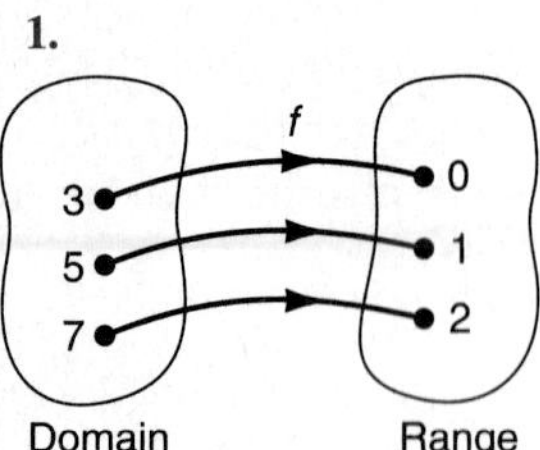

2.

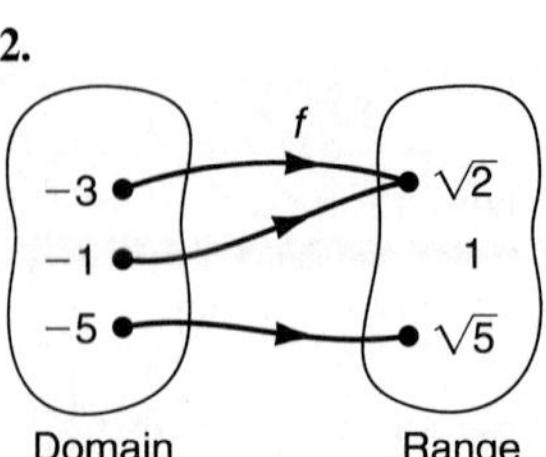

3.

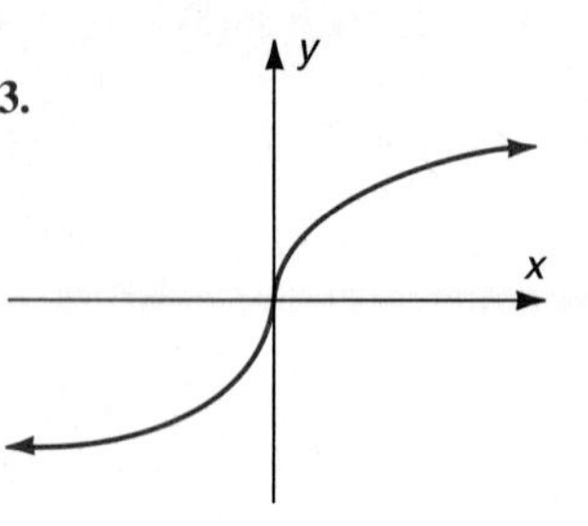

4.

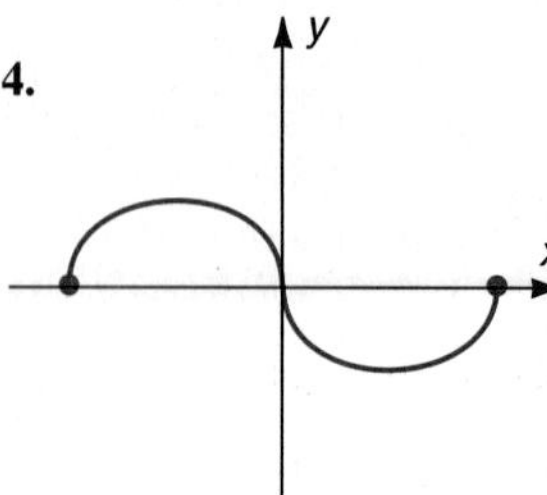

In Problems 5 through 8, which graphs represent one-to-one functions? See Example 3.

5.

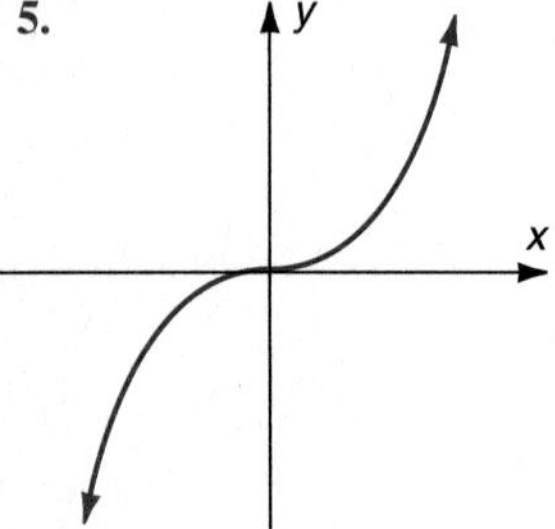

6.

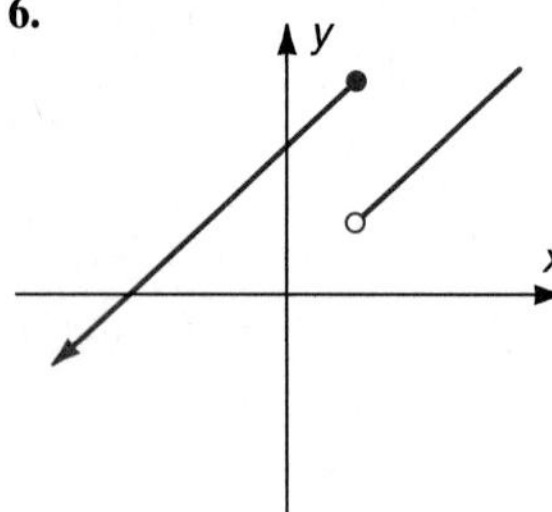

7.

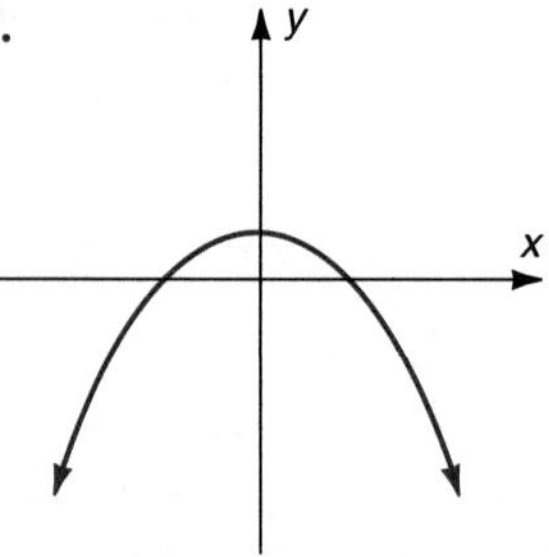

8.

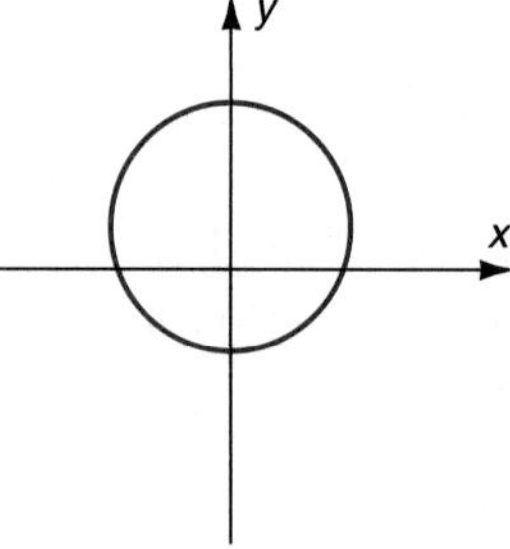

In Problems 9 through 12, use the graph of h shown in Figure 4.81. See Example 4.

9. Is h a function?

10. Is h one-to-one?

11. Sketch the graph of h^{-1}.

12. Find $h(2)$, $h(4)$, $h^{-1}(1)$, $h^{-1}(0)$, $h^{-1}(\sqrt{2})$.

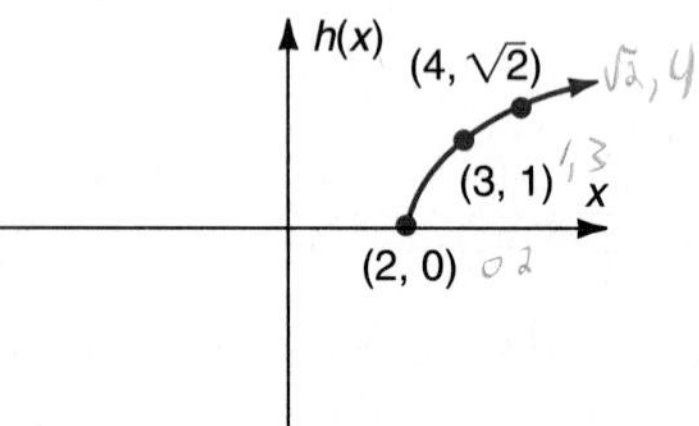

Fig. 4.81

In Problems 13 and 14, verify that the members of each pair of functions are inverses. See Example 5.

13. $f(x) = \sqrt[3]{2x + 1}$; $g(x) = \dfrac{x^3 - 1}{2}$

14. $f(x) = \dfrac{x}{x + 2}$; $g(x) = \dfrac{2x}{1 - x}$

In Problems 15 through 18, find $f^{-1}(x)$. For Problems 15 and 16, see Example 6.

15. $f(x) = 3x + 7$

16. $f(x) = \dfrac{2}{3}x - 5$

For Problems 17 and 18, see Example 7.

17. $f(x) = \sqrt{x + 3}$ **18.** $f(x) = \sqrt{x - 4}$

Practice Exercises

In Problems 19 through 30, sketch the graph of each function and determine whether or not each function is one-to-one.

19. $f(x) = 2x + 1$ **20.** $h(x) = -5$ **21.** $f(x) = x^2 - 1$

22. $f(x) = 3 - x^2$ **23.** $g(x) = (x + 1)^2 + 1$ **24.** $g(x) = (x + 2)^2$

25. $h(x) = |x + 1|$ **26.** $f(x) = |x| - 3$ **27.** $f(x) = \sqrt{x + 2}$

28. $g(x) = \sqrt{x - 5}$ **29.** $h(x) = \begin{cases} x + 2 & \text{if } x \le -1 \\ 2 - x & \text{if } x > -1 \end{cases}$ **30.** $s(x) = \begin{cases} x & \text{if } x < 0 \\ 3 & \text{if } x \ge 0 \end{cases}$

In Problems 31 through 38, find $f^{-1}(x)$. Sketch f and f^{-1} on the same set of axes. Give the domain and range for both f and f^{-1}.

31. $f(x) = x + 3$ **32.** $f(x) = 3x$ **33.** $f(x) = \dfrac{x + 1}{x}$

34. $f(x) = \dfrac{x}{x - 2}$ **35.** $f(x) = \sqrt{x + 7}$ **36.** $f(x) = \sqrt{x - 5}$

37. $f(x) = x^2 + 1;\ x \ge 0$ **38.** $f(x) = 1 - x^2;\ x \ge 0$

In Problems 39 through 52, determine if each function is one-to-one and, if so, find its inverse.

39. $f(x) = 5x - 9$ **40.** $g(x) = -7x - 1$ **41.** $h(t) = \dfrac{t - 1}{t + 1}$

42. $S(x) = \dfrac{1 - x}{1 + x}$ **43.** $G(x) = 1 - x^2$ **44.** $F(z) = \dfrac{1}{z^2}$

45. $T(r) = |r^2 - 1|$ **46.** $v(x) = |4x^2 - 9|$ **47.** $\Delta(x) = x^4 - 16$

48. $\Gamma(t) = \dfrac{t^2}{1 - t}$ **49.** $U(x) = x^2 - 4x + 4;\ x \le 2$ **50.** $V(s) = 2|s - 1|;\ s \le 1$

51. $k(\theta) = \sqrt{\theta^3}$ **52.** $H(x) = \sqrt{8x^3 - 1}$

In Problems 53 through 58, determine if each function is one-to-one and, if so, sketch its inverse.

53.

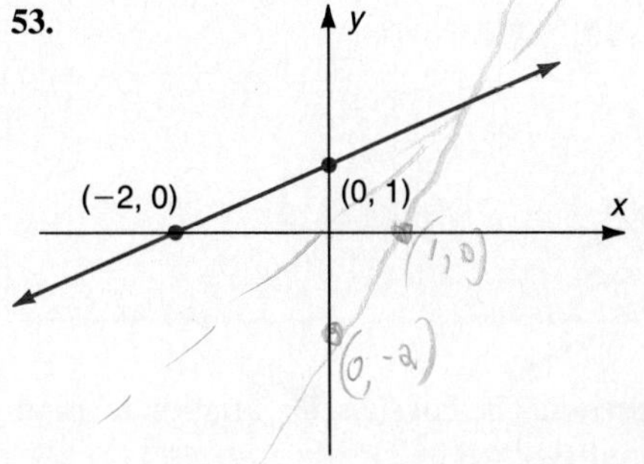

54.

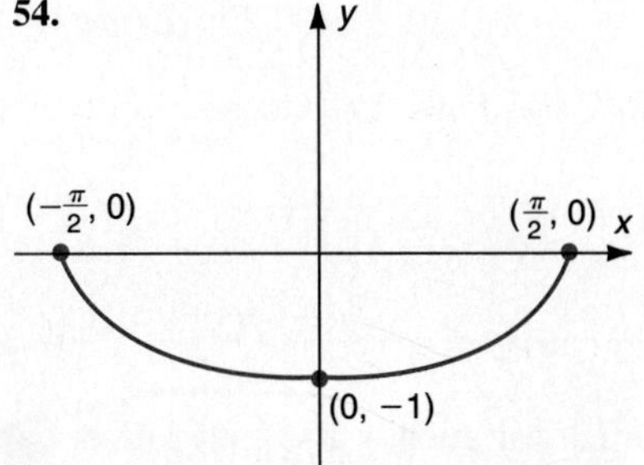

55.

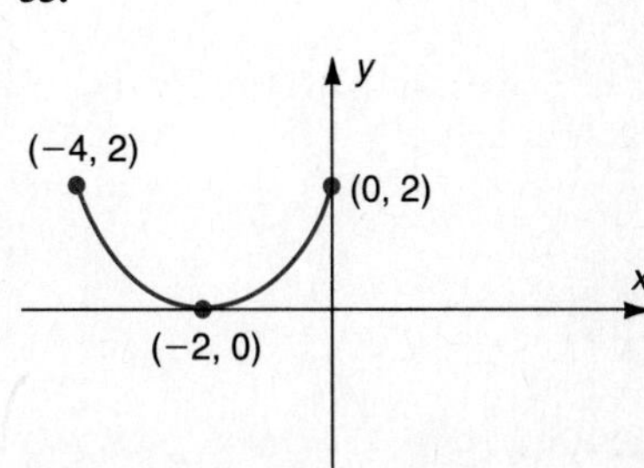

56.

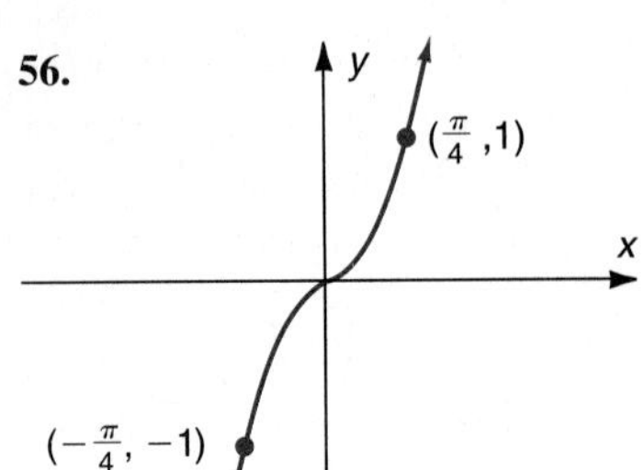

57.

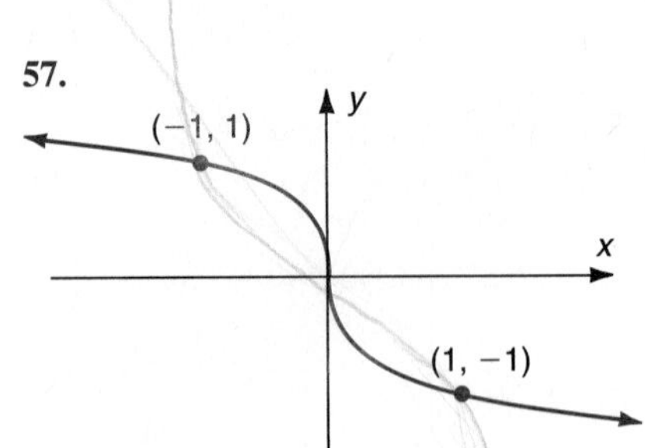

58.

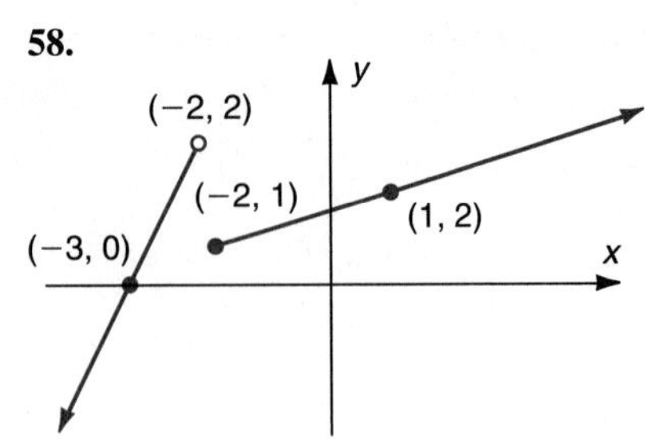

59. Prove that an even function with more than one number in its domain *cannot* be one-to-one.

60. Is every odd function one-to-one? Prove the conjecture or provide a counterexample.

61. Notice that all our examples of one-to-one functions have been *increasing* functions or *decreasing* functions. Prove that a function that is either increasing throughout its domain or decreasing throughout its domain is one-to-one.

Challenge Problems

62. Find a function f such that $f(x) = f^{-1}(x)$.

63. Let $f(x) = 2x + 1$ and $g(x) = x - 5$. Find $(f \circ g)^{-1}(x)$ and $(g^{-1} \circ f^{-1})(x)$.

64. Suppose f and g are one-to-one functions. Prove that $(f \circ g)^{-1}(x) = (g^{-1} \circ f^{-1})(x)$.

65. Prove that if a function is an increasing function, then its inverse is also an increasing function.

66. The converse of the theorem proved in Problem 61 *is not true*. Find a one-to-one function whose domain is R that is neither increasing on R nor decreasing on R.

IN YOUR OWN WORDS . . .

67. What is a one-to-one function?

68. If f is a one-to-one function, describe f^{-1}.

4.6 VARIATION

Variation describes certain relationships that may exist between variables. For example, the formula $F = ma$ relates the force F on a body of mass m to its acceleration a. Notice that as the force *increases,* the acceleration *increases,* and as the force *decreases* the acceleration *decreases*. We say F **varies directly** as a.

> **Definition: Direct Variation**
>
> y varies directly as x means that
>
> $$y = kx$$
>
> where k is a nonzero constant.

Direct variation is also called **direct proportion.** The constant in variation relationships is called the **constant of proportionality.**

EXAMPLE 1 Hooke's Law states that the force required to stretch a spring is directly proportional to the distance stretched from the natural length of the spring. If a force of 10 pounds is required to stretch a spring 6 inches, what force is required to stretch the spring 1 foot?

Solution This is a problem in direct variation. If F is the force and D is the distance stretched,

$$F = kD$$

Variation problems usually have a two-part solution. First, the constant of proportionality is calculated, and then the resulting formula is used to complete the problem. In this problem we are told that a force of 10 pounds stretches the spring $\frac{1}{2}$ foot; that is,

$$10 = k \cdot \frac{1}{2}$$

$$20 = k$$

Now, we have the value of k (called the spring constant), and we can write the formula

$$F = 20d$$

Next, we use the formula to complete the problem.

$$F = 20 \cdot 1 = 20$$

It requires a force of 20 pounds to stretch the spring 1 foot. □

The average rate of an object traveling a fixed distance is related to the time needed for the trip by the relationship, rate = distance/time. Notice that as time *increases,* rate *decreases*. We say R **varies inversely** as T.

Definition: Inverse Variation

y varies inversely as x means that

$$y = \frac{k}{x}$$

where k is a nonzero constant.

Inverse variation is also called **inverse proportion.**

EXAMPLE 2 The intensity of illumination varies inversely as the square of the distance. If the intensity of a particular light is 120 candlepower at a distance of 9 feet from the light, what is the intensity 25 feet from the light?

Solution If I is the intensity and d is the distance, then since I varies inversely as d^2, we have

$$I = \frac{k}{d^2}$$

First we find k. Since I is 120 when d is 9,

$$120 = \frac{k}{9^2}$$

$$k = 9720$$

Now the formula for this particular light is

$$I = \frac{9720}{d^2}$$

Therefore, when the distance is 25 feet, the intensity is

$$I = \frac{9720}{25^2} \approx 15.55$$

At 25 feet the intensity is approximately 15.55 candlepower. □

Definition: Joint Variation

y **varies jointly** as w and x means that

$$y = kwx$$

where k is a nonzero constant.

We often say that y is **jointly proportional** to w and x. Joint variation is sometimes called **combined variation.**

EXAMPLE 3 The power that can be safely transmitted by a rotating shaft is jointly proportional to the speed of rotation and the cube of the diameter of the shaft. If a particular shaft 2 inches in diameter can transmit 71 horsepower at 1200 revolutions per minute, by how much should the diameter of the shaft be increased in order to transmit 100 horsepower at 1000 revolutions per minute?

Solution Let P be the power in horsepower, s the shaft speed in revolutions per minute, and d the diameter of the shaft in inches.

$$P = ksd^3$$

We find k.

$$71 = k(1200)(2^3)$$

$$k = \frac{71}{1200 \cdot 8}$$

Therefore the formula for this particular shaft material is

$$P = \frac{71sd^3}{9600}$$

Finally, we solve for d when P is 100 and s is 1000.

$$100 = \frac{71 \cdot 1000d^3}{9600}$$

$$\frac{960}{71} = d^3$$

$$d = \sqrt[3]{\frac{960}{71}} \approx 2.38$$

The diameter of the shaft should be increased to at least 2.38 inches. □

Often, the forms of variation are combined in a single mathematical model.

EXAMPLE 4 Newton's Law of Gravitation states that the force between two particles varies jointly with their masses and inversely as the square of the distance between them. What is the effect on the gravitational force between two bodies if their masses are doubled and the distance between them is tripled?

Solution Let the original masses of the two bodies be m_1 and m_2 and let d be the distance between them. Then the gravitational force F is given by the formula

$$F = \frac{km_1m_2}{d^2}$$

If F_n is the gravitational force when the masses are doubled and the distance between them is tripled, then

$$F_n = \frac{k(2m_1)(2m_2)}{(3d)^2}$$

$$= \frac{4km_1m_2}{9d^2}$$

$$= \frac{4}{9}\frac{km_1m_2}{d^2} = \frac{4}{9}F$$

Doubling the masses and tripling the distance between them reduces the gravitational force by a factor of $\frac{4}{9}$. □

The numerical value of the constant of proportionality will depend on the units of measure of the variables.

EXAMPLE 5 The period of a pendulum varies directly as the square root of its length.

(a) Suppose a pendulum of length 30 centimeters has a period of 1.1 seconds. What is the constant of proportionality to the nearest thousandth?

(b) What is the period of a pendulum 35.4 centimeters long?

(c) If the length of the pendulum in part (a) is measured in *inches* instead of *centimeters,* the period will not change. Taking 1 inch to be 2.54 centimeters, find the constant of proportionality to the nearest thousandth using inches as the unit of length.

(d) What is the period of a pendulum 10 inches long?

Solutions Let T be the period of a pendulum and L be its length. We are given that

$$T = k\sqrt{L}$$

(a) If T is 1.1 seconds and L is 30 centimeters, we have

$$1.1 = k\sqrt{30}$$

$$k = \frac{1.1}{\sqrt{30}} \approx 0.201$$

(b) Since

$$T = k\sqrt{L}$$

and L is 35.4 centimeters,

$$T \approx 1.195$$

A pendulum 35.4 centimeters long has a period of approximately 1.195 seconds.

(c) If T is 1.1 seconds and L is 30/2.54 inches, we have

$$1.1 = k_2\sqrt{\frac{30}{2.54}}$$

$$k_2 = 1.1\sqrt{\frac{2.54}{30}} \approx 0.320$$

(d) In this case, L is measured in inches. Therefore,

$$T \approx 1.012$$

A pendulum 10 inches long has a period of approximately 1.012 seconds. □

PROBLEM SET 4.6

Warm-ups

For Problems 1 and 2, see Example 1.

1. V varies directly as the square of w. If V is 100 when w is 5, what is V when w is 6?

2. If a force of 71 pounds stretches a spring 2.5 inches, how far, to the nearest thousandth, will a force of 100 pounds stretch the same spring?

For Problems 3 and 4, see Example 2.

3. Suppose R varies inversely with the cube root of Z. If R is 11 when Z is 125, what is R when Z is 8?

4. If the intensity of a particular light is approximately 0.331 candlepower at 10 feet, what is its approximate intensity at 2 feet to the nearest thousandth of a candlepower?

For Problems 5 and 6, see Example 3.

5. s is jointly proportional to t and the square root of g. If s is 10 when t is 3 and g is $\frac{25}{4}$, what is s when t is 7 and g is 36?

6. The volume of a cylinder varies jointly with its height and the square of its radius. If a cylinder 22 centimeters in height with a radius of 4.88 centimeters has approximate volume of 1645.93 cubic centimeters, what is the approximate volume of a cylinder 44 centimeters in height with a radius of 2.44 centimeters?

For Problems 7 and 8, see Example 4.

7. What is the effect on the gravitational force between two particles if the distance between them is halved?

8. The resistance of a wire varies directly as its length and inversely as the square of its diameter. If 50 feet of wire with a diameter of 0.01 inch has a resistance of 5 ohms, what is the resistance of 100 feet of the same kind of wire if its diameter is 0.02 inch?

For Problems 9 and 10, see Example 5.

9. The distance a stone falls when dropped off a cliff is directly proportional to the square of time. If a stone falls 144.9 feet in 3 seconds, what is the constant of proportionality? If the stone falls 61.55 meters in 3 seconds, what is the constant of proportionality? How far will the stone have fallen after 2 seconds in meters? In feet?

10. The gravitational attraction between two bodies varies jointly with their masses and inversely with the square of the distance between them. The constant of proportionality is called the **gravitational constant.** If the gravitational force of attraction between two bowling balls, each with mass 0.5 slugs, 2 feet apart is 4.26×10^{-8} pounds, what is the numerical value of the gravitational constant in the English system? In the metric system, the mass of each of the balls is 0.0342 kilogram, 1 foot is 0.305 meter, and the force of attraction is 2.10×10^{-12} newton. What is the numerical value of the gravitational constant in the metric system? Give approximations to three significant digits.

Practice Exercises

In Problems 11 through 18, write a formula expressing each statement.

11. M varies directly as the cube of t.

12. G varies directly as the square root of p.

13. U varies inversely as the square of L.
14. b varies inversely as the fourth power of r.
15. x varies directly as u and inversely as the square of s.
16. Q varies directly as the square of P and inversely as q.
17. I is jointly proportional to w_1 and w_2, and inversely proportional to the square root of t.
18. y is jointly proportional to h_0 and the cube of L, and inversely proportional to h_{av}.

In Problems 19 through 26, express each statement as a formula and determine the constant of proportionality.

19. Z varies directly as the square of d; Z is 405 when d is 9.
20. L varies directly as the fifth power of α; L is 8 when α is 2.
21. R varies inversely as the cube of n; R is $\frac{3}{5}$ when n is $\frac{3}{2}$.
22. E varies inversely as the square root of x; E is 0.12 when x is 0.0289.
23. θ varies directly as the square of t and inversely as w; θ is $\frac{1}{2}$ when t is 5 and w is 25.
24. ϕ varies directly as m and inversely as the square root of ω; ϕ is 11 when m is 14 and ω is 14,161.
25. A is jointly proportional to a_1 and the fourth power of c, and inversely proportional to a_2; A is 1 when a_1 is 0.07, a_2 is 2.1, and c is 5.
26. R_{av} is jointly proportional to r_1 and r_2, and inversely proportional to the three-fifths power of ρ; R_{av} was 14 when r_1 was 9, r_2 was 12, and ρ was 243.
27. The period of a pendulum is directly proportional to the square root of its length. If a pendulum 4 feet long has a period of 2 seconds, how long is the period of a pendulum 2 feet long?
28. Hooke's Law also states that the force necessary to compress a spring is directly proportional to the distance compressed. If a force of 2 newtons compresses a spring 7 centimeters, what force is required to compress the same spring 10 centimeters?
29. The rate of rotation of gears meshed together varies inversely as the number of teeth. If a gear with 36 teeth makes 2400 revolutions per minute, what is the rate of rotation of a 24-tooth gear meshed with it?
30. The volume of a sphere varies as the cube of its diameter. If a sphere of diameter 1 parsec has a volume of 18.14 cubic light-years, what is the volume of a sphere of diameter 3 parsecs?
31. The weight of a body in space varies inversely with the square of the distance from the center of the earth. If an astronaut weighed 170 pounds on earth, what would his weight be, to the nearest pound, if stationary 500 miles above the earth? Assume the radius of the earth is 3960 miles.
32. The time necessary to prepare a certain field for professional football is directly proportional to the number of days since it was last prepared and inversely proportional to the number of people available for grounds work. If it takes 2 days to prepare the field when the time since last preparation has been 7 days and there are 8 people available for work, how long will it take with 12 workers if it's been 2 weeks since the last preparation?
33. The number of watts that a resistor in an electrical circuit can dissipate varies jointly as the resistance and the square of the current. If a given 15-ohm resistor dissipates 25 watts when the current is 2.5 amperes, how many watts will it dissipate if the current is increased to 4.8 amperes?
34. The Ray Warren Corporation has found that its sales volume for barber scissors is directly proportional to the size of the sales force and inversely proportional to the price of the scissors. If 77,625 pairs were sold in 1992 when there were 54 salespeople and the price was $22.50 a pair, what is the sales forecast for this year with 50 salespeople and a price of $18.75 a pair?
35. The resistance to fluid flow in a small tube is directly proportional to the length of the tube and inversely proportional to the fourth power of the diameter. If an artery of diameter d has resistance R to the flow of blood, what happens to the resistance if plaque buildup reduces the diameter to one-half d?

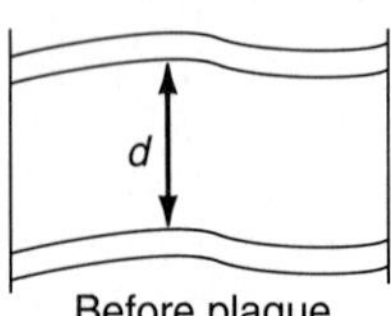

Before plaque

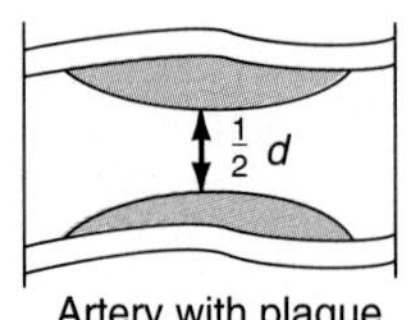

Artery with plaque

36. The maximum safe load that can be applied to a horizontal beam of rectangular cross section varies jointly with the width of the beam and the square of its depth, and inversely with its length from support to support.

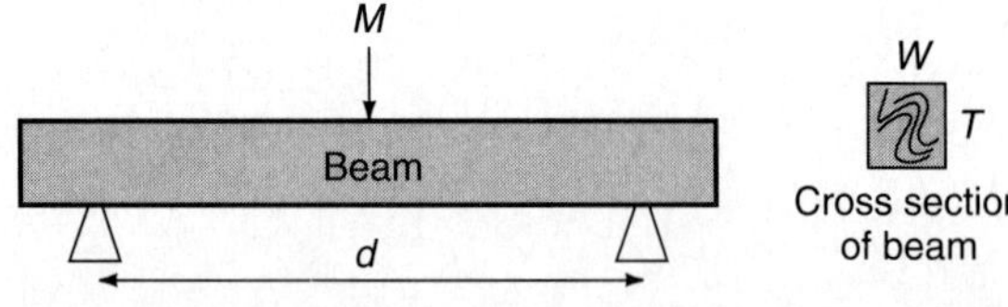

Suppose M is the maximum load that can be applied to the beam in the figure at the bottom of page 300. What will the maximum load be if the supports are moved to be twice as far apart and the depth of the beam is doubled?

Challenge Problems

37. The maximum load on a column made of fine-grain hard maple varies jointly as the area of the cross section and the square of the least dimension of its cross section, and inversely as the square of its length. If such a column 25 feet long and 1 foot square in cross section can carry a maximum load of 28,149 pounds, what is the maximum load, to the nearest pound, that can be carried by a 20-foot-long column of the same material but circular in cross section with a diameter of 1.5 feet?

38. Kepler's Third Law of Planetary Motion states that the square of the time it takes a planet to make one revolution around the sun is directly proportional to the cube of the average distance of the planet from the sun. Assume the average distance of the earth from the sun is 93 million miles and it takes 365 days for the earth to make one orbit. If the average distance of Jupiter from the sun is 484 million miles, what is the length of the Jupiter year to the nearest earth-day?

IN YOUR OWN WORDS . . .

39. Compare direct, inverse, and joint variation.

40. Suppose A varies directly as the cube of k. Explain what happens to A if k is tripled.

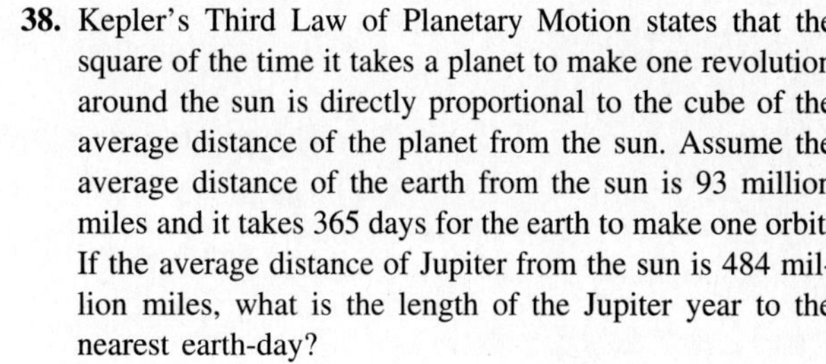

CHAPTER SUMMARY

GLOSSARY

A **polynomial function of degree n:** A function of the form $f(x) = a_n x^n + a_{n-1}x^{n-1} + \cdots + a_1 x + a_0$, where each coeffficient is a real number and n is a whole number. The given form is called **standard form.**

A **rational function:** A function of the form $f(x) = \dfrac{g(x)}{h(x)}$, where $g(x)$ and $h(x)$ are polynomials.

y **varies directly** as x: $y = kx$, where k is a nonzero constant.

y **varies inversely** as x: $y = \dfrac{k}{x}$, where k is a nonzero constant.

y **varies jointly** as w and x: $y = kwx$, where k is a nonzero constant.

LINEAR FUNCTIONS

A function of the form $f(x) = mx + b$, where m and b are real numbers, is called a **linear function.**

The graph of a linear function is a line.

QUADRATIC FUNCTIONS

A function of the form $f(x) = ax^2 + bx + c$; where a, b, and c are real numbers and $a \neq 0$, is called a **quadratic function.**

The graph of a quadratic function is a parabola.

ABSOLUTE VALUE FUNCTIONS

A function in the form of $f(x) = a|x - h| + k$ is called an **absolute value function.** Its graph is a V with vertex at the point (h, k). The V opens upward if a is positive, and downward if a is negative.

TRANSLATIONS

If f is a function and $c > 0$, then

The graph of $y = f(x) + c$ is the graph of $y = f(x)$ shifted *up* c units.
The graph of $y = f(x) - c$ is the graph of $y = f(x)$ shifted *down* c units.
The graph of $y = f(x + c)$ is the graph of $y = f(x)$ shifted c units to the *left*.
The graph of $y = f(x - c)$ is the graph of $y = f(x)$ shifted c units to the *right*.

SKETCHING POLYNOMIAL FUNCTIONS

1. Find the x-intercepts.
2. Determine the end behavior of the graph from the degree of the polynomial and whether the leading coefficient is positive or negative.
3. If an x-intercept was obtained from a linear factor to an *odd* power, the graph will *cross* the x-axis at that intercept.
4. If an x-intercept was obtained from a linear factor to an *even* power, the graph will *just touch* the x-axis at that intercept.

FINDING VERTICAL ASYMPTOTES OF RATIONAL FUNCTIONS

Suppose $f(x) = \dfrac{g(x)}{h(x)}$ is a rational function in reduced form.

If $h(a) = 0$, then $x = a$ will be a vertical asymptote.

FINDING HORIZONTAL ASYMPTOTES OF RATIONAL FUNCTIONS

$f(x) = \dfrac{a_m x^m + a_{m-1}x^{m-1} + \cdots + a_1 x + a_0}{b_n x^n + b_{n-1}x^{n-1} + \cdots + b_1 x + b_0}$; $a_m \neq 0$, $b_n \neq 0$ is a rational function in lowest terms.

1. If $m < n$, then $y = 0$ is a horizontal asymptote of the graph of f.
2. If $m = n$, then $y = \dfrac{a_m}{b_n}$ is a horizontal asymptote of the graph of f.
3. If $m > n$, there is no horizontal asymptote.

SKETCHING RATIONAL FUNCTIONS

1. Find any vertical asymptotes.
2. Find any horizontal asymptote. Determine if the graph intersects any horizontal asymptotes.
3. Find x- and y-intercepts.
4. Plot several points, selecting one from each interval as determined by the x-intercepts and the vertical asymptotes.
5. Sketch the graph as a smooth curve over the domain of the function.

OPERATIONS WITH FUNCTIONS

$(f+g)(x) = f(x) + g(x)$ Sum

$(f-g)(x) = f(x) - g(x)$ Difference

$(fg)(x) = f(x) \cdot g(x)$ Product

$\left(\frac{f}{g}\right)(x) = \frac{f(x)}{g(x)};\ g(x) \neq 0$ Quotient

COMPOSITION OF FUNCTIONS

If f and g are functions, then $(f \circ g)(x) = f(g(x))$ for all x in the domain of g such that $g(x)$ is in the domain of f.

$f \circ g$ is called the composition of f with g or f composed with g.

ONE-TO-ONE FUNCTION

A function f is said to be one-to-one if every number in the range is paired with exactly one number in the domain. That is, for x and y in the domain of f, $x = y$ if and only if $f(x) = f(y)$.

HORIZONTAL LINE TEST

If each horizontal line intersects the graph of a function in at most one point, then the function is one-to-one.

DEFINITION OF f^{-1}

If f is a one-to-one function, then f^{-1} is the function such that $f(f^{-1}(x)) = x$ for all x in the domain of f^{-1} and $f^{-1}(f(x)) = x$ for all x in the domain of f.

f^{-1} is called the inverse of f.

REVIEW PROBLEMS

Sketch the graph of each function. Label all intercepts, asymptotes, and the vertex of parabolas and absolute value functions. Determine if each function is one-to-one.

1. $f(x) = -2x$

2. $g(x) = 2x - 3$

3. $f(x) = |x + 1|$

4. $g(x) = 2|x|$

5. $f(x) = 4x^2 + 1$

6. $f(x) = 3 - x^2$

7. $f(x) = x^2 - 2;\ x > 1$

8. $g(x) = \frac{1}{2}x + 3;\ x \geq 2$

9. $f(x) = [\![x + 2]\!]$

10. $f(x) = [\![x]\!] - 1$

11. $f(x) = x(x + 3)(x - 3)$

12. $f(x) = x(x + 3)(x - 1)$

13. $s(x) = \frac{-1}{(x + 2)^2}$

14. $f(x) = \frac{-1}{x - 2}$

15. $h(x) = |2x| + 2$

16. $h(x) = -|2x + 1| - 1$

17. $h(x) = \sqrt{x + 4}$

18. $f(x) = \sqrt{x - 6}$

19. $g(x) = x^2 - 2x - 4$

20. $h(x) = x^2 + 6x + 8$

21. $f(x) = x^5 - 4x^3$

22. $f(x) = x^3 - x^2 - 2x$

23. $f(x) = \frac{x}{3x - 1}$

24. $f(x) = \frac{2x}{x + 1}$

25. $f(x) = -3;\ x < -1$

26. $g(x) = 2x - 3;\ x \leq 0$

27. $g(x) = 3x^2 - 6x + 4$

28. $h(x) = 2x^2 - 12x + 8$

29. $f(x) = (2 + 3x)(x - 1)(x - 3)$

30. $f(x) = (7 - 3x)^2(2x + 5)$

31. $f(x) = -x^2 + 2x - 3$

32. $f(x) = -2x^2 + 8x - 5$

33. $g(x) = \sqrt{x} - 4$

34. $h(x) = \sqrt{x} + 3$

35. $f(x) = \dfrac{4}{x + 2}$

36. $g(x) = \dfrac{-3}{x - 2}$

37. $f(x) = |x - 2| + 3$

38. $g(x) = -|x + 1| - 2$

39. $f(x) = x^2(2x - 3)^2$

40. $f(x) = x(5x + 2)^2$

In Problems 41 through 50, consider the following functions:

$$f(x) = \begin{cases} x + 1 & \text{if } x < 2 \\ x^2 & \text{if } x \geq 2 \end{cases} \qquad g(x) = \begin{cases} x^2 - 1 & \text{if } x \leq -2 \\ -2 & \text{if } x \geq 2 \end{cases}$$

$$\Theta(x) = \begin{cases} |x - 2| & \text{if } x < 1 \\ |x + 3| & \text{if } x \geq 1 \end{cases} \qquad \Omega(x) = \begin{cases} \sqrt{1 - x} & \text{if } x < 1 \\ 2x - 3 & \text{if } x \geq 1 \end{cases}$$

41. Find $f(-5)$.
42. Find $f(5)$.
43. Find $g(0)$.
44. Find $g(3)$.
45. Find $\Theta(-1)$.
46. Find $\Theta(5)$.
47. Find $\Omega(1)$.
48. Find $\Omega(-2)$.
49. Graph f.
50. Graph Θ.

In Problems 51 through 56, sketch the graph of each function. Figure 4.82 shows the graph of the function F.

51. $f(x) = F(x - 1)$
52. $g(x) = F(x) - 1$
53. $U(x) = -F(x)$
54. $V(x) = F(-x)$
55. $h(x) = 2F(x)$
56. $k(x) = F(-x - 2) + 1$

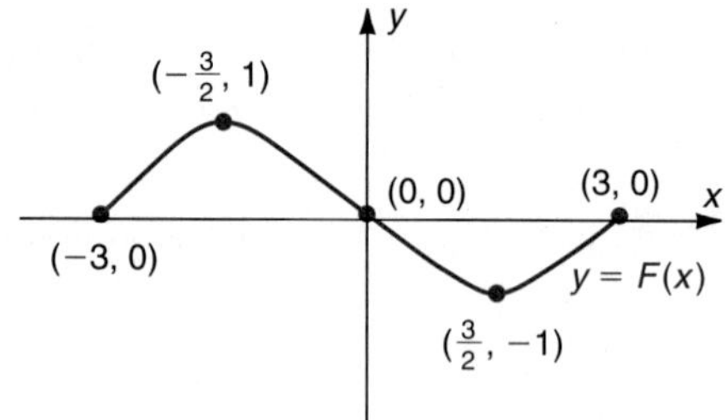

Fig. 4.82

In Problems 57 and 58, find $(f + g)(x)$, $(f - g)(x)$, $(fg)(x)$, $\left(\dfrac{f}{g}\right)(x)$ *and give the domain of each function.*

57. $f(x) = 3x - 2;\ g(x) = \dfrac{3}{2}x$

58. $f(x) = \sqrt{x};\ g(x) = \sqrt{x + 1}$

In Problems 59 through 64, let $f(x) = \sqrt{2x + 1}$ *and* $g(x) = 1 - x^2$*. Find the value of each.*

59. $(g - f)(0)$
60. $(gf)\left(\dfrac{3}{2}\right)$
61. $(f \circ g)(1)$
62. $(g \circ f)(0)$
63. $(f \circ f)(1)$
64. $(g \circ g)(1)$

In Problems 65 and 66, find $(f \circ g)(x)$ *and* $(g \circ f)(x)$ *and give the domain of each composite function.*

65. $f(x) = \dfrac{x}{3};\ g(x) = x^2 + 4$

66. $f(t) = t^2;\ g(t) = \sqrt{t - 1}$

In Problems 67 and 68, decompose the composite function into two functions.

67. $h(x) = (x + 2)^4$

68. $h(x) = \sqrt[3]{x + 2}$

In Problems 69 and 70, find a partial fraction decomposition of each function.

69. $f(t) = \dfrac{t + 11}{(t - 1)(t + 3)}$

70. $g(x) = \dfrac{3x^2 + x - 6}{x^3 + 2x^2}$

In Problems 71 through 76, find $f^{-1}(x)$. Sketch f and f^{-1} on the same set of axes. Give the domain and range for both f and f^{-1}.

71. $f(x) = x + 2$

72. $f(x) = \dfrac{x}{x - 3}$

73. $f(x) = \sqrt{x - 3}$

74. $f(x) = 2 - x^2;\ x \geq 0$

75. A stone is thrown in a still pond making a circular ripple whose radius is increasing with time according to the function $r(t) = \frac{9}{2}t$, where t is seconds from the time of the splash and r is in feet. The function $A(r) = \pi r^2$ gives the area of a circle of radius r. Form the function $A \circ r$. What is $(A \circ r)(4)$? What is the meaning of $A \circ r$?

76. The daily cost in dollars of operating a television assembly line is given by the function

$$C(x) = 10x^{2/3} + 411;\ 1 \leq x \leq 100$$

where x is the number of television sets assembled in a day. Suppose the number of television sets assembled in h hours is given by the function $x(h) = 8h;\ 0 \leq h \leq 12$. What does the function $C \circ x$ give? How much does it cost to operate the assembly line on a day when the assembly line operates 8 hours? What if it only operates 1 hour?

LET'S NOT FORGET . . .

77. Which form is more useful? $f(x) = \dfrac{x^3 - 4x}{x^2 - 2x - 15} = \dfrac{x(x - 2)(x + 2)}{(x - 5)(x + 3)}$

(a) If you are graphing the function f?
(b) If you are beginning a partial fraction decomposition?

78. Watch the role of negative signs. Graph the three functions.
(a) $f(x) = \sqrt{x}$
(b) $g(x) = -\sqrt{x}$
(c) $h(x) = \sqrt{-x}$.

79. From memory. Match each function in the left column with the best description of its graph in the right column.

1. $f(x) = 2x - 6$	**a.** A parabola opening downward
2. $g(x) = \sqrt{x - 3}$	**b.** A parabola opening upward
3. $h(x) = 2x^2 - 11x$	**c.** A line
4. $F(x) = \dfrac{x + 1}{x - 1}$	**d.** The upper half of a parabola
5. $G(x) = 13 - 4x - x^2$	**e.** Graph with a vertical asymptote

80. With a calculator. Given $f(t) = \dfrac{(3t - 2)(t - 1)^3}{0.12345(t^4 - 2)}$. Approximate $f(\pi)$ to five significant digits.

CHAPTER 5

Polynomial Equations

CONNECTIONS

Polynomial equations have fascinated mathematicians since the dawn of history. The ancient Babylonians were able to solve quadratic equations for real roots. However, it wasn't until the sixteenth century before much progress was made on higher-degree polynomial equations.

Around 1540, a tangled tale unfolded. It seems an Italian professor of mathematics, Scipione del Ferro, discovered how to solve certain cubic equations, but failed to publish his findings. Fior, his student, was aware of his findings, but he also did not publish. About this time, Niccolo Fontana, known as Tartaglia (the stammerer), claimed publicly to be able to solve cubic equations of the form $x^3 + mx = n$. Since Fior could solve such equations, he arranged for a public contest to be held, each contestant to provide 30 questions to the other. Just before the contest, Tartaglia discovered how to solve cubic equations of the form $x^3 + px^2 = n$. Therefore, Tartaglia solved all of Fior's questions in the contest and Fior solved none of Tartaglia's.

Upon hearing of the matter, Girolamo Cardano—a well-known mathematician, physician, astrologer, and scoundrel—invited Tartaglia to his home where he tricked him into revealing his solution, while promising never to reveal it. Shortly thereafter, Cardano published his monumental mathematics book which included Tartaglia's solution to cubic equations. Ferrari, a student of Cardano, discovered a method of solving the general fourth-degree polynomial equation.

In the nineteenth century, a Norwegian mathematician, Niels Abel, proved that a general solution to polynomial equations of degree higher than four is not possible. In 1831 a young French mathematician, Evariste Galois, using methods of abstract algebra, classified those higher-degree equations that can be solved. Galois was unable to develop his theory fully because he was killed in a duel in 1832 at the age of 21.

5.1 SYNTHETIC DIVISION

We have seen that quadratic equations as well as polynomial equations of degree higher than 2 arise as mathematical models for applications. In Section 2.4 we solved quadratic equations and in Section 2.6 we solved certain polynomial equations of degree higher than 2 by factoring. In general, we can not solve polynomial equations by factoring. Finding exact solutions of polynomial equations of degree 1, 2, 3, and 4 can be done with formulas such as the Quadratic Formula which produces solutions of second-degree equations. The formulas for third- and fourth-degree equations are complicated. With computers and graphing calculators it is easy to approximate solutions to equations. However, even with technology, knowing as much as possible about the solutions before finding them saves time and effort. In this chapter we will study polynomial equations with emphasis on finding rational and nonreal complex solutions and approximating real irrational solutions.

To continue the study of polynomial equations we need to develop another tool. We saw how to divide polynomials in Section 1.3. Polynomial division is very important when the divisor is of the form $x - r$, for some number r. Long division, with such a divisor, can be condensed into a revised procedure called **synthetic division.**

Let's examine a typical long division display when the divisor is of the form $x - r$.

$$
\begin{array}{r|l}
 & 2x^2 + 5x - 3 \\
\hline
x - 3 & 2x^3 - x^2 - 18x + 16 \\
 & \underline{2x^3 - 6x^2} \\
 & \qquad 5x^2 - 18x \\
 & \qquad \underline{5x^2 - 15x} \\
 & \qquad\qquad -3x + 16 \\
 & \qquad\qquad \underline{-3x + 9} \\
 & \qquad\qquad\qquad 7
\end{array}
$$

Notice that all the arithmetic is done with the *coefficients.* The various powers of the variable merely maintained the structure. So, if we carefully align the coefficients, we can remove the variables completely.

$$
\begin{array}{r|l}
 & 2 + 5 - \ \ 3 \\
\hline
1 - 3 & 2 - 1 - 18 + 16 \\
 & 2 - 6 \\
\cline{2-2}
 & \quad\ \ 5 - 18 \\
 & \quad\ \ 5 - 15 \\
\cline{2-2}
 & \qquad\ \ - \ \ 3 + 16 \\
 & \qquad\ \ - \ \ 3 + \ \ 9 \\
\cline{2-2}
 & \qquad\qquad\quad\ \ 7
\end{array}
$$

Next we see that many numbers in the display are repeated in different locations according to a definite pattern. We can eliminate the numbers circled above without loss of information. If we remove the circled numbers and then push the display up to make it more compact, we have

$$
\begin{array}{r|r}
 & 2 + 5 - \ \ 3 \\
\hline
1 - 3 & 2 - 1 - 18 + 16 \\
 & - 6 - 15 + \ \ 9 \\
\cline{2-2}
 & 7
\end{array}
$$

Notice that the essence of the entire long division procedure is still present. Since we use synthetic division only when the divisor is of the form $x - r$, we can eliminate the leading 1 in the divisor. Also note that if we change the sign on the -3 in the divisor, it will change all the signs in the third row, but then we can *add instead of subtracting.*

$$
\begin{array}{r|r}
 & 2 + 5 - \ \ 3 \\
\hline
3 & 2 - 1 - 18 + 16 \\
 & + 6 + 15 - \ \ 9 \\
\cline{2-2}
 & 7
\end{array}
$$

Now, if we move the quotient down to the line with the remainder and remove the unnecessary signs, we will finally have the same division problem as in the example written in the synthetic division form.

$$
\begin{array}{l r|rrrr l}
\text{Divisor, } x - 3 & 3 & 2 & -1 & -18 & 16 & \text{Dividend, } 2x^3 - x^2 - 18x + 16 \\
 & & & 6 & 15 & -9 & \\
\cline{3-6}
 & & 2 & 5 & -3 & 7 &
\end{array}
$$

Quotient, $2x^2 + 5x - 3$ Remainder, 7

EXAMPLE 1 Use synthetic division to divide $3t^4 - 2t^3 - 11t^2 + 6t - 2$ by $t - 2$.

Solution We put the 2 from the divisor, $t - 2$, and the coefficients from the dividend into the display. Then we bring down the 3.

$$\begin{array}{r|rrrrr} 2 & 3 & -2 & -11 & 6 & -2 \\ & & & & & \\ \hline & 3 & & & & \end{array}$$

Next, we multiply the 2 by the 3 and write the result under the -2 and *add.*

$$\begin{array}{r|rrrrr} 2 & 3 & -2 & -11 & 6 & -2 \\ & & 6 & & & \\ \hline & 3 & 4 & & & \end{array}$$

We multiply the 2 by the 4 and write the result under the -11 and add. Then we continue this procedure all the way to the end.

$$\begin{array}{r|rrrrr} 2 & 3 & -2 & -11 & 6 & -2 \\ & & 6 & 8 & -6 & 0 \\ \hline & 3 & 4 & -3 & 0 & -2 \end{array}$$

We now can read the result directly off the display. $3t^4 - 2t^3 - 11t^2 + 6t - 2$ divided by $t - 2$ is $3t^3 + 4t^2 - 3t$ with a remainder of -2. □

Two other ways to write the result of Example 1 are

$$\frac{3t^4 - 2t^3 - 11t^2 + 6t - 2}{t - 2} = 3t^3 + 4t^2 - 3t + \frac{-2}{t - 2}$$

and

$$3t^4 - 2t^3 - 11t^2 + 6t - 2 = (t - 2)(3t^3 + 4t^2 - 3t) - 2$$

EXAMPLE 2 Perform the indicated division: $\dfrac{y^4 - 12y^2 + 2}{y - 4}$.

Solution Since the position of the coefficients in synthetic division is *very important,* we must insert zeros for the coefficients of y^3 and y in the dividend. Then we proceed as before.

$$\begin{array}{r|rrrrr} 4 & 1 & 0 & -12 & 0 & 2 \\ & & 4 & 16 & 16 & 64 \\ \hline & 1 & 4 & 4 & 16 & 66 \end{array}$$

$$\frac{y^4 - 12y^2 + 2}{y - 4} = y^3 + 4y^2 + 4y + 16 + \frac{66}{y - 4}$$ □

EXAMPLE 3 Use synthetic division to find the remainder when $x^3 + 5x^2 - 18$ is divided by $x + 3$.

Solution The procedure for synthetic division is based on division by $x - r$. Therefore, in this problem we must think of $x + 3$ as $x - (-3)$ and insert -3 in the display. We must also insert 0 as the coefficient of the first degree-term in the dividend.

$$\begin{array}{r|rrrr} -3 & 1 & 5 & 0 & -18 \\ & & -3 & -6 & 18 \\ \hline & 1 & 2 & -6 & 0 \end{array}$$

When $x^3 + 5x^2 - 18$ is divided by $x + 3$, the remainder is 0. □

Synthetic division also works with complex numbers.

EXAMPLE 4 Use synthetic division to divide.

$$\frac{x^3 - 2x^2 + x - 2}{x - i}$$

Solution Since the divisor in $\frac{x^3 - 2x^2 + x - 2}{x - i}$ is $x - i$, we use i as the divisor. The procedure is the same as with real numbers.

We must, however, add and multiply complex numbers.

$$\begin{array}{r|rrrr} i & 1 & -2 & 1 & -2 \\ & & i & (-1 - 2i) & 2 \\ \hline & 1 & -2 + i & -2i & 0 \end{array}$$

$$\frac{x^3 - 2x^2 + x - 2}{x - i} = x^2 + (-2 + i)x - 2i$$ □

The result of dividing the polynomial $P(x)$ by $x - r$ can be recorded in two different ways:

$$\frac{P(x)}{x - r} = Q(x) + \frac{R}{x - r}$$

and

$$P(x) = Q(x) \cdot (x - r) + R$$

Notice that if the remainder is 0, then the divisor is a factor of the polynomial.

EXAMPLE 5 Use synthetic division to determine if $x + 4$ is a factor of $x^3 + 3x^2 - x + 8$.

Solution $x + 4$ is a factor of $x^3 + 3x^2 - x + 8$ if the remainder is 0 when $x^3 + 3x^2 - x + 8$ is divided by $x + 4$.

$$\begin{array}{r|rrrr} -4 & 1 & 3 & -1 & 8 \\ & & -4 & 4 & -12 \\ \hline & 1 & -1 & 3 & -4 \end{array}$$

The remainder is -4. Thus, $x + 4$ is not a factor of $x^3 + 3x^2 - x + 8$. □

PROBLEM SET 5.1

Warm-ups

In Problems 1 through 4, use synthetic division to divide. See Examples 1 and 2.

1. $\dfrac{2x^3 - x^2 - 3x - 4}{x - 2}$

2. $\dfrac{t^4 - t^3 + t^2 - t + 1}{t - 1}$

3. $(x^4 - 2x^2 - 5x + 3) \div (x - 2)$

4. $(3w^3 - 12w^2 - 14w - 6) \div (w - 5)$

In Problems 5 through 8, use synthetic division to find the remainder. See Example 3.

5. $\dfrac{x^4 - 2x^2 + 2}{x - 3}$

6. $(R^4 + 1) \div (R + 1)$

7. $\dfrac{2x + 1 + 3x^3 + 4x^2}{x + 2}$

8. $(t^5 - 1) \div (t - 1)$

In Problems 9 and 10, use synthetic division to divide. See Example 4.

9. $\dfrac{x^4 - 2x^2 + x - 2}{x - i}$

10. $\dfrac{x^2 - 4x + 5}{x - (2 - i)}$

In Problems 11 and 12, use synthetic division to determine if $x + 1$ is a factor of the given polynomial. See Example 5.

11. $3x^5 + x^3 - 2x^2 - 5$

12. $x^6 + 2x^5 - 2x^4 + x^3 - x^2 + x - 1$

Practice Exercises

In Problems 13 through 26, use synthetic division to perform the indicated divisions, writing the results in the form: Polynomial = quotient · divisor + remainder

13. $(x^3 + 3x^2 + 3x + 1) \div (x - 1)$

14. $(w^3 - 2w^2 - 2w - 1)/(w - 1)$

15. $\dfrac{q^3 + 3q^2 - 4q - 12}{q - 2}$

16. $\dfrac{2x^3 - 7x^2 - x + 12}{x - 3}$

17. $\dfrac{t^3 - 5t^2 + 3t + 10}{t - 3}$

18. $\dfrac{2x^3 - 6x^2 - x + 7}{x - 2}$

19. $\dfrac{2x^3 + 2x^2 + 5x + 5}{x + 1}$

20. $\dfrac{t^3 + 4t^2 + 5t + 2}{t + 2}$

21. $\dfrac{3k^3 + 7k^2 - 25k + 10}{k + 4}$

22. $\dfrac{5x^3 + 11x^2 - 11x + 7}{x + 3}$

23. $\dfrac{x^6 - 2^6}{x - 2}$

24. $\dfrac{s^5 - 3^5}{s - 3}$

25. $\dfrac{x^4 + 2x^2 + 1}{x - i}$

26. $\dfrac{x^4 - 2x^3 + x^2 + 2x - 2}{x - 1 + i}$

In Problems 27 through 40, use synthetic division to find the remainder.

27. $\dfrac{3w^3 - 15w + 10}{w - 2}$

28. $\dfrac{2x^3 - 5x^2 + 5}{x - 3}$

29. $(7x^3 + 10x^2 + 5) \div (x + 1)$

30. $(10r^3 - 37r + 2) \div (r + 2)$

31. $\dfrac{s^4 - 5s^3 - s^2 + 12s + 4}{s - 2}$

32. $\dfrac{2x^4 - 7x^3 - x^2 + 10x - 8}{x - 1}$

33. $\dfrac{4x^4 + 12x^3 - 2x^2 - 3x + 10}{x + 3}$

34. $\dfrac{3g^4 + 11g^3 - 4g^2 + g - 10}{g + 4}$

35. $\dfrac{t^4 - 4t^2 - 100}{t - 6}$

36. $\dfrac{2x^4 - 11x^3 + 100}{x - 5}$

37. $(2x^4 + 15x + 7) \div (x + 2)$

38. $(t^4 + 20t - 11) \div (t + 3)$

39. $\dfrac{3w^6 - 3w^5 + 2w^2 - 2}{w - 1}$

40. $\dfrac{x^7 - 8x^4 - x + 12}{x - 2}$

In Problems 41 through 44, determine if $x + 1$ is a factor of the given polynomial.

41. $x^8 + 1$ **42.** $x^9 + 1$ **43.** $x^7 - 1$ **44.** $x^6 - 1$

In Problems 45 through 48, determine if $x - 1$ is a factor of the given polynomial.

45. $x^8 + 1$ **46.** $x^9 + 1$ **47.** $x^7 - 1$ **48.** $x^6 - 1$

In Problems 49 and 50, factor each polynomial if possible.

49. $x^5 + 1$ **50.** $x^5 - 1$

In Problems 51 through 56, perform the indicated division.

51. $(2x^4 + 3x^3 - 6x^2 + 1) \div \left(x - \dfrac{1}{2}\right)$

52. $(6x^4 + 8x^3 + 3x^2 + x + 1) \div \left(x + \dfrac{1}{3}\right)$

53. $(3z^4 + 7z^3 + 3z + 2) \div \left(z + \dfrac{1}{3}\right)$

54. $(3x^4 - 7x^3 - 2x + 1) \div \left(x - \dfrac{1}{3}\right)$

55. $(x^3 - 3x^2 + x + 1)/(x - 1 + \sqrt{2})$

56. $(x^3 - 3x^2 - x + 3)/(x - 1 - \sqrt{3})$

57. What number must k be so that $x - 5$ is a factor of $2x^3 - 7x^2 + kx + 5$?

58. What number must k be so that $x + 1$ is a factor of $x^3 + kx^2 - 2x + 3$?

Challenge Problems

In Problems 59 and 60, use synthetic division to find the quotient and remainder.

59. $\dfrac{2x^4 + x^3 - 5x^2 + 4x - 1}{2x - 1}$ (*Hint:* Write the divisor as $2\left(x - \dfrac{1}{2}\right)$, divide by $\left(x - \dfrac{1}{2}\right)$ using synthetic division, and complete the division by dividing the quotient and the remainder by 2.)

60. $\dfrac{6x^4 - 2x^3 - x^2 + 8x + 1}{3x + 2}$ (See hint for Problem 59.)

In Problems 61 through 64, perform the indicated division and assume k is a real number.

61. $\dfrac{x^4 - k^4}{x + k}$ **62.** $\dfrac{x^4 + k^4}{x - k}$ **63.** $\dfrac{x^3}{x - k}$ **64.** $\dfrac{x^3}{x + k}$

65. Use synthetic division to perform the indicated division.

(a) $(2x^3 - x^2 - 13x - 6) \div [(x - 3)(x + 2)]$

(b) $\dfrac{x^4 + x^3 - x - 1}{(x - 1)(x + 1)}$

IN YOUR OWN WORDS . . .

66. Describe the process of synthetic division.

67. Explain how synthetic division can be used to factor a polynomial.

5.2 ZEROS OF POLYNOMIAL EQUATIONS; THE FACTOR AND REMAINDER THEOREMS

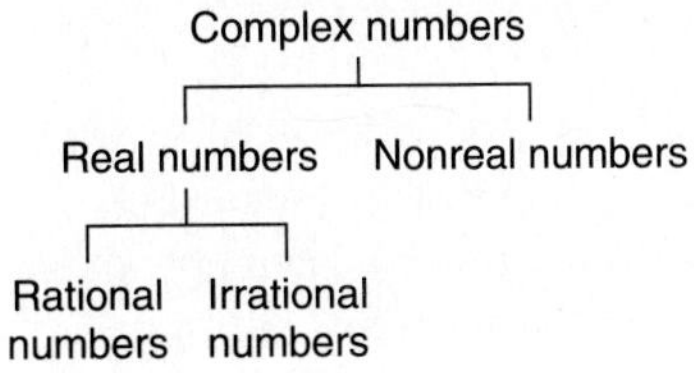

Complex numbers
Real numbers Nonreal numbers
Rational numbers Irrational numbers

Fig. 5.1

Since we are operating in the system of complex numbers, all solutions to polynomial equations are complex numbers. The theorems we will develop in this chapter make it necessary for us to distinguish complex numbers as real numbers or nonreal numbers. Likewise, real numbers are either rational or irrational. Figure 5.1 shows this relationship. For example, $3 + 2i$ is a nonreal complex number, while $\sqrt{2}$ is a real, irrational, complex number.

Zeros of Polynomial Functions

The main thrust of this chapter is solving polynomial equations. We do this by considering **polynomial functions.** We will use the following ideas throughout the chapter. (n is a nonnegative integer, and $a_n \neq 0$.)

Polynomial of degree n: $a_nx^n + a_{n-1}x^{n-1} + \cdots + a_1x + a_0$

Polynomial equation: $a_nx^n + a_{n-1}x^{n-1} + \cdots + a_1x + a_0 = 0$

Polynomial function: $P(x) = a_nx^n + a_{n-1}x^{n-1} + \cdots + a_1x + a_0$

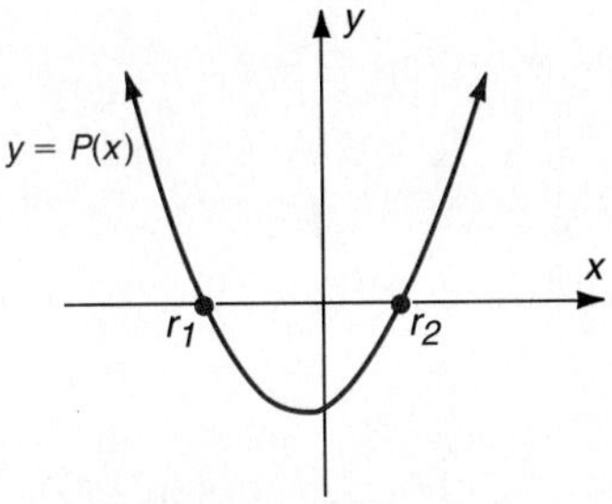

A **zero** of a polynomial is a solution of the polynomial equation. That is, if P is a polynomial and $P(r) = 0$, then r is a zero of P. Solving the polynomial equation

$$a_nx^n + a_{n-1}x^{n-1} + \cdots + a_1x + a_0 = 0$$

is equivalent to finding the zeros of the polynomial P. These zeros may be nonreal complex numbers.

Notice in Figure 5.2 that a *real* zero of a polynomial equation P is an x-intercept of the graph of P.

y
y = P(x)
x
r_1 r_2

r_1 and r_2 are real zeros of P.

r_1 and r_2 are the x-intercepts of the graph of P.

Fig. 5.2

EXAMPLE 1 Find the zeros of each polynomial equation and interpret them graphically.

(a) $f(x) = x^4 - 2x^3 - 2x^2$ (b) $g(x) = 2x^3 + 3x^2 - 18x - 27$
(c) $h(t) = t^4 + 64$

Solutions (a) $f(x) = x^4 - 2x^3 - 2x^2$

We must find the numbers x such that $f(x) = 0$. That is, we must solve the equation

$x^4 - 2x^3 - 2x^2 = 0$

$x^2(x^2 - 2x - 2) = 0$ Factor.

$x^2 = 0$ or $x^2 - 3x - 2 = 0$ Property of Zero Products

$$x = \frac{2 \pm \sqrt{(-2)^2 - 4(1)(-2)}}{2 \cdot 1}$$ Quadratic Formula

$$x = \frac{2 \pm \sqrt{4 + 8}}{2} = 1 \pm \sqrt{3}$$

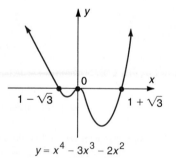

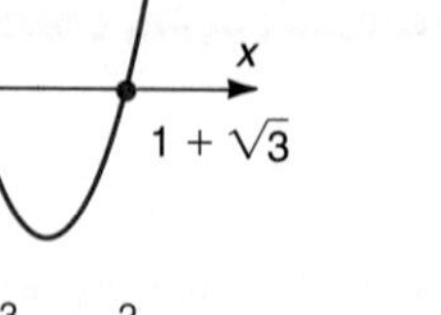

$y = x^4 - 3x^3 - 2x^2$

Fig. 5.3

So, the zeros of this polynomial equation are 0, $1 - \sqrt{3}$, and $1 + \sqrt{3}$. Notice that 0 is a zero of multiplicity 2. The graph of f is shown in Figure 5.3.

Inspecting the graph in Figure 5.3, we confirm that the x-intercepts of the graph are the zeros of the function. Notice that all the zeros are *real numbers.* Also, notice that the graph touches, but does not cross, the x-axis at the double zero.

(b) $g(x) = 2x^3 + 3x^2 - 18x - 27$

We form the polynomial equation.

$2x^3 + 3x^2 - 18x - 27 = 0$

Since we have a four-term polynomial to factor, we try grouping.

$x^2(2x + 3) - 9(2x + 3) = 0$

$(2x + 3)(x^2 - 9) = 0$ Factor.

$(2x + 3)(x + 3)(x - 3) = 0$ Factor completely.

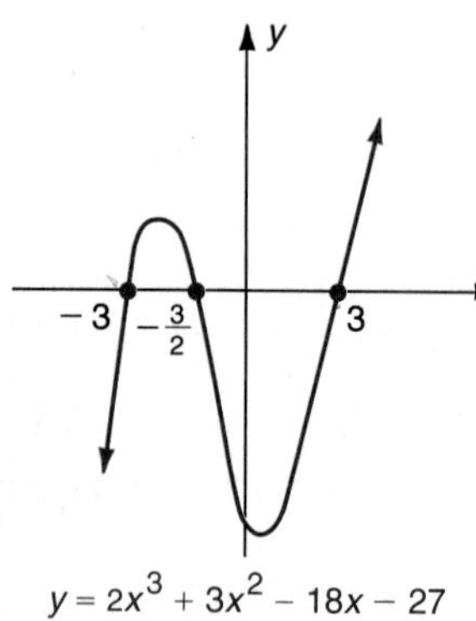

$y = 2x^3 + 3x^2 - 18x - 27$

Fig. 5.4

And we see the zeros are $-\dfrac{3}{2}$, -3, and 3.

Again we see from Figure 5.4 that the graph crosses the x-axis exactly at the zeros of the function.

(c) $h(t) = t^4 + 64$

Form the polynomial equation.

$t^4 + 64 = 0$

At first glance, it looks like the Square Root Property will do the job. However, it leads to expressions like $\sqrt{8i}$ which we are not prepared to handle. (Note that the i is *inside* the square root symbol.) But we *can* factor this by completing the square.

$$t^4 + 16t^2 + 64 - 16t^2 = 0$$ Add and subtract $16t^2$.

The first three terms are a perfect square.

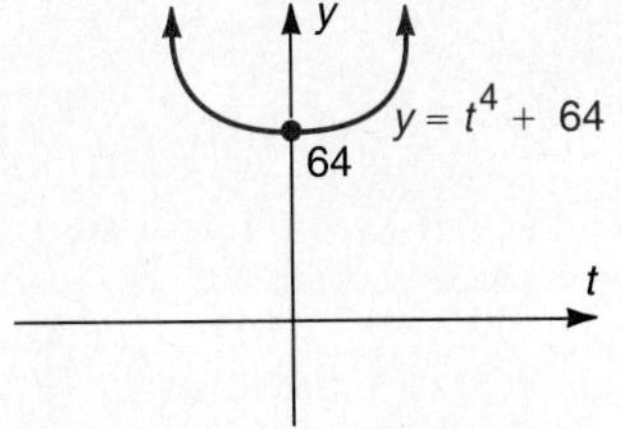

Notice that the graph never touches the t axis and that the function has no *real* zeros.

Fig. 5.5

$$(t^2 + 8)^2 - 16t^2 = 0$$

$$[(t^2 + 8) - 4t][(t^2 + 8) + 4t] = 0 \qquad \text{Difference of two squares}$$

$$(t^2 - 4t + 8)(t^2 + 4t + 8) = 0$$

$$t^2 - 4t + 8 = 0 \quad \text{or} \quad t^2 + 4t + 8 = 0 \qquad \text{Property of Zero Products}$$

$$t = \frac{4 \pm \sqrt{16 - 32}}{2} \quad \text{or} \quad t = \frac{-4 \pm \sqrt{16 - 32}}{2} \qquad \text{Quadratic Formula}$$

$$t = 2 \pm 2i \quad \text{or} \quad t = -2 \pm 2i$$

The zeros are $2 + 2i$, $2 - 2i$, $-2 + 2i$, and $-2 - 2i$ (Figure 5.5). □

Division Algorithm

The polynomial equations in Example 1 were all special in that they could be factored without too much effort. To continue the study of polynomial equations we need to apply what we learned about the division of polynomials. We know that

$$\frac{2x^4 - 3x^3 + 11x - 7}{x^2 - x + 2} = 2x^2 - x - 5 + \frac{8x + 3}{x^2 - x + 2}$$

Notice that the answer is written in the form

$$\frac{\text{Polynomial}}{\text{Divisor}} = \text{quotient} + \frac{\text{remainder}}{\text{divisor}}$$

That is,

$$\frac{P(x)}{D(x)} = Q(x) + \frac{R(x)}{D(x)}$$

If we rewrite the above results by multiplying both sides by the divisor, we have

$$2x^4 - 3x^3 + 11x - 7 = (x^2 - x + 2)(2x^2 - x - 5) + 8x + 3$$

which is of the form

$$\text{Polynomial} = \text{divisor} \cdot \text{quotient} + \text{remainder}$$

That is,

$$P(x) = D(x) \cdot Q(x) + R(x)$$

This is, in fact, a statement of the **division algorithm,** which we give without proof.

Theorem: The Division Algorithm

If $P(x)$ and $D(x)$ are polynomials with the degree of $D(x)$ less than or equal to the degree of $P(x)$, and $D(x)$ is not the zero polynomial, then there are unique polynomials $Q(x)$ and $R(x)$ such that

$$P(x) = D(x) \cdot Q(x) + R(x)$$

The degree of $R(x)$ is less than the degree of $D(x)$ or $R(x)$ is the zero polynomial. $D(x)$ is called the divisor, $Q(x)$ the quotient, and $R(x)$ the remainder.

This result is most useful in the special case when $D(x)$ is of the form $x - r$, for some number r. In this case the division algorithm takes the form

$$P(x) = (x - r)Q(x) + R$$

where R is a number.

The Remainder Theorem

If the polynomial $P(x)$ is divided by $x - r$, then the remainder is $P(r)$.

Proof:

$$\begin{aligned} P(x) &= (x - r)Q(x) + R && \text{Division Algorithm} \\ P(r) &= (r - r)Q(r) + R && \text{Substitution} \\ P(r) &= R \end{aligned}$$

Thus, the value of the function when x is r is the *remainder* when $P(x)$ is divided by $x - r$.

The Remainder Theorem gives us another way to evaluate a polynomial.

EXAMPLE 2 If $P(x) = 7x^4 - 19x^3 - 8x^2 - 1$, use synthetic division to find $P(3)$.

Solution Rather than evaluating the polynomial by direct substitution, we use synthetic division to divide $7x^4 - 19x^3 - 8x^2 - 1$ by $x - 3$.

$$\begin{array}{r|rrrrr} 3 & 7 & -19 & -8 & 0 & -1 \\ & & 21 & 6 & -6 & -18 \\ \hline & 7 & 2 & -2 & -6 & -19 \end{array}$$

The Remainder Theorem says that $P(3) = -19$. □

When a polynomial is divided by $x - r$ and the remainder is 0, then $x - r$ is a factor of the polynomial.

The Factor Theorem

The polynomial $P(x)$ has $x - r$ as a factor if and only if $P(r) = 0$.

Proof: We first prove that if $P(r) = 0$, then $x - r$ is a factor of $P(x)$.

$P(x) = (x - r) \cdot Q(x) + R$ Division Algorithm

$P(r) = R$ Remainder Theorem

Since $P(r) = 0$, R must be 0.

So, $P(x) = (x - r) \cdot Q(x)$. Thus, $x - r$ is a factor of $P(x)$.

Now we prove that if $x - r$ is a factor of $P(x)$, then $P(r) = 0$.

We can write $P(x)$ as the product of $x - r$ and some polynomial $T(x)$.

$P(x) = (x - r)T(x)$ $x - r$ is a factor of $P(x)$.

$P(r) = (r - r)T(r)$

$P(r) = 0$

Thus, we have shown that $x - r$ is a factor of the polynomial $P(x)$ if and only if $P(r) = 0$.

The Factor Theorem says that if a polynomial has $x - r$ as a *factor*, then r is a *root* of the polynomial equation, and conversely, if r is a *root*, then $x - r$ is a *factor* of the polynomial. Notice that $x - r$ is a *linear* factor. So, knowing a linear factor is equivalent to knowing a root.

EXAMPLE 3 Use the Factor Theorem to show that $x - 3$ is a factor of $x^3 - 4x^2 + 3x$.

Solution Let $P(x) = x^3 - 4x^2 + 3x$.
Then $P(3) = 27 - 36 + 9 = 0$.
$x - 3$ is a factor of $P(x)$. Factor Theorem □

EXAMPLE 4 Find a polynomial equation of degree 5 with roots of -1, 2, and 5 and no others.

Solution The Factor Theorem tells us that we have factors of $x + 1$, $x - 2$, and $x - 5$. Since the polynomial is fifth degree, two of these must be roots of multiplicity 2. With no more information, any two of the given roots could be double roots. One such equation is $(x + 1)^2(x - 2)^2(x - 5) = 0$. Another would be $(x + 1)(x - 2)^2(x - 5)^2 = 0$. We could also make other polynomial equations satisfying the given information by changing the leading coefficient. Thus, $a(x + 1)^2(x - 2)^2(x - 5) = 0$, where a is any real number, is such an equation. □

EXAMPLE 5 Find a second-degree polynomial function with zeros of $\frac{1}{2}$ and -1 and integer coefficients.

Solution Since $\frac{1}{2}$ and -1 are zeros, the polynomial will have factors of $x - \frac{1}{2}$ and $x + 1$. So, $f(x) = \left(x - \frac{1}{2}\right)(x + 1)$ is a second-degree polynomial function with zeros of $\frac{1}{2}$ and -1. But, we want the integer coefficients. If we multiply the factors by 2, we do not change the zeros and the coefficients will be integers.

$$f(x) = 2\left(x - \frac{1}{2}\right)(x + 1)$$

$$f(x) = (2x - 1)(x + 1)$$

$$f(x) = 2x^2 + x - 1$$

□

An important use of the Factor Theorem is to use a *known root* to reduce the degree of the polynomial in a polynomial equation.

EXAMPLE 6 Solve the equation $2x^3 + 7x - 9 = 0$ given that 1 is a solution.

Solution Since 1 is a solution, we know that the polynomial has $x - 1$ as a factor. Thus

$$2x^3 + 7x - 9 = (x - 1)Q(x)$$

for some polynomial $Q(x)$. But by the pattern in synthetic division we know that $Q(x)$ must be of second degree, and we know how to solve all quadratic equations. We use synthetic division to find $Q(x)$.

$$\begin{array}{r|rrrr} 1 & 2 & 0 & 7 & -9 \\ & & 2 & 2 & 9 \\ \hline & 2 & 2 & 9 & 0 \end{array}$$

Now we have $Q(x)$. It is $2x^2 + 2x + 9$. Therefore,

$$2x^3 + 7x - 9 = (x - 1)(2x^2 + 2x + 9)$$

We see by inspection that 1 is a solution of the original equation (we knew that!). Any additional solutions must come from the quadratic equation,

$$2x^2 + 2x + 9 = 0 \qquad \text{Property of Zero Products}$$

Using the quadratic formula,

$$x = \frac{-2 \pm \sqrt{4 - 4 \cdot 2 \cdot 9}}{2 \cdot 2}$$

$$= \frac{-2 \pm 2\sqrt{17}i}{4} = -\frac{1}{2} \pm \frac{\sqrt{17}}{2}i$$

$$\left\{1, -\frac{1}{2} \pm \frac{\sqrt{17}}{2}i\right\} \qquad \text{Solution set}$$

□

EXAMPLE 7 Find the zeros of the polynomial function $F(t) = 4t^4 - 16t^3 + 13t^2 + 15t - 18$ given that -1 and 2 are zeros.

Solution We are given that $F(-1) = 0$ and $F(2) = 0$. So, we can use the Factor Theorem twice. First, since -1 is a zero, $t - (-1) = t + 1$ is a factor. By synthetic division,

$$\begin{array}{r|rrrrr} -1 & 4 & -16 & 13 & 15 & -18 \\ & & -4 & 20 & -33 & 18 \\ \hline & 4 & -20 & 33 & -18 & 0 \end{array}$$

Therefore,

$$F(t) = (t + 1)(4t^3 - 20t^2 + 33t - 18) \qquad \text{Factor Theorem}$$

Next, we use synthetic division to divide out the other zero. We are given that $F(2) = 0$, so $F(t)$ has $t - 2$ as a factor by the Factor Theorem. Certainly it is not a factor of $t + 1$, so it must be a factor of $4t^3 - 20t^2 + 33t - 18$.

$$\begin{array}{r|rrrr} 2 & 4 & -20 & 33 & -18 \\ & & 8 & -24 & 18 \\ \hline & 4 & -12 & 9 & 0 \end{array}$$

Now we know

$$F(t) = (t + 1)(t - 2)(4t^2 - 12t + 9) \qquad \text{Factor Theorem}$$

The quadratic polynomial can be factored by inspection.

$$F(t) = (t + 1)(t - 2)(2t - 3)^2$$

Therefore, the zeros of F are -1, 2, and $\frac{3}{2}$. The zero $\frac{3}{2}$ has multiplicity 2. □

In Example 7, note how the two synthetic divisions *fit together,* the second on the end of the first.

$$\begin{array}{r|rrrrr} -1 & 4 & -16 & 13 & 15 & -18 \\ & & -4 & 20 & -33 & 18 \\ \hline 2 & 4 & -20 & 33 & -18 & 0 \\ & & 8 & -24 & 18 & \\ \hline & 4 & -12 & 9 & 0 & \end{array}$$

This is a nice feature of the synthetic division pattern when dividing out more than one root.

With the division algorithm, the Factor Theorem, and the Remainder Theorem we have seen that the following ideas are equivalent.

Four Equivalent Statements about a Polynomial $P(x)$

1. r is a root of the polynomial equation $P(x) = 0$.

2. $P(r) = 0$. That is, r is a zero of the polynomial P.

> 3. $x - r$ is a factor of $P(x)$.
> 4. The remainder when $P(x)$ is divided by $x - r$ is 0.

Also important but not equivalent to the statements in the box is the fact that the x-intercepts of the graph of P are the real zeros of the function P.

PROBLEM SET 5.2

Warm-ups

In Problems 1 through 4, find the zeros of each polynomial equation and sketch the graph illustrating the real zeros, if any. See Example 1.

1. $f(x) = x^3 - x^2 - 6x$
2. $g(x) = x^3 - 2x^2 - x$
3. $F(t) = 2t^3 - 5t^2 + 2t - 5$
4. $G(z) = z^4 - 16$

In Problems 5 through 8, use synthetic division to find the indicated function value. See Example 2.

5. $P(x) = x^4 - 6x^3 - 11x^2 - 9x + 3$; $P(2)$
6. $P(y) = 7y^5 - 16y^4 - 9y^2 + 11$; $P(3)$
7. $P(t) = t^6 + 11t^3 + 10t^2 - 4t$; $P(-2)$
8. $P(x) = 2x^4 - 15x^3 - 31x^2 + 36x - 7$; $P(-9)$

In Problems 9 and 10, show that $x - 4$ is a factor of each polynomial. See Example 3.

9. $2x^3 - 5x^2 - 11x - 4$
10. $x^4 - 14x^2 - 32$

In Problems 11 and 12, find a polynomial equation satisfying the given information. See Example 4.

11. Roots of -3, 0, and 2 only; fourth degree
12. Double root of 4; leading coefficient 3

In Problems 13 and 14, find a polynomial function with integer coefficients satisfying the given information. See Example 5.

13. Double root of $\frac{1}{2}$; degree 2
14. Roots of $\frac{1}{2}$ and $\frac{1}{3}$; degree 3

In Problems 15 and 16, solve each equation using the given information. See Example 6.

15. $x^3 + x^2 - 3x + 1 = 0$; given that 1 is a solution
16. $t^3 - 2t + 4 = 0$; given that -2 is a solution

In Problems 17 and 18, find the zeros of each polynomial function using the given information. See Example 7.

17. $f(x) = x^4 + 2x^3 - 13x^2 - 14x + 24$; given that 1 and -2 are zeros
18. $g(r) = 2r^5 + r^4 - 21r^3 + 2r^2 + 52r - 24$; given that -3 and 2 are zeros, with 2 a zero of multiplicity 2

Practice Exercises

In Problems 19 through 22, factor each polynomial into linear factors using the given information.

19. $f(z) = z^3 + 5z^2 + 7z + 3$; given that -3 is a zero
20. $g(x) = 2x^3 + 6x^2 + 5x + 2$; given that -2 is a zero
21. $h(x) = x^4 + 2x^3 - 10x^2 - 17x + 6$; given that -2 and 3 are zeros
22. $k(s) = s^4 - 3s^3 + 5s^2 - s - 10$; given that -1 and 2 are zeros

In Problems 23 through 26, find the zeros of each polynomial function using the given information.

23. $F(x) = x^4 - 9x^3 + 16x^2 + 15x + 25$; given that 5 is a zero of multiplicity 2

24. $G(x) = x^4 + 7x^3 + 7x^2 - 24x - 16$; given that -4 is a zero of multiplicity 2

25. $H(w) = w^5 + 5w^4 - w^3 - 5w^2 - 2w - 10$; given that -5, $\sqrt{2}$, and $-\sqrt{2}$ are zeros

26. $P(k) = k^5 - 7k^4 + k^3 - 7k^2 - 12k + 84$; given that 7, $\sqrt{3}$, and $-\sqrt{3}$ are zeros

In Problems 27 through 34, solve each equation using the given information.

27. $x^3 - 6x^2 + 7x + 4 = 0$; given that 4 is a solution

28. $r^3 + 5r^2 + 4r - 6 = 0$; given that -3 is a root

29. $p^5 - 10p^4 - 12p^3 + 10p^2 + 11p = 0$; given that -1 is a root of multiplicity 2

30. $2x^5 - 19x^4 + 57x^3 - 56x^2 + 16x = 0$; given that 4 is a root of multiplicity 2

31. $x^6 - 3x^4 - 2x^3 + 4x^2 + 4x = 4$; given that 1 is a root of multiplicity 2 and $\pm\sqrt{2}$ are roots

32. $w^6 + 8w^5 + 14w^4 - 16w^3 - 35w^2 - 24w = 48$; given that -4 is a root of multiplicity 2 and $\pm\sqrt{3}$ are roots

33. $9t^4 - 12t^3 + 13t^2 - 12t + 4 = 0$; given that $\frac{2}{3}$ is a root of multiplicity 2

34. $4x^4 + 16x^3 + 29x^2 + 33x + 18 = 0$; given that $-\frac{3}{2}$ is a root of multiplicity 2

35. Use the graphs of the functions in Figure 5.6 to determine the number and the multiplicity of the real zeros for each function.

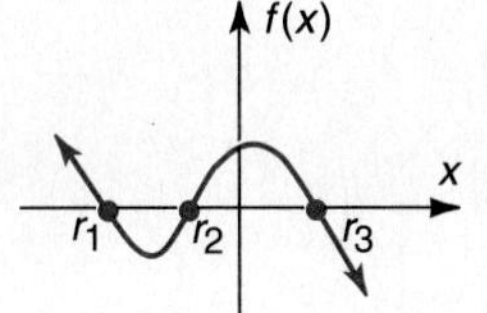

$f(x) = ax3 + bx2 + cx + d$

(a)

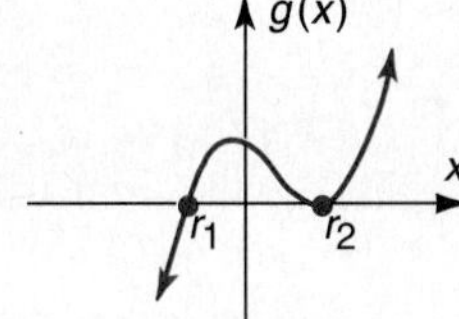

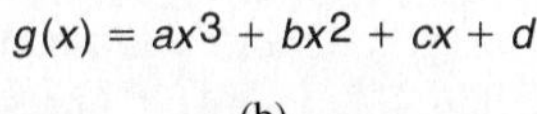

(b)

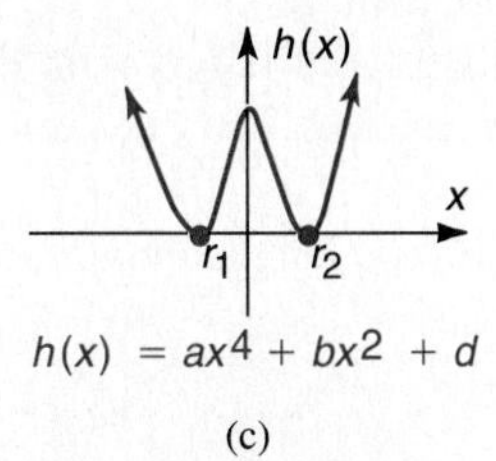

$h(x) = ax4 + bx2 + d$

(c)

Fig. 5.6

36. How many real zeros does the function $f(x) = a(x - h)^3 + k$ have?

37. How many real zeros can a second-degree polynomial have? Under what circumstances will the function have no real zeros?

38. How many real zeros can third- and fourth-degree polynomial functions have?

39. The graph of the function $f(x) = 4x^3 - 11x^2 + 6x + 1$ is sketched in Figure 5.7. Point A is called a relative maximum, and point B is called a relative minimum. Use synthetic division to guess the coordinates of points A and B. (*Hint:* The x-coordinates are rational numbers.)

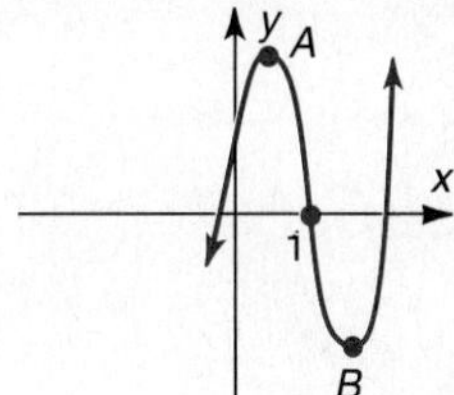

Fig. 5.7

40. Evaluate $3x^5 - x^4 - 10x^3 + 9x^2 + 16x + 2$ when x is $-\frac{5}{3}$.

41. If $f(x) = 6x^5 - 2x^4 + 3x^3 - x + 1$, find $f\left(\frac{3}{2}\right)$.

42. Use synthetic division to show that i is a zero of $2t^4 - 3t^3 + 6t^2 - 3t + 4$.

43. Evaluate $z^5 - z^4 + z^3 + 2z^2 + 2$ when z is $1 + i$.

Challenge Problems

44. Show that every polynomial function with real coefficients of odd degree has at least one real zero.

45. Show that every polynomial function with real coefficients of even degree has an even number of real zeros if a zero of multiplicity k is counted k times.

IN YOUR OWN WORDS . . .

46. What is the connection between the Factor Theorem and the Remainder Theorem and the zeros of polynomials?

47. Describe two methods for evaluating a polynomial.

48. Explain the relationship between the multiplicity of a real zero of a function and the graph of the function.

5.3 THE FUNDAMENTAL THEOREM OF ALGEBRA; RATIONAL ZEROS

It is now time to face an important question, Does every polynomial equation have zeros? The answer to this question lies in a theorem with the imposing name, the Fundamental Theorem of Algebra. It was first proved by Carl Friedrich Gauss (1777–1855). We omit the proof.

The Fundamental Theorem of Algebra

Every polynomial equation of degree $n \geq 1$ has at least one complex zero.

We have noticed from the study of quadratic equations that even though all the coefficients of a polynomial are real numbers, the solutions of the polynomial equation may not be real numbers. The Fundamental Theorem merely states that every nonconstant polynomial equation has some kind of zero.

The Fundamental Theorem guarantees that every polynomial equation of degree greater than or equal to 1 has *at least* one zero. The question remains, How many zeros should we expect?

Let f be a polynomial equation of degree $n \geq 1$. By the Fundamental Theorem f has a zero. Call it r_1. Therefore, $f(r_1) = 0$ and

$$f(x) = (x - r_1)g(x) \qquad \text{Factor Theorem}$$

Now g must be a polynomial equation of degree $n - 1$. Either g is of zero degree and f has the form

$$f(x) = (x - r_1)k$$

where k is a nonzero constant, or g has degree greater than zero and therefore has a zero by the Fundamental Theorem. If g is of zero degree, then f is a first-degree polynomial equation with *exactly* one zero, r_1. If the degree of g is greater than 0, then

$$f(x) = (x - r_1)(x - r_2)h(x) \qquad \text{Factor Theorem}$$

where r_2 is the guaranteed zero of g and h is a polynomial equation of degree $n - 2$.

Repeating the argument, either h is of zero degree and f is a second-degree polynomial equation with *exactly* two zeros or f has the form

$$f(x) = (x - r_1)(x - r_2)(x - r_3)j(x) \qquad \text{Factor Theorem}$$

where r_3 is the guaranteed zero of the polynomial function h and j is a polynomial function of degree $n - 3$. As the degree of the remaining polynomial function decreases by 1 on every step, this procedure will end after a finite number of steps. This leads to the Linear Factors Theorem.

The Linear Factors Theorem

If f is a polynomial equation of degree $n \geq 1$, then $f(x)$ can be expressed as the product of n linear factors,

$$f(x) = a_n(x - r_1)(x - r_2) \cdots (x - r_n)$$

where a_n is the leading coefficient of f and $a_n \neq 0$.

Some or all of the r_i's in the Linear Factors Theorem may not be real numbers, and some may be repeated.

EXAMPLE 1 Write each polynomial in the form $a_n(x - r_1)(x - r_2) \cdots (x - r_n)$.

(a) $5x^2 + 2x - 3$ (b) $x^2 + x + 1$

Solutions (a) $5x^2 + 2x - 3 = (5x - 3)(x + 1)$ Factor.

Each factor must be of the form $x - r_i$. So, we must factor out the 5 in the first factor.

$$= 5\left(x - \frac{3}{5}\right)[x - (-1)]$$

(b) We use the Quadratic Formula to find the zeros of $x^2 + x + 1$.

$$x = \frac{-1 \pm \sqrt{-3}}{2}$$

$$= \frac{-1 \pm \sqrt{3}i}{2}$$

So, $x^2 + x + 1 = \left(x - \frac{-1 + \sqrt{3}i}{2}\right)\left(x - \frac{-1 - \sqrt{3}i}{2}\right)$ □

By the Factor Theorem, each r_i is a root of the polynomial equation $f(x) = 0$.

If a root of multiplicity k is counted k times, then we have exactly n roots. There are no other roots because if r is any number not equal to an r_i, then $f(r) = a_n(r - r_1)(r - r_2)\cdots(r - r_n) \neq 0$. Thus an nth-degree polynomial equation has exactly n complex roots, some of which may be repeated.

Theorem: Number of Roots of an nth-Degree Polynomial Equation

Every polynomial equation of degree $n \geq 1$ has exactly n roots if a root of multiplicity k is counted k times.

EXAMPLE 2 Find a polynomial function F with zeros -1, 0, and 3 and with 0 a zero of multiplicity 3 and $F(1) = 8$. Write $F(x)$ in standard form.

Solution The expressions $[x - (-1)]$, $(x - 0)$, and $(x - 3)$ all must be factors of $F(x)$, with $(x - 0)$ occurring three times. Clearly,

$$F(x) = a_n(x + 1)(x - 0)^3(x - 3)$$

is such a function.

$$F(x) = a_n x^3(x + 1)(x - 3)$$

To find a_n, we use $F(1) = 8$.

$$F(1) = a_n 1^3(1 + 1)(1 - 3) = -4a_n$$

$$8 = -4a_n$$

$$a_n = -2$$

$$F(x) = -2x^5 + 4x^4 + 6x^3 \qquad \text{Standard form}$$

□

EXAMPLE 3 Find a polynomial equation G with zeros 1 and $2 \pm i$ and leading coefficient 4. Write $G(t)$ in standard form.

Solution We want a polynomial equation in the variable t with three linear factors and a leading coefficient of 4. From the Linear Factors Theorem we see that

$$G(t) = 4(t - 1)[t - (2 + i)][t - (2 - i)]$$

is such an equation.

$$= 4(t - 1)(t - 2 - i)(t - 2 + i)$$

$$= 4(t - 1)[(t - 2)^2 - i^2] \qquad \text{Difference of two squares}$$

$$= 4(t - 1)(t^2 - 4t + 5)$$

$$= 4(t^3 - 5t^2 + 9t - 5)$$

$$G(t) = 4t^3 - 20t^2 + 36t - 20 \qquad \text{Standard form}$$

□

Rational Zeros

Now suppose a polynomial equation $f(x) = a_nx^n + a_{n-1}x^{n-1} + \cdots + a_1x + a_0$ has all integer coefficients and a *rational* zero r. That is, r is a rational number and $f(r) = 0$. Since r is a rational number, we can write $r = \frac{p}{q}$, where p and q are integers *that do not have a common factor* (other than ± 1). Therefore, $f\left(\frac{p}{q}\right) = 0$.

So,

$$a_n\left(\frac{p}{q}\right)^n + a_{n-1}\left(\frac{p}{q}\right)^{n-1} + \cdots + a_1\left(\frac{p}{q}\right) + a_0 = 0$$

or

$$a_n\frac{p^n}{q^n} + a_{n-1}\frac{p^{n-1}}{q^{n-1}} + \cdots + a_1\frac{p}{q} + a_0 = 0$$

Multiply by q^n.

$$a_np^n + a_{n-1}p^{n-1}q + \cdots + a_1pq^{n-1} + a_0q^n = 0$$

From the boxed equation we can get two interesting relationships. First subtract a_0q^n from both sides and factor p from the left side.

$$p(a_np^{n-1} + a_{n-1}p^{n-2} + \cdots + a_1q^{n-1}) = -a_0q^n$$

As the integer p is a factor of the left side, it must be a factor of the right side. But as p and q have no common factors, p must be a factor of the integer a_0. Next, subtract a_np^n from both sides of the boxed equation and factor q from the left side.

$$q(a_{n-1}p^{n-1} + \cdots + a_1pq^{n-2} + a_0q^{n-1}) = -a_np^n$$

By a similar argument, q must be a factor of the integer a_n. This discussion suggests the following theorem.

The Rational Zero Theorem

If the polynomial equation $f(x) = a_nx^n + a_{n-1}x^{n-1} + \cdots + a_1x + a_0$ has all integer coefficients and $r = \frac{p}{q}$ is a rational zero of f (p and q have no common prime factors; $a_n \neq 0$; $a_0 \neq 0$), then

1. p is a factor of the constant term a_0
2. q is a factor of the leading coefficient a_n.

Notice that this theorem does not tell us how many rational zeros a polynomial function has, or if it has any at all. It does limit the number of rational zeros possible.

EXAMPLE 4 Find the zeros of the function $f(x) = 2x^3 - 3x^2 - 4x - 1$ and sketch the graph of the function.

Solution Since the polynomial $2x^3 - 3x^2 - 4x - 1$ does not factor by grouping and the degree is higher than 2, we hope it has a rational zero. If there is a rational zero, the Rational Zero Theorem tells us it can be written in the form $\frac{p}{q}$, where p is a factor of the constant term -1 and q is a factor of the leading coefficient 2. Let's look at the possibilities.

Factors of the constant term: ± 1

Factors of the leading coefficient: $\pm 1, \pm 2$

Possibilities for $\frac{p}{q}$: $\pm 1, \pm \frac{1}{2}$

By substitution we see that neither 1 nor -1 is a zero. So the search is limited to $\pm \frac{1}{2}$. We will use synthetic division.

$$
\begin{array}{r|rrrr}
\frac{1}{2} & 2 & -3 & -4 & -1 \\
 & & 1 & -1 & -\frac{5}{2} \\
\hline
 & 2 & -2 & -5 & -\frac{7}{2}
\end{array}
$$

Since the remainder is not 0, $\frac{1}{2}$ is not a zero. Next we try $-\frac{1}{2}$.

$$
\begin{array}{r|rrrr}
-\frac{1}{2} & 2 & -3 & -4 & -1 \\
 & & -1 & 2 & 1 \\
\hline
 & 2 & -4 & -2 & 0
\end{array}
$$

We're in luck. Since the remainder is 0, we know that $-\frac{1}{2}$ is a zero of the polynomial function. Furthermore, from the synthetic division we know that $f(x)$ can be factored to

$$f(x) = \left(x + \frac{1}{2}\right)(2x^2 - 4x - 2)$$

or

$$f(x) = (2x + 1)(x^2 - 2x - 1)$$

Since the remaining factor is quadratic, we can find other zeros with the Quadratic Formula.

$$x = \frac{2 \pm \sqrt{4 - 4 \cdot 1(-1)}}{2 \cdot 1}$$

$$= \frac{2 \pm \sqrt{8}}{2} = 1 \pm \sqrt{2}$$

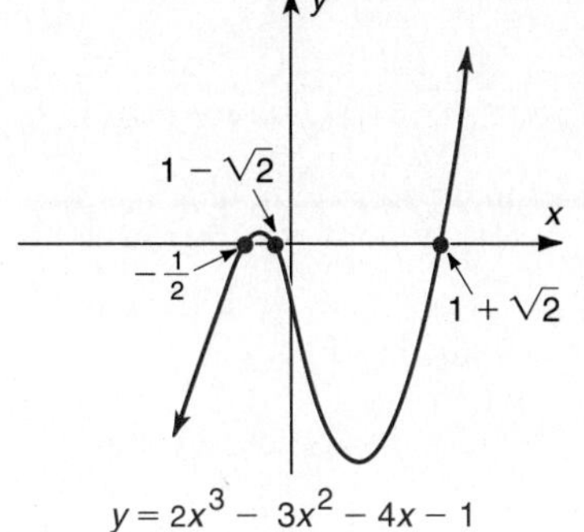

Fig. 5.8

The three zeros of f are $-\frac{1}{2}$ and $1 \pm \sqrt{2}$. The graph in Figure 5.8 illustrates the function and its zeros. Notice that with this scale on the graph the zeros at $-\frac{1}{2}$ and $1 - \sqrt{2}$ (≈ -0.414) are almost indistinguishable. □

EXAMPLE 5 Solve $t^4 - 4t^3 + 5t^2 - 4t + 4 = 0$

Solution The solutions of the given equation are the zeros of the polynomial equation

$$f(t) = t^4 - 4t^3 + 5t^2 - 4t + 4$$

If we can find at least two rational zeros, then the problem is reduced to solving a second-degree equation. The possibilities for rational zeros are

Factors of the constant term: $\pm 1, \pm 2, \pm 4$

Factors of the leading coefficient: ± 1

Possibilities for $\frac{p}{q}$: $\pm 1, \pm 2, \pm 4$

After eliminating ± 1, we try synthetic division with 2. When 2 turns out to be a root, we try 2 *again.* It may be a multiple root. The two divisions follow.

$$\begin{array}{r|rrrrr} 2 & 1 & -4 & 5 & -4 & 4 \\ & & 2 & -4 & 2 & -4 \\ \hline 2 & 1 & -2 & 1 & -2 & 0 \\ & & 2 & 0 & 2 & \\ \hline & 1 & 0 & 1 & 0 & \end{array}$$

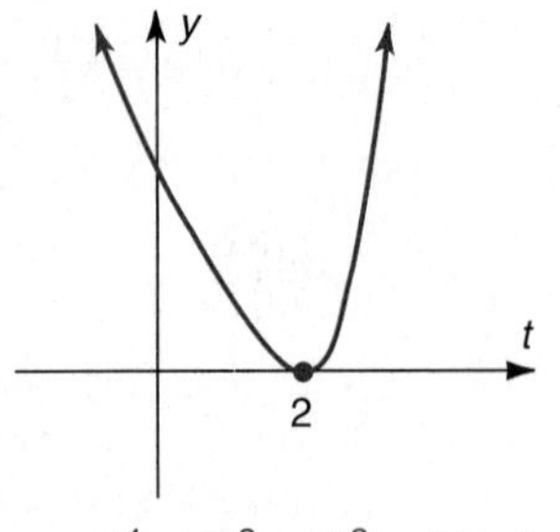

$y = t^4 - 4t^3 + 5t^2 - 4t + 4$

Fig. 5.9

And we have discovered that 2 *is* a multiple root. We rewrite the original equation in factored form. The graph of f is shown in Figure 5.9.

$$(t - 2)^2(t^2 + 1) = 0$$

$$\{2, \pm i\} \qquad \text{Solution set}$$

□

EXAMPLE 6 Find the rational zeros of $g(t) = 2t^4 - t^3 + 3t^2 - t + 1$.

Solution The possibilities for rational zeros are given by the following list.

Factors of the constant term: ± 1

Factors of the leading coefficient: $\pm 1, \pm 2$

Possibilities for $\dfrac{p}{q}$: $\pm 1, \pm \dfrac{1}{2}$

Neither 1 nor -1 is a zero. We show synthetic division for the tests of $\pm \dfrac{1}{2}$.

$$\begin{array}{r|rrrrr} -\frac{1}{2} & 2 & -1 & 3 & -1 & 1 \\ & & -1 & 1 & -2 & \frac{3}{2} \\ \hline & 2 & -2 & 4 & -3 & \frac{5}{2} \end{array}$$

$$\begin{array}{r|rrrrr} \frac{1}{2} & 2 & -1 & 3 & -1 & 1 \\ & & 1 & 0 & \frac{3}{4} & \frac{1}{4} \\ \hline & 2 & 0 & 3 & \frac{1}{2} & \frac{5}{4} \end{array}$$

So we conclude that g has no rational zeros. □

Bounds on Real Zeros

A number U is said to be an **upper bound** for the set of *real* zeros of a function if U is greater than or equal to every number in the set. A number L is said to be a **lower bound** for the set of *real* zeros of a function if L is less than or equal to every number in the set. Neither the upper bound nor the lower bound is unique. The synthetic division display can often help find upper or lower bound. We state the Upper and Lower Bound Theorem without proof. Problem 51 gives some insight into the proof.

Upper and Lower Bound Theorem

Suppose f is a polynomial function with real coefficients and a positive leading coefficient.

1. If $f(x)$ is divided by $x - U$, where $U > 0$, using synthetic division, and the bottom row of the display contains *no negative numbers,* then U is an upper bound for the real zeros of f.
2. If $f(x)$ is divided by $x - L$, where $L < 0$, using synthetic division, and the numbers in the bottom row of the display *alternate in sign,* then L is a lower bound for the real zeros of f. (Treat 0s in the bottom row as either positive or negative, as needed.)

The Upper and Lower Bound Theorem limits the size of *real* zeros of a function. If r is a zero of a function, then $L \leq r \leq U$. Since rational numbers are real numbers, the theorem also limits the choices of rational zeros.

EXAMPLE 7 Find the rational zeros of $F(x) = x^4 + 9x^2 - 2x + 24$.

Solution The possibilities for rational zeros are given by the following list.

Factors of the constant term: $\pm 1, \pm 2, \pm 3, \pm 4, \pm 6, \pm 8, \pm 12, \pm 24$

Factors of the leading coefficient: ± 1

Possibilities for $\dfrac{p}{q}$: $\pm 1, \pm 2, \pm 3, \pm 4, \pm 6, \pm 8, \pm 12, \pm 24$

It seems that we will have to perform 16 synthetic divisions before we can conclude F has no rational zeros. We start with 1.

$$\begin{array}{r|rrrrr} 1 & 1 & 0 & 9 & -2 & 24 \\ & & 1 & 1 & 10 & 8 \\ \hline & 1 & 1 & 10 & 8 & 32 \end{array}$$

We notice that 1 is not a zero. However, we also note that the bottom row in the display *contains no negative numbers.* Therefore, by the Upper and Lower Bound Theorem, 1 is an *upper bound* for the zeros of F, and it is unnecessary to check 2, 3, 4, 6, 8, 12, and 24 because they are all greater than 1. Let's try -1.

$$\begin{array}{r|rrrrr} -1 & 1 & 0 & 9 & -2 & 24 \\ & & -1 & 1 & -10 & 12 \\ \hline & 1 & -1 & 10 & -12 & 36 \end{array}$$

Now we notice that -1 is also not a zero. In addition, this time the signs on the bottom row *alternate*. So, by the Upper and Lower Bound Theorem, -1 is a *lower bound* for the zeros of F. It is unnecessary to check $-2, -3, -4, -6, -8, -12$, or -24 because they are all less than -1.

Thus, we learned that all the real zeros of F are between -1 and 1, but neither -1 nor 1 is a zero. Since there are no other possibilities for rational roots in this interval, we conclude that F has no rational zeros. □

PROBLEM SET 5.3

Warm-ups

In Problems 1 through 4 write each polynomial in the form $a_n(x - r_1)(x - r_2)\cdots(x - r_n)$. See Example 1.

1. $(3x + 2)(x - 5)(2x - 3)$ **2.** $9x^2 - 5$ **3.** $x^2 - 2x + 2$ **4.** $x^2 + 3x + 9$

For Problems 5 through 8, see Examples 2 and 3.

5. Find a polynomial function f having zeros 1 and 2, with 1 a zero of multiplicity 3 and $f(0) = 5$.

6. Find a polynomial function g having zeros -4, 0, and 2, with a leading coefficient of 3. Write $g(x)$ in standard form.

7. Find a polynomial function F having zeros -2 and -3, each of multiplicity 2, and 3 for its leading coefficient.

8. Find a polynomial function G having zeros $-2 \pm i$ and $\pm i$, with i and $-i$ zeros of multiplicity 2 and leading coefficient -1. Write $G(t)$ in standard form.

In Problems 9 and 10, find the zeros of each function and sketch the graph. See Example 4.

9. $f(x) = 3x^3 - 19x^2 + 27x - 7$ **10.** $g(t) = t^4 - 8t^2 - 8t$

In Problems 11 through 14, solve each equation. See Example 5.

11. $6x^3 + 5x^2 - 2x - 1 = 0$ **12.** $2x^4 + 2x^3 - 7x^2 - 8x - 4 = 0$
13. $8z^4 + 4z^3 - 24z^2 - 28z = 8$ **14.** $y^4 + 2y^3 + 3y^2 + 4y + 2 = 0$

In Problems 15 through 18, find the rational zeros of each function, if any. See Examples 6 and 7.

15. $f(x) = 2x^4 + 2x^3 + 3x^2 - x - 2$ **16.** $g(x) = 2x^3 - 11x^2 + 7x - 10$
17. $h(u) = u^4 - 2u^3 + 3u^2 + 6u - 18$ **18.** $k(t) = t^5 + 3t^4 + 12t^3 + 18t^2 + 32t + 24$

Practice Exercises

In Problems 19 through 36, solve each equation.

19. $x^3 - 4x^2 + x + 6 = 0$ **20.** $p^3 - 8p^2 + 21p - 18 = 0$ **21.** $12s^3 + 8s^2 - s - 1 = 0$
22. $18x^3 - 21x^2 + 8x - 1 = 0$ **23.** $2x^3 + 3x^2 - 5x = 6$ **24.** $3y^3 + 4y^2 + 5y = 6$
25. $2z^4 - 15z^3 + 34z^2 - 15z - 18 = 0$ **26.** $2x^4 - 9x^3 + 6x^2 + 11x - 6 = 0$ **27.** $2x^4 + 5x^3 - 2x^2 - 7x + 2 = 0$
28. $2t^4 - t^3 - 3t^2 - 5t = 2$ **29.** $12v^4 + 14v^3 - 6v^2 - 3v + 1 = 0$ **30.** $12x^4 + 16x^3 - x^2 - 17x = 10$

31. $x^4 + \frac{1}{6}x^3 - \frac{17}{6}x^2 + \frac{7}{6}x = 1$

32. $k^4 - \frac{1}{2}k^3 - \frac{4}{3}k^2 + \frac{1}{6}k + \frac{2}{3} = 0$

33. $18T^4 - 9T^3 - 17T^2 = -4T - 4$

34. $12x^4 - 44x^3 + 55x^2 = 30x - 9$

35. $x^5 - 5x^4 + 10x^3 - 10x^2 + 5x - 1 = 0$

36. $z^6 + 6z^5 + 15z^4 + 20z^3 + 15z^2 + 6z + 1 = 0$

In Problems 37 and 38, a function and its graph are given (Figures 5.10 and 5.11). Find the zeros of each function.

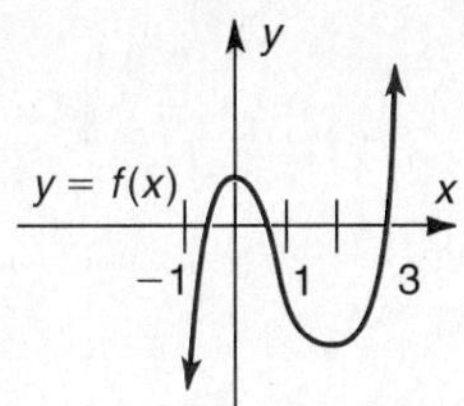

Fig. 5.10

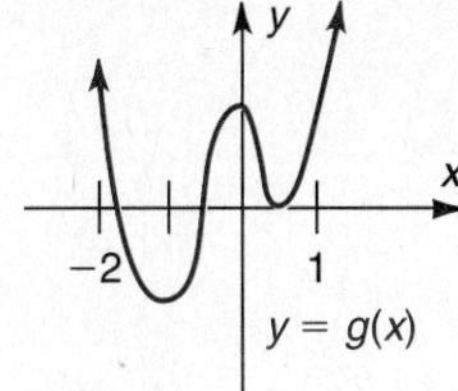

Fig. 5.11

37. $f(x) = 6x^5 - 19x^4 + 7x^3 - 13x^2 + x + 6$

38. $g(x) = 144x^6 + 264x^5 + 289x^4 + 469x^3 + 12x^2 - 118x + 20$

39. A small rectangular box is twice as wide as it is high and 8 inches longer than it is high. What are its dimensions if its volume is 198 cubic inches?

40. A rectangular pool has a constant depth and holds 96 cubic yards of water. What are its dimensions if its width is three times its depth and it is 6 yards longer than its depth?

41. The lengths of the sides of a right triangle are all integers. If the length of the hypotenuse is 4 feet longer than the length of one leg and the area of the triangle is 24 square feet, what are its dimensions?

42. A packaging company wishes to manufacture rectangular boxes without tops by cutting squares from each corner of rectangular sheets of cardboard and then bending up the flaps. If the sheets of cardboard are 16 inches by 21 inches and the volume of the boxes is to be 450 cubic inches, what size square should be cut from each corner?

43. The sum of the squares of the first n natural numbers is determined by the function

$$S(n) = \frac{n(n+1)(2n+1)}{6}$$

How many natural numbers would give a sum of squares of 385?

44. The sum of the cubes of the first n natural numbers is determined by the function

$$C(n) = \frac{n^2(n+1)^2}{4}$$

How many natural numbers would give a sum of cubes of 1296?

45. A manufacturer of baseball souvenirs predicts revenue by the function $R(x) = x^3 - 6x^2 + 10x$, where x is the number of thousands of souvenirs sold. How many souvenirs must be sold to generate \$10,608 in revenue?

46. The total cost of producing and marketing a product is predicted by the function $T(x) = 3x^3 + 190x^2 - 800x$, where x is the number of thousands of units produced. How many units must be produced to make the total cost \$40,875?

In Problems 47 through 50, solve each inequality.

47. $2x^3 + x^2 - 5x + 2 \leq 0$

48. $3x^3 - 7x^2 - 7x + 3 \geq 0$

49. $x^4 + 2x^3 + 4x^2 + 6x + 3 > 0$

50. $2x^5 - 3x^4 - 2x + 3 < 0$

Challenge Problems

51. Show that 3 is an upper bound for the zeros of $f(x) = x^4 - 2x^3 - x^2 - 18$ by finding a polynomial $Q(x)$ such that $f(x) = (x - 3)Q(x)$. Then show that $Q(x) > 0$ for $x > 3$. Use a similar argument to show that -2 is a lower bound for the zeros of f.

52. Show that $\sqrt[3]{-1 + \sqrt{2}} + \sqrt[3]{-1 - \sqrt{2}}$ is a root of $x^3 + 3x + 2 = 0$. Approximate $\sqrt[3]{-1 + \sqrt{2}} + \sqrt[3]{-1 - \sqrt{2}}$ to three decimal places.

For Graphing Calculators

Graph each polynomial function and use ZOOM *to count the number of zeros. Try to determine how many have multiplicity 2. How many nonreal complex zeros does each have?*

53. $\frac{4}{3}x^5 - 2x^4 - 4x^3 + 6x^2 + 3x - \frac{9}{2}$

54. $x^6 - 3x^4 + 4$

55. $1 - 2x - x^2 + x^3 + 3x^4 + x^5 - 2x^6 - x^7$

IN YOUR OWN WORDS . . .

56. Why is the Fundamental Theorem of Algebra important in the study of polynomials?

57. Discuss the Upper and Lower Bound Theorem and its uses.

58. Explain the relationship between the zeros of the function f and the function g if $f(x) = 2x^3 - 3x^2 - 2x + 3$ and $g(x) = x^3 - \frac{3}{2}x^2 - x + \frac{3}{2}$. Make a generalization about the relationship.

59. The Rational Zero Theorem requires that the constant term be nonzero. Explain how this theorem could be used to find the rational zeros of a polynomial with a constant term of zero.

5.4 DESCARTES' RULE OF SIGNS; APPROXIMATING REAL ZEROS

Thus far we have discovered a trial-and-error method guaranteed to find all the *rational* zeros of a polynomial function and a theorem that will help find bounds for real zeros. Let's see what we can do about the *irrational* zeros. We start with a rule that will allow us to count all the *real* zeros. If we can count all the real zeros, then the Rational Zero Theorem will allow us to find out how many irrational zeros there are and, we hope, the Upper and Lower Bound Theorem will limit their size. We state Descartes' Rule of Signs without proof.

Descartes' Rule of Signs

Suppose f is a polynomial function of degree $n \geq 1$ with nonzero real coefficients written in standard form.

1. The number of *positive real* zeros of f is equal to the number of changes of sign in $f(x)$, or less than that number by an *even* integer.
2. The number of *negative real* zeros is given by the same rule applied to $f(-x)$.

A change of sign occurs when two successive coefficients have opposite signs. For example, the polynomial function $f(x) = x^4 + 2x^3 - 3x^2 - 4x - 5$ has one change of sign, while the function $g(x) = x^3 - 2x^2 + 3x - 4$ has three changes of sign. In Descartes' Rule of Signs, a zero of multiplicity k *is counted* k *times.*

EXAMPLE 1 What does Descartes' Rule of Signs tell us about the zeros of each of the following functions?

(a) $f(x) = x^4 + x^3 - 6x^2 - 14x - 12$ (b) $g(t) = 3t^4 + 9t^2 + 11$

(c) $h(x) = x^{10} - 7$ (d) $F(x) = x^4 + 9x^2 - 2x + 24$

Solutions (a) $f(x) = x^4 + x^3 - 6x^2 - 14x - 12$

The function has four zeros because the degree is 4. (Some may be repeated.) Since there is one change of sign in $f(x)$, f has exactly one positive zero.

$$f(-x) = x^4 - x^3 - 6x^2 + 14x - 12$$

There are three changes of sign in $f(-x)$, so f has *either* three negative zeros *or* one negative zero. Zero is the only real number that is neither positive nor negative. Since 0 is not a zero of f, we can conclude that the zeros are positive, negative, or nonreal complex. Putting this information together in a table will help describe the possibilities for the zeros.

NUMBER OF POSITIVE REAL	NUMBER OF NEGATIVE REAL	NUMBER OF NONREAL COMPLEX	TOTAL NUMBER
1	3	0	4
1	1	2	4

We conclude that f has one positive zero and three negative zeros, or f has one positive zero, one negative zero, and two complex (not real) zeros.

(b) $g(t) = 3t^4 + 9t^2 + 11$

The function has four zeros.
Since there are no changes of sign in $g(t)$, g has no positive zeros.

$$g(-t) = 3t^4 + 9t^2 + 11$$

There are no changes of sign in $g(-t)$. So g has no negative zeros.
We conclude that g has four nonreal complex zeros because 0 is not a zero of g.

(c) $h(x) = x^{10} - 7$

The function has 10 zeros.
There is one change of sign in $h(x)$, so h has exactly one positive zero.

$$h(-x) = x^{10} - 7$$

Also, there is one change of sign in $h(-x)$, so h has exactly one negative zero. Thus, h has one positive zero, one negative zero, and eight nonreal complex zeros.

(d) $F(x) = x^4 + 9x^2 - 2x + 24$

The function has four zeros.
There are two changes of sign in $F(x)$, so F has either two positive zeros or no positive zeros.

$$F(-x) = x^4 + 9x^2 + 2x + 24$$

There are no changes of sign in $F(-x)$, so F has no negative zeros.
Therefore, either F has two positive zeros and two nonreal complex zeros, or it has four nonreal complex zeros. (In Section 5.3, Example 7, we learned that *if* F has any real zeros, they are irrational and lie between -1 and 1.) □

Approximating Irrational Zeros

We now introduce a theorem that will allow us to approximate irrational real zeros. It says that if a polynomial function value is *positive* at one number and *negative* at another, then the function value is 0 at some number in between the two numbers.

The Intermediate Value Theorem for Polynomials

If f is a polynomial function and p and q are numbers such that $f(p) > 0$ and $f(q) < 0$, then there is a real number k, between p and q, such that $f(k) = 0$.

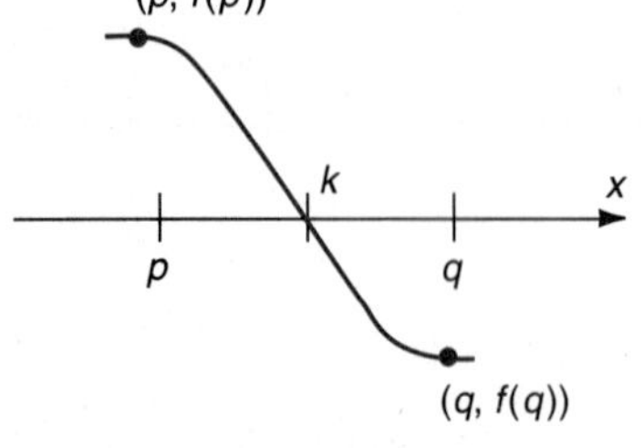

Fig. 5.12

The proof of this theorem will have to await calculus, but it certainly seems right, as seen by the graph in Figure 5.12. (Remember, the graphs of polynomial functions have no breaks.)

If we know there is a real zero of f between p and q, we can approximate it to any desired accuracy by a **method of successive approximations.** There are several such methods. They involve selecting a number between p and q, say s, and evaluating $f(s)$. If $f(s) = 0$, s is a zero of f and we are through. If $f(s) \neq 0$, then it is either positive or negative and we repeat the procedure with s and p or s and q, whichever is a positive-negative pair. Of course, a calculator is essential for easy calculation.

The first numerical method we will use is called **bisection.** We select the number *halfway* between the number that gives a positive value and the one that gives a negative value. We find it by adding the two numbers together and dividing by 2.

EXAMPLE 2 Approximate any real roots of the equation $t^3 - 2t^2 - 2 = 0$ to the nearest hundredth using the method of bisection.

Solution We know that we will have three roots in all, but some may be complex, not real. By Descartes' Rule of Signs we see that the function $f(t) = t^3 - 2t^2 - 2$ has exactly one positive zero and no negative zeros. So, we are looking for exactly one positive real root for the equation. Possible rational zeros are ± 1 and ± 2. However, none of these is a zero. There is one positive irrational zero. Note that

$$f(0) = -2$$
$$f(1) = -3$$
$$f(2) = -2$$
$$f(3) = 7$$

So, by the Intermediate Value Theorem for polynomials, there is a zero of f between 2 and 3 since f is negative at 2 and positive at 3. We evaluate f at the number halfway between 2 and 3.

$$f(2.5) = 1.125$$

Now, the same theorem tells us the zero is between 2 and 2.5. Again, we split the difference.

$$f(2.25) \approx -0.734$$

We continue in the same manner. The procedure ends when we have an approximation with the desired accuracy. As we want an answer correct to the nearest hundredth, we continue until we have the zero trapped between two approximations that agree to the nearest hundredth. Therefore, we can round our approximations to the nearest thousandth as we go.

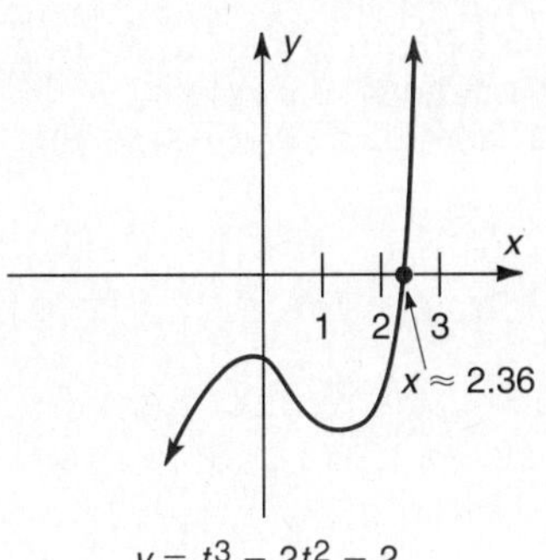

$y = t^3 - 2t^2 - 2$

Fig. 5.13

$$f(2.375) \approx 0.115$$
$$f(2.313) \approx -0.325$$
$$f(2.344) \approx -0.109$$
$$f(2.360) \approx 0.005$$
$$f(2.352) \approx -0.053$$
$$f(2.356) \approx -0.024$$

Now, as the zero is between 2.356 and 2.360, we conclude that the zero is 2.36 to the nearest hundredth. The real root of the equation is approximately 2.36.

In Figure 5.13 a sketch of the graph of the function $f(t) = t^3 - 2t^2 - 2$ shows there is a real zero between 2 and 3. □

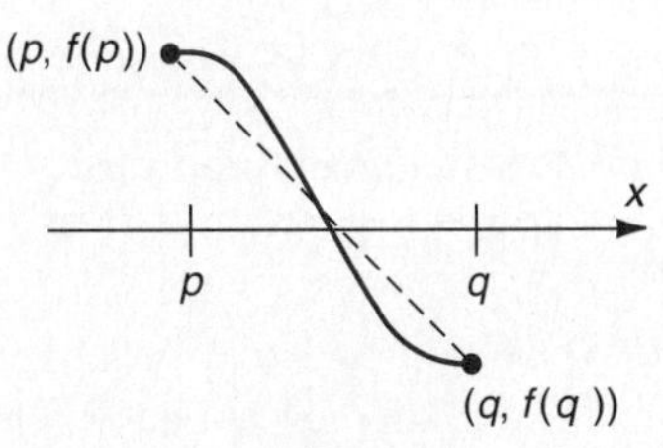

Fig. 5.14

The next method of successive approximations is called **linear interpolation.** It is a bit more complicated, but it usually takes fewer trials. The idea is to get a much better guess as to where the zero is than simply taking the midpoint. The idea is shown in Figure 5.14. Notice that the graphed function is positive at p and negative at q. That is, $f(p) > 0$ and $f(q) < 0$. Since there is a great deal of regularity in the graphs of polynomial functions, the x-intercept of the *line connecting* $(p, f(p))$ *and* $(q, f(q))$ is usually closer to the actual zero than the value obtained by simply choosing the midpoint. It can be shown (Problem 37) that the x-intercept of the line connecting $(p, f(p))$ and $(q, f(q))$ is given by

$$x = p - \frac{p - q}{f(p) - f(q)} \cdot f(p);\ f(p) \neq f(q)$$

EXAMPLE 3 There is a zero of $f(x) = x^3 + x - 3$ between 1 and 2. Use linear interpolation to locate this zero to the nearest hundredth.

Solution First we note that $f(1) = -1$ and $f(2) = 7$, so there is, in fact, a zero of f between 1 and 2. We approximate this zero with the linear interpolation formula. For convenience, let p be 2, let q be 1, and call the first approximation x_1.

$$x = p - \frac{p - q}{f(p) - f(q)} \cdot f(p) \qquad \text{Linear Interpolation Formula}$$

$$x_1 = 2 - \frac{2 - 1}{f(2) - f(1)} \cdot f(2) = 2 - \frac{2 - 1}{7 - (-1)} \cdot 7$$

$$x_1 = 1.125$$

Next we find $f(1.125)$. It is, to the nearest thousandth, -0.4512. As this value is negative, we replace q with 1.125 and calculate x_2.

$$x_2 = 2 - \frac{2 - 1.125}{7 - (-0.4512)} \cdot 7 \approx 1.178$$

$f(1.178) \approx -0.1874$, so we replace q with 1.1784 and calculate x_3.

$$x_3 = 2 - \frac{2 - 1.1784}{7 - (-0.1874)} \cdot 7 \approx 1.200$$

$f(1.200) \approx -0.072$, so we replace q with 1.200.

$$x_4 = 2 - \frac{2 - 1.200}{7 - (-0.072)} \cdot 7 \approx 1.208$$

$f(1.208) \approx -0.029$

$$x_5 = 2 - \frac{2 - 1.208}{7 - (-0.029)} \cdot 7 \approx 1.211$$

Since x_4 and x_5 are both 1.21 to the nearest hundredth, we have found the desired approximation. The graph of f in Figure 5.15 illustrates this zero. □

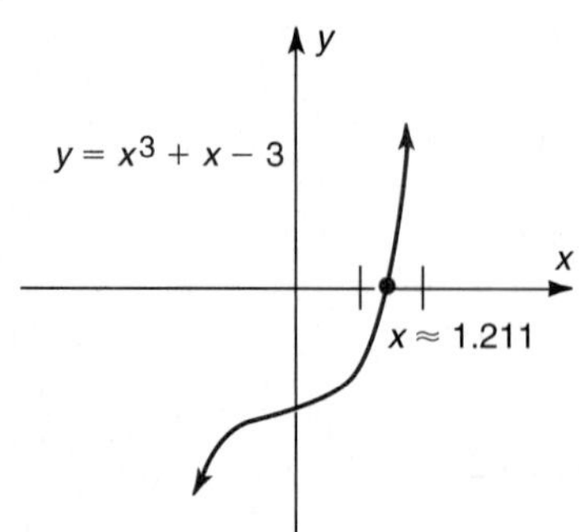

Fig. 5.15

EXAMPLE 4 Solve $x^4 - 2x^3 - 3x^2 - 6 = 0$, giving any approximate solutions to the nearest hundredth.

Solution By Descartes' Rule of Signs we expect one positive solution and one negative solution. A sketch of the graph in Figure 5.16 confirms this. Using the Rational Root Theorem we check for any rational solutions. The possibilities are

Factors of the numerator: $\pm 1, \pm 2, \pm 3, \pm 6$

Factors of the denominator: ± 1

Possible rational roots: $\pm 1, \pm 2, \pm 3, \pm 6$

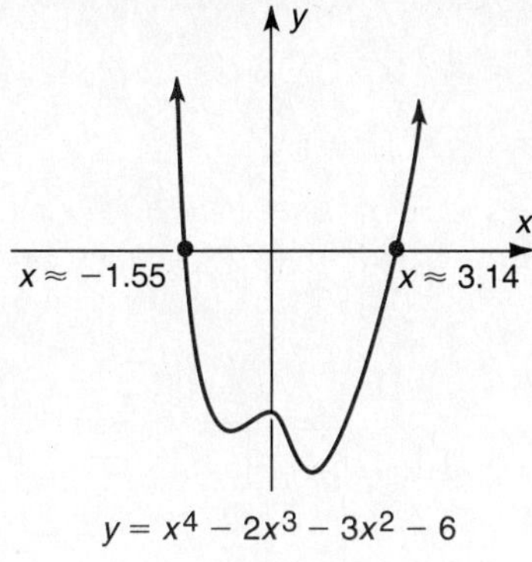

Fig. 5.16

With synthetic division we discover there are no rational roots. The test of 3 is shown.

$$\begin{array}{r|rrrrr} 3 & 1 & -2 & -3 & 0 & -6 \\ & & 3 & 3 & 0 & 0 \\ \hline & 1 & 1 & 0 & 0 & -6 \end{array}$$

If we let f be the polynomial function, by the Remainder Theorem we see $f(3) = -6$. (See synthetic division above.) Since $f(4) = 74$, there is a zero between 3 and 4. We search with linear interpolation, taking p to be 4 and q to be 3.

$$x = p - \frac{p - q}{f(p) - f(q)} \cdot f(p)$$

$$x_1 = 4 - \frac{4 - 3}{f(4) - f(3)} \cdot f(4) = 4 - \frac{4 - 3}{74 - (-6)} \cdot 74$$

$$x_1 \approx 3.075$$

$$f(3.075) \approx -3.110$$

$$x_2 = 4 - \frac{4 - 3.075}{74 - (-3.110)} \cdot 74 \approx 3.112$$

$$f(3.112) \approx -1.540$$

$$x_3 = 4 - \frac{4 - 3.112}{74 - (-1.540)} \cdot 74 \approx 3.130$$

$$f(3.130) \approx -0.740$$

$$x_4 = 4 - \frac{4 - 3.130}{74 - (-0.740)} \cdot 74 \approx 3.138$$

$$f(3.138) \approx -0.377$$

$$x_5 = 4 - \frac{4 - 3.138}{74 - (-0.377)} \cdot 74 \approx 3.142$$

Therefore, the positive solution is 3.14 to the nearest hundredth.

To search for the negative root we note that $f(-2) = 14$, while $f(-1) = -6$. We let p be -1 and q be -2.

$$x_1 = -1 - \frac{-1 - (-2)}{-6 - 14}(-6) \approx -1.300$$

$$f(-1.300) \approx -3.820$$

As this value of f is *negative,* we calculate x_2 with $p = -1.300$. Remember, we always stay between positive and negative functional values.

$$x_2 = -1.300 - \frac{-1.300 - (-2)}{-3.820 - 14}(-3.820) \approx -1.450$$

$$f(-1.450) \approx -1.790$$

$$x_3 = -1.450 - \frac{-1.450 - (-2)}{-1.790 - 14}(-1.790) \approx -1.512$$

$$f(-1.512) \approx -0.719$$

$$x_4 = -1.512 - \frac{-1.512 - (-2)}{-0.719 - 14}(-0.719) \approx -1.536$$

$$f(-1.536) \approx -0.264$$

$$x_5 = -1.536 - \frac{-1.536 - (-2)}{-0.264 - 14}(-0.264) \approx -1.546$$

$$f(-1.546) \approx -0.107$$

$$x_6 = -1.546 - \frac{-1.546 - (-2)}{-0.107 - 14}(-0.107) \approx -1.547$$

The negative solution is -1.55 to the nearest hundredth.

The only real solutions to the equations are -1.55 and 3.14 to the nearest hundredth. □

GRAPHING CALCULATOR BOX

Approximating Zeros

The graphing calculator can take most of the thinking and all the work from the job of finding the real zeros of polynomials (and most other functions). Since we know the real zeros occur at the x-intercepts of the graph, we simply graph and use [ZOOM] and [TRACE] to approximate the x-intercepts to whatever degree of accuracy is desired.

Technique and Accuracy

Let's find the real zeros of the polynomial function $f(x) = 10x^3 - 11x^2 - 20x - 6$ to the nearest thousandth. First, we graph it in the default window.

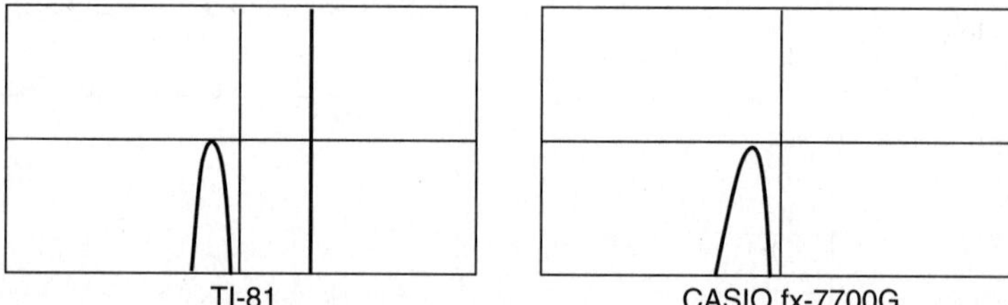

TI-81 CASIO fx-7700G

Since we know the general shape of a cubic, we expect another x-intercept in the CASIO display. A range change in x to $[-10, 10]$ yields a display similar to the TI-81. We see an intercept between 2 and 2.5 and an unknown situation between -1 and 0.

We find the easy zero first. Change the range to $[2, 2.5]$ by $[-1, 1]$ and regraph. The [TRACE] feature does not work well with nearly vertical lines. To enlarge our view and flatten the graph to a more suitable angle, use the "box" feature of ZOOM to make a box as indicated.

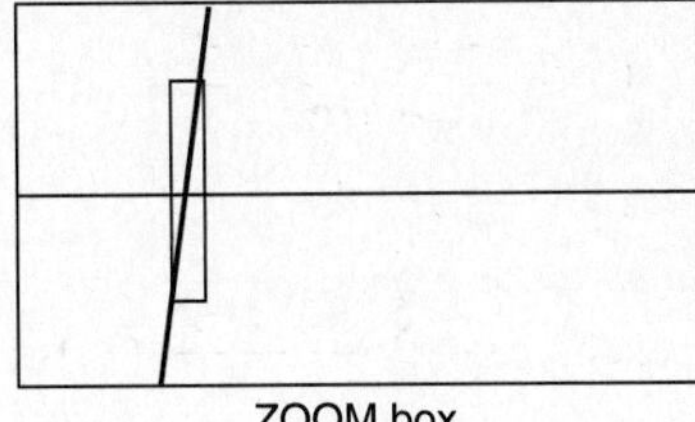
ZOOM box

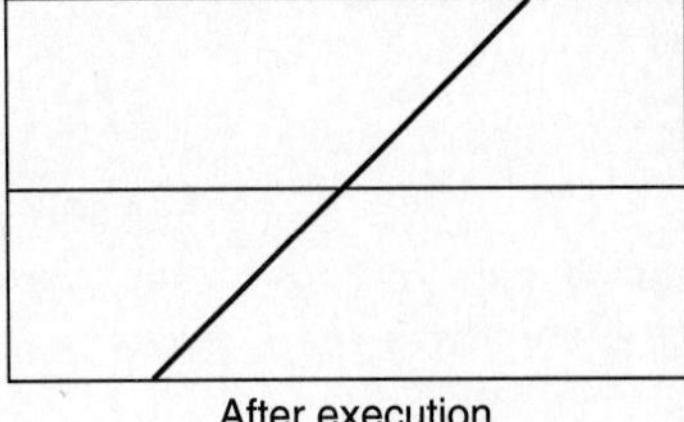
After execution

Now we try TRACE. We move the trace mark along the graph until the y-coordinate changes from positive to negative (or negative to positive) in one press of the arrow. Note the x-coordinates.

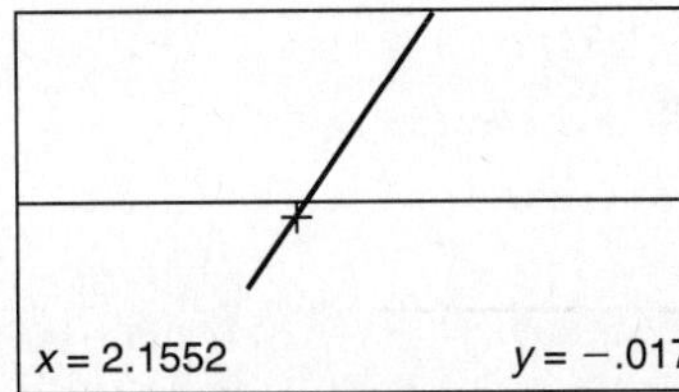

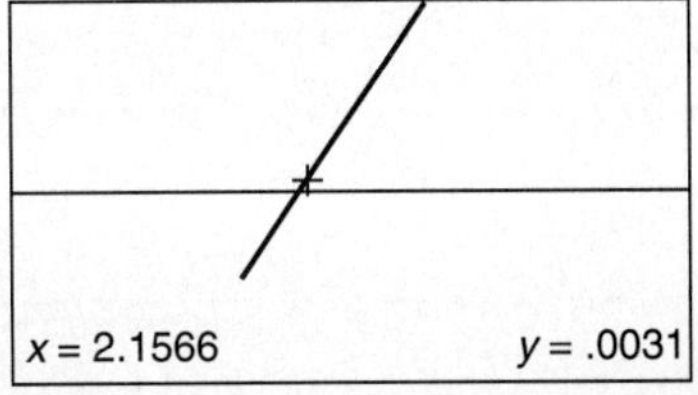

Now we know our zero to the nearest thousandth will be 2.155 or 2.156. But which will it be? One more zoom should settle the matter. Again, we make the box longer in the vertical direction to help the trace.

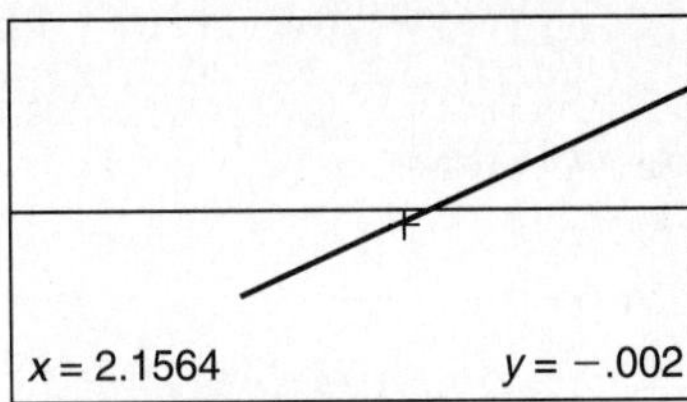

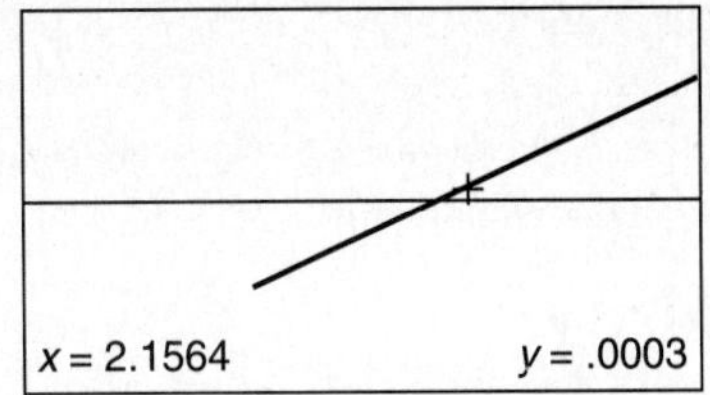

This zero is 2.156 to the nearest thousandth.

Now to investigate the unknown situation between -1 and 0 we change the range to $[-1, 0]$ by $[-1, 1]$ and regraph. It is still not clear! We zoom as indicated.

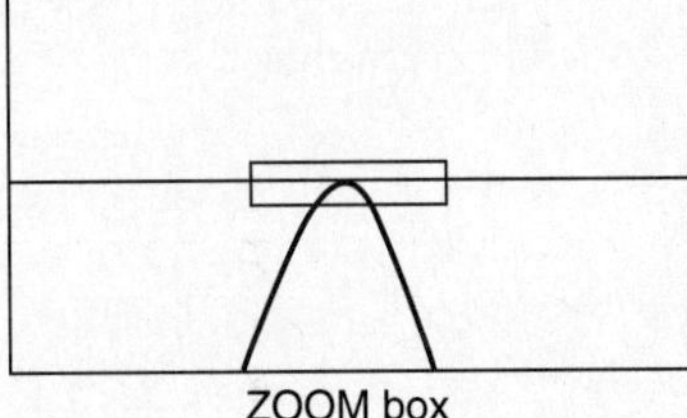
ZOOM box

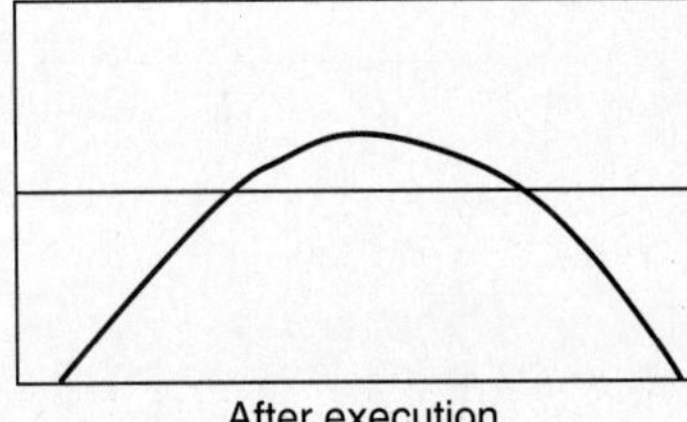
After execution

Now the situation is clear. There are two more zeros. Using the same zoom-and-trace technique, we find them to be -0.556 and -0.500 to the nearest thousandth.

PROBLEM SET 5.4

Warm-ups

In Problems 1 through 4, what does Descartes' Rule of Signs tell about the real zeros of each function? See Example 1.

1. $f(x) = x^4 - 2x^3 - 3x^2 - 8x + 2$
2. $g(t) = 5t^4 - 3t^3 - t^2 + 9t - 16$
3. $P(x) = x^4 - 6x^2 + 2x - 11$
4. $Q(x) = 3x^6 - x - 1$

In Problems 5 and 6, approximate any real roots of the given equation to the nearest hundredth using the method of bisection. See Example 2.

5. $x^3 + x - 7 = 0$
6. $t^4 + t^2 + 3t = 4$

In Problems 7 and 8, approximate any real roots of the given equation to the nearest hundredth using the method of linear interpolation. See Examples 3 and 4.

7. $5x^3 - x^2 + 3 = 0$
8. $x^3 - 3x^2 - 4x + 6 = 0$

Practice Exercises

In Problems 9 through 18, determine the possible number of positive and negative zeros of each function.

9. $f(x) = 2x^3 + 7$
10. $f(t) = 5t^4 + 11$
11. $F(q) = 7q^5 - 2q$
12. $F(x) = x^8 - 6x^3$
13. $g(s) = 3s^3 - 2s^2 + s - 16$
14. $g(x) = -x^3 + 7x^2 - 2x + 11$
15. $h(x) = x^3 - 10x^2 - 3x + 1$
16. $h(w) = w^3 + 6w^2 - 19w + 2$
17. $G(x) = 6x^3 + 4x^2 + 2x + 1$
18. $G(y) = 3y^8 + 2y^6 + y^4 + 2y^2$

In Problems 19 through 24, sketch the graph of the function and approximate to the nearest hundredth any real zeros using the method of bisection.

19. $f(x) = 3x^3 - 6x - 5$
20. $F(z) = 4z^3 - 5z + 4$
21. $g(t) = t^3 - 2t^2 + 5t - 9$
22. $G(x) = x^3 + 2x^2 - 5x + 10$
23. $h(x) = -2x^3 + 2x^2 + 5x - 1$
24. $H(y) = -y^3 - 2y^2 + 5y - 1$

In Problems 25 through 30. Use the method of linear interpolation to approximate any irrational roots to the nearest hundredth.

25. $s^4 + s^3 - s^2 + 3s + 4 = 0$
26. $x^4 - x^3 - x^2 + 4x - 3 = 0$
27. $2x^4 - 5x^3 + 3x^2 + 6x - 9 = 0$
28. $2t^4 - t^3 - 6t^2 + 4t + 6 = 0$
29. $6y^5 + 12y^4 + 4y^3 - 5y^2 = 4y + 1$
30. $4x^5 - 16x^4 + 17x^3 = 5x^2 - 8x + 4$

31. The outer radius of the spherical shell shown in the figure is 1 centimeter less than twice the inner radius. Approximate to the nearest hundredth of a centimeter the length of the inner radius if the volume of the shell is 202π cubic centimeters.

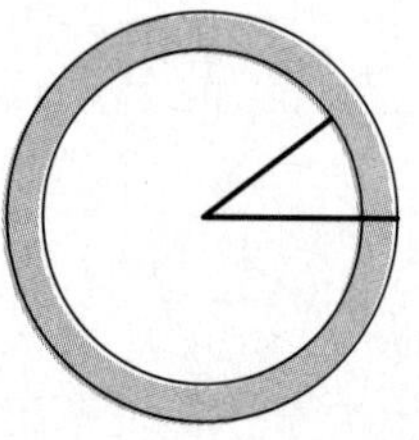

32. Dr. Silver wishes to build a 50-gallon fish tank for her dental office. As shown in the figure, the height of the tank is to be 2 inches more than the width and the length is to be twice the height. Approximate to the nearest hundredth of an inch the dimensions of the tank. Assume that 1 gallon ≈ 230 cubic inches.

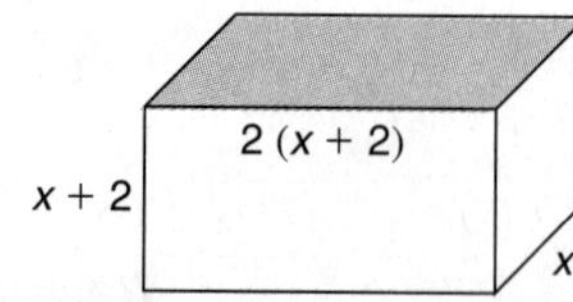

Challenge Problems

33. The graph of a function g is shown in the figure. If $g(x) = x^3 + ax + b$, approximate the real root of g to the nearest tenth.

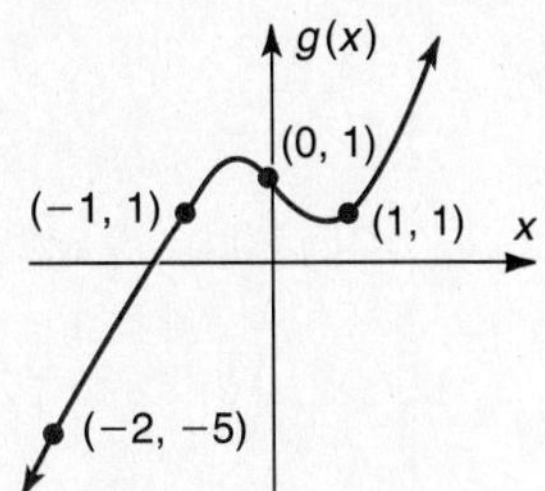

In Problems 34 and 35, each function has an irrational zero between 0 and 1. Approximate it to the nearest ten-thousandth.

34. $f(x) = x^3 - 3x^2 + 1$

35. $g(x) = x^3 - 3x + 1$

36. Use the Rational Root Theorem and Descartes' Rule of Signs to show that $\sqrt{2}$ is an irrational real number. (*Hint:* Solve the equation $x^2 - 2 = 0$ using the Square Root Property and then reexamine the equation for real roots and rational roots.)

37. Show that the x-intercept of the line containing the points $(p, f(p))$ and $(q, f(q))$, when $f(p) \neq f(q)$, is given by

$$x_0 = p - \frac{p - q}{f(p) - f(q)} \cdot f(p)$$

IN YOUR OWN WORDS . . .

38. Explain how the method of bisection is used to approximate the irrational real roots of an equation.

39. What is the idea behind the method of linear interpolation, and what advantage does it have over the method of bisection?

40. When will $f(x) = f(-x)$? What does this say about the zeros of f?

5.5 COMPLEX ZEROS

As we are well aware from our knowledge of quadratic equations and our previous study in this chapter, polynomial equations often have nonreal complex solutions. For example, the equation $x^4 + x^3 - 6x^2 - 14x - 12 = 0$ has -2, 3, *and* $1 \pm i$ as its solutions. Note that the two complex solutions, $1 + i$ and $1 - i$, are *complex conjugates.* Do nonreal complex solutions always occur in conjugate pairs? Let us examine this idea.

Complex Conjugates

Suppose z and w are complex numbers. Let's consider $\bar{z} + \bar{w}$ and $\overline{z + w}$. Suppose $z = a + bi$ and $w = c + di$. Then $\bar{z} = a - bi$ and $\bar{w} = c - di$. Therefore, we have

$$\overline{z + w} = \overline{(a + bi) + (c + di)} = \overline{(a + c) + (b + d)i} = (a + c) - (b + d)i$$
$$= a - bi + c - di = \bar{z} + \bar{w}$$

We have shown that the conjugate of a sum is the sum of the conjugates. The remaining properties listed below may be shown in a similar manner.

Properties of Complex Conjugates

For z and w any complex numbers and k any real number,

1. $\overline{z + w} = \bar{z} + \bar{w}$ Conjugate of a sum
2. $\overline{zw} = \bar{z}\bar{w}$ Conjugate of a product
3. $\overline{z^n} = \bar{z}^n$ Conjugate of a power
4. $\bar{k} = k$ Conjugate of a real number

If $z = a + bi$

5. $z + \bar{z} = 2a$
6. $z\bar{z} = a^2 + b^2$

Now suppose the complex number z is a solution of the polynomial equation

$$a_n x^n + a_{n-1}x^{n-1} + \cdots + a_1 x + a_0 = 0$$

and further suppose that the coefficients are all *real* numbers. Then,

$$a_n z^n + a_{n-1}z^{n-1} + \cdots + a_1 z + a_0 = 0$$

If complex numbers are equal, their conjugates are equal. So,

$$\overline{a_n z^n + a_{n-1}z^{n-1} + \cdots + a_1 z + a_0} = \bar{0}$$

By the properties of complex conjugates,

$$\overline{a_n z^n} + \overline{a_{n-1}z^{n-1}} + \cdots + \overline{a_1 z} + \bar{a}_0 = \bar{0}$$

$$\bar{a}_n \overline{z^n} + \bar{a}_{n-1}\overline{z^{n-1}} + \cdots + \bar{a}_1 \bar{z} + \bar{a}_0 = \bar{0}$$

$$\bar{a}_n \bar{z}^n + \bar{a}_{n-1}\bar{z}^{n-1} + \cdots + \bar{a}_1 \bar{z} + \bar{a}_0 = \bar{0}$$

However, the a_i's and 0 are *real* numbers, which means that each $\bar{a}_i = a_i$. So,

$$a_n \bar{z}^n + a_{n-1}\bar{z}^{n-1} + \cdots + a_1 \bar{z} + a_0 = 0$$

and we see that $\bar{z}$, the *conjugate* of the solution z, is also a solution of the polynomial equation. Notice that for this argument to be valid, the coefficients *must be real numbers.*

The Complex Zero Theorem

If the coefficients of polynomial function P are all real numbers and z is a nonreal zero of P, then $\bar{z}$ is also a zero of P.

Thus, nonreal complex zeros occur in pairs. That is, if $a + bi$ is a zero, then $a - bi$ is also a zero.

EXAMPLE 1 Find the zeros of $f(t) = t^4 - 3t^2 + 6t - 2$ given that $1 + i$ is a zero of f.

Solution Since f has all real coefficients and $1 + i$ is a zero, its conjugate $1 - i$ must also be a zero. We use synthetic division to factor out the known zeros.

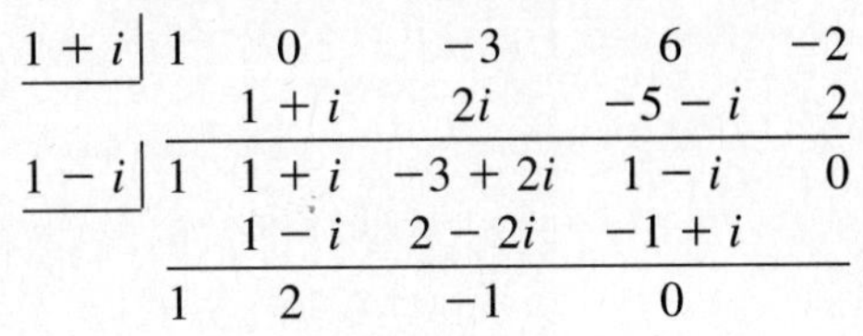

$1+i$	1	0	-3	6	-2
		$1+i$	$2i$	$-5-i$	2
$1-i$	1	$1+i$	$-3+2i$	$1-i$	0
		$1-i$	$2-2i$	$-1+i$	
	1	2	-1	0	

Thus, $f(t) = [t - (1 + i)][t - (1 - i)](t^2 + 2t - 1)$. The Quadratic Formula can be used to find the other two zeros.

$$t = \frac{-2 \pm \sqrt{4 - 4(-1)}}{2} = -1 \pm \sqrt{2}$$

Fig. 5.17

The zeros of f are $1 + i$, $1 - i$, $-1 + \sqrt{2}$, and $-1 - \sqrt{2}$. Figure 5.17 shows a sketch of the graph of f in Example 1. □

EXAMPLE 2 Write a fourth-degree polynomial equation with real coefficients that has $2 + 3i$ and $-i$ as two of its solutions. Write the equation with the variable x and in standard form.

Solution Since we want real coefficients, $2 - 3i$ and i must also be roots, as they are the conjugates of the two known solutions. Therefore, an equation of the desired type is

$$[x - (2 - 3i)][x - (2 + 3i)](x - i)(x + i) = 0$$

$$(x^2 - 4x + 13)(x^2 + 1) = 0$$

$$x^4 - 4x^3 + 14x^2 - 4x + 13 = 0$$ □

Factoring Polynomials Over the Real Numbers

There are times when we wish to factor a polynomial and are limited to real number coefficients. This is called **factoring over the reals.** What kind of factors can we expect? The zeros of a polynomial are either real or nonreal. Suppose P is a polynomial function with real coefficients and k is a real zero of P. Then $x - k$ is a **linear factor** of $P(x)$ with real coefficients. Now suppose z is a nonreal complex zero of P. Then $\bar{z}$ is also a zero of P. Therefore, two of the factors of $P(x)$ are $x - z$ and $x - \bar{z}$. Notice their product.

$$(x - z)(x - \bar{z}) = x^2 - (z + \bar{z})x + z\bar{z}$$

Since $z + \bar{z}$ and $z\bar{z}$ are both *real numbers,* the product of the factors $x - z$ and $x - \bar{z}$ is a quadratic polynomial *with real coefficients!* Therefore, we conclude that a polynomial with real coefficients can be factored *over the reals* into linear and second-degree factors.

Theorem: Factoring Polynomials Over the Real Numbers

If P is a polynomial with real coefficients, then $P(x) = a_n(x - r_0)\cdots(x - r_m)(x^2 + b_0x + c_0)\cdots(x^2 + b_nx + c_n)$, where all the coefficients are real numbers.

EXAMPLE 3 Factor $y^4 - 4y^3 + 26y^2 + 12y - 87$ over the reals, given that $2 + 5i$ is a zero.

Solution Since $2 + 5i$ is a zero, $2 - 5i$ is a zero. Therefore, $[y - (2 - 5i)]$ and $[y - (2 + 5i)]$ are factors. Thus their product, $[y - (2 - 5i)][y - (2 + 5i)] = y^2 - 4y + 29$, is a factor. Notice that this second-degree factor has *real* coefficients. We use long division to reduce the original polynomial.

$$\begin{array}{r|l} & y^2 - 3 \\ \hline y^2 - 4y + 29 & y^4 - 4y^3 + 26y^2 + 12y - 87 \\ & \underline{y^4 - 4y^3 + 29y^2} \\ & \quad -3y^2 + 12y - 87 \\ & \quad \underline{-3y^2 + 12y - 87} \\ & \qquad 0 \end{array}$$

Thus another factor is $y^2 - 3$. However, this can be factored further over the reals to $(y - \sqrt{3})(y + \sqrt{3})$.
Factoring over the reals results in

$$y^4 - 4y^3 + 26y^2 + 12y - 87 = (y^2 - 4y + 29)(y - \sqrt{3})(y + \sqrt{3}) \quad \square$$

Strategy for Finding Zeros of Polynomial Functions

We have studied several theorems that help us find the zeros of a polynomial function. We end this chapter with a strategy to follow when trying to find the zeros of a polynomial function.

1. Try to factor the polynomial.

2. Use Descartes' Rule of Signs to determine the possibilities for negative, positive, and nonreal complex roots.
3. Look for rational zeros. Keep the Upper and Lower Bound Theorem in mind.
4. Use the Intermediate Value Theorem and either bisection, linear interpolation or a graphing calculator to approximate the irrational solutions.

PROBLEM SET 5.5

Warm-ups

In Problems 1 through 4, find the zeros of each polynomial equation using the given information. See Example 1.

1. $f(x) = x^4 + 4x^3 - x^2 - 6x + 18$; given that $1 - i$ is a zero
2. $g(t) = t^4 + 2t^3 - 3t^2 + 2t - 4$; given that i is a zero
3. $h(x) = 2x^4 - 3x^3 + 3x^2 + 77x - 39$; given that $2 - 3i$ is a zero
4. $p(s) = s^4 - 6s^3 + 18s^2 - 102s + 289$; given that $-1 + 4i$ is a zero

In Problems 5 and 6, write polynomials according to the given specifications. See Example 2.

5. Write a third-degree polynomial equation with real coefficients that has 5 and $1 - 2i$ as two of its roots. Write the equation in standard form with the variable x.
6. Write a fourth-degree polynomial equation with real coefficients that has $3 - 2i$ and $-2 + i$ as two of its zeros. Write the polynomial in standard form with the variable t.

In Problems 7 and 8, factor each polynomial over the reals. See Example 3.

7. $x^4 - 4x^3 + 3x^2 - 4x + 2$; given that $-i$ is a zero
8. $t^4 - 4t^3 + 5t^2 - 2t - 2$; given that $1 + i$ is a zero

Practice Exercises

In Problems 9 through 20, solve each equation.

9. $2x^4 + 4x^3 + 9x^2 + 16x + 4 = 0$; one root is $-2i$.
10. $w^4 + 4w^3 + 10w^2 + 36w + 9 = 0$; one root is $3i$.
11. $t^4 + t^3 + 13t + 5 = 0$; one root is $1 + 2i$.
12. $x^4 - 9x^3 + 30x^2 - 45x + 25 = 0$; one root is $2 - i$.
13. $2x^5 - 9x^4 + 19x^3 - 39x^2 + 44x - 12 = 0$; $2i$ and $\frac{3}{2}$ are roots.
14. $3z^5 + 4z^4 + 5z^3 + 2z - 4 = 0$; $\frac{2}{3}$ and $-1 - i$ are roots.
15. $k^5 - 2k^4 - 9k + 18 = 0$; $\sqrt{3}i$ and 2 are roots.
16. $2x^5 + 5x^4 - 50x - 125 = 0$; $-\sqrt{5}i$ and $-\frac{5}{2}$ are roots.
17. $6x^5 - 8x^4 + 7x^3 + 16x^2 = 5x + 6$; $1 - \sqrt{2}i$ and $-\frac{2}{3}$ are roots.
18. $9S^5 + 33S^4 + 45S^3 = 43S^2 + 34S - 14$; $-2 + \sqrt{3}i$ and $\frac{1}{3}$ are roots.
19. $4y^5 - 9y^4 - 2y^3 + 57y^2 - 182y + 42 = 0$; $1 + \sqrt{5}i$ is a root, and there is a rational root.
20. $2x^5 - 6x^4 - 17x^3 + 75x^2 + 21x - 99 = 0$; $3 - \sqrt{2}i$ is a root, and there is a rational root.

In Problems 21 through 24, factor each polynomial over the reals.

21. $x^4 - 4x^3 + 2x^2 + 20x - 35$; given that $(x - 2 - \sqrt{3}i)$ is a factor
22. $t^4 - t^3 - t^2 + 2t - 2$; given that $\left(t - \frac{1}{2} + \frac{\sqrt{3}}{2}i\right)$ is a factor

23. $2z^4 + 13z^3 + 27z^2 + 5z - 11$

24. $3x^4 - 8x^3 + x^2 + 44x - 28$

25. Prove that the conjugate of the product of two complex numbers is equal to the product of the conjugates of the two numbers.

26. Prove that the conjugate of a real number is the real number.

27. Prove that the conjugate of the cube of a complex number is the cube of the conjugate of the complex number.

28. Prove that the conjugate of the conjugate of a complex number is the complex number.

Challenge Problems

29. Two polynomials are equal if and only if coefficients of ith-degree terms are equal. If r_1, r_2, and r_3 are zeros of $f(x) = ax^3 + bx^2 + cx + d$, show that $r_1 + r_2 + r_3 = -\frac{b}{a}$.

30. Centuries ago mathematicians knew that a cubic equation such as $x^3 + px^2 + qx + r = 0$ could be changed into $t^3 + at + b = 0$ by substituting $x = t - \frac{p}{3}$, thus eliminating the squared term. Show that this is true and express a and b in terms of p, q, and r. Use this idea to solve $x^3 + 3x^2 - x - 3 = 0$.

IN YOUR OWN WORDS . . .

31. What is the importance of the Complex Zero Theorem?

32. Explain the relationship between the Linear Factors Theorem and the Factoring Over the Real Numbers Theorem.

CHAPTER SUMMARY

GLOSSARY

Polynomial of degree n: An expression of the form $a_nx^n + a_{n-1}x^{n-1} + \cdots + a_1x + a_0$.

Polynomial equation: An equation in the form $a_nx^n + a_{n-1}x^{n-1} + \cdots + a_1x + a_0 = 0$.

Polynomial function: A function of the form $P(x) = a_nx^n + a_{n-1}x^{n-1} + \cdots + a_1x + a_0$.

Zero of a polynomial: A solution of the polynomial equation $P(x) = 0$.

Upper bound for a set of real numbers: U is said to be an upper bound for a set of real numbers if U is *greater than* or *equal to* every number in the set.

Lower bound for a set of real numbers: L is said to be an lower bound for a set of real numbers if L is *less than* or *equal to* every number in the set.

Method of bisection: A method of approximating polynomial zeros by successively dividing intervals in half.

Method of linear interpolation: A method of approximating zeros of polynomial functions by finding the x-intercept of an associated line.

THE DIVISION ALGORITHM

If $P(x)$ and $D(x)$ are polynomials with the degree of $D(x)$ less than or equal to the degree of $P(x)$, and $D(x)$ not the zero polynomial, then there are unique polynomials $Q(x)$ and $R(x)$ such that

$$P(x) = D(x) \cdot Q(x) + R(x)$$

and the degree of $R(x)$ is less than the degree of $D(x)$ or $R(x)$ is the zero polynomial.

THE REMAINDER THEOREM

If the polynomial $P(x)$ is divided by $x - r$, then the remainder is $P(r)$.

THE FACTOR THEOREM

The polynomial $P(x)$ has $x - r$ as a factor if and only if $P(r) = 0$.

FOUR EQUIVALENT POLYNOMIAL STATEMENTS

1. r is a root of the polynomial equation $P(x) = 0$.
2. $P(r) = 0$. That is, r is a zero of the function P.
3. $x - r$ is a factor of $P(x)$.
4. The remainder when $P(x)$ is divided by $x - r$ is 0.

THE FUNDAMENTAL THEOREM OF ALGEBRA

Every polynomial equation of degree ≥ 1 has at least one complex zero.

THE LINEAR FACTORS THEOREM

If P is a polynomial equation of degree $n \geq 1$, then $P(x)$ can be expressed as the product of n linear factors, $P(x) = a_n(x - r_1)(x - r_2)\cdots(x - r_n)$, where a_n is the leading coefficient of P and $a_n \neq 0$.

THE NUMBER OF ROOTS OF AN *n*th DEGREE POLYNOMIAL EQUATION

Every polynomial equation of degree $n \geq 1$ has exactly n roots if a root of multiplicity k is counted k times.

THE RATIONAL ZERO THEOREM

If the polynomial function $F(x) = a_nx^n + a_{n-1}x^{n-1} + \cdots + a_1x + a_0$ has all integer coefficients and $r = \dfrac{p}{q}$ is a rational zero of F (p and q have no common prime factors; $a_n \neq 0$; $a_0 \neq 0$), then

1. p is a factor of the constant term a_0.
2. q is a factor of the leading coefficient a_n.

THE UPPER AND LOWER BOUND THEOREM

Suppose P is a polynomial function with real coefficients and a positive leading coefficient.

1. If $P(x)$ is divided by $x - U$, where $U > 0$, using synthetic division and the bottom row of the display *contains no negative numbers,* then U is an upper bound for the zeros of P.
2. If $P(x)$ is divided by $x - L$, where $L < 0$, using synthetic division and the numbers in the bottom row of the display *alternate in sign,* then L is a lower bound for the zeros of P. (Treat 0s in the bottom row as either positive or negative, as needed.)

DESCARTES' RULE OF SIGNS

Suppose P is a nonconstant polynomial function with nonzero real coefficients written in standard form.

1. The number of *positive* real zeros of P is equal to the number of changes of sign in $P(x)$, or less than that number by an *even* integer.
2. The number of *negative* real zeros is given by the same rule with $P(-x)$.

THE INTERMEDIATE VALUE THEOREM FOR POLYNOMIALS

If F is a polynomial function and p and q are numbers such that $F(p) > 0$ and $F(q) < 0$, then there is a real number k, between p and q, such that $F(k) = 0$.

THE LINEAR INTERPOLATION FORMULA

The x-intercept of the line connecting $[p, f(p)]$ and $[q, f(q)]$ is given by

$$x = p - \frac{p-q}{f(p) - f(q)} \cdot f(p);\ f(p) \neq f(q)$$

PROPERTIES OF COMPLEX CONJUGATES

For z and w any complex numbers and k any real number,

1. $\overline{z+w} = \bar{z} + \bar{w}$ Conjugate of a sum
2. $\overline{zw} = \bar{z}\bar{w}$ Conjugate of a product
3. $\overline{z^n} = \bar{z}^n$ Conjugate of a power
4. $\bar{k} = k$ Conjugate of a real number

If $z = a + bi$

5. $z + \bar{z} = 2a$
6. $z\bar{z} = a^2 + b^2$

THE COMPLEX ZERO THEOREM

If the coefficients of polynomial function P are all real numbers and z is a nonreal zero of P, then $\bar{z}$ is also a zero of P.

FACTORING POLYNOMIALS OVER THE REAL NUMBERS

If P is a polynomial with real coefficients, then

$$P(x) = a_n(x - r_0)\cdots(x - r_m)(x^2 + b_0x + c_0)\cdots(x^2 + b_nx + c_n)$$

where all the coefficients are real numbers.

REVIEW PROBLEMS

In Problems 1 through 4, use synthetic division to perform the indicated operation.

1. $\dfrac{3s^3 + 4s^2 - s + 7}{s + 1}$

2. $\dfrac{5x^4 - 11x^3 + 4x^2 - 7x + 6}{x - 2}$

3. $(x^4 + 2x^3 + 9x + 1) \div (x + 3)$

4. $(2k^3 + 5k^2 - 5k + 1) \div \left(k - \frac{1}{2}\right)$

In Problems 5 through 8, use synthetic division to find the remainder.

5. $\dfrac{3w^3 - 14w^2 + 7w + 9}{w - 4}$

6. $\dfrac{x^4 + 7x^3 - 30x - 20}{x + 3}$

7. $(6x^4 - 7x^3 + 11x^2 - 12x + 5) \div \left(x - \frac{2}{3}\right)$

8. $(y^3 - y^2 + y - 2) \div (y - i)$

In Problems 9 through 12, evaluate each polynomial for the given value of the variable.

9. $x^3 - 4x^2 - 15x + 20$; when x is 6

10. $x^4 + 12x^3 - 15x^2 - 18x - 30$; when x is -7

11. $t^5 + 7t^4 - 100$; when t is -3

12. $10z^4 - 50z^3 + 38z^2 + 30$; when z is 4

In Problems 13 through 16, determine the number of possible positive and negative zeros of each function.

13. $f(x) = 2x^4 + 3x^3 - 4x^2 - x + 7$

14. $g(x) = 5x^4 - x^3 + 2x^2 - 3x + 11$

15. $h(t) = t^5 - t^2$

16. $T(v) = v^8 - v^7 + v - 1$

In Problems 17 through 20, find the zeros of each function using the given information.

17. $f(t) = t^3 - 5t^2 + 5t + 3$; given that 3 is a zero

18. $g(x) = 2x^3 + 4x^2 + x + 2$; given that -2 is a zero

19. $h(x) = x^4 + 3x^3 - 7x^2 - 7x + 2$; given that -1 and 2 are zeros

20. $k(x) = 2x^4 + x^3 - 35x^2 - 85x - 75$; given that -3 and 5 are zeros

In Problems 21 through 30, solve each equation.

21. $x^3 - 4x^2 + 6x = 4$; given that 2 is a solution

22. $x^5 + 3x^4 - x^3 - 5x^2 - 4x + 6 = 0$; given that -3 and 1 are roots and 1 has multiplicity 2

23. $r^4 + 4r^3 + 3r^2 + 4r + 2 = 0$; given that i is a root

24. $2y^3 + 3y^2 - 5y - 6 = 0$

25. $6b^4 + 25b^3 + 26b^2 = 4b + 8$

26. $4x^5 + 33x^4 + 108x^3 + 169x^2 + 116x + 20 = 0$

27. $x^6 - 5x^5 + 6x^4 - 16x^2 + 80x = 96$

28. $4t^5 - 4t^4 + t^3 - 32t^2 + 32t = 8$

29. $2g^5 - 3g^4 - 6g^3 + 21g^2 - 22g + 6 = 0$; given that $1 + i$ is a solution

30. $x^4 - 8x^3 + 26x^2 - 40x + 25 = 0$; given that $2 - i$ is a solution

In Problems 31 and 32, approximate the real root of each equation to the nearest hundredth.

31. $2x^3 - 3x + 2 = 0$

32. $5 - 3x^2 - x^3 = 0$

In Problems 33 and 34, factor each polynomial over the reals.

33. $x^4 + 2x^3 - x^2 - 8x - 12$; given that $x + 1 + \sqrt{2}i$ is a factor

34. $3x^4 - 3x^3 + 4x^2 - 8x + 8$; given that $-\frac{1}{2} - \frac{\sqrt{7}}{2}i$ is a zero of the polynomial

35. Use the graph of $f(x) = x^3 - 2x^2 + x - 2$ shown in Figure 5.18 to solve the equation $x^3 - 2x^2 + x - 2 = 0$.

36. Use the graph of $g(x) = 6x^4 - x^3 + 53x^2 - 9x - 9$ shown in Figure 5.19 to solve the equation $x^4 - \frac{1}{6}x^3 + \frac{53}{6}x^2 - \frac{3}{2}x - \frac{3}{2} = 0$.

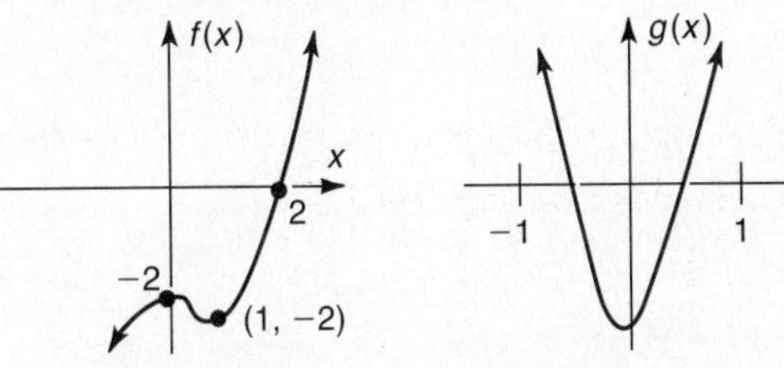

Fig. 5.18

Fig. 5.19

37. Find a third-degree polynomial function with integer coefficients and zeros of $\frac{1}{4}$ and i.

38. Find a fifth-degree polynomial function with integer coefficients and zeros of $\frac{1}{2}$, $\frac{1}{5}$, and 2, and no other zeros.

39. The height of a rectangular metal box is 1 inch less than the width, and the length is twice the width. If the volume of the box is 1800 cubic inches, find its dimensions.

40. Revenue from sales of commemorative buttons can be predicted by the function $R(x) = x^3 - 4x^2 + 8x$, where x is the number of hundreds of buttons sold. How many buttons must be sold to generate \$203 in revenue?

41. The trough shown in Figure 5.20 has a volume of $144\sqrt{3}$ cubic feet. The ends are equilateral triangles with sides of length s feet, and the length of the trough is $(3s - 2)$ feet. The volume of the trough can be calculated by multiplying the area of one of the triangles by the length of the trough. Find s.

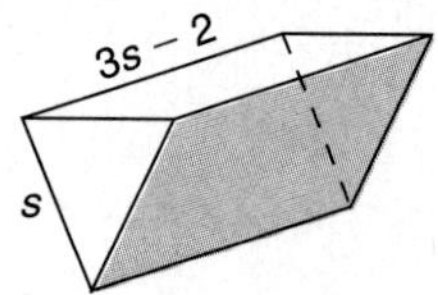

Fig. 5.20

42. A rectangular box has a width of 5 meters, a height of 12 meters, and a length of 13 meters. If each dimension is increased by the same number of meters, the volume is increased by 690 cubic meters. By how many meters is each dimension increased?

LET'S NOT FORGET . . .

43. Which form is more useful? $(x^2 + 1)(x^2 - 1) = 0$ or $(x - i)(x + i)(x - 1)(x + 1) = 0$ if the problem is to solve the equation.

44. Watch the role of negative signs.
   (a) Graph $f(x) = -(x - 2)(x + 2)(x + 4)$ and find the zeros of the function.
   (b) How many negative real zeros does $f(x) = x^4 - 2x^3 - x^2 - x + 2$ have?

45. From memory. How are the possible rational zeros of a polynomial function with integer coefficients determined?

46. With a calculator. Approximate the real zero of $f(x) = x^3 - x - 2$ to the nearest hundredth.

CHAPTER 6

Exponential and Logarithmic Functions

CONNECTIONS

Would anyone give serious consideration to a job that paid one penny for the first day of each month, two cents for the second day, four cents for the third day, and so on, doubling each day? At first, it sounds like a ridiculous pay scale. However, if the job didn't involve immoral conduct or require too much physical labor, a mathematician would snap it up! She or he would realize that things that *double* periodically may start very slowly but tend to grow dramatically later. In fact, this job would pay over a million dollars on the 28th day of each month, making a total of $2,684,354.55 for February, the shortest month of the year! This is an example of exponential growth, which, along with exponential decay, is the first of two major mathematical topics in this chapter.

The most notable improvement in arithmetic in the sixteenth and seventeenth centuries was the invention of logarithms. John Napier, a Scotsman, is credited with the development of logarithms while making calculations about astrological problems. An English mathematician, Henry Briggs, is credited with much of the development of logarithms and constructing extensive logarithm tables. Logarithms remained the major tool for extensive calculations until the development of modern computing machinery in the middle of the twentieth century. Although we no longer use logarithms for computing, they still have great theoretical importance and are the second major mathematical topic of this chapter.

6.1 EXPONENTIAL FUNCTIONS

Power functions such as $f(x) = x^2$ and $g(x) = x^3$ represent a special kind of polynomial function which we studied in Section 4.2. In such functions, the base is a variable and the exponent is a constant. In this section we will examine another class of important functions whose rules look much like those for a power function. However, the base will be a constant and the exponent a variable. For example, $f(x) = 2^x$ and $g(x) = \left(\frac{1}{2}\right)^x$. Such functions are called **exponential functions.**

Let's consider first $f(x) = 2^x$. Two questions immediately come to mind. Is this really a function? If so, what is its domain? Our exponent definitions clearly specify the value of 2^r if r is a rational number. So $f(x) = 2^x$ *is a function* whose domain is *at least* all rational numbers. In fact, calculus shows that b^x is a real number if x is a real number and b is a *positive* real number. Thus the domain of $f(x) = 2^x$ is all real numbers.

If x is an integer, we evaluate 2^x easily. For example, $2^4 = 16$ and $2^{-6} = \frac{1}{64}$. A calculator can be used to approximate other values of 2^x.

EXAMPLE 1 Approximate each exponential to six significant digits.

(a) $2^{\sqrt{3}}$ (b) $7^{3/5}$ (c) $\left(\frac{1}{5}\right)^{\pi}$ (d) $\pi^{-\pi}$

Solutions

(a) $2^{\sqrt{3}}$

The keystrokes [2] [yˣ] [3] [√] [=] yield a display like **3.321997085**.

$2^{\sqrt{3}} \approx 3.32200$

(b) $7^{3/5}$

The keystrokes [7] [yˣ] [(] [3] [÷] [5] [)] [=] give a display like **3.21409585**.

$7^{3/5} \approx 3.21410$

(c) $\left(\frac{1}{5}\right)^{\pi} \approx 0.00636973$

(d) $\pi^{-\pi} \approx 0.0274257$ □

Graphing Exponential Functions

Now we return to the function $f(x) = 2^x$. To help in graphing the function f, we make a table of some coordinates for $y = 2^x$, plot the points, and connect them with a smooth curve (Figure 6.1).

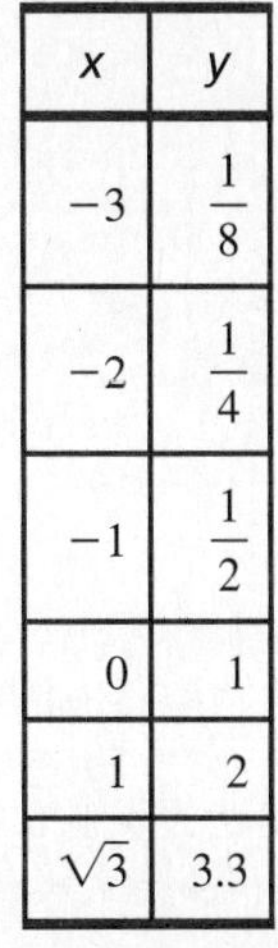

x	y
-3	$\frac{1}{8}$
-2	$\frac{1}{4}$
-1	$\frac{1}{2}$
0	1
1	2
$\sqrt{3}$	3.3

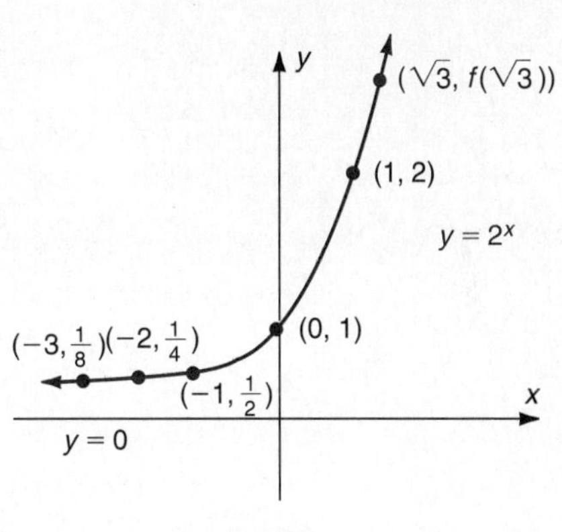

Fig. 6.1

As x gets smaller, the graph gets closer and closer to the x-axis, but it never touches it. That is, $f(x) \to 0$ as $x \to -\infty$. Thus the line $y = 0$ (the x-axis) is a horizontal asymptote for the graph. As x increases, y also increases. In fact, y will get larger and larger as x increases. We see that the *range* of this function is all positive real numbers.

> **Exponential Function**
>
> If x and b are real numbers, where $b > 0$ and $b \neq 1$, the function
>
> $$f(x) = b^x$$
>
> is called an **exponential function.**
> Its domain is all real numbers. Its range is $(0, +\infty)$.

We exclude $b = 1$ from the definition because the graph of $f(x) = 1$ is a horizontal line.

EXAMPLE 2 Sketch the graph of each function and state its domain and range.

(a) $g(x) = 10^x$ (b) $h(x) = \left(\frac{1}{2}\right)^x$

Solutions (a) $g(x) = 10^x$

Again, we make a table and sketch the graph (Figure 6.2).

x	y
-2	$\frac{1}{100}$
-1	$\frac{1}{10}$
0	1
1	10
2	100

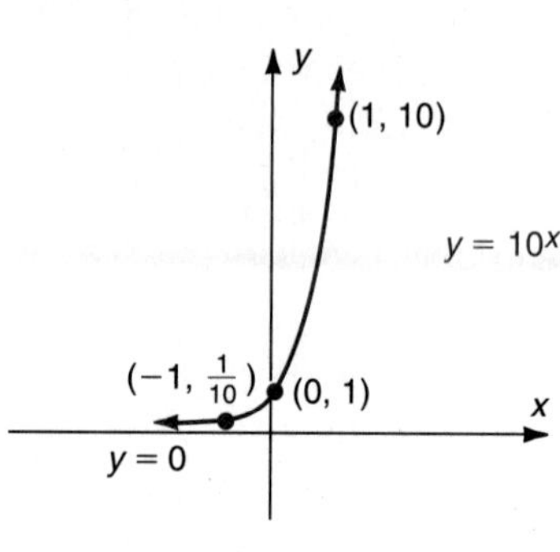

Fig. 6.2

The domain is R, and the range is $(0, +\infty)$

(b) $h(x) = \left(\dfrac{1}{2}\right)^x$

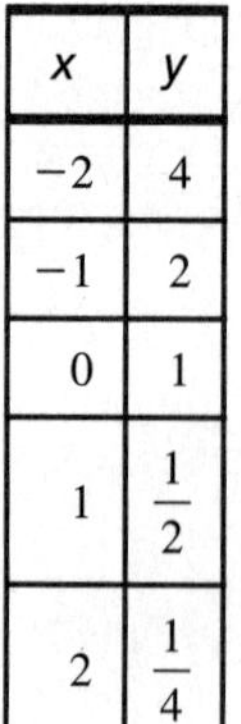

x	y
-2	4
-1	2
0	1
1	$\frac{1}{2}$
2	$\frac{1}{4}$

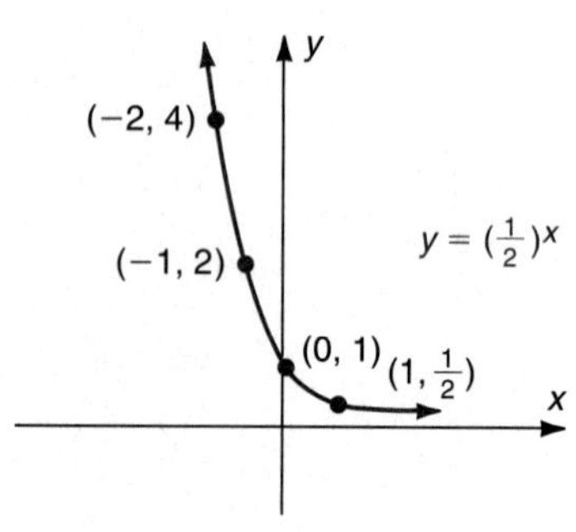

Fig. 6.3

The domain is R, and the range is $(0, +\infty)$. Notice that $\left(\dfrac{1}{2}\right)^x = 2^{-x}$. Thus, $h(x) = \left(\dfrac{1}{2}\right)^x = 2^{-x}$, which is the reflection of $y = 2^x$ about the y-axis (Figure 6.3). □

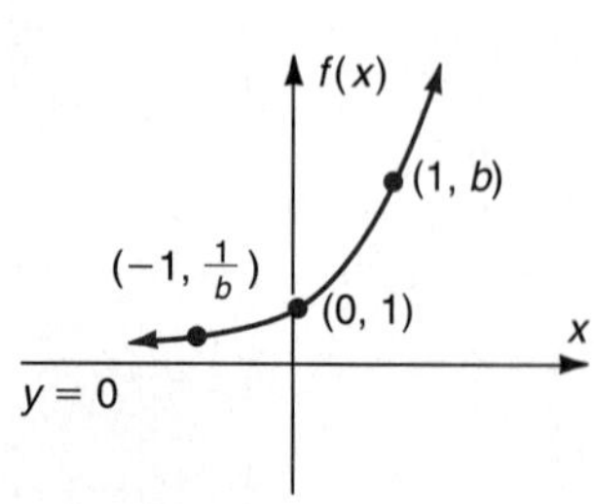

$f(x) = b^x$, $b > 1$

Increasing exponential function

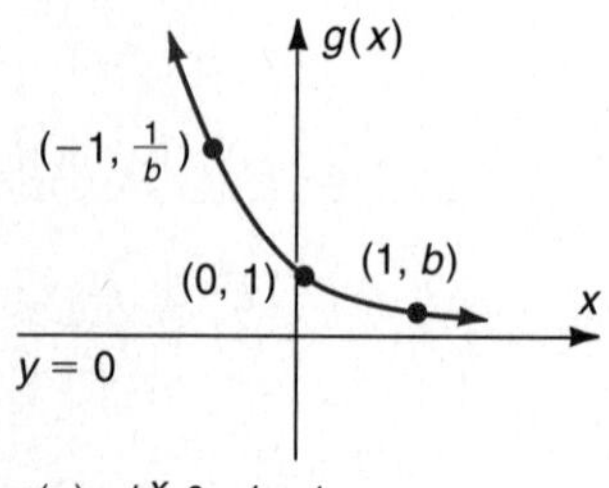

$g(x) = b^x$, $0 < b < 1$

Decreasing exponential function

Fig. 6.4

Example 2 illustrates two types of exponential functions, increasing and decreasing. If $b > 1$, the function $f(x) = b^x$ is an *increasing function,* and if $0 < b < 1$, it is a *decreasing function*. See Figure 6.4.

The Number e

One particular base for exponential functions occurs so often naturally in applications that it has a special name. It is an irrational number, like π. We call it e. A key marked e^x can be found on every scientific calculator. Look for it above the [ln] key.

To find the value of e on a calculator, press [1] [e^x] for most scientific calculators and [e^x] [1] [ENTER] for function-first calculators.

$$e \approx 2.718281828$$

The number e is sometimes defined using the following:

$$\left(1 + \frac{1}{n}\right)^n \to e \text{ as } n \to +\infty.$$

n	***Approximation of*** $\left(1 + \frac{1}{n}\right)^n$
100	2.704813829
1000	2.716923932
10,000	2.718145927
1,000,000	2.718280469

EXAMPLE 3 Approximate each exponential to six significant digits.

(a) e^4 (b) $e^{\sqrt{2}}$ (c) e^{-1}

Solutions (a) $e^4 \approx 54.5982$ Keystrokes for a scientific calculator: [4] [e^x]

For a function-first calculator: [e^x] [4] [ENTER]

(b) $e^{\sqrt{2}} \approx 4.11325$

(c) $e^{-1} \approx 0.367879$ Keystrokes for a scientific calculator: [1] [+/−] [e^x]

For a function-first calculator: [e^x] [(−)] [1] [ENTER]

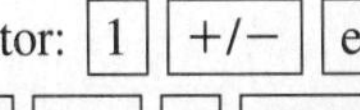

The Natural Exponential Function

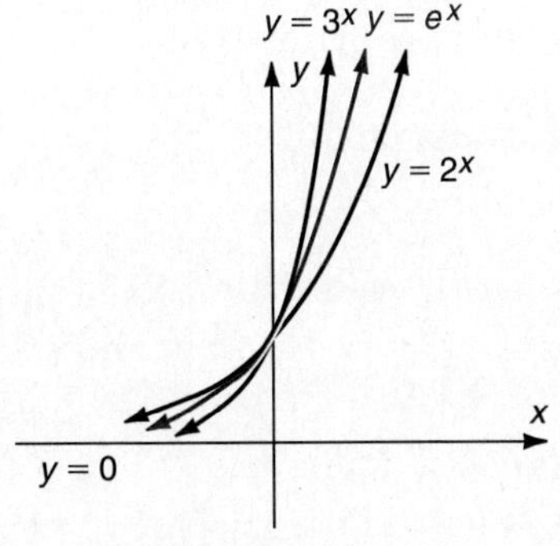

Fig. 6.5

We call the exponential function $f(x) = e^x$ the **natural exponential function.** It is used in physics, engineering, business, and other fields. Because $e > 1$, the graph of the natural exponential function is increasing. The graphs of $y = e^x$, $y = 2^x$, and $y = 3^x$ are shown in Figure 6.5.

Translations

The graphs of exponential functions can be translated in the same manner as the graphs of other functions.

EXAMPLE 4 Graph each of the following pairs of functions on the same set of axes.

(a) $f(x) = e^x$ and $g(x) = e^x + 2$ (b) $f(x) = e^x$ and $g(x) = e^x - 1$

Solutions (a) $f(x) = e^x$ and $g(x) = e^x + 2$

First we graph $y = e^x$. Then, on the same set of axes, we graph $y = e^x + 2$ as shown in Figure 6.6. Notice that the graph of $g(x) = e^x + 2$ is the graph of $f(x) = e^x$ shifted upward 2 units. Note that the horizontal asymptote becomes the line $y = 2$. The domain of g is R, and its range is $(2, +\infty)$.

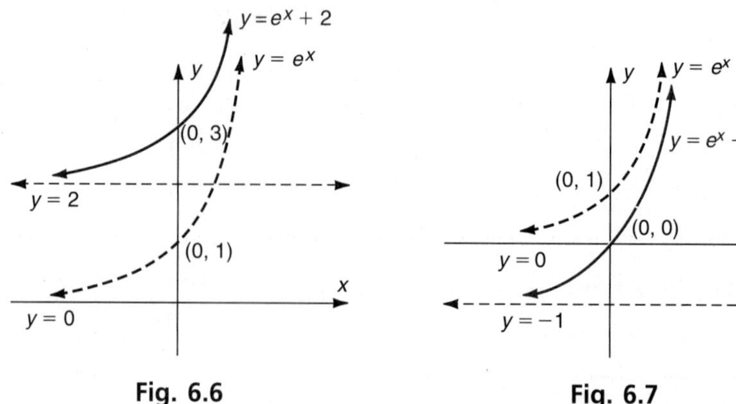

Fig. 6.6 **Fig. 6.7**

(b) $f(x) = e^x$ and $g(x) = e^x - 1$

We graph $y = e^x - 1$ and $y = e^x$ on the same set of coordinate axes. See Figure 6.7. Notice that the shift is 1 unit downward and that the horizontal asymptote is $y = -1$. Here the domain of g is R, and its range is $(-1, +\infty)$. □

EXAMPLE 5 Graph each of the following pairs of functions on the same set of axes.

(a) $f(x) = 3^x$ and $g(x) = 3^{x-2}$ (b) $f(x) = 3^x$ and $g(x) = 3^{x+1}$

Solutions (a) $f(x) = 3^x$ and $g(x) = 3^{x-2}$

First we graph $y = 3^x$. Then, on the same axes, we graph $y = 3^{x-2}$. Notice in Figure 6.8 that the graph of $g(x) = 3^{x-2}$ is the graph of $f(x) = 3^x$ shifted right 2 units. The domain of g is R, and its range is $(0, +\infty)$.

(b) $f(x) = 3^x$ and $g(x) = 3^{x+1}$

This time we graph $y = 3^{x+1}$ on the same set of axes as the graph of $y = 3^x$ (Figure 6.9). The shift is to the left 1 unit. The domain of g is R, and its range is $(0, +\infty)$. □

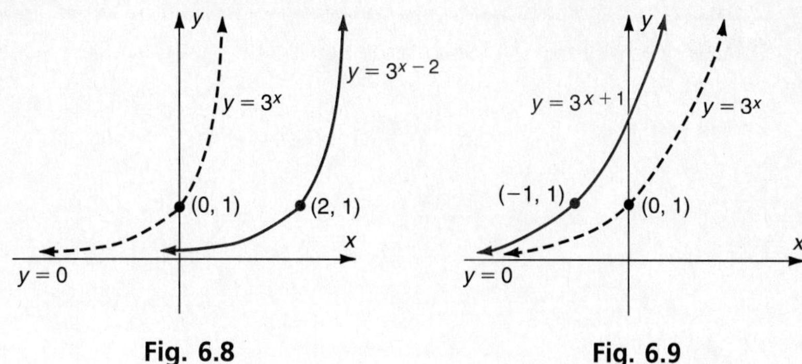

Fig. 6.8

Fig. 6.9

We have just seen that $y = b^x + k$ is the graph of an exponential function *shifted* k units up or down, and $y = b^{x-h}$ is the graph *shifted* h units right or left. In general, $y = b^{x-h} + k$ is the graph of $y = b^x$ shifted h units right or left and k units up or down.

EXAMPLE 6 Sketch the graph of $y = \left(\frac{1}{4}\right)^{x-2} + 1$.

Solution This graph is the graph of $y = \left(\frac{1}{4}\right)^x$ shifted right 2 units and up 1 unit as shown in Figure 6.10. □

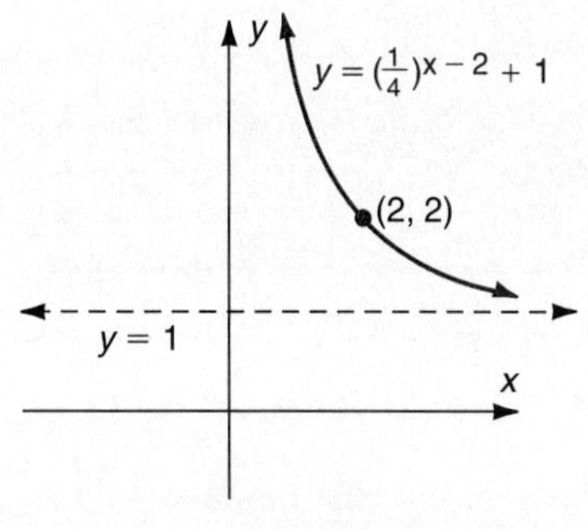

Fig. 6.10

Solving Some Exponential Equations and Inequalities

The horizontal line test indicates that an exponential function is a one-to-one function. Because of this, we have the following result.

One-to-One Property of Exponential Functions

For $b > 0$ and $b \neq 1$,

if $b^x = b^y$, then $x = y$

The One-to-One Property allows us to equate exponents when the bases are the same. This property provides a tool to use when solving certain equations and inequalities.

EXAMPLE 7 Solve each equation.

(a) $3^x = 81$ (b) $4^x = \frac{1}{8}$ (c) $5^{x-7} = 125$

Solutions (a) $3^x = 81$

$3^x = 3^4$

$x = 4$ One-to-One Property of Exponential Functions

$\{4\}$

(b) $4^x = \frac{1}{8}$

$(2^2)^x = 2^{-3}$ $4 = 2^2$ and $2^{-3} = \frac{1}{8}$

$2^{2x} = 2^{-3}$ Power of a power

$2x = -3$ One-to-One Property

$x = -\frac{3}{2}$

$\left\{-\frac{3}{2}\right\}$

(c) $5^{x-7} = 125 = 5^3$

$x - 7 = 3$ One-to-One Property

$x = 10$

$\{10\}$

□

EXAMPLE 8 Solve the inequality $\left(\frac{1}{3}\right)^{2x} > 9$.

Solution Since this is a nonlinear inequality, we use the method of boundary numbers. First, we find the boundary numbers.

$\left(\frac{1}{3}\right)^{2x} = 9$ Associated equation

$(3^{-1})^{2x} = 3^2$

$3^{-2x} = 3^2$ Power of a power

$-2x = 2$ One-to-One Property

$x = -1$ The boundary number is -1.

Test -2: $\left(\frac{1}{3}\right)^{-4} = 3^4 = 81$. Since $81 > 9$ is true, the left region in Figure 6.11 is in the solution set.

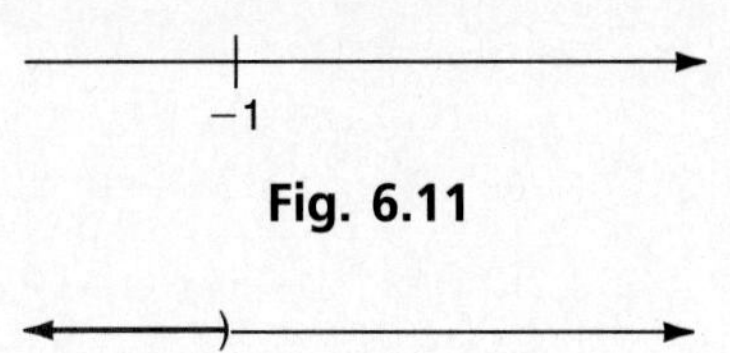

Fig. 6.11

Fig. 6.12

Test 0: $\left(\frac{1}{3}\right)^0 = 1$. Since $1 > 9$ is false, the right region is *not* in the solution set. The boundary number is not included as equality is not allowed. Figure 6.12 is a picture of the solution set.

$$(-\infty, -1) \qquad \text{Solution set}$$ □

Inequalities of the form $f(x) < g(x)$ can sometimes be solved by looking at the graphs of $y = f(x)$ and $y = g(x)$.

EXAMPLE 9 The inequality $\left(\frac{1}{2}\right)^x < 3^x$ is of the form $f(x) < g(x)$, where f and g are exponential functions. Solve the inequality by examining the graphs of $y = f(x)$ and $y = g(x)$.

Solution The graphs of $y = \left(\frac{1}{2}\right)^x$ and $y = 3^x$ are shown in Figure 6.13. Notice in Figure 6.13 that the graph of $y = \left(\frac{1}{2}\right)^x$ is below the graph of $y = 3^x$ when x is positive. Thus, the solution set for $\left(\frac{1}{2}\right)^x < 3^x$ is $(0, +\infty)$. □

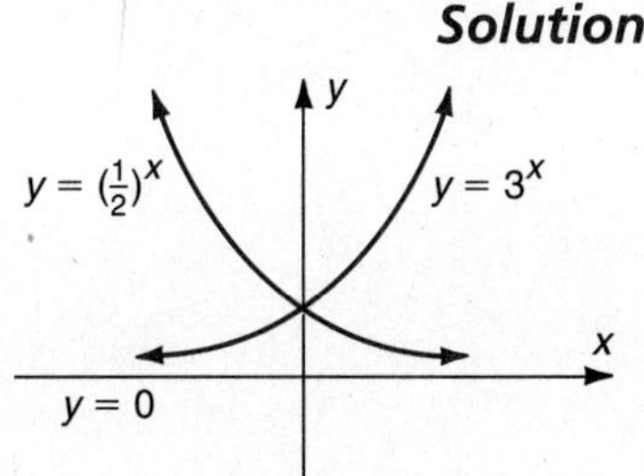

Fig. 6.13

PROBLEM SET 6.1

Warm-ups

In Problems 1 through 4, approximate each exponential to five significant digits. See Example 1.

1. $3^{\sqrt{2}}$
2. $13^{7/6}$
3. $\pi^{\pi/4}$
4. $(\sqrt{11})^{\sqrt{11}}$

In Problems 5 and 6, sketch the graph of each function and state its domain and range. See Example 2.

5. $f(x) = 5^x$
6. $g(x) = \left(\frac{1}{5}\right)^x$

In Problems 7 and 8 approximate each exponential to five significant digits. See Example 3.

7. e^{-2}
8. e^{π}

For Problems 9 and 10, see Examples 4, 5, and 6.

9. Graph the function $f(x) = 6^x$. On the same axes graph the functions g and h, where $g(x) = 6^x + 2$ and $h(x) = 6^{x+2}$, by *shifting* the graph of f.
10. Graph the function $F(x) = 4^x$. On the same axes graph the functions G and H, where $G(x) = 4^{x-3} - 1$ and $H(x) = 4^{x+3} + 1$, by *shifting* the graph of F.

In Problems 11 through 14, solve each equation. See Example 7.

11. $2^x = 128$ **12.** $5^x = \frac{1}{625}$ **13.** $4^{x-7} = 256$ **14.** $9^{-x} = 27$

In Problems 15 and 16, solve each inequality. See Example 8.

15. $16 < 8^x$ **16.** $25^{2x+1} - 125 \leq 0$

In Problems 17 and 18, each inequality is of the form $f(x) < g(x)$, where f and g are exponential functions. Solve each inequality by examining the graphs of $y = f(x)$ and $y = g(x)$. See Example 9.

17. $\left(\frac{1}{3}\right)^x \geq 2^x$ **18.** $5^x < 10^x$

Practice Exercises

In Problems 19 through 26, determine which expression is larger.

19. $5^{0.7}$ or $7^{0.5}$ **20.** $7^{3/4}$ or $19^{9/4}$ **21.** $26^{\sqrt{5}}$ or $10^{\sqrt{10}}$ **22.** $(\sqrt{22})^{4/9}$ or $(\sqrt{97})^{23/15}$

23. $(\sqrt[3]{47})^{\sqrt{7}}$ or $(\sqrt[4]{183})^{\sqrt{2}}$ **24.** e^{π} or π^e **25.** $\left(1 + \frac{1}{500}\right)^{500}$ or e **26.** $\left(1 + \frac{0.09}{4}\right)^4$ or $e^{0.09}$

In Problems 27 through 58, solve each equation or inequality.

27. $3^x = 243$ **28.** $6^x = 216$ **29.** $11^{-x} = 1331$ **30.** $2^{-x} = 1024$

31. $5^{2x-1} = 625$ **32.** $7^{2-x} = \frac{1}{49}$ **33.** $8^x = 32$ **34.** $\frac{1}{27} = 9^x$

35. $4^x < 128$ **36.** $9^x > 27$ **37.** $125^{1-2x} = 5^x$ **38.** $16^x = 2^{3x-1}$

39. $x^3 = 8$ **40.** $x^2 = 1$ **41.** $6^{x+2} \geq 216^x$ **42.** $343^{1-x} \leq 7^x$

43. $64^{-2x} = 32^{1-3x}$ **44.** $27^{3x+1} = 81^{2x-7}$ **45.** $(2 - x)^{1/2} = 5$ **46.** $(x - 7)^{1/3} = 1$

47. $3^x \cdot 9^x = 3^7 \cdot 27^{-x}$ **48.** $\sqrt{2} \cdot 16^{-x} = \left(\frac{1}{2}\right)^x \cdot 32^x$ **49.** $5^{x^2} = \frac{1}{625}$ **50.** $5^{-x^2} = \frac{1}{5}$

51. $8^{|x|} = 64$ **52.** $\left(\frac{1}{2}\right)^{|x|} = \frac{1}{32}$ **53.** $e^x - 1 = 0$ **54.** $e^x - e^3 = 0$

55. $2^{2x} - 6 \cdot 2^x + 8 = 0$ **56.** $3^{2x} - 12 \cdot 3^x + 27 = 0$ **57.** $x^2e^x - e^x = 0$ **58.** $xe^x - xe^2 = 0$

In Problems 59 through 80, sketch the graph of each function and state its domain and range.

59. $f(x) = 4^x$ **60.** $g(x) = \left(\frac{1}{3}\right)^x$ **61.** $h(x) = 4^{-x}$ **62.** $k(x) = \left(\frac{1}{3}\right)^{-x}$

63. $u(x) = 4^x + 1$ **64.** $v(x) = \left(\frac{1}{3}\right)^x + 2$ **65.** $F(x) = 4^{x-1}$ **66.** $G(x) = \left(\frac{1}{3}\right)^{x-2}$

67. $H(x) = 4^{x+1} - 2$ **68.** $K(x) = \left(\frac{1}{3}\right)^{x+2} - 1$ **69.** $U(x) = -4^x$ **70.** $V(x) = -\left(\frac{1}{3}\right)^x$

71. $f(x) = e^x + 2$ **72.** $f(x) = e^{x+2}$ **73.** $g(x) = e^{-x}$ **74.** $g(x) = e^{-x} - 1$

75. $h(x) = 2^{|x|}$ **76.** $h(x) = 2^{-|x|}$ **77.** $F(x) = 2^{x^2}$ **78.** $F(x) = 2^{-x^2}$

79. $G(x) = e^{2x}$ **80.** $G(x) = e^{x/2}$

In Problems 81 through 84, solve each inequality graphically.

81. $3^x \geq \left(\frac{1}{2}\right)^{x-2} + 1$ **82.** $\left(\frac{1}{2}\right)^x + 9 > 10^{x+1}$ **83.** $5^{x+2} < 3^{x+2}$ **84.** $3^{x-4} + 1 \leq \left(\frac{1}{2}\right)^{x-5}$

Challenge Problems

The hyperbolic sine and hyperbolic cosine are two interesting functions that arise in calculus. They are defined by

$$\sinh(x) = \frac{e^x - e^{-x}}{2} \qquad \cosh(x) = \frac{e^x + e^{-x}}{2}$$

Hyperbolic sine Hyperbolic cosine

85. Find each value.
(a) $\sinh(0)$ (b) $\cosh(0)$
(c) $\sinh(1)$ (d) $\cosh(1)$

86. Approximate each value to the nearest thousandth.
(a) $\sinh(0.5)$ (b) $\sinh(5)$
(c) $\cosh(0.5)$ (d) $\cosh(5)$

87. Write $[\cosh(x)]^2 - [\sinh(x)]^2$ in a simpler form.

88. A catenary is the curve formed by a flexible cable hanging from two points under its own weight. Power cables and telephone lines hang in this shape. The equation of a catenary is

$$y = k\cosh\left(\frac{x}{k}\right)$$

where k is a positive constant. Make a careful graph of a catenary on the interval $[-2, 2]$. Assume k has value 2.

89. Sketch the graph of $y = \frac{1}{\sqrt{2\pi}}e^{-x^2/2}$. This curve is the bell-shaped or standard normal curve used in statistics.

IN YOUR OWN WORDS . . .

90. What is an exponential function?

91. What conclusions can be drawn about the domain and range of exponential functions?

92. Discuss the graphs of $y = e^x$, $y = e^{-x}$, $y = -e^x$.

6.2 APPLICATIONS OF EXPONENTIAL FUNCTIONS

The graphs of exponential functions are either increasing or decreasing. Real-life situations modeled by an increasing exponential function are examples of exponential growth, while applications modeled by a decreasing exponential function are examples of exponential decay. See Figure 6.14.

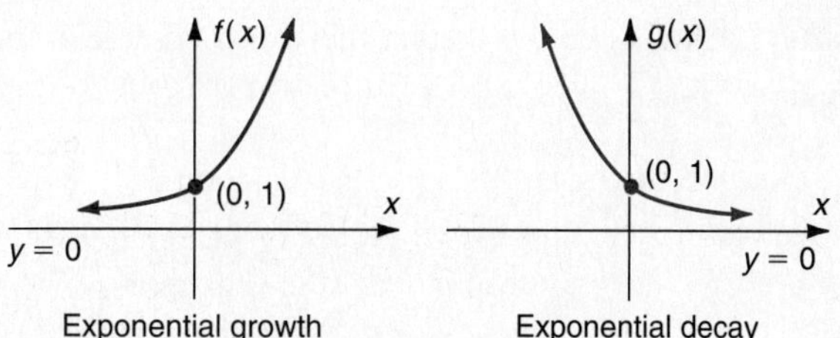

Fig. 6.14

The exponential model is usually of the form $g(x) = a \cdot b^{kx}$, where a and k are constants.

EXAMPLE 1 Sketch the graph of each function.

(a) $f(x) = 2 \cdot e^{3x}$ (Figure 6.15) (b) $g(x) = 3 \cdot e^{-x/2}$ (Figure 6.16)

Solutions (a)

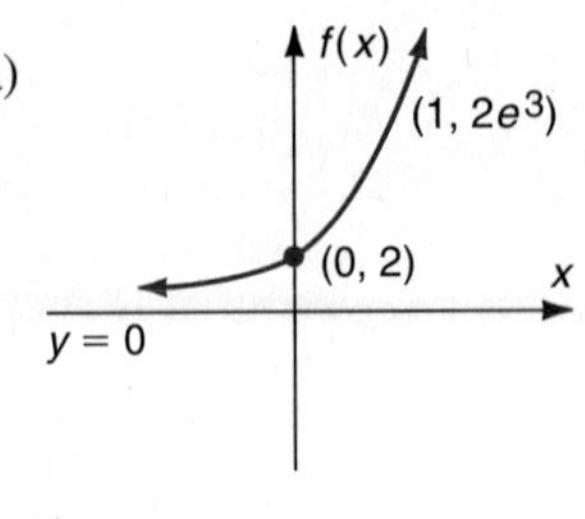

Fig. 6.15

(b)

g(x)
(−1, 3e^(1/2))
(0, 3)
(1, 3e^(−1/2))
x
y = 0

Fig. 6.16 □

Exponential Growth

Population growth is an example of exponential growth. In 1990, the world population was about 5.3 billion, with a growth rate of 1.7% each year. If this growth rate continues, we can express the world population as a function of time.

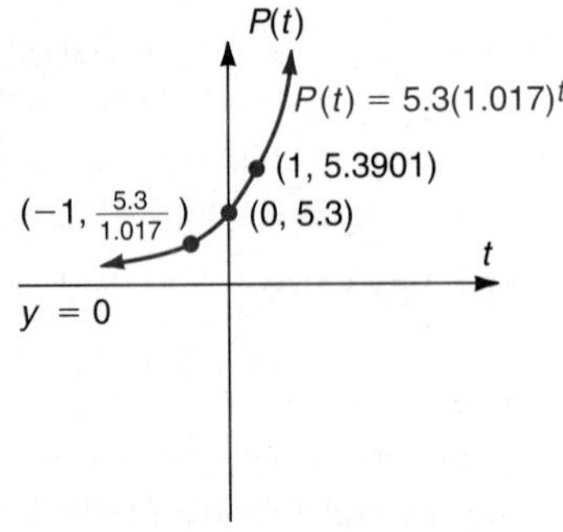

Fig. 6.17

Number of Years After 1990	*Population in Billions*
1	$5.3 + 0.017(5.3) = 5.3(1 + 0.017)$
2	$5.3(1 + 0.017) + 0.017 \cdot 5.3(1 + 0.017) = 5.3(1 + 0.017)^2$
3	$5.3(1 + 0.017)^2 + 0.017 \cdot 5.3(1 + 0.017)^2 = 5.3(1 + 0.017)^3$
t	$5.3(1 + 0.017)^t$

Thus, $P(t) = 5.3(1.017)^t$ is a function that expresses world population in billions t years after 1990. The graph of P as shown in Figure 6.17 depicts the exponential growth rate of $P(t)$ as t increases.

EXAMPLE 2 Predict the world population in 2000 to the nearest tenth of a billion. (Assume a 1.7% increase each year.)

Solution We use the function $P(t) = 5.3(1.017)^t$ to predict the world population in the year 2000. Since 2000 is 10 years after 1990, t is 10.

$$P(10) = 5.3(1.017)^{10}$$

$$P(10) \approx 6.273146$$

The world population in 2000 will be approximately 6.3 billion. □

EXAMPLE 3 If the population of Greenwood, Indiana, at the end of any year can be predicted by

$$P(y) = 24{,}000\left(\frac{6}{5}\right)^{y/10}$$

where y is the number of years after 1990, what is the expected population of Greenwood in the 2000 census? What is the expected population in 1995?

Solution We need to find $P(10)$ for the expected population in the 2000 census.

$P(10) = 24{,}000\left(\frac{6}{5}\right)^{10/10} = 24000 \cdot \frac{6}{5} = 28800$. No calculator required!

For 1995, y is 5 and $P(5) = 24000\left(\frac{6}{5}\right)^{5/10}$.

Press [6] [÷] [5] [=] [y^x] [.5] [×] [24000] [=] and read **26290.6827** on the display.

A population of 28,800 is expected at the 2000 census, and 26,290 is the predicted 1995 population. □

A situation where the rate of change of a quantity is proportional to the amount present is modeled by an exponential function. For example, a population that grows under ideal conditions has the characteristic that its rate of increase is proportional to the existing population. If P_0 is the population when t is 0, then the population at time t is

$$P(t) = P_0 e^{kt}; \text{ where } k \text{ is a constant}$$

EXAMPLE 4 In a certain bacteria culture the rate of growth is proportional to the number of bacteria present. If the initial population is 5000 bacteria and the population doubles every 20 minutes, the population in t minutes is given by $P(t) = 5000e^{0.035t}$. What will the population be in 1 hour?

Solution The amount of bacteria is an exponential function of time. P_0 is 5000, and k is 0.035. Since P is a function of t in *minutes*, we are to find $P(60)$.

$$P(60) = 5000e^{(0.035)(60)}$$

This computation is easily done with a scientific calculator. Key in

[5000] [×] [(] [.035] [×] [60] [)] [e^x] [=] and read **40830.85**.

$P(60) \approx 40830.85$

However, we cannot count that 0.85 bacteria!
The population will be 40,830 bacteria in 1 hour. □

Exponential Decay

The decay of radioactive substances is an example of exponential decay. The *half-life* of a substance is the amount of time it takes for half of the substance to decay. If 100 grams of a substance with a half-life of 10 years is present initially, we can express the amount of the substance remaining after t years as a function of time.

Number of Years	*Amount of Substance Remaining*
10	$100\left(\frac{1}{2}\right)$
20	$100\left(\frac{1}{2}\right)\cdot\frac{1}{2} = 100\left(\frac{1}{2}\right)^2$
30	$100\left(\frac{1}{2}\right)^2\cdot\frac{1}{2} = 100\left(\frac{1}{2}\right)^3$
t	$100\left(\frac{1}{2}\right)^{t/10} = 100\cdot 2^{-t/10}$

Thus, $A(t) = 100 \cdot 2^{-t/10}$ expresses the amount of the radioactive substance remaining after t years. In general, if h is the half-life measured in the same units as t, the amount of the radioactive substance present at time t is given by

$$A(t) = A_0 2^{-t/h}$$

EXAMPLE 5 Dr. Willard Libby (1908–1980) was awarded the 1960 Nobel Prize in chemistry for developing a technique for determining the age of death of once-living matter by radioactive dating. Simply stated, during its lifetime a tree accumulates regular amounts of natural carbon from the atmosphere. At its death, this accumulation stops. However, naturally occurring carbon contains a trace of radioactive carbon 14 (^{14}C) which decays with a half-life of 5700 years. This allows us to date organic materials containing carbon with great accuracy. Suppose an archaeologist knows that 1 milligram of ^{14}C was in a sample that died 5000 years ago. How much ^{14}C would she expect to find in the sample today?

Solution Using the formula $A(t) = A_0 2^{-t/h}$, where A_0 is 1, h is 5700, and t is 5000, we have

$$A(5000) = 1 \cdot 2^{-5000/5700} \approx 0.5444$$

She should expect to find 0.5444 milligram of ^{14}C in the sample. □

Compound Interest

Exponential functions are also indispensable for calculating interest. One important formula gives the compound interest on an investment or loan. If P dollars is invested at an annual interest rate of r compounded n times a year, then $A(t)$, the amount of dollars in the account after t years, is given by

$$A(t) = P\left(1 + \frac{r}{n}\right)^{nt}$$

The table below shows the dollars accumulated after 1 year if \$2500 is invested at 9% for different interest periods.

Interest Period	***Dollars in Account After 1 Year***
Annually ($n = 1$)	$2500(1 + 0.09)^1 = 2725$
Quarterly ($n = 4$)	$2500\left(1 + \frac{0.09}{4}\right)^4 \approx 2732.71$
Monthly ($n = 12$)	$2500\left(1 + \frac{0.09}{12}\right)^{12} \approx 2734.52$
Weekly ($n = 52$)	$2500\left(1 + \frac{0.09}{52}\right)^{52} \approx 2735.22$
Daily ($n = 365$)	$2500\left(1 + \frac{0.09}{365}\right)^{365} \approx 2735.41$
Hourly ($n = 8760$)	$2500\left(1 + \frac{0.09}{8760}\right)^{8760} \approx 2735.43$

It appears that the amount accumulated approaches some value as n increases without bound. We say the interest is **compounded continuously** if n increases without bound. The value of $2500\left(1 + \frac{0.09}{n}\right)^n \to 2500e^{0.09}$ as $n \to +\infty$. If the compounding is done continuously, the formula becomes

$$A(t) = Pe^{rt}$$

In business applications, Pe^{rt} is used sometimes to approximate $P\left(1 + \frac{r}{n}\right)^{nt}$.

Compound Interest Formulas

If P dollars is invested at an annual interest rate of r compounded n times a year, then $A(t)$, the amount of dollars in the account after t years, is given by

$$A(t) = P\left(1 + \frac{r}{n}\right)^{nt}$$

If the compounding is continuous, the formula becomes

$$A(t) = Pe^{rt}$$

EXAMPLE 6 Fred and Jane Sheffey have just invested \$10,000 in a money market account at 7.65% interest. How much will they have in this account in 5 years if the interest is

(a) Compounded quarterly? (b) Compounded continuously?

Solutions (a) We use the compound interest formula $A(t) = P\left(1 + \frac{r}{n}\right)^{nt}$, when P is 10000, r is 0.0765, n is 4, and t is 5.

$$A(5) = 10000\left(1 + \frac{0.0765}{4}\right)^{(4)(5)} \approx 14606.599$$

Fred and Jane should have \$14,606.60 in their account at the end of 5 years with interest compounded quarterly.

(b) Now we use the continuous formula $A(t) = Pe^{rt}$, when P is again 10000, r is 0.0765, and t is 5.

$$A(5) = 10000e^{(0.0765)(5)} \approx 14659.449$$

With continuous compounding, they should have \$14,659.45 at the end of 5 years. □

PROBLEM SET 6.2

Warm-ups

In Problems 1 through 4, sketch the graph of each function. See Example 1.

1. $f(x) = 3 \cdot e^{x/2}$

2. $g(x) = \frac{1}{2} \cdot e^{-3x}$

3. $h(x) = 10 \cdot 3^{-2x}$

4. $k(x) = 2 \cdot 3^{x/3}$

For Problems 5 and 6, see Example 2.

5. The number of registered participants at the Annual Mathematics Conference has increased by 12% each year since 1970. If 250 mathematicians participated in the 1970 conference, express the number of registered participants at the Annual Mathematics Conference as a function of the number of years since 1970.

6. Use the function in Problem 5 to predict the number of registered participants at the Annual Mathematics Conference in 1997.

For Problems 7 and 8, see Example 3.

7. What is the expected population of Greenwood, Indiana, in the 2020 census?

8. What is the expected population of Greenwood, Indiana, in the 2010 census?

For problems 9 and 10, see Example 4.

9. In the bacteria culture what will the population be in $2\frac{1}{2}$ hours?

10. What will the population be in $\frac{1}{2}$ hour?

For Problems 11 and 12, see Example 5.

11. Suppose an archaeologist knows there was 0.05 milligrams of ^{14}C in an organic sample when it died 4500 years ago. How much ^{14}C should he expect to find today?

12. Suppose there was 0.5 milligrams of ^{14}C in an organic sample when it died 2000 years ago. How much ^{14}C is there in the sample today?

For Problems 13 and 14, see Example 6.

13. Antonia Lopez currently has \$67,894 in a savings account that pays 6.22% annually compounded monthly. If she neither deposits nor withdraws from the account, how much money will she have in 7 years? How much more would she have if the interest is compounded continuously?

14. Answer the questions in Problem 13 for an interest rate of 8.25%.

Practice Exercises

15. Kevin wishes to deposit \$3000 for a period of 2 years for a special purpose. After visiting several banks and investigating his options, he has selected the following four plans:
A. 6.5% interest per year compounded continuously
B. 6.55% interest per year compounded monthly
C. 6.6% interest per year compounded quarterly
D. 6.65% per year simple interest
What should Kevin do? How much will he lose if he chooses the worst plan?

In Problems 16 through 18, Wei Hung trades his business for some rental property and \$200,000 which he deposits in an account paying 7.4% compounded daily.

16. How much money will the account contain in 2 years?

17. How much more interest would Wei earn in 2 years if interest were compounded continuously?

18. How much interest would Wei lose in 2 years if interest were compounded monthly?

19. The half-life of radium is approximately 1600 years. How much of a 200-milligram sample will be left after 100 years?

20. The Natick Radiation Laboratory was initially charged with 2.5 million curies of cobalt 60 (^{60}Co) in July 1962. If the half-life of ^{60}Co is 5.2 years, how many curies remain in July 1996?

21. The population of Lincolnton t years after 1986 is modeled by the function

$$P(t) = 1522e^{0.035t}$$

What is the expected population at the end of the year 2000?

22. In Problem 21, what would the expected population be if k were doubled, that is, if $P(t) = 1522e^{0.07t}$?

23. *Escherichia coli* bacteria double in number every 20 minutes. If 5000 *E. coli* bacteria were present in a sample initially, express the number of bacteria in the sample as a function of time. By approximately how many bacteria has the number of bacteria present increased between 1 and 2 hours? Between 3 and 4 hours? Explain both answers in terms of the graph of the function.

24. An antibiotic causes the number of bacteria in an infection to decrease by one-eighth every hour. If 1.8×10^{42} bacteria were present when the antibiotic was administered, express the number of bacteria present as a function of time. Compare the approximate number of bacteria present 24 hours after the antibiotic is administered, the approximate number present after 36 hours, and the approximate number present after 48 hours. What does this indicate about the infection?

25. The atmospheric pressure in pounds per square inch x miles above sea level is given by the function

$$P(x) = 14.7e^{-0.21x}$$

Approximate the atmospheric pressure on top of Mount Ranier which is 14,410 feet above sea level.

26. The resale value of certain computers after t years is predicted by the function

$$R(t) = 3500e^{-0.2t}$$

Approximate the resale value after 3 years. After 5 years.

27. As $n \to +\infty$, $\left(1 + \frac{1}{n}\right)^n \to e$. Compare the value of e^r to the value of $\left(1 + \frac{r}{n}\right)^n$ when $r = 0.08$. First approximate $e^{0.08}$ and then fill in the table below. Make approximations to as many decimals as possible.

n	$\left(1+\frac{0.08}{n}\right)^n$
1	
10	
100	
1000	
10,000	
1,000,000	

As $n \to +\infty$, what appears to be happening to $\left(1+\frac{0.08}{n}\right)^n$? What seems to be true about the value of e^r and of $\left(1+\frac{r}{n}\right)^n$?

28. If $n \to +\infty$, show that $P\left(1+\frac{r}{n}\right)^{nt} \approx Pe^{rt}$.

29. The bacterial growth curve for bacteria in a liquid medium can be divided into four phases. The first phase is the *lag phase* where little population increase takes place. The second phase is the *growth phase*. Eventually the number of deaths balances the number of new cells, and the *stationary phase* is reached. If fresh medium is not added, the number of deaths exceeds the number of new cells and the *decline phase* begins. The following function approximates the number of bacteria present after x hours.

$$G(x) = \begin{cases} 1732 & \text{if } 0 < x \leq 0.5 \\ 1000 \cdot 3^x & \text{if } 0.5 < x \leq 3.2 \\ 33{,}635 & \text{if } 3.2 < x \leq 5.8 \\ 121{,}890\left(\frac{1}{5}\right)^{x-5} & \text{if } x > 5.8 \end{cases}$$

Sketch the graph of G and describe each of the phases in terms of the graph.

30. John has just received his degree from DeKalb College. He has two job offers. The first pays \$50 the first month plus a 10% raise each month thereafter, and the second pays \$500 the first month plus a \$50 raise each month thereafter. For each job offer, write a function that expresses his monthly salary as a function of the number of months worked. Graph each function. Which job offer will give a higher salary after 3 years? After 5 years?

31. Susan has also just received her degree from DeKalb College. She has two job offers. The first pays \$1000 the first month with a 1% raise each month thereafter, and the second pays \$750 the first month with a 1.5% raise each month thereafter. For each job offer, write a function that expresses her monthly salary as a function of the number of months worked. Which job offer will give a higher salary after 3 years? After 5 years?

Challenge Problems

32. The average rate of change of $f(x)$ with respect to x over an interval $[a, b]$ is defined as

$$\text{Average rate of change} = \frac{f(b) - f(a)}{b - a}$$

Calculate the average rate of change over the indicated interval for each function.

	$f(x) = 2x$	$g(x) = x^2$	$h(x) = 2^x$
[1, 3]			
[3, 6]			
[8, 12]			

IN YOUR OWN WORDS . . .

33. Explain what is meant by the phrase *exponential growth.*

6.3 LOGARITHMIC FUNCTIONS

Since an exponential function is a one-to-one function, it has an inverse. To find a rule for the inverse of $f(x) = b^x$, we follow the procedure developed in Section 4.5.

$y = b^x$ Replace $f(x)$ with y.

$x = b^y$ Interchange x and y.

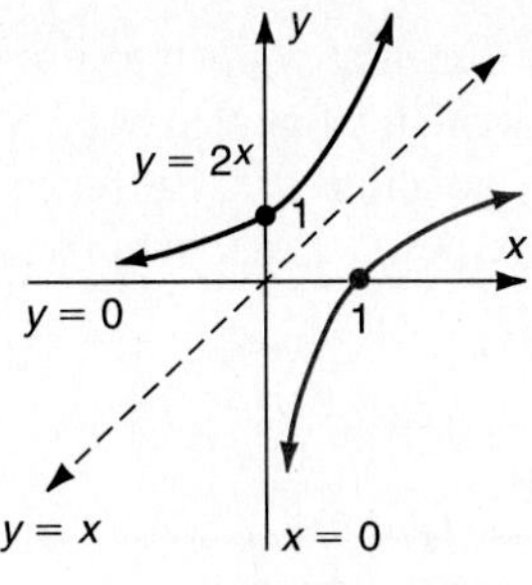

y = 2^x and its inverse

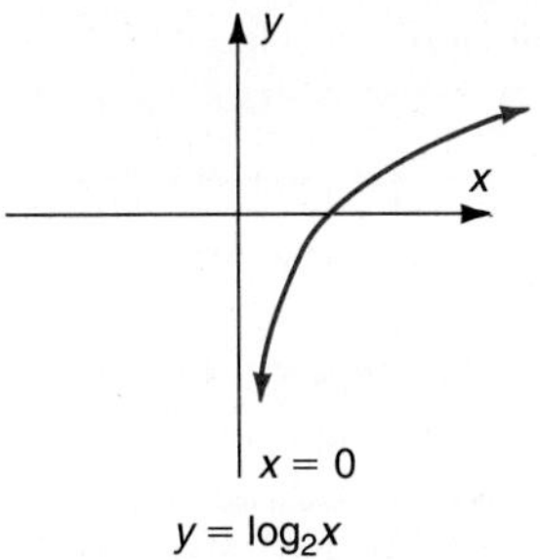

y = log₂x

Fig. 6.18

The next step is to solve the equation $x = b^y$ for y. We don't have the tools to solve this equation for y, but since f has an inverse we know such a function exists. The name for this function is $\log_b$. So continuing the procedure, we write

$$y = \log_b x \qquad \text{Solve for } y.$$

$$f^{-1}(x) = \log_b x \qquad \text{Replace } y \text{ with } f^{-1}(x).$$

We read $\log_b x$ as "logarithm of x to the base b" or simply "log to the base b of x." The symbol $\log_b$ is the name of a function just as f, g, h, and f^{-1} are names of functions.

Since logarithmic and exponential functions are inverses of each other, we can sketch the graph of $y = \log_2 x$ because we know the graph of $y = 2^x$. Using the idea that the graphs of inverse functions are reflections about the line $y = x$, we see the shape of the graph of $y = \log_2 x$ as shown in Figure 6.18. Notice in Figure 6.18 that the graph of $y = \log_2 x$ has the y-axis as a vertical asymptote and crosses the x-axis at 1. Also, we see from the graph that the function $f(x) = \log_2 x$ has all positive real numbers as its domain and all real numbers as its range.

We used 2 as the base, but we could have used any base $b > 0$ and $b \neq 1$. The function $f(x) = \log_b x$ is the inverse of the function $g(x) = b^x$.

Logarithmic Function

If x and b are positive real numbers, with $b \neq 1$, the function

$$f(x) = \log_b x$$

is called a **logarithmic function** to the base b.
Its domain is $(0, +\infty)$. Its range is R.

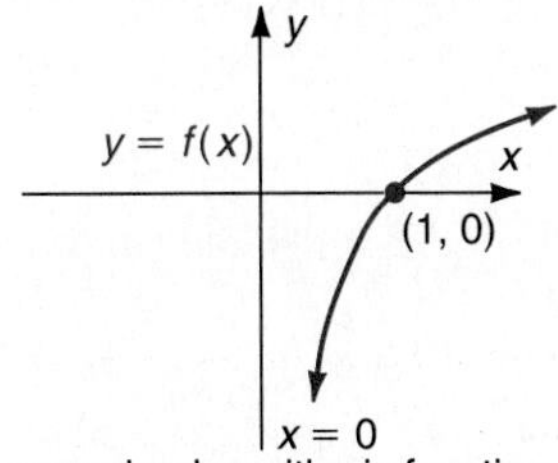

Increasing logarithmic function

$f(x) = \log_b x;\ b > 1$

Logarithmic functions are increasing or decreasing just like exponential functions. See Figure 6.19.

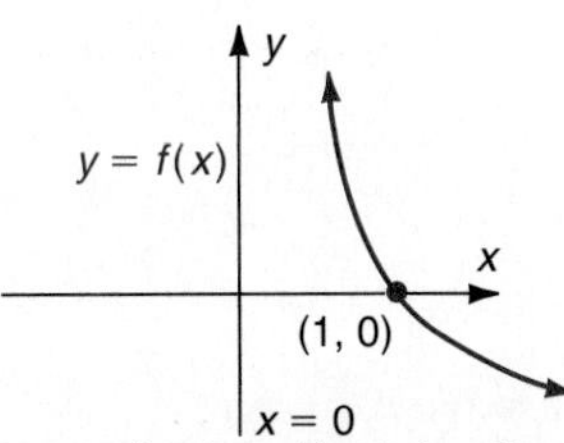

Decreasing logarithmic function

$f(x) = \log_b x;\ \ 0 < b < 1$

Fig. 6.19

Fundamental Equivalence

Since $g(x) = b^x$ and $f(x) = \log_b x$ are inverses of each other, the statements $y = \log_b x$ and $x = b^y$ are equivalent. This equivalence is important in the further study of algebra and should be memorized.

Fundamental Equivalence Between Logarithms and Exponentials

If b and x are positive real numbers with $b \neq 1$, then

$$y = \log_b x \text{ is equivalent to } b^y = x$$

Logarithmic form   Exponential form

One important observation we can make from the fundamental equivalence is that logarithms are not really new, they are just exponents written in another way. Since we will use this equivalence often, we will use the notation $\Leftrightarrow$ to mean "is equivalent to" or "if and only if." That is, if b and x are positive real numbers with $b \neq 1$,

$$y = \log_b x \Leftrightarrow b^y = x$$

EXAMPLE 1 Change $5 = \log_2 32$ to exponential form.

Solution $5 = \log_2 32 \Leftrightarrow 2^5 = 32$ Fundamental equivalence □

EXAMPLE 2 Change $3^4 = 81$ to logarithmic form.

Solution $3^4 = 81 \Leftrightarrow 4 = \log_3 81$ Fundamental equivalence □

EXAMPLE 3 Evaluate the following logarithms.

(a) $\log_8 64$ (b) $\log_4 \dfrac{1}{16}$ (c) $\log_6(-3)$ (d) $\log_{1/3} 27$

Solutions (a) $\log_8 64$

We let $y = \log_8 64$. Now all we need to do is find y. But, by the fundamental equivalence,

$y = \log_8 64 \Leftrightarrow 8^y = 64$

$8^y = 8^2$

$y = 2$ One-to-One Property of exponential functions

Therefore, $\log_8 64 = 2$.

(b) $y = \log_4 \dfrac{1}{16} \Leftrightarrow 4^y = \dfrac{1}{16}$ Fundamental equivalence

$4^y = \dfrac{1}{4^2}$

$4^y = 4^{-2}$ Property of negative exponent

$y = -2$ One-to-One Property of exponential functions

$\log_4 \dfrac{1}{16} = -2$

(c) The domain of $\log_b x$ is all *positive* real numbers. Therefore, -3 is *not in the domain* of $\log_6 x$.

$\log_6(-3)$ *is undefined.*

(d) $y = \log_{1/3} 27 \Leftrightarrow \left(\frac{1}{3}\right)^y = 27$ Fundamental equivalence

$$\frac{1}{3^y} = 3^3$$

$$3^{-y} = 3^3$$

$$-y = 3$$ One-to-One Property of exponential functions

$$y = -3$$

$\log_{1/3} 27 = -3$ □

Common Logs and Natural Logs

Logarithms were developed during the latter part of the sixteenth century by a Scots mathematician, John Napier. Napier and an English mathematician, Henry Briggs, determined that the most useful base for logarithms was 10, as 10 is the base for our number system. For nearly 400 years these base 10 logarithms were used for scientific and navigational computations. The slide rule, which was an indispensable tool of physics and chemistry students until the development of the hand-held calculator, is based on a logarithmic scale. Base 10 logarithms are called **common logarithms** and are usually written omitting the base. That is, $\log x$ means $\log_{10} x$. However, as will be seen in calculus, logarithms arise *naturally* in nature and mathematical development. These logarithms are to the base e and are called **natural logarithms.** We abbreviate natural logarithms by writing $\ln x$. In other words, $\ln x$ means $\log_e x$.

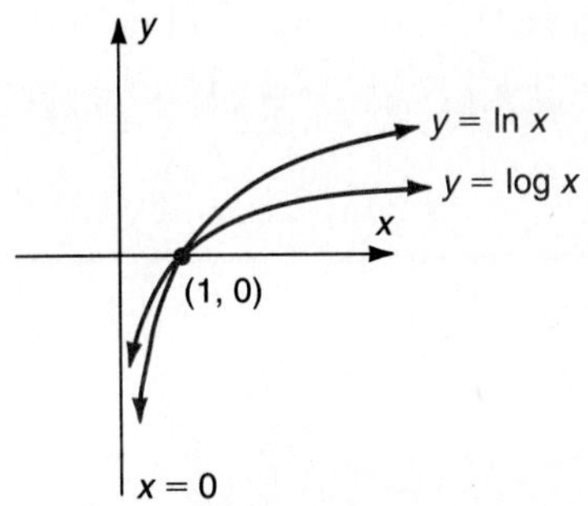

Fig. 6.20

Common Logs and Natural Logs

The following two abbreviations are widely used.

Common logarithms: $\log x = \log_{10} x$

Natural logarithms: $\ln x = \log_e x$

The graphs of $y = \log x$ and $y = \ln x$ are shown in Figure 6.20.

EXAMPLE 4 Find each common or natural logarithm.

(a) $\log 100$ (b) $\log 0.0001$ (c) $\ln e$ (d) $\ln \sqrt{e}$

Solutions (a) $\log 100 = y \Leftrightarrow 10^y = 100$ Fundamental equivalence

$$10^y = 10^2$$

$$y = 2$$ One-to-one function

$\log 100 = 2$

(b) $\log 0.0001 = y \Leftrightarrow 10^y = 0.0001$ Fundamental equivalence

$10^y = 10^{-4}$

$y = -4$ One-to-one function

$\log 0.0001 = -4$

(c) $\ln e = y \Leftrightarrow e^y = e$ Fundamental equivalence

$y = 1$ One-to-one function

$\ln e = 1$

(d) $\ln \sqrt{e} = y \Leftrightarrow e^y = \sqrt{e}$ Fundamental equivalence

$e^y = e^{1/2}$

$y = \frac{1}{2}$ One-to-one function

$\ln \sqrt{e} = \frac{1}{2}$

□

Finding Common and Natural Logs with a Calculator

The scientific calculator has replaced logarithms and the slide rule as computational tools. However, logarithmic functions express many relationships in business and science and are used in mathematics. Both common and natural logarithms can be found on a scientific calculator. Look for a key marked [log] or [log x] for evaluating the *common logarithm* of a number, and a key marked [ln] or [ln x] for finding the *natural logarithm*.

EXAMPLE 5 Find each common logarithm to six significant digits.

(a) $\log 2$ (b) $\log \frac{1}{3}$

Solutions (a) log 2 does not simplify like the integer powers of 10. We approximate it on a scientific calculator. Press [2] followed by [log], and the display should show something like [0.301029996]. If this sequence of keystrokes does not work, try [log] followed by [2] [ENTER] or [EXE].

$\log 2 \approx 0.301030$

(b) Use the following sequence of keystrokes.

[(] [1] [÷] [3] [)] [log] for most scientific calculators.

[log] [(] [1] [÷] [3] [)] [ENTER] for graphing calculators and function-first calculators.

$$\log \frac{1}{3} \approx -0.477121$$

□

EXAMPLE 6 Find each natural logarithm to six significant digits.

(a) $\ln 2$ (b) $\ln \frac{2}{7}$

Solutions (a) ln 2 also must be approximated on a calculator. Try [2] followed by [ln] (or [ln] followed by [2] [ENTER]) and read **0.693147181** on the display.

$\ln 2 \approx 0.693147$

(b) [(] [2] [÷] [7] [)] [ln] Scientific calculator

[ln] [(] [2] [÷] [7] [)] [ENTER] Function-first calculator

$$\ln \frac{2}{7} \approx -1.25276$$

□

Solving Logarithmic Equations

Sometimes it is necessary to find a number, given its logarithm. For this we use the fundamental equivalence and a calculator, if necessary.

EXAMPLE 7 Solve each equation. Write approximations to five significant digits.

(a) $\log x = 5$ (b) $\log x = -3$ (c) $\log x = 3.99231$ (d) $\ln x = -2.6671$

Solutions (a) $\log x = 5 \Longleftrightarrow x = 10^5$ Solution set Fundamental equivalence

$\{100000\}$ Solution set

(b) $\log x = -3 \Longleftrightarrow x = 10^{-3}$ Fundamental equivalence

$\{0.001\}$

(c) $\log x = 3.99231 \Longleftrightarrow x = 10^{3.99231}$ Fundamental equivalence

Now we need a calculator. The 10^x key gives the *inverse* of the log key and is usually a second function. Enter [3.99231] followed by [10^x] and read **9824.48966** on the display. (For function-first calculators, enter [10^x] first.)

$x \approx 9824.5$

(d) $\ln x = -2.6671 \Longleftrightarrow x = e^{-2.6671}$ Fundamental equivalence

The e^x key gives the *inverse* of the ln key, and it is usually a second function.

$x \approx 0.069453$

□

We can use the fundamental equivalence to solve other simple logarithmic equations.

EXAMPLE 8 Solve each equation.

(a) $\log_5 x = 2$ (b) $\log_6 216 = x$ (c) $\log_x 7 = \frac{1}{2}$

Solutions

(a) $\log_5 x = 2 \Leftrightarrow 5^2 = x$ — Fundamental equivalence
$x = 25$
$\{25\}$ — Solution set

(b) $\log_6 216 = x \Leftrightarrow 6^x = 216$ — Fundamental equivalence
$6^x = 6^3$
$x = 3$ — One-to-one function
$\{3\}$ — Solution set

(c) $\log_x 7 = \frac{1}{2} \Leftrightarrow x^{1/2} = 7$ — Fundamental equivalence
$\sqrt{x} = 7$ — Definition of $\frac{1}{2}$ exponent
$x = 49$ — Square both sides.
$\{49\}$ — Solution set. □

Graphing Logarithmic Functions

Because a logarithmic function is the inverse of an exponential function, we have a general idea of its graph. However, we can use the fundamental equivalence to plot as many points as we desire.

EXAMPLE 9 Graph the function $f(x) = \log_3 x$.

Solution We graph the equation $y = f(x)$.

$$y = \log_3 x \Leftrightarrow 3^y = x \quad \text{Fundamental equivalence}$$

Therefore, the graph of $y = \log_3 x$ is the same as the graph of $x = 3^y$. Now we can make a table by choosing a value for y first and calculating the corresponding value of x. (As we make the sketch we remember that the y-axis is an asymptote for the graph.) See Figure 6.21.

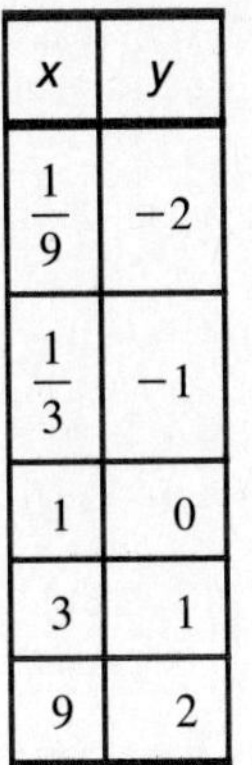

x	y
$\frac{1}{9}$	-2
$\frac{1}{3}$	-1
1	0
3	1
9	2

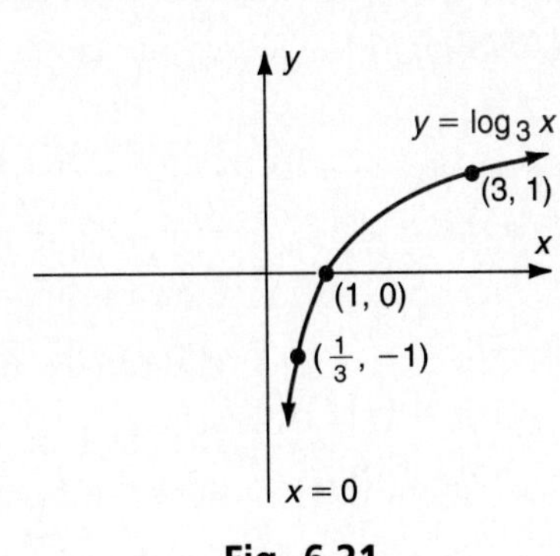

Fig. 6.21

□

The graphs of logarithmic functions can be shifted and reflected in the same manner as those of the other functions we have seen.

EXAMPLE 10 Sketch the graph of each function by shifting the graph of $y = \log_2 x$.

(a) $f(x) = \log_2 x + 1$ (b) $g(x) = \log_2 x - 2$
(c) $h(x) = \log_2(x + 1)$ (d) $j(x) = \log_2(x - 2)$

Solutions The graph of $y = \log_2 x$ is shown in Figure 6.22.

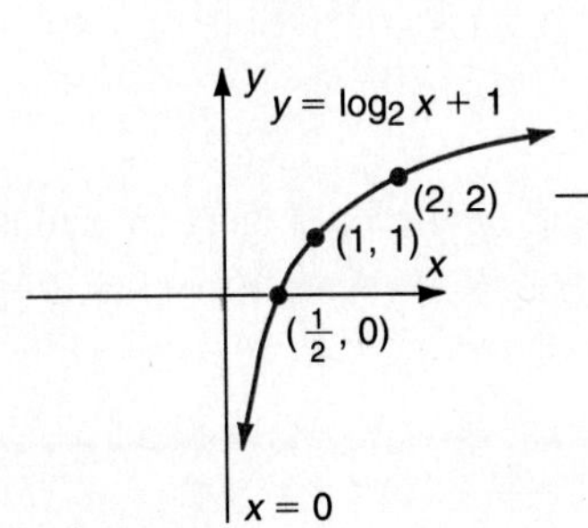

Fig. 6.22

(a) $f(x) = \log_2 x + 1$
The graph is a shift of $y = \log_2 x$ one unit upward as shown in Figure 6.23.

(b) $g(x) = \log_2 x - 2$
The graph is a shift of $y = \log_2 x$ two units *downward* as shown in Figure 6.24.

(c) $h(x) = \log_2(x + 1)$
The graph is a shift of $y = \log_2 x$ one unit to the left. See Figure 6.25.

(d) $j(x) = \log_2(x - 2)$
The graph is a shift of $y = \log_2 x$ two units to the *right*. See Figure 6.26.

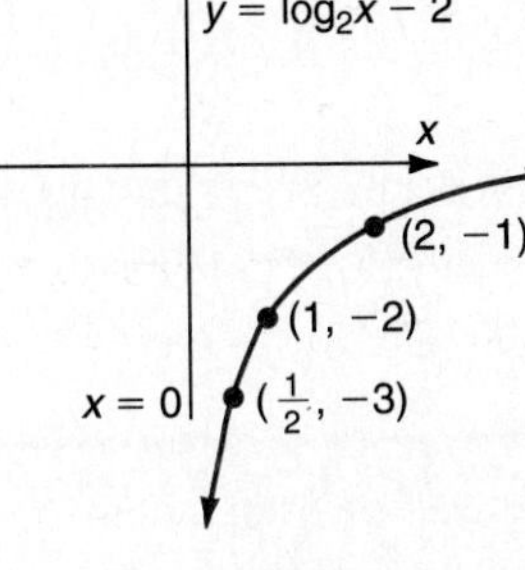

Fig. 6.23

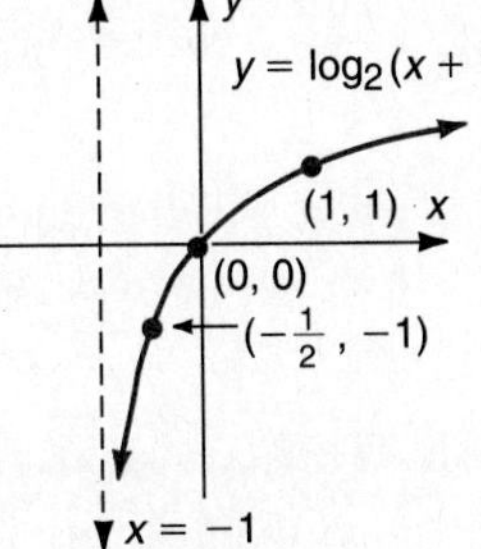

Fig. 6.24

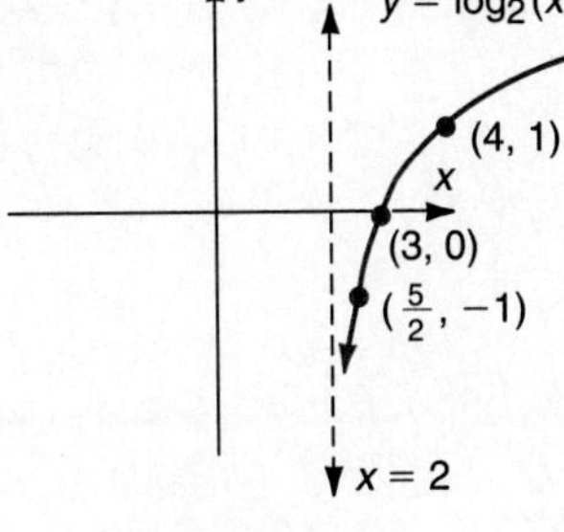

Fig. 6.25

Fig. 6.26

□

EXAMPLE 11 Sketch the graph of each function and give the domain and range for each.

(a) $f(x) = \ln(-x)$ (b) $g(x) = -\ln x$ (c) $h(x) = \ln|x|$

Solutions (a) The graph of $f(x) = \ln(-x)$ is the graph of $y = \ln x$ reflected about the y-axis as shown in Figure 6.27.

The domain of g is $(-\infty, 0)$, and the range is $(-\infty, +\infty)$.

(b) The graph of $g(x) = -\ln x$ is the graph of $y = \ln x$ reflected about the x-axis as shown in Figure 6.28.

The domain of g is $(0, +\infty)$, and the range is $(-\infty, +\infty)$.

(c) Since $|x| = \begin{cases} x & \text{if } x \geq 0 \\ -x & \text{if } x < 0 \end{cases}$, we can write h as $h(x) = \begin{cases} \ln x & \text{if } x \geq 0 \\ \ln(-x) & \text{if } x < 0 \end{cases}$. Notice that we can use both positive and negative numbers for x. See Figure 6.29. The domain of h is $(-\infty, 0) \cup (0, +\infty)$, and the range is $(-\infty, +\infty)$.

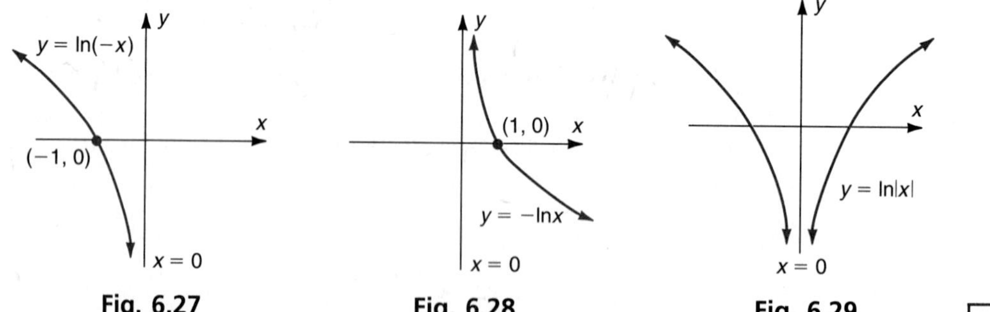

Fig. 6.27 Fig. 6.28 Fig. 6.29 □

Applications

Loudness of sound is usually measured in decibels. Because the human ear can detect an incredible range of loudness, the level of sound is given by a logarithmic function,

$$\beta = 10 \log_{10} \frac{I}{I_0}$$

where β is the loudness in decibels, I is the intensity in watts per square meter, and I_0 is the intensity of the threshold of hearing, taken to be 10^{-12} watts per square meter.

EXAMPLE 12 Find the loudness in decibels of each sound.

(a) A rustle of leaves with intensity 10^{-11} watts per square meter

(b) Busy street traffic with intensity 10^{-5} watts per square meter

(c) Ordinary conversation with intensity 3×10^{-6} watts per square meter

Solutions (a) A rustle of leaves with intensity 10^{-11} watts per square meter

$$\beta = 10 \log_{10} \frac{I}{I_0}$$

where $I = 10^{-11}$ watts per square meter and $I_0 = 10^{-12}$ watts per square meter

$$\beta = 10 \log_{10} \frac{10^{-11}}{10^{-12}} \qquad \text{Substitution}$$

$$= 10 \log_{10} 10^{1} \qquad \text{Division property of exponents}$$

$$\beta = 10 \cdot 1 = 10 \qquad \log_{10} 10 = 1$$

A rustle of leaves is about 10 decibels.

(b) Busy street traffic with intensity 10^{-5} watts per square meter.

$$\beta = 10 \log_{10} \frac{I}{I_0}$$

$$\beta = 10 \log_{10} \frac{10^{-5}}{10^{-12}} \qquad \text{Substitution}$$

$$= 10 \log_{10} 10^{7} \qquad \text{Division Property of exponents}$$

$$y = \log_{10} 10^{7} \Longleftrightarrow 10^{y} = 10^{7}$$

So $y = 7$.

$$\beta = 10 \cdot 7 = 70 \qquad \log_{10} 10^{7} = 7$$

Busy street traffic is about 70 decibels.

(c) Ordinary conversation with intensity 3×10^{-6} watts per square meter

$$\beta = 10 \log \frac{I}{I_0}$$

$$\beta = 10 \log \frac{3 \times 10^{-6}}{10^{-12}} \qquad \text{Substitution}$$

$$= 10 \log(3 \times 10^{6}) \qquad \text{Property of exponents}$$

As 3 is not a simple power of 10, the fundamental equivalence will not help here. However, a calculator will. The scientific calculator has a key to simplify entering numbers written in scientific notation, such as the 3×10^{6} in this problem. It is a key labeled EXP or EE. Its use is illustrated in this problem. For scientific calculators, enter the keystrokes 10 × 3 EXP 6 log and read 64.77121255 on the display.

$$\beta \approx 64.8$$

Ordinary conversation is approximately 65 decibels. □

In chemistry, the pH of a solution is defined to be $-\log[H^+]$, where $[H^+]$ is the hydrogen ion concentration in the solution.

EXAMPLE 13 Find the hydrogen ion concentration in a sample of lemon juice whose pH is 3.42.

Solution

$\text{pH} = -\log[H^+]$	Formula
$3.42 = -\log[H^+]$	Substitution
$-3.42 = \log[H^+] \Longleftrightarrow [H^+] = 10^{-3.42}$	Fundamental equivalence
$[H^+] \approx 0.00038$	

□

PROBLEM SET 6.3

Warm-ups

In Problems 1 through 6, change each to exponential form. See Example 1.

1. $3 = \log_5 125$

2. $\log_8 8 = 1$

3. $\log_{1/2} \dfrac{1}{16} = 4$

4. $\log_{1/3} 243 = -5$

5. $-1 = \log_6 \dfrac{1}{6}$

6. $k = \log_b w$

In Problems 7 through 12, change each to logarithmic form. See Example 2.

7. $8 = 2^3$

8. $625 = 5^4$

9. $10^4 = 10000$

10. $\dfrac{1}{81} = 3^{-4}$

11. $64 = \left(\dfrac{1}{4}\right)^{-3}$

12. $r = p^q$

In Problems 13 through 18, evaluate each logarithm. See Example 3.

13. $\log_7 49$

14. $\log_{10} 100000$

15. $\log_{11}(-11)$

16. $\log_9 \dfrac{1}{729}$

17. $\log_{1/5} 125$

18. $\log_{1/2} \dfrac{1}{128}$

In Problems 19 through 21, evaluate each logarithm. See Example 4.

19. $\log 0.1$

20. $\ln \dfrac{1}{e}$

21. $\log 10000$

In Problems 22 through 27, approximate each logarithm to five significant digits. See Examples 5 and 6.

22. $\log 346.7$

23. $\log 0.08763$

24. $\log \sqrt{23.56}$

25. $\ln 23.5$

26. $\ln \dfrac{2}{7}$

27. $\ln 0.007^3$

In Problems 28 through 33, solve each equation. Write approximations to four significant digits. See Example 7.

28. $\log x = 3$

29. $\ln x = -2.4415$

30. $\log x = -0.043771$

31. $\ln x = e$

32. $\log_3 x = 3$

33. $\log x = 0.001$

In Problems 34 through 39, solve each equation. See Example 8.

34. $\log_7 x = 3$

35. $\log_x 1024 = 5$

36. $\log_{1/5} 125 = x$

37. $\log_2(x + 1) = 4$

38. $\log_3(9x) = 81$

39. $\log_{1/3} \dfrac{1}{27} = 3x - 1$

In Problems 40 through 42, sketch the graph of each function by plotting points. See Example 9.

40. $F(x) = \log_5 x$

41. $G(x) = \log_{1/2} x$

42. $H(x) = \log_7 x$

In Problems 43 through 45, sketch each graph by shifting the graph of $y = \ln x$. See Example 10.

43. $f(x) = \ln x - 2$

44. $g(x) = \ln(x - 1)$

45. $h(x) = \ln(x + 2) + 1$

In Problems 46 through 48, sketch the graph of each function and give the domain and range for each. See Example 11.

46. $f(x) = \log(-x)$

47. $g(x) = \ln(x + 1) - 2$

48. $h(x) = \ln(-x) + 1$

For Problems 49 and 50, see Example 12.

49. Using the formula for the loudness of sound, find the loudness in decibels of a siren where the intensity is 10^{-2} watt per square meter.

50. What is the loudness in decibels of a radio when the intensity is 2×10^{-8} watt per square meter?

For Problems 51 and 52, see Example 13.

51. Find the hydrogen ion concentration in a sample of pure water whose pH is 7.

52. Find the hydrogen ion concentration in a sample of a window cleaning solution whose pH is 8.28.

Practice Exercises

In Problems 53 through 72, evaluate each logarithm. Write approximations to five significant digits.

53. $\log_7 \dfrac{1}{49}$

54. $\log_8 \dfrac{1}{512}$

55. $\log_{1/3} 27$

56. $\log_{1/4} 4$

57. $\log 21$

58. $\log 0.034$

59. $\ln 0.65$

60. $\ln 19$

61. $\ln 15^2$

62. $\ln(0.91)^3$

63. $(\ln 15)^2$

64. $[\ln(0.91)]^3$

65. $2 \ln 15$

66. $3 \ln(0.91)$

67. $\log 5 + \log 3$

68. $\log 6 - \log 2$

69. $\log(5 + 3)$

70. $\log(6 - 2)$

71. $\log(5 \cdot 3)$

72. $\log \dfrac{6}{2}$

In Problems 73 through 84, solve each equation. Write approximations to five significant digits.

73. $\log_4 x = 3$

74. $\log_5 t = -1$

75. $\log_b 343 = 3$

76. $\log_y 121 = -2$

77. $\ln x = 3$

78. $\ln w = 0.08$

79. $\log x = 2.34451$

80. $\log s = -1.18335$

81. $\log_4(1 - x) = 2$

82. $\log_5(2x + 1) = -1$

83. $\log(x^2 + 1) = 1$

84. $\log(t^2 - 21) = 2$

In Problems 85 and 86, sketch the graph of each function.

85. $F(x) = \log_4 x$

86. $G(x) = \log_6 x$

In Problems 87 through 92, use the results of Problems 85 and 86 to sketch the graph of each function. Give the domain and range for each function.

87. $f(x) = \log_4 x - 1$

88. $g(x) = \log_6 x + 2$

89. $h(x) = \log_4(x - 2)$

90. $k(x) = \log_6(x + 1)$

91. $u(x) = \log_4(x + 1) + 2$

92. $v(x) = \log_6(x - 2) - 1$

93. The threshold of pain is considered to be a sound intensity of 1 watt per square meter. How many decibels is this?

94. What is the loudness in decibels of an F-16 taking off 100 feet away if the intensity of the sound is 11 watts per square meter?

95. What is the hydrogen ion concentration in a sample of vinegar whose pH is 5.1?

96. What is the hydrogen ion concentration in a soil sample whose pH is 6.4?

In Problems 97 through 100, use graphing to solve each inequality.

97. $\log_3 x < 0$

98. $\log_3 x < 1$

99. $\log_2 x < -1$

100. $\log_2 x > 1$

Challenge Problems

The Greek astronomers Hipparchus and Ptolemy called the brightest stars *first magnitude,* and the faintest *sixth magnitude.* The invention of the telescope brought many more stars into view and made a mathematical definition of their magnitude necessary. The formula used is

$$M = \sqrt[5]{100}(\log I_0 - \log I)$$

where I is the intensity of light from the star and I_0 is the intensity of light from a star of magnitude 0. The smaller the magnitude, the brighter the star. Sirius, the brightest star in the sky, has magnitude -1.45, while the planet Venus can reach magnitude -4.4. The Hubble space telescope can observe stars as faint as the twenty-fifth magnitude. In the following problems, take the measure of I_0 to be 2.45×10^{-6}.

101. What is the measure of the intensity of light from Sirius?

102. What is the measure of the intensity of light from a twenty-fifth-magnitude star?

103. How much brighter is a first-magnitude star than a sixth-magnitude star?

IN YOUR OWN WORDS . . .

104. What is the logarithm of a number?

105. Explain the relationship between the graphs of $y = \ln x$, $y = \ln(-x)$, and $y = -\ln x$.

6.4 PROPERTIES OF LOGARITHMS

From the fundamental equivalence between logarithms and exponentials,

$$y = \log_b x \Longleftrightarrow b^y = x$$

we know that logarithms are actually exponents. As such, they inherit the properties of exponents, and we use this concept to investigate the properties of logarithms. Here, and in the rest of this section, we will assume that the base b is a positive real number not equal to 1.

Suppose $p = \log_b x$ and $q = \log_b y$. Then by the fundamental equivalence,

$$p = \log_b x \Leftrightarrow b^p = x$$

and

$$q = \log_b y \Leftrightarrow b^q = y$$

Thus,

$$b^p b^q = xy \qquad \text{Multiplication}$$

$$b^{p+q} = xy \qquad \text{Property of exponents}$$

$$b^{p+q} = xy \Leftrightarrow p + q = \log_b(xy) \qquad \text{Fundamental equivalence}$$

Therefore, $\log_b(xy) = \log_b x + \log_b y$, and we have shown the first of several properties of logarithms.

Log of a Product

If x and y are both positive real numbers, then

$$\log_b(xy) = \log_b x + \log_b y$$

The logarithm of a product is the sum of the logarithms of the factors.

In the same manner we show another property.

$$\frac{b^p}{b^q} = \frac{x}{y} \qquad \text{Division}$$

$$b^{p-q} = \frac{x}{y} \qquad \text{Property of exponents}$$

$$b^{p-q} = \frac{x}{y} \Leftrightarrow p - q = \log_b \frac{x}{y} \qquad \text{Fundamental equivalence}$$

Log of a Quotient

If x and y are both positive real numbers, then

$$\log_b \frac{x}{y} = \log_b x - \log_b y$$

The logarithm of a quotient is the logarithm of the numerator minus the logarithm of the denominator.

With the same technique, another important property can be shown. The proof is an exercise in the problem set.

Log of a Power

If x and r are real numbers with x positive, then

$$\log_b x^r = r \log_b x$$

The logarithm of a number to a power is the power times the logarithm of the number.

Notice that for any appropriate base,

$y = \log_b 1 \Longleftrightarrow b^y = 1$	Fundamental equivalence
$b^y = b^0$	$b^0 = 1.$
$y = 0$	One-to-one property of exponential functions

Log of 1

$$\log_b 1 = 0$$

The logarithm (to *any* allowable base) of 1 is 0.

The log of 1 and the log of a quotient lead to the following result.

Log of a Reciprocal

If x is a positive real number, then

$$\log_b \frac{1}{x} = -\log_b x$$

The logarithm of the reciprocal of a number is the opposite of the logarithm of the number.

Also, for any appropriate base,

$y = \log_b b \Longleftrightarrow b^y = b$	Fundamental equivalence
$b^y = b^1$	
$y = 1$	One-to-one Property of exponential functions

Log of the Base

$$\log_b b = 1$$

The logarithm of the base is 1.

Combining the log of the base with the log of a power gives us another property.

$$\log_b(b^x) = x \cdot \log_b b = x \cdot 1 = x$$

Log of the Base to a Power

$$\log_b(b^x) = x$$

Another property is useful to know.

$$p = \log_b x \Longleftrightarrow b^p = x \qquad \text{Fundamental equivalence}$$

Chaining these facts in a different way gives

$$x = b^p = b^{\log_b x}$$

Base to a Log with the Same Base

$$b^{\log_b x} = x$$

Although we proved the last two properties without using the fact that exponential and logarithmic functions are inverses, both properties are a statement of the fact that

$$f(f^{-1}(x)) = x \text{ for } x \text{ in the domain of } f^{-1}.$$

$$f^{-1}(f(x)) = x \text{ for } x \text{ in the domain of } f.$$

The proof of these properties using this idea is left as an exercise.

In summary, the basic properties of logarithms are as follows.

Properties of Logarithms

If x, y, and r are real numbers, with x and y both positive, then

1. $\log_b(xy) = \log_b x + \log_b y$ Log of a product
2. $\log_b \dfrac{x}{y} = \log_b x - \log_b y$ Log of a quotient
3. $\log_b x^r = r \log_b x$ Log of a power
4. $\log_b \dfrac{1}{x} = -\log_b x$ Log of a reciprocal

5. $\log_b 1 = 0$	Log of 1
6. $\log_b b = 1$	Log of the base
7. $\log_b(b^x) = x$	Log of a base to a power
8. $b^{\log_b x} = x$	Base to a log with the same base

Remember that these properties apply for any base. The natural logarithm is used so often that we list the properties again with base e. They look somewhat different, especially properties 7 and 8.

1. $\ln(xy) = \ln x + \ln y$	Log of a product
2. $\ln \dfrac{x}{y} = \ln x - \ln y$	Log of a quotient
3. $\ln x^r = r \ln x$	Log of a power
4. $\ln \dfrac{1}{x} = -\ln x$	Log of a reciprocal
5. $\ln 1 = 0$	Log of 1
6. $\ln e = 1$	Log of a base
7. $\ln e^x = x$	Log of a base to a power
8. $e^{\ln x} = x$	Base to a log with the same base

EXAMPLE 1 Use the properties of logarithms to write each logarithm in terms of the logarithms of the positive numbers A, B, and C.

(a) $\log_b(ABC)$ (b) $\log_b \dfrac{AB}{C}$ (c) $\log_b \dfrac{A}{BC}$

(d) $\log_b(A^2B^3)$ (e) $\log_b \dfrac{1}{A^4}$ (f) $\log_b(A + BC)$

Solutions (a) $\log_b(ABC) = \log_b A + \log_b(BC)$ Log of a product

$= \log_b A + \log_b B + \log_b C$ Log of a product

(b) $\log_b \dfrac{AB}{C} = \log_b AB - \log_b C$ Log of a quotient

$= \log_b A + \log_b B - \log_b C$ Log of a product

(c) $\log_b \dfrac{A}{BC} = \log_b A - \log_b BC$ Log of a quotient

$= \log_b A - (\log_b B + \log_b C)$ Log of a product

$= \log_b A - \log_b B - \log_b C$

(d) $\log_b(A^2B^3) = \log_b A^2 + \log_b B^3$ Log of a product

$= 2 \log_b A + 3 \log_b B$ Log of a power

(e) $\log_b \frac{1}{A^4} = -\log_b A^4$ Log of a reciprocal

$= -4 \log_b A$ Log of a power

(f) $\log_b(A + BC)$ *cannot be simplified.* There is no simple property for simplifying the logarithm of a sum or difference. □

EXAMPLE 2 Write each expression as a single logarithm with a coefficient of 1.

(a) $\ln 2 + 2 \ln 3$ (b) $2 \ln 6 - \ln 12$ (c) $\frac{1}{2} \ln 8 - 2 \ln 4 + 3 \ln y$

Solutions (a) $\ln 2 + 2 \ln 3 = \ln 2 + \ln 3^2$ Log of a power

$= \ln 2 \cdot 3^2$ Log of a product

$= \ln 18$

(b) $2 \ln 6 - \ln 12 = \ln 6^2 - \ln 12$ Log of a power

$= \ln \frac{6^2}{12}$ Log of a quotient

$= \ln 3$

(c) $\frac{1}{2} \ln 8 - 2 \ln 4 + 3 \ln y = \ln 8^{1/2} - \ln 4^2 + \ln y^3$ Log of a power

$= \ln \frac{\sqrt{8}}{16} + \ln y^3$ Log of a quotient

$= \ln \frac{2\sqrt{2}}{16} \cdot y^3$ Log of a product

$= \ln \frac{\sqrt{2}y^3}{8}$ □

EXAMPLE 3 Suppose $\log_b 2 = 0.5$ and $\log_b 3 = 0.8$. Evaluate each logarithm.

(a) $\log_b 6$ (b) $\log_b 9$ (c) $\log_b \frac{8}{27}$ (d) $\log_b \frac{b}{\sqrt{3}}$

Solutions (a) We try to express $\log_b 6$ in terms of the two given logs, $\log_b 2$ and $\log_b 3$.

$\log_b 6 = \log_b 2 \cdot 3$ Factor

$= \log_b 2 + \log_b 3$ Log of a product

$= 0.5 + 0.8$ Given values

$= 1.3$

(b) $\log_b 9 = \log_b 3^2$

$= 2 \log_b 3$ Log of a power

$= 2(0.8)$

$= 1.6$

(c) $\log_b \dfrac{8}{27} = \log_b 8 - \log_b 27$ Log of a quotient

$= \log_b 2^3 - \log_b 3^3$

$= 3 \log_b 2 - 3 \log_b 3$ Log of a power

$= 3(0.5) - 3(0.8)$ Given values

$= -0.9$

(d) $\log_b \dfrac{b}{\sqrt{3}} = \log_b b - \log_b \sqrt{3}$ Log of a quotient

$= 1 - \log_b \sqrt{3}$ Log of the base

$= 1 - \log_b 3^{1/2}$ Fractional exponent

$= 1 - \dfrac{1}{2} \log_b 3$ Log of a power

$= 1 - \dfrac{1}{2}(0.8)$ Given value

$= 0.6$ □

EXAMPLE 4 Find the value of each expression.

(a) $b^{\log_b 7}$ (b) $e^{2\ln 3}$

Solutions (a) $b^{\log_b 7} = 7$ Base to a Log with the Same Base Property

(b) The same property can be used to find the value of $e^{2\ln 3}$. However, we must use the Log of a Power Property first.

$e^{2\ln 3} = e^{\ln 3^2}$ Log of a Power Property

$= 3^2$ Base to a Log with the Same Base Property

$= 9$ □

Change of Base

It is often convenient to change the base of a logarithm from one number to another. Suppose we have the expression $\log_b x$ and would like to rewrite it in terms of the base c, where, of course, $c > 0$ and $c \neq 1$.

$$y = \log_b x \iff b^y = x \qquad \text{Fundamental equivalence}$$

Thus, b^y and x represent the same positive number. Therefore, $\log_c b^y$ must be equal to $\log_c x$.

$$\log_c b^y = \log_c x$$

$$y \log_c b = \log_c x \qquad \text{Log of a power}$$

Solve for y. Since $b \neq 1$, we can divide both sides by $\log_c b$.

$$y = \frac{\log_c x}{\log_c b}$$

But $y = \log_b x$, so we have

$$\log_b x = \frac{\log_c x}{\log_c b}$$

We have developed the following formula.

Change of Base Formula

If b and c are both positive numbers not equal to 1, then

$$\log_b x = \frac{\log_c x}{\log_c b} \quad \text{for all } x > 0.$$

EXAMPLE 5 Write $\log_2 100$ as a base 10 logarithm.

Solution

$$\log_2 100 = \frac{\log 100}{\log 2} \qquad \text{Change of base formula}$$

$$= \frac{\log 10^2}{\log 2} = \frac{2 \log 10}{\log 2} \qquad \text{Log of a power}$$

$$= \frac{2 \cdot 1}{\log 2} \qquad \text{Log of the base}$$

$$\log_2 100 = \frac{2}{\log 2}$$

□

EXAMPLE 6 Approximate each logarithm to four significant digits.

(a) $\log_6 11$ (b) $\log_{1/3} 44$

Solutions (a) Since a calculator has logarithms only to the bases 10 and e, we must convert to one of those bases. Here we convert to base 10.

$$\log_6 11 = \frac{\log 11}{\log 6} \qquad \text{Change of base formula}$$

Using the $\boxed{\log}$ key on a scientific calculator we obtain

$$\log_6 11 \approx 1.338$$

(b) This time we will use the $\boxed{\ln}$ key. (It doesn't make any difference!)

$$\log_{1/3} 44 = \frac{\ln 44}{\ln \frac{1}{3}} \qquad \text{Change of base formula}$$

$$= \frac{\ln 44}{-\ln 3} \qquad \text{Log of a reciprocal}$$

$$\log_{1/3} 44 \approx -3.445$$

□

PROBLEM SET 6.4

In this problem set assume all variables represent positive real numbers.

Warm-ups

In Problems 1 through 6, use the properties of logarithms to write each in terms of the logarithms of p, q, and r. See Example 1.

1. $\log_b pqr^2$ **2.** $\log_6 \frac{p^2}{q^3r^4}$ **3.** $\ln p\sqrt{q}$

4. $\log_3 \frac{\sqrt[3]{p}}{qr}$ **5.** $\log_b \sqrt[3]{p^2}$ **6.** $\ln \frac{e^2}{\sqrt{p}}$

In Problems 7 through 12, write each expression as a single logarithm with a coefficient of 1. See Example 2.

7. $2 \log_b 3 + \log_b 10$ **8.** $3 \ln 2 - 2 \ln 6$ **9.** $5 \ln 2 - 3 \ln x - \ln 3$

10. $\frac{1}{2} \log_b 27 - \frac{2}{3} \log_b 8$ **11.** $\log_b 12 - 2(\log_b 3 - \log_b 6)$ **12.** $2\left(\frac{1}{3} \log_b x - \frac{2}{5} \log_b y\right)$

In Problems 13 through 18, evaluate each logarithm given that $\log_b 3 = 0.9$ and $\log_b 5 = 1.5$. See Example 3.

13. $\log_b 15$ **14.** $\log_b 25$ **15.** $\log_b \frac{5}{9}$

16. $\log_b \sqrt{3}$ **17.** $\log_b \sqrt{75}$ **18.** $\log_b \sqrt[5]{5b}$

In Problems 19 through 21, find the value of each expression. See Example 4.

19. $6^{-2\log_6 9}$ **20.** $e^{\ln 10}$ **21.** $10^{\ln e}$

In Problems 22 through 24, write each logarithm as a base 10 logarithm. See Example 5.

22. $\log_3 10$ **23.** $\log_7 0.001$ **24.** $\log_b 10^n$

In Problems 25 through 27, approximate each logarithm to five significant digits. See Example 6.

25. $\log_5 43$

26. $\log_{1/4} 7$

27. $\log_{14} \frac{4}{17}$

Practice Exercises

In Problems 28 through 39, evaluate each expression.

28. $\ln e$

29. $\ln 1$

30. $\ln e^3$

31. $\ln \sqrt[3]{e}$

32. $\ln \frac{1}{e^2}$

33. $\ln e^n$

34. $\ln \frac{1}{e}$

35. $\ln \frac{1}{\sqrt{e}}$

36. $e^{\ln 1}$

37. $e^{\ln \sqrt{2}}$

38. $e^{2 \ln 3}$

39. $e^{-\ln 2}$

In Problems 40 through 45, each problem contains a pair of tempting "look alikes." Find the value of each expression given that $\log_k 2 = 0.4$ and $\log_k 5 = 0.9$. Notice how each pair contains different ideas.

40. $\log_k 10$; $(\log_k 2)(\log_k 5)$

41. $\log_k \frac{2}{5}$; $\frac{\log_k 2}{\log_k 5}$

42. $\log_k 2^3$; $(\log_k 2)^3$

43. $\log_k \sqrt{2}$; $\sqrt{\log_k 2}$

44. $\log_k \frac{1}{5}$; $\frac{1}{\log_k 5}$

45. $\log_k 5$; $\log_5 k$

In Problems 46 through 57, write each expression as a single logarithm with a coefficient of 1.

46. $1 - 2 \log_b 5$

47. $2 - \log_b 8$

48. $\log x - 2 \log 5 + 3 \log y$

49. $2 \log t - \log s + 3 \log 2$

50. $\frac{2}{3} \ln 27 + \frac{1}{3} \ln 8$

51. $\frac{3}{4} \ln 16 + \frac{1}{2} \ln 9$

52. $6 \ln t - 2(\ln u - 3 \ln v)$

53. $3(2 \ln y + \ln z) - 4 \ln x$

54. $\log(2x + 3) - 2 \log(x + 1)$

55. $2 \log(t - 1) - \log(t + 1)$

56. $2 \ln x - \ln(x + 1) + \ln(x + 2)$

57. $\ln(t + 1) - \ln(t - 1) - \ln(t + 2)$

In Problems 58 through 63, evaluate each logarithm to the nearest ten-thousandth.

58. $\log_5 2$

59. $\log_7 5$

60. $\frac{1}{3} \log_{11} 17$

61. $\frac{2}{5} \log_8 127$

62. $(\log_3 7)\left(\log_3 \frac{3}{8}\right)$

63. $\frac{2 \log_6 3}{3 \log_6 2}$

64. Prove the Log of a Power Property. That is, if x and r are real numbers with x positive, then prove

$$\log_b x^r = r \log_b x$$

65. If $f(x) = \log_b x$, use the fact that

$f(f^{-1}(x)) = x$ for x in the domain of f^{-1}.

$f^{-1}(f(x)) = x$ for x in the domain of f.

to show that $\log_b(b^x) = x$ (Log of Base to a Power Property) and that $b^{\log_b x} = x$ (Base to a Log with Same Base Property).

66. Prove that for a positive number q and all x, $q^x = e^{x \ln q}$.

67. Prove that $\log_2 x = \frac{1}{\log_x 2}$ when x is positive and not equal to 1.

Challenge Problems

68. Graph $y = \log x^2$ and $y = 2 \log x$. Are the graphs the same? Does this contradict the Log of a Power Property?

69. Graph $y = \ln ex$.

70. Graph $y = \log_x 10$.

IN YOUR OWN WORDS . . .

71. How is the change of base formula useful?

72. Explain how the Log of a Product, Log of a Quotient, and Log of a Power Properties act like properties of exponents.

6.5 EXPONENTIAL AND LOGARITHMIC EQUATIONS

We have already solved some simple types of both exponential and logarithmic equations. In this section we examine more general kinds of such equations. Often, logarithmic equations can be solved by converting to exponential form with the fundamental equivalence.

Solving Logarithmic Equations

EXAMPLE 1 Solve each equation.

(a) $\log_3 x^2 = 4$ (b) $(\log_3 x)^2 = 4$ (c) $\log_5 |x| = 2$

Solutions (a) $\log_3 x^2 = 4 \Longleftrightarrow 3^4 = x^2$ Fundamental equivalence

$81 = x^2$

$\pm 9 = x$ Square Root Property

$\{\pm 9\}$ Solution set

(b) $(\log_3 x)^2 = 4$

Notice the important difference between this equation and the one in part (a).

$\log_3 x = \pm 2$ Square Root Property

We now have two equations to solve

$\log_3 x = 2 \Longleftrightarrow 3^2 = x$ and $\log_3 x = -2 \Longleftrightarrow 3^{-2} = x$

$9 = x$ $\qquad\qquad \dfrac{1}{9} = x$

$\left\{\dfrac{1}{9}, 9\right\}$

(c) $\log_5 |x| = 2 \Longleftrightarrow 5^2 = |x|$ Fundamental equivalence

$25 = |x|$

$x = 25$ or $x = -25$

$\{\pm 25\}$

□

Since both exponential functions and logarithmic functions are one-to-one, they each have the powerful property $f(p) = f(q)$ if and only if $p = q$.

One-to-One Property of Exponential and Logarithmic Functions

$$b^x = b^y \Longleftrightarrow x = y$$

for *all* real numbers x, y

and

$$\log_b x = \log_b y \Longleftrightarrow x = y$$

for all positive numbers x, y and $b > 0$ and $b \neq 1$.

These properties prove to be very useful in solving exponential and logarithmic equations. If $x = y$ and we use the One-to-One Property of logarithmic functions, we mean "take the log of both sides" as we write $\log_b x = \log_b y$.

Consider the logarithmic equation $\log(x + 1) + \log x = \log 2$.

$$\log(x + 1) + \log x = \log 2$$

$$\log(x + 1)x = \log 2 \qquad \text{Log of a product}$$

$$(x + 1)x = 2 \qquad \text{One-to-One Property of logarithms}$$

$$x^2 + x - 2 = 0 \qquad \text{Standard form}$$

$$(x - 1)(x + 2) = 0 \qquad \text{Factor.}$$

$$x = 1 \quad \text{or} \quad x = -2 \qquad \text{Property of Zero Products}$$

It is important to note that in logarithmic equations such as this, we *must* check our answers. Notice that in this case -2 will not check because -2 *is not* in the domain of either $\log x$ or $\log(x + 1)$; thus it *cannot* be a solution of the given equation.

$$\{1\} \qquad \text{Solution set}$$

□

EXAMPLE 2 Solve each equation.

(a) $\ln(x + 3) - \ln(x - 1) = \ln x$ (b) $\log(t - 1) - 1 = \log(t + 2)$

Solutions (a) $\ln(x + 3) - \ln(x - 1) = \ln x$

$$\ln \frac{x + 3}{x - 1} = \ln x \qquad \text{Log of a quotient}$$

However, as $\ln x$ is a one-to-one function, we can write

$$\frac{x+3}{x-1} = x$$

$$x + 3 = x^2 - x$$

$$x^2 - 2x - 3 = 0 \qquad \text{Standard form}$$

$$(x-3)(x+1) = 0 \qquad \text{Factor.}$$

$$x = 3 \quad \text{or} \quad x = -1$$

Since -1 does not check (not in the domain of $\ln x$), it is not in the solution set.

$$\{3\} \qquad \text{Solution set}$$

(b) $\log(t-1) - 1 = \log(t+2)$

Notice that the *base* of the logarithms in this problem is 10. Therefore, we can write 1 as log 10.

$$\log(t-1) - \log 10 = \log(t+2) \qquad \text{Log of the base}$$

$$\log\frac{t-1}{10} = \log(t+2) \qquad \text{Log of a quotient}$$

$$\frac{t-1}{10} = t + 2 \qquad \text{One-to-One Property of logarithmic function}$$

$$t - 1 = 10t + 20$$

$$-21 = 9t$$

$$t = -\frac{7}{3}$$

However, $-\frac{7}{3}$ is not in the domain of $\log(t-1)$ so it does not check.

$$\varnothing \qquad \text{Solution set}$$

□

Solving General Exponential Equations

Equations like $3^x = 11$ represent a common form of exponential equation. If the right side were a known power of 3, we could solve this equation by equating exponents. Even though 11 is not a known power of 3, we can still approximate the solution by taking the logarithm of both sides. We choose logarithm to the base 10 so we can use a calculator. The natural logarithm would work equally well.

$$3^x = 11$$

$$\log 3^x = \log 11 \qquad \text{One-to-One Property of logarithmic function}$$

$$x \log 3 = \log 11 \qquad \text{Log of a power}$$

$$x = \frac{\log 11}{\log 3} \qquad \text{Divide both sides by log 3.}$$

The keystrokes [11] [log] [÷] [3] [log] [=] should give [2.182658339] on a scientific calculator.

$$x \approx 2.1827$$

EXAMPLE 3 Approximate solutions to five significant digits.

(a) $10^x = 16$ (b) $23^t = 3$ (c) $e^{x^2} = 38$ (d) $3^{x+2} = 7^{2x}$

Solutions (a) $10^x = 16$

Because the 10 appears as a base, we select logarithms to the base 10.

$\log 10^x = \log 16$	Log of both sides (One-to-One Property)
$x \log 10 = \log 16$	Log of a power
$x \cdot 1 = \log 16$	Log of the base
$x \approx 1.2041$	

(b) $23^t = 3$

$\log 23^t = \log 3$	Log of both sides
$t \log 23 = \log 3$	Log of a power
$t = \dfrac{\log 3}{\log 23}$	Divide both sides by log 23.
$t \approx 0.35038$	

(c) $e^{x^2} = 38$

As the base here is e, we choose natural logs.

$\ln e^{x^2} = \ln 38$	Natural log of both sides
$x^2 \ln e = \ln 38$	Log of a power
$x^2 \cdot 1 = \ln 38$	Log of the base
$x^2 = \ln 38$	
$x = \pm\sqrt{\ln 38}$	Square Root Property
$x \approx \pm 1.9072$	

(d) $3^{x+2} = 7^{2x}$

$\ln 3^{x+2} = \ln 7^{2x}$	Natural log of both sides
$(x + 2)\ln 3 = 2x \ln 7$	Log of a power
$x \ln 3 + 2 \ln 3 = 2x \ln 7$	Distributive Property
$2 \ln 3 = 2x \ln 7 - x \ln 3$	
$2 \ln 3 = (2 \ln 7 - \ln 3)x$	
$\dfrac{2 \ln 3}{2 \ln 7 - \ln 3} = x$	

We should be very careful to hold the denominator together with parentheses when keying this into a calculator.

$$x \approx 0.78663$$

□

GRAPHING CALCULATOR BOX

Approximating Solutions of Equations

The graphing calculator is a powerful tool for finding approximate solutions of difficult equations. Suppose x_1 is a solution of the equation $f(x) = g(x)$. Then the numbers $f(x_1)$ and $g(x_1)$ are equal. Now consider the graphs of f and g. The point $(x_1, f(x_1))$ is on the graph of f, and the point $(x_1, g(x_1))$ is on the graph of g. However, since $f(x_1)$ and $g(x_1)$ are the same number, these two points are the same! Thus, solutions of the equation $f(x) = g(x)$ occur at points where the graphs of f and g intersect.

Let's approximate the solutions to the equation $e^x = 3 - 2x$ to four significant digits. First, we graph $y = e^x$ and $y = 3 - 2x$ on the *same screen*. Start with the default window. (To enter e^x on either calculator, press [e^x] then [X|T] or [X,θ,T]. Do *not* use [^] or [x^y].)

TI-81

Use Y_1 and Y_2 for e^x and $3 - 2x$ making sure both equal signs are darkened; then graph.

CASIO fx-7700G

Use [F MEM] to store the two functions and use the multistatement connector [↵] to graph them on the same screen.

Now use box and trace to find the intersections to the desired degree of accuracy.

Answer: One solution; approximately 0.5942.

Sometimes it is convenient to write the equation $f(x) = g(x)$ in the form $f(x) - g(x) = 0$, then find where the graph of $y = f(x) - g(x)$ touches the x-axis. Let's use this technique to solve the equation $e^{-x/10} = \log x$. First, we write the equation in the form $e^{-x/10} - \log x = 0$, then we graph $y = e^{-x/10} - \log x$. (Enter $e^{-x/10}$ by pressing [e^x] [(] [(−)] [X|T] [÷] [10] [)] on either calculator.) The graph crosses the x-axis around 4. Now, we use box and trace to find the point of intersection to four significant digits.

Answer: One solution; approximately 4.404.

PROBLEM SET 6.5

Warm-ups

In Problems 1 through 6, solve each equation. See Example 1.

1. $\log_3 y^2 = 4$

2. $(\log_3 y)^2 = 4$

3. $\log_3 |2z| = 3$

4. $\log x^2 = 2$

5. $(\log x)^2 = 9$

6. $\ln |2x - 3| = 1$

In Problems 7 through 10, solve each equation. See Example 2.

7. $\ln x + \ln(x - 1) = \ln 6$
8. $\log(t + 8) - \log t = \log(t - 1)$
9. $\log y + \log(2y + 1) = 1$
10. $\ln(x + 1) - \ln(2x - 1) = \ln(x - 1)$

In Problems 11 through 16, approximate the solutions to five decimal places. See Example 3.

11. $10^x = 7$
12. $6^x = 5$
13. $e^{t-1} = 6$
14. $51^y = 9.116388$
15. $\sqrt{37} = 4^{x/3}$
16. $5^{2x-3} = 8^{x+1}$

Practice Exercises

In Problems 17 through 60, solve each equation. Give all approximations to four significant digits.

17. $\log_2 z^2 = 6$
18. $\log_5 x^3 = 6$
19. $\log_5 2 + \log_5 x = 2$
20. $\log_4 5 - 3 = \log_4 t$
21. $4^y = 8^{y+3}$
22. $27^x = 9^{x-7}$
23. $11^x = 24$
24. $8^x = 41$
25. $e^{2x} = \sqrt{3}$
26. $e^{x/2} = 10$
27. $\ln(2x - 7) = \ln x$
28. $\log(3s - 8) = \log s$
29. $\ln(2x - 7) = \ln x + \ln 3$
30. $\log(3s - 8) = \log s + \log 5$
31. $2 \log t = \log(2 - t)$
32. $2 \ln x = \ln(6 + x)$
33. $e^{2\ln x} = 4$
34. $e^{-\ln x} = 2$
35. $\log 10^{x^2} = 1$
36. $\log_2 2^{x-1} = 3$
37. $(\sqrt{13})^z = 15^{z+1}$
38. $7^{x+3} = (\sqrt{10})^x$
39. $\log_6 |2x - 1| = 2$
40. $\log_5 |2 - 5x| = 1$
41. $\log(s - 3) = 1 - \log s$
42. $\log t + \log(t - 9) = 1$
43. $\left(\frac{2}{3}\right)^t = 500$
44. $\left(\frac{3}{4}\right)^t = 0.34$
45. $\ln 2 + 2 \ln x = \ln(3x + 5)$
46. $2 \ln x + \ln 3 = \ln(4 - x)$
47. $\log(2r - 3) = 2 \log r - \log(r - 2)$
48. $\log w - \log 2 = \log(w + 1)$
49. $10e^{-2t} = 40$
50. $30 = 10e^{-3t}$
51. $e^{-x} = 6^x$
52. $\frac{e^x}{2} = e^{2x-1}$
53. $\ln(z + 1) = 2 \ln(z - 1)$
54. $2 \ln v = \ln(2 - v) + \ln(4 - v)$
55. $\frac{2^x - 1}{2^x + 1} = 2^x$
56. $\frac{e^x + 1}{e^x - 1} = e^x$
57. $\ln(\ln x) = 0$
58. $\log(\log x) = 0$
59. $\log x^2 = (\log x)^2$
60. $\ln x^2 = (\ln x)^2$

Challenge Problems

In Problems 61 through 63, solve each equation.

61. $x^x = x$
62. $(\ln x)^x = 1$
63. $(\ln x)^{-1} = \ln(x^{-1})$

In the challenge problems of Section 6.1, we defined the hyperbolic sine and hyperbolic cosine.

$$\sinh(x) = \frac{e^x - e^{-x}}{2} \quad \textit{and} \quad \cosh(x) = \frac{e^x + e^{-x}}{2}$$

In Problems 64 through 66, solve each equation.

64. $\cosh(x) = 1$ (*Hint:* Multiply both sides by $2e^x$ and look for a quadratic equation in the variable e^x.)

65. $\sinh(x) = 2$
66. $2 \sinh(x) = \cosh(x)$

For Graphing Calculators

In Problems 67 through 74, approximate all solutions of each equation to four significant digits.

67. $3^x = 4 - x^2$

68. $2^{-x} = x^3$

69. $e^x = x + 5$

70. $10^x = -4 - x$

71. $\log(x - 1) = x - 5$

72. $\ln(2x + 1) = 2x - 1$

73. $\log_3 x = e^{-x}$ (*Hint:* Use change of base formula.)

74. $\log_{15} x^2 - x^2 = 0$

IN YOUR OWN WORDS . . .

75. Why is it necessary to learn about logarithmic and exponential equations?

76. When is taking the log of both sides of an equation useful?

6.6 APPLICATIONS OF LOGARITHMIC FUNCTIONS

Exponential and logarithmic equations occur in a wide variety of mathematical models in business, engineering, and the sciences. Such diverse things as compound interest, radioactive decay, and population growth lead to exponential equations, while the intensity of earthquakes, loudness of sound, and depreciation of equipment lead to logarithmic equations.

Compound Interest

In Section 6.2 we learned that if P dollars are invested at an annual interest rate of r compounded n times a year, then the amount of dollars in the account after t years, $A(t)$, is given by the compound interest formula

$$A(t) = P\left(1 + \frac{r}{n}\right)^{nt}$$

We solved problems using this formula to find the amount of money accumulated. In this section we can find t by taking the log of both sides of the equation.

EXAMPLE 1 John Deduck deposits the proceeds of the sale of a small plot of land in a savings account that pays 7.5% annual interest compounded monthly. How long will it take for this investment to double?

Solution We wish to find t so that $A(t) = 2P$. We are given that $r = 0.075$ and $n = 12$.

$$A(t) = P\left(1 + \frac{r}{n}\right)^{nt} \qquad \text{Compound interest formula}$$

$$2P = P\left(1 + \frac{0.075}{12}\right)^{12t} \qquad \text{Substitution}$$

$$2 = \left(1 + \frac{0.075}{12}\right)^{12t} \qquad \text{Divide both sides by } P.$$

$$\log 2 = \log\left(1 + \frac{0.075}{12}\right)^{12t} \qquad \text{Log of both sides}$$

$$\log 2 = 12t \log\left(1 + \frac{0.075}{12}\right) \qquad \text{Log of a power}$$

$$\frac{\log 2}{12 \log\left(1 + \frac{0.075}{12}\right)} = t$$

$$t \approx 9.2708$$

It will take John about 9.27 years to double this money. □

Radioactive Decay

Radioactive materials decay into another form in an exponential manner. A quantity A_0 of a radioactive substance will decay to $A(t)$ of the substance in t years by the relationship

$$A(t) = A_0 e^{kt}$$

where k is a constant that depends on the radioactive material; k is called the decay constant.

EXAMPLE 2 Suppose a new radioactive isotope of iridium is discovered on Mars. A small sample is isolated, and after 1 month only half of it remains. What is the value to four significant digits of the decay constant of this new isotope?

Solution In the formula

$$A(t) = A_0 e^{kt}$$

we are given that $A(t)$ is $\frac{1}{2}A_0$ when t is $\frac{1}{12}$ years.

$$\frac{1}{2}A_0 = A_0 e^{k(1/12)} \qquad \text{Substitution}$$

$$\frac{1}{2} = e^{k/12} \qquad \text{Divide both sides by } A_0.$$

$$\ln \frac{1}{2} = \ln e^{k/12} \qquad \text{Natural log of both sides}$$

$$-\ln 2 = \frac{k}{12} \ln e \qquad \text{Log of a reciprocal and log of a power}$$

$$-\ln 2 = \frac{k}{12} \qquad \text{Log of the base}$$

$$k = -12 \ln 2$$

$$k \approx -8.317766$$

The decay constant for this newly discovered isotope is approximately -8.318. □

Population Growth

Population growth can be modeled by the function

$$P(t) = P_0 e^{kt}$$

where P_0 is the initial population, $P(t)$ is the population after t years, and k is a constant.

EXAMPLE 3 Suppose the population of Jacksonville, Florida, is predicted by $P(t) = P_0 e^{kt}$. If the population in the 1980 census was 541,000 and it increased to 704,000 in the 1990 census, what can we expect it to be in the 2010 census?

Solution Our strategy in a problem like this is to use the two given population figures to approximate the growth constant and then approximate the 2010 population.

$$P(t) = P_0 e^{kt}$$

$$P(10) = P_0 e^{k(10)} \qquad \text{After 10 years}$$

$$704000 = 541000 e^{10k} \qquad \text{Substitution}$$

$$\frac{704}{541} = e^{10k}$$

$$\ln \frac{704}{541} = \ln e^{10k} \qquad \text{Natural logs}$$

$$\ln \frac{704}{541} = 10k \ln e \qquad \text{Log of a power}$$

$$\ln \frac{704}{541} = 10k \qquad \text{Log of the base}$$

$$k = \frac{1}{10} \ln \frac{704}{541}$$

$$k \approx 0.0263359077$$

At this point the approximation for k should be on the calculator display. *Leave it there or store it!* Now that we have k, we can calculate the 2010 population.

$$P(t) = P_0 e^{kt} \qquad \text{Formula for population growth}$$

$$P(30) = 541000 e^{k(30)} \qquad \text{Substitution (2010 is 30 years after 1980.)}$$

The value for k should still be on the calculator display. The following keystrokes for a scientific calculator complete the calculation.

[×] [30] [=] [e^x] [×] [541000] [=]

$$P(30) \approx 1192129$$

The population in the 2010 census should be approximately 1,192,000. □

Richter Scale

The Richter scale is a well-known measure of the destructive power of an earthquake. It was developed in 1935 by the American geologist Charles Richter. The Richter number is given by the following formula.

$$R = \log \frac{I}{I_0}$$

where I is the intensity of the earthquake and I_0 is the minimum intensity that can be felt. I_0 is usually taken to be 1. Thus the Richter number simplifies to $R = \log I$. An earthquake with a Richter number of 6 will cause considerable damage to ordinary buildings. A Richter 7 earthquake is a major earthquake; about 10 occur each year. Richter 8 measures a great earthquake. These occur about every 5 to 10 years.

EXAMPLE 4 At 5 o'clock on the evening of October 17, 1989, a powerful earthquake with epicenter at Loma Prieta shook the San Francisco Bay area, devastating the city and killing 60 people. It measured 7.1 on the Richter scale. How much more powerful was the famous San Francisco earthquake of 1906 which measured 8.3 on the Richter scale?

Solution First, we find the intensity of each earthquake.

$$R = \log I \Longleftrightarrow I = 10^R \qquad \text{Fundamental equivalence}$$

For the 1906 earthquake,

$$I = 10^{8.3}$$

$$I \approx 199526231$$

For the 1989 earthquake,

$$I = 10^{7.1}$$

$$I \approx 12589254$$

Therefore, the 1906 earthquake was $\dfrac{199526231}{12589254}$ or about 16 times more powerful than the 1989 quake. □

Depreciation

The value of equipment generally declines as time passes. Of course, such things as antiques and old restored cars are notable exceptions. However, in business, the value of a piece of equipment is usually depreciated according to an established formula for tax purposes. Often the double-declining balance method is used. If $V(t)$ is the value of the piece of equipment after t years, C is the original cost, and n is the life expectancy in years, the double-declining balance formula gives

$$V(t) = C\left(1 - \frac{2}{n}\right)^t$$

EXAMPLE 5 When new, a punch press costs \$32,000 and has a life expectancy of 25 years. According to the double-declining balance method,

(a) What is the value after 10 years?
(b) What is its age when its value is \$16,000?

Solutions (a) We apply the formula directly.

$$V(t) = C\left(1 - \frac{2}{n}\right)^t \qquad \text{Double-declining balance formula}$$

$$V(10) = 32000\left(1 - \frac{2}{25}\right)^{10} \qquad \text{Substitution}$$

$$V(10) \approx 13900.43$$

The value after 10 years is approximately \$13,900.

(b) To find the age given the value, we substitute 16000 for $V(t)$ and find t.

$$16000 = 32000\left(1 - \frac{2}{25}\right)^t$$

$$\frac{1}{2} = \left(1 - \frac{2}{25}\right)^t$$

$$\log \frac{1}{2} = \log\left(1 - \frac{2}{25}\right)^t \qquad \text{Log of both sides}$$

$$-\log 2 = t \log\left(1 - \frac{2}{25}\right) \qquad \text{Log of a reciprocal and log of a power}$$

$$\frac{-\log 2}{\log\left(1 - \frac{2}{25}\right)} = t$$

$$t \approx 8.31295$$

The press is about 8.3 years old. □

PROBLEM SET 6.6

Warm-ups

For Problems 1 and 2, see Example 1.

1. How long will it take for John Deduck's investment to triple?

2. How long will it take for John Deduck's investment to quadruple?

For Problems 3 and 4, see Example 2.

3. Strontium 90 (^{90}Sr) has a half-life of 28 years. How much of a 55-milligram sample will remain after 5 years?

4. The half-life of plutonium 230 (^{230}Pu) is approximately 24,500 years. How long will it take for 10 grains of a certain sample of ^{230}Pu to decay to 9 grains?

For Problems 5 and 6, see Example 3.

5. A colony of bacteria grows according to the relationship $P(t) = P_0e^{kt}$. If the colony has 2500 members at noon and 2600 members at 4 P.M., what will be the expected population at midnight? (t is in hours.)

6. How long will it take the bacteria in Problem 5 to double?

For Problems 7 and 8, see Example 4.

7. Compare the Mexican earthquake of 1985 which measured 7.8 on the Richter scale with the San Francisco earthquake of 1989 which measured 7.1. *See Example 4.*

8. Compare the earthquake of 1985 in Indonesia which measured 6.8 on the Richter scale with the San Francisco earthquake of 1989 which measured 7.1.

For Problems 9 and 10, see Example 5.

9. Use the double-declining balance formula to determine when a piece of equipment that costs \$100,000 and has an expected life of 30 years will be worth \$60,000.

10. According to the double-declining balance method, how long will it take a piece of equipment with an expected lifetime of 40 years to lose half its value?

Practice Exercises

11. Jim invests \$22,000 of his retirement savings in a CD that pays 6.85% annual interest compounded monthly. When will the CD be worth \$35,000?

12. A money market account pays 6.9% compounded daily. How long will it take for \$12,000 to grow to \$15,000 in this account?

13. In the study of audio systems, the **gain** β (in decibels) is defined by

$$\beta = 10 \log \frac{P_o}{P_i}$$

where P_i is the power input to the system and P_o is the power output. Suppose a stereo amplifier sends 100 watts of power to the speakers from an input of 2.5×10^{-3} watt. What is its gain?

14. What is the gain of an amplifier that accepts a signal of 7.22×10^{-4} watt and sends out a signal of 78 watts?

15. The half-life of radium is approximately 1690 years. What percent of the amount present today will be present in 50 years? In 500 years?

16. Gloria Levert is considering buying a double-socket linear trip-felder for her business. The machine costs \$31,560 and should last 20 years. She is worried about how fast it will depreciate. Use the double-declining balance method to figure when the machine will be worth \$10,000.

17. Newton's Law of Cooling states that the temperature of an object will change at a rate proportional to the difference in the temperature of the object and the temperature of the surrounding medium. That results in the formula

$$Q(t) = Q_m + (Q_o - Q_m)e^{-kt}$$

where Q_m is the temperature of the medium, Q_o is the initial temperature of the object, and k is a constant. Suppose the county coroner examines the body of a recent murder victim and finds the temperature of the body to be 80°F at 9:00 P.M. One hour later she finds the temperature to be 75°F. Assuming the victim's body temperature was 98.6°F when he was murdered and the room was maintained at a constant temperature of 68°F, approximately what was the time of death?

18. If a cup of coffee at 84°C is on a desk in a room maintained at 20°C and in 10 minutes it measures 78°C, find its temperature after 100 minutes.

19. The simple formula for population growth assumes ideal conditions always remain in effect. Several models have been proposed that are more realistic. One of the better models is the **logistics equation.**

$$Q(t) = \frac{Q_{\max}Q_0}{Q_0 + (Q_{\max} - Q_0)e^{-kQ_{\max}t}}$$

where $Q_{\max}$ is the maximum supportable population, Q_0 is the population when t is 0, and k is a constant. If 2000 individual microbes are in a culture that can support 10,000 individuals, and 1 hour later there are 2250 microbes, according to the logistics model, how many individual microbes will there be in the culture after 10 hours?

20. Suppose the population of Jacksonville, Florida, as seen in Example 3, is given by the logistics equation. What will the population be 2010? ($Q_{\max} = 2{,}000{,}000$)

21. A specific object falls from rest in a medium that offers resistance proportional to the square of the velocity. The distance traveled in feet after t seconds is given by

$$y(t) = 2\ln\frac{e^{4t} + e^{-4t}}{2}$$

How far will the object have fallen after 1 second? After 2 seconds? After 10 seconds?

22. If the object in Problem 21 is dropped from 100 feet, when will it strike the ground?

23. Express the time required for a continuously compounded investment to double as a function of the interest rate.

24. Express the time required for an investment compounded monthly to triple as a function of the interest rate.

25. At what rate compounded continuously must money be invested to double in 5 years?

26. At what rate compounded continuously must money be invested to triple in 10 years?

27. In Section 6.2 we learned that the amount of a radioactive substance with a half-life of h present at time t is predicted by the function $A(t) = A_0 \cdot 2^{-t/h}$, where A_0 is the amount present when $t = 0$. Bones were estimated to contain half of the ^{14}C found in living bones. How old are the bones? (The half-life of ^{14}C is 5700 years.)

28. If the bones in Problem 27 contained 95% of the ^{14}C found in living bones, how old would the bones be?

Challenge Problems

29. Express the doubling time for interest compounded annually as a function of the interest rate. Graph this function. Compare the doubling time for rates of 1% to 20% with rates of 20% to 50%.

IN YOUR OWN WORDS . . .

30. When is the technique of taking the logarithm of both sides of an equation used?

CHAPTER SUMMARY

GLOSSARY

Exponential function: A function of the form $f(x) = b^x$, where $b > 0$ and $b \neq 1$.

The real number e: An important naturally occurring constant. Like π, it is an irrational number. $e \approx 2.718281828$.

Natural Exponential Function: The function defined by $f(x) = e^x$.

Half-life: The time required for half of a given sample of radioactive material to decay.

Logarithmic function: The function f, given by

$f(x) = \log_b x$, is the *inverse* of the exponential function $g(x) = b^x$.

Common logarithm: A logarithm function with base 10. That is, $\log x = \log_{10} x$.

Natural logarithm: A logarithm function with base e. That is, $\ln x = \log_e x$.

EXPONENTIAL GROWTH AND DECAY

In exponential growth or decay, $A(t)$, the amount present at time t, is given by

$$A(t) = A_0 e^{kt}$$

where A_0 is the amount present when time is 0 and k is a constant.

COMPOUND INTEREST FORMULAS

If P dollars are invested at an annual interest rate of r, then $A(t)$, the amount of dollars in the account after t years, is given by one of the following.

1. If the interest is compounded n times a year,

$$A(t) = P\left(1 + \frac{r}{n}\right)^{nt}$$

2. If the interest is compounded continuously,

$$A(t) = Pe^{rt}$$

FUNDAMENTAL EQUIVALENCE BETWEEN LOGARITHMS AND EXPONENTIALS

If b and x are positive real numbers with $b \neq 1$, then

$$y = \log_b x \Longleftrightarrow b^y = x$$

PROPERTIES OF LOGARITHMS

If x, y, and r are real numbers with x and y both positive, then

1. $\log_b(xy) = \log_b x + \log_b y$	Log of a product
2. $\log_b \frac{x}{y} = \log_b x - \log_b y$	Log of a quotient
3. $\log_b x^r = r \log_b x$	Log of a power
4. $\log_b \frac{1}{x} = -\log_b x$	Log of a reciprocal
5. $\log_b 1 = 0$	Log of 1
6. $\log_b b = 1$	Log of the base
7. $\log_b b^x = x$	Log of the base to a power
8. $b^{\log_b x} = x$	Base to a log with the same base

CHANGE OF BASE FORMULA

If b and c are both positive numbers not equal to 1, then

$$\log_b x = \frac{\log_c x}{\log_c b}; \text{ for all } x > 0$$

ONE-TO-ONE PROPERTY OF EXPONENTIAL AND LOGARITHMIC FUNCTIONS

For *all* real numbers x, y, and $b > 0$; $b \neq 1$

$$b^x = b^y \iff x = y$$

For all *positive* numbers x, y, and $b > 0$; $b \neq 1$

$$\log_b x = \log_b y \iff x = y$$

REVIEW PROBLEMS

In Problems 1 through 10, evaluate each expression. Write approximations to five significant digits.

1. $5^{0.9}$
2. $e^{3/8}$
3. $e^{\sqrt{5}}$
4. $(\sqrt[3]{31})^{\sqrt{5}}$
5. $\log 28$
6. $\log 0.055$
7. $\ln 0.86$
8. $\ln 19^3$
9. $\log_5 7$
10. $\frac{3}{5}\log_7 122$

In Problems 11 through 18, solve each equation or inequality.

11. $3^x = 81$
12. $12^{-x} = 1728$
13. $5^{2x+1} = 625$
14. $9^x = 27$
15. $3^x < 243$
16. $6^{x-3} \geq 216^{x+1}$
17. $625^{1-2x} = 5^{3x}$
18. $32^{-2x} = 128^{1-3x}$

In Problems 19 through 26, write each expression as a single logarithm with a coefficient of 1.

19. $\log_b 23 - 3\log_b 4$
20. $2\log_6 10 + 5\log_6 2$
21. $\log x - 2\log 6 + 5\log y$
22. $\frac{1}{3}\ln 27 + \frac{2}{3}\ln 8$
23. $4\ln t - 3(\ln u - 2\ln v)$
24. $\log(2x - 5) - 2\log(x + 3)$
25. $3\ln x - \ln(x + 2) + \ln(x + 1)$
26. $2\ln x - 3\ln y + \frac{1}{2}\ln z^3 - \ln 2w$

In Problems 27 through 46, solve each equation. Write approximations to five significant digits.

27. $\log_3 x = 4$
28. $\log_y 125 = -3$
29. $\log_3 z^2 = 6$
30. $\log_4 2 + \log_4 x = 2$
31. $3^y = 27^{y+3}$
32. $16^x = 64^{x-2}$
33. $\log z = -8$
34. $\ln w = -3$
35. $\log_6(3 - x) = 2$
36. $\log(k^2 - 69) = 2$
37. $\ln(3x - 4) = \ln x$
38. $\ln(3x - 5) = \ln x + \ln 4$
39. $2\log t = \log(42 - t)$
40. $\log|2x - 3| = 2$
41. $(\sqrt{17})^z = 12^{z+1}$
42. $8^x = 24$
43. $\log(s - 3) = 1 - \log s$
44. $\ln 2 + 2\ln x = \ln(x + 6)$
45. $e^{-2x} = 5^x$
46. $\ln(2z + 1) = 2\ln(z - 1)$

In Problems 47 through 56, sketch the graph of each function and state its domain and range.

47. $f(x) = 5^x$
48. $h(x) = 5^{-x}$
49. $u(x) = 5^x + 1$
50. $F(x) = 5^{x-1}$
51. $H(x) = 5^{x+1} - 2$
52. $U(x) = -5^x$
53. $F(x) = \log_5 x$
54. $f(x) = \log_5 x - 1$
55. $h(x) = \log_5(x - 2)$
56. $u(x) = \log_5(x + 1) + 2$

In Problems 57 through 60, solve each inequality graphically.

57. $\ln(x - 2) < 1$
58. $\log_2 x \geq -1$
59. $\ln x \geq 1$
60. $\log(x + 1) < 0$

61. The population of a city can be predicted by the model

$$P(t) = P_0 e^{kt}$$

where P_0 is the initial population and $P(t)$ is the population at time t. If the population increased from 10,000 to 20,000 in 25 years, how long will it take the population to double again?

62. Find the half-life of a radioactive isotope if 5% of the isotope decomposes in 100 years.

63. The number of bacteria in a growing culture is given by the function

$$N(t) = 100 \cdot 3^t$$

where t is the time in hours and $N(t)$ is the number of bacteria in the culture. How many bacteria were present initially? How many bacteria are present after 2 hours? After 20 minutes?

64. If \$5000 is invested at a rate of 12.25% compounded monthly, how much will be in the account after 10 years?

65. If a money market account pays 6.9% compounded daily, how long will it take for \$12,000 to grow to \$15,000?

66. Which is best: 9% simple interest, 8.75% annual interest compounded semiannually, or 8.5% annual interest compounded continuously?

67. Atmospheric pressure in pounds per square inch x miles above sea level is given by the function

$$P(x) = 14.7e^{-0.21x}$$

At what height above sea level is the atmospheric pressure one-fourth that at sea level?

68. The absolute ceiling for certain aircraft is 15,000 meters. What is the atmospheric pressure at this altitude? (1 meter ≈ 3.28 feet.)

69. Paying off a debt in a given length of time by equal periodic payments which include interest is called *amortizing the debt*. The following amortization formula shows the relationship among the amount of the loan (A), the payment (P), the number of payments or periods (n), and the interest rate per period (r).

$$P = A \cdot \frac{r}{1 - (1 + r)^{-n}}$$

Warren bought a television set for \$1220 and agreed to pay for it in 18 equal monthly installments at $1\frac{1}{2}$% interest per month. How much are Warren's monthly payments? How much interest did Warren pay?

70. An *annuity* is a series of equal periodic deposits. If the deposits are made at the end of each time period, the annuity is called *ordinary*. The *future value* of an annuity is the sum of all payments plus all interest earned. The following formula gives the future value (S) of an ordinary annuity consisting of n equal deposits (or periods) of R dollars each with interest rate of i per period.

$$S = R \cdot \frac{(1 + i)^n - 1}{i}$$

JoAnn deposited \$150 per month into an annuity for 15 years. If the interest rate was 8.75%, what is the value of the annuity? How much interest did JoAnn earn?

LET'S NOT FORGET . . .

71. Which form is better when solving the equation?
(a) $3^x = 5$ or $\log_3 5 = x$
(b) $3^x = 9$ or $\log_3 9 = x$

72. Watch the role of negative signs. If $f(x) = \ln x$, what is the domain of $f(-x)$?

73. If $f(x) = e^x$, sketch the graph of $y = f(-x)$ and $y = -f(x)$.

74. From memory. Refer to the list at left. Match each expression on the left with an expression of the right if possible.

(1) $\ln AB$	(a) $(\ln A)(\ln B)$
(2) $\ln \frac{A}{B}$	(b) $\ln A + \ln B$
(3) $\ln A^n$	(c) $\ln A - \ln B$
(4) $\ln(A + B)$	(d) $n \cdot \ln A$
(5) $\ln(A - B)$	(e) $\frac{\ln A}{\ln B}$

75. With a calculator. Approximate $\log_3 7$ to four significant digits.

76. Approximate $e^{2.5}$ to four significant digits.

CHAPTER 7

Conics

CONNECTIONS

It is remarkable how often yesterday's theoretic mathematics becomes today's workaday science. One of the greatest of these connections occurred between Appollonius, "The Great Geometer," and Johannes Kepler, a seventeenth century German astronomer. In the third century B.C., Appollonius studied the figures formed when a cone is sliced by a plane. These purely imaginary "conic sections" lay dormant until Kepler, some 1800 years later, connected them to the motions of the planets about the sun. This important discovery led Newton to create modern celestial mechanics and form the basis of today's physics.

Conic sections occur everywhere in our daily lives. Lenses in telescopes, reflectors in flashlights, and cables on a suspension bridge are all parabolas. Comets travel in orbits that are ellipses or hyperbolas. The amazing accuracy obtained by our spacecraft when they are millions of miles from their guidance laboratory on Earth is largely because their orbits are conic sections.

In the first three sections of Chapter 7 we investigate the conic sections in a modern setting. Then in the last section we show the connection between the conic sections and the general quadratic equation in two variables.

7.1 THE PARABOLA

As early as the fourth century, the Greeks defined figures formed by the intersection of a plane with a right circular cone. Parabolas, ellipses (including circles), and hyperbolas are all formed by the intersection of a plane and a right circular cone with two nappes. Thus, they are called **conic sections** or **conics.** Figure 7.1 illustrates some possibilities for obtaining these curves.

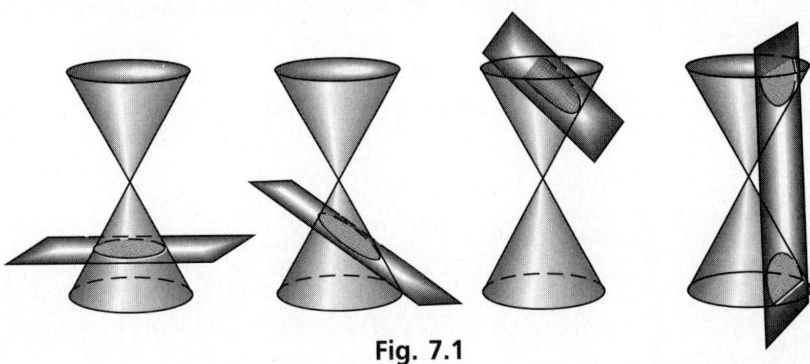

Fig. 7.1

In Sections 3.3 and 3.4 we defined the circle and the parabola as the locus of a set of points. In addition, we have seen that each of these figures is the graph of a second-degree equation. Out interest in these graphs stems from their relationship to the general second-degree equation in two variables,

$$Ax^2 + Bxy + Cy^2 + Dx + Ey + F = 0$$

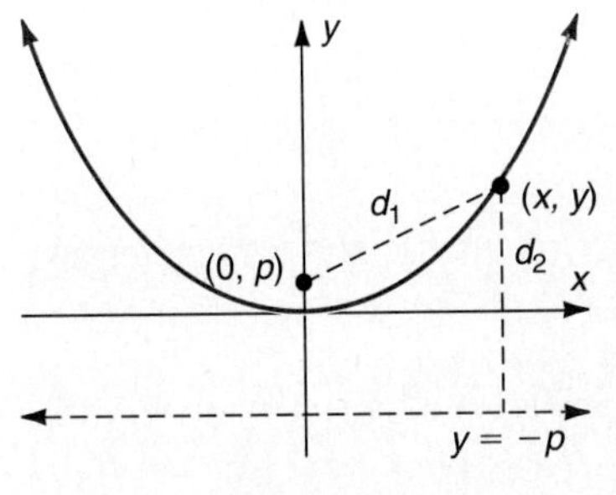

Fig. 7.2

where A, B, and C are not all zero. Except for some special cases, the graphs of these equations turn out to be parabolas, circles, ellipses, or hyperbolas. In this section we study the parabola, with emphasis on the focus and directrix. In Sections 7.2 and 7.3 we study the ellipse and the hyperbola.

A **parabola** is the set of points in a plane that are equidistant from a fixed point *(focus)* and a fixed line *(directrix)*. Suppose that p is positive, the focus is the point $(0, p)$, and the directrix is the line $y = -p$ as shown in Figure 7.2. Notice that the origin is a point on the parabola because it is equidistant from the point $(0, p)$ and the line $y = -p$.

Since each point (x, y) on the parabola is equidistant from the focus and the directrix, we have $d_1 = d_2$. The distance d_1 is the distance between (x, y) and $(0, p)$. So, using the distance formula,

$$d_1 = \sqrt{(x - 0)^2 + (y - p)^2}$$

The distance d_2 is $y + p$.

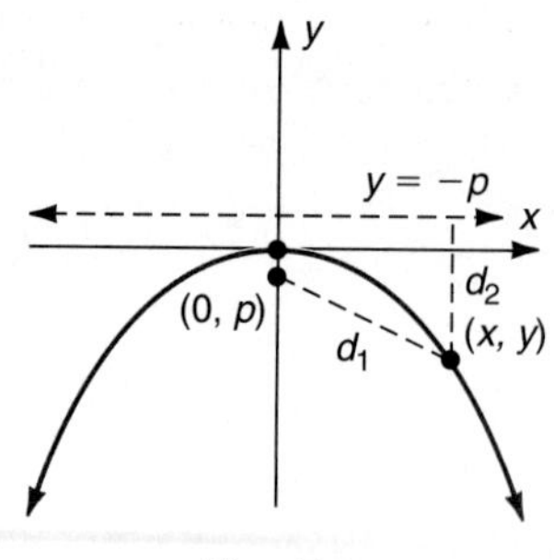

Fig. 7.3

$$d_1 = d_2$$
$$\sqrt{(x-0)^2 + (y-p)^2} = y + p$$
$$x^2 + (y-p)^2 = (y+p)^2 \qquad \text{Square both sides.}$$
$$x^2 + y^2 - 2py + p^2 = y^2 + 2py + p^2$$
$$x^2 = 4py$$

Thus, the equation of a parabola is $x^2 = 4py$.

If p is negative, the parabola opens downward as shown in Figure 7.3.

If the focus is placed at $(p, 0)$, then the equation of the resulting parabola will be $y^2 = 4px$ and the parabola will open to the left or the right. The proof of this result is left as an exercise. Notice that when p is positive, the parabola opens either upward or to the right, and when p is negative, the parabola opens either downward or to the left.

If the parabola is translated to a position where the vertex is at (h, k), then the equation will have the form $(x - h)^2 = 4p(y - k)$ for vertical parabolas and $(y - k)^2 = 4p(x - h)$ for horizontal parabolas. Figure 7.4 shows four possibilities of such translated parabolas where p is the directed distance from the vertex to the focus.

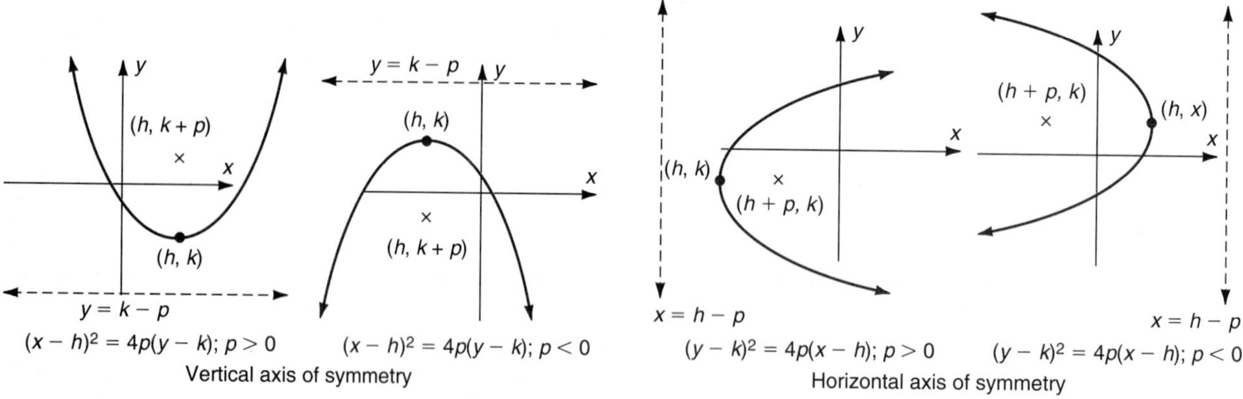

Fig. 7.4

Equations of Parabolas ($p \neq 0$)

Vertical Axis of Symmetry	***Horizontal Axis of Symmetry***
$(x - h)^2 = 4p(y - k)$	$(y - k)^2 = 4p(x - h)$
Vertex: (h, k)	Vertex: (h, k)
Focus: $(h, k + p)$	Focus: $(h + p, k)$
Directrix: $y = k - p$	Directrix: $x = h - p$

EXAMPLE 1 Find the focus and the equation of the directrix for each parabola.

(a) $y^2 = -6x$ (b) $(x - 1)^2 = 4(y + 3)$

Solutions (a) The graph of $y^2 = -6x$ is shown in Figure 7.5.

If we write the equation in the form $y^2 = 4px$, we can find the value of p. Since we are given $y^2 = -6x$, we have

$$4p = -6$$

$$p = -\frac{3}{2}$$

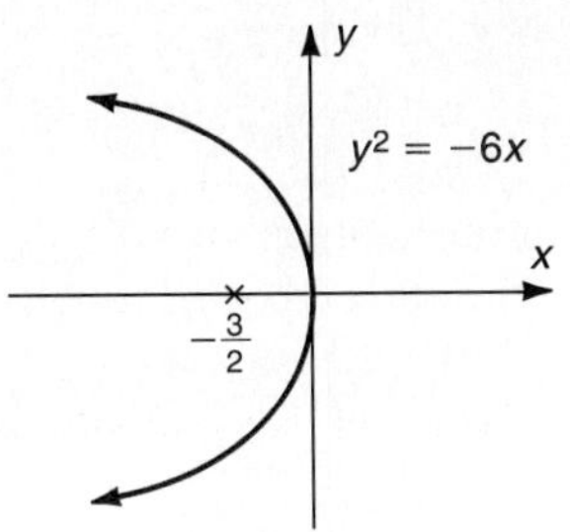

Fig. 7.5

The focus is on the x-axis since the parabola opens to the left. Since the focus is p units from the vertex, the focus is $\left(-\frac{3}{2}, 0\right)$. The equation of the directrix is $x = \frac{3}{2}$.

(b) The graph of $(x - 1)^2 = 4(y + 3)$ is shown in Figure 7.6.

The vertex has been translated to the point $(1, -3)$. Matching the form of $(x - h)^2 = 4p(y - k)$ gives $4p = 4$ or $p = 1$. Since the focus is 1 unit from the vertex, the focus is $(1, -2)$ and the equation of the directrix is $y = -4$. □

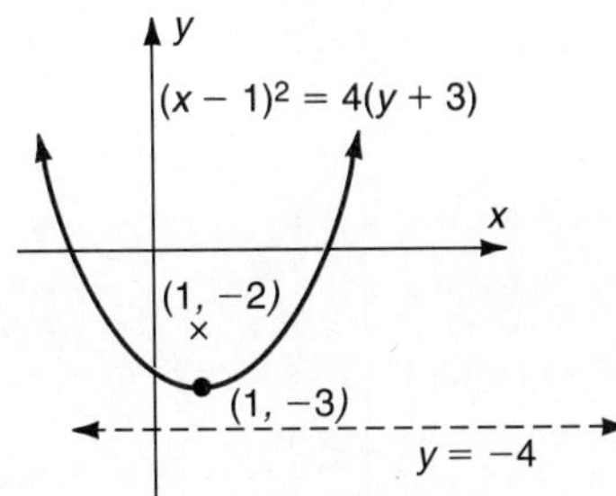

Fig. 7.6

EXAMPLE 2 Write an equation of the parabola with focus $(0, 2)$ and vertex at $(3, 2)$.

Solution If we locate the focus and the vertex, as shown in Figure 7.7, we see that the graph of such a parabola must open to the left. Thus, p is negative. The distance between the points $(0, 2)$ and $(3, 2)$ is 3. Thus, p is -3. The equation of the parabola is

$$(y - 2)^2 = 4(-3)(x - 3)$$

$$(y - 2)^2 = -12(x - 3)$$ □

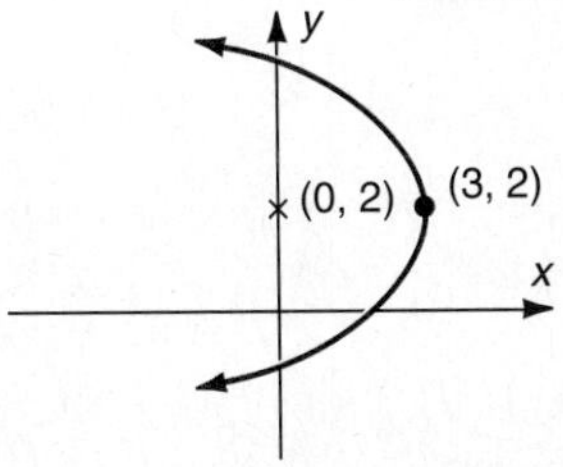

Fig. 7.7

Sometimes we must complete the square to find the focus or vertex.

EXAMPLE 3 Find the focus and an equation of the directrix for the parabola with equation $y = \frac{1}{2}x^2 + 2x + 3$.

Solution If we write the equation in the form $(x - h)^2 = 4p(y - k)$, then we know that the focus is $(h, k + p)$ and an equation of the directrix is $y = k - p$. Completing the square will allow us to write the equation in this form.

$$y - 3 = \frac{1}{2}x^2 + 2x \qquad \text{Subtract 3 from both sides.}$$

$$y - 3 = \frac{1}{2}(x^2 + 4x) \qquad \text{Factor.}$$

$$y - 3 = \frac{1}{2}(x^2 + 4x + 4 - 4) \qquad \text{Complete the square.}$$

$$y - 3 = \frac{1}{2}(x + 2)^2 - 2$$

$$y - 1 = \frac{1}{2}(x + 2)^2 \qquad \text{Add 2 to both sides.}$$

$$2(y - 1) = (x + 2)^2 \qquad \text{Multiply both sides by 2.}$$

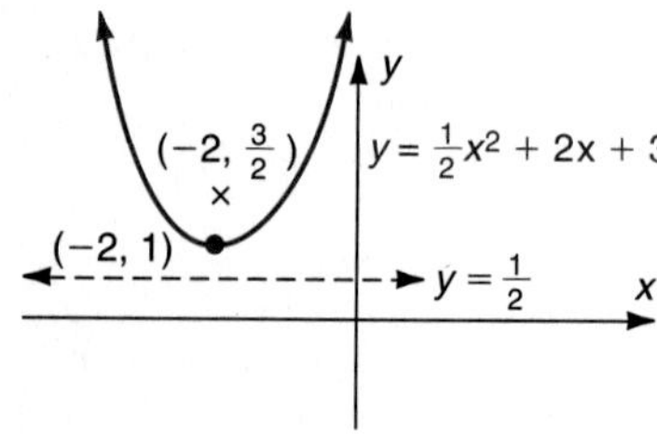

Fig. 7.8

$(x + 2)^2 = 2(y - 1)$ is the form $(x - h)^2 = 4p(y - k)$, where h is -2, k is 1, and p is $\frac{1}{2}$. Thus, the focus $\left(-2, \frac{3}{2}\right)$, and the equation of the directrix is $y = \frac{1}{2}$. The graph is shown in Figure 7.8. □

Applications of conics often involve the focus.

EXAMPLE 4 Flashlights are built so that the light bulb is placed at the focus of a parabolic reflector in the flashlight. This allows light to be reflected off the reflector parallel to the axis of the parabola. (See Figure 7.9.)

If the equation of the parabola is $y^2 = 2x$, where should the light bulb be placed so that light is reflected parallel to the axis of the parabola?

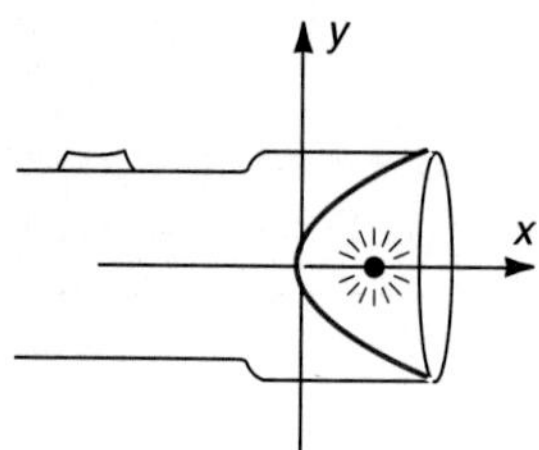

Fig. 7.9

Solution The bulb should be placed at the focus of the parabola. Since $4p = 2$, $p = \frac{1}{2}$. The focus is $\frac{1}{2}$ unit to the right of the vertex. The light bulb should be placed $\frac{1}{2}$ unit to the right of the vertex of the parabola. □

PROBLEM SET 7.1

Warm-ups

In Problems 1 through 4, find the focus and an equation of the directrix of each parabola. See Example 1.

1. $x^2 = 8y$ **2.** $(y + 1)^2 = -7(x - 2)$ **3.** $(x - 2)^2 = -6(y + 1)$ **4.** $y^2 = 5(x - 3)$

In Problems 5 through 8, find an equation of the parabola with the given conditions. See Example 2.

5. Vertex at $(1, 0)$ and focus at $(3, 0)$

6. Directrix $x = -2$ and focus at $(0, 0)$

7. Directrix $y = -5$ and focus at $(0, 5)$

8. Vertex at $(0, 0)$ and directrix $2y = 3$

In Problems 9 through 12, find the focus and an equation of the directrix for the parabola with the given equation. See Example 3.

9. $y = 2x^2 - 4x$ **10.** $y = -\frac{1}{2}x^2 + 2x - 1$ **11.** $x = -3y^2 + 3y - 2$ **12.** $x = 3y^2 + 1$

For Problems 13 and 14, see Example 4.

13. A parabolic searchlight has a bulb at the focus that is 4 inches from the vertex. If the depth of the searchlight is 9 inches, find the diameter of the searchlight.

14. A satellite dish with a parabolic cross section has a radius of 6 feet and is 4 feet deep at its center. How far is the focus from the center?

Practice Exercises

In Problems 15 through 20, find the focus and an equation of the directrix of the parabola with the given equation.

15. $x = \frac{1}{2}y^2$ **16.** $y = -3x^2$ **17.** $y = x^2 - 4x + 5$ **18.** $x = y^2 - 2y + 3$

19. $x = -2y^2 + 4y$ **20.** $y = 4x^2 - 8x$

In Problems 21 through 24, find an equation of the parabola shown.

21.

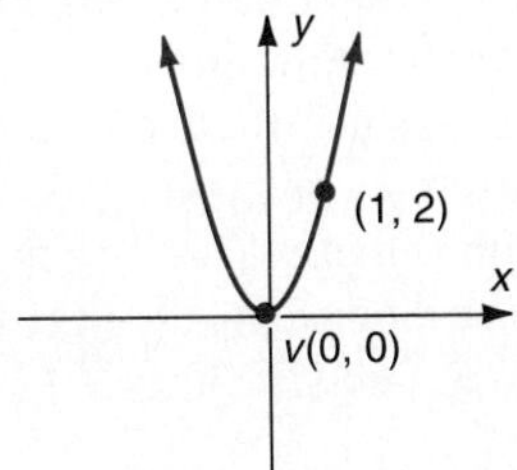

22.

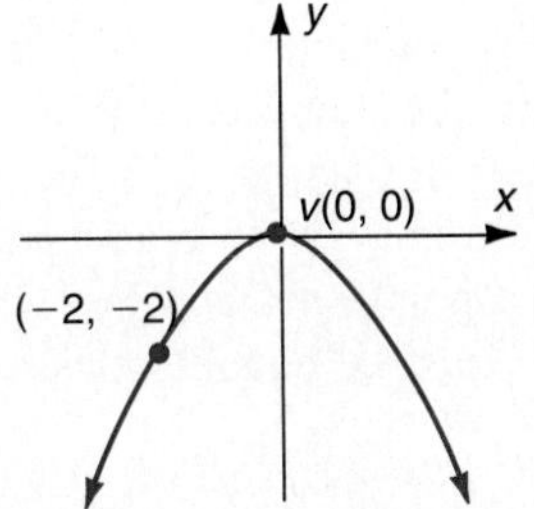

23.

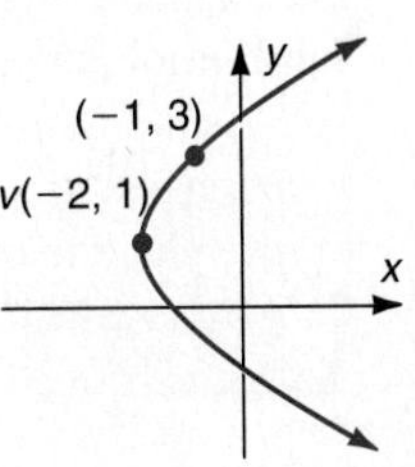

24.

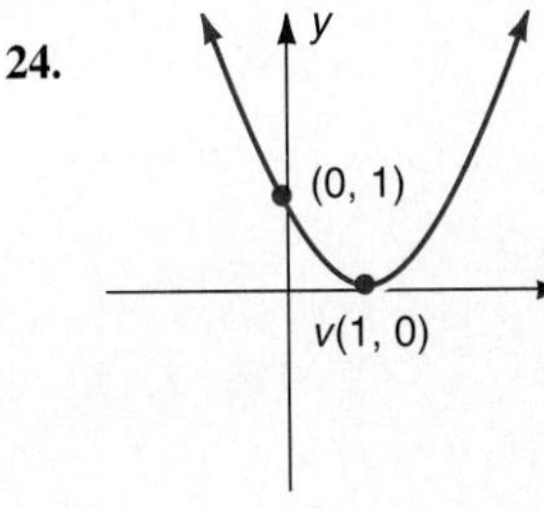

25. Find an equation of all parabolas with vertex $(2, 5)$ and containing the point $(-2, 0)$.

26. Find an equation of all parabolas with vertex $(-1, 0)$ and containing the point $(1, 3)$.

27. Find the focus and an equation of the directrix of two parabolas with x-intercept of -1 and 3.

28. Find the focus and directrix of a parabola with a y-intercept of 2 and containing the point $(2, -1)$.

29. The towers of a suspension bridge are 300 meters apart and 40 meters above the road as shown in the figure. If the

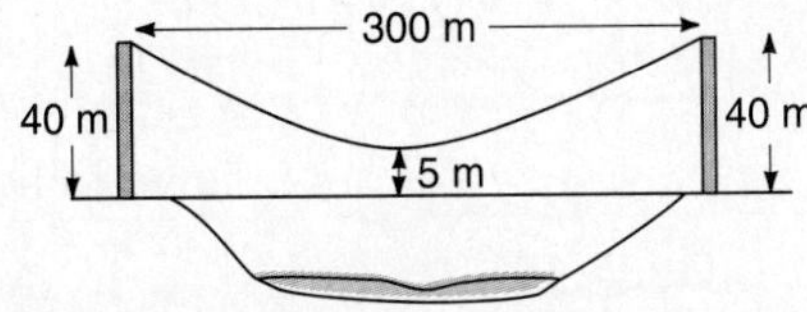

cables connecting the towers hang in the shape of a parabola with vertex 5 meters above the road, find the height of the cable above the road at a point half way between the tower and the vertex.

30. A telescope has a mirror that is parabolic and constructed so that the focus is 10 meters from the vertex. If the distance across the top of the mirror is 1.5 meters, how deep is the mirror at its center?

31. A train overpass that is parabolic in shape must be constructed over a four-lane highway 16 meters wide. Specifications indicate that the overpass must touch the ground 4 meters beyond the highway and be at least 16 meters high over the highway. What is the clearance of the overpass?

32. A parabolic arch is 15 feet high and 40 feet wide. At what height is the arch 20 feet wide?

Challenge Problems

33. The *latus rectum* of a parabola is the line segment with end points on the parabola passing through the focus perpendicular to the axis of the parabola. Prove that the length of the latus rectum is $|4p|$.

IN YOUR OWN WORDS . . .

34. Explain the relationship among the focus, directrix, and vertex of a parabola.

7.2 THE ELLIPSE

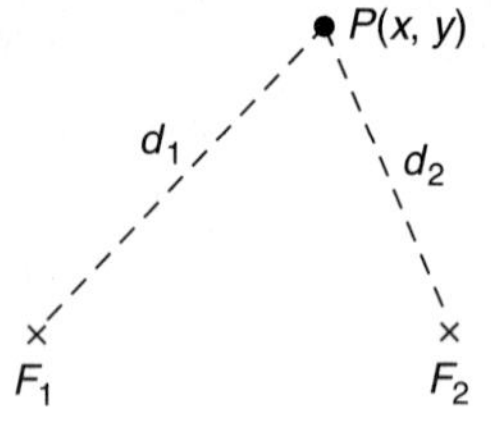

Fig. 7.10

An **ellipse** is the set of all points in a plane the sum of whose distances from two fixed points is constant as shown in Figure 7.10. The two fixed points are called **foci.**

Imagine a string attached to a pencil at point P and held fixed by thumbtacks at F_1 and F_2. The ellipse is the set of points (x,y) such that $d_1 + d_2 = k$, where k is a constant. The pencil traces an ellipse as shown in Figure 7.11.

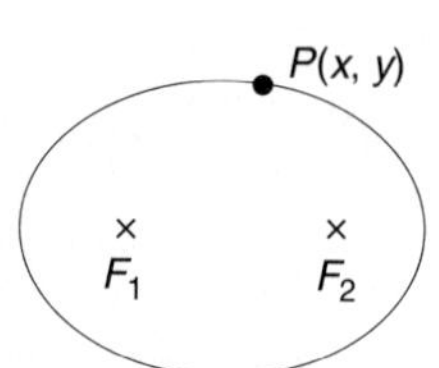

Fig. 7.11

As shown in Figure 7.12, we place a set of axes on the ellipse so that the foci F_1 and F_2 are on the x-axis and the origin is halfway between them at the **center** of the ellipse.

The points V_1 and V_2 are called **vertices.** The segment $\overline{V_1V_2}$ is called the **major axis,** and the segment $\overline{M_1M_2}$ is the **minor axis.** The segment $\overline{OV_1}$ or $\overline{OV_2}$ is called the **semimajor axis,** and the segment $\overline{OM_1}$ or $\overline{OM_2}$ is called the **semiminor axis.**

As in Figure 7.12, let the coordinates of the foci be $(c, 0)$ and $(-c, 0)$ and let the x-intercepts be $\pm a$, with $c > 0$ and $a > 0$. The distance from F_1 to a point P on the ellipse plus the distance from F_2 to the same point P on the ellipse is a constant by definition. Since V_2 is a point on the ellipse, the distance from F_1 to V_2 plus the distance from F_2 to V_2 must be equal to this constant.

$$\overline{F_1V_2} + \overline{F_2V_2} = (a + c) + (a - c) = 2a$$

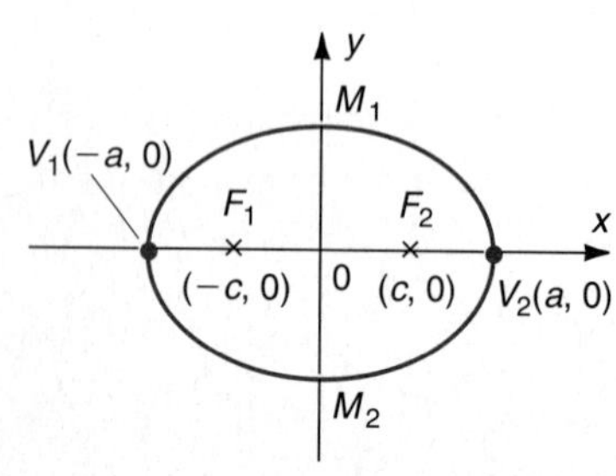

Fig. 7.12

Thus, the sum of the distances from F_1 to P and from F_2 to P must be $2a$. So, we have

$$F_1P + F_2P = 2a$$

$$\sqrt{(x + c)^2 + y^2} + \sqrt{(x - c)^2 + y^2} = 2a$$

If we isolate a radical and square both sides of this equation and simplify, we obtain

$$\sqrt{(x + c)^2 + y^2} = 2a - \sqrt{(x - c)^2 + y^2}$$

$$(x + c)^2 + y^2 = 4a^2 - 4a\sqrt{(x - c)^2 + y^2} + (x - c)^2 + y^2$$
$$x^2 + 2xc + c^2 + y^2 = 4a^2 - 4a\sqrt{(x - c)^2 + y^2} + x^2 - 2xc + c^2 + y^2$$
$$4xc - 4a^2 = -4a\sqrt{(x - c)^2 + y^2}$$
$$cx - a^2 = -a\sqrt{(x - c)^2 + y^2}$$

We square both sides again.

$$c^2x^2 - 2cxa^2 + a^4 = a^2((x^2 - 2xc + c^2) + y^2)$$
$$c^2x^2 - 2cxa^2 + a^4 = a^2x^2 - 2a^2xc + a^2c^2 + a^2y^2$$
$$a^4 - a^2c^2 = a^2x^2 - c^2x^2 + a^2y^2$$
$$a^2(a^2 - c^2) = (a^2 - c^2)x^2 + a^2y^2$$

If we divide both sides by $a^2(a^2 - c^2)$, we have

$$1 = \frac{x^2}{a^2} + \frac{y^2}{(a^2 - c^2)}$$

Notice that $a > c$. So, $a^2 - c^2 > 0$. So, we can let $b^2 = a^2 - c^2$ (which means that $a > b$). The equation becomes

$$1 = \frac{x^2}{a^2} + \frac{y^2}{b^2}$$

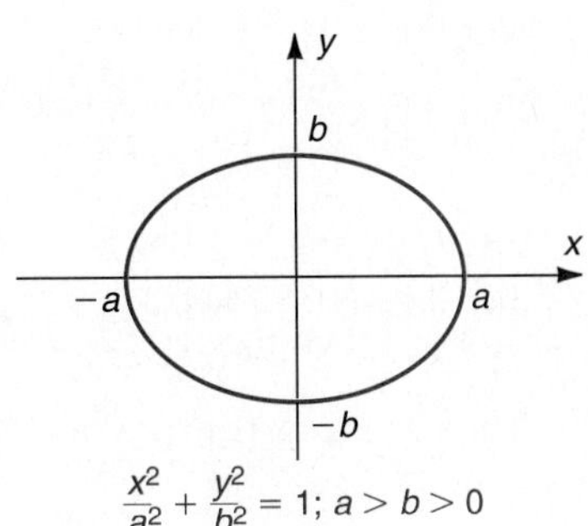

$\frac{x^2}{a^2} + \frac{y^2}{b^2} = 1;\ a > b > 0$

Fig. 7.13

We can find the y-intercepts for this ellipse. If x is 0,

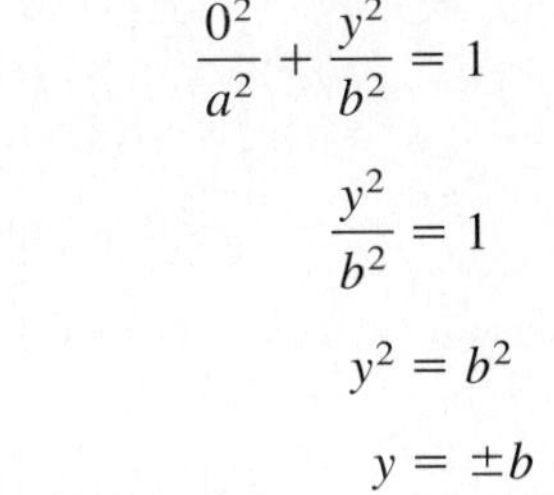

$$\frac{0^2}{a^2} + \frac{y^2}{b^2} = 1$$
$$\frac{y^2}{b^2} = 1$$
$$y^2 = b^2$$
$$y = \pm b$$

Thus, the x-intercepts are $\pm a$ and the y-intercepts are $\pm b$. The graph of $\frac{x^2}{a^2} + \frac{y^2}{b^2} = 1$, where $a > b$, is shown in Figure 7.13.

If the coordinates of the foci are $(0, -c)$ and $(0, c)$, then the equation of the ellipse is also $\frac{x^2}{a^2} + \frac{y^2}{b^2} = 1$, where $b > a$ and $b^2 = a^2 + c^2$ as shown in Figure 7.14.

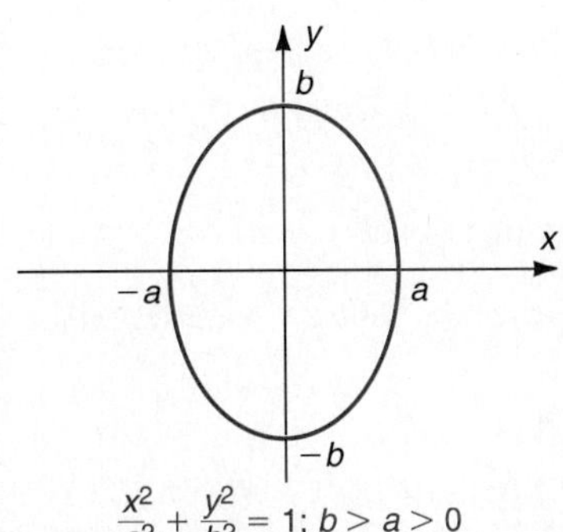

$\frac{x^2}{a^2} + \frac{y^2}{b^2} = 1;\ b > a > 0$

Fig. 7.14

The Ellipse Centered at the Origin

An ellipse centered at the origin with x-intercepts $\pm a$ and y-intercepts $\pm b$ is the graph of the equation

$$\frac{x^2}{a^2} + \frac{y^2}{b^2} = 1$$

If $a > b > 0$, the foci are on the x-axis.
If $b > a > 0$, the foci are on the y-axis.

If $a > b$, the major axis is on the x-axis. If $b > a$, the major axis is on the y-axis.

The x-intercepts $\pm a$ and the y-intercepts $\pm b$ are sufficient to sketch the graph of the equation $\frac{x^2}{a^2} + \frac{y^2}{b^2} = 1$. Plot these four points and connect them with a smooth oval shape.

EXAMPLE 1 Sketch the graph of $4x^2 + 9y^2 = 36$.

Solution If we divide both sides of the equation by 36, the equation will be of the form $\frac{x^2}{a^2} + \frac{y^2}{b^2} = 1$.

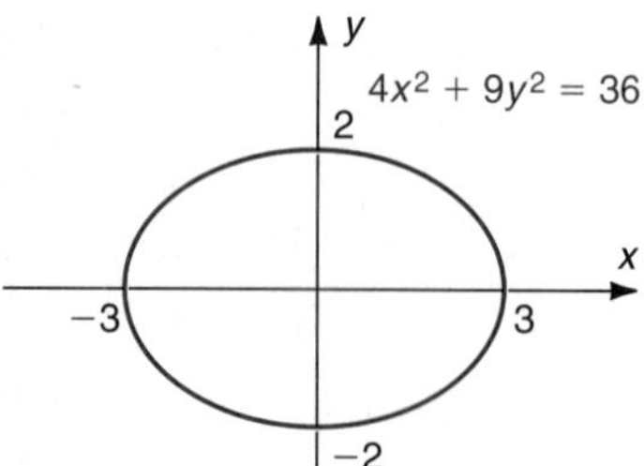

Fig. 7.15

$$4x^2 + 9y^2 = 36$$

$$\frac{x^2}{9} + \frac{y^2}{4} = 1$$

$$\frac{x^2}{3^2} + \frac{y^2}{2^2} = 1$$

So, the graph is an ellipse with $a = 3$ and $b = 2$. The x-intercepts are ± 3, and the y-intercepts are ± 2. Since $a > b$, the major axis is on the x-axis as shown in Figure 7.15. □

EXAMPLE 2 Sketch the graph of $\frac{x^2}{16} + \frac{y^2}{25} = 1$.

Solution The equation is of the form $\frac{x^2}{a^2} + \frac{y^2}{b^2} = 1$. So, the graph is an ellipse with $a = 4$ and $b = 5$. The x-intercepts are ± 4, and the y-intercepts are ± 5. Since $b > a$, the major axis is on the y-axis. See Figure 7.16.

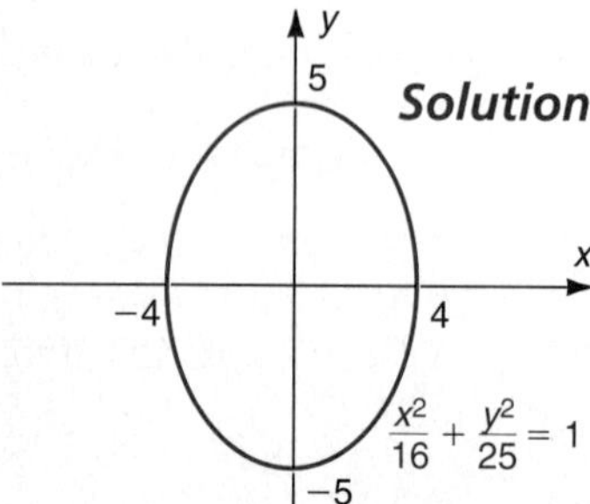

Fig. 7.16

EXAMPLE 3 Sketch the graph of $4x^2 + 9y^2 = 9$.

Solution We divide both sides of the equation by 9 so that the equation will be in the form $\frac{x^2}{a^2} + \frac{y^2}{b^2} = 1$.

$$\frac{4x^2}{9} + \frac{y^2}{1} = 1$$

To match the form $\frac{x^2}{a^2} + \frac{y^2}{b^2} = 1$, we must rewrite the equation so that the coefficient of x^2 is 1.

$$\frac{x^2}{\frac{9}{4}} + \frac{y^2}{1} = 1 \qquad \text{Divide numerator and denominator by 4.}$$

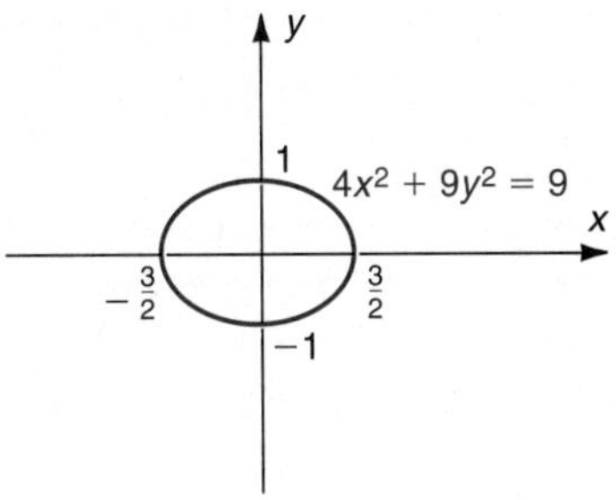

The graph is an ellipse with $a = \frac{3}{2}$ and $b = 1$ as shown in Figure 7.17. □

Fig. 7.17

As in the parabola and the circle, if x is replaced with $x - h$ and y and $y - k$, the ellipse will be translated so that it is centered at the point (h, k) instead of at the origin.

The Ellipse

The graph of the equation

$$\frac{(x-h)^2}{a^2} + \frac{(y-k)^2}{b^2} = 1$$

is the graph of $\frac{x^2}{a^2} + \frac{y^2}{b^2} = 1$ shifted so that its center is at the point (h, k).

EXAMPLE 4 Sketch the graph of $\frac{(x+2)^2}{4} + \frac{(y+1)^2}{9} = 1$.

Solution The graph will be an ellipse centered at $(-2, -1)$. An easy way to draw the graph is to copy a set of axes with its origin at the point $(-2, -1)$ and then graph the ellipse $\frac{x^2}{4} + \frac{y^2}{9} = 1$. Locate the center $(-2, -1)$. Then plot four points on the ellipse by moving left and right 2 units and up and down 3 units from the center. See Figure 7.18.

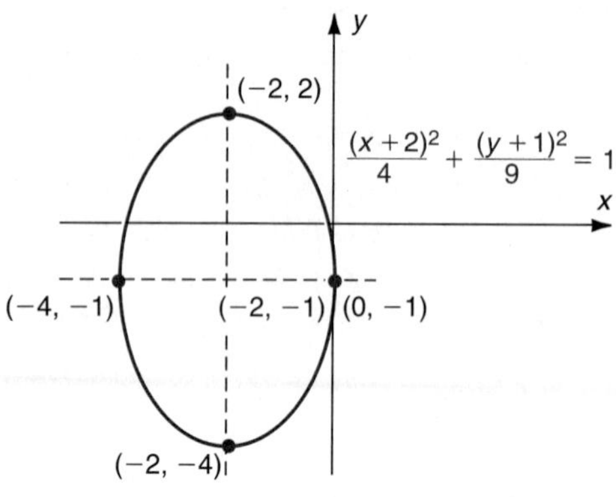

Fig. 7.18

□

EXAMPLE 5 Sketch the graph of $4x^2 + y^2 + 24x - 4y + 24 = 0$.

Solution We must complete the square in x and in y so that we can find the center and vertices of the graph.

$$4x^2 + y^2 + 24x - 4y + 24 = 0$$

Group the x-terms and the y-terms and subtract 24 from both sides.

$$(4x^2 + 24x) + (y^2 - 4y) = -24$$

The coefficient of x^2 must be 1 to complete the square. So, we factor 4 from the x-terms.

$$4(x^2 + 6x) + (y^2 - 4y) = -24$$

Complete the square in x and in y.

$$4(x^2 + 6x + 9 - 9) + (y^2 - 4y + 4 - 4) = -24$$

$$4(x^2 + 6x + 9) - 36 + (y^2 - 4y + 4) - 4 = -24$$

$$4(x + 3)^2 + (y - 2)^2 = 16$$

$$\frac{(x + 3)^2}{4} + \frac{(y - 2)^2}{16} = 1$$

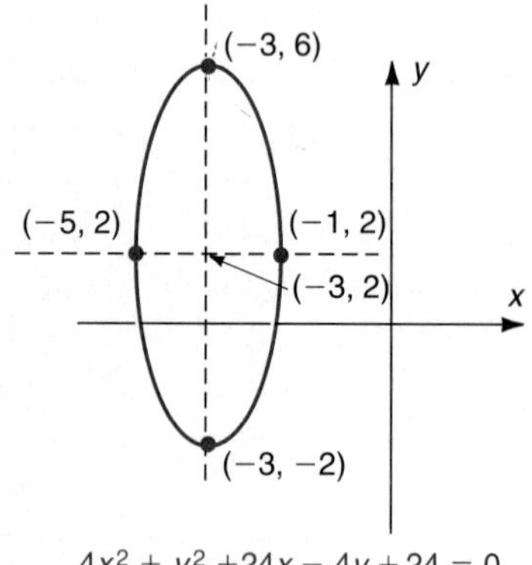

$4x^2 + y^2 + 24x - 4y + 24 = 0$

Fig. 7.19

The graph is an ellipse centered at $(-3, 2)$ as shown in Figure 7.19. □

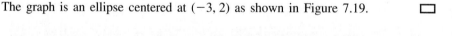

Equation of the Ellipse

$$\frac{(x - h)^2}{a^2} + \frac{(y - k)^2}{b^2} = 1$$

Major Axis Parallel to the x-Axis	*Major Axis Parallel to the y-Axis*
$a > b > 0$	$b > a > 0$
Center: (h, k)	Center: (h, k)
Vertices: $(h + a, k)$ and $(h - a, k)$	Vertices: $(h, k + b)$ and $(h, k - b)$
Foci: $(h + c, k)$ and $(h - c, k)$	Foci: $(h, k + c)$ and $(h, k - c)$

(Length of semimajor axis)^2 = (length of semiminor axis)^2 + c^2

EXAMPLE 6 Find an equation of the ellipse with center at $(-2, 4)$, foci at $(1, 4)$ and $(-5, 4)$, and minor axis of length 8.

Solution Graph the ellipse first (Figure 7.20). The major axis is parallel to the x-axis. Since the minor axis is 8 units, $2b = 8$ or $b = 4$. Since the foci are 3 units away from the center, $c = 3$.

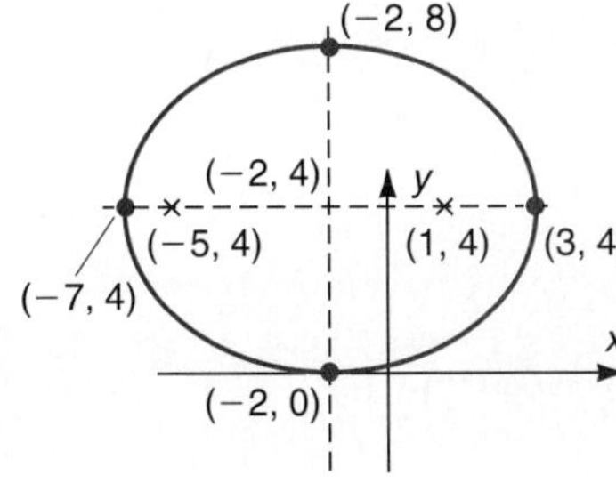

Fig. 7.20

$$\text{(Length of semimajor axis)}^2 = \text{(length of semiminor axis)}^2 + c^2$$

$$a^2 = b^2 + c^2$$

$$a^2 = 4^2 + 3^2$$

$$a^2 = 25$$

The equation of the ellipse is $\dfrac{(x + 2)^2}{25} + \dfrac{(y - 4)^2}{16} = 1.$ □

EXAMPLE 7 Find the foci of the graph of $\dfrac{x^2}{25} + \dfrac{y^2}{16} = 1.$

Solution The equation tells us that $a^2 = 25$ and $b^2 = 16$. Since $a > b$, the foci are $(-c, 0)$ and $(c, 0)$.

$$\text{(Length of semimajor axis)}^2 = \text{(length of semiminor axis)}^2 + c^2$$

$$a^2 = b^2 + c^2$$

$$25 = 16 + c^2$$

$$9 = c^2$$

$$\pm 3 = c$$

Thus, the foci are $(3, 0)$ and $(-3, 0)$. □

Ellipses have the property that a ray from one focus can be reflected through the other focus. The rotunda of the Capitol in Washington, D.C., has an elliptical ceiling. Sound waves that reach the ceiling at one focus are reflected to the other focus, making a "whispering gallery." Another such gallery is at the Mormon Tabernacle in Salt Lake City, Utah.

PROBLEM SET 7.2

Warm-ups

In Problems 1 through 14, sketch the graph of each equation.

For Problems 1 through 4, see Examples 1 and 2.

1. $9x^2 + 4y^2 = 36$

2. $4x^2 + 25y^2 = 100$

3. $\dfrac{x^2}{9} + \dfrac{y^2}{16} = 1$

4. $\dfrac{x^2}{4} + \dfrac{y^2}{36} = 1$

For Problems 5 and 6, see Example 3.

5. $9x^2 + 4y^2 = 4$

6. $16x^2 + 25y^2 = 4$

For Problems 7 through 10, see Example 4.

7. $\dfrac{(x-1)^2}{9} + \dfrac{(y-2)^2}{4} = 1$

8. $\dfrac{(x-2)^2}{16} + \dfrac{(y-1)^2}{4} = 1$

9. $\dfrac{(x+1)^2}{9} + \dfrac{(y+2)^2}{4} = 1$

10. $\dfrac{(x+2)^2}{16} + \dfrac{(y+1)^2}{4} = 1$

For Problems 11 through 14, see Example 5.

11. $4x^2 + y^2 - 24x + 4y + 24 = 0$

12. $4x^2 + 9y^2 + 32x + 36y + 64 = 0$

13. $9x^2 + 4y^2 - 18x - 8y - 23 = 0$

14. $x^2 + 9y^2 - 4x + 36y + 31 = 0$

In Problems 15 and 16, find an equation of the ellipse with the given conditions. See Example 6.

15. Center at the origin, foci at $(-4, 0)$ and $(4, 0)$, and length of minor axis 6

16. Foci at $(2, -8)$ and $(2, 4)$, center at $(2, -2)$, and length of minor axis $4\sqrt{3}$

In Problems 17 and 18, find the foci of each graph. See Example 7.

17. $\dfrac{x^2}{9} + \dfrac{y^2}{4} = 1$

18. $\dfrac{(x-2)^2}{16} + \dfrac{y^2}{9} = 1$

Practice Exercises

In Problems 19 through 44, sketch the graph of each equation.

19. $25x^2 + 9y^2 = 225$

20. $4x^2 + 64y^2 = 256$

21. $4x^2 + 25y^2 = 25$

22. $9x^2 + 16y^2 = 9$

23. $\dfrac{x^2}{16} + \dfrac{y^2}{36} = 1$

24. $\dfrac{x^2}{9} + \dfrac{y^2}{25} = 1$

25. $\dfrac{x^2}{9} + y^2 = 1$

26. $x^2 + \dfrac{y^2}{4} = 1$

27. $\dfrac{(x-1)^2}{16} + \dfrac{(y+2)^2}{9} = 1$

28. $\dfrac{(x+2)^2}{4} + \dfrac{(y-1)^2}{9} = 1$

29. $\dfrac{(x+2)^2}{25} + \dfrac{y^2}{4} = 1$

30. $\dfrac{x^2}{9} + \dfrac{(y-1)^2}{16} = 1$

31. $4(x-1)^2 + (y-1)^2 = 4$

32. $(x+1)^2 + 9(y-2)^2 = 9$

33. $2x^2 + y^2 = 8$

34. $x^2 + 3y^2 = 12$

35. $4(x+1)^2 + 32(y-1)^2 = 8$

36. $2(x-2)^2 + 4(y-1)^2 = 8$

37. $25x^2 + 9y^2 - 50x - 200 = 0$

38. $4x^2 + 9y^2 - 32x + 36y + 64 = 0$

39. $4x^2 + 9y^2 + 8x - 18y - 23 = 0$

40. $9x^2 + y^2 - 36x + 4y + 31 = 0$

41. $9x^2 + y^2 - 4y - 5 = 0$

42. $x^2 + 9y^2 + 2x - 8 = 0$

43. $4x^2 + 5y^2 + 8x - 20y + 4 = 0$

44. $5x^2 + 3y^2 + 20x - 6y + 8 = 0$

45. Johannes Kepler, a German astronomer, published three laws of planetary motion in the 1600s. Kepler's first law says that the planets move in elliptical orbits with the sun as a focus as shown in the figure. The point of the orbit closest to the sun is called its *perihelion,* and the point of the orbit farthest from the sun is called its *aphelion.* These occur at the vertices. If the length of the major axis of the earth's orbit around the sun is 185,920,000 miles and if the sun is 1,500,000 miles from the center of the earth's elliptical orbit, find the earth's perihelion and aphelion. Explain why the distance from the earth to the sun is usually given as 93,000,000 miles.

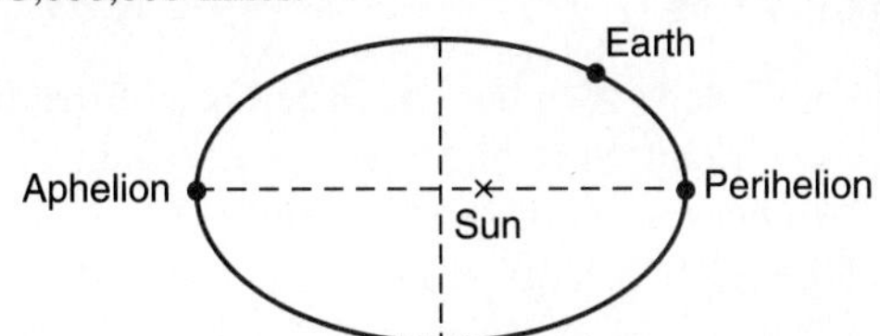

46. The moon revolves around the earth in an elliptical orbit as shown in the figure. The vertices of the ellipse are the points closest and farthest away from the earth. The point of the orbit closest to the earth is 221,456 miles *(perigee)* and the point farthest from the earth is 252,711 miles *(apogee).* How far from the center of the orbit is the earth?

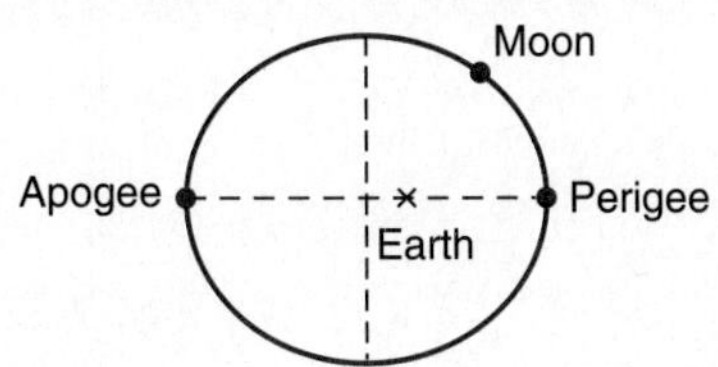

47. An arch of a bridge is in the shape of a semiellipse with a span of 48 feet and a height of 10 feet as shown in the figure. How high is the arch at a point 12 feet from its center? Is the height of the arch at the foci greater than or less than the height of the arch 12 feet from its center?

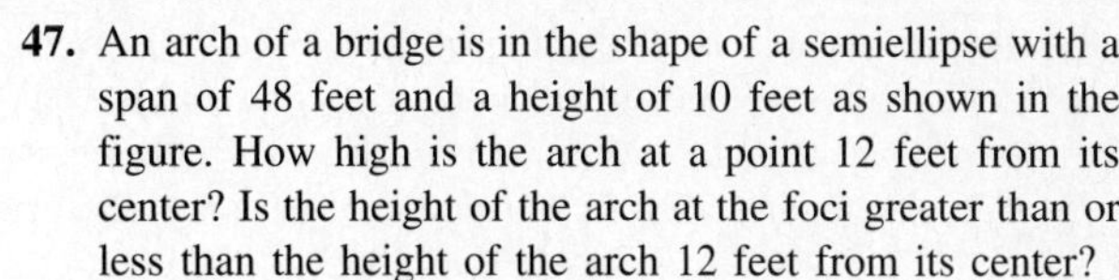

48. A track is to be built in the shape of an ellipse around a ballfield that is 120 yards by 60 yards as shown in the figure. Find the foci for the ellipse.

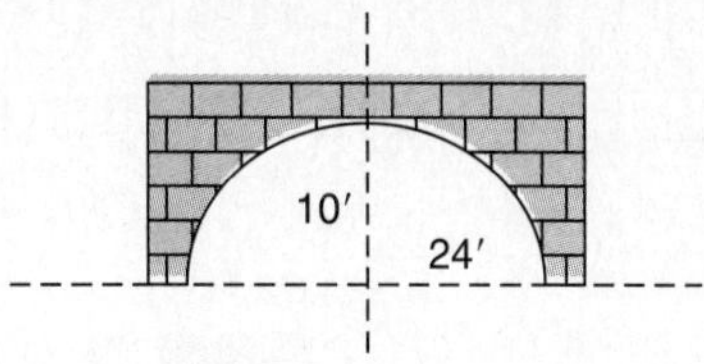

49. A circle is a special case of an ellipse. Graph $\dfrac{(x-1)^2}{9} + \dfrac{(y+1)^2}{9} = 1$ and explain why the statement is true.

50. Show that the equation $\dfrac{(x-h)^2}{a^2} + \dfrac{(y-k)^2}{b^2} = 1$ can be written in the form

$$Ax^2 + Bxy + Cy^2 + Dx + Ey + F = 0$$

51. Make up an equation whose graph is an ellipse that does not intersect the graph of $y - 1 = -4x^2$.

Challenge Problems

In Problems 52 through 55, sketch the graph of each equation. (Hint: Square both sides of the equation.)

52. $y = \sqrt{1 - \dfrac{x^2}{4}}$ **53.** $y = -\sqrt{1 - \dfrac{x^2}{4}}$ **54.** $x = -\sqrt{1 - \dfrac{y^2}{9}}$ **55.** $x = \sqrt{1 - \dfrac{y^2}{9}}$

56. Find an equation for the top half of an ellipse centered at (2, 3) with a semimajor axis 4 units in length and a semiminor axis 3 units in length.

57. Find an equation for the right half of the graph of $\dfrac{x^2}{4} + \dfrac{(y-1)^2}{2} = 1$.

58. Graph each equation.

(a) $\dfrac{x^2}{16} + \dfrac{y^2}{3} = 1$ (b) $\dfrac{x^2}{9} + \dfrac{y^2}{3} = 1$ (c) $\dfrac{x^2}{4} + \dfrac{y^2}{3} = 1$

(d) $\dfrac{x^2}{3.5} + \dfrac{y^2}{3} = 1$ (e) $\dfrac{x^2}{3.1} + \dfrac{y^2}{3} = 1$ (f) $\dfrac{x^2}{3.01} + \dfrac{y^2}{3} = 1$

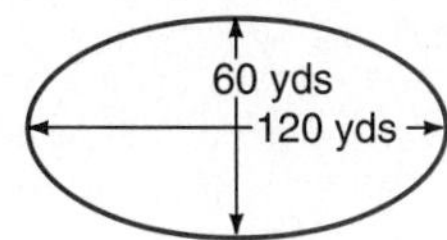

Describe what is happening to the graphs. The *eccentricity* e of an ellipse is defined as the ratio of c to a, where c is the distance from the center to the focus and a is the length of the semimajor axis. Calculate the eccentricity of each ellipse in parts (a) through (f). What is the largest and smallest eccentricity of an ellipse such as those in parts (a) through (f)? Describe what is happening to the graphs in terms of eccentricity.

IN YOUR OWN WORDS . . .

59. Explain how to graph $\dfrac{(x-h)^2}{a^2} + \dfrac{(y-k)^2}{b^2} = 1$.

60. When is completing the square necessary when graphing ellipses?

7.3 THE HYPERBOLA

A **hyperbola** is the set of points in a plane such that the difference of the distances from two fixed points is a positive constant. Sketching these points results in a shape like that shown in Figure 7.21. We place a set of axes so that the two fixed points, called **foci,** are on the x-axis and the origin is halfway between them at the **center** of the hyperbola.

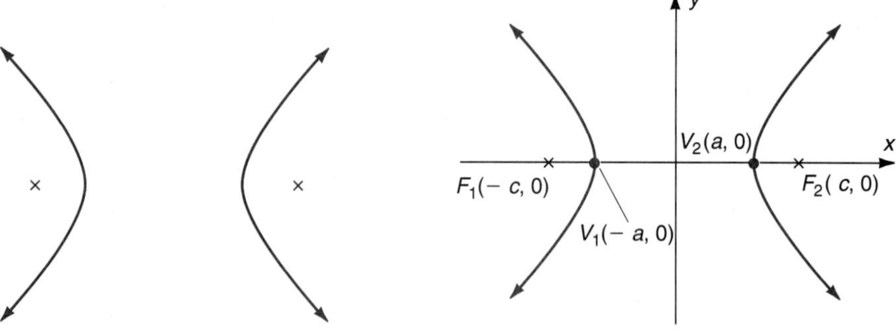

Fig. 7.21 **Fig. 7.22**

In Figure 7.22, the points V_1 and V_2 are called **vertices.** The segment $\overline{V_1V_2}$ is called the **transverse axis.** As in Figure 7.22, let the coordinates of the foci be $(c, 0)$ and $(-c, 0)$ and let the x-intercepts be $\pm a$, with $c > a > 0$. The difference in the distances from F_1 to P on the hyperbola and from F_2 to the same point P on the hyperbola is a positive constant by definition. Since V_2 is a point on the hyperbola, the distance from F_1 to V_2 minus the distance from F_2 to V_2 must be equal to this constant.

$$\overline{F_1V_2} - \overline{F_2V_2} = (a + c) - (c - a) = 2a$$

Thus, the difference in the distances from F_1 to P and from F_2 to P must be $2a$. So, we have

$$|\overline{F_1P} - \overline{F_2P}| = 2a \qquad \text{Absolute value ensures the constant is positive.}$$

Equivalently

$$\sqrt{(x + c)^2 + y^2} - \sqrt{(x - c)^2 + y^2} = \pm 2a$$

If we isolate a radical and square both sides of this equation and simplify, we obtain

$$\sqrt{(x+c)^2+y^2} = \pm 2a + \sqrt{(x-c)^2+y^2}$$

$$(x+c)^2+y^2 = 4a^2 \pm 4a\sqrt{(x-c)^2+y^2} + (x-c)^2+y^2$$

$$x^2+2xc+c^2+y^2 = 4a^2 \pm 4a\sqrt{(x-c)^2+y^2} + x^2-2xc+c^2+y^2$$

$$4xc-4a^2 = \pm 4a\sqrt{(x-c)^2+y^2}$$

$$cx-a^2 = \pm a\sqrt{(x-c)^2+y^2}$$

We square both sides again.

$$c^2x^2 - 2cxa^2 + a^4 = a^2[(x^2-2xc+c^2)+y^2]$$

$$c^2x^2 - 2cxa^2 + a^4 = a^2x^2 - 2a^2xc + a^2c^2 + a^2y^2$$

$$c^2x^2 + a^4 = a^2x^2 + a^2c^2 + a^2y^2$$

$$c^2x^2 - a^2x^2 - a^2y^2 = a^2c^2 - a^4$$

$$(c^2-a^2)x^2 - a^2y^2 = a^2(c^2-a^2)$$

If we divide both sides by $a^2(c^2-a^2)$, we have

$$\frac{x^2}{a^2} - \frac{y^2}{c^2-a^2} = 1$$

Since $c > a$, $c^2 - a^2 > 0$. So, we can let $b^2 = c^2 - a^2$. The equation becomes

$$\frac{x^2}{a^2} - \frac{y^2}{b^2} = 1$$

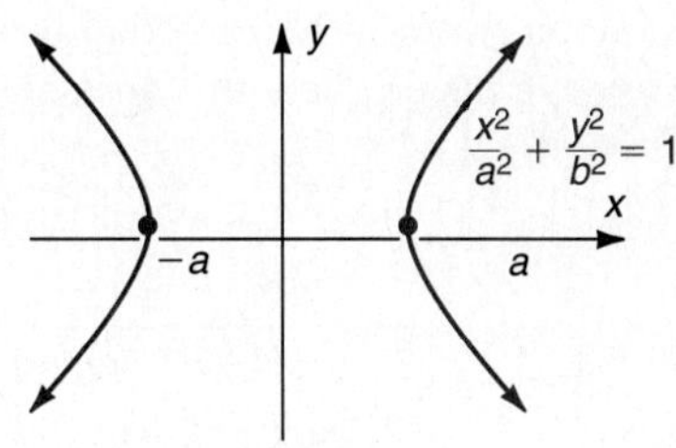

Fig. 7.23

We can find the y-intercepts for this hyperbola. If x is 0,

$$\frac{0^2}{a^2} - \frac{y^2}{b^2} = 1$$

$$\frac{y^2}{b^2} = -1$$

$$y^2 = -b^2$$

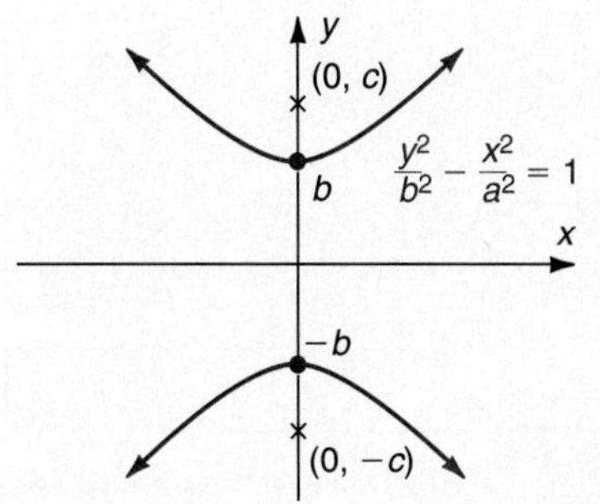

Fig. 7.24

Since $b > 0$, this equation has no real solutions. So, the graph has no y-intercepts as shown in Figure 7.23.

If the foci are $(0, -c)$ and $(0, c)$, then the equation of the hyperbola is $\frac{y^2}{b^2} - \frac{x^2}{a^2} = 1$, where $\pm b$ are the y-intercepts as shown in Figure 7.24.

The hyperbola has asymptotes. If we solve the equation $\frac{x^2}{a^2} - \frac{y^2}{b^2} = 1$ for y, we can gain some insight into why this is true.

$$\frac{x^2}{a^2} - \frac{y^2}{b^2} = 1$$

$$-\frac{y^2}{b^2} = 1 - \frac{x^2}{a^2} \qquad \text{Subtract } \frac{x^2}{a^2} \text{ from both sides.}$$

$$\frac{y^2}{b^2} = \frac{x^2}{a^2} - 1 \qquad \text{Multiply both sides by } -1.$$

$$y^2 = \frac{x^2b^2}{a^2} - b^2 \qquad \text{Multiply both sides by } b^2.$$

$$y = \pm\sqrt{\frac{b^2}{a^2}x^2 - b^2} \qquad \text{Square Root Property}$$

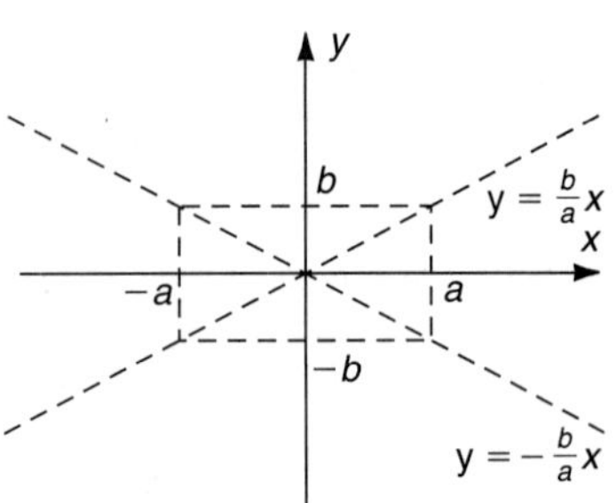

Fig. 7.25

As x grows *very* large, the number b^2 becomes negligible. Thus, as $x \to +\infty$, $y \to \pm\sqrt{\frac{b^2}{a^2}x^2}$. Therefore, $y \to \pm\frac{b}{a}x$. This means that the lines $y = \pm\frac{b}{a}x$ are asymptotes of the hyperbola. The farther the hyperbola is from the origin, the closer it gets to the asymptotes. However, the hyperbola never intersects the asymptotes.

We use the values of a and b to sketch a hyperbola by drawing some construction lines. First, we draw the rectangle shown in Figure 7.25 using the values of a and b. Next, we draw the diagonals of the rectangle, extending them as far as necessary. The two diagonals are **asymptotes** of the hyperbola. The vertices will be either $(0, \pm b)$ or $(\pm a, 0)$. Using these construction lines as guides, we now sketch the hyperbola, either left-right or up-down, depending on the form of the equation.

Notice that the asymptotes are lines that pass through the origin with slopes of $\frac{b}{a}$ and $-\frac{b}{a}$.

The Hyperbola Centered at the Origin

The graph of the equation $\frac{x^2}{a^2} - \frac{y^2}{b^2} = 1$ is a hyperbola centered at the origin with x-intercepts $\pm a$.

The graph of the equation $\frac{y^2}{b^2} - \frac{x^2}{a^2} = 1$ is a hyperbola centered at the origin with y-intercepts $\pm b$.

The asymptotes for these hyperbolas have equations $y = \pm\frac{b}{a}x$.

EXAMPLE 1 Sketch the graph of each equation.

(a) $x^2 - 4y^2 = 16$ (b) $9y^2 - 4x^2 = 36$.

Solutions (a) $x^2 - 4y^2 = 16$

$$\frac{x^2}{16} - \frac{y^2}{4} = 1 \qquad \text{Divide both sides by 16.}$$

The graph is a hyperbola with x-intercepts of ± 4. We use $a = 4$ and $b = 2$ to sketch the asymptotes as shown in Figure 7.26.

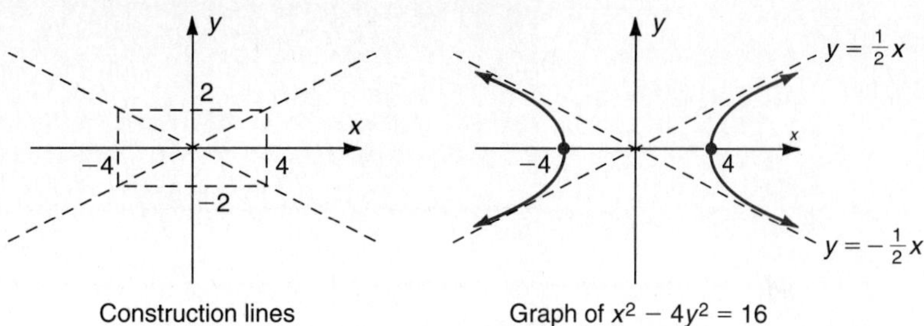

Fig. 7.26

(b) $9y^2 - 4x^2 = 36$

$$\frac{y^2}{4} - \frac{x^2}{9} = 1 \qquad \text{Divide both sides by 36.}$$

The graph is a hyperbola with y-intercepts of ± 2 as shown in Figure 7.27. Using $a = 3$ and $b = 2$, sketch the asymptotes.

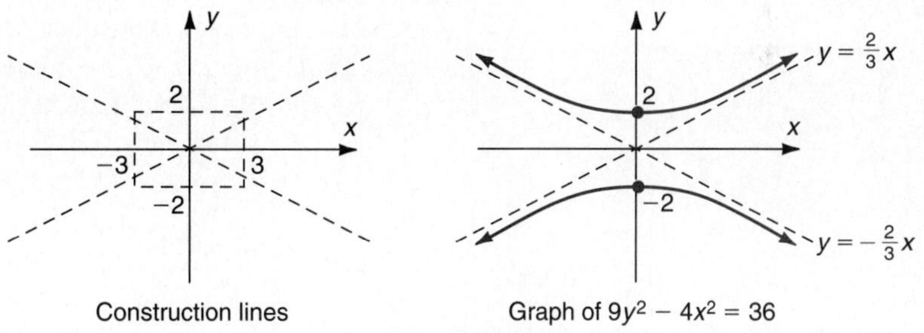

Fig. 7.27

Like the ellipse, if x is replaced with $x - h$ and $y - k$, the hyperbola will be centered at the point (h, k) instead of at the origin.

Hyperbolas

The graph of the equation

$$\frac{(x-h)^2}{a^2} - \frac{(y-k)^2}{b^2} = 1 \qquad \text{for } a > 0 \text{ and } b > 0$$

is the graph of the hyperbola $\frac{x^2}{a^2} - \frac{y^2}{b^2} = 1$ shifted so that its center is at the point (h, k).

> The graph of the equation
>
> $$\frac{(y-k)^2}{b^2} - \frac{(x-h)^2}{a^2} = 1 \qquad \text{for } a > 0 \text{ and } b > 0$$
>
> is the graph of the hyperbola $\frac{y^2}{b^2} - \frac{x^2}{a^2} = 1$ shifted so that its center is at the point (h, k).
>
> The asymptotes of these hyperbolas have equations $y - k = \pm\frac{b}{a}(x - h)$.

EXAMPLE 2 Sketch the graph of each equation.

(a) $4(x-2)^2 - 25(y+1)^2 = 100$ (b) $(y-2)^2 - \frac{x^2}{4} = 1$

Solutions (a) $4(x-2)^2 - 25(y+1)^2 = 100$

$$\frac{(x-2)^2}{25} - \frac{(y+1)^2}{4} = 1 \qquad \text{Divide both sides by 100.}$$

The graph is a hyperbola centered at $(2, -1)$. To sketch the graph, draw in a set of axes with its origin at $(2, -1)$. Use the new axes to sketch the asymptotes using $a = 5$ and $b = 2$. See Figures 7.28 and 7.29.

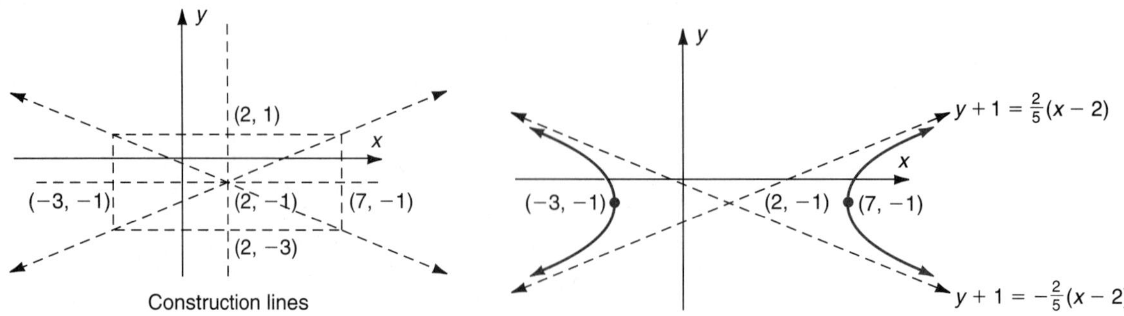

Fig. 7.28 **Fig. 7.29**

(b) $(y-2)^2 - \frac{x^2}{4} = 1$

This graph will be a hyperbola centered at $(0, 2)$. Using $a = 2$ and $b = 1$, we can sketch the graph. See Figure 7.30.

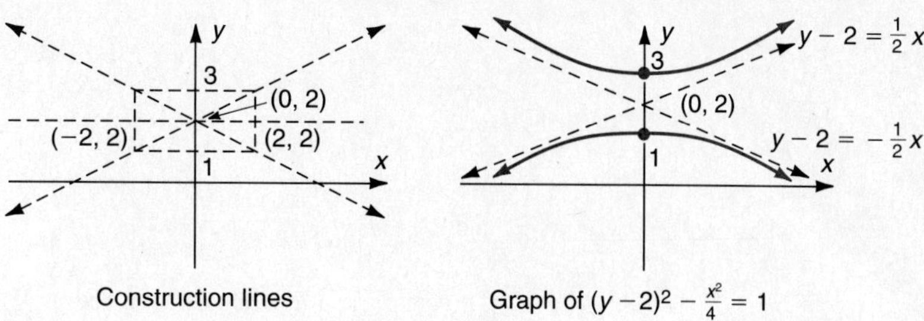

Fig. 7.30

EXAMPLE 3 Sketch the graph of each equation.

(a) $x^2 - 4y^2 - 2x = 3$ (b) $4y^2 - x^2 + 2x + 8y - 1 = 0$

Solutions (a) We must complete the square in x to determine the center of the hyperbola.

$$x^2 - 2x - 4y^2 = 3$$

$$(x^2 - 2x + 1 - 1) - 4y^2 = 3 \qquad \text{Complete the square.}$$

$$(x - 1)^2 - 1 - 4y^2 = 3$$

$$(x - 1)^2 - 4y^2 = 4$$

$$\frac{(x - 1)^2}{4} - y^2 = 1$$

The center of the hyperbola is (1, 0). Draw the construction lines from the center using $a = 2$ and $b = 1$. See Figure 7.31.

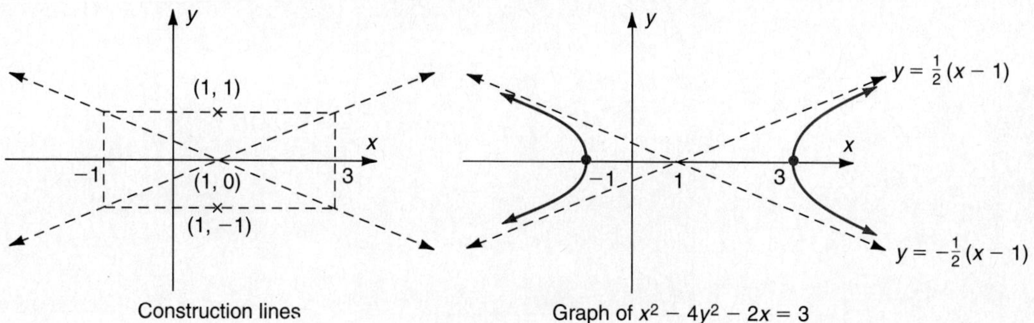

Fig. 7.31

(b) $4y^2 - x^2 + 2x + 8y - 1 = 0$
We must complete the square in both x and y. Be careful with the negative sign in front of the second parentheses.

$$4(y^2 + 2y) - (x^2 - 2x) = 1 \qquad \text{Group } x \text{ and } y \text{ terms.}$$

$$4(y^2 + 2y + 1 - 1) - (x^2 - 2x + 1 - 1) = 1 \qquad \text{Complete the square.}$$

$$4(y + 1)^2 - (x - 1)^2 = 4$$

$$(y + 1)^2 - \frac{(x - 1)^2}{4} = 1$$

The graph is a hyperbola centered at $(1, -1)$ with $a = 2$ and $b = 1$ as shown in Figure 7.32.

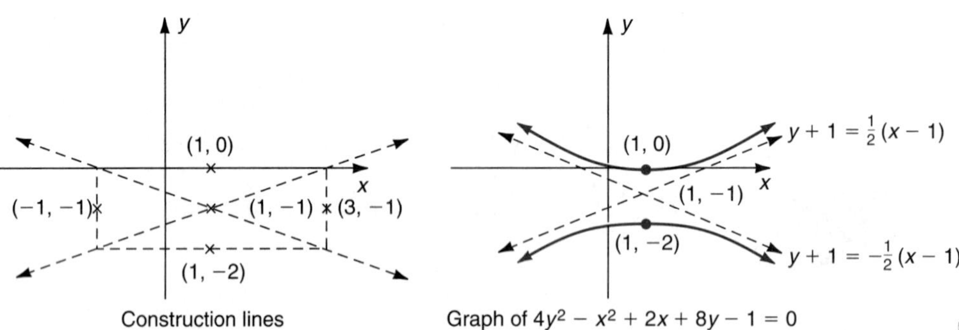

Fig. 7.32

Equations of Hyperbolas

For positive a, b, and c.

Transverse Axis Parallel to the x-Axis

$$\frac{(x - h)^2}{a^2} - \frac{(y - k)^2}{b^2} = 1$$

Center: (h, k)

Vertices: $(h + a, k)$ and $(h - a, k)$

Foci: $(h + c, k)$ and $(h - c, k)$

$a^2 + b^2 = c^2$

Transverse Axis Parallel to the y-Axis

$$\frac{(y - k)^2}{b^2} - \frac{(x - h)^2}{a^2} = 1$$

Center: (h, k)

Vertices: $(h, k + b)$ and $(h, k - b)$

Foci: $(h, k + c)$ and $(h, k - c)$

$a^2 + b^2 = c^2$

EXAMPLE 4 Find the foci of the graph of $\dfrac{x^2}{9} - \dfrac{y^2}{16} = 1$.

Solution $a^2 = 9$ and $b^2 = 16$ from the equation. So,

Fig. 7.33

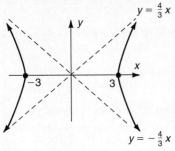

$$a^2 + b^2 = c^2$$
$$9 + 16 = c^2$$
$$25 = c^2$$
$$\pm 5 = c$$

The graph shown in Figure 7.33 is useful in understanding where the foci are. Since the foci are on the x-axis, they are $(-5, 0)$ and $(5, 0)$. □

LORAN is a system for long-range navigation. Two transmitting stations act as foci of a hyperbola. A navigator figures the difference in times between signals sent by the stations (T_1 and T_2) and calculates the difference in the distances from his ship to the stations. Thus, the ship is on the hyperbola determined by the transmitting stations. The process is repeated for another pair of transmitters (S_1 and S_2) to determine another hyperbola. The position of the navigator is the intersection of the two hyperbolas. See Figure 7.34.

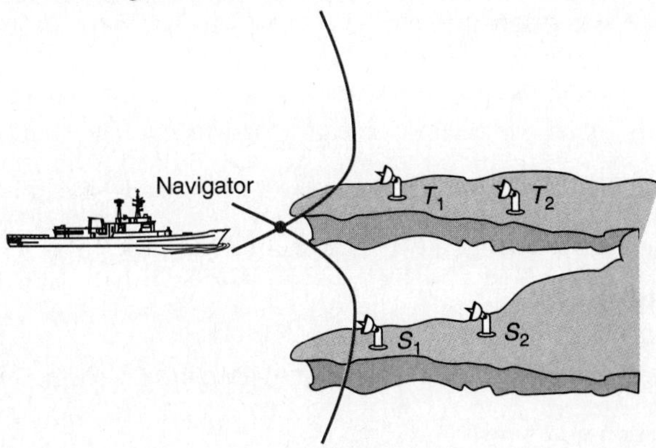

Fig. 7.34

GRAPHING CALCULATOR BOX

Graphing the Conic Sections

The conic sections may be graphed in the same manner as circles—first the "top," then the "bottom."

The Ellipse

Let's graph the ellipse, $\frac{(x+3)^2}{5} + \frac{(y-5)^2}{7} = 1$. To enter the equation, we must first solve for y.

$$7(x+3)^2 + 5(y-5)^2 = 35$$
$$5(y-5)^2 = 35 - 7(x+3)^2$$

$$(y-5)^2 = 7 - \frac{7}{5}(x+3)^2$$

$$y = 5 \pm \sqrt{7 - \frac{7}{5}(x+3)^2}$$

We use the + of ± to graph the top of the ellipse and the − to graph the bottom. Since the ellipse is centered at $(-3, 5)$ and has semimajor axis of length $\sqrt{7}$, set the range to $[-7, 1]$ by $[1, 9]$.

TI-81

Enter one equation in Y1 and the other in Y2. Make sure that both equal signs are dark.

CASIO fx-7700G

Enter one equation in the function memory [F MEM] as f1 and the other as f2.

Press [GRAPH] [F2] [1] [↵] [GRAPH] [F2] [2] [EXE]. Note the [↵] *multistatement* key that connects the two graphs.

Something about the graph appears wrong. The ellipse should be longer in the y direction. As we discussed when graphing circles, we need to "square" the window. The "square" command (zoom menu) will square up the TI-81. With the CASIO the difference in the x-coordinates should be $\frac{3}{2}$ the difference in the y-coordinates. Therefore, a change in the x-range to $[-9, 3]$ will do the trick.

The Hyperbola

We graph the hyperbola in a manner similar to the ellipse. Consider $\frac{(y-5)^2}{7} - \frac{(x-2)^2}{3} = 1$. Solving for y yields the two equations,

$$y = 5 \pm \sqrt{7 + \frac{7}{3}(x-2)^2}$$

A window of $[-10, 16]$ by $[-5, 15]$ shows the important features of the graph. It can be squared up by changing the x-range to $[-12, 18]$.

Graphing parabolas with the graphing calculator was discussed in Chapter 3.

PROBLEM SET 7.3

Warm-ups

In Problems 1 through 4, sketch the graph of each hyperbola. See Example 1.

1. $9x^2 - 4y^2 = 36$

2. $4y^2 - 25x^2 = 100$

3. $\frac{x^2}{9} - \frac{y^2}{16} = 1$

4. $\frac{y^2}{4} - \frac{x^2}{36} = 1$

In Problems 5 through 8, sketch the graph of each hyperbola. See Example 2.

5. $\frac{(x-1)^2}{9} - \frac{(y-2)^2}{4} = 1$

6. $\frac{(x-2)^2}{16} - \frac{(y-1)^2}{4} = 1$

7. $\frac{(y+1)^2}{9} - \frac{(x+2)^2}{4} = 1$

8. $\frac{(y+2)^2}{16} - \frac{(x+1)^2}{4} = 1$

In Problems 9 through 12, sketch the graph of each hyperbola. See Example 3.

9. $x^2 + 2x - 4y^2 = 3$

10. $x^2 - y^2 - 2x + 2y - 4 = 0$

11. $4y^2 - x^2 + 2x - 16y + 11 = 0$

12. $9y^2 - 36x^2 + 72x = 0$

In Problems 13 through 16, find the foci of each graph. See Example 4.

13. $\frac{x^2}{9} - \frac{y^2}{4} = 1$

14. $\frac{x^2}{16} - \frac{y^2}{9} = 1$

15. $\frac{y^2}{9} - \frac{x^2}{4} = 1$

16. $\frac{y^2}{16} - \frac{x^2}{9} = 1$

Practice Exercises

In Problems 17 through 80, sketch the graph of each equation.

17. $25x^2 - 9y^2 = 225$

18. $4x^2 - 64y^2 = 256$

19. $4y^2 - 25x^2 = 25$

20. $9y^2 - 16x^2 = 9$

21. $\frac{x^2}{16} - \frac{y^2}{36} = 1$

22. $\frac{x^2}{9} - \frac{y^2}{25} = 1$

23. $\frac{y^2}{9} - x^2 = 1$

24. $y^2 - \frac{x^2}{4} = 1$

25. $4(x-2)^2 - 9(y+1)^2 = 36$

26. $\frac{(y+2)^2}{25} - \frac{(x-1)^2}{4} = 1$

27. $\frac{(y-1)^2}{4} - (x+2)^2 = 1$

28. $\frac{x^2}{9} - \frac{(y+1)^2}{4} = 1$

29. $\frac{(x+1)^2}{16} - \frac{(y+2)^2}{9} = 1$

30. $\frac{(x+2)^2}{4} - \frac{(y-1)^2}{9} = 1$

31. $\frac{(x+2)^2}{25} - \frac{y^2}{4} = 1$

32. $\frac{x^2}{9} - \frac{(y-1)^2}{16} = 1$

33. $4(x-1)^2 - (y-1)^2 = -4$

34. $(x+1)^2 - 9(y-2)^2 = -9$

35. $y^2 - x^2 = -1$

36. $x^2 - y^2 = -1$

37. $x^2 - 2y^2 = 8$

38. $y^2 - 3x^2 = 12$

39. $12(x-2)^2 - 16y^2 = 4$

40. $x^2 - 18(y+1)^2 = 9$

41. $x^2 - y^2 - 6x - 4y - 4 = 0$

42. $9x^2 - 36y^2 - 72y = 0$

43. $4x^2 - y^2 - 24x + 4y + 24 = 0$

44. $4y^2 - 9x^2 + 36x + 32y + 64 = 0$

45. $9y^2 - 4x^2 - 8x - 18y - 31 = 0$

46. $x^2 - 9y^2 - 4x + 36y - 41 = 0$

47. $y^2 - x^2 - 4x + 6y - 4 = 0$

48. $5y^2 - 4x^2 + 8x + 20y + 36 = 0$

49. $9x^2 - y^2 + 4y - 13 = 0$

50. $x^2 - 9y^2 + 2x - 8 = 0$

51. $4x^2 - 5y^2 + 8x - 20y - 36 = 0$

52. $4x^2 - 5y^2 + 8x + 10y - 21 = 0$

53. $4x^2 - 9y^2 - 4x - 6y - 36 = 0$

54. $4x^2 - 36y^2 + 12x + 36y - 36 = 0$

In Problems 55 through 60, write an equation of the hyperbola satisfying the given conditions.

55. Vertices $(-3, 3)$ and $(-3, -3)$; one focus at $(-3, 5)$

56. Vertices $(4, 4)$ and $(-4, 4)$; one focus at $(5, 4)$

57. Center at origin; one vertex at $(0, 3)$; one focus at $(0, -5)$

58. Center at origin; one vertex at $(4, 0)$; one focus at $(-5, 0)$

59. Focus at $(4, 0)$; vertex at $(2, 0)$; center at $(0, 0)$

60. Center at $(-1, 3)$; $a = \sqrt{2}$; $b = \sqrt{3}$; transverse axis parallel to x-axis

Challenge Problems

61. Describe the graphs of $\frac{x^2}{a^2} - \frac{y^2}{b^2} = 1$ and $\frac{x^2}{a^2} - \frac{y^2}{b^2} = -1$.
These graphs are called **conjugate hyperbolas.**

62. The graph of each of the following equations is a degenerate hyperbola. Sketch the graph of each.
(a) $x^2 - 4y^2 = 0$
(b) $\frac{(x-2)^2}{4} - \frac{(y+1)^2}{9} = 0$

In Problems 63 through 66, sketch the graph of each equation.

63. $y = \sqrt{1 + x^2}$

64. $y = -\sqrt{1 + x^2}$

65. $x = -\sqrt{1 + y^2}$

66. $x = \sqrt{1 + y^2}$

67. Find an equation of the top half of the graph of $\dfrac{(x-1)^2}{4} - \dfrac{(y+1)^2}{9} = 1.$

68. Find an equation of the top half of the graph of $\dfrac{(y+1)^2}{9} - \dfrac{(x-1)^2}{4} = 1.$

IN YOUR OWN WORDS . . .

69. Explain how to graph $\dfrac{(x-h)^2}{a^2} - \dfrac{(y-k)^2}{b^2} = 1.$

7.4 THE GENERAL SECOND-DEGREE EQUATION IN TWO VARIABLES

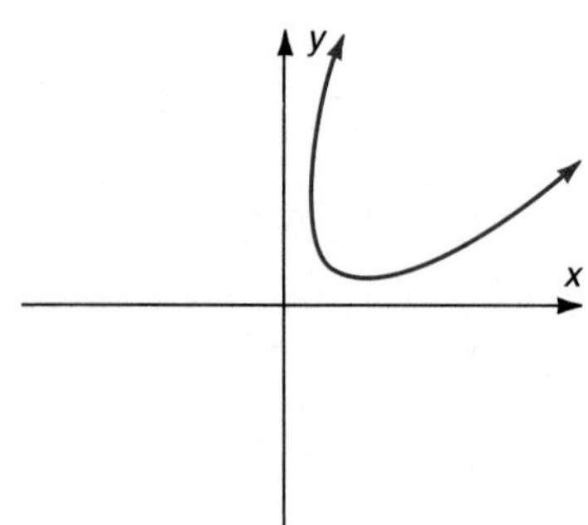

Fig. 7.35

In Sections 7.1, 7.2, and 7.3 we studied parabolas, ellipses, and hyperbolas as conic sections. Each of the conics is the graph of a second-degree equation in two variables. The general second-degree equation in two variables is

$$Ax^2 + Bxy + Cy^2 + Dx + Ey + F = 0$$

where A, B, and C are not all 0. The examples we examined did not contain an xy-term. The graphs of two-variable quadratic equations with an xy-term are conics that have been turned through angles other than 90° to the axes. For example, the parabola in Figure 7.35 has been shifted and *rotated.*

We have seen that conics translated from the origin can be graphed on a new set of axes parallel to the original axes. Translations result when x is replaced by $x - h$ and y is replaced by $y - k$ in the equations for parabolas, ellipses, and hyperbolas at the origin. In a similar manner second-degree equations with an xy-term can be changed into equations with no xy-term and graphed on a new set of axes formed by *rotating* the original axes. The process requires trigonometry.

A simple example of a second-degree equation in two variables that contains an xy-term is

$$xy = 1$$

The graph of $xy = 1$ is a hyperbola. If we solve for y, the equation becomes $y = \dfrac{1}{x}$. The graph is shown in Figure 7.36. If we make a new set of axes by rotating the xy-axes 45° to form the $x'y'$-axes and if we let

$$x = \frac{1}{\sqrt{2}}x' - \frac{1}{\sqrt{2}}y' \qquad \text{and} \qquad y = \frac{1}{\sqrt{2}}x' + \frac{1}{\sqrt{2}}y'$$

then the equation $xy = 1$ becomes

$$\frac{(x')^2}{2} - \frac{(y')^2}{2} = 1$$

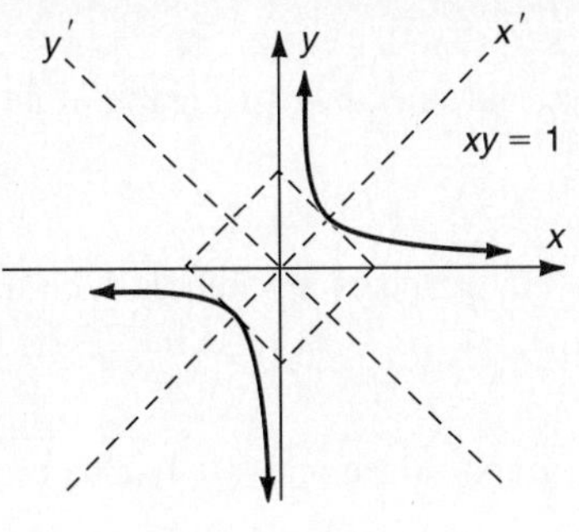

Fig. 7.36

with respect to the rotated $x'y'$-axes. We can graph this equation on the $x'y'$-axes as we did in Section 7.3. See Figure 7.36.

The coefficients of the general second-degree equation in two variables with no xy-term can tell us what the graph of the equation will be. The general equation in standard form is

$$Ax^2 + Bxy + Cy^2 + Dx + Ey + F = 0$$

If $B = 0$, the equation has no xy-term and it becomes

$$Ax^2 + Cy^2 + Dx + Ey + F = 0$$

The Graph of $Ax^2 + Cy^2 + Dx + Ey + F = 0$
(A and C not both 0)

1. If A or C is zero, the graph is a parabola.
   (a) The axis of symmetry is horizontal if $A = 0$.
   (b) The axis of symmetry is vertical if $C = 0$.
2. If $A = C$, the graph is a circle.
3. If A and C are both positive or both negative, the graph is an ellipse.
4. If A and C are opposite in sign, the graph is a hyperbola.
5. There are certain degenerate forms that *may* result from these equations such as lines, points, and no graph at all.

Some examples of the degenerate forms mentioned above are

Equation	*Graph*
$x^2 + y^2 = 0$	The point (0, 0)
$x^2 - 4 = 0$	Two parallel lines, $x = 2$ and $x = -2$
$x^2 - y^2 = 0$	Two intersecting lines, $y = \pm x$
$x^2 + y^2 + 1 = 0$	No graph, the empty set

EXAMPLE 1 Identify each of the following as the graph of a parabola, circle, ellipse, or hyperbola. (None are degenerate.)

(a) $3x^2 - 4y^2 - 7x + 8y + 14 = 0$ (b) $5x^2 + 4y^2 + x - 10 = 0$

(c) $5x^2 + 4y + x - 10 = 0$ (d) $-3x^2 - 3y^2 + 6x + 5y - 11 = 0$

Solutions (a) $3x^2 - 4y^2 - 7x + 8y + 14 = 0$
The coefficients of x^2 and y^2 are opposite in sign. The graph is a hyperbola.

(b) $5x^2 + 4y^2 + x - 10 = 0$
The coefficients of x^2 and y^2 are both positive and unequal. The graph is an ellipse.

(c) $5x^2 + 4y + x - 10 = 0$
There is no y^2 term. This means that C is 0. The graph is a parabola with a vertical axis of symmetry.

(d) $-3x^2 - 3y^2 + 6x + 5y - 11 = 0$
The coefficients of x^2 and y^2 are equal. The graph is a circle. (It is also an ellipse.) □

PROBLEM SET 7.4

Warm-ups

In Problems 1 through 4, determine whether the graph of the equation is a parabola, a circle, an ellipse, or a hyperbola. See Example 1.

1. $y^2 - 6x = 0$

2. $x^2 + 3y^2 - 6x + 6y = 0$

3. $-3x^2 - 3y^2 - 6x - 5y = 11$

4. $x^2 + 3x - 4y = 16 + 3y^2$

Practice Exercises

In Problems 5 through 20, determine whether the graph of the equation is a parabola, a circle, an ellipse, a hyperbola, or not a conic. Indicate whether the axis of symmetry is parallel to the x-axis or the y-axis for each parabola.

5. $x^2 - y^2 + 3x - y - 4 = 0$

6. $x^2 + 4y = 6$

7. $5x^2 + 5y^2 - 10x + 10y = 0$

8. $3x^2 + 2y^2 + 14y = 0$

9. $15y = 32x^2$

10. $(y - 1)^2 = x$

11. $3x + 5y = 9$

12. $y = 2x + 8$

13. $18x^2 + 1 = 24y^2$

14. $x^2 = y^2 + 1$

15. $-3x^2 - 6y^2 = -3$

16. $3x^2 - y^2 = -1$

17. $y = 6$

18. $x + 3 = 0$

19. $x^2 - 3x + 4y = 16 + 3y^2$

20. $12x^2 + 12x - 11y - 7 = 0$

Challenge Problems

In Problems 21 and 22, sketch the graph of each conic.

21. $x^2 + xy + y^2 = 1$ $\left(\textit{Hint:}\right.$ Make $x'y'$-axes by rotating the xy-axes 45° and rewrite the equation by substituting $x = \frac{1}{\sqrt{2}}x' - \frac{1}{\sqrt{2}}y'$ and $y = \frac{1}{\sqrt{2}}x' + \frac{1}{\sqrt{2}}y'.\left.\right)$

22. $x^2 + \sqrt{3}xy = 1$ $\left(\textit{Hint:}\right.$ Make $x'y'$-axes by rotating the xy-axes 30° and rewrite the equation by substituting $x = \frac{\sqrt{3}}{2}x' - \frac{1}{2}y'$ and $y = \frac{1}{2}x' + \frac{\sqrt{3}}{2}y'.\left.\right)$

IN YOUR OWN WORDS . . .

23. What is the graph of the general second-degree equation in two variables?

CHAPTER SUMMARY

GLOSSARY

Parabola: The set of points in a plane that are equidistant from a fixed point **(focus)** and a fixed line **(directrix).**

Ellipse: The set of all points in a plane the sum of whose distances from two fixed points is constant. The two fixed points are called **foci.**

Hyperbola: The set of points in a plane such that the difference of the distances from two fixed points is a positive constant.

THE PARABOLA ($p \neq 0$)

Vertical Axis of Symmetry	*Horizontal Axis of Symmetry*
$(x - h)^2 = 4p(y - k)$	$(y - k)^2 = 4p(x - h)$
Vertex: (h, k)	Vertex: (h, k)
Focus: $(h, k + p)$	Focus: $(h + p, k)$
Directrix: $y = k - p$	Directrix: $x = h - p$

THE ELLIPSE

The graph of the equation $\dfrac{(x-h)^2}{a^2} + \dfrac{(y-k)^2}{b^2} = 1$ is the graph of $\dfrac{x^2}{a^2} + \dfrac{y^2}{b^2} = 1$ shifted so that its center is at the point (h, k).

Major Axis Parallel to the x-Axis	*Major Axis Parallel to the y-Axis*
Center: (h, k)	Center: (h, k)
Vertices: $(h + a, k)$ and $(h - a, k)$	Vertices: $(h, k + b)$ and $(h, k - b)$
Foci: $(h + c, k)$ and $(h - c, k)$	Foci: $(h, k + c)$ and $(h, k - c)$

(Length of semimajor axis)^2 = (length of semiminor axis)$^2 + c^2$

THE HYPERBOLA

The graph of the equation $\dfrac{(x-h)^2}{a^2} - \dfrac{(y-k)^2}{b^2} = 1$ is the graph of the hyperbola $\dfrac{x^2}{a^2} - \dfrac{y^2}{b^2} = 1$ shifted so that its center is at the point (h, k). The graph of the equation, $\dfrac{(y-k)^2}{b^2} - \dfrac{(x-h)^2}{a^2} = 1$ is the graph of the hyperbola $\dfrac{y^2}{b^2} - \dfrac{x^2}{a^2} = 1$ shifted so that its center is at the point (h, k). The asymptotes of these hyperbolas have equations $y - k = \pm\dfrac{b}{a}(x - h)$.

Transverse Axis Parallel to the x-Axis

$$\frac{(x-h)^2}{a^2} - \frac{(y-k)^2}{b^2} = 1$$

Center: (h, k)

Vertices: $(h + a, k)$ and $(h - a, k)$

Foci: $(h + c, k)$ and $(h - c, k)$

$a^2 + b^2 = c^2$

Transverse Axis Parallel to the y-Axis

$$\frac{(y-k)^2}{b^2} - \frac{(x-h)^2}{a^2} = 1$$

Center: (h, k)

Vertices: $(h, k + b)$ and $(h, k - b)$

Foci: $(h, k + c)$ and $(h, k - c)$

$a^2 + b^2 = c^2$

THE GRAPH OF $Ax^2 + Cy^2 + Dx + Ey + F = 0$ (A and C not both 0)

1. If A or C is zero, the graph is a parabola.
   (a) The axis of symmetry is horizontal if $A = 0$.
   (b) The axis of symmetry is vertical if $C = 0$.
2. If $A = C$, the graph is a circle.
3. If A and C are both positive or both negative, the graph is an ellipse.
4. If A and C are opposite in sign, the graph is a hyperbola.
5. There are certain degenerate forms that *may* result from these equations such as lines, points, and no graph at all.

REVIEW PROBLEMS

In Problems 1 through 24, sketch the graph of each equation.

1. $3x^2 + y^2 = 3$

2. $x^2 + (y - 1)^2 = 0$

3. $y = (x + 2)^2$

4. $x^2 = y^2 + 3$

5. $3x + 2 = 0$

6. $x^2 + 2x + y^2 = 0$

7. $y = 2x^2 + 8x + 9$

8. $x = 2y^2 + 2y + 3$

9. $x^2 + 4y^2 = 0$

10. $x^2 + 2x + y^2 + 6y = 12$

11. $y^2 = 4 - x^2$

12. $x^2 = 3 - y^2$

13. $y = \frac{1}{3}x^2$

14. $x^2 + 4x + y^2 + 4y - 12 = 0$

15. $y = 2x - x^2$

16. $x = 3 - 4y - y^2$

17. $x^2 - y^2 = 9$

18. $y^2 - x^2 = 25$

19. $\frac{(x-2)^2}{9} + (y + 3)^2 = 1$

20. $(x - 2)^2 + \frac{(y+1)^2}{4} = 1$

21. $4x^2 + y^2 - 2y - 3 = 0$

22. $x^2 - 6x + 16y^2 - 7 = 0$

23. $x^2 - 4y^2 - 8y - 8 = 0$

24. $x^2 - 4y^2 + 4x = 0$

In Problems 25 through 30, write an equation for the indicated conic.

25. Ellipse with center at the origin, vertices at $(-3, 0)$ and $(3, 0)$, and length of major axis 10 units

26. Hyperbola with center at the origin, vertices at $(-3, 0)$ and $(3, 0)$, and one focus at $(5, 0)$

27. Hyperbola with center at $(2, -1)$, vertices at $(2, 4)$ and $(2, -6)$, and one focus at $(2, 6)$

28. Ellipse with foci at $(-8, 2)$ and $(4, 2)$, center at $(-2, 2)$, and length of minor axis 8.

29. Parabola with vertex at $(1, -1)$ and focus at $(1, 1)$.

30. Parabola with focus at $(-4, 0)$ and equation of directrix $x = 4$.

31. A suspension footbridge 30 meters wide is supported by cables that hang in the shape of a parabola. The ends of the cable are 7 meters above the footpath, and the center of the cable is 3.5 meters above the footpath. Find the focus of the parabola. (See Figure 7.37.)

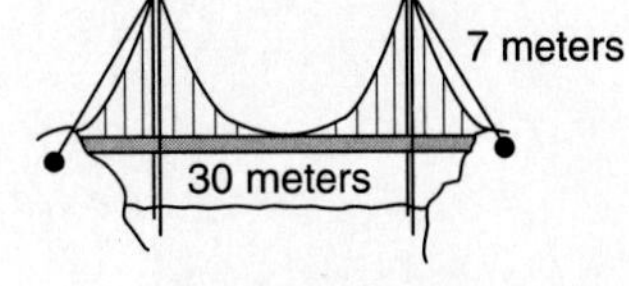

Fig. 7.37

32. The orbit of a satellite circling the earth is elliptical with the center of the earth as one focus. If the perigee of the orbit is 300 kilometers and the apogee is 500 kilometers, find an equation of the orbit. Assume the diameter of the earth is 12,740 kilometers. (See Figure 7.38.)

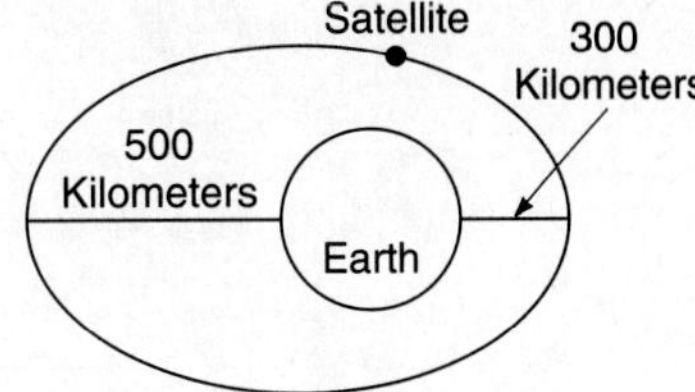

Fig. 7.38

LET'S NOT FORGET . . .

33. Which form is more useful? $\dfrac{(x-1)^2}{4} + \dfrac{y^2}{9} = 1$ or $9x^2 + 4y^2 - 18x - 27 = 0$ if the problem is to graph the equation.

34. Watch the role of negative signs. Graph $x^2 - y^2 = -1$.

35. From memory. When will the graph of $Ax^2 + Cy^2 + Dx + Ey + F = 0$ be
    (a) A parabola?
    (b) A hyperbola?
    (c) An ellipse?
    (d) A circle?

36. Approximate with a calculator. Approximate to the nearest thousandth the radius and the vertices of the graph of $\dfrac{(x-\pi)^2}{7} + \dfrac{(y-2)^2}{5} = 1$.

CHAPTER 8

Systems of Equations and Inequalities

CONNECTIONS

Engineers, economists, and scientists are continuously developing mathematical models of how things work. For example, an aeronautical engineer makes a model that describes how an airplane flies by defining the relationships between various parts of the airplane. These real world models usually require more than one equation in more than one variable. Such a model is called a system of equations.

Today, computers and calculators have increased the importance of systems by their ability to carry out an immense number of calculations in fractions of a second. Therefore, it is important to develop methodical techniques for solving systems that can be programmed in computers.

This chapter develops the mathematical theory behind solving systems of equations and inequalities. We will take a simplistic approach to real world problems in order to learn the techniques and procedures for formulating models and solving systems.

8.1 SYSTEMS OF EQUATIONS

Many times the mathematical model for a real-life situation is better stated with two equations and two variables. For example, Alex must replace a hot water heater. Brand A costs \$300 to purchase and \$10 a month to operate, while Brand B costs \$150 and \$15 a month to operate. Which brand should Alex buy?

Let m be the number of months after purchase and C be the total amount Alex will spend over m months. Then, if he buys brand A, his total cost after m months will be

$$C = 10m + 300$$

However, if he buys brand B his total cost will be

$$C = 15m + 150$$

We now have a problem made up of two equations with two variables. Before solving Alex's problem, we consider the general idea of models composed of more than one equation.

A **system of equations** is a set of two or more equations linked together and considered simultaneously. Systems of equations arise naturally when finding the points of intersection of graphs. The mathematical model needed to solve many problems in fields such as business, physics, chemistry, engineering, social sciences, biology, and meteorology is often a system of equations.

In this section we focus on systems of two equations with two variables, like those describing the problem of Alex and his choice of hot water heaters. Another example of a system of two equations in two variables is

$$\begin{cases} x^2 + y^2 = 4 \\ y = x + 2 \end{cases}$$

The brace { indicates that the two equations are linked together and that x and y represent the same number in both equations. A **solution** to a system of two equations in two variables is an ordered pair that satisfies *both* equations.

The **solution set** of a system consists of all solutions. To **solve** a system means to find the solution set. Two systems with the same solution set are called **equivalent.** A system with at least one solution is called **consistent,** while a system with no solutions is called **inconsistent.** A system with an infinite number of solutions is called **dependent.** In this chapter we will find real solutions for systems of equations.

A system that contains only linear equations is called a **linear** system. A system that contains at least one nonlinear equation is called a **nonlinear** system.

EXAMPLE 1 Determine whether each system is linear or nonlinear.

(a) $\begin{cases} x + y = 2 \\ y = 3x + 5 \end{cases}$ (b) $\begin{cases} x^2 + y^2 = 1 \\ 2x + 3y = 7 \end{cases}$ (c) $\begin{cases} x + y = 0 \\ y = \sqrt{x} + 1 \end{cases}$

Solutions (a) Both equations are first-degree or linear. The system is linear.

(b) The first equation is second-degree. The system is nonlinear.

(c) The second equation is nonlinear. The system is nonlinear. □

We will examine three methods for solving systems of equations: **graphing, substitution,** and **elimination.** The method of *graphing* involves graphing the equations and then looking for points of intersection. However, since guessing and estimation may be involved, graphing may not be the method of choice. *Substitution* is the general method and may be used to solve both linear and nonlinear systems. *Elimination* works well for linear systems, but may not work at all for nonlinear systems.

Graphing

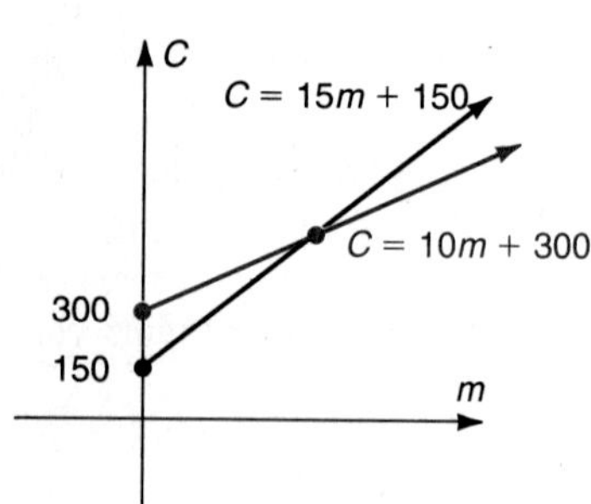

Fig. 8.1

We look at graphing first by helping Alex with his water heater problem. Recall that we had the two equations

$$\begin{cases} C = 10m + 300 \\ C = 15m + 150 \end{cases}$$

where C is the total cost at the end of m months. The graph of each of these equations is shown in Figure 8.1. The graph tells us that the two lines intersect in a point. Since the graph of each equation is a picture of the solution set of that equation, the point of intersection represents *a solution to both equations*. That is, the total cost is the same after a certain number of months. Notice from the graph that after the point of intersection, brand B costs more than brand A.

How can we find the coordinates of the point of intersection? We could make a guess by looking at the graph. Suppose that our guess is (30, 600). This means that (30, 600) is a solution to both equations. We check this guess.

$C = 10m + 300$	$C = 15m + 150$
If m is 30,	If m is 30,
$C = 10(30) + 300$	$C = 15(30) + 150$
$C = 600$	$C = 600$

Thus, (30, 600) is a solution to both equations. The point of intersection is (30, 600).

So, after 30 months it will cost $600 to own either brand. If the hot water heater will last longer than 30 months, Alex should buy brand A.

EXAMPLE 2 Graph each pair of equations and guess the point(s) of intersection, if any.

(a) $\begin{cases} x^2 + y^2 = 4 \\ y = -\dfrac{1}{2}x^2 + 2 \end{cases}$ (b) $\begin{cases} y = 2x^2 - 1 \\ y = \sqrt{x} \end{cases}$

(c) $\begin{cases} x = y^2 + 1 \\ (x+1)^2 + y^2 = 1 \end{cases}$ (d) $\begin{cases} y = 2 \\ (x-1)^2 + (y-1)^2 = 1 \end{cases}$

Solutions

(a) The graph of the first equation is a circle, and the second is a parabola as shown in Figure 8.2.

$$\begin{cases} x^2 + y^2 = 4 \\ y = -\dfrac{1}{2}x^2 + 2 \end{cases}$$

There are three points of intersection: (0, 2), (−2, 0), and (2, 0). Each ordered pair is a solution of both equations. Thus, the solution set is {(0, 2), (−2, 0), (2, 0)}.

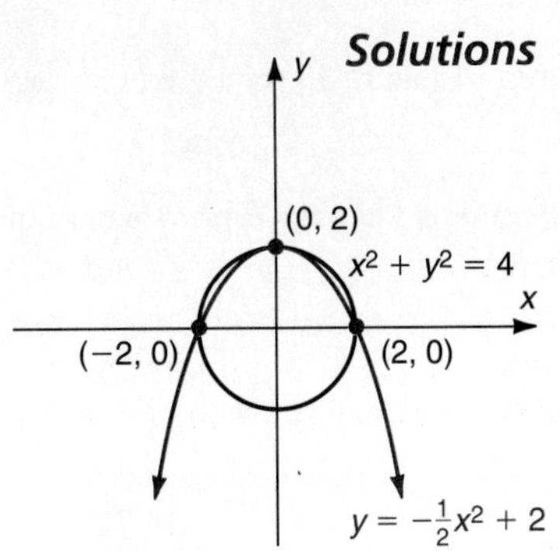

Fig. 8.2

(b) $\begin{cases} y = 2x^2 - 1 \\ y = \sqrt{x} \end{cases}$

There is one point of intersection. It is (1, 1). The solution set is {(1, 1)}. See Figure 8.3.

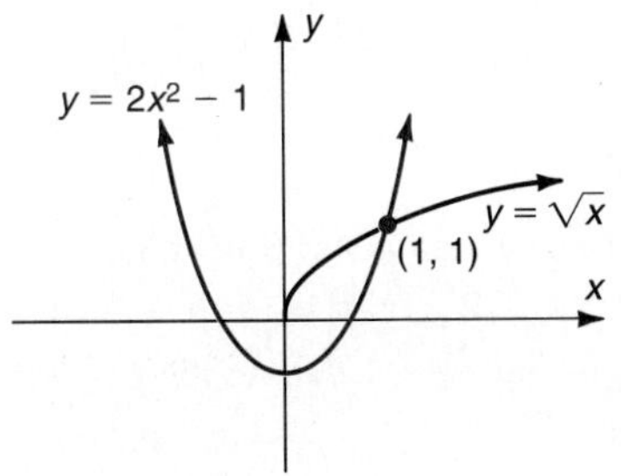

Fig. 8.3

(c) $\begin{cases} x = y^2 + 1 \\ (x+1)^2 + y^2 = 1 \end{cases}$

There are no points of intersection. The solution set is empty, ∅. See Figure 8.4.

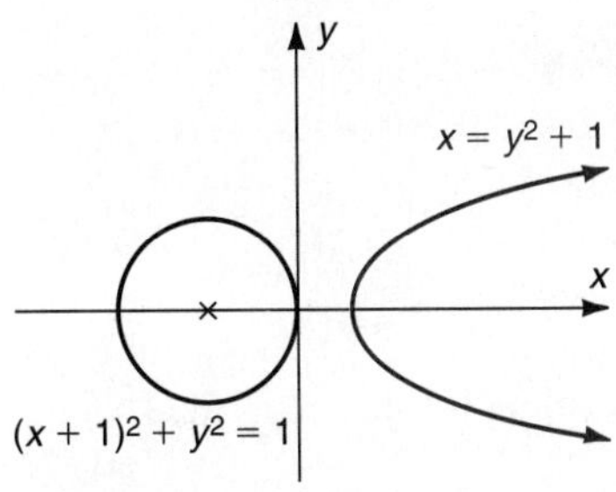

Fig. 8.4

(d) $\begin{cases} y = 2 \\ (x-1)^2 + (y-1)^2 = 1 \end{cases}$

There is one point of intersection. It is (1, 2). The solution set is {(1, 2)} as shown on Figure 8.5. □

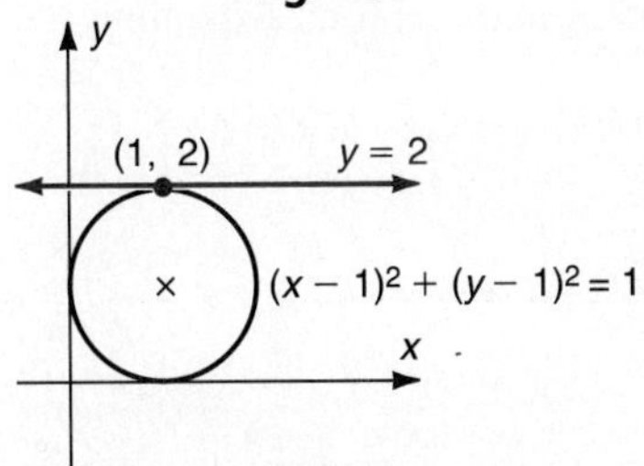

Fig. 8.5

Substitution

The second method we shall discuss is the method of substitution.

Method of Substitution

1. Solve one of the equations for one of the variables.
2. Substitute this expression for the same variable in the other equation.
3. Solve the resulting equation.
4. Substitute each solution back into step 1 to find the other variable.
5. Write the solution set.

EXAMPLE 3 Find the solution set for the system by substitution.

$$\begin{matrix}\boxed{1} \\ \boxed{2}\end{matrix}\begin{cases} x^2 + y^2 = 4 \\ x \;\; - y \;\; = 2 \end{cases}$$

(The equations are numbered $\boxed{1}$ and $\boxed{2}$ for convenience.)

Solution We examine the system to solve for the most convenient variable. Either x or y in equation $\boxed{2}$ is suitable. Solve equation $\boxed{2}$ for x.

$$\boxed{3} \quad x = y + 2$$

Now we substitute this into equation $\boxed{1}$ and solve.

$$(y + 2)^2 + y^2 = 4$$

$$y^2 + 4y + 4 + y^2 = 4 \qquad \text{Square } (y + 2).$$

$$2y^2 + 4y = 0 \qquad \text{Standard form}$$

$$2y(y + 2) = 0 \qquad \text{Factor.}$$

$$y = 0 \qquad \text{or} \qquad y = -2$$

Now we must find the number x that corresponds to each number y. Substitute each value of y back into $\boxed{3}$.

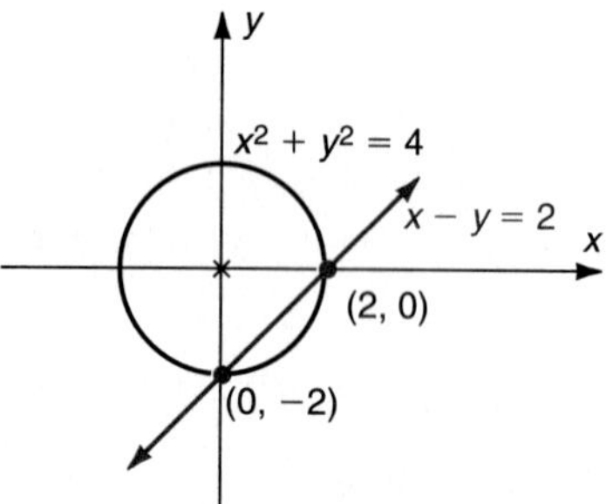

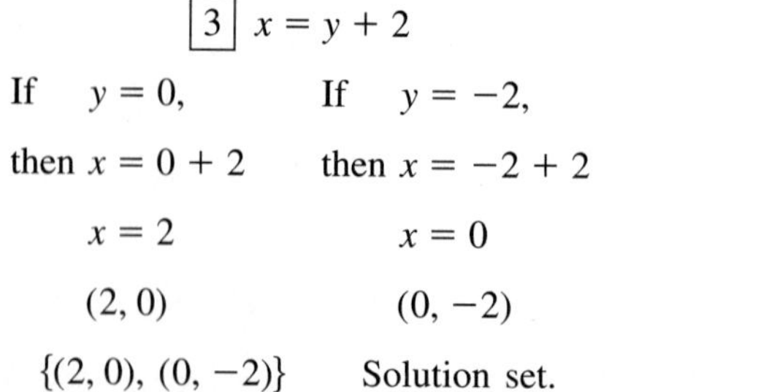

$$\boxed{3} \quad x = y + 2$$

If $y = 0$,	If $y = -2$,
then $x = 0 + 2$	then $x = -2 + 2$
$x = 2$	$x = 0$
$(2, 0)$	$(0, -2)$

$\{(2, 0), (0, -2)\}$ Solution set. □

Fig. 8.6

The graphs of the equations in Example 3 are shown in Figure 8.6. Notice that the points of intersection are the ordered pairs in the solution set.

EXAMPLE 4 Find the solution set for each system by substitution and graph the equations.

(a) $\begin{cases} 3x - 6y = 1 \\ -x + 2y = 2 \end{cases}$ (b) $\begin{cases} y = x^2 + 2 \\ y = x + 1 \end{cases}$

Solutions (a) $\begin{matrix}\boxed{1} \\ \boxed{2}\end{matrix}\begin{cases} 3x - 6y = 1 \\ -x + 2y = 2 \end{cases}$

We solve equation $\boxed{2}$ for x.

$$-x = 2 - 2y$$

$$x = -2 + 2y$$

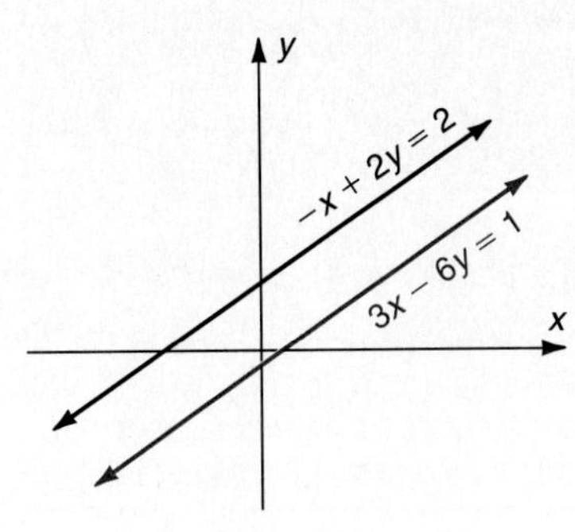

Fig. 8.7

Putting this into equation $\boxed{1}$ gives

$$3(-2 + 2y) - 6y = 1$$

$$-6 + 6y - 6y = 1$$

$$-6 = 1$$

What has happened? The variable y is gone, and we are left with $-6 = 1$, a false statement! The graph of the two equations (Figure 8.7) demonstrates the problem.

Since the slope of each line is $\frac{1}{2}$, they are parallel. There is no point common to both lines. The system is *inconsistent*. The solution set is the empty set, for which we use the symbol $\varnothing$.

(b) $\begin{cases} y = x^2 + 2 \\ y = x \;\; + 1 \end{cases}$

We replace y in the first equation with $x + 1$.

$$x + 1 = x^2 + 2$$

$$0 = x^2 - x + 1 \qquad \text{Standard form}$$

This quadratic equation cannot be solved by factoring. Substituting in the quadratic formula,

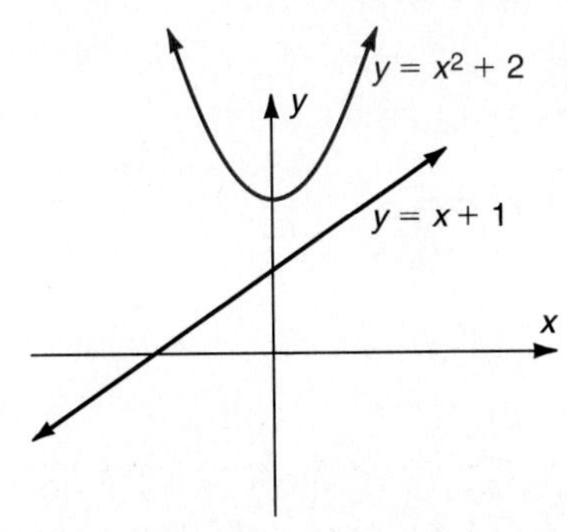

Fig. 8.8

$$x = \frac{-(-1) \pm \sqrt{(-1)^2 - 4(1)(1)}}{2}$$

$$= \frac{1 \pm \sqrt{-3}}{2}$$

There are two complex solutions to the equation, which means that there are no real number solutions to the system of equations. The graphs (Figure 8.8) show us what is happening.

The line and the parabola do not intersect. The system is *inconsistent*. The solution set is the empty set $\varnothing$. □

Elimination

The method of elimination stems from the addition property of equality that allows us to add two equations together. This method for solving systems of equations works particularly well for linear systems but may not work at all in nonlinear cases.

Method of Elimination

1. Multiply both sides of each equation by a suitable real number so that one of the variables will be eliminated by addition of the equations. (This step may not be necessary.)
2. Add the equations and solve the resulting equation.
3. Substitute the solution found in step 2 into one of the original equations and solve this equation.
4. Write the solution set.

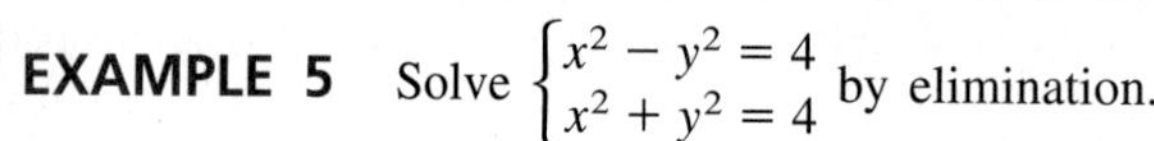

EXAMPLE 5 Solve $\begin{cases} x^2 - y^2 = 4 \\ x^2 + y^2 = 4 \end{cases}$ by elimination.

Solution If we add the equations, we will eliminate the variable y.

$$\begin{cases} x^2 - y^2 = 4 \\ x^2 + y^2 = 4 \end{cases}$$

$$2x^2 = 8$$

$$x^2 = 4$$

$$x = \pm 2$$

We substitute each of these x numbers into either of the original equations. Using the second equation,

If x is 2, then	If x is -2, then
$2^2 + y^2 = 4$	$(-2)^2 + y^2 = 4$
$y^2 = 0$	$y^2 = 0$
$y = 0$	$y = 0$
$(2, 0)$	$(-2, 0)$

Both solutions check in the original equations. The solution set is $\{(2, 0), (-2, 0)\}$. □

EXAMPLE 6 Solve each system by elimination. Graph the equations.

(a) $\begin{cases} 5x - 3y = 13 \\ 2x + y = 3 \end{cases}$ (b) $\begin{cases} 3x + 7y = 15 \\ 2x + 5y = 11 \end{cases}$

Solutions (a) If we add these equations together, we get the equation $7x - 2y = 16$, which doesn't help. But note what happens if we multiply both sides of the second equation by 3 and then add. The new system is equivalent to the original system because $2x + y = 3$ is equivalent to $6x + 3y = 9$.

$$\begin{cases} 5x - 3y = 13 \\ 6x + 3y = 9 \end{cases}$$

$$11x = 22$$

$$x = 2$$

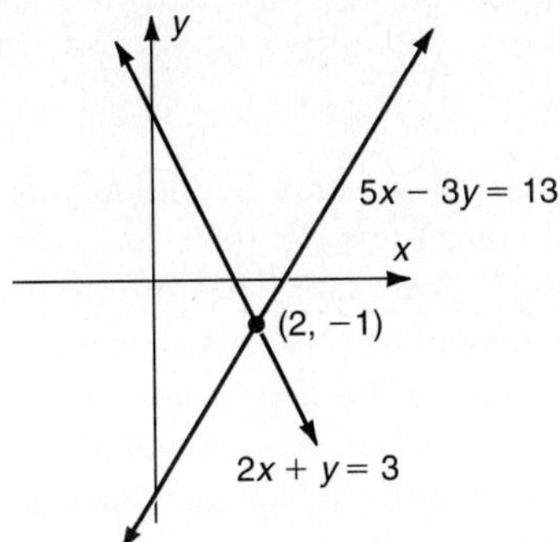

Fig. 8.9

We now find y by substituting into the second equation.

$$2(2) + y = 3$$

$$4 + y = 3$$

$$y = -1$$

The solution set is $\{(2, -1)\}$. The graphs (Figure 8.9) are lines that intersect at $(2, -1)$.

(b) $\begin{cases} 3x + 7y = 15 \\ 2x + 5y = 11 \end{cases}$

The first step is to "fix the equations up" so that the elimination step will work. One way to do this is to multiply both sides of the first equation by -2, and both sides of the second equation by 3.

$$\begin{cases} -6x - 14y = -30 \\ 6x + 15y = 33 \end{cases}$$

This system is equivalent to the first, but notice what happens if we add the two equations together.

$$\begin{cases} -6x - 14y = -30 \\ 6x + 15y = 33 \end{cases}$$

$$y = 3$$

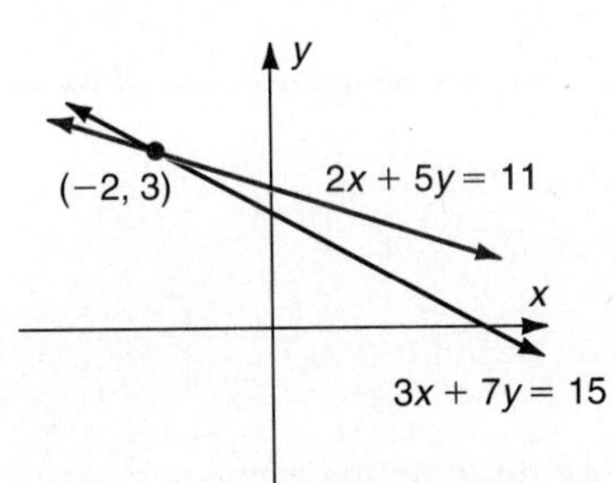

Fig. 8.10

Substituting 3 for y in the second equation gives

$$2x + 5(3) = 11$$

$$2x + 15 = 11$$

$$2x = -4$$

$$x = -2$$

And we have found the only solution, $(-2, 3)$. The solution set is $\{(-2, 3)\}$. The situation is illustrated in Figure 8.10. □

From Example 6, we see that it is sometimes necessary to multiply both sides of one or both of the equations by suitable numbers for the elimination step to work.

EXAMPLE 7 Solve $\begin{cases} x - y = 3 \\ 2x - 2y = 6 \end{cases}$

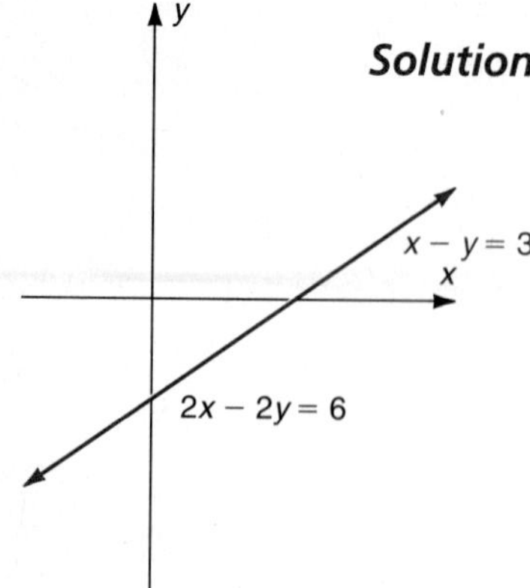

Fig. 8.11

Solution If we multiply the first equation by -2 and add, look what happens.

$$\begin{cases} -2x + 2y = -6 \\ 2x - 2y = 6 \end{cases}$$

$$0 = 0$$

Notice that the second equation is obtained from the first equation by multiplying both sides by 2. This means that the graph of each equation *is the same line*. The intersection of the two lines is in fact the line itself. Thus, there are an infinite number of points of intersection, hence an infinite number of solutions to the system. Figure 8.11 illustrates the situation. Every point on the line represents an ordered pair that is a solution to the system. We could write the solution set as $\{(x, y) \mid x - y = 3\}$. However, we often write the solution set in another manner. Solve either one of the equations for either x or y. Let's solve the first equation for x.

$$x - y = 3$$

$$x = 3 + y$$

Therefore, the solution set is $\{(3 + y, y) \mid y \text{ is a real number}\}$. □

Systems with an infinite number of solutions, such as the one illustrated in Example 7, are called *dependent*. Rather than use the letter x or y in the solution sets of dependent systems, we often employ a letter not used for variables in the system, usually s or t. In this case we write the solution set as $\{(3 + s, s) \mid s \in R\}$. s is called a **parameter.**

EXAMPLE 8 Solve $\begin{cases} \dfrac{1}{x} - \dfrac{3}{y} = 4 \\ \dfrac{2}{x} + \dfrac{1}{y} = 1 \end{cases}$

Solution We can eliminate the variable y from the system if we multiply the second equation by 3 and add.

$$\begin{cases} \dfrac{1}{x} - \dfrac{3}{y} = 4 \\ \dfrac{6}{x} + \dfrac{3}{y} = 3 \end{cases}$$

$$\frac{7}{x} = 7$$

$$7 = 7x \qquad \text{Clear fractions.}$$

$$1 = x$$

Substitute 1 for x in the first equation to find y.

$$1 - \frac{3}{y} = 4$$

$$y - 3 = 4y \qquad \text{Clear fractions.}$$

$$-3 = 3y$$

$$-1 = y$$

Write the solution set: $\{(1, -1)\}$ □

It is a good idea to examine the system to decide whether substitution or elimination should be used.

EXAMPLE 9 Determine whether substitution, elimination, or both can be used to solve each system.

(a) $\begin{cases} x^2 + y^2 = 4 \\ x + y = 7 \end{cases}$ (b) $\begin{cases} 3x - 2y = 4 \\ x - y = 0 \end{cases}$

(c) $\begin{cases} x^2 + y^2 = 2 \\ xy = 1 \end{cases}$ (d) $\begin{cases} 2x^2 + y^2 = 1 \\ x^2 - 2y^2 = 1 \end{cases}$

Solutions (a) Only substitution will work. We must solve the second equation for either x or y.

(b) This is a linear system. Both elimination and substitution will work.

(c) Only substitution will work. We can solve the second equation for x or y.

(d) Both substitution and elimination will work. □

Applications with Two Variables

Many word problems become much simpler if we use more than one variable.

EXAMPLE 10 Find the dimensions of a rectangle if its area is 250 square feet and its perimeter is 70 feet.

Solution Let L and W be the length and width of the rectangle in feet. (See Figure 8.12.)

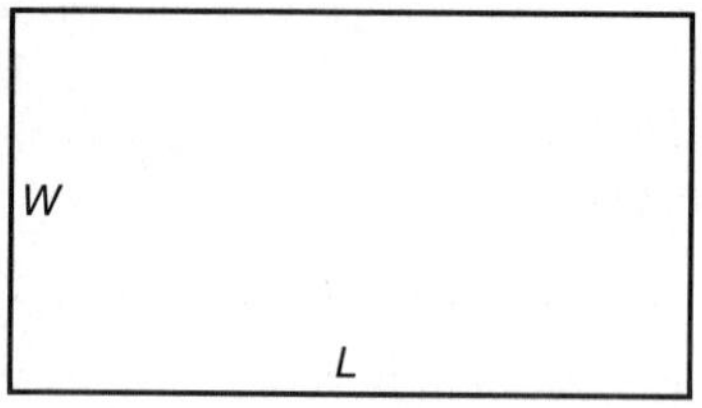

W

L

Fig. 8.12

The area is 250. Thus, we form the equation $LW = 250$.
Since the perimeter is 70, we have $2L + 2W = 70$ or $L + W = 35$.
We now have a system of equations.

$$\begin{cases} LW = 250 \\ L + W = 35 \end{cases}$$

To solve the system, we will use substitution and solve the second equation for L.

$$L = 35 - W$$

Substituting into the first equation,

$$(35 - W)W = 250$$

$$35W - W^2 = 250 \qquad \text{Distributive Property}$$

$$0 = W^2 - 35W + 250 \qquad \text{Standard form}$$

$$0 = (W - 25)(W - 10) \qquad \text{Factor.}$$

$$W = 25 \quad \text{or} \quad W = 10$$

If W is 25, then L is 10, and if W is 10, L is 25. In either case, the dimensions are 10 feet by 25 feet. □

EXAMPLE 11 Carolyn bought 3 pounds of apples and 2 pounds of pears from Sam's Market for \$4.05. Anne bought 1 pound of apples and 4 pounds of pears for \$4.65. What is the price per pound Sam charges for his apples and pears?

Solution Let A be the price per pound of apples, and P be the price per pound of pears. Then we have the system of two equations,

$$\begin{cases} 3A + 2P = 4.05 \\ A + 4P = 4.65 \end{cases}$$

Using elimination, we multiply the top equation by -2 and then add.

$$\begin{cases} -6A - 4P = -8.10 \\ A + 4P = 4.65 \end{cases}$$

$$-5A = -3.45$$

$$A = 0.69$$

Substituting 0.69 for A in the second equation gives

$$0.69 + 4P = 4.65$$

$$4P = 4.65 - 0.69$$

$$4P = 3.96$$

$$P = 0.99$$

Sam's Market charges \$0.69 per pound for apples and \$0.99 per pound for pears. □

Distance-rate-time problems can often be simplified by using more than one variable; particularly "upstream-downstream" problems.

EXAMPLE 12 Levy's boat takes 1 hour to travel 5 miles upstream to Long Bridge, but only 20 minutes to return. What is the speed of the current in the river and what is the average speed of Levy's boat in still water?

Solution We let x be the average speed (in miles per hour) of the boat in still water and y be the speed (in miles per hour) of the current. Thus Levy's speed upstream is $x - y$, and downstream is $x + y$.

Now we make our distance-rate-time table for each part of the trip.

	DISTANCE	RATE	TIME
Upstream	5	$x - y$	1
Downstream	5	$x + y$	$\frac{1}{3}$

We form two equations using the relationship $D = RT$.

$$5 = (x - y)1$$

$$5 = (x + y)\frac{1}{3}$$

Multiply the second equation through by 3 and write the system in the usual form.

$$\begin{cases} x - y = 5 \\ x + y = 15 \end{cases}$$

Add the equations together to find x.

$$2x = 20$$

$$x = 10$$

Substitute into the second equation

$$10 + y = 15$$

$$y = 5$$

Levy's boat averages 10 miles per hour in still water, and the current in the river is 5 miles per hour. □

"Working together-working separately" problems also lend themselves to a several-variable approach.

EXAMPLE 13 Library assistants Barney and Vontella, working together, can clear the stacks in 6 hours. One day Barney is called away after working 3 hours, and it takes Vontella four more hours to complete the job. How long does it take each to do the job working alone?

Solution Let B be the number of hours it would take Barney to do the job working alone. Let V be the number of hours it would take Vontella to do the job working alone.

If it takes Barney B hours to do the job alone, then in 1 hour Barney must do $\frac{1}{B}$ of the job. If it takes Vontella V hours to do the job alone, she must do $\frac{1}{V}$ of the job in 1 hour. Since $\frac{1}{6}$ of the job is done in 1 hour when they both are working, we form the equation

$$\frac{1}{B} + \frac{1}{V} = \frac{1}{6}$$

When they actually do the job, Barney works 3 hours, thus doing $3\left(\frac{1}{B}\right)$ of the job, while Vontella works 7 hours doing $7\left(\frac{1}{V}\right)$ of the job. But, since they do one job, we have the equation

$$\frac{3}{B} + \frac{7}{V} = 1$$

We solve the system $\begin{cases} \frac{1}{B} + \frac{1}{V} = \frac{1}{6} \\ \frac{3}{B} + \frac{7}{V} = 1 \end{cases}$ by elimination.

$$\begin{cases} \frac{-3}{B} - \frac{3}{V} = \frac{-3}{6} \\ \frac{3}{B} + \frac{7}{V} = 1 \end{cases} \qquad \text{Multiply the first equation by } -3.$$

$$\frac{4}{V} = \frac{1}{2} \qquad \text{Add.}$$

$$V = 8$$

Substitute 8 for V in the first equation to find B.

$$\frac{1}{B} + \frac{1}{8} = \frac{1}{6}$$

$$24 + 3B = 4B \qquad \text{Clear fractions.}$$

$$24 = B$$

It would take Vontella 8 hours to do the job alone, and it would take Barney 24 hours alone. □

PROBLEM SET 8.1

Warm-ups

In Problems 1 through 4, determine whether each system is linear or nonlinear. See Example 1.

1. $\begin{cases} x^2 + y = 0 \\ y = x + 3 \end{cases}$

2. $\begin{cases} 3x - 4y = 9 \\ x + 5y = 7 \end{cases}$

3. $\begin{cases} \dfrac{2}{x} - \dfrac{1}{y} = 2 \\ \dfrac{1}{x} + \dfrac{1}{y} = 3 \end{cases}$

4. $\begin{cases} y = 3x \\ y = \sqrt{x - 1} \end{cases}$

In Problems 5 through 8, graph each system and guess the point(s) of intersection, if any. See Example 2.

5. $\begin{cases} x + y = 2 \\ y = \sqrt{x} \end{cases}$

6. $\begin{cases} y = (x - 1)^2 \\ x = y^2 + 1 \end{cases}$

7. $\begin{cases} y = x \\ (x + 1)^2 + (y - 1)^2 = 1 \end{cases}$

8. $\begin{cases} x - y = 2 \\ x = y^2 \end{cases}$

In Problems 9 through 12, find the solution set for the system by the method of substitution. Graph each system. For Problems 9 and 10, see Example 3.

9. $\begin{cases} x^2 + y^2 = 9 \\ x - y = -3 \end{cases}$

10. $\begin{cases} x^2 - 2x + y^2 = 0 \\ x + y = 2 \end{cases}$

For Problems 11 and 12, see Example 4.

11. $\begin{cases} x + y = 4 \\ y = 5 - x \end{cases}$

12. $\begin{cases} x^2 + y = 0 \\ x + y = 2 \end{cases}$

In Problems 13 through 20, solve each system by the method of elimination. For Problems 13 and 14, see Example 5.

13. $\begin{cases} x^2 - y^2 = 8 \\ x^2 + y^2 = 8 \end{cases}$

14. $\begin{cases} x^2 - y^2 = 4 \\ x + y^2 = 2 \end{cases}$

For Problems 15 and 16, see Example 6.

15. $\begin{cases} 3x - y = -6 \\ 2x + 3y = -4 \end{cases}$

16. $\begin{cases} 5x - 4y = 0 \\ 3x - 8y = -7 \end{cases}$

For Problems 17 and 18, see Example 7.

17. $\begin{cases} 3x - y = -6 \\ 2y = 6x + 12 \end{cases}$

18. $\begin{cases} 5x - 4y = 0 \\ x = \dfrac{4}{5}y \end{cases}$

For Problems 19 and 20, see Example 8.

19. $\begin{cases} \dfrac{3}{x} - \dfrac{2}{y} = 5 \\ \dfrac{2}{x} + \dfrac{1}{y} = 1 \end{cases}$

20. $\begin{cases} \dfrac{1}{x} + \dfrac{2}{y} = 3 \\ \dfrac{2}{x} - \dfrac{1}{y} = 1 \end{cases}$

In Problems 21 through 24, determine whether substitution, elimination, or both can be used to solve each system. See Example 9.

21. $\begin{cases} x^2 - y^2 = 4 \\ x + y = 0 \end{cases}$

22. $\begin{cases} 5x - y = 4 \\ 3x - 7y = 0 \end{cases}$

23. $\begin{cases} x^2 + y^2 = 9 \\ xy = 1 \end{cases}$

24. $\begin{cases} x^2 + 3y^2 = 6 \\ 2x^2 + 3y^2 = 1 \end{cases}$

For Problems 25 and 26, see Example 10.

25. A rectangular search zone in the desert of Iraq has an area of 96 square miles and a perimeter of 40 miles. Find the dimensions of the search zone.

26. The winner of the poster contest at Mainstreet Elementary School is a rectangular poster that is 5 inches longer than it is wide. If its area is 300 square inches, find its dimensions.

For Problems 27 and 28, see Example 11.

27. Barbara bought 3 heads of lettuce and 4 pounds of tomatoes at Kroger for $6.03. Fred bought 2 heads of lettuce and 3 pounds of tomatoes for $4.35 at the same store. What is the price of a head of lettuce and a pound of tomatoes at Kroger?

28. Sid bought 3 pounds of grapes and 6 lemons for $3.57 at Circle K. Ilene bought 2 pounds of grapes and 5 lemons at the same store and paid $2.48. How much are lemons and grapes at Circle K?

For Problems 29 and 30, see Example 12.

29. Ingavald drove his new boat 48 miles downstream to his friend's cabin in 3 hours. The current was swift, and he traveled only 20 miles in 5 hours on the return trip. What is the average speed of Inky's boat in still water and what is the speed of the current?

30. Flying with the wind, Margie Clark travels from San Bernardino to Flagstaff in 2 hours. The return trip, against the wind, takes $2\frac{1}{2}$ hours. The distance between San Bernardino and Flagstaff is 500 kilometers. What is the average speed of the plane in still air and what is the wind speed?

For Problems 31 and 32, see Example 13.

31. A small swimming pool can be filled with water by two pipes in 8 hours. If the pool is filled in 10 hours when one of the pipes is shut off after 4 hours, how long will it take each of the pipes to fill the pool alone?

32. Frank and Grace, working together, can paint a room in 4 hours. One morning Frank painted alone for 3 hours and left for the day. Grace arrived that afternoon and spent 6 hours finishing the job. How long does it take each of them, working alone, to paint the room?

Practice Exercises

In Problems 33 through 48, find the solution set.

33. $\begin{cases} 2x - y = 10 \\ x + 2y = -5 \end{cases}$

34. $\begin{cases} 5x - y = 12 \\ 3x + 2y = 15 \end{cases}$

35. $\begin{cases} \dfrac{2}{3}x + 8y = \dfrac{6}{7} \\ \dfrac{3}{4}x + 9y = 3 \end{cases}$

36. $\begin{cases} \dfrac{2}{3}x - \dfrac{2}{5}y = 1 \\ \dfrac{5}{2}x - \dfrac{3}{2}y = 2 \end{cases}$

37. $\begin{cases} 3x - y = 4 \\ y = 3x - 4 \end{cases}$

38. $\begin{cases} y = 4 - x \\ 2x + 2y = 8 \end{cases}$

39. $\begin{cases} x^2 - 2y^2 = 4 \\ 2x^2 + y^2 = 8 \end{cases}$

40. $\begin{cases} 3x^2 - y^2 = -1 \\ x^2 + 2y^2 = 9 \end{cases}$

41. $\begin{cases} y = 2 - x^2 \\ 3x + 4y = 12 \end{cases}$

42. $\begin{cases} x = y^2 - 3 \\ x + y = -1 \end{cases}$

43. $\begin{cases} x + xy = 1 \\ xy = 4 \end{cases}$

44. $\begin{cases} x^2 - xy = -2 \\ 4xy = 5 \end{cases}$

45. $\begin{cases} \dfrac{3}{x} + \dfrac{5}{y} = -\dfrac{1}{6} \\ \dfrac{2}{x} + \dfrac{4}{y} = \dfrac{1}{4} \end{cases}$

46. $\begin{cases} \dfrac{7}{x} - \dfrac{5}{y} = -\dfrac{3}{4} \\ \dfrac{3}{x} - \dfrac{2}{y} = \dfrac{5}{4} \end{cases}$

47. $\begin{cases} x - y = 0 \\ y = \sqrt{x + 2} \end{cases}$

48. $\begin{cases} x + y = 0 \\ x = \sqrt{2 - y} \end{cases}$

In Problems 49 through 58, graph each system and find all points of intersection.

49. $\begin{cases} x^2 + y^2 = 9 \\ x - y = 3 \end{cases}$

50. $\begin{cases} x^2 + y^2 = 1 \\ x - y = 1 \end{cases}$

51. $\begin{cases} \dfrac{1}{2}x + \dfrac{3}{4}y = \dfrac{7}{4} \\ \dfrac{1}{3}x - \dfrac{1}{6}y = \dfrac{1}{2} \end{cases}$

52. $\begin{cases} \dfrac{1}{5}x - \dfrac{2}{3}y = -\dfrac{1}{15} \\ \dfrac{3}{4}x - \dfrac{5}{6}y = \dfrac{17}{12} \end{cases}$

53. $\begin{cases} x = y^2 - 2y \\ x = -y^2 + 4y \end{cases}$

54. $\begin{cases} y = x^2 - 4x + 4 \\ y = -x^2 + 2x \end{cases}$

55. $\begin{cases} y = x^2 - 4 \\ 2x^2 - 2y = 8 \end{cases}$

56. $\begin{cases} y = x^2 + 3 \\ 2y - 2x^2 = 6 \end{cases}$

57. $\begin{cases} y = x^3 \\ y - 4x = 0 \end{cases}$

58. $\begin{cases} y = -4x \\ y = x(x - 3)(x + 2) \end{cases}$

59. The sum of the squares of two numbers is 9. If the difference between the two numbers is 3, find the numbers.

60. Bonnie knows of two consecutive integers such that the sum of the larger number and twice the smaller number is -5. Find the two numbers.

61. Two cars start from towns 360 miles apart and travel toward each other. One travels 8 miles per hour faster than the other. If they met after 3 hours, what was the average speed of each car?

62. It takes a barge 8 hours to travel 24 miles upstream and 8 hours to travel 88 miles downstream. What is the average speed of the barge in still water, and what is the speed of the current?

63. Joyce Garrett bought 2 cans of Penn tennis balls and 1 can of Wilson tennis balls for \$9.00 at the pro shop. Art Abling was in the pro shop the same day and bought 1 can of Penn and 4 cans of Wilson for \$13.95. Find the price of a can of Penn and a can of Wilson tennis balls.

64. John Worth works at the local used bookstore after school during the week and on weekends. He makes more per hour on the weekends. The first week he worked 40 hours during the week and 10 hours on the weekend, and his check was \$199. The second week he worked 30 hours during the week and 12 hours on the weekend to bring home \$169.50. How much pay per hour does John receive both during the week and on the weekend?

65. Marion wishes to mix a 17% sugar solution with a 30% sugar solution to obtain 26 liters of a 24% sugar solution. How much of each should she mix?

66. Gloria Hitchcock wishes to mix a 5% ammonia solution with an 8% ammonia solution to obtain 12 quarts of a 6% ammonia solution. How much of each should she mix?

67. The perimeter of a rectangular deck is 54 feet. The width of the deck is 3 feet less than twice the length. What are the dimensions of the deck?

68. Madelyn is fencing a rectangular play area of 3750 square feet for Demon and Rosie. If she bought 250 feet of chain link fence to use, what are the dimensions of the play area?

Challenge Problems

69. Find all numbers k so that the system $\begin{cases} 5x + 3y = 5 \\ x - 3y = k \end{cases}$ has one solution.

70. For what numbers a and b will the system $\begin{cases} ax + y = p \\ bx + y = q \end{cases}$ have a solution?

71. Discuss the possible relationships between the three lines in the system $\begin{cases} ax + by = 0 \\ cx + dy = 0 \\ ex + fy = 0 \end{cases}$ and determine how many solutions the system could have.

For Graphing Calculators

In Problems 72 through 75, approximate to the nearest ten-thousandth the real solutions of each system.

72. $\begin{cases} y = x^2 \\ x = (y-4)^2 - 2 \end{cases}$

73. $\begin{cases} y = x^6 - 2x^3 + 1 \\ y = 1 - x^2 \end{cases}$

74. $\begin{cases} y = \ln x \\ y = (x-3)^2 + 1 \end{cases}$

75. $\begin{cases} y = e^x \\ y = \frac{1}{3}x^3 + \frac{3}{2}x^2 + 2x + 1 \end{cases}$

76. The supply and demand equations for a small company that prints posters are:
Supply: $y = x^2 + 4$
Demand: $y = 8 + 9x - x^3$
where x is the price in dollars and y is the number in thousands of posters. At what price will supply equal demand?

77. Dr. McComb ordered a drug that affects the number of Type A bacteria and Type B bacteria in the bloodstream to be given to a patient intravenously. Another medication must be administered precisely when the number of Type A bacteria present is equal to the number of Type B bacteria present. If the number of Type A bacteria present after t hours is given by the equation $N = 3{,}000e^{2t}$ and the number of Type B bacteria present after t hours is given by the equation $N = 5{,}000e^{-0.15t}$, after how many minutes and seconds will the number of Type A bacteria be equal to the number of Type B bacteria?

IN YOUR OWN WORDS . . .

78. Describe the method of substitution and give an example of a system that requires substitution to solve.

79. Describe the method of elimination and give an example of a system that could be solved by elimination.

80. Make up a word problem that could be solved using a system of equations.

8.2 LINEAR SYSTEMS

In this section we focus on solving systems of *linear* equations. We have already solved linear systems with two equations and two variables by substitution and elimination. We will extend the method of elimination to larger systems of linear equations.

Mathematical models often result in linear equations in more than two variables. For example, $x + y + z = 4$ is a linear equation in three variables. A solution to the equation is an ordered triple that makes the equation a true statement. The ordered triples $(5, 1, -2)$ and $(0, -2, 6)$ are solutions. $x_1 + 3x_2 - x_3 + 4x_4 = 0$ is a linear equation in four variables. One solution is $(0, 1, -1, -1)$.

Graphing

In the last section we found the geometric interpretation of equations in two variables to be very useful. It would be nice to extend those ideas to equations with more than two variables. The *ideas* extend quite nicely. However, as three dimensions are necessary for three variables and four dimensions are needed for four variables, the *drawings* become very difficult to downright impossible!

Let's look at the three-variable case, where the solutions are ordered triples

(x, y, z) of real numbers. To graph such an ordered triple we need *three* axes; an x-axis, a y-axis, and a z-axis. By convention, we put the x-axis and the y-axis in a *horizontal* plane perpendicular to each other, with the z-axis vertical. The usual three-dimensional coordinate system is shown in Figure 8.13.

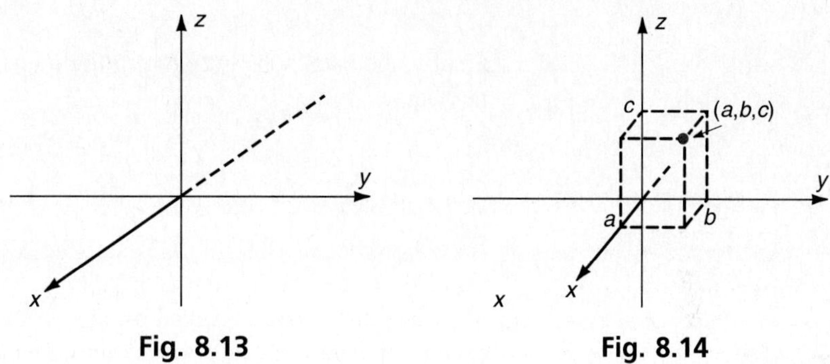

Fig. 8.13

Fig. 8.14

The coordinates are located on the appropriate axes in the usual manner. The point associated with the ordered triple (a, b, c) is located as shown in Figure 8.14.

EXAMPLE 1 Graph each point.

(a) $(1, 5, 3)$ (b) $(4, 2, -3)$

Solutions

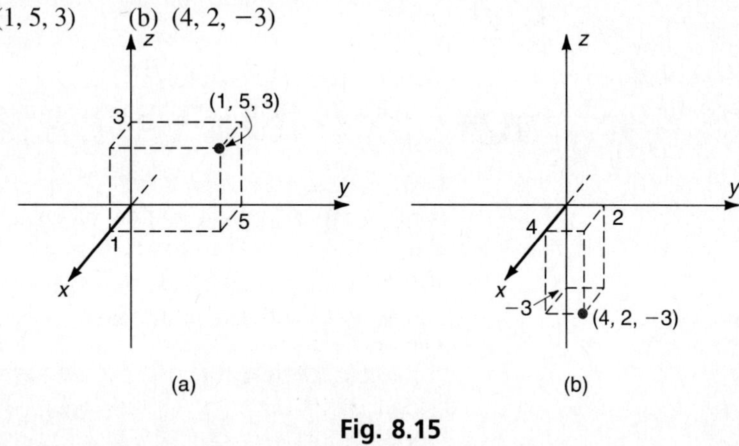

Fig. 8.15

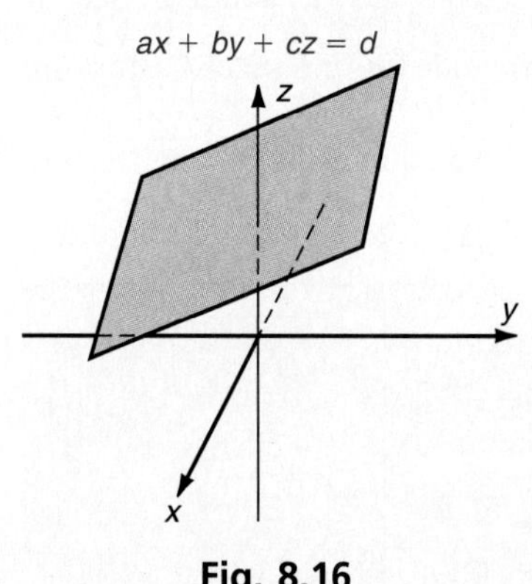

Fig. 8.16

The graph of a linear equation in three variables is a *plane* in the three-dimensional space formed by the coordinate axes as shown in Figure 8.16.

Gaussian Elimination

The basic strategy involved in Gaussian elimination is to change the original system into an equivalent system that is easier to solve. Consider the system

$$\begin{cases} x + y - z = 3 \\ \quad\;\; y - z = 2 \\ \qquad\;\; 2z = 6 \end{cases}$$

We can see from the third equation that $z = 3$. If we substitute 3 for z in the second equation, we can find y. This process is called **back-substitution.**

$$y - 3 = 2$$
$$y = 5$$

Finally, the back-substitution of 5 for y and 3 for z in the first equation gives us the number x.

$$x + 5 - 3 = 3$$
$$x = 1$$

Now the ordered triple (1, 5, 3) satisfies all three of the given equations. Thus the solution set for the system is {(1, 5, 3)}.

Back-substitution worked because certain coefficients were zero. In the second equation x is missing, and both x and y are missing in the third equation. A system so arranged is in **triangular form.**

How can we change a system to triangular form? Let's look at some examples of equivalent systems with two variables to get some ideas.

1. $\begin{cases} x + 2y = 0 \\ 3x - y = 7 \end{cases}$ is equivalent to $\begin{cases} 3x - y = 7 \\ x + 2y = 0 \end{cases}$. Each has a solution set of $\{(2, -1)\}$. Notice that the equations have simply been interchanged.

2. $\begin{cases} 2x - y = 1 \\ x + 2y = 3 \end{cases}$ is equivalent to $\begin{cases} 4x - 2y = 2 \\ x + 2y = 3 \end{cases}$. Each has a solution set of $\{(1, 1)\}$. Multiplying the first equation by 2 produces an equivalent equation.

3. $\begin{cases} x + y = 1 \\ x - 2y = 4 \end{cases}$ is equivalent to $\begin{cases} x + y = 1 \\ 3x = 6 \end{cases}$. Each has a solution set of $\{(2, -1)\}$. Two times the first equation has been added to the second equation.

These ideas are the basis for changing systems into equivalent systems and are commonly called **elementary operations.**

Elementary Operations

An equivalent system of equations will be produced if any of the following operations are performed.

1. Interchange any two equations.
2. Multiply both sides of any equation by the same nonzero real number.
3. Add a multiple of any equation to any other equation.

EXAMPLE 2 Solve $\begin{cases} x + y - z = 3 \\ -y + z = 2. \\ y + z = 6 \end{cases}$

Solution The system will be in triangular form if we can eliminate y from the third equation. This can be done if we add the second equation to the third equation using elementary operations.

$$\begin{cases} x + y - z = 3 \\ -y + z = 2 \\ 2z = 8 \end{cases}$$

Now we see that $z = 4$, and by back-substitution,

$$-y + 4 = 2$$

$$2 = y$$

Back-substitution into the first equation gives

$$x + 2 - 4 = 3$$

$$x = 5$$

Since the ordered triple $(5, 2, 4)$ satisfies all three equations in the system, the solution set is $\{(5, 2, 4)\}$. □

EXAMPLE 3 Solve $\begin{cases} x - y + z = 3 \\ 2x + y - z = 0 \\ y + z = -2 \end{cases}$.

Solution We number the equations for convenience in the discussion.

$$\begin{array}{l} \boxed{1} \\ \boxed{2} \\ \boxed{3} \end{array} \begin{cases} x - y + z = 3 \\ 2x + y - z = 0 \\ y + z = -2 \end{cases}$$

To eliminate x in equation $\boxed{2}$, we add -2 times equation $\boxed{1}$ to equation $\boxed{2}$.

$$\begin{array}{l} \boxed{1} \\ \boxed{2} \\ \boxed{3} \end{array} \begin{cases} x - y + z = 3 \\ 3y - 3z = -6 \\ y + z = -2 \end{cases}$$

Notice that we can make smaller coefficients in equation $\boxed{2}$ if we divide both sides by 3 $\left(\text{or multiply by } \frac{1}{3}\right)$.

$$\begin{array}{l} \boxed{1} \\ \boxed{2} \\ \boxed{3} \end{array} \begin{cases} x - y + z = 3 \\ \phantom{x - {}} y - z = -2 \\ \phantom{x - {}} y + z = -2 \end{cases}$$

To eliminate y in equation $\boxed{3}$, we add -1 times equation $\boxed{2}$ to equation $\boxed{3}$.

$$\begin{array}{l} \boxed{1} \\ \boxed{2} \\ \boxed{3} \end{array} \begin{cases} x - y + z = 3 \\ \phantom{x - {}} y - z = -2 \\ \phantom{x - y + {}} 2z = 0 \end{cases}$$

Thus, $z = 0$. Back-substituting into equation $\boxed{2}$ gives $y = -2$. Continuing back-substitution into equation $\boxed{1}$, we have

$$x + 2 + 0 = 3$$

$$x = 1$$

The solution set is $\{(1, -2, 0)\}$. ☐

Notice in the last elimination step of Example 3 that we could have simply added equation $\boxed{2}$ to equation $\boxed{3}$ to get y directly.

$$\begin{array}{l} \boxed{1} \\ \boxed{2} \\ \boxed{3} \\ \end{array} \begin{cases} x - y + z = 3 \\ y - z = -2 \\ y + z = -2 \end{cases}$$

$$2y = -4$$

The systematic approach of writing the system in triangular form may not always be the shortest approach. However, it will lead to the solution set.

EXAMPLE 4 Solve $\begin{cases} x + y + 2z = 1 \\ 3x + 2y - z = 9. \\ 2x - y + z = 2 \end{cases}$

Solution

$$\begin{array}{l} \boxed{1} \\ \boxed{2} \\ \boxed{3} \end{array} \begin{cases} x + y + 2z = 1 \\ 3x + 2y - z = 9 \\ 2x - y + z = 2 \end{cases}$$

We eliminate x in equation $\boxed{2}$ by multiplying equation $\boxed{1}$ by -3 and adding to equation $\boxed{2}$.

$$\begin{array}{l} \boxed{1} \\ \boxed{2} \\ \boxed{3} \end{array} \begin{cases} x + y + 2z = 1 \\ -y - 7z = 6 \\ 2x - y + z = 2 \end{cases}$$

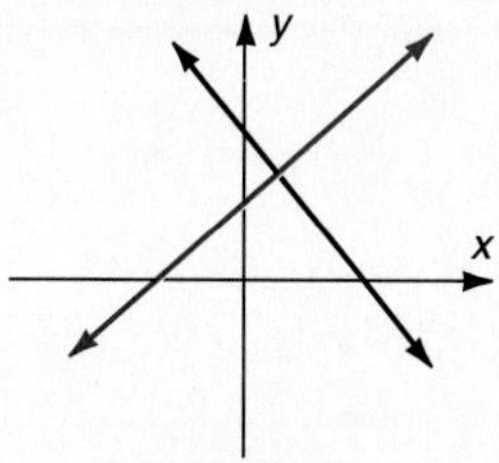

One solution
Two lines intersecting at a point
Consistent system

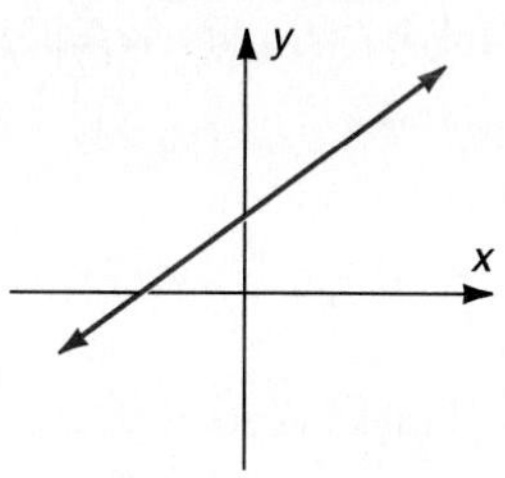

Infinitely many solutions
Two coincident lines
Consistent, dependent system

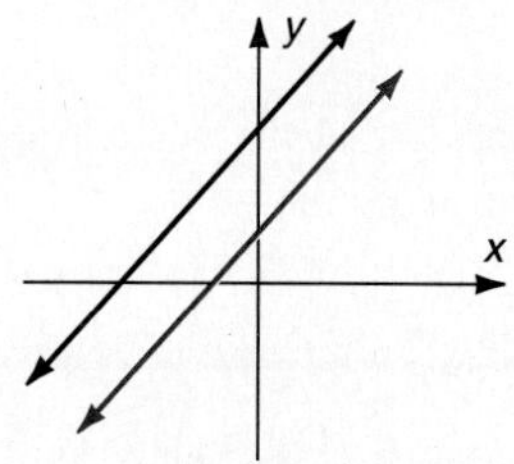

No solutions
Two parallel lines
Inconsistent system

Fig. 8.17

We eliminate x in equation $\boxed{3}$ by multiplying equation $\boxed{1}$ by -2 and adding to equation $\boxed{3}$.

$$\begin{array}{c} \boxed{1} \\ \boxed{2} \\ \boxed{3} \end{array} \begin{cases} x + y + 2z = 1 \\ \quad -y - 7z = 6 \\ \quad -3y - 3z = 0 \end{cases}$$

To finish the triangular form, we multiply equation $\boxed{2}$ by -3 and add to equation $\boxed{3}$.

$$\begin{array}{c} \boxed{1} \\ \boxed{2} \\ \boxed{3} \end{array} \begin{cases} x + y + 2z = 1 \\ \quad -y - 7z = 6 \\ \quad\quad 18z = -18 \end{cases}$$

Now $z = -1$. Back-substitution gives $y = 1$, and then $x = 2$.
The solution set is $\{(2, 1, -1)\}$. ▭

The systems we have examined thus far have all had one solution. Do we always get exactly one solution for a linear system? Let's consider the possibilities that arise when solving a system of two linear equations in two variables. Since the graph of such equations is a line, there are only three cases shown in Figure 8.17.

As seen earlier, the graph of an equation in three variables is a plane in three-dimensional space. The solutions of systems of equations in three variables are illustrated by the various ways that planes can intersect. Three (of many) possible situations are shown in Figure 8.18. As you can imagine, we seldom graph equations with more than two variables. With four or more variables, we find it difficult to even imagine a graph.

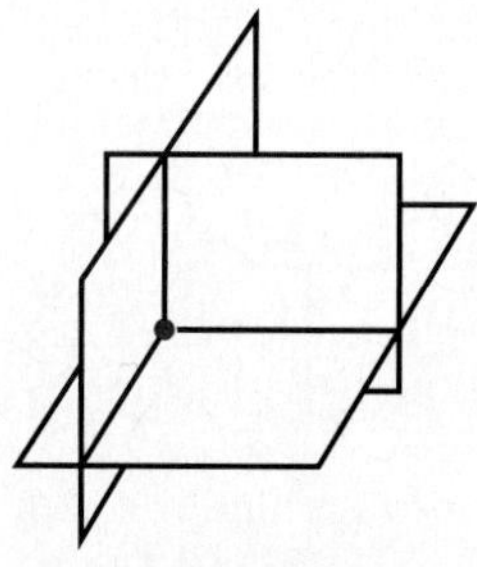
One solution
three planes intersecting
at a point
Consistent system

Infinitely many solutions
three planes intersecting in a line
Consistent, dependent
system

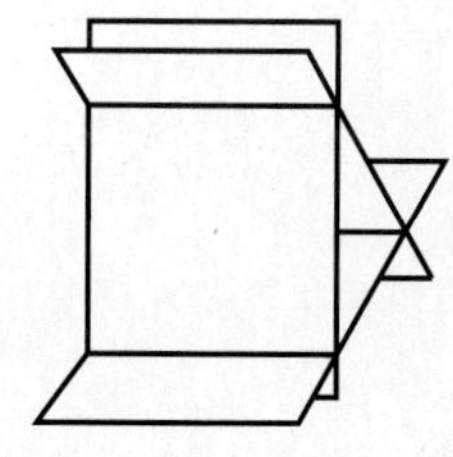
No solutions
three planes with no common
intersection
Inconsistent system

Fig. 8.18

Every system of linear equations has no solution, one solution, or infinitely many solutions.

EXAMPLE 5 Solve $\begin{cases} 2x + y - z = 2 \\ y - z = 0. \\ x - y + z = 2 \end{cases}$

Solution We begin by interchanging the first and third equations.

$$\begin{cases} x - y + z = 2 \\ y - z = 0 \\ 2x + y - z = 2 \end{cases}$$

We eliminate x from the (new) third equation by multiplying the first equation by -2 and adding to the third equation.

$$\begin{cases} x - y + z = 2 \\ y - z = 0 \\ 3y - 3z = -2 \end{cases}$$

Now we eliminate y from the third equation by multiplying the second equation by -3 and adding to the third equation.

$$\begin{cases} x - y + z = 2 \\ y - z = 0 \\ 0 = -2 \end{cases}$$

The third equation, $0 = -2$, is a false statement indicating that the system has no solution. The solution set is the empty set. The system is inconsistent. □

EXAMPLE 6 Solve $\begin{cases} x + 2y + 3z = 0 \\ x + y + 4z = 4. \\ y - z = -4 \end{cases}$

Solution We can eliminate x from the second equation by subtracting the first equation from the second.

$$\begin{cases} x + 2y + 3z = 0 \\ -y + z = 4 \\ y - z = -4 \end{cases}$$

Add the second equation to the third to eliminate y.

$$\begin{cases} x + 2y + 3z = 0 \\ -y + z = 4 \\ 0 = 0 \end{cases}$$

The third equation, $0 = 0$, is a true statement for all numbers x, y, and z. To write the solution set, we solve the second equation for y.

$$y = z - 4$$

Using back-substitution in the first equation, we can solve for x in terms of z.

$$x + 2(z - 4) + 3z = 0$$
$$x + 2z - 8 + 3z = 0$$
$$x = 8 - 5z$$

Therefore, $(8 - 5z, z - 4, z)$ is a solution *for any number z!* This system has an infinite number of solutions. Written with the parameter t, the solution set is $\{(8 - 5t, t - 4, t) \mid t \in R\}$. □

Notice that the solution set in Example 6 contains an infinite number of solutions, one for each real number t. For example, taking 1 for t gives the ordered triple $(3, -3, 1)$ which satisfies all three of the given equations.

EXAMPLE 7 Find an equation for the parabola with a vertical axis of symmetry that contains the three points $(1, 2)$, $(-1, 8)$, $(0, 4)$.

Solution A parabola with a vertical axis of symmetry has an equation of the form

$$y = ax^2 + bx + c$$

Substituting the given ordered pairs into this equation results in a system of equations. Using the point $(1, 2)$, we substitute 1 for x and 2 for y in $y = ax^2 + bx + c$ to get the equation

$$2 = a(1)^2 + b(1) + c$$

Likewise, the other equations come from using the other two points.

$$8 = a(-1)^2 + b(-1) + c$$
$$4 = a(0)^2 + b(0) + c$$

Therefore, we have three equations in the three unknowns, a, b, and c.

$$\begin{cases} a + b + c = 2 \\ a - b + c = 8 \\ \qquad\quad c = 4 \end{cases}$$

If we subtract the second equation from the first, we form an equivalent system.

$$\begin{cases} a + b + c = 2 \\ \qquad 2b = -6 \\ \qquad\quad c = 4 \end{cases}$$

Thus, we see that $c = 4$, $b = -3$, and $a = 1$.
An equation of the desired parabola is $y = x^2 - 3x + 4$. □

Systems Where the Number of Equations Differs from the Number of Variables

Thus far, we have considered only **square systems,** that is, systems with the same number of variables as equations. A system with fewer equations than variables is said to be **underspecified,** since there is insufficient information to fully evaluate all the variables. An underspecified system is either dependent or inconsistent. A system that has more equations than it has variables is said to be **overspecified,** as there is more information than is needed. An overspecified system can be consistent, dependent, or inconsistent.

EXAMPLE 8 Solve each system.

(a) $\begin{cases} x + y + z = 1 \\ x - y - z = 2 \end{cases}$ (b) $\begin{cases} x - 4y = -5 \\ 2x - 3y = 0 \\ 3x + 2y = 13 \end{cases}$

Solutions (a) $\begin{cases} x + y + z = 1 \\ x - y - z = 2 \end{cases}$

As this system has three variables but only two equations, it is underspecified. Adding the equations together yields

$$2x = 3$$

$$x = \frac{3}{2}$$

Putting this result into either original equation gives

$$y + z = -\frac{1}{2}$$

$$y = -\frac{1}{2} - z$$

$$\left\{\left(\frac{3}{2}, -\frac{1}{2} - t, t\right) \middle| t \in R\right\} \qquad \text{Solution set}$$

(b) $\begin{cases} x - 4y = -5 \\ 2x - 3y = 0 \\ 3x + 2y = 13 \end{cases}$

Here we have two variables and three equations. It is overspecified. We put it into triangular form. Multiply the first equation by -2 and add to the second.

$$\begin{cases} x - 4y = -5 \\ 5y = 10 \\ 3x + 2y = 13 \end{cases}$$

Multiply the first equation by -3 and add to the third.

$$\begin{cases} x - 4y = -5 \\ 5y = 10 \\ 14y = 28 \end{cases}$$

At this point we can see that the last two equations are the same. We have $y = 2$ and, by back-substitution, $x = 3$. These check in all three equations. The system is consistent, and the solution set is $\{(3, 2)\}$. □

Notice what happens if we continue to put the system in part (b) of Example 9 in triangular form.

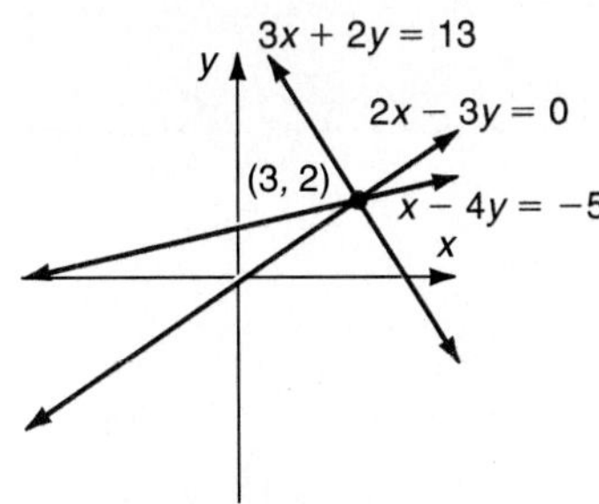

Overspecified, consistent system

Fig. 8.19

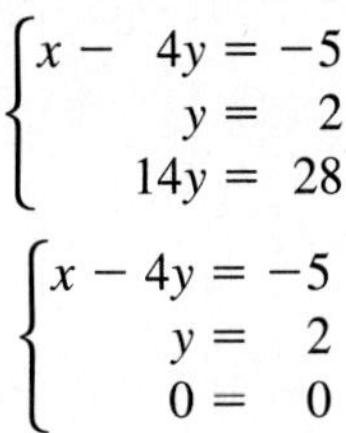

$$\begin{cases} x - 4y = -5 \\ y = 2 \\ 14y = 28 \end{cases}$$

$$\begin{cases} x - 4y = -5 \\ y = 2 \\ 0 = 0 \end{cases}$$

Note that although the variables drop out of the last equation, leaving $0 = 0$, the system *is not dependent*. With more equations than variables, this will always happen to a *consistent* system. If we change the constant term in the last equation, notice the result.

$$\begin{cases} x - 4y = -5 \\ 2x - 3y = 0 \\ 3x + 2y = 27 \end{cases}$$

Overspecified, inconsistent system

Fig. 8.20

Reduced to triangular form it becomes

$$\begin{cases} x - 4y = -5 \\ y = 2 \\ 0 = 1 \end{cases}$$

This system is inconsistent.

It is instructive to graph all the equations of the system in part (b) of Example 8 (Figure 8.19) and those of the inconsistent system (Figure 8.20).

PROBLEM SET 8.2

Warm-ups

In Problems 1 through 4, graph each ordered triple. See Example 1.

1. $(1, 2, 3)$ **2.** $(2, 1, -3)$ **3.** $(3, -2, 1)$ **4.** $(-1, 3, 2)$

In Problems 5 through 14, find the solution set for each system. For Problems 5 and 6, see Example 2.

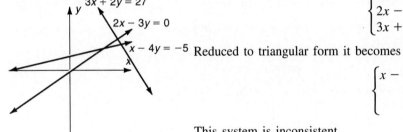

5. $\begin{cases} x + y - z = -1 \\ 2y - 3z = 3 \\ 2z = -2 \end{cases}$

6. $\begin{cases} x - y - z = 1 \\ y + z = 6 \\ -y + z = 4 \end{cases}$

For Problems 7 and 8, see Example 3.

7. $\begin{cases} x - y + z = -2 \\ 2x - y - z = -3 \\ -2y + z = -1 \end{cases}$

8. $\begin{cases} x + y + z = 7 \\ 3x + 4y - z = -5 \\ 3y + z = 0 \end{cases}$

For Problems 9 and 10, see Example 4.

9. $\begin{cases} x + y + 2z = 7 \\ 2x + y - z = 5 \\ 3x - y + z = 15 \end{cases}$

10. $\begin{cases} x - y + 5z = 1 \\ 5x - 4y + z = -16 \\ -2x + y - 3z = 2 \end{cases}$

For Problems 11 and 12, see Example 5.

11. $\begin{cases} x + 2y - z = -3 \\ 2x + 5y + z = 6 \\ y + 3z = 4 \end{cases}$

12. $\begin{cases} -x + 2y + z = 3 \\ -2x + 5y - z = 0 \\ y - 3z = 6 \end{cases}$

For Problems 13 and 14, see Example 6.

13. $\begin{cases} x + 3y - z = 2 \\ 2x + 7y - z = 3 \\ y + z = -1 \end{cases}$

14. $\begin{cases} x + y - z = 0 \\ x - y + z = 0 \\ y - z = 0 \end{cases}$

In Problems 15 and 16, find an equation of the parabola with a vertical axis of symmetry that contains the three points given. See Example 7.

15. (1, 1); (2, 0); (0, 4)

16. (−1, 3); (2, −6); (0, 2)

In Problems 17 through 22, solve each system. See Example 8.

17. $\begin{cases} x + 2y + z = 4 \\ y - 3z = 2 \end{cases}$

18. $\begin{cases} x + 2y + 3z = 7 \\ x = 5 \end{cases}$

19. $\begin{cases} x + 2y + z = 0 \\ 2x + y - z = 3 \end{cases}$

20. $\begin{cases} x_1 - 2x_2 - 3x_3 = 6 \\ 3x_1 + x_2 + 2x_3 = 0 \end{cases}$

21. $\begin{cases} x + 3y = 2 \\ 2x + 7y = 5 \\ x + y = 0 \end{cases}$

22. $\begin{cases} 2x + 3y = 1 \\ 3x + 4y = 0 \\ 4x + 5y = 1 \end{cases}$

Practice Exercises

In Problems 23 through 44, solve each system.

23. $\begin{cases} x - 2y = 5 \\ 3x + y = 4 \end{cases}$

24. $\begin{cases} 3x + y = 4 \\ x - 4y = 0 \end{cases}$

25. $\begin{cases} \frac{1}{2}x - \frac{1}{3}y = 0 \\ \frac{2}{5}x + \frac{3}{5}y = 0 \end{cases}$

26. $\begin{cases} \frac{2}{3}x + \frac{1}{6}y = 0 \\ \frac{1}{2}x - \frac{1}{4}y = 0 \end{cases}$

27. $\begin{cases} x + y = 4 \\ \frac{1}{2}x + \frac{1}{2}y = 0 \end{cases}$

28. $\begin{cases} \frac{1}{3}x - \frac{3}{4}y = 1 \\ 4x - 9y = 6 \end{cases}$

29. $\begin{cases} x + 6y = 1 \\ \frac{1}{2}x + 3y = \frac{1}{2} \end{cases}$

30. $\begin{cases} \frac{1}{3}x + \frac{3}{2}y = 0 \\ \frac{2}{7}x + \frac{9}{7}y = 0 \end{cases}$

31. $\begin{cases} x_1 + 2x_2 + x_3 = 3 \\ x_1 + 3x_2 - x_3 = -4 \\ x_1 - 2x_2 - 2x_3 = -2 \end{cases}$

32. $\begin{cases} x_1 + 2x_2 + x_3 = 0 \\ x_1 + 2x_2 + 2x_3 = -2 \\ x_1 - 2x_2 - x_3 = 0 \end{cases}$

33. $\begin{cases} x - y + 3z = 1 \\ -x + 2y - 2z = 1 \\ x - 3y + z = 2 \end{cases}$

34. $\begin{cases} x - y + 4z = -7 \\ 4x - 5y + 7z = 20 \\ -2x + 7y - 2z = -13 \end{cases}$

35. $\begin{cases} x + y + z = 17 \\ x - y = 6 \\ x - z = 7 \end{cases}$

36. $\begin{cases} x + y + z = 26 \\ y - z = -3 \\ y - x = -8 \end{cases}$

37. $\begin{cases} x + 2z = 28 \\ y + 10z = -45 \\ x - 2y + 4z = -36 \end{cases}$

38. $\begin{cases} x + y + 2z = 2 \\ 2x - 2y + 2z = 3 \\ x + y = 3 \end{cases}$

39. $\begin{cases} x_1 + 2x_2 + x_3 = 7 \\ x_1 - x_2 - 2x_3 = 4 \\ x_1 + x_2 = 4 \end{cases}$

40. $\begin{cases} x_1 - 5x_2 + 3x_3 = 9 \\ 2x_1 - 9x_2 + x_3 = 3 \\ x_2 - 5x_3 = -15 \end{cases}$

41. $\begin{cases} x + 2y - z = 0 \\ y + z = 2 \end{cases}$

42. $\begin{cases} y + 3z = 7 \\ x = 5 \end{cases}$

43. $\begin{cases} 5x - y + 3z + w = -2 \\ 3y - 2z - w = 2 \\ z + w = 3 \\ 8w = 16 \end{cases}$

44. $\begin{cases} x + y + z + w = 0 \\ y + z + w = -4 \\ z + w = 3 \\ x + w = 7 \end{cases}$

45. The sum of three numbers is 64. One of the numbers is twice the average of the other two. The difference between the other two is 6. What are the three numbers?

46. Find three numbers such that the first plus twice the second plus three times the third is 57, if three times the first plus five times the third is 89 and the third is the average of the other two.

47. Amanda, Mimi, and Kristen made several trips to the concession stand at the local basketball game. Amanda bought 3 small and 2 large drinks and 1 hotdog for \$5.45. Mimi bought 1 small drink and 2 large drinks with 1 hotdog for \$4.15, and Kristen bought 1 large drink and 2 hotdogs for \$4.00. What was the price of hotdogs and drinks at the concession stand?

48. The Boyds went soup shopping independently at the local grocery for their trip to the mountains. Spencer bought 5 cans of tomato and 3 cans of mushroom for \$5.99, while Lisa bought 2 cans of chicken noodle and 1 can of mushroom for \$2.36. If Jennifer bought 2 cans of tomato and 3 cans of chicken noodle for \$4.37, what is the price of tomato soup, mushroom soup, and chicken noodle soup at the Boyd's grocery?

49. Pipe A and pipe B, working together with the drain closed, can fill a tank in 12 minutes. With pipe A closed and the drain open, it takes pipe B 40 minutes to fill the tank. With pipe B closed and the drain open, it takes pipe A 2 hours to fill the tank. How long does it take the open drain to empty a full tank?

50. Write an equation for the parabola with a vertical axis of symmetry that passes through the points (0, 2), (2, 4), and (1, 2).

51. Is it possible to draw a parabola with a horizontal axis of symmetry through the three points in Problem 50?

52. Write an equation for the parabola with a vertical axis of symmetry that contains the points (0,4), (1, 5), and (−2, 8).

53. Is it possible to draw a parabola with a horizontal axis of symmetry through the three points in Problem 52?

54. Write an equation of the circle that contains the points (1,2), (−1, 0), and (1, −2). (*Hint:* The equation of a circle can be written in the form $x^2 + y^2 + ax + by + c = 0$.)

55. Write an equation for the circle that contains the points (5, −1), (3, 1), and (1, −1).

Challenge Problems

56. Given any three points, can we always draw a parabola through the points?

57. For what numbers k will the system $\begin{cases} x + y + z = 4 \\ y + z = 8 \\ kz = 10 \end{cases}$ have one solution? What is that solution?

58. A system of equations in which all the constant terms are zero is called a **homogeneous system.**
(a) Solve the homogeneous system in column 2 graphically.

$$\begin{cases} x + y = 0 \\ 3x - y = 0 \end{cases}$$

(b) Can a homogeneous system have no solution?
(c) Can a homogeneous system have an infinite number of solutions?
(d) Write a theorem about homogeneous systems of equations.

IN YOUR OWN WORDS . . .

59. Explain what *triangular form* means in relation to a system of equations. Why is it so named?

60. Compare graphical methods with algebraic methods of solving systems with two variables. Which is easier? Which is more reliable? What if the system has more than two variables?

8.3 THE METHOD OF THE AUGMENTED MATRIX

Consider the following system of linear equations.

$$\begin{cases} 2x + 3y = 5 \\ 2x - 3y = 11 \end{cases}$$

In a few steps we find the solution set $\{(4, -1)\}$.
Next consider the system

$$\begin{cases} 2s + 3t = 5 \\ 2s - 3t = 11 \end{cases}$$

Notice that it has the same coefficients and the same solution set. Whether the variables are called x and y, or s and t, or even θ and ϕ, the solutions are exactly the same. It's the *coefficients* that determine the solutions of a system. If we remove everything except the coefficients in the above system, we have a rectangular array of numbers

$$\begin{matrix} 2 & 3 & 5 \\ 2 & -3 & 11 \end{matrix}$$

that still contains the essence of the problem. In this section we exploit this property to develop a technique for solving systems of linear equations.

Matrices

A **matrix** is a rectangular array of numbers usually written inside brackets or parentheses. For example,

$$\begin{pmatrix} 1 & 2 & 3 & 4 \\ 5 & 6 & 7 & 8 \\ 9 & 10 & 11 & 12 \end{pmatrix}$$

The numbers in the array are called **elements** or **entries.** We identify the position of elements in a matrix by a row and a column.

$$\begin{array}{r c} & \text{4 Columns} \\ & \downarrow \quad \downarrow \quad \downarrow \quad \downarrow \\ \begin{array}{r} \rightarrow \\ \text{3 Rows} \rightarrow \\ \rightarrow \end{array} & \begin{pmatrix} 1 & 2 & 3 & 4 \\ 5 & 6 & 7 & 8 \\ 9 & 10 & 11 & 12 \end{pmatrix} \end{array}$$

The number 9 in this matrix is in row 3 column 1. We identify the **size** (or **dimension**) of a matrix by the number of rows followed by the number of columns (rows always before columns). The matrix above has 3 rows and 4 columns, so it is a 3×4 matrix.

$$\begin{matrix} 2 \times 3 & 2 \times 2 & 3 \times 2 & 3 \times 1 \\ \begin{pmatrix} 1 & 2 & 3 \\ 4 & 5 & 6 \end{pmatrix} & \begin{pmatrix} 1 & 0 \\ 0 & 1 \end{pmatrix} & \begin{pmatrix} 1 & 0 \\ 3 & 2 \\ 5 & -1 \end{pmatrix} & \begin{pmatrix} -3 \\ 2 \\ -6 \end{pmatrix} \end{matrix}$$

We solved systems of linear equations by Gaussian elimination in the last section. Much copying of equations was required to produce equivalent systems of equations. If we record our work carefully, we can carry out the same steps without writing the variables.

Suppose we have a system of three linear equations in three variables written in the indicated form.

$$\begin{cases} Ax + By + Cz = P \\ Dx + Ey + Fz = Q \\ Gx + Hy + Iz = R \end{cases}$$

If we copy the *coefficients* of the variables into a matrix in the same position that they occur in the system, we form the matrix

$$\begin{pmatrix} A & B & C \\ D & E & F \\ G & H & I \end{pmatrix}$$

This matrix is called the **matrix of coefficients** or **coefficient matrix** of the system.

Augmented Matrices

If we add a column on the right side of the matrix containing the constant terms of the system, we have the **augmented matrix** of the system.

$$\left(\begin{array}{ccc|c} A & B & C & P \\ D & E & F & Q \\ G & H & I & R \end{array}\right)$$

A line is often placed in the matrix to separate the matrix of coefficients from the constants. It is not part of the matrix and is thus not required. Two augmented matrices are **equivalent** if the systems they represent are equivalent.

EXAMPLE 1 Write the augmented matrix of the system $\begin{cases} 4x + y - z = 3 \\ x + 2y + z = 0. \\ 2x - y + 3z = -5 \end{cases}$

Solution We write the matrix by inspection.

$$\left(\begin{array}{ccc|c} 4 & 1 & -1 & 3 \\ 1 & 2 & 1 & 0 \\ 2 & -1 & 3 & -5 \end{array}\right)$$

□

EXAMPLE 2 Write the augmented matrix of the system $\begin{cases} 7x + 2y - z = 3 \\ x + z = 0. \\ y + 3z = 5 \end{cases}$

Solution First, we write the system inserting zeros as coefficients where variables are missing.

$$\begin{cases} 7x + 2y - z = 3 \\ x + 0y + z = 0 \\ 0x + y + 3z = 5 \end{cases}$$

Now we write the augmented matrix.

$$\left(\begin{array}{ccc|c} 7 & 2 & -1 & 3 \\ 1 & 0 & 1 & 0 \\ 0 & 1 & 3 & 5 \end{array}\right)$$

□

EXAMPLE 3 Write a system of equations that has $\left(\begin{array}{cc|c} 1 & 2 & 3 \\ 4 & 5 & 6 \end{array}\right)$ as its augmented matrix.

Solution If we use x and y as variables, the given matrix yields the system $\begin{cases} x + 2y = 3 \\ 4x + 5y = 6 \end{cases}$.

□

Solving Systems Using Augmented Matrices

Consider the following steps in solving the given system and notice the augmented matrix at each step.

System of equations — Augmented matrix

$$\begin{cases} x + 2y - z = -1 \\ x + y + z = 2 \\ y - 3z = -5 \end{cases} \qquad \left(\begin{array}{ccc|c} 1 & 2 & -1 & -1 \\ 1 & 1 & 1 & 2 \\ 0 & 1 & -3 & -5 \end{array}\right)$$

Subtract the second equation from the first.

$$\begin{cases} x + 2y - z = -1 \\ y - 2z = -3 \\ y - 3z = -5 \end{cases} \qquad \left(\begin{array}{ccc|c} 1 & 2 & -1 & -1 \\ 0 & 1 & -2 & -3 \\ 0 & 1 & -3 & -5 \end{array}\right)$$

Subtract the third equation from the second equation.

$$\begin{cases} x + 2y - z = -1 \\ y - 2z = -3 \\ z = 2 \end{cases} \qquad \left(\begin{array}{ccc|c} 1 & 2 & -1 & -1 \\ 0 & 1 & -2 & -3 \\ 0 & 0 & 1 & 2 \end{array}\right)$$

The system is in triangular form at this point. Notice that the missing variables show up in the augmented matrix as zeros. The augmented matrix is said to be in *triangular form* if the system it represents is in triangular form. Notice the 1s on the diagonal of the coefficient matrix. These 1s are called **leading ones.** Notice also that each entry below a leading 1 is zero. Such a matrix is said to be in *row echelon form.*

Row Echelon Form

A matrix having the following properties is said to be in **row echelon form.**

1. All rows that are completely zeros are grouped at the bottom of the matrix.
2. If a row is not all zeros, then the first nonzero entry is a 1.
3. Each leading 1 is located to the *right* of the leading 1 of the row immediately above it.

If, in addition to the above, each column that contains a leading 1 has zeros *everywhere else,* then the matrix is said to be in **reduced row echelon form.**

The idea of the method of augmented matrices is to form the augmented matrix of the system to be solved and then write a series of equivalent matrices, ending with a matrix in row echelon form. The elementary operations that we used in Section 8.2 are now called *elementary row operations.*

Elementary Row Operations

The following operations when performed on an augmented matrix will produce an equivalent matrix.

1. Interchange two rows.
2. Multiply the numbers in any row by a nonzero number.
3. Add a multiple of one row to another row.

To produce row echelon form or reduced row echelon form, it is best to work each column in turn from left to right in the augmented matrix. The strategy is to get the leading 1s first and then make the entries below the leading 1 in the column zero (and above the leading 1 for *reduced* row echelon form). The next example will illustrate the idea.

EXAMPLE 4 Solve the system $\begin{cases} x + 2y + z = -1 \\ 2x + y - z = -5. \\ 3x + y + 4z = -2 \end{cases}$

Solution Form the augmented matrix for the system.

$$\left(\begin{array}{ccc|c} 1 & 2 & 1 & -1 \\ 2 & 1 & -1 & -5 \\ 3 & 1 & 4 & -2 \end{array}\right)$$

Now, to find an equivalent matrix in row echelon form, we notice that there is a 1 in row 1 column 1. We will get zeros in the rest of column 1.

$$\left(\begin{array}{ccc|c} 1 & 2 & 1 & -1 \\ 2 & 1 & -1 & -5 \\ 3 & 1 & 4 & -2 \end{array}\right) \xrightarrow[\text{and add to row 2.}]{\text{Multiply row 1 by } -2} \left(\begin{array}{ccc|c} 1 & 2 & 1 & -1 \\ 0 & -3 & -3 & -3 \\ 3 & 1 & 4 & -2 \end{array}\right)$$

$$\left(\begin{array}{ccc|c} 1 & 2 & 1 & -1 \\ 0 & -3 & -3 & -3 \\ 3 & 1 & 4 & -2 \end{array}\right) \xrightarrow[\text{and add to row 3.}]{\text{Multiply row 1 by } -3} \left(\begin{array}{ccc|c} 1 & 2 & 1 & -1 \\ 0 & -3 & -3 & -3 \\ 0 & -5 & 1 & 1 \end{array}\right)$$

Now we will move to the second column and get a leading 1 in row 2. We divide row 2 by -3.

$$\left(\begin{array}{ccc|c} 1 & 2 & 1 & -1 \\ 0 & -3 & -3 & -3 \\ 0 & -5 & 1 & 1 \end{array}\right) \xrightarrow[\text{Multiply row 2 by } -\frac{1}{3}.]{} \left(\begin{array}{ccc|c} 1 & 2 & 1 & -1 \\ 0 & 1 & 1 & 1 \\ 0 & -5 & 1 & 1 \end{array}\right)$$

The next step is to get a zero under the leading 1. We multiply row 2 by 5 and add to row 3.

$$\left(\begin{array}{ccc|c} 1 & 2 & 1 & -1 \\ 0 & 1 & 1 & 1 \\ 0 & -5 & 1 & 1 \end{array}\right) \xrightarrow[\text{and add to row 3.}]{\text{Multiply row 2 by 5}} \left(\begin{array}{ccc|c} 1 & 2 & 1 & -1 \\ 0 & 1 & 1 & 1 \\ 0 & 0 & 6 & 6 \end{array}\right)$$

Now we move to the third column and get a leading 1 in row 3.

$$\left(\begin{array}{ccc|c} 1 & 2 & 1 & -1 \\ 0 & 1 & 1 & 1 \\ 0 & 0 & 6 & 6 \end{array}\right) \xrightarrow[\text{Multiply row 3 by } \frac{1}{6}.]{} \left(\begin{array}{ccc|c} 1 & 2 & 1 & -1 \\ 0 & 1 & 1 & 1 \\ 0 & 0 & 1 & 1 \end{array}\right)$$

Now the augmented matrix is in row echelon form. We write the system that is equivalent to this matrix.

$$\begin{cases} x + 2y + z = -1 \\ \quad\;\; y + z = 1 \\ \qquad\quad z = 1 \end{cases}$$

Using back-substitution, we find the solution set $\{(-2, 0, 1)\}$. □

The next example points out some useful variations that arise with augmented matrices.

EXAMPLE 5 Each matrix is an augmented matrix for a system of linear equations. Find the solution set for each system.

(a) $\left(\begin{array}{ccc|c} 1 & 0 & 0 & 2 \\ 0 & 1 & 0 & 0 \\ 0 & 0 & 1 & -1 \end{array}\right)$ (b) $\left(\begin{array}{ccc|c} 1 & 2 & -1 & 3 \\ 0 & 1 & 1 & 2 \\ 0 & 0 & 1 & 5 \end{array}\right)$ (c) $\left(\begin{array}{ccc|c} 1 & 1 & 1 & 0 \\ 0 & 2 & -1 & 1 \\ 0 & 0 & 3 & 3 \end{array}\right)$

Solutions (a) $\left(\begin{array}{ccc|c} 1 & 0 & 0 & 2 \\ 0 & 1 & 0 & 0 \\ 0 & 0 & 1 & -1 \end{array}\right)$

This matrix is in *reduced* row echelon form. Notice that we can write the solution set directly by inspection. It is $\{(2, 0, -1)\}$. The solutions can be found without back-substitution because there are also zeros *above* the leading 1s. This is the advantage of the reduced row echelon form.

(b) $\left(\begin{array}{ccc|c} 1 & 2 & -1 & 3 \\ 0 & 1 & 1 & 2 \\ 0 & 0 & 1 & 5 \end{array}\right)$

The matrix is in row echelon form with zeros below leading 1s. We see from the third row that $z = 5$. Using back-substitution in the second row, we have $y = -3$. Using the first row, we have $x = 14$. The solution set is $\{(14, -3, 5)\}$.

(c) $\left(\begin{array}{ccc|c} 1 & 1 & 1 & 0 \\ 0 & 2 & -1 & 1 \\ 0 & 0 & 3 & 3 \end{array}\right)$

Notice that this matrix is not in row echelon form because the leading 1s are missing. However, we can still use back-substitution to find the solution set. The third row gives $z = 1$. Substituting into the second row gives $y = 1$, and into the first row gives $x = -2$. The solution set is $\{(-2, 1, 1)\}$. □

Each of the augmented matrices in Example 5 illustrates a way of using an augmented matrix to solve a system of linear equations. It is a matter of preference as to which form to use.

The Method of the Augmented Matrix

1. Write the augmented matrix.
2. Using elementary row operations, change the matrix to row echelon form. It is important to make the changes in the following order.
   (a) Move all completely zero rows to the bottom of the matrix.
   (b) Get the number 1 in row 1 column 1.

(c) Get zeros in column 1 under the leading 1 in row 1.
(d) Get a 1 in row 2 column 2 if possible. If not, go to the next column.
(e) Get zeros under the leading 1 formed in step (d).
(f) Continue steps similar to steps (a), (d), and (e) until the matrix is in row echelon form.

3. Find each variable in turn.

4. Write the solution set.

In steps 2c and 2e, also get zeros *above* the leading 1 for *reduced* row echelon form.

EXAMPLE 6 Find the solution set for each system using the augmented matrix method.

$$\text{(a)} \begin{cases} 3x - y + 4z = 0 \\ x - y + z = 0 \\ 4x - 2y + 5z = 2 \end{cases} \qquad \text{(b)} \begin{cases} x - y = 4 \\ y + z = 3 \\ x + z = 7 \end{cases}$$

Solutions (a) We form the augmented matrix.

$$\left(\begin{array}{rrr|r} 3 & -1 & 4 & 0 \\ 1 & -1 & 1 & 0 \\ 4 & -2 & 5 & 2 \end{array}\right) \xrightarrow[\text{Interchange row 1 and row 2.}]{} \left(\begin{array}{rrr|r} 1 & -1 & 1 & 0 \\ 3 & -1 & 4 & 0 \\ 4 & -2 & 5 & 2 \end{array}\right)$$

$$\left(\begin{array}{rrr|r} 1 & -1 & 1 & 0 \\ 3 & -1 & 4 & 0 \\ 4 & -2 & 5 & 2 \end{array}\right) \xrightarrow[\text{Multiply row 1 by } -3 \text{ and add to row 2.}]{} \left(\begin{array}{rrr|r} 1 & -1 & 1 & 0 \\ 0 & 2 & 1 & 0 \\ 4 & -2 & 5 & 2 \end{array}\right)$$

$$\left(\begin{array}{rrr|r} 1 & -1 & 1 & 0 \\ 0 & 2 & 1 & 0 \\ 4 & -2 & 5 & 2 \end{array}\right) \xrightarrow[\text{Multiply row 1 by } -4 \text{ and add to row 3.}]{} \left(\begin{array}{rrr|r} 1 & -1 & 1 & 0 \\ 0 & 2 & 1 & 0 \\ 0 & 2 & 1 & 2 \end{array}\right)$$

Rather than make a leading 1 in column 2, we notice that rows 2 and 3 are alike except for the constants. Notice what happens when row 3 is subtracted from row 2.

$$\left(\begin{array}{rrr|r} 1 & -1 & 1 & 0 \\ 0 & 2 & 1 & 0 \\ 0 & 2 & 1 & 2 \end{array}\right) \xrightarrow[\text{Subtract row 2 from row 3.}]{} \left(\begin{array}{rrr|r} 1 & -1 & 1 & 0 \\ 0 & 2 & 1 & 0 \\ 0 & 0 & 0 & 2 \end{array}\right)$$

The last row says that $0x + 0y + 0z = 2$ or $0 = 2$. This false statement indicates that there are no solutions to the system. The system is inconsistent. The solution set is $\varnothing$.

(b) We must be careful to insert zeros in forming the augmented matrix for

$$\begin{cases} x - y = 4 \\ y + z = 3. \\ x + z = 7 \end{cases}$$

$$\left(\begin{array}{ccc|c} 1 & -1 & 0 & 4 \\ 0 & 1 & 1 & 3 \\ 1 & 0 & 1 & 7 \end{array}\right) \xrightarrow[\text{from the third row.}]{\text{Subtract the first row}} \left(\begin{array}{ccc|c} 1 & -1 & 0 & 4 \\ 0 & 1 & 1 & 3 \\ 0 & 1 & 1 & 3 \end{array}\right)$$

Notice what happens if we subtract the second row from the third row.

$$\left(\begin{array}{ccc|c} 1 & -1 & 0 & 4 \\ 0 & 1 & 1 & 3 \\ 0 & 1 & 1 & 3 \end{array}\right) \xrightarrow[\text{from the third row.}]{\text{Subtract the second row}} \left(\begin{array}{ccc|c} 1 & -1 & 0 & 4 \\ 0 & 1 & 1 & 3 \\ 0 & 0 & 0 & 0 \end{array}\right)$$

The system that is equivalent to the original system is $\begin{cases} x - y = 4 \\ y + z = 3. \\ 0 = 0 \end{cases}$

The last equation says that $0 = 0$. This indicates that the system is dependent. To write the solution set, solve the second equation for y.

$$y + z = 3$$

$$y = 3 - z$$

Solving the first equation for x gives $x = 4 + y$. Now substitute $3 - z$ for y, and we have

$$x = 4 + 3 - z$$

$$x = 7 - z$$

The solution set is $\{(7 - t, 3 - t, t) \mid t \in R\}$. □

PROBLEM SET 8.3

Warm-ups

In Problems 1 and 2, write the augmented matrix of each system. See Examples 1 and 2.

1. $\begin{cases} 4x + y = 7 \\ x + 2y + 3z = -8 \\ 2x + 3z = -1 \end{cases}$

2. $\begin{cases} 4x + 2y - z - 6w = 3 \\ x + y - z + 5w = -1 \\ x + z - 2w = -1 \\ 3x - 2y + z + w = 0 \end{cases}$

In Problems 3 and 4, write a system of equations that has the given matrix as its augmented matrix. See Example 3.

3. $\begin{pmatrix} 1 & 0 & -3 \\ -4 & 5 & 8 \end{pmatrix}$

4. $\begin{pmatrix} 3 & -5 & 1 & 9 \\ 1 & -2 & 4 & 0 \\ 0 & 7 & 1 & -2 \end{pmatrix}$

In Problems 5 and 6, solve the system using the augmented matrix. See Example 4.

5. $\begin{cases} 2x + 3y - 2z = 2 \\ x - y + 2z = 5 \\ 2y + z = 7 \end{cases}$

6. $\begin{cases} 3x + 2y = -1 \\ x + 2y - 2z = 0 \\ y - 4z = -1 \end{cases}$

In Problems 7 through 10, the matrix given is an augmented matrix for a system of linear equations. Find the solution set for each system. See Example 5.

7. $\begin{pmatrix} 1 & 0 & 0 & 7 \\ 1 & 1 & 0 & -4 \\ 0 & 1 & 1 & -1 \end{pmatrix}$ **8.** $\begin{pmatrix} 1 & -2 & 1 & 0 \\ 0 & 1 & 1 & 7 \\ 0 & 0 & 1 & 3 \end{pmatrix}$ **9.** $\begin{pmatrix} 1 & 0 & 0 \\ 0 & 1 & 0 \end{pmatrix}$ **10.** $\begin{pmatrix} 1 & 1 & 3 \\ 1 & 0 & -1 \end{pmatrix}$

In Problems 11 through 14, find the solution set for each system using the augmented matrix method. See Example 6.

11. $\begin{cases} x + y = 2 \\ x + z = 3 \\ y - z = -1 \end{cases}$ **12.** $\begin{cases} 2x + y + z = 5 \\ x - y + z = -1 \\ 3x + 2z = 4 \end{cases}$ **13.** $\begin{cases} 3x - y + z = 4 \\ y + z = 1 \\ 3x + 2z = 5 \end{cases}$ **14.** $\begin{cases} x + y + z = 0 \\ 2x + y - z = 3 \\ y + 3z = 0 \end{cases}$

Practice Exercises

In Problems 15 through 20, an augmented matrix for a system of linear equations is given. Find the solution set for each system.

15. $\begin{pmatrix} 1 & 2 & -1 & 4 \\ 0 & 1 & 0 & 3 \\ 0 & 0 & 0 & 5 \end{pmatrix}$ **16.** $\begin{pmatrix} 1 & 0 & 1 & 0 \\ 0 & 1 & 1 & 0 \\ 0 & 0 & 1 & 0 \end{pmatrix}$ **17.** $\begin{pmatrix} 2 & 3 & -4 \\ 0 & 1 & 3 \end{pmatrix}$ **18.** $\begin{pmatrix} 1 & 1 & 1 \\ 0 & 0 & 0 \end{pmatrix}$

19. $\begin{pmatrix} 1 & 2 & 3 & 1 \\ 0 & 1 & -1 & 4 \\ 0 & 0 & 2 & 0 \end{pmatrix}$ **20.** $\begin{pmatrix} 1 & 2 & -1 & 4 \\ 0 & 1 & 1 & 3 \end{pmatrix}$

In Problems 21 through 42, find the solution set using an augmented matrix.

21. $\begin{cases} x_1 + 3x_2 = 4 \\ 3x_1 + 8x_2 = 2 \end{cases}$ **22.** $\begin{cases} 3x_1 - 2x_2 = 7 \\ x_1 - x_2 = 4 \end{cases}$ **23.** $\begin{cases} 2x - 3y = 6 \\ 3x + 2y = 2 \end{cases}$ **24.** $\begin{cases} 5x - 10y = 2 \\ 3x - 4y = 6 \end{cases}$

25. $\begin{cases} \frac{1}{2}x - \frac{1}{2}y = 3 \\ 3x - 3y = 0 \end{cases}$ **26.** $\begin{cases} 5x + 5y = 3 \\ \frac{1}{2}x + \frac{1}{2}y = \frac{1}{5} \end{cases}$ **27.** $\begin{cases} 3x + 7y = 0 \\ x + \frac{7}{3}y = 0 \end{cases}$ **28.** $\begin{cases} x - 2y = 6 \\ \frac{1}{3}x - \frac{2}{3}y = 2 \end{cases}$

29. $\begin{cases} x - 3y + 6z = -3 \\ 3y - 3z = 2 \\ y + 9z = 4 \end{cases}$ **30.** $\begin{cases} x - y + z = 1 \\ x - 2y - z = 2 \\ 4y + z = -1 \end{cases}$ **31.** $\begin{cases} x + y + z = 2 \\ 2x + y - z = -3 \\ x - 3y - z = 4 \end{cases}$ **32.** $\begin{cases} x + 2y - z = 3 \\ 2x + 5y - 4z = 3 \\ -x + y = 8 \end{cases}$

33. $\begin{cases} x_1 - 2x_2 + x_3 = 3 \\ 2x_1 + 3x_2 - x_3 = 4 \\ 3x_1 + x_2 = 0 \end{cases}$ **34.** $\begin{cases} x_1 - 3x_2 + x_3 = 11 \\ 3x_1 + 10x_2 = 4 \\ 4x_1 + 7x_2 + x_3 = 8 \end{cases}$ **35.** $\begin{cases} x + 3y - z = -2 \\ 2x - 5y + z = 3 \\ 3x - 2y = 1 \end{cases}$ **36.** $\begin{cases} x + z = 4 \\ y + z = 6 \\ 3x - 3y = -6 \end{cases}$

37. $\begin{cases} x + y + z = 0 \\ 2x + y + 3z = 0 \\ 3x + 4y + z = 0 \end{cases}$ **38.** $\begin{cases} x - y - z = 0 \\ 2x - y + z = 0 \\ -3x + 2y - z = 0 \end{cases}$ **39.** $\begin{cases} x + y + z = 0 \\ 2x + y + 3z = 0 \\ x + 2z = 0 \end{cases}$ **40.** $\begin{cases} x - y - z = 0 \\ 2x - y + z = 0 \\ -x + y = 0 \end{cases}$

41. $\begin{cases} x + y + z = 4 \\ y + z + w = 6 \\ x + y + w = 5 \\ x + z + w = 9 \end{cases}$ **42.** $\begin{cases} x + 2y - w = -9 \\ 2y - 2z - 2w = 5 \\ x + 4y + z = 4 \\ z + w = -2 \end{cases}$

43. The sum of four integers is 19. The sum of the first and second numbers is 1, while the sum of the second and third numbers is the fourth number. The third is 3 more than the second. Find the numbers.

44. The sum of four numbers is 11. Twice the first plus the third is 2, while three times the first added to twice the third is 5. If three times the second plus twice the fourth is 17, what are the four numbers?

Challenge Problems

45. In the following augmented matrix, how can a 1 in row 1 column 1 be obtained without introducing fractions?

$$\left(\begin{array}{ccc|c} 2 & 3 & 4 & 1 \\ 5 & 1 & -1 & 4 \\ 3 & 7 & 2 & 0 \end{array}\right)$$

IN YOUR OWN WORDS . . .

46. Explain what an augmented matrix is and how it is used.
47. Compare solving a system by writing the augmented matrix in *row echelon form* and writing the augmented matrix in *reduced row echelon form.* Which is more difficult? Which is more convenient?

8.4 MATRIX ALGEBRA

Using an augmented matrix to solve a system of linear equations is but one use of matrices. In this section we examine operations with matrices and learn about some properties of matrices.

We have seen that a matrix is a rectangular array of numbers with its size specified by indicating the number of rows followed by the number of columns. Capital letters are commonly used to name matrices. It is convenient to use lowercase letters with two subscripts to denote the entries, the first to designate the row and the second to designate the column. For example, a general 2×3 matrix named A could be designated by

$$A = \begin{pmatrix} a_{11} & a_{12} & a_{13} \\ a_{21} & a_{22} & a_{23} \end{pmatrix}$$

A matrix with the same number of rows as columns is called a **square** matrix. For example, $B = \begin{pmatrix} b_{11} & b_{12} & b_{13} \\ b_{21} & b_{22} & b_{23} \\ b_{31} & b_{32} & b_{33} \end{pmatrix}$ is a 3×3 square matrix. The entries b_{11}, b_{22}, and b_{33} form the **main diagonal.**

Equality

We begin our further study of matrices by defining equality of two matrices.

Definition: Equal Matrices

Two **matrices are equal** if and only if they are the same size and corresponding entries are equal.

That is, $A = B$ if and only if $a_{ij} = b_{ij}$ for all appropriate i and j.

EXAMPLE 1 If $\begin{pmatrix} 1 & 2 \\ 3 & 4 \end{pmatrix} = \begin{pmatrix} x+y & 2 \\ 3 & 2x \end{pmatrix}$, find x and y.

Solution The two matrices can be equal only if corresponding entries are equal. This means that $x + y = 1$ and $2x = 4$. So x is 2 and y is -1. □

We perform operations of addition, subtraction, and multiplication with matrices. Arithmetic with matrices is patterned somewhat on arithmetic with real numbers.

Addition

Definition: Addition of Matrices

If A and B are matrices of the same size, then $A + B$ is the matrix formed by adding corresponding entries in A and B.

That is, $C = A + B$ if and only if $c_{ij} = a_{ij} + b_{ij}$ for all appropriate i and j.

To add two matrices, both matrices must be the same size and the sum will also be the same size.

EXAMPLE 2 Perform the indicated additions.

(a) $\begin{pmatrix} 1 & -1 \\ -2 & 3 \end{pmatrix} + \begin{pmatrix} -2 & 3 \\ 4 & 7 \end{pmatrix}$ (b) $\begin{pmatrix} 2 & 3 & 4 \\ 5 & 6 & 7 \end{pmatrix} + \begin{pmatrix} -2 & -3 & -4 \\ -5 & -6 & -7 \end{pmatrix}$

(c) $\begin{pmatrix} 1 & 2 \\ 0 & -1 \end{pmatrix} + \begin{pmatrix} 1 & 3 & -4 \\ -5 & 6 & 7 \end{pmatrix}$

Solutions (a) $\begin{pmatrix} 1 & -1 \\ -2 & 3 \end{pmatrix} + \begin{pmatrix} -2 & 3 \\ 4 & 7 \end{pmatrix} = \begin{pmatrix} -1 & 2 \\ 2 & 10 \end{pmatrix}$

(b) $\begin{pmatrix} 2 & 3 & 4 \\ 5 & 6 & 7 \end{pmatrix} + \begin{pmatrix} -2 & -3 & -4 \\ -5 & -6 & -7 \end{pmatrix} = \begin{pmatrix} 0 & 0 & 0 \\ 0 & 0 & 0 \end{pmatrix}$

(c) $\begin{pmatrix} 1 & 2 \\ 0 & -1 \end{pmatrix} + \begin{pmatrix} 1 & 3 & -4 \\ -5 & 6 & 7 \end{pmatrix}$ The matrices are not the same size. The addition is not defined. □

A matrix that has all entries as zeros is called a **zero matrix.** Because $A + 0 = A = 0 + A$ when the addition is defined, a zero matrix is called an **additive identity.**

Scalar Multiplication

Two kinds of multiplication are defined with matrices—the product of two matrices and the product of a real number and a matrix. We consider the product of a real number and a matrix first. Real numbers in this product are called **scalars.** Thus, the name given to this multiplication is *scalar multiplication.*

Definition: Scalar Multiplication

If A is a matrix and c a scalar, then cA is the matrix formed by multiplying each entry in A by c.

EXAMPLE 3 If $A = \begin{pmatrix} 1 & -1 \\ 2 & 3 \\ -3 & 4 \end{pmatrix}$, find each of the following.

(a) $2A$ (b) $-2A$ (c) $(kl)A$

Solutions (a) $2A = 2\begin{pmatrix} 1 & -1 \\ 2 & 3 \\ -3 & 4 \end{pmatrix} = \begin{pmatrix} 2 & -2 \\ 4 & 6 \\ -6 & 8 \end{pmatrix}$ Each entry in A is multiplied by 2.

(b) $-2A = -2\begin{pmatrix} 1 & -1 \\ 2 & 3 \\ -3 & 4 \end{pmatrix} = \begin{pmatrix} -2 & 2 \\ -4 & -6 \\ 6 & -8 \end{pmatrix}$

(c) $(kl)A = (kl)\begin{pmatrix} 1 & -1 \\ 2 & 3 \\ -3 & 4 \end{pmatrix} = \begin{pmatrix} kl & -kl \\ 2kl & 3kl \\ -3kl & 4kl \end{pmatrix}$ □

Before we can define subtraction, we must define the opposite of a matrix. Using the matrix in Example 3, $A = \begin{pmatrix} 1 & -1 \\ 2 & 3 \\ -3 & 4 \end{pmatrix}$, notice that

$$A + (-1)A = \begin{pmatrix} 1 & -1 \\ 2 & 3 \\ -3 & 4 \end{pmatrix} + \begin{pmatrix} -1 & 1 \\ -2 & -3 \\ 3 & -4 \end{pmatrix}$$

$$= \begin{pmatrix} 0 & 0 \\ 0 & 0 \\ 0 & 0 \end{pmatrix}$$

The sum of the two matrices is the additive identity. That means that $(-1)A$ is the opposite of A.

> **Definition: Opposite of a Matrix**
>
> If A is a matrix, then $-A = (-1)A$.
>
> $-A$ is called the **opposite** of A.

Subtraction

The opposite of a matrix leads to the definition of subtraction.

> **Definition: Subtraction of Matrices**
>
> If A and B are matrices of the same size, then $A - B = A + (-B)$.

That is, subtracting a matrix is the same as adding its opposite.

EXAMPLE 4 Perform the indicated operations given that $A = \begin{pmatrix} 2 & -1 \\ 0 & 4 \end{pmatrix}$ and $B = \begin{pmatrix} 1 & 5 \\ -3 & 2 \end{pmatrix}$.

(a) $A - B$ (b) $B - A$ (c) $2A - 3B$

Solutions (a) $A - B = \begin{pmatrix} 2 & -1 \\ 0 & 4 \end{pmatrix} - \begin{pmatrix} 1 & 5 \\ -3 & 2 \end{pmatrix} = \begin{pmatrix} 1 & -6 \\ 3 & 2 \end{pmatrix}$

(b) $B - A = \begin{pmatrix} 1 & 5 \\ -3 & 2 \end{pmatrix} - \begin{pmatrix} 2 & -1 \\ 0 & 4 \end{pmatrix} = \begin{pmatrix} -1 & 6 \\ -3 & -2 \end{pmatrix}$

(c) $2A - 3B = 2\begin{pmatrix} 2 & -1 \\ 0 & 4 \end{pmatrix} - 3\begin{pmatrix} 1 & 5 \\ -3 & 2 \end{pmatrix} = \begin{pmatrix} 4 & -2 \\ 0 & 8 \end{pmatrix} - \begin{pmatrix} 3 & 15 \\ -9 & 6 \end{pmatrix}$

$= \begin{pmatrix} 1 & -17 \\ 9 & 2 \end{pmatrix}$ □

Matrix Multiplication

Now we consider the product of two matrices. The definition of matrix multiplication was developed from applications of matrices to engineering and business problems. The obvious idea, multiplying entry by entry, proved to be not very useful. Those applications of matrix multiplication have motivated the definition we use today.

Suppose that $A \cdot B = C$. The entries in C are obtained from the entries in A and B. An entry c_{ij} in C is calculated by multiplying the ith row of A by the jth column of B. How can we multiply a row by a column? Let's examine the product of the two matrices A and B in Figure 8.21. Let C be their product. The entry c_{32} in the product is the result of multiplying the third row of the first matrix by the second column of the second matrix as follows.

$$\underset{A}{\begin{pmatrix} \square & \square \\ \square & \square \\ 5 & 4 \end{pmatrix}} \cdot \underset{B}{\begin{pmatrix} \square & 3 & \square & \square \\ \square & -1 & \square & \square \end{pmatrix}} = \underset{C}{\begin{pmatrix} \square & \square & \square & \square \\ \square & \square & \square & \square \\ \square & ? & \square & \square \end{pmatrix}}$$

Fig. 8.21

A natural idea is to multiply each entry in the row by the corresponding entry in the column. We must end up with one number. So, we add these products.

$$5(3) + 4(-1) = 15 - 4 = 11$$

$$\begin{pmatrix} \square & \square \\ \square & \square \\ 5 & 4 \end{pmatrix} \cdot \begin{pmatrix} \square & 3 & \square & \square \\ \square & -1 & \square & \square \end{pmatrix} = \begin{pmatrix} \square & \square & \square & \square \\ \square & \square & \square & \square \\ \square & 11 & \square & \square \end{pmatrix}$$

This definition places some requirements on the size of A, B, and C. The number of entries in row 3 of matrix A must match the number of entries in column 2 of matrix B. This means that the number of *columns* in A must be the same as the number of *rows* in B. Also, the size of C is determined by the number of rows in A and the number of columns in B.

Matrix:	A	$\cdot$ B	$=$	C
Dimension:	3×2	2×4		3×4

That is, to be **compatible for multiplication,** the dimensions of two matrices must fit the pattern in Figure 8.22.

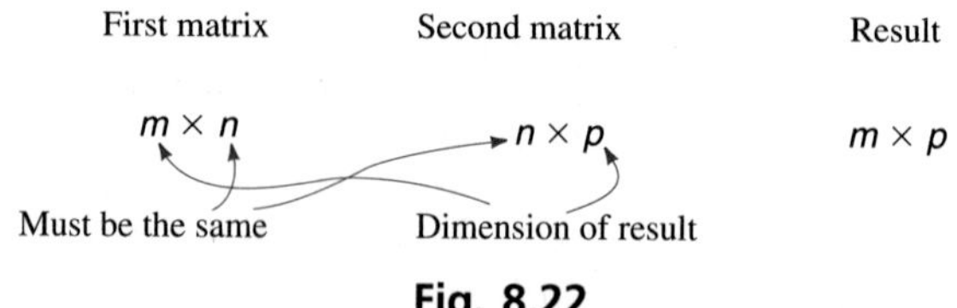

Fig. 8.22

This discussion leads us to the definition of multiplication of matrices.

> **Definition: Multiplication of Matrices**
>
> If A is an $m \times n$ matrix and B is an $n \times p$ matrix, then AB is an $m \times p$ matrix whose entries are computed as follows: The entry in the ith row and the jth column of AB is the sum of products of the corresponding entries in the ith row of A and jth column of B.

The number of columns in the first factor must be the same as the number of rows in the second factor in order for the multiplication to be defined.

EXAMPLE 5 Find the product $\begin{pmatrix} 1 & -1 \\ 2 & 3 \\ -3 & 4 \end{pmatrix} \cdot \begin{pmatrix} 5 & 0 \\ -6 & 7 \end{pmatrix}$.

Solution The size of the first matrix is 3×2, and of the second 2×2. The multiplication is defined, and the product will be of size 3×2.

$$\begin{pmatrix} 1 & -1 \\ 2 & 3 \\ -3 & 4 \end{pmatrix} \cdot \begin{pmatrix} 5 & 0 \\ -6 & 7 \end{pmatrix} = \begin{pmatrix} \square & \square \\ \square & \square \\ \square & \square \end{pmatrix}$$

Let's compute the entry in row 2 column 1. The rule indicates that this entry is the sum of the products of corresponding entries from the second row of the first matrix and the first column of the second matrix.

$$\begin{pmatrix} 1 & -1 \\ 2 & 3 \\ -3 & 4 \end{pmatrix} \cdot \begin{pmatrix} 5 & 0 \\ -6 & 7 \end{pmatrix} = \begin{pmatrix} \square & \square \\ -8 & \square \\ \square & \square \end{pmatrix} \qquad \text{Calculations:} \quad 2(5) + 3(-6)$$

The entry in row 1 column 1 is the sum of the products of entries from the first row and first column.

$$\begin{pmatrix} 1 & -1 \\ 2 & 3 \\ -3 & 4 \end{pmatrix} \cdot \begin{pmatrix} 5 & 0 \\ -6 & 7 \end{pmatrix} = \begin{pmatrix} 11 & \square \\ -8 & \square \\ \square & \square \end{pmatrix} \qquad \text{Calculations: } 1(5) + (-1)(-6)$$

The entry in row 1 column 2 is the sum of the products of the entries from row 1 and column 2.

$$\begin{pmatrix} 1 & -1 \\ 2 & 3 \\ -3 & 4 \end{pmatrix} \cdot \begin{pmatrix} 5 & 0 \\ -6 & 7 \end{pmatrix} = \begin{pmatrix} 11 & -7 \\ -8 & \square \\ \square & \square \end{pmatrix} \qquad \text{Calculations: } 1(0) + (-1)(7)$$

We continue to find each entry in the product.

$$\begin{pmatrix} 1 & -1 \\ 2 & 3 \\ -3 & 4 \end{pmatrix} \cdot \begin{pmatrix} 5 & 0 \\ -6 & 7 \end{pmatrix} = \begin{pmatrix} 11 & -7 \\ -8 & 21 \\ \square & \square \end{pmatrix} \qquad \text{Calculations: } 2(0) + 3(7)$$

$$\begin{pmatrix} 1 & -1 \\ 2 & 3 \\ -3 & 4 \end{pmatrix} \cdot \begin{pmatrix} 5 & 0 \\ -6 & 7 \end{pmatrix} = \begin{pmatrix} 11 & -7 \\ -8 & 21 \\ -39 & \square \end{pmatrix} \qquad \text{Calculations: } (-3)(5) + 4(-6)$$

$$\begin{pmatrix} 1 & -1 \\ 2 & 3 \\ -3 & 4 \end{pmatrix} \cdot \begin{pmatrix} 5 & 0 \\ -6 & 7 \end{pmatrix} = \begin{pmatrix} 11 & -7 \\ -8 & 21 \\ -39 & 28 \end{pmatrix} \qquad \text{Calculations: } (-3)(0) + 4(7)$$

□

Suppose we want to find the product $\begin{pmatrix} 5 & 0 \\ -6 & 7 \end{pmatrix} \cdot \begin{pmatrix} 1 & -1 \\ 2 & 3 \\ -3 & 4 \end{pmatrix}$. Is it defined?

The sizes, 2×2 and 3×2, don't match up. So this product is undefined.

However, the product in the other order $\begin{pmatrix} 1 & -1 \\ 2 & 3 \\ -3 & 4 \end{pmatrix} \cdot \begin{pmatrix} 5 & 0 \\ -6 & 7 \end{pmatrix}$ is defined, and we found it in Example 5.

This one case demonstrates that matrix multiplication is not commutative. That is, AB *does not* necessarily equal BA, when A and B are matrices.

We conclude this section on matrices with a summary of properties of matrices.

Properties of Matrices

Matrices are indicated with capital letters, and scalars with lowercase letters. Assume that the sizes of the matrices are appropriate for the operation to be performed.

1. $A + B = B + A$ — Commutative Property for addition

2. $A + (B + C) = (A + B) + C$ — Associative Property for addition

3. $A(BC) = (AB)C$	Associative Property for multiplication
4. $A(B + C) = AB + AC$	Distributive Property
5. $(B + C)A = BA + CA$	Distributive Property
6. $a(B + C) = aB + aC$	
7. $(a + b)C = aC + bC$	
8. $(ab)C = a(bC)$	
9. $a(BC) = (aB)C = B(aC)$	

The commutative property *does not* hold for multiplication.

GRAPHING CALCULATOR BOX

Operations with Matrices

Entering a Matrix

Let's enter the matrix $A = \begin{pmatrix} 1 & -2 \\ 0 & -4 \end{pmatrix}$.

TI-81

Three matrices of dimension no bigger than 6×6 can be defined using the [MATRX] key. Press [MATRX] to select the matrix menu. Press the right arrow to highlight EDIT. Choose A by pressing [1]. The dimensions of the matrix are shown on the top line. We must set the dimensions as 2×2. Press [2] [ENTER] to set the number of rows and then [2] [ENTER] to set the number of columns. Enter each matrix element, row by row, pressing [ENTER] after each entry.

[1] [ENTER] [(−)] [2] [ENTER] [0] [ENTER] [(−)] [4] [ENTER]

Press [QUIT] to exit the matrix menu. Notice the matrix identifiers [*A*], [*B*], and [*C*] just above the number keys, [1], [2], and [3]. The matrix A can be displayed by pressing [[A]] [ENTER].

CASIO fx-7700G

Two matrices of dimension no bigger than 9×9 can be defined using the matrix mode. Press [MODE] [0] to select the matrix mode. Choose *A* by pressing [F1]. Pressing [F6] take us to the dimension menu. [F1] selects dimension. Use the arrows to move the cursor and press [2] [EXE] to set the number of rows and [2] [EXE] to set the number of columns. Now *A* has dimension 2×2 and *A* is displayed. Enter each element by entering the number followed by [EXE]. The editing arrows may be used if necessary. The number at the bottom of the screen is the value in the entry at the location of the cursor.

Four menus are available in this mode. One menu allows A, B, or C to be displayed and allows addition, subtraction, and multiplication to be performed. Displaying A or B calls up a separate menu for A and B which allows us to set dimensions and work with a single matrix. The dimension menu is accessed after displaying the matrix and pressing [F6]. From the dimension menu, [PRE] calls up the matrix status line.

To exit the matrix mode press [MODE] or [MODE] and select another mode.

Performing Operations

Dimensions must be appropriate to perform any operation. Enter $B = \begin{pmatrix} 1 & 0 \\ 0 & 1 \end{pmatrix}$.

Let's find the scalar product $4A$, the sum $A + B$, the difference $A - B$, and the matrix product AB.

TI-81

The result of performing an operation is stored in [ANS].

[4] [[A]] [ENTER]	gives $4A$.
[[A]] [+] [[B]] [ENTER]	gives $A + B$.
[[A]] [−] [[B]] [ENTER]	gives $A - B$.
[[A]] [×] [[B]] [ENTER]	gives AB.

To find $B - A$ or BA, just enter the matrices in the desired order.

CASIO fx-7700G

Set the calculator in matrix mode and select +, −, or × to add, subtract or multiply. Notice that the result of an operation is stored in C and the dimensions of C are set accordingly.

[4] [F1]	yields $C = 4A$.	(A must be displayed.)
[F3]	yields $C = A + B$.	
[F4]	yields $C = A - B$.	
[F5]	yields $C = AB$.	

To find $B - A$ or BA, we must exchange A and B using the $A \rightleftarrows B$ option. With A or B displayed, press [F5]. Now subtract with [F4] or multiply with [F5].

PROBLEM SET 8.4

Warm-ups

In Problems 1 and 2, find x and y so that the matrices are equal. See Example 1.

1. $\begin{pmatrix} 1 & 2 \\ 3 & 4 \end{pmatrix} = \begin{pmatrix} x - y & 2 \\ 3 & x + y \end{pmatrix}$

2. $\begin{pmatrix} 1 & 0 \\ 0 & 1 \end{pmatrix} = \begin{pmatrix} x - y & 0 \\ 0 & x + y \end{pmatrix}$

In Problems 3 through 8, perform the operations indicated. See Examples 2, 3, and 4.

3. $2\begin{pmatrix} 1 & -4 \\ -5 & 0 \end{pmatrix} + 4\begin{pmatrix} -1 & 7 \\ 0 & 3 \end{pmatrix}$

4. $\begin{pmatrix} 2 & 3 & 4 \\ 5 & 6 & 7 \end{pmatrix} - 3\begin{pmatrix} 1 & -3 & 2 \\ 7 & -1 & 1 \end{pmatrix}$

5. $3\begin{pmatrix} 1 & 2 \\ 0 & -1 \end{pmatrix} - \begin{pmatrix} 1 & 3 & -4 \\ -5 & 6 & 7 \end{pmatrix}$

6. $c\begin{pmatrix} 1 & -1 \\ 2 & 3 \\ -3 & 4 \end{pmatrix}$

7. $3a\begin{pmatrix} 2 & -1 \\ 0 & 4 \end{pmatrix} + 2a\begin{pmatrix} 1 & 5 \\ -3 & 2 \end{pmatrix}$

8. $\begin{pmatrix} 1 & -1 \\ 2 & 3 \\ -3 & 4 \end{pmatrix} - \begin{pmatrix} 5 & 0 \\ -6 & 7 \end{pmatrix}$

In Problems 9 through 12, find each product if it is defined. See Example 5.

9. $\begin{pmatrix} 2 & -1 \\ 0 & 1 \\ -3 & 4 \end{pmatrix} \cdot \begin{pmatrix} 7 & 8 \\ -6 & 5 \end{pmatrix}$

10. $\begin{pmatrix} 1 & -3 \\ 0 & 1 \end{pmatrix} \cdot \begin{pmatrix} 2 & 4 \\ 7 & 5 \end{pmatrix}$

11. $\begin{pmatrix} 7 & 8 \\ -6 & 5 \end{pmatrix} \cdot \begin{pmatrix} 2 & -1 \\ 0 & 1 \\ -3 & 4 \end{pmatrix}$

12. $\begin{pmatrix} 2 & 4 \\ 7 & 5 \end{pmatrix} \cdot \begin{pmatrix} 1 & -3 \\ 0 & 1 \end{pmatrix}$

Practice Exercises

In Problems 13 through 34, use the following matrices to perform the indicated operations.

$A = \begin{pmatrix} 2 & -1 \\ 1 & -1 \end{pmatrix}$; $B = \begin{pmatrix} 3 & 0 & 1 \\ -2 & 1 & 2 \end{pmatrix}$; $C = \begin{pmatrix} 1 \\ 4 \\ 5 \end{pmatrix}$; $D = \begin{pmatrix} 1 & -1 & 2 \\ 0 & 1 & -1 \\ 2 & 0 & 1 \\ -1 & 0 & 2 \end{pmatrix}$;

$E = (1 \quad -3 \quad 2)$; $F = \begin{pmatrix} -3 & 4 \\ 5 & 1 \end{pmatrix}$; $G = \begin{pmatrix} 1 & -1 & 3 \\ 0 & 2 & -1 \\ 2 & 1 & 0 \end{pmatrix}$.

13. $2A - 3F$
14. $3F - 2A$
15. AB
16. BA
17. BD
18. DG
19. A^2 $(A \cdot A)$
20. F^2
21. $AB + B$
22. $FB - 2A$
23. CE
24. EC
25. DC
26. CD
27. GB
28. BG
29. $A^2 - F^2$
30. $(A + F)^2$
31. $(A + F)(A - F)$
32. $A^2 + 2AF + F^2$
33. A^3
34. F^3

In Problems 35 through 44, find x_1 and x_2 so that each equation is true. Use $X = \begin{pmatrix} x_1 \\ x_2 \end{pmatrix}$.

$A = \begin{pmatrix} 1 & 3 \\ -1 & 2 \end{pmatrix}$, *and* $B = \begin{pmatrix} 3 \\ 7 \end{pmatrix}$, $I = \begin{pmatrix} 1 & 0 \\ 0 & 1 \end{pmatrix}$ *and* $O = \begin{pmatrix} 0 \\ 0 \end{pmatrix}$.

35. $X = AB$
36. $X = A(-B)$
37. $X = 2A(3B)$
38. $X = 3A(-2B)$
39. $2X = B$
40. $2X = AB$
41. $AX = B$
42. $AX = O$
43. $(A - 2I)X = O$
44. $(A - 2I)X = B$

In Problems 45 through 48, use $A = \begin{pmatrix} a & b \\ c & d \end{pmatrix}$, $B = \begin{pmatrix} e & f \\ g & h \end{pmatrix}$, *and* $C = \begin{pmatrix} i & j \\ k & l \end{pmatrix}$ *to show that the property listed is true for 2×2 matrices.*

45. Commutative property for addition
46. $C(A + B) = CA + CB$
47. $k(A + B) = kA + kB$
48. Associative property for addition

49. Find the following products and make a rule for multiplication.

(a) $\begin{pmatrix} 1 & 0 & 0 \\ 0 & 1 & 0 \\ 0 & 0 & 1 \end{pmatrix} \cdot \begin{pmatrix} a & b & c \\ d & e & f \\ g & h & i \end{pmatrix}$ (b) $\begin{pmatrix} a & b & c \\ d & e & f \\ g & h & i \end{pmatrix} \cdot \begin{pmatrix} 1 & 0 & 0 \\ 0 & 1 & 0 \\ 0 & 0 & 1 \end{pmatrix}$

Challenge Problems

50. Find nonzero matrices A and B, both 2×2, so that $AB = 0$.

51. Find a 2×2 matrix A such that $A^2 = A$. Matrices with this property are called **idempotent.**

52. If $AB = AC$, does $B = C$? (A, B, and C are appropriate-sized matrices.)

53. When we introduced matrix multiplication, we mentioned that the obvious scheme of multiplying entry by entry did not prove to be useful. In general, this is the case. However, in certain fields of mathematics research, such a definition *has* proved to be useful. It is called the **Hadamard product.** We denote it $A \circ B$. Find nonzero 2×2 matrices A and B such that $A \circ B = AB$.

For Graphing Calculators

54. The Hennessy Company sells four models of cars at its five dealerships. The inventory of models at each store is given by the matrix

Models

$$A = \begin{bmatrix} 12 & 8 & 7 & 6 \\ 22 & 11 & 3 & 9 \\ 7 & 2 & 11 & 5 \\ 10 & 7 & 6 & 9 \\ 3 & 7 & 15 & 13 \end{bmatrix}$$

	X	Y	Z	W	
	12	8	7	6	⟵ Dealership 1
	22	11	3	9	⟵ Dealership 2
A =	7	2	11	5	⟵ Dealership 3
	10	7	6	9	⟵ Dealership 4
	3	7	15	13	⟵ Dealership 5

Wholesale and retail values for each model are given in matrix B with wholesale prices in column 1 and retail prices in column 2.

	12,000	15,000	⟵ X
B =	7,999	10,900	⟵ Y
	10,800	14,500	⟵ Z
	18,650	22,500	⟵ W

(a) Find AB. What do the entries represent?
(b) What is the wholesale value of the cars at Dealership 1?
(c) What is the retail value of the cars at Dealership 3?
(d) What is the wholesale value of all the cars?
(e) If both wholesale and retail values are increased by 3%, find a new matrix B showing the new wholesale and retail values of each model.

55. The River Rice Company blends brown rice and white rice to make their four best selling blends. The number of grams of protein, carbohydrates, and fat per ounce in each rice are given in matrix A. The number of ounces of each rice used in a 32 ounce package of each blend is given in matrix B.

	Brown	***White***	
	2.7	3	Protein
A =	23.6	35.5	Carbohydrates
	0.9	0.1	Fat

Blends

	1	2	3	4	
B =	5	20	10	17	Brown
	27	12	22	15	White

(a) Find AB. What do the entries in AB represent?
(b) How many grams of protein in Blend 1?
(c) How many grams of fat in Blend 4?

IN YOUR OWN WORDS . . .

56. Explain some similarities in operations with matrices and operations with real numbers.

57. What are some differences in operations with matrices and operations with real numbers.

8.5 THE INVERSE OF A SQUARE MATRIX

We have seen that matrix arithmetic resembles arithmetic with real numbers. We continue our study of matrices by looking at the inverse of a square matrix.

Identity Matrices

Consider the matrices

$A = \begin{pmatrix} a & b & c \\ d & e & f \\ g & h & i \end{pmatrix}$ and $I = \begin{pmatrix} 1 & 0 & 0 \\ 0 & 1 & 0 \\ 0 & 0 & 1 \end{pmatrix}$. The matrix I has a special property that is illustrated as follows.

$$AI = \begin{pmatrix} a & b & c \\ d & e & f \\ g & h & i \end{pmatrix} \cdot \begin{pmatrix} 1 & 0 & 0 \\ 0 & 1 & 0 \\ 0 & 0 & 1 \end{pmatrix} = \begin{pmatrix} a & b & c \\ d & e & f \\ g & h & i \end{pmatrix} = A$$

$$IA = \begin{pmatrix} 1 & 0 & 0 \\ 0 & 1 & 0 \\ 0 & 0 & 1 \end{pmatrix} \cdot \begin{pmatrix} a & b & c \\ d & e & f \\ g & h & i \end{pmatrix} = \begin{pmatrix} a & b & c \\ d & e & f \\ g & h & i \end{pmatrix} = A$$

Thus $AI = A = IA$. The matrix I is the multiplicative identity for 3×3 matrices. Such a square matrix is called an **identity matrix.** There are identity matrices for all sizes of square matrices.

Multiplicative Inverse

In real numbers the multiplicative identity is the number 1. Because $5 \cdot \frac{1}{5} = 1$ we call $\frac{1}{5}$ the multiplicative inverse (or simply inverse) of 5, and 5 the inverse of $\frac{1}{5}$.

Consider the matrices A and B, where $A = \begin{pmatrix} 1 & -1 & 0 \\ 3 & 0 & 2 \\ -1 & 0 & -1 \end{pmatrix}$ and $B = \begin{pmatrix} 0 & 1 & 2 \\ -1 & 1 & 2 \\ 0 & -1 & -3 \end{pmatrix}$. Notice that

$$AB = \begin{pmatrix} 1 & -1 & 0 \\ 3 & 0 & 2 \\ -1 & 0 & -1 \end{pmatrix} \cdot \begin{pmatrix} 0 & 1 & 2 \\ -1 & 1 & 2 \\ 0 & -1 & -3 \end{pmatrix} = \begin{pmatrix} 1 & 0 & 0 \\ 0 & 1 & 0 \\ 0 & 0 & 1 \end{pmatrix}$$

and

$$BA = \begin{pmatrix} 0 & 1 & 2 \\ -1 & 1 & 2 \\ 0 & -1 & -3 \end{pmatrix} \cdot \begin{pmatrix} 1 & -1 & 0 \\ 3 & 0 & 2 \\ -1 & 0 & -1 \end{pmatrix} = \begin{pmatrix} 1 & 0 & 0 \\ 0 & 1 & 0 \\ 0 & 0 & 1 \end{pmatrix}$$

In the same manner as the real numbers 5 and $\frac{1}{5}$, we see that $AB = I$. If A and B are square matrices of the same size and $AB = BA = I$, we say that A and B are **inverses** of each other. We usually write A^{-1} for the inverse of A, and B^{-1} for the inverse of B. We see that

$$AA^{-1} = A^{-1}A = I$$

Inverse of a Square Matrix

Let A be a square matrix.

If there is a square matrix A^{-1} such that

$$AA^{-1} = A^{-1}A = I$$

then A^{-1} is called the inverse of A.

It can be shown that if A^{-1} exists, it is unique. The size of A, A^{-1}, and I is the same. A matrix that has an inverse in called **invertible** or **nonsingular.** A matrix that does not have an inverse is called **singular.**

EXAMPLE 1 Find A^{-1} if $A = \begin{pmatrix} 1 & -1 \\ -2 & 1 \end{pmatrix}$.

Solution Let $A^{-1} = \begin{pmatrix} x & y \\ z & w \end{pmatrix}$. The product, AA^{-1}, must be the identity.

$$AA^{-1} = I$$

$$\begin{pmatrix} 1 & -1 \\ -2 & 1 \end{pmatrix}\begin{pmatrix} x & y \\ z & w \end{pmatrix} = \begin{pmatrix} 1 & 0 \\ 0 & 1 \end{pmatrix}$$

We multiply the matrices on the left.

$$\begin{pmatrix} x - z & y - w \\ -2x + z & -2y + w \end{pmatrix} = \begin{pmatrix} 1 & 0 \\ 0 & 1 \end{pmatrix}$$

This is a statement that two matrices are equal. Thus, corresponding entries are equal.

$$x - z = 1$$
$$y - w = 0$$
$$-2x + z = 0$$
$$-2y + w = 1$$

These four equations yield two systems of equations.

$$\begin{cases} x - z = 1 \\ -2x + z = 0 \end{cases} \qquad \begin{cases} y - w = 0 \\ -2y + w = 1 \end{cases}$$

Both of these systems can be solved by elimination to give $x = -1$, $y = -1$, $z = -2$, and $w = -1$. Thus, $A^{-1} = \begin{pmatrix} -1 & -1 \\ -2 & -1 \end{pmatrix}$. This result can be checked by verifying that $AA^{-1} = I = A^{-1}A$. □

Finding the inverse in Example 1 depended on a system of linear equations having a solution. However, not all systems of this type have solutions. This implies that not all square matrices have inverses. If the inverse of a matrix exists, we can always mimic the procedure used in Example 1 to find the inverse. However, the systems that arise in this procedure allow us to carry out all the steps at once by adjoining the identity matrix to the matrix A. We begin with the matrix $(A \mid I)$ and, using elementary row operations, change it into $(I \mid A^{-1})$. This transformation can always be done if the inverse exists. Notice how the process works in the next example.

EXAMPLE 2 Find B^{-1} if $B = \begin{pmatrix} 4 & -1 \\ -3 & 1 \end{pmatrix}$.

Solution The idea is to first form the matrix $(B \mid I)$ by annexing the identity matrix to the matrix B.

$$(B \mid I) = \left(\begin{array}{cc|cc} 4 & -1 & 1 & 0 \\ -3 & 1 & 0 & 1 \end{array}\right)$$

We must change B into I, the identity using elementary row operations. Our first step is to get a 1 in row 1 column 1.

$$\left(\begin{array}{cc|cc} 4 & -1 & 1 & 0 \\ -3 & 1 & 0 & 1 \end{array}\right) \xrightarrow[\text{Add row 2 to row 1.}]{} \left(\begin{array}{cc|cc} 1 & 0 & 1 & 1 \\ -3 & 1 & 0 & 1 \end{array}\right)$$

We get a 0 under the leading 1.

$$\left(\begin{array}{cc|cc} 1 & 0 & 1 & 1 \\ -3 & 1 & 0 & 1 \end{array}\right) \xrightarrow[\text{Multiply row 1 by 3 and add to row 2.}]{} \left(\begin{array}{cc|cc} 1 & 0 & 1 & 1 \\ 0 & 1 & 3 & 4 \end{array}\right)$$

Now we have $(I \mid B^{-1})$, where

$$B^{-1} = \begin{pmatrix} 1 & 1 \\ 3 & 4 \end{pmatrix}$$

□

EXAMPLE 3 Find the inverse of $\begin{pmatrix} 1 & 0 & 0 \\ -3 & 1 & 0 \\ 0 & -1 & 2 \end{pmatrix}$.

Solution Our strategy is to annex the identity matrix and then transform it using elementary row operations.

$$\left(\begin{array}{rrr|rrr} 1 & 0 & 0 & 1 & 0 & 0 \\ -3 & 1 & 0 & 0 & 1 & 0 \\ 0 & -1 & 2 & 0 & 0 & 1 \end{array}\right) \xrightarrow[\text{and add to row 2.}]{\text{Multiply row 1 by 3}} \left(\begin{array}{rrr|rrr} 1 & 0 & 0 & 1 & 0 & 0 \\ 0 & 1 & 0 & 3 & 1 & 0 \\ 0 & -1 & 2 & 0 & 0 & 1 \end{array}\right)$$

$$\left(\begin{array}{rrr|rrr} 1 & 0 & 0 & 1 & 0 & 0 \\ 0 & 1 & 0 & 3 & 1 & 0 \\ 0 & -1 & 2 & 0 & 0 & 1 \end{array}\right) \xrightarrow[\text{Add row 2 to row 3.}]{} \left(\begin{array}{rrr|rrr} 1 & 0 & 0 & 1 & 0 & 0 \\ 0 & 1 & 0 & 3 & 1 & 0 \\ 0 & 0 & 2 & 3 & 1 & 1 \end{array}\right)$$

$$\left(\begin{array}{rrr|rrr} 1 & 0 & 0 & 1 & 0 & 0 \\ 0 & 1 & 0 & 3 & 1 & 0 \\ 0 & 0 & 2 & 3 & 1 & 1 \end{array}\right) \xrightarrow[\text{Multiply row 3 by } \frac{1}{2}]{} \left(\begin{array}{rrr|rrr} 1 & 0 & 0 & 1 & 0 & 0 \\ 0 & 1 & 0 & 3 & 1 & 0 \\ 0 & 0 & 1 & \frac{3}{2} & \frac{1}{2} & \frac{1}{2} \end{array}\right)$$

The inverse of $\begin{pmatrix} 1 & 0 & 0 \\ -3 & 1 & 0 \\ 0 & -1 & 2 \end{pmatrix}$ is $\begin{pmatrix} 1 & 0 & 0 \\ 3 & 1 & 0 \\ \frac{3}{2} & \frac{1}{2} & \frac{1}{2} \end{pmatrix}$.

□

Not all square matrices have inverses. In using the procedure in the previous example, when it is impossible to change the given matrix into the identity matrix, the given matrix does not have an inverse. It is singular. Notice how this happens.

EXAMPLE 4 Find the inverse of $\begin{pmatrix} 1 & 1 & 3 \\ 2 & 2 & 1 \\ 3 & 3 & 4 \end{pmatrix}$.

Solution We annex the identity matrix to the given matrix.

$$\left(\begin{array}{rrr|rrr} 1 & 1 & 3 & 1 & 0 & 0 \\ 2 & 2 & 1 & 0 & 1 & 0 \\ 3 & 3 & 4 & 0 & 0 & 1 \end{array}\right) \xrightarrow[\text{and add to row 2.}]{\text{Multiply row 1 by } -2} \left(\begin{array}{rrr|rrr} 1 & 1 & 3 & 1 & 0 & 0 \\ 0 & 0 & -5 & -2 & 1 & 0 \\ 3 & 3 & 4 & 0 & 0 & 1 \end{array}\right)$$

$$\left(\begin{array}{rrr|rrr} 1 & 1 & 3 & 1 & 0 & 0 \\ 0 & 0 & -5 & -2 & 1 & 0 \\ 3 & 3 & 4 & 0 & 0 & 1 \end{array}\right) \xrightarrow[\text{and add to row 3.}]{\text{Multiply row 1 by } -3} \left(\begin{array}{rrr|rrr} 1 & 1 & 3 & 1 & 0 & 0 \\ 0 & 0 & -5 & -2 & 1 & 0 \\ 0 & 0 & -5 & -3 & 0 & 1 \end{array}\right)$$

At this point, we try to get a 1 in row 2 column 2. This is impossible to do and leave the first column intact. We conclude that the given matrix has no inverse. It is *singular.* □

Solving Systems with Matrix Algebra

Consider the following system of equations.

$$\begin{cases} 4x - y = 5 \\ -3x + y = -3 \end{cases}$$

If we call the matrix of coefficients A, then $A = \begin{pmatrix} 4 & -1 \\ -3 & 1 \end{pmatrix}$. Now let $X = \begin{pmatrix} x \\ y \end{pmatrix}$ and $B = \begin{pmatrix} 5 \\ -3 \end{pmatrix}$. Note the matrix equation

$$A \cdot X = B$$

$$\underset{2 \times 2}{\begin{pmatrix} 4 & -1 \\ -3 & 1 \end{pmatrix}} \cdot \underset{2 \times 1}{\begin{pmatrix} x \\ y \end{pmatrix}} = \underset{2 \times 1}{\begin{pmatrix} 5 \\ -3 \end{pmatrix}}$$

If we multiply the matrices on the left side, we have

$$\begin{pmatrix} 4x - y \\ -3x + y \end{pmatrix} = \begin{pmatrix} 5 \\ -3 \end{pmatrix}$$

Two matrices are equal if the corresponding entries are equal. So, we see that the original system is equivalent to the matrix equation $AX = B$.

EXAMPLE 5 Write the matrix equation that corresponds to the system, $\begin{cases} 3x + y - z = 7 \\ x + z = 8. \\ y + 5z = 0 \end{cases}$

Solution We write the coefficient matrix by inspection and form the matrix equation.

$$\begin{pmatrix} 3 & 1 & -1 \\ 1 & 0 & 1 \\ 0 & 1 & 5 \end{pmatrix} \cdot \begin{pmatrix} x \\ y \\ z \end{pmatrix} = \begin{pmatrix} 7 \\ 8 \\ 0 \end{pmatrix}$$

□

Now notice the following. If A is the matrix of coefficients and A^{-1} exists, we can solve the matrix equation $AX = B$ for X.

$$AX = B$$

$$A^{-1}(AX) = A^{-1}B \qquad \text{Multiply both sides by } A^{-1} \text{ on the } \textit{left}.$$

$$(A^{-1}A)X = A^{-1}B \qquad \text{Associative Property}$$
$$IX = A^{-1}B$$
$$X = A^{-1}B$$

Therefore, we can solve a system that has a solution by using the inverse of the coefficient matrix. The next example illustrates this idea.

EXAMPLE 6 Solve $\begin{cases} 4x - y = 5 \\ -3x + y = -3 \end{cases}$ using an inverse matrix.

Solution The solution to the system can be written as the matrix equation $X = A^{-1}B$, where A is the matrix of coefficients and B is the matrix of constants.

By inspection,

$$A = \begin{pmatrix} 4 & -1 \\ -3 & 1 \end{pmatrix} \text{ and } B = \begin{pmatrix} 5 \\ -3 \end{pmatrix}$$

We calculated A^{-1} in Example 2.

$$A^{-1} = \begin{pmatrix} 1 & 1 \\ 3 & 4 \end{pmatrix}$$
$$\begin{pmatrix} x \\ y \end{pmatrix} = \begin{pmatrix} 1 & 1 \\ 3 & 4 \end{pmatrix} \cdot \begin{pmatrix} 5 \\ -3 \end{pmatrix}$$
$$\begin{pmatrix} x \\ y \end{pmatrix} = \begin{pmatrix} 2 \\ 3 \end{pmatrix}$$

Thus $x = 2$ and $y = 3$. The solution set is $\{(2, 3)\}$. □

GRAPHING CALCULATOR BOX

Finding the Inverse of a Matrix

Enter $A = \begin{pmatrix} 2 & 3 & 0 \\ 1 & 1 & -2 \\ -1 & 0 & 1 \end{pmatrix}$.

Finding the Inverse of A

TI-81

[A] x^{-1} ENTER gives A^{-1}. (It is stored in ANS.) As a check, press ANS × [A] ENTER and note that the 3 × 3 identity matrix is the result.

CASIO fx-7700G

In matrix mode, press F4 and A^{-1} is displayed in C. To check, copy C into B by selecting $C \to B$ (press F2). Return to the matrix status line with PRE, then select matrix multiplication with F5. The 3×3 identity matrix is the result.

Solving Systems by Matrix Inversion

Consider the system of equations,

$$\begin{cases} x + y + 2z = 1 \\ \quad\;\; y + 3z = -1 \\ 3x \qquad + 2z = 1 \end{cases}$$

This is of the form, $AX = B$ where $A = \begin{pmatrix} 1 & 1 & 2 \\ 0 & 1 & 3 \\ 3 & 0 & 2 \end{pmatrix}$, $X = \begin{pmatrix} x \\ y \\ z \end{pmatrix}$, and $B = \begin{pmatrix} 1 \\ -1 \\ 1 \end{pmatrix}$

If A^{-1} exists, then $X = A^{-1}B$.
Therefore, compute $A^{-1}B$, and then read x, y and z from the result.

TI-81

Enter A and B. Return to the computational screen with QUIT, then press [A] x^{-1} × [B] ENTER and read $x = 1$, $y = 2$, and $z = -1$ from the resulting matrix.

CASIO fx-7700G

Enter A and B, then press PRE F1 F4 to find A^{-1}. Store the result back into A with F1. Return to the matrix status line with PRE and press F5 to multiply A (which now holds A^{-1}) times B. Read $x = 1$, $y = 2$, and $z = -1$ from the resulting matrix.

PROBLEM SET 8.5

Warm-ups

In Problems 1 through 8, find the inverse of each matrix if it exists. See Examples 1, 2, 3, and 4.

1. $\begin{pmatrix} 1 & -1 \\ -4 & -3 \end{pmatrix}$ **2.** $\begin{pmatrix} 1 & -3 \\ -4 & 5 \end{pmatrix}$ **3.** $\begin{pmatrix} 2 & -2 \\ -4 & 1 \end{pmatrix}$ **4.** $\begin{pmatrix} 3 & 3 \\ -2 & 1 \end{pmatrix}$

5. $\begin{pmatrix} 1 & 1 & -1 \\ -3 & 2 & -1 \\ 3 & -3 & 2 \end{pmatrix}$ **6.** $\begin{pmatrix} 3 & 2 & 2 \\ 2 & 2 & 2 \\ -4 & 4 & 3 \end{pmatrix}$ **7.** $\begin{pmatrix} 2 & -1 \\ -2 & 1 \end{pmatrix}$ **8.** $\begin{pmatrix} 3 & -4 \\ -3 & 4 \end{pmatrix}$

In Problems 9 through 12, write the matrix equation that corresponds to the system. See Example 5.

9. $\begin{cases} 5x - 7y = 1 \\ 3x + 8y = 0 \end{cases}$ **10.** $\begin{cases} x - y = 5 \\ -3x + 4y = 1 \end{cases}$ **11.** $\begin{cases} x + 2y - z = 11 \\ 3x + 2z = 5 \\ y - 5z = 0 \end{cases}$ **12.** $\begin{cases} x + y - z = 1 \\ x + y = 1 \\ y + z = 1 \end{cases}$

In Problems 13 through 18, solve each system using the inverse of the coefficient matrix. See Example 6. (The inverse for each coefficient matrix was found in Problems 1 through 6.)

13. $\begin{cases} x - y = -1 \\ -4x - 3y = 3 \end{cases}$

14. $\begin{cases} x - 3y = 5 \\ -4x + 5y = -3 \end{cases}$

15. $\begin{cases} 2x - 2y = 0 \\ -4x + y = -3 \end{cases}$

16. $\begin{cases} 3x + 3y = 7 \\ -2x + y = -3 \end{cases}$

17. $\begin{cases} x + y - z = 0 \\ -3x - 2y - z = -2 \\ 3x - 3y + 2z = 3 \end{cases}$

18. $\begin{cases} 3x + 2y + 2z = 3 \\ 2x + 2y + 2z = 2 \\ -4x + 4y + 3z = -3 \end{cases}$

Practice Exercises

In Problems 19 through 30, find the inverse of each matrix if it exists.

19. $\begin{pmatrix} 4 & 0 \\ 0 & 2 \end{pmatrix}$

20. $\begin{pmatrix} 1 & 0 \\ 0 & 1 \end{pmatrix}$

21. $\begin{pmatrix} 1 & 1 \\ 4 & 5 \end{pmatrix}$

22. $\begin{pmatrix} 2 & 2 \\ -3 & -3 \end{pmatrix}$

23. $\begin{pmatrix} 2 & 3 \\ 5 & 6 \end{pmatrix}$

24. $\begin{pmatrix} 0 & -1 \\ -1 & 0 \end{pmatrix}$

25. $\begin{pmatrix} 1 & 0 & 0 \\ 0 & 1 & 0 \\ 0 & 0 & 1 \end{pmatrix}$

26. $\begin{pmatrix} 1 & 1 & 2 \\ 0 & 1 & 3 \\ 3 & 0 & -2 \end{pmatrix}$

27. $\begin{pmatrix} 3 & 0 & 0 \\ 0 & 4 & 0 \\ 0 & 0 & 5 \end{pmatrix}$

28. $\begin{pmatrix} 0 & 0 & -3 \\ 0 & -2 & 0 \\ 4 & 0 & 1 \end{pmatrix}$

29. $\begin{pmatrix} 1 & -2 & -1 & -2 \\ 3 & -5 & -2 & -3 \\ 2 & 5 & -2 & -5 \\ -1 & 4 & 4 & 11 \end{pmatrix}$

30. $\begin{pmatrix} 1 & 1 & 1 & 2 \\ 1 & 0 & 1 & 0 \\ 0 & 1 & 0 & 1 \\ 1 & 0 & 0 & 1 \end{pmatrix}$

In Problems 31 through 34, use the inverse found in Problem 21 to solve each system.

31. $\begin{cases} x + y = 3 \\ 4x + 5y = 6 \end{cases}$

32. $\begin{cases} x + y = 5 \\ 4x + 5y = 1 \end{cases}$

33. $\begin{cases} x + y = 1 \\ 4x + 5y = 0 \end{cases}$

34. $\begin{cases} x + y = 0 \\ 4x + 5y = 0 \end{cases}$

In Problems 35 and 36, use the inverse found in Problem 26 to solve each system.

35. $\begin{cases} x + y + 2z = 7 \\ y + 3z = -2 \\ 3x - 2z = 0 \end{cases}$

36. $\begin{cases} x + y + 2z = 12 \\ y + 3z = 8 \\ 3x - 2z = -3 \end{cases}$

Challenge Problems

37. Use the inverse found in Problem 26 to solve the system

$$\begin{cases} x + y + 2z = a \\ y + 3z = b. \\ 3x - 2z = c \end{cases}$$

38. If A is a 2×3 matrix, is it possible to find a matrix I such that $AI = A$ and $IA = A$?

Problems 39 and 40, use the Hadamard product. See Section 8.4, Problem 53.

39. Find the general family of Hadamard identity matrices. Denote a Hadamard identity I_h.

40. What matrices are Hadamard-invertible? That is, under what conditions on A does a matrix B exist such that $A \circ B = I_h$?

For Graphing Calculators

41. The Azar Toy Company makes three small toys, X, Y and Z. Labor, material, and miscellaneous costs for making each toy are given in matrix A.

$$\begin{array}{cc} & \begin{array}{ccc} X & Y & Z \end{array} \\ A = & \begin{bmatrix} 0.89 & 0.89 & 1.19 \\ 0.59 & 0.79 & 0.99 \\ 0.79 & 0.89 & 1.29 \end{bmatrix} \end{array} \begin{array}{l} \longleftarrow \text{Labor} \\ \longleftarrow \text{Material} \\ \longleftarrow \text{Miscellaneous} \end{array}$$

Three plants reported these costs.

	Plant 1	Plant 2	Plant 3
Labor	\$1,466.96	\$1,210.62	\$1,407.86
Material	1,111.16	971.32	1,079.74
Miscellaneous	1,404.76	1,227.02	1,352.84

How many of each kind of toy did each plant make?

42. Is there a cubic equation of the form $y = ax^3 + bx^2 + cx + d$ that contains the four points $(-1, -2)$, $(1, 2)$, $(4, 2)$, and $(5, 3)$? If so, find it.

IN YOUR OWN WORDS . . .

43. Explain how a system of two equations in two variables can be written as a matrix equation.

44. Explain why a nonsquare matrix cannot have an inverse.

45. Explain what the inverse of a matrix is.

8.6 DETERMINANTS AND CRAMER'S RULE

To each square matrix we assign a number called its **determinant.** We indicate the determinant with a bar on each side of the array of numbers in the matrix. For example, if $A = \begin{pmatrix} 1 & 2 & 0 \\ 3 & -1 & 6 \\ 21 & 5 & 13 \end{pmatrix}$, then the determinant of A is $\begin{vmatrix} 1 & 2 & 0 \\ 3 & -1 & 6 \\ 21 & 5 & 13 \end{vmatrix}$. We refer to the determinant of A as $\det A$ or $|A|$. The number of rows (or columns) of the matrix is called the **order** of the determinant. The example above is a third-order determinant. The following examples show a second-order and a fourth-order determinant.

$$\begin{vmatrix} 1 & 3 \\ 0 & 1 \end{vmatrix} \qquad \begin{vmatrix} 1 & 0 & 2 & 1 \\ 5 & 1 & 0 & 0 \\ 8 & 0 & 0 & 2 \\ 1 & 0 & 0 & 1 \end{vmatrix}$$

Evaluating Determinants

A determinant represents a number just like $\sqrt{4}$ represents a number. To evaluate any determinant, we must first define a second-order determinant.

Value of a Second-Order Determinant

$$\begin{vmatrix} a & b \\ c & d \end{vmatrix} = ad - bc$$

EXAMPLE 1 Evaluate the following determinants.

(a) $\begin{vmatrix} 1 & 2 \\ 3 & 4 \end{vmatrix}$ (b) $\begin{vmatrix} 2 & 0 \\ -4 & 1 \end{vmatrix}$ (c) $\begin{vmatrix} 3 & -5 \\ 4 & -2 \end{vmatrix}$ (d) $\begin{vmatrix} -1 & 7 \\ 3 & -6 \end{vmatrix}$

Solutions (a) $\begin{vmatrix} 1 & 2 \\ 3 & 4 \end{vmatrix} = 1(4) - 2(3) = -2$

(b) $\begin{vmatrix} 2 & 0 \\ -4 & 1 \end{vmatrix} = 2(1) - 0(-4) = 2$

(c) $\begin{vmatrix} 3 & -5 \\ 4 & -2 \end{vmatrix} = 3(-2) - (-5)(4) = -6 + 20 = 14$

(d) $\begin{vmatrix} -1 & 7 \\ 3 & -6 \end{vmatrix} = (-1)(-6) - 7(3) = 6 - 21 = -15$ □

A determinant of order higher than 2 is evaluated by reducing it in successive steps until it is a sum of second-order determinants. In order to do this we need two more definitions.

The **minor** of entry a_{ij} in a determinant is the determinant that remains when the ith row and jth column are deleted. It is convenient to let M_{ij} be the minor of entry a_{ij}. For example, the minor of the entry b in $\begin{vmatrix} a & b & c \\ d & e & f \\ g & h & i \end{vmatrix}$ is indicated as $M_{12} = \begin{vmatrix} d & f \\ g & i \end{vmatrix}$.

The number $(-1)^{i+j}M_{ij}$ is called the **cofactor** of an element a_{ij}. Notice that $(-1)^{i+j}$ is either 1 or -1 depending on whether $i + j$ is odd or even. We can verify that a quick way to determine the + or − sign is to use the array of signs shown in Figure 8.23.

$$\begin{vmatrix} + & - & + \\ - & + & - \\ + & - & + \end{vmatrix} \qquad \begin{vmatrix} + & - & + & - \\ - & + & - & + \\ + & - & + & - \\ - & + & - & + \end{vmatrix} \qquad \begin{vmatrix} + & - & + & - & + \\ - & + & - & + & - \\ + & - & + & - & + \\ - & + & - & + & - \\ + & - & + & - & + \end{vmatrix}$$

Third order Fourth order Fifth order

Sign of cofactor of elements by position

Fig. 8.23

Notice that the signs alternate like the squares on a checkerboard, with a + in the upper, left-hand location.

EXAMPLE 2 Write the cofactor of f in the determinant $\begin{vmatrix} a & b & c \\ d & e & f \\ g & h & i \end{vmatrix}$.

Solution The minor of f is the determinant that remains when we delete the row and column of f. (The second row and the third column).

$$\begin{vmatrix} a & b & c \\ d & e & f \\ g & h & i \end{vmatrix}$$

Minor of $f = \begin{vmatrix} a & b \\ g & h \end{vmatrix}$.

The position of f in the array of signs is

$$\begin{vmatrix} + & - & + \\ - & + & - \\ + & - & + \end{vmatrix}$$

The cofactor of f is the minor with that sign.

Cofactor of $f = -\begin{vmatrix} a & b \\ g & h \end{vmatrix}$ □

Now we are in a position to evaluate a determinant of order higher than 2. We select *any* row or column, multiply every element in that row or column by its cofactor, and sum these products. The number obtained is the value of the determinant.

To Evaluate a Determinant of Order Higher Than 2

1. Select any row or column.
2. For each entry in the chosen row (or column), multiply the element by its cofactor.
3. The value of the determinant is the sum of the products found in step 2.

EXAMPLE 3 Evaluate the following determinant by expanding about the first row.

$$\begin{vmatrix} 2 & 4 & -3 \\ 1 & 1 & -4 \\ -2 & 5 & 0 \end{vmatrix}$$

Solution The minors of the entries in the first row are

$$\begin{vmatrix} 1 & -4 \\ 5 & 0 \end{vmatrix} \qquad \begin{vmatrix} 1 & -4 \\ -2 & 0 \end{vmatrix} \qquad \begin{vmatrix} 1 & 1 \\ -2 & 5 \end{vmatrix}$$

First entry Second entry Third entry

The cofactors of the entries in the first row are

$$+\begin{vmatrix} 1 & -4 \\ 5 & 0 \end{vmatrix} \qquad -\begin{vmatrix} 1 & -4 \\ -2 & 0 \end{vmatrix} \qquad +\begin{vmatrix} 1 & 1 \\ -2 & 5 \end{vmatrix}$$

The product of each entry and its cofactor is

$$2\begin{vmatrix} 1 & -4 \\ 5 & 0 \end{vmatrix} \qquad -4\begin{vmatrix} 1 & -4 \\ -2 & 0 \end{vmatrix} \qquad -3\begin{vmatrix} 1 & 1 \\ -2 & 5 \end{vmatrix}$$

Thus, the value of the determinant is given by

$$\begin{aligned}
\begin{vmatrix} 2 & 4 & -3 \\ 1 & 1 & -4 \\ -2 & 5 & 0 \end{vmatrix} &= 2\begin{vmatrix} 1 & -4 \\ 5 & 0 \end{vmatrix} - 4\begin{vmatrix} 1 & -4 \\ -2 & 0 \end{vmatrix} - 3\begin{vmatrix} 1 & 1 \\ -2 & 5 \end{vmatrix} \\
&= 2[0 - (-20)] - 4(0 - 8) - 3[5 - (-2)] \\
&= 2(20) - 4(-8) - 3(7) \\
&= 40 + 32 - 21 \\
&= 51
\end{aligned}$$

The astonishing thing is that *the same result* is obtained no matter which row or column is selected! Notice that the determinant would have been easier to evaluate by expanding about the third row.

$$\begin{aligned}
\begin{vmatrix} 2 & 4 & -3 \\ 1 & 1 & -4 \\ -2 & 5 & 0 \end{vmatrix} &= -2\begin{vmatrix} 4 & -3 \\ 1 & -4 \end{vmatrix} - 5\begin{vmatrix} 2 & -3 \\ 1 & -4 \end{vmatrix} + 0\begin{vmatrix} 2 & 4 \\ 1 & 1 \end{vmatrix} \\
&= -2(-16 + 3) - 5(-8 + 3) + 0 \\
&= -2(-13) - 5(-5) \\
&= 26 + 25 = 51
\end{aligned}$$

□

EXAMPLE 4 Evaluate each determinant.

(a) $\begin{vmatrix} 2 & 0 & -5 \\ -3 & 1 & 4 \\ 2 & 0 & 3 \end{vmatrix}$ (b) $\begin{vmatrix} 1 & 1 & 0 & 2 \\ 0 & 3 & 2 & 1 \\ 0 & -1 & 5 & 0 \\ 0 & 3 & 0 & 1 \end{vmatrix}$

Solutions (a) We expand about the second column.

$$\begin{vmatrix} 2 & 0 & -5 \\ -3 & 1 & 4 \\ 2 & 0 & 3 \end{vmatrix} = 0 \cdot C_{12} + 1\begin{vmatrix} 2 & -5 \\ 2 & 3 \end{vmatrix} + 0 \cdot C_{32}$$

$$= 6 + 10$$

$$= 16$$

(b) It is best to use the first column for expansion.

$$\begin{vmatrix} 1 & 1 & 0 & 2 \\ 0 & 3 & 2 & 1 \\ 0 & -1 & 5 & 0 \\ 0 & 3 & 0 & 1 \end{vmatrix} = 1\begin{vmatrix} 3 & 2 & 1 \\ -1 & 5 & 0 \\ 3 & 0 & 1 \end{vmatrix}$$

Expand about the second row.

$$= -(-1)\begin{vmatrix} 2 & 1 \\ 0 & 1 \end{vmatrix} + 5\begin{vmatrix} 3 & 1 \\ 3 & 1 \end{vmatrix}$$

$$= 2 + 5(0)$$

$$= 2$$

□

Determinants are an important structure found throughout mathematics. We will explore two applications.

Invertibility of a Matrix

In Section 8.5 we learned that some square matrices have inverses and some do not. Those that have inverses are called *invertible* (or *nonsingular*) and proved to be important in the solution of matrix equations. A square matrix that does not have an inverse is called *singular*. Just as every real number has a multiplicative inverse unless it is zero, every square matrix has an inverse unless *its determinant* is zero.

Invertibility of a Matrix

Let A be a square matrix.

A^{-1} exists if and only if $\det A \neq 0$

EXAMPLE 5 Determine which of the following matrices have inverses.

(a) $A = \begin{pmatrix} 1 & 0 & 1 \\ 1 & 1 & 0 \\ 0 & 1 & 1 \end{pmatrix}$ (b) $B = \begin{pmatrix} 1 & 0 & 2 \\ 0 & 4 & 0 \\ 3 & 7 & 6 \end{pmatrix}$ (c) $C = \begin{pmatrix} 2 & 1 & 3 \\ 0 & 5 & 7 \end{pmatrix}$

Solutions (a) $\det A = 2$
Since $\det A \neq 0$, A^{-1} exists. A has an inverse.

(b) $\det B = 0$
Since $\det B = 0$, B is singular and does not have an inverse.

(c) A matrix must be square to have an inverse. C does not have an inverse.

□

Cramer's Rule

Cramer's Rule is an application of determinants for solving certain systems of linear equations. Cramer's Rule can be used whenever the determinant of the matrix of coefficients is not zero. There must be the same number of linear equations as there are variables.

Cramer's Rule for a System of Two Linear Equations in Two Variables

The solution to the system of equations $\begin{cases} ax + by = p \\ cx + dy = q \end{cases}$ can be found using,

$$x = \frac{\begin{vmatrix} p & b \\ q & d \end{vmatrix}}{\begin{vmatrix} a & b \\ c & d \end{vmatrix}} \text{ and } y = \frac{\begin{vmatrix} a & p \\ c & q \end{vmatrix}}{\begin{vmatrix} a & b \\ c & d \end{vmatrix}}, \text{ provided } \begin{vmatrix} a & b \\ c & d \end{vmatrix} \neq 0.$$

Notice that the determinant in the denominator is made up of the coefficients of the variables of the original equation in their same positions. We call it the *determinant of coefficients* and designate it D. The determinant in the numerator of the formula for x is the determinant of coefficients with the x-column replaced by the column of constant terms. We designate it D_x, and D_y, the denominator in the formula for y, is formed in a similar manner. Using this notation, we can state Cramer's Rule as follows

$$x = \frac{D_x}{D} \qquad \text{and} \qquad y = \frac{D_y}{D} \qquad \text{provided } D \neq 0$$

If the determinant of coefficients is zero, the system is either inconsistent or dependent, and Cramer's Rule will not work. Solve the system by elimination or substitution.

EXAMPLE 6 Use Cramer's Rule to solve the system $\begin{cases} 4x - y = 2 \\ 3x + y = 5 \end{cases}$.

Solution

$$x = \frac{\begin{vmatrix} 2 & -1 \\ 5 & 1 \end{vmatrix}}{\begin{vmatrix} 4 & -1 \\ 3 & 1 \end{vmatrix}} = \frac{2-(-5)}{4-(-3)} = \frac{7}{7} = 1$$

$$y = \frac{\begin{vmatrix} 4 & 2 \\ 3 & 5 \end{vmatrix}}{\begin{vmatrix} 4 & -1 \\ 3 & 1 \end{vmatrix}} = \frac{4(5)-3(2)}{7} = \frac{14}{7} = 2$$

Notice that the denominator is the same in the calculation of both variables. Using this observation saves a little work. The solution set is $\{(1, 2)\}$. □

Cramer's Rule for Three Equations in Three Variables

The solution for the system of equations $\begin{cases} ax + by + cx = p \\ dx + ey + fz = q \\ gx + hy + iz = r \end{cases}$ can be found using

$$x = \frac{\begin{vmatrix} p & b & c \\ q & e & f \\ r & h & i \end{vmatrix}}{\begin{vmatrix} a & b & c \\ d & e & f \\ g & h & i \end{vmatrix}}, \quad y = \frac{\begin{vmatrix} a & p & c \\ d & q & f \\ g & r & i \end{vmatrix}}{\begin{vmatrix} a & b & c \\ d & e & f \\ g & h & i \end{vmatrix}}, \text{ and } z = \frac{\begin{vmatrix} a & b & p \\ d & e & q \\ g & h & r \end{vmatrix}}{\begin{vmatrix} a & b & c \\ d & e & f \\ g & h & i \end{vmatrix}},$$

provided that $\begin{vmatrix} a & b & c \\ d & e & f \\ g & h & i \end{vmatrix} \neq 0.$

Again, D is the determinant of coefficients of the system, and the determinants D_x, D_y, and now D_z are formed in a manner similar to those of the two-dimensional case.

EXAMPLE 7 Use Cramer's Rule to solve the following system.

$$\begin{cases} x - y + z = -3 \\ \quad x + z = -1 \\ \quad y + 2z = 2 \end{cases}$$

Solution

$$D = \begin{vmatrix} 1 & -1 & 1 \\ 1 & 0 & 1 \\ 0 & 1 & 2 \end{vmatrix} = -1\begin{vmatrix} -1 & 1 \\ 1 & 2 \end{vmatrix} + 0 - 1\begin{vmatrix} 1 & -1 \\ 0 & 1 \end{vmatrix} = (-1)(-3) - 1(1) = 2$$

$$D_x = \begin{vmatrix} -3 & -1 & 1 \\ -1 & 0 & 1 \\ 2 & 1 & 2 \end{vmatrix} = 1\begin{vmatrix} -1 & 1 \\ 1 & 2 \end{vmatrix} + 0 - 1\begin{vmatrix} -3 & -1 \\ 2 & 1 \end{vmatrix} = -3 - 1(-1) = -2$$

$$D_y = \begin{vmatrix} 1 & -3 & 1 \\ 1 & -1 & 1 \\ 0 & 2 & 2 \end{vmatrix} = 0 - 2\begin{vmatrix} 1 & 1 \\ 1 & 1 \end{vmatrix} + 2\begin{vmatrix} 1 & -3 \\ 1 & -1 \end{vmatrix} = 4$$

$$D_z = \begin{vmatrix} 1 & -1 & -3 \\ 1 & 0 & -1 \\ 0 & 1 & 2 \end{vmatrix} = -1\begin{vmatrix} -1 & -3 \\ 1 & 2 \end{vmatrix} + 0 - (-1)\begin{vmatrix} 1 & -1 \\ 0 & 1 \end{vmatrix} = 0$$

$$x = \frac{D_x}{D} = \frac{-2}{2} = -1$$

$$y = \frac{D_y}{D} = \frac{4}{2} = 2$$

$$z = \frac{D_z}{D} = \frac{0}{2} = 0$$

Therefore the solution set is $\{(-1, 2, 0)\}$. □

Systems with more than three variables can be solved with Cramer's Rule. There must be as many equations as there are variables, and all the equations must be linear. First, the system should be written in the form shown above for the three-variable case, being sure to insert zeros where necessary. The various determinants used in Cramer's Rule can then be formed by inspection.

GRAPHING CALCULATOR BOX

Finding the Determinant of a Matrix

Let's find $\begin{vmatrix} 2 & 3 & 0 \\ 1 & 1 & -2 \\ -1 & 0 & 1 \end{vmatrix}$. Enter the matrix as matrix A.

TI-81

[MATRX] [5] [[A]] [ENTER] gives det A = 5. (It is stored in ANS.)

CASIO fx-7700G

In matrix mode with A displayed, press [F3] and det A = 5 is displayed.

A General Method for Solving a System of Equations

Let's examine a technique for solving square systems (same number of equations as variables) on a graphing calculator. Let's solve the system,

$$\begin{cases} x + y + z + w = 2 \\ x + y + 2z = -1 \\ 2y + 3z - 3w = -9 \\ 2x + 3y + 2w = 6 \end{cases}$$

First, enter the matrix of coefficients in matrix A and evaluate its determinant. We find that det A = 4. Since det A is not 0, A^{-1} exists. We can solve the system by calculating $A^{-1}B$. The solution is (1, 0, −1, 2).

Now let's solve the system, $\begin{cases} x + y - z = -8 \\ 4x + 5y - 6z = -2 \\ 2x + 3y - 4z = 14 \end{cases}$. We enter the matrix of coefficients as matrix A and evaluate its determinant. We find det A = 0. The matrix of coefficients has no inverse. We now solve the system by the method of the augmented matrix. The TI-81 allows us to perform elementary row operations and reduce the augmented matrix.

The Augmented Matrix Method

TI-81

Return to the matrix A in [MATRX] EDIT. Change its dimensions to 3 × 4. Notice that a column of zeros was added to the existing matrix of coefficients. Simply arrow down to entry 1,4 and enter −8. Then enter −2 in entry 2,4 and 14 in entry 3,4. Press [QUIT] to return to the computational screen, then [[A]] [ENTER] to examine the new matrix A. Notice that it is the augmented matrix for the system. To multiply row 1 by −4 and add it to row 2 we press, [MATRX] [4] for the proper command and see [∗ Row+(] displayed. We enter the constant −4, the matrix A, the row being used 1, and the row affected 2, with the statement −4, A, 1, 2. To accomplish this, we press [(−)] [4] [ALPHA] [,] [[A]] [ALPHA] [,] [1] [ALPHA] [,] [2] [)] [ENTER]. The modified matrix is displayed. (It is also saved in ANS.) To get a 0 in the first entry in row 3, press [MATRX] [4] followed by

[(−)] [2] [ALPHA] [,] [ANS] [ALPHA] [,] [1] [ALPHA] [,] [3] [)] [ENTER].

We now have the augmented matrix,

$$\begin{pmatrix} 1 & 1 & -1 & -8 \\ 0 & 1 & -2 & 30 \\ 0 & 1 & -2 & 30 \end{pmatrix}$$

The given system is dependent. The solutions in parametric form are (−38 − t, 30 + 2t, t).

The other two elementary row operations are available on the TI-81. To interchange rows 2 and 3 of matrix A we press [MATRX] [1] followed by [[A]] [ALPHA] [,] [2] [ALPHA] [,] [3] [)]. To multiply row 2 of matrix A by 5 we press [MATRX] [3] followed by [5] [ALPHA] [,] [[A]] [ALPHA] [,] [2] [)].

PROBLEM SET 8.6

Warm-ups

In Problems 1 through 4, evaluate each determinant. See Example 1.

1. $\begin{vmatrix} 1 & -3 \\ 3 & -4 \end{vmatrix}$ **2.** $\begin{vmatrix} -2 & 0 \\ -11 & 10 \end{vmatrix}$ **3.** $\begin{vmatrix} -3 & -1 \\ -4 & -7 \end{vmatrix}$ **4.** $\begin{vmatrix} -2 & 4 \\ 3 & -6 \end{vmatrix}$

In Problems 5 through 8, write the cofactor of the entry given in the determinant $\begin{vmatrix} a & b & c \\ d & e & f \\ g & h & i \end{vmatrix}$. *See Example 2.*

5. d **6.** g **7.** a **8.** e

In Problems 9 and 10, evaluate each determinant by expanding about the first row. See Example 3.

9. $\begin{vmatrix} 2 & 4 & -3 \\ 1 & 1 & -4 \\ -2 & 5 & 0 \end{vmatrix}$ **10.** $\begin{vmatrix} -1 & 5 & -7 \\ 0 & 1 & 0 \\ 3 & 2 & 1 \end{vmatrix}$

In Problems 11 through 14, choose an appropriate row or column to use for expansion and evaluate each determinant. See Example 4.

11. $\begin{vmatrix} 1 & 1 & -5 \\ -3 & 1 & 4 \\ 2 & 0 & 3 \end{vmatrix}$ **12.** $\begin{vmatrix} 1 & 2 & -1 \\ 0 & 1 & 0 \\ 4 & 0 & 1 \end{vmatrix}$ **13.** $\begin{vmatrix} 1 & 2 & 0 & 2 \\ 0 & 1 & 2 & 1 \\ 0 & -1 & 5 & 0 \\ 1 & 3 & 0 & 1 \end{vmatrix}$ **14.** $\begin{vmatrix} 0 & 1 & 0 & 1 \\ 1 & 1 & 2 & 1 \\ 0 & 0 & 1 & 0 \\ 2 & 3 & 4 & 1 \end{vmatrix}$

In Problems 15 through 18, determine which matrices have inverses. See Example 5.

15. $\begin{pmatrix} 3 & -6 \\ -4 & -8 \end{pmatrix}$ **16.** $\begin{pmatrix} -2 & -4 \\ -5 & -10 \end{pmatrix}$ **17.** $\begin{pmatrix} 1 & 1 & -3 \\ -1 & 1 & 0 \\ 2 & 2 & -6 \end{pmatrix}$ **18.** $\begin{pmatrix} 1 & 3 & -2 \\ 0 & 1 & 0 \\ 1 & 0 & -1 \end{pmatrix}$

In Problems 19 through 22, use Cramer's Rule to solve each system. See Examples 6 and 7.

19. $\begin{cases} 3x - y = -2 \\ 3x + 2y = 5 \end{cases}$ **20.** $\begin{cases} 4x + 2y = 0 \\ 3x + y = -5 \end{cases}$ **21.** $\begin{cases} x - y + z = 0 \\ x \quad\;\; + z = 0 \\ \quad\;\; y + 2z = 0 \end{cases}$ **22.** $\begin{cases} x - y + z = 0 \\ x + 2y + z = -2 \\ \quad\;\; y + 2z = 0 \end{cases}$

Practice Exercises

In Problems 23 and 24, B, C, and D are obtained from A with one elementary row operation. (a) Determine what operation was used in each case. (b) Find det A, det B, det C, and det D. (c) Guess what effect an elementary row operation has on the value of a determinant.

23. $A = \begin{pmatrix} 1 & 2 \\ 3 & 4 \end{pmatrix}$; $B = \begin{pmatrix} 3 & 4 \\ 1 & 2 \end{pmatrix}$; $C = \begin{pmatrix} 5 & 10 \\ 3 & 4 \end{pmatrix}$; $D = \begin{pmatrix} 1 & 2 \\ 0 & -2 \end{pmatrix}$

24. $A = \begin{pmatrix} 1 & 2 & 3 \\ 4 & 5 & 6 \\ 0 & 1 & 0 \end{pmatrix}$; $B = \begin{pmatrix} 0 & 1 & 0 \\ 4 & 5 & 6 \\ 1 & 2 & 3 \end{pmatrix}$; $C = \begin{pmatrix} 3 & 6 & 9 \\ 4 & 5 & 6 \\ 0 & 1 & 0 \end{pmatrix}$; $D = \begin{pmatrix} 1 & 2 & 3 \\ 0 & -3 & -6 \\ 0 & 1 & 0 \end{pmatrix}$

25. Evaluate $\begin{vmatrix} a & b & c \\ 0 & d & e \\ 0 & 0 & f \end{vmatrix}$. Make a conjecture and prove it.

26. What is the value of the determinant of a matrix with a row of zeros? Prove the conjecture.

27. What is the value of the determinant of a matrix that has two identical rows or columns?

In Problems 28 through 35, use the ideas in Problems 23 through 27 in evaluating each determinant.

28. $\begin{vmatrix} 5 & 0 \\ 0 & 7 \end{vmatrix}$

29. $\begin{vmatrix} 1 & 0 & 0 \\ 5 & 3 & 0 \\ 7 & 0 & 8 \end{vmatrix}$

30. $\begin{vmatrix} 1 & 2 & 3 \\ 50 & 1 & 7 \\ 1 & 2 & 3 \end{vmatrix}$

31. $\begin{vmatrix} 1 & -5 & 7 \\ 11 & -9 & 13 \\ 0 & 0 & 0 \end{vmatrix}$

32. $\begin{vmatrix} -1 & 3 & 5 \\ 4 & 7 & 3 \\ -5 & 4 & -2 \end{vmatrix}$

33. $\begin{vmatrix} 3 & 2 & 5 \\ 0 & 1 & -7 \\ -3 & -2 & 1 \end{vmatrix}$

34. $\begin{vmatrix} 1 & 0 & 0 & 0 \\ 0 & 11 & 0 & 0 \\ 2 & 0 & 3 & 0 \\ -5 & 0 & 0 & -1 \end{vmatrix}$

35. $\begin{vmatrix} 4 & 0 & 0 & 2 \\ 0 & 5 & 0 & 0 \\ 6 & 0 & 3 & 0 \\ -1 & 0 & 0 & -1 \end{vmatrix}$

In Problems 36 through 45, solve each system by Cramer's Rule.

36. $\begin{cases} 4x - 7y = -5 \\ 3x + 9y = -18 \end{cases}$

37. $\begin{cases} 11x + 12y = 0 \\ 3x + 2y = -14 \end{cases}$

38. $\begin{cases} 2x_1 - x_2 + x_3 = 0 \\ x_1 + 2x_2 - x_3 = 0 \\ -x_1 + x_2 + x_3 = 0 \end{cases}$

39. $\begin{cases} x - 2y + z = 6 \\ x + y = -2 \\ 2x + y + 2z = 0 \end{cases}$

40. $\begin{cases} 2x_1 - x_2 = 2 \\ x_1 + 2x_2 - x_3 = 3 \\ -x_1 + x_2 + x_3 = 0 \end{cases}$

41. $\begin{cases} 2x_2 + x_3 = -1 \\ x_1 + x_2 + x_3 = 2 \\ 2x_1 + x_2 = 0 \end{cases}$

42. $\begin{cases} 2x_1 + 5x_2 + 2x_3 = 9 \\ x_1 + 3x_2 - x_3 = 0 \\ 2x_1 + 3x_2 - 3x_3 = 1 \end{cases}$

43. $\begin{cases} x + 2y + 2z = 3 \\ 2x + 3y + 6z = 2 \\ -x + y + z = 0 \end{cases}$

44. $\begin{cases} x_1 + x_2 + x_3 + x_4 = -1 \\ x_2 + x_3 + x_4 = -3 \\ x_3 + x_4 = -2 \\ x_1 + x_2 + x_3 = 2 \end{cases}$

45. $\begin{cases} x_1 + x_2 + x_3 = 0 \\ -x_1 + x_3 + x_4 = 0 \\ 3x_1 - x_2 = 1 \\ 5x_2 + 2x_3 = 4 \end{cases}$

Challenge Problems

In Problems 46 through 48, evaluate each determinant if $\begin{vmatrix} a & b \\ c & d \end{vmatrix} = 5.$

46. $\begin{vmatrix} c & d \\ a & b \end{vmatrix}$

47. $\begin{vmatrix} 2a & 2b \\ c & d \end{vmatrix}$

48. $\begin{vmatrix} a & b \\ c - a & d - b \end{vmatrix}$

In Problems 49 and 50, consider the general system $\begin{cases} a_{11}x_1 + a_{12}x_2 = b_1 \\ a_{21}x_1 + a_{22}x_2 = b_2 \end{cases}$.

49. Show that $x_1 = \dfrac{a_{22}b_1 - a_{12}b_2}{a_{11}a_{22} - a_{12}a_{21}}$, providing $a_{11}a_{22} \neq a_{12}a_{21}$.

50. Show that $x_2 = \dfrac{a_{11}b_2 - a_{21}b_1}{a_{11}a_{22} + a_{12}a_{21}}$, providing $a_{11}a_{22} \neq a_{12}a_{21}$.

IN YOUR OWN WORDS . . .

51. Explain Cramer's Rule without using the formula.

8.7 SYSTEMS OF INEQUALITIES AND LINEAR PROGRAMMING

Inequalities in Two Variables

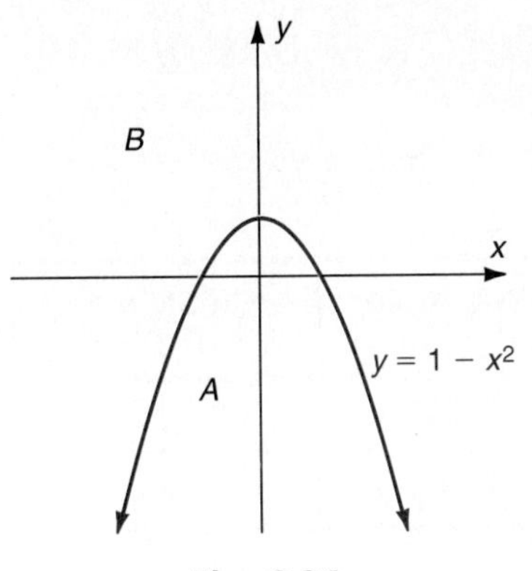

Fig. 8.24

In Chapter 2 we studied inequalities in one variable. In this section we extend the method of boundary numbers so that we can graph inequalities in two variables. The **graph** of an inequality in two variables consists of all ordered pairs that make the inequality a true statement. An inequality in two variables, like an equation in two variables, requires a coordinate system to display its graph.

Let's look at the graph of the inequality $y \leq 1 - x^2$. The graph is the collection of all ordered pairs (x, y) that make the statement $y \leq 1 - x^2$ true.

Using a technique similar to the method of boundary numbers, we graph the equation $y = 1 - x^2$.

This parabola is called a **boundary curve.** It divides the plane into two regions, marked A and B in Figure 8.24.

We test region A by selecting any point in the interior of region A, say $(0, 0)$, and determining whether it makes the statement of the inequality true or false. Since $0 \leq 1$ is a true statement, every point in region A is in the solution set. Likewise we pick any point in region B, say $(2,1)$, and see that the statement $1 \leq -3$ is false. Thus no point in region B is in the solution set. The parabola itself is in the solution set because the original statement, $y \leq 1 - x^2$, includes equality. The solution set is graphed in Figure 8.25.

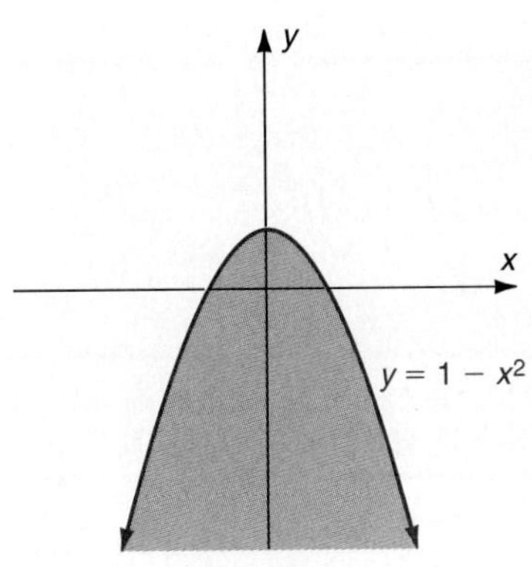

Fig. 8.25

A Procedure for Graphing an Inequality in Two Variables

1. Graph the boundary curve.
   (a) Draw a solid curve if equality is included.
   (b) Draw a dashed curve if equality is not included.
2. Determine which region(s) formed by the curve makes the inequality true by testing with one point from inside each region.
3. Shade the region(s) that make the inequality true.

EXAMPLE 1 Graph the solution set for $x - 2y > 0$.

Solution First we draw the boundary line $x - 2y = 0$ (See Figure 8.26.) We draw a dashed line because equality is not included. Test a point from each region in $x - 2y > 0$.

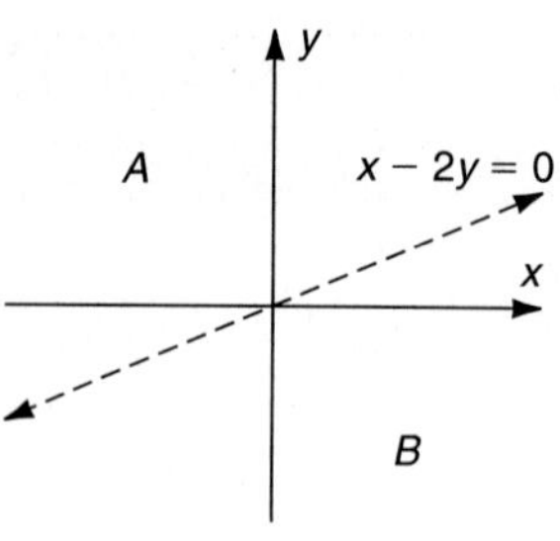

Fig. 8.26

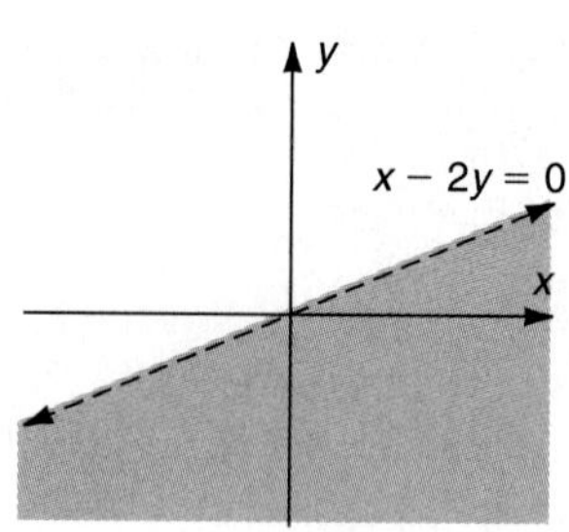

Fig. 8.27

REGION	TEST POINT	STATEMENT: $x - 2y > 0$	TRUTH OF STATEMENT	REGION IN SOLUTION SET?
A	$(-1, 2)$	$-1 - 4 > 0$	False	No
B	$(1, -1)$	$1 + 2 > 0$	True	Yes

We shade region B. See Figure 8.27. □

EXAMPLE 2 Graph the solution set for $x^2 + y^2 \leq 4$.

Solution We graph the boundary $x^2 + y^2 = 4$. It will be a solid circle because equality is included. See Figure 8.28.

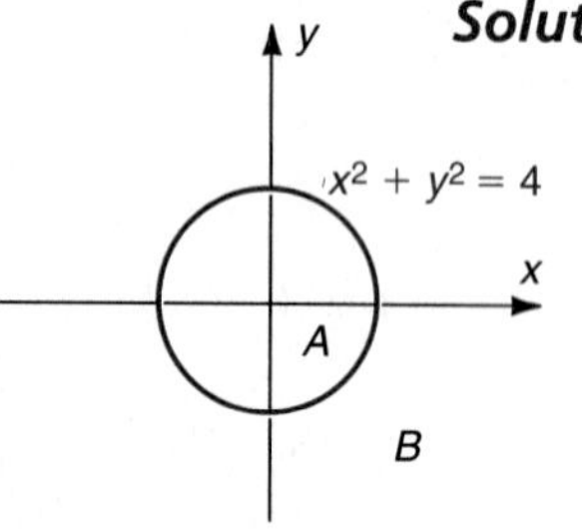

Fig. 8.28

REGION	TEST POINT	STATEMENT: $x^2 + y^2 \leq 4$	TRUTH OF STATEMENT	REGION IN SOLUTION SET?
A	$(0, 0)$	$0 \leq 4$	True	Yes
B	$(3, 0)$	$9 + 0 \leq 4$	False	No

We shade region A. See Figure 8.29. □

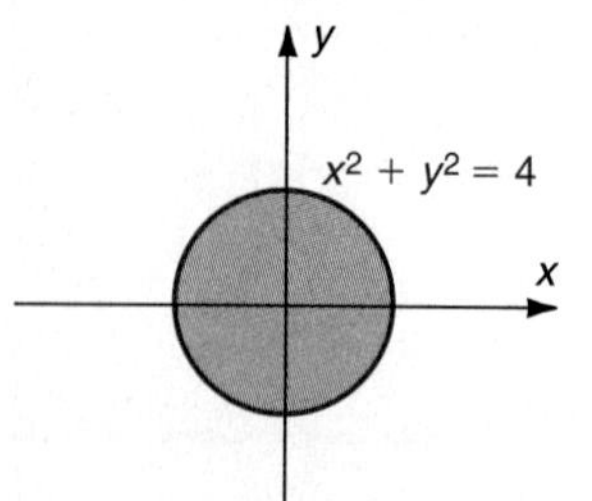

Fig. 8.29

Systems of Inequalities in Two Variables

When we consider two or more linear inequalities simultaneously, we form a **system of inequalities.** In general, the solution set of such a system is a region in the plane where *all* the inequalities are true.

EXAMPLE 3 Graph the solution set of the following system of inequalities: $\begin{cases} x - y < 0 \\ x + y \geq 1 \end{cases}$.

Solution First we graph the inequality $x - y < 0$. See Figure 8.30.

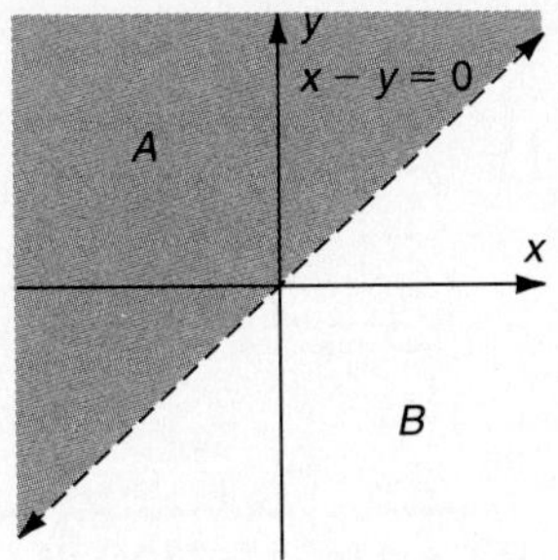

Fig. 8.30

REGION	TEST POINT	STATEMENT: $x - y < 0$	TRUTH OF STATEMENT	REGION IN SOLUTION SET?
A	(0, 1)	$0 - 1 < 0$	True	Yes
B	(1, 0)	$1 - 0 < 0$	False	No

So, we see that the first inequality is true for all points to the left of the line. However, the solution set of the *system* is the set of points where *both* inequalities are true. Next, we sketch the graph of $x + y \geq 1$. See Figure 8.31.

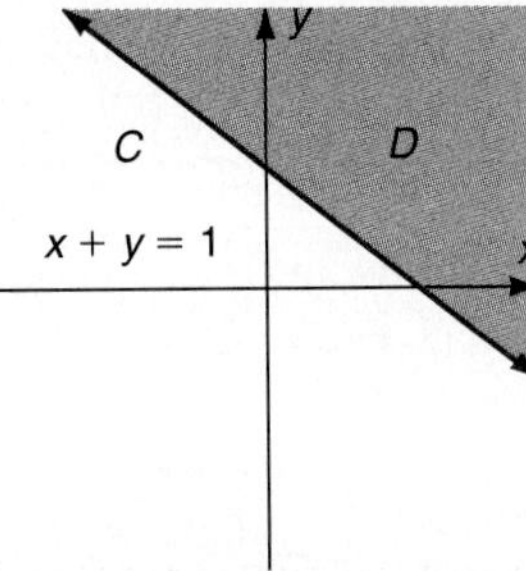

Fig. 8.31

REGION	TEST POINT	STATEMENT: $x + y \geq 1$	TRUTH OF STATEMENT	REGION IN SOLUTION SET?
C	(0, 0)	$0 + 0 \geq 1$	False	No
D	(2, 0)	$2 + 0 \geq 1$	True	Yes

The second inequality is true for all points to the right of the line. We are looking for points that satisfy *both* inequalities. Notice the result (Figure 8.32) if we graph both solution sets on the same coordinate system.

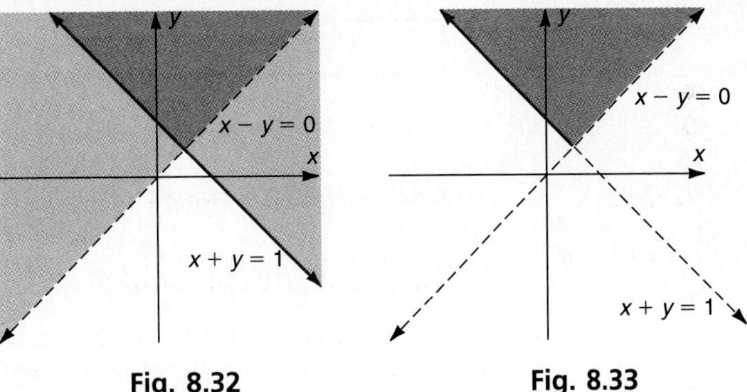

Fig. 8.32

Fig. 8.33

The intersection of the two solution sets is the solution set for the system. Notice that the lower segment of the solid boundary line is *not* in the solution set of the inequality $x - y < 0$ and therefore must be changed from solid to dashed in Figure 8.33, the sketch of the solution set of the system. □

A Procedure for Graphing the Solution Set of a System of Inequalities

1. Graph each inequality on the same axes, lightly shading its solution set.
2. Darken the *intersection* of the lightly shaded regions.
3. Change any portion of any solid boundary curve not in the intersection of the solutions from solid to dashed.

EXAMPLE 4 Graph the solution set of the system of inequalities $\begin{cases} y \geq x^2 - 1 \\ y \leq 2x - 1 \end{cases}$.

Solution First we graph the solution set for $y \geq x^2 - 1$, lightly shading its solution set as shown in Figure 8.34.

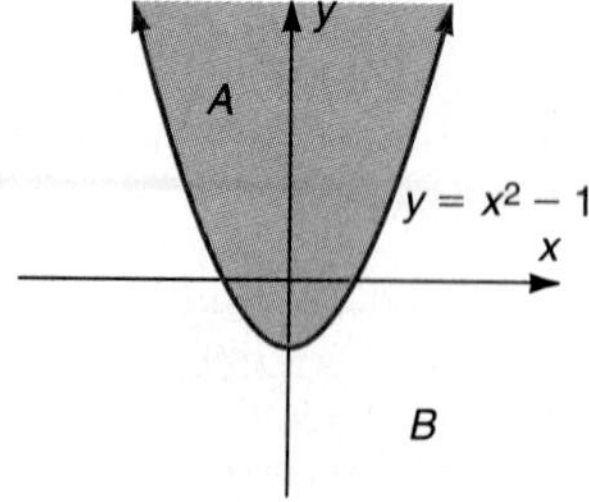

Fig. 8.34

REGION	TEST POINT	STATEMENT: $y \geq x^2 - 1$	TRUTH OF STATEMENT	REGION IN SOLUTION SET?
A	(0, 0)	$0 \geq 0 - 1$	True	Yes
B	(0, −2)	$-2 \geq 0 - 1$	False	No

Now we graph the solution set for $y \leq 2x - 1$, again lightly shading the solution set. See Figure 8.35.

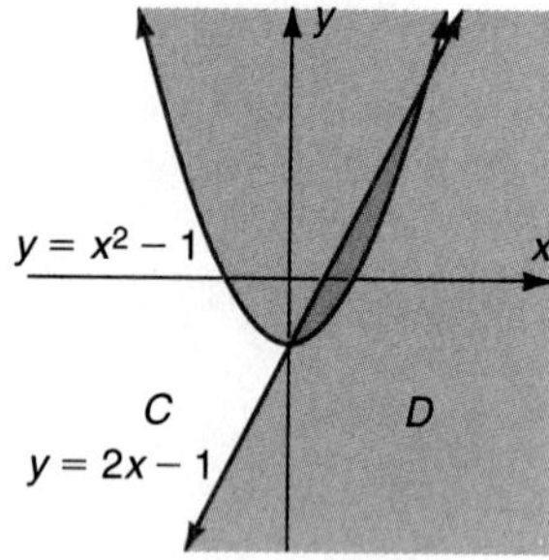

Fig. 8.35

REGION	TEST POINT	STATEMENT: $y \leq 2x - 1$	TRUTH OF STATEMENT	REGION IN SOLUTION SET?
C	(0, 0)	$0 \leq -1$	False	No
D	(0, −2)	$-2 \leq -1$	True	Yes

Finally, in Figure 8.36 we graph the solution set of the system. □

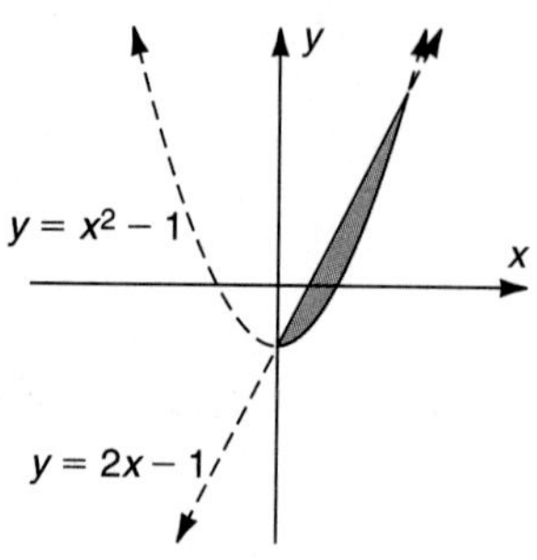

Fig. 8.36

Linear Programming

In management science, business, and engineering, systems of linear inequalities provide the basis for **linear programming** problems. Such problems require finding optimum values such as a minimum cost, a minimum drug dosage, a maximum number of calories consumed, or a maximum profit. The mathematical model for such problems consists of a system of linear inequalities and a function that calculates the quantity to be minimized or maximized. The inequalities in the system are called **constraints.** The solution set for the system of inequalities is called the **feasible region,** and the function to be optimized is called the **objective function.** The points of intersection of the lines in the system are called **vertices.**

The Fundamental Theorem of Linear Programming

If the objective function has an optimum value, it must occur at one of the vertices of the feasible region.

EXAMPLE 5 Find the maximum value of P if $P = 4x + 3y$ subject to the constraints

$$x \geq 0$$
$$y \geq 0$$
$$y \leq 6$$
$$y \geq x$$
$$y \leq 2x.$$

Solution In the graph in Figure 8.37, the feasible region is shaded and the vertices of the feasible region are labeled. By the Fundamental Theorem of Linear Programming, the maximum value of P must occur at one of the vertices.

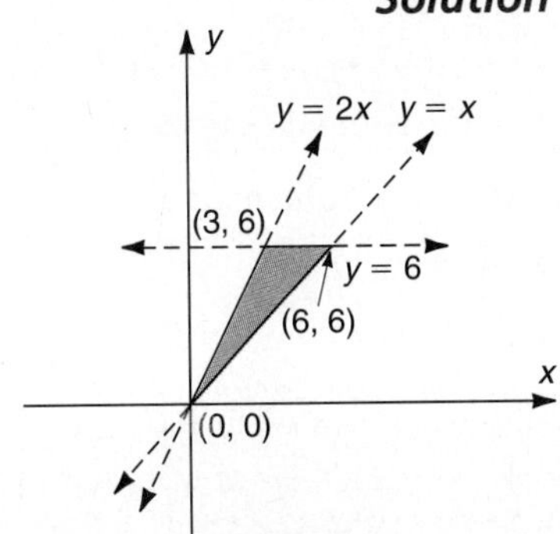

Fig. 8.37

Vertex	$P = 4x + 3y$
(0, 0)	$P = 0$
(3, 6)	$P = 4(3) + 3(6) = 12 + 18 = 30$
(6, 6)	$P = 4(6) + 3(6) = 24 + 18 = 42$

The maximum value of P subject to the constraints is 42. □

EXAMPLE 6 Find the minimum value of $C = x + 2y$ subject to the constraints

$$x \geq 0$$
$$y \geq 0$$
$$2x + 3y \geq 18$$
$$x + y \geq 7$$

Solution In the graph in Figure 8.38, the feasible region is shaded and the vertices of the feasible region are labeled. The minimum value of C must occur at one of the vertices.

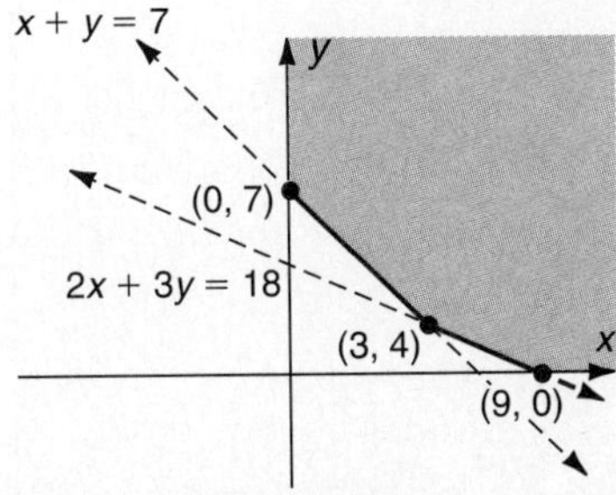

Fig. 8.38

Vertex	$C = x + 2y$
(9, 0)	$C = 9$
(3, 4)	$C = 3 + 2(4) = 11$
(0, 7)	$C = 14$

The minimum value of C subject to the constraints is 9. □

The feasible region in Example 5 is said to be **bounded,** while the feasible region in Example 6 is **unbounded.**

EXAMPLE 7 Emily and Lyle decorate T-shirts and sweatshirts to sell at the Olympics. Emily can finish a T-shirt in 2 hours, while Lyle takes only 1 hour. Sweatshirts take each of them 3 hours to decorate. Emily can work no more than 12 hours per week on shirts, and Lyle can work no more than 9 hours a week. They make \$8 profit on each T-shirt and \$15 profit on each sweatshirt. How many of each kind of shirt must they sell in a week to maximize their profit?

Solution Let x be the number of T-shirts and y be the number of sweatshirts made in a week. Let P be the profit made in a week. Since x T-shirts will yield $8x$ profit and y sweatshirts will yield $15y$ profit, the profit is given by $P = 8x + 15y$.

Objective function: $P = 8x + 15y$

Constraints:		
	$x \geq 0$	Number of T-shirts must be nonnegative.
	$y \geq 0$	Number of sweatshirts must be nonnegative.
	$2x + 3y \leq 12$	Emily's working time
	$x + 3y \leq 9$	Lyle's working time

Fig. 8.39

The feasible region is shaded in Figure 8.39, and the vertices are labeled. The maximum profit will occur at a vertex.

Vertex	$P = 8x + 15y$
(6, 0)	$P = 8(6) + 0 = 48$
(0, 3)	$P = 0 + 15(3) = 45$
(3, 2)	$P = 8(3) + 15(2) = 24 + 30 = 54$

The maximum profit is \$54. Emily and Lyle must sell three T-shirts and two sweatshirts to obtain a maximum weekly profit. □

GRAPHING CALCULATOR BOX

Graphing Inequalities in Two Variables

Graphing One Inequality in Two Variables

Let's graph the inequality $y \leq 2x - x^2$. Set the [RANGE] at the default window.

TI-81

Make sure that all graphs in the [Y=] menu are turned off.
[DRAW] (third row above PRGM) gives a menu that will SHADE a region above one graph and below another. We must identify the lower graph first followed by the upper graph. If the inequality is solved for y,

then greater than indicates above the graph and less than indicates below the graph. To graph a single inequality in two variables, we must enter both a lower and an upper graph. So, to graph $y \le 2x - x^2$, the parabola is the upper graph and we must choose a lower graph, such as $y = -5$. We will see SHADE(-5, $2x - x^2$) on the display as we enter the instructions.

[DRAW] [7] [−5] [ALPHA] [,] [2] [X|T] [−] [X|T] [x²] [)] [ENTER]

To clear the DRAW screen, press [DRAW] [ENTER] [ENTER].

CASIO fx-7700G

We must change to inequality mode. Press [MODE] [÷].

Enter [GRAPH] [F4] and notice that $y \le$ appears on the screen.

Now enter $2x - x^2$. Press [EXE] and the graph is drawn.

Graphing a System of Inequalities in Two Variables

Let's graph the solution set of $\begin{cases} y \le 2x - x^2 \\ y > \quad -x \end{cases}$.

TI-81

Make sure that all graphs in the [Y=] menu are turned off.
The TI-81 will graph this system because one inequality has less than and one has greater than. This makes it possible to specify a lower and an upper function.
$y = -x$ is the lower function and $y = 2x - x^2$ is the upper function.

[DRAW] [7] [(−)] [X|T] [ALPHA] [,] [2] [X|T] [−] [X|T] [x²] [)] [ENTER]

However, if both inequalities are less than or both greater than, it is diffidult to shade the solution set. In such cases, graph both equations at the same time and figure out which region to shade.

CASIO fx-7700G

In the inequality mode, just connect the two inequalities. Enter

[GRAPH] [F4] [2] [X,θ,T] [−] [X,θ,T] [x²] [↵] [GRAPH] [F1] [−] [X,θ,T] [EXE].

PROBLEM SET 8.7

Warm-ups

In Problems 1 through 4, graph the solution set for each inequality. See Examples 1 and 2.

1. $x - y > 3$

2. $y \le x^2 + 3$

3. $x^2 + y^2 \le 16$

4. $y > 3x - 2$

In Problems 5 through 8, graph the solution set of each system of inequalities. See Examples 3 and 4.

5. $\begin{cases} 2x - y \ge 0 \\ x + y < -1 \end{cases}$

6. $\begin{cases} x + y \ge 4 \\ 2x - y \ge 3 \end{cases}$

7. $\begin{cases} y \le -2x + 1 \\ y < 1 - x^2 \end{cases}$

8. $\begin{cases} x^2 + y^2 \le 4 \\ x > y^2 \end{cases}$

In Problems 9 through 12, find the indicated optimum value of F subject to the given constraints. See Examples 5 and 6.

9. Maximize F if $F = 10x - 12y$
Constraints: $x \geq 0$
$y \geq 0$
$y \leq 6$
$y \geq x$
$y \leq 2x$

10. Maximize F if $F = 6x - 3y$
Constraints: $x \geq 0$
$y \geq 0$
$x \leq 10$
$y \leq 12$
$x + y \geq 16$

11. Minimize F if $F = 20x + 30y$
Constraints: $x \geq 0$
$y \geq 0$
$3x + 2y \leq 30$
$x + 2y \geq 14$

12. Minimize F if $F = 2x + 5y$
Constraints: $x \geq 0$
$y \geq 0$
$2x + y \geq 8$
$x + 2y \geq 10$

For Problems 13 and 14, see Example 7.

13. Bud, an American short-hair cat, eats both dry and canned cat food. Dry cat food costs \$0.45 per ounce, and canned costs \$1.09 per ounce. An ounce of dry cat food contains 4 units of protein and 1 unit of fat, while an ounce of canned food contains 6 units of protein and 3 units of fat. The minimum daily requirements of protein and fat are 160 units and 70 units, respectively. How many ounces of each kind of cat food must Bud eat to obtain the minimum daily requirements and so that his owner spends the least amount of money?

14. HI, Inc., makes scientific and graphing calculators. Workers take 1 hour to make a scientific calculator and 3 hours to make a graphing calculator. The cost of making a scientific calculator is \$15, and the cost of making a graphing calculator is \$90. The company has available 1200 hours of labor and \$27,000 for manufacturing costs. The profit on a scientific calculator is \$4, and on a graphing calculator it is \$10. What should the production schedule be to maximize profit?

Practice Exercises

In Problems 15 through 30, graph the solution set of each inequality.

15. $y > 2x + 1$

16. $y < x$

17. $2x + 3y \geq 6$

18. $x + 2y \leq 4$

19. $y \leq |x|$

20. $y > |x + 1|$

21. $y \geq (x - 2)^2$

22. $y < (x + 2)^2$

23. $(x - 2)^2 + y^2 \leq 4$

24. $x^2 + (y - 2)^2 \leq 4$

25. $y < \sqrt{x - 1}$

26. $y > \sqrt{x}$

27. $y \leq 2^{x+1}$

28. $y > 2^x$

29. $y > \ln x$

30. $y < \ln x + 1$

In Problems 31 through 52, graph the solution set of each system of inequalities.

31. $\begin{cases} x + y \leq 1 \\ y \leq x + 5 \end{cases}$

32. $\begin{cases} 2x + 3y \geq 2 \\ x - y \leq 1 \end{cases}$

33. $\begin{cases} y > 4x + 2 \\ x \geq -1 \end{cases}$

34. $\begin{cases} y < -2x + 3 \\ y \leq 5 \end{cases}$

35. $\begin{cases} x - y \leq 1 \\ y > x \end{cases}$

36. $\begin{cases} x + y > 2 \\ y \geq -x \end{cases}$

37. $\begin{cases} x + 2y \leq 0 \\ x + 2y \geq 4 \end{cases}$

38. $\begin{cases} 2x - y \leq 4 \\ 2x - y \geq 0 \end{cases}$

39. $\begin{cases} y > (x - 1)^2 + 2 \\ y \leq x + 3 \end{cases}$

40. $\begin{cases} y \leq (x + 1)^2 + 2 \\ y > -x + 3 \end{cases}$

41. $\begin{cases} y \leq 1 - x^2 \\ y > x^2 - 1 \\ x \geq 0 \\ y \geq 0 \end{cases}$

42. $\begin{cases} y > 9 - x^2 \\ y \geq x^2 - 9 \\ x \geq 0 \\ y \geq 0 \end{cases}$

43. $\begin{cases} x + y \leq 1 \\ (x - 1)^2 + (y + 1)^2 \leq 1 \end{cases}$

44. $\begin{cases} x + y \geq 1 \\ (x - 1)^2 + (y - 1)^2 < 1 \end{cases}$

45. $\begin{cases} y \le 4 - x^2 \\ y \le -2x + 5 \\ x \ge 0 \\ y \ge 0 \end{cases}$

46. $\begin{cases} y \le x^2 + 4 \\ y \le 2x + 3 \\ x \ge 0 \\ y \ge 0 \end{cases}$

47. $\begin{cases} y \le 4 - x^2 \\ y \ge -2x + 5 \end{cases}$

48. $\begin{cases} y > x^2 + 4 \\ y \ge 2x + 3 \end{cases}$

49. $\begin{cases} x^2 + y^2 \ge 4 \\ x^2 + y^2 < 9 \\ x \ge 0 \\ y \ge 0 \end{cases}$

50. $\begin{cases} x^2 + y^2 \le 4 \\ x^2 + y^2 < 9 \\ x \ge 0 \\ y \ge 0 \end{cases}$

51. $\begin{cases} y \ge 3 \\ y \le 7 - x \\ 2x - 3y \ge -11 \end{cases}$

52. $\begin{cases} x > 3 \\ x + y \le 13 \\ y > 3x - 11 \end{cases}$

In Problems 53 through 58, find the minimum and maximum value of the given objective function subject to the given constraints.

53. Objective function: $C = 6x + 9y$
Constraints: $x \ge 0$
$y \ge 0$
$2x + y \le 12$
$x + 2y \ge 6$

54. Objective function: $C = 20x + 10y$
Constraints: $x \ge 0$
$y \ge 0$
$9x + 3y \le 18$
$6x + 4y \ge 24$

55. Objective function: $C = 12x + 3y$
Constraints: $x \ge 0$
$y \ge 0$
$x + 2y \ge 8$
$2x + y \ge 10$
$5x + 4y \le 40$

56. Objective function: $C = 20x + 7y$
Constraints: $x \ge 0$
$y \ge 0$
$x + y \ge 7$
$8x + 3y \ge 12$
$12x + 7y \le 84$

57. Objective function: $C = 2x + 5y$
Constraints: $x \ge 0$
$y \ge 0$
$4x + 3y \le 44$
$x + y \ge 9$
$2x + 5y \ge 36$

58. Objective function: $C = x + 15y$
Constraints: $x \ge 0$
$y \ge 0$
$x + y \ge 6$
$-x + 2y \le 14$
$4x + y \le 34$

59. Lady Grace plans to invest her \$60,000 royalty check in AAA bonds and AA bonds that yield 6% and 10%, respectively. She must invest at least twice as much in AAA bonds as in AA bonds, and she must invest no more than half in either kind of bond. How much should she invest in each to maximize her interest?

60. Louis makes designer cotton dresses and skirts trimmed in silk. He has 90 yards of cotton and 30 yards of silk. A dress requires 6 yards of cotton and 2 yards of silk, while a skirt requires 4 yards of cotton and $\frac{1}{2}$ yard of silk. How many of each should he make to maximize income if a dress sells for \$120 and a skirt for \$65?

Challenge Problems

Graph the solution set for each system.

61. $\begin{cases} y < nx; \ n < 0 \\ y < mx; \ m > 0 \end{cases}$

62. $\begin{cases} y \ge nx + b; \ n < 0, b > 0 \\ y \le mx + b; \ m > 0, b > 0 \end{cases}$

63. $\begin{cases} x < b; \ b < 0 \\ y < a; \ a > 0 \end{cases}$

64. $\begin{cases} x < b; \ b > 0 \\ y > a; \ a < 0 \end{cases}$

IN YOUR OWN WORDS . . .

65. Describe the graph of $y \le ax^2$; $a > 0$.

66. Describe the graph of $x^2 + y^2 > r^2$, $r > 0$.

8.8 USING GRAPHS TO SOLVE EQUATIONS IN ONE VARIABLE

Often, in the solution of real problems in engineering, business, and the sciences, equations arise that are not solvable with our usual collection of tools and tricks. For example,

$$\ln x = \sqrt[3]{x}$$

is a difficult equation to solve using the techniques we have developed. Various methods of approximation are used to solve such equations. In this section we investigate one such technique.

Suppose (h, k) is a solution of the system of equations

$$\begin{cases} y = f(x) \\ y = g(x) \end{cases}$$

Then we have $f(h) = k$ and $g(h) = k$. Therefore, *h is a solution* of the equation

$$f(x) = g(x)$$

Conversely, suppose h is a solution of the equation

$$f(x) = g(x)$$

Then $f(h) = g(h) = k'$, for some real number k'. Thus, the ordered pair *(h, k') is a solution* of the system of equations

$$\begin{cases} y = f(x) \\ y = g(x) \end{cases}$$

We have just shown that it is possible to find the real solutions of an equation in one variable by finding the x-coordinates of the solutions of a particular system of equations.

Theorem: Equations in One Variable and Systems of Two Equations

h is a real solution of the equation $f(x) = g(x)$ if and only if (h, k) is a solution of the system

$$\begin{cases} y = f(x) \\ y = g(x) \end{cases}$$

for some real number k.

Thus, solving the equation $f(x) = g(x)$ is equivalent to solving the system

$$\begin{cases} y = f(x) \\ y = g(x) \end{cases}$$

Have we gained anything? If $f(x) = g(x)$ is truly difficult, the system will certainly be difficult also. However, we learned in Section 8.1 that the real solutions of a system of two equations in two variables can be estimated from their graphs. Since the graph of the equation $y = f(x)$ is the graph of f, and the graph of the equation $y = g(x)$ is the graph of g, we have the following result.

Graphic Solution of an Equation in One Variable

The real solutions of the equation $f(x) = g(x)$ are the x-coordinates of the intersections of the graphs of f and g.

Therefore, a technique that can be used to determine the number of real solutions of an equation in one variable is to graph the left-hand side and the right-hand side of the equation and count the number of intersections. Notice how this works in the next example.

EXAMPLE 1 Determine how many real solutions the equation $\ln x = \dfrac{1}{x-2}$ has.

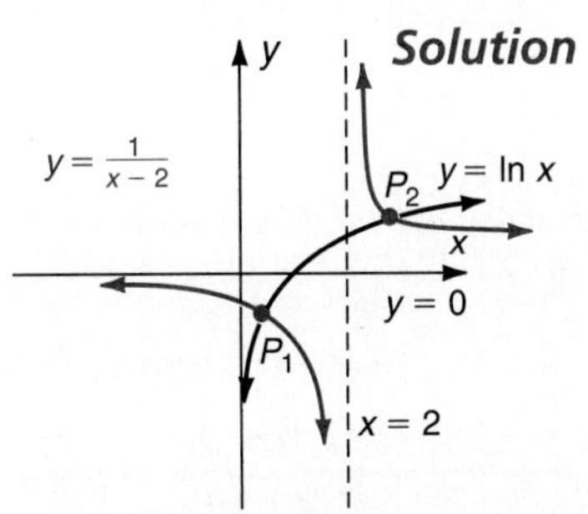

Fig. 8.40

Solution We graph $y = \ln x$ and $y = \dfrac{1}{x-2}$ on the same coordinate system and count the times the graphs intersect. The graphs intersect two times: at P_1 and P_2 as shown in Figure 8.40. There are two real solutions to the equation $\ln x = \dfrac{1}{x-2}$. □

If we wanted to find the solutions in Example 1, we could draw the graphs very carefully and approximate the x-coordinates of the points of intersection.

EXAMPLE 2 Find the real solutions of $2^x = 1 - x$.

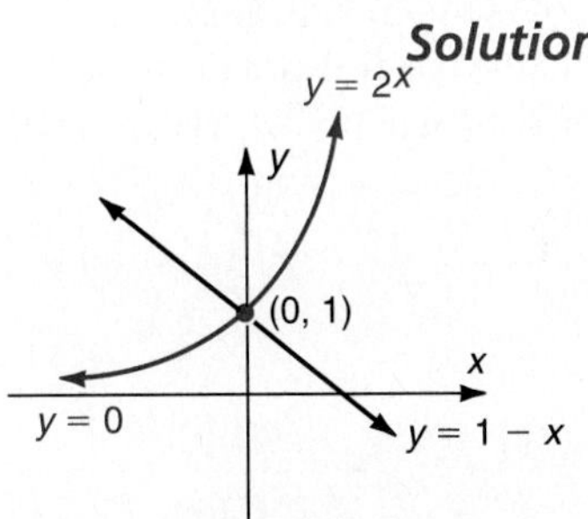

Fig. 8.41

Solution We graph both sides and look for points of intersection (Figure 8.41). The graphs intersect at one point and it *appears* to be (0, 1). We check 0 (the x-coordinate) in the original equation and see that it is in fact a solution. The only real solution is 0. □

EXAMPLE 3 Find the real solutions of $x^2 + \log x - 1 = 0$.

Solution This is certainly an unusual equation to try to factor! Furthermore, using the graphical method we find that the left side is not convenient to graph. But since we know the graphs of $\log x$ and $1 - x^2$, we rearrange the given equation.

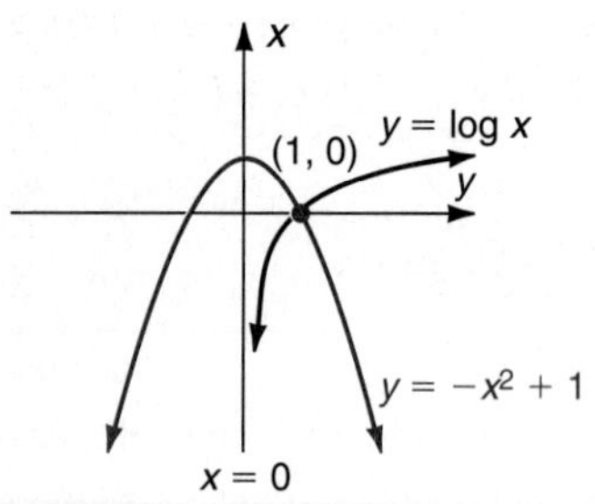

Fig. 8.42

$$\log x = -x^2 + 1$$

Then we graph both sides (Figure 8.42). The apparent intersection is (1, 0), so we check 1 in the original equation and find it to be a solution. The only real solution is 1. □

When using the graphing technique, we seldom find solutions as convenient as those in Examples 2 and 3. Usually a careful graph and approximation are required.

EXAMPLE 4 Approximate the real solution(s) of $\ln x = |x - 3|$ to the nearest tenth.

Solution We draw a careful graph of both sides (Figure 8.43) and look for points of intersection. There are clearly two points of intersection, and we estimate their x-coordinates to be 2.2 and 4.5. The approximate real solutions to the equation are 2.2 and 4.5 to the nearest tenth. □

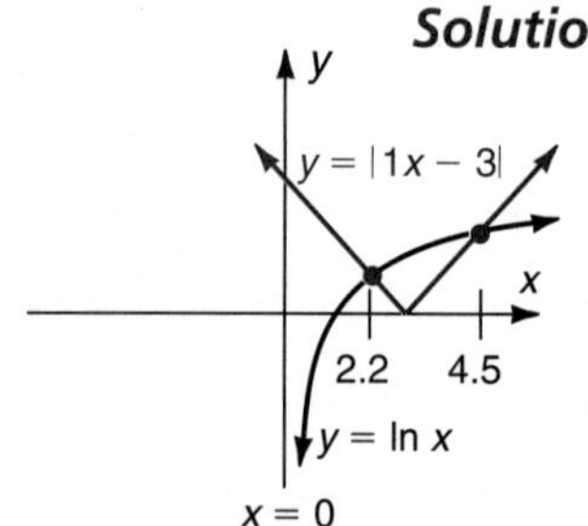

Fig. 8.43

EXAMPLE 5 Approximate the real solutions of the $x^2 + 1 = 2^x$.

Solution We see no direct way to attack this equation, so we graph both $y = x^2 + 1$ and $y = 2^x$, as shown in Figure 8.44, and look for points of intersection. Inspecting the graphs, it *appears* that they intersect in two points. The two points apparently are (0, 1) and (1, 2). So we check the two x-coordinates in the original equation and see that they *are* solutions. However, this example points out a danger in this approach.

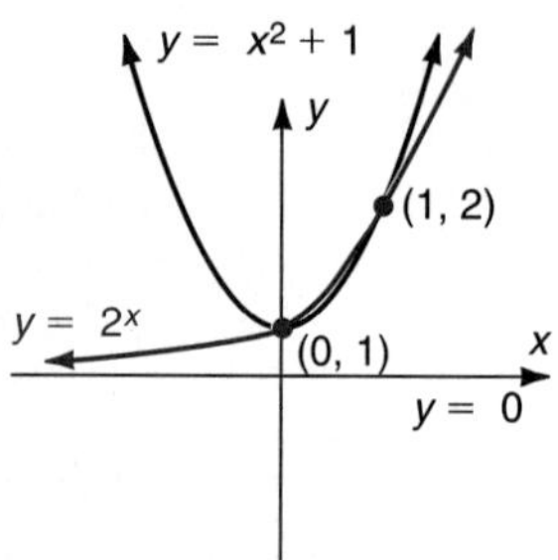

Fig. 8.44

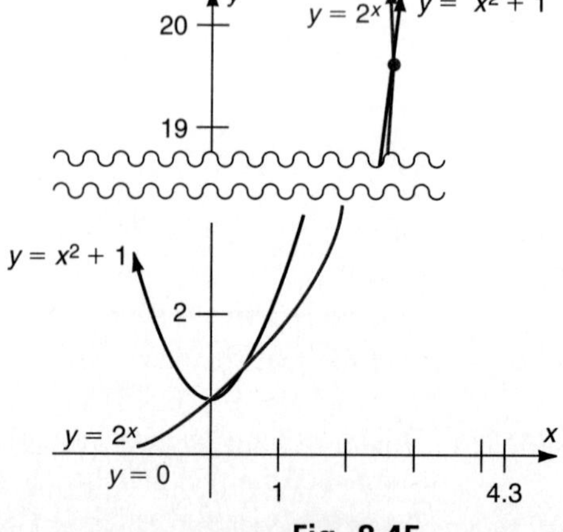

Fig. 8.45

We tend to investigate the graphs only near the origin. Could these graphs intersect again elsewhere? If we check the graphs between 4 and 5 on the x-axis and for larger y-coordinates (Figures 8.45), we see that the graphs *do intersect again!* The x-coordinate of this intersection is approximately 4.3. The equation $x^2 + 1 = 2^x$ has three real solutions. They are 0, 1, and approximately 4.3. □

PROBLEM SET 8.8

Warm-ups

In Problems 1 through 4, determine how many real solutions each equation has. See Example 1.

In Problems 1 and 2, use the graph provided.

1. $x^3 - 9x = x - x^3$ **2.** $16 - x^4 = |4 - x^2|$ **3.** $x^3 = \sqrt{x + 1}$ **4.** $\dfrac{1}{x} = 3 - x^2$

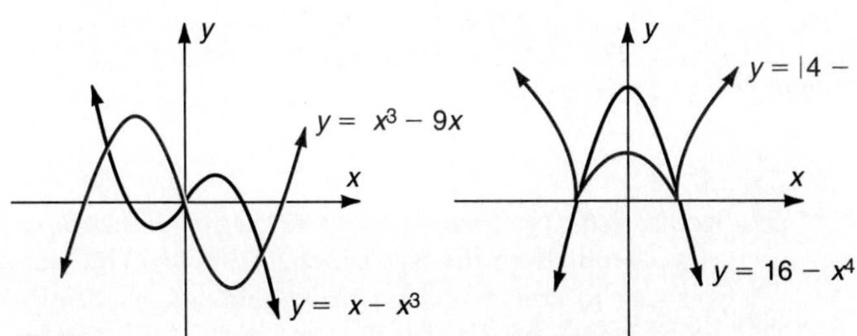

In Problems 5 and 6, find the real solutions of each equation. See Example 2.

5. $2|x| = x^2$ **6.** $3^x = 3 - 2(x - 1)^2$

In Problems 7 and 8, find the real solutions of each equation. See Example 3.

7. $2^{-x} + (x + 1)^2 - 2 = 0$ **8.** $\sqrt{x - 1} - |x - 1| = 0$

In Problems 9 and 10, approximate each real solution to the nearest tenth. See Example 4.

9. $\ln x = -x^2$ **10.** $\dfrac{1}{x - 1} = e^x$

In Problems 11 and 12, find all real solutions. Give approximations to the nearest tenth. See Example 5.

11. $\sqrt{x} - 1 = \ln x$ (two solutions) **12.** $(x + 1)^2 = e^x$ (three solutions)

Practice Exercises

In Problems 13 through 16, use the graphs of $y = x^3 - 4x$ and $y = x(x - 2)^2(x + 2)$ shown in Figure 8.46 to find the real solutions.

13. $x^3 - 4x = x(x - 2)^2(x + 2)$ **14.** $x^3 - 4x = 0$

15. $x(x - 2)^2(x + 2) = 0$ **16.** $\begin{cases} y = x^3 - 4x \\ y = x(x - 2)^2(x + 2) \end{cases}$

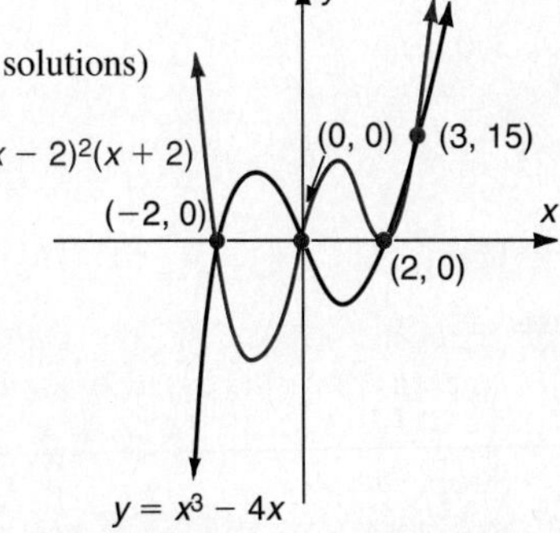

Fig. 8.46

In Problems 17 through 22, use the graphs of the functions f, g, and h shown in Figure 8.47 to find all real solutions to the indicated equation, system, or inequality.

17. $\begin{cases} y = f(x) \\ y = g(x) \end{cases}$.

18. $\begin{cases} y = f(x) \\ y = h(x) \end{cases}$.

19. $f(x) = h(x)$.

20. $f(x) = g(x)$.

21. $f(x) \geq g(x)$.

22. $f(x) < h(x)$.

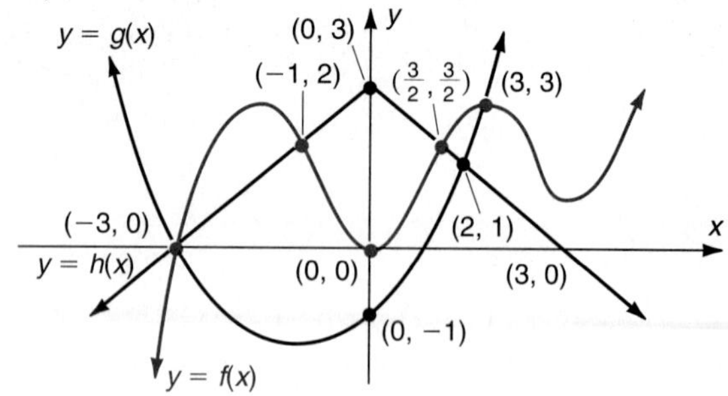

Fig. 8.47

In Problems 23 through 26, use the methods of this section to find the real solutions of each equation.

23. $x^4 = (x - 1)^3 + 1$

24. $x^3 + 1 = (x + 1)^2$

25. $x^3 + 1 = x^2 + x$

26. $-x^3 = (x - 1)^2 - 1$

In Problems 27 through 30, find all real solutions to each equation. Give approximations to the nearest tenth.

27. $2^x = 3 - x^2$

28. $(x - 1)^3 = \ln x$

29. $\ln x = -x^3$

30. $2^{x-1} = \sqrt{x - 1}$

31. The Machine Tools Division of Lunar Effects, Inc., had sales of \$50 million last year and forecasts sales to increase at a rate of 10% yearly in the foreseeable future. That is, $S_m = 50(1.1)^t$, where S_m is in millions and t is in years. The newest division of Lunar Effects in the Laser Cutter Division which sold \$10 million last year and forecasts a 20% annual increase, or $S_l = 10(1.2)^t$.

(a) Graph S_m and S_l on the same set of axes using the horizontal axis as t (in years).

(b) Estimate from the graph when Laser Cutter sales will catch up with Machine Tools sales? ($S_l = S_m$.)

(c) Approximate to the nearest tenth the solution of $S_l = S_m$ using the methods of Section 6.5.

(d) Suppose Laser Cutter sales actually increase at a rate of 25% yearly. When will they catch up?

32. Jorge Zapata, president of Zapata Industries, wishes to increase profit from the Kitchenware Division. The vice-president in charge of sales claims that his plan (an increase in the sales force) will yield yearly profits for this division of $P_1(t) = 100t^2 + 1200$, where P is in dollars and t is in years. The vice-president in charge of computer resources says that her plan (computer and software upgrades) will give yearly profits of $P_2(t) = 200e^{t/2} + 960$. Assume that the cost of implementing each plan is about the same.

(a) Graph the profit from each plan on the same set of axes.

(b) Which plan is the better short-range solution? Which is the better long-range solution?

(c) When are they even?

Challenge Problems

In Problems 33 and 34, find all real solutions to each equation. Give approximations to the nearest tenth.

33. $e^x = x^4 - 2x^2 + 1$

34. $\ln x = \sqrt{x - 1}$

IN YOUR OWN WORDS . . .

35. Explain the relationship between a system of equations and the graphs of the equations in the system.

36. Explain how solving a system of equations can provide the real solution(s) to an equation in one variable.

CHAPTER SUMMARY

GLOSSARY

System of equations: Two or more equations considered simultaneously.

Solution to a system of equations in two variables: An ordered pair that is a solution to all equations in the system. The **solution set** consists of all solutions.

Solve a system: Find the solution set.

Consistent system: A system with at least one solution.

Inconsistent system: A system with no solutions.

Dependent system: A system with an infinite number of solutions.

Equivalent systems: Two systems with the same solution set.

Linear system: A system that contains only linear equations.

Nonlinear system: A system that contains at least one nonlinear equation.

Square system: A system with the same number of variables as equations.

Matrix: A rectangular array of numbers, usually written inside brackets or parentheses.

Elements or **entries:** The numbers in a matrix.

Dimension or size of a matrix: The number of rows followed by the number of columns in the matrix.

Square matrix: A matrix that has the same number of rows as columns. The entries that have the same row and column number form the **main diagonal.**

Zero matrix: A matrix that has all entries as zeros. Because $A + 0 = A = 0 + A$ when the addition is defined, a zero matrix is called an **additive identity.**

Identity matrix: A square matrix with ones down the main diagonal and zeros elsewhere.

Inverse of a matrix: If A is a square matrix and there is a matrix A^{-1} such that $AA^{-1} = A^{-1}A = I$, where I is an identity matrix, then A^{-1} is called the inverse of A.

Singular matrix: A matrix that does not have an inverse.

Nonsingular matrix: A matrix that has an inverse.

Matrix of coefficients: The matrix formed by the coefficients in a system of linear equations.

Augmented matrix: The matrix formed by annexing the column of constant terms to the matrix of coefficients.

Determinant: A number assigned to each square matrix.

SOLVING SYSTEMS OF EQUATIONS

Method of Substitution

1. Solve one of the equations for one of the variables.
2. Substitute this expression for the same variable in the other equation.
3. Solve the resulting equation.
4. Substitute each value back into step 1 to find the other variable.
5. Write the solution set.

Method of Elimination

1. Multiply both sides of each equation by a suitable real number so that one of the variables will be eliminated by addition of the equations. (This step may not be necessary).

2. Add the equations and solve the resulting equation.
3. Substitute the value found in step 2 into one of the original equations and solve this equation.
4. Write the solution set.

ELEMENTARY OPERATIONS

An equivalent system of equations will be produced if any of the following operations is performed.

1. Interchange any two equations.
2. Multiply both sides of any equation by the same nonzero real number.
3. Add a multiple of any equation to any other equation.

ELEMENTARY ROW OPERATIONS

The following operations when performed on an augmented matrix will produce an equivalent matrix.

1. Interchange two rows.
2. Multiply the numbers of any row by a nonzero number.
3. Add a multiple of any row to any other row.

THE METHOD OF THE AUGMENTED MATRIX

1. Write the augmented matrix.
2. Using elementary row operations, change the matrix to row echelon form. It is important to make the changes in the following order.
   (a) Move all completely zero rows to the bottom of the matrix.
   (b) Get the number 1 in row 1, column 1.
   (c) Get zeros in column 1 under the leading 1 in row 1.
   (d) Get 1 in row 2, column 2 if possible. If not, go to the next column.
   (e) Get zeros under the leading 1 formed in step (d).
   (f) Continue steps similar to (d) and (e) until the matrix is in row echelon form.
3. Find each variable in turn.
4. Write the solution set.

EQUAL MATRICES

Two matrices are equal if and only if they are the same size and corresponding entries are equal. That is, $A = B$, if and only if $a_{ij} = b_{ij}$ for all appropriate i and j.

ADDITION OF MATRICES

If A and B are matrices of the same size, then $A + B$ is the matrix formed by adding corresponding entries in A and B. That is, $C = A + B$ if and only if $c_{ij} = a_{ij} + b_{ij}$ for all appropriate i and j.

SCALAR MULTIPLICATION

If A is a matrix and c a scalar, then cA is the matrix formed by multiplying each entry in A by c.

OPPOSITE OF A MATRIX If A is a matrix, then $-A = (-1)A$.

SUBTRACTION OF MATRICES If A and B are matrices of the same size, then $A - B = A + (-B)$.

MULTIPLICATION OF MATRICES If A is an $m \times n$ matrix and B is an $n \times p$ matrix, then AB is an $m \times p$ matrix whose entries are computed as follows: The entry in the ith row and jth column of AB is the sum of products of the corresponding entries in the ith row of A and jth column of B.

PROPERTIES OF MATRICES Matrices are indicated with capital letters, and scalars are lowercase letters. Assume the sizes of the matrices are appropriate for the operation to be performed.

1. $A + B = B + A$ Commutative Property for addition
2. $A + (B + C) = (A + B) + C$ Associative Property for addition
3. $A(BC) = (AB)C$ Associative Property for multiplication
4. $A(B + C) = AB + AC$ Distributive Property
5. $(B + C)A = BA + CA$ Distributive Property
6. $a(B + C) = aB + aC$
7. $(a + b)C = aC + bC$
8. $(ab)C = a(bC)$
9. $a(BC) = (aB)C = B(aC)$

The commutative property *does not* hold for multiplication of matrices.

INVERSE OF A SQUARE MATRIX Let A be a square matrix. If there is a square matrix A^{-1} such that $AA^{-1} = A^{-1}A = I$, then A^{-1} is called the inverse of A.

INVERTIBILITY OF A MATRIX Let A be a square matrix. A^{-1} exists if and only if $\det A \neq 0$.

VALUE OF A SECOND-ORDER DETERMINANT

$$\begin{vmatrix} a & b \\ c & d \end{vmatrix} = ad - bc$$

TO EVALUATE A DETERMINANT OF ORDER HIGHER THAN 2

1. Select a row or column to expand about.
2. For each entry in the chosen row (or column), multiply the element by its cofactor.
3. The value of the determinant is the sum of the products found in step 2.

CRAMER'S RULE FOR A SYSTEM OF TWO EQUATIONS IN TWO VARIABLES

The solution to the system of equations $\begin{cases} ax + by = p \\ cx + dy = q \end{cases}$ can be found using

$$x = \frac{\begin{vmatrix} p & b \\ q & d \end{vmatrix}}{\begin{vmatrix} a & b \\ c & d \end{vmatrix}} \text{ and } y = \frac{\begin{vmatrix} a & p \\ c & q \end{vmatrix}}{\begin{vmatrix} a & b \\ c & d \end{vmatrix}}, \text{ provided } \begin{vmatrix} a & b \\ c & d \end{vmatrix} \neq 0.$$

CRAMER'S RULE FOR THREE EQUATIONS IN THREE VARIABLES

The solution for the system of equations $\begin{cases} ax + by + cz = p \\ dx + ey + fz = q \\ gx + hy + iz = r \end{cases}$ can be found using

$$x = \frac{\begin{vmatrix} p & b & c \\ q & e & f \\ r & h & i \end{vmatrix}}{\begin{vmatrix} a & b & c \\ d & e & f \\ g & h & i \end{vmatrix}}, \; y = \frac{\begin{vmatrix} a & p & c \\ d & q & f \\ g & r & i \end{vmatrix}}{\begin{vmatrix} a & b & c \\ d & e & f \\ g & h & i \end{vmatrix}}, \text{ and } z = \frac{\begin{vmatrix} a & b & p \\ d & e & q \\ g & h & r \end{vmatrix}}{\begin{vmatrix} a & b & c \\ d & e & f \\ g & h & i \end{vmatrix}},$$

$$\text{provided } \begin{vmatrix} a & b & c \\ d & e & f \\ g & h & i \end{vmatrix} \neq 0.$$

A PROCEDURE FOR GRAPHING AN INEQUALITY IN TWO VARIABLES

1. Graph the boundary curve.
   (a) Draw a solid curve if equality is included.
   (b) Draw a dashed curve if equality is not included.
2. Determine which region(s) formed by the curve makes the inequality true by testing with one point from inside each region.
3. Shade the region(s) that make the inequality true.

A PROCEDURE FOR GRAPHING THE SOLUTION SET OF A SYSTEM OF INEQUALITIES

1. Graph each inequality on the same coordinate system, lightly shading its solution set.
2. Darken the *intersection* of the lightly shaded regions.
3. Change any portion of any solid boundary curve not in the intersection of the solutions from solid to dashed.

THE FUNDAMENTAL THEOREM OF LINEAR PROGRAMMING

If the objective function has an optimum value, it must occur at one of the vertices of the feasible region.

EQUATIONS IN ONE VARIABLE AND SYSTEMS OF EQUATIONS

h is a solution of the equation $f(x) = g(x)$ if and only if (h, k) is a solution of the system

$$\begin{cases} y = f(x) \\ y = g(x) \end{cases} \quad \text{for some real number } k$$

GRAPHIC SOLUTION OF AN EQUATION IN ONE VARIABLE

The real solutions of the equation $f(x) = g(x)$ are the x-coordinates of the intersections of the graphs of f and g.

REVIEW PROBLEMS

In Problems 1 through 8, find the solution set.

1. $\begin{cases} x^2 - y^2 = 4 \\ x + y^2 = 8 \end{cases}$

2. $\begin{cases} x + 3y = 12 \\ y = \sqrt{x-2} \end{cases}$

3. $\begin{cases} (x-2)^2 + (y-5)^2 = 4 \\ x + y = 9 \end{cases}$

4. $\begin{cases} x^2 - y^2 = 9 \\ x^2 + y^2 = 1 \end{cases}$

5. $\begin{cases} y = |x| \\ x = 2 - y^2 \end{cases}$

6. $\begin{cases} 3x_1 + 2x_2 = 7 \\ 2x_1 - 3x_2 = 11 \end{cases}$

7. $\begin{cases} \dfrac{1}{x} - \dfrac{3}{y} = \dfrac{9}{10} \\ \dfrac{2}{x} - \dfrac{1}{y} = -\dfrac{6}{5} \end{cases}$

8. $\begin{cases} x^2 + y^2 = 4 \\ (x+5)^2 + y^2 = 9 \end{cases}$

In Problems 9 through 12, find the solution set and graph the system.

9. $\begin{cases} 3x_1 + x_2 = 0 \\ 2x_2 = -6x_1 \end{cases}$

10. $\begin{cases} x + y^2 = 2 \\ x = y^2 - 2 \end{cases}$

11. $\begin{cases} y = 3x - 7 \\ 3x - y = 0 \end{cases}$

12. $\begin{cases} \dfrac{1}{3}x - y = 4 \\ \dfrac{1}{2}x - \dfrac{2}{3}y = 2 \end{cases}$

In Problems 13 through 24, find the solution set.

13. $\begin{cases} x + y + z = 1 \\ x + 3z = 5 \\ 2x + y - z = 0 \end{cases}$

14. $\begin{cases} 2x - y - z = -7 \\ x + y + z = 1 \\ 3x + 2y - 2z = 4 \end{cases}$

15. $\begin{cases} x_1 + x_2 + x_3 = 0 \\ 2x_1 - x_2 - x_3 = 0 \\ 3x_1 - x_2 + 2x_3 = 0 \end{cases}$

16. $\begin{cases} 3x + 6y + 3z = 9 \\ x - 2y = 5 \\ 3y + z = -5 \end{cases}$

17. $\begin{cases} 3x - y + 2z = 11 \\ y - 3z = 4 \\ 3y - 6z = 8 \end{cases}$

18. $\begin{cases} x + y = 9 \\ y + z = 7 \\ x + z = 8 \end{cases}$

19. $\begin{cases} 2x + 3y + 4z = 18 \\ 4x - 3y - 2z = 4 \\ x + y + z = 6 \end{cases}$

20. $\begin{cases} x + 3y + 2z = 0 \\ y - z = 0 \end{cases}$

21. $\begin{cases} x - y + z = -5 \\ 2x + 3y = 0 \\ 3x - z = 5 \end{cases}$

22. $\begin{cases} 2x - y - z = -3 \\ x + 2y + 3z = -10 \\ 3x - y - 2z = -5 \end{cases}$

23. $\begin{cases} x_1 + x_2 = -1 \\ x_2 + x_3 = -1 \\ x_3 + x_4 = -1 \\ x_1 + x_4 = -1 \end{cases}$

24. $\begin{cases} x + y + z + w = 4 \\ x - y - z + w = 0 \\ x + y + z - w = 2 \\ x + y - z - w = 0 \end{cases}$

In Problems 25 through 34, graph the solution set.

25. $y \le x$

26. $x^2 + y^2 \le 9$

27. $y < (x+1)^2$

28. $x + 2y > 0$

29. $\begin{cases} x + 2y \le 0 \\ y < 4 - x^2 \end{cases}$

30. $\begin{cases} x^2 + y^2 \le 9 \\ x^2 + y^2 > 4 \end{cases}$

31. $\begin{cases} y < (x-1)^2 \\ y \ge x - 1 \end{cases}$

32. $\begin{cases} x > (y+1)^2 - 2 \\ y > x - 1 \end{cases}$

33. $\begin{cases} 3x + 2y \le 6 \\ x + \dfrac{2}{3}y > 1 \end{cases}$

34. $\begin{cases} x + 2y \le 0 \\ 2x - y \ge 0 \end{cases}$

Each matrix in Problems 35 through 40 is the augmented matrix for a system of linear equations. In each case, find the solution set for the system.

35. $\begin{pmatrix} 1 & 0 & 2 \\ 0 & 1 & 3 \end{pmatrix}$

36. $\begin{pmatrix} 1 & 0 & 3 \\ 0 & 0 & 2 \end{pmatrix}$

37. $\begin{pmatrix} 1 & 0 & 0 & 4 \\ 0 & 1 & 1 & 5 \\ 0 & 0 & 1 & 6 \end{pmatrix}$ **38.** $\begin{pmatrix} 1 & 1 & 1 & 3 \\ 0 & 0 & 1 & 2 \\ 0 & 0 & 0 & 0 \end{pmatrix}$ **39.** $\begin{pmatrix} 1 & 1 & 1 & -2 \\ 1 & 0 & 1 & 3 \\ 0 & 0 & 2 & 6 \end{pmatrix}$ **40.** $\begin{pmatrix} 1 & 1 & 1 & 0 \\ 0 & 1 & 2 & 3 \\ 0 & 0 & 0 & 0 \end{pmatrix}$

In Problems 41 and 42, write the system in the form $AX = B$, where A, X, and B are appropriate matrices.

41. $\begin{cases} 3x - 2y = 4 \\ x + y = 0 \end{cases}$

42. $\begin{cases} 2x - y + z = 5 \\ x + 3y = 0 \\ 3x - z = 5 \end{cases}$

43. Given $\begin{vmatrix} a & b \\ c & d \end{vmatrix} = 4$, $\begin{vmatrix} a & e \\ c & f \end{vmatrix} = 8$, $\begin{vmatrix} e & b \\ f & d \end{vmatrix} = -12$, find the solution set of the system $\begin{cases} ax + by = e \\ cx + dy = f \end{cases}$.

In Problems 44 through 55, find the indicated matrix or determinant given that $A = \begin{pmatrix} 2 & 3 \\ -1 & 5 \end{pmatrix}$,

$B = \begin{pmatrix} 3 \\ -7 \end{pmatrix}$, $C = \begin{pmatrix} 2 & -1 & 0 \\ 4 & 1 & 3 \end{pmatrix}$, $D = \begin{pmatrix} 1 & 0 & -2 \\ 2 & 1 & 0 \\ 1 & 0 & -3 \end{pmatrix}$.

44. A^{-1}
45. $|D|$
46. $A^2 - 2A$
47. AB
48. AC
49. D^{-1}
50. CD
51. X if $AX = B$
52. X if $3X = C$
53. $|A|$
54. $|3A|$
55. $AC - 3C$

In Problems 56 and 57, use determinants to determine whether the given matrix has an inverse.

56. $\begin{pmatrix} 1 & 0 & 1 \\ 3 & 2 & 5 \\ 8 & 0 & 7 \end{pmatrix}$

57. $\begin{pmatrix} -3 & 0 & -4 \\ -3 & 4 & 0 \\ 0 & 1 & 1 \end{pmatrix}$

In Problems 58 through 61, find the minimum and maximum values of the given objective function subject to the given constraints.

58. Objective function: $P = x + 7y$
Constraints: $x \geq 0$
$y \geq 0$
$2x + y \leq 10$
$x + 2y \leq 4$

59. Objective function: $P = 2x + y$
Constraints: $x \geq 0$
$y \geq 0$
$4x + 3y \leq 12$
$6x + 3y \leq 18$

60. Objective function: $C = 12x + 2y$
Constraints: $x \geq 0$
$y \geq 0$
$2x + y \geq 8$
$x + 2y \geq 10$
$5x + 4y \leq 40$

61. Objective function: $C = 10x + 7y$
Constraints: $x \geq 0$
$y \geq 0$
$x + y \geq 5$
$6x + 3y \geq 9$
$12x + 7y \leq 84$

In Problems 62 and 63, find all solutions. Give approximations to the nearest tenth.

62. $\log x + 1 = x^2$

63. $3^x = x + 2$

64. Metro Pool and Chemical Company provides year-round service for swimming pools. Treatment involves using sesque, chlorine, and a stabilizer. The sesque balances the alkalinity, and the stabilizer holds the chlorine in the water. A well-kept average-sized private pool requires 5 pounds of stabilizer, 3 pounds of chlorine, and $\frac{1}{4}$ pound of sesque, while a neglected pool requires 6 pounds of stabilizer, 9 pounds of chlorine, and $\frac{3}{4}$ pound of sesque. No more than 12 pounds of chlorine, 22 pounds of stabilizer, and 2 pounds of sesque can be used at one time. If chlorine costs \$1.65 per pound, sesque \$2.00 per pound, and stabilizer \$1.09 per pound, how many well-kept pools and how many neglected pools should Metro service to minimize the cost of these chemicals?

65. The area of a rectangular deck is 384 square feet, and its perimeter is 112 feet. Find the dimensions of the deck.

66. The sum of the squares of two numbers is 346, and the sum of the two numbers is 26. Find the two numbers.

67. Tickets to the ballet were \$12.50 for matinees and \$18.00 for evening performances. The total attendance was 2650. If the total receipts were \$40,913, how many attended evening performances?

68. A 576-mile plane trip takes 4 hours with the wind and 6 hours against the wind. Find the speed of the plane and the speed of the wind.

LET'S NOT FORGET . . .

69. Which form is more useful?

$$\text{(a)} \begin{pmatrix} 2 & 0 & 1 & 5 \\ 1 & 1 & 0 & -1 \\ 1 & 1 & 1 & 0 \end{pmatrix} \quad \text{or} \quad \text{(b)} \begin{pmatrix} 1 & 1 & 1 & 0 \\ 0 & 1 & \frac{1}{2} & -\frac{5}{2} \\ 0 & 0 & 1 & -3 \end{pmatrix}$$

when solving a system of equations by the method of the augmented matrix.

70. Watch the role of negative signs. Evaluate

$$\begin{vmatrix} 0 & 0 & 0 & 1 \\ 0 & 1 & 0 & 2 \\ -1 & 2 & -1 & 3 \\ -1 & -1 & 3 & 4 \end{vmatrix}$$

71. From memory. What is a *singular* matrix? Is there an easy test? Is either matrix singular?

$$\text{a)} \begin{pmatrix} 1 & 0 & 2 \\ 0 & 2 & 5 \\ 4 & 0 & 7 \end{pmatrix} \qquad \text{b)} \begin{pmatrix} -3 & 6 & -4 \\ -2 & 4 & -7 \\ 0 & 0 & -5 \end{pmatrix}$$

72. With a calculator. Approximate the solution(s) of $\log x = -x^2$ to the three significant digits.

CHAPTER 9

Additional Topics From Algebra

CONNECTIONS

In our final chapter, we develop some topics that are natural extensions of the work we have been doing. In particular, we will investigate some important connections to calculus.

We begin by developing summation notation, a compact way of writing sums containing a large number of terms. We follow with an introduction to mathematical induction which provides the structure for proofs involving the natural numbers. Then we examine $(a + b)^n$ when n is a natural number, culminating in a proof of the binomial theorem by mathematical induction.

We conclude the chapter by introducing sequences and series, counting and probability. These are topics that become more and more important as our mathematical maturity develops.

9.1 SUMMATION NOTATION AND FINITE SUMS

This section will deal with sums that contain a great number of terms. Since these terms can be unwieldy to write, we introduce a very compact notation called **summation notation.**

Consider the sum $a_1 + a_2 + a_3 + a_4 + a_5 + a_6 + a_7 + a_8 + a_9$. We use summation notation to abbreviate it as follows.

$$\sum_{j=1}^{9} a_j = a_1 + a_2 + a_3 + a_4 + a_5 + a_6 + a_7 + a_8 + a_9$$

The symbol Σ is the capital Greek letter sigma. Therefore, this is often called **sigma notation.** The assignment $j = 1$, under the sigma, is the first value of the variable j to be used in the sum. The integer 9, at the top of the sigma, is the final value of the variable j to be used in the sum. There is a term in the sum for every integer value of j from 1 to 9.

Summation Notation

For integers m and n with $m \leq n$,

$$\sum_{j=m}^{n} A_j = A_m + A_{m+1} + A_{m+2} + \cdots + A_n$$

j is called the **index of summation,** m is called the **lower limit,** and n is called the **upper limit.**

EXAMPLE 1 Evaluate the following sums.

(a) $\displaystyle\sum_{j=1}^{4} 2j$ (b) $\displaystyle\sum_{j=3}^{6} (2 + j)$ (c) $\displaystyle\sum_{k=1}^{4} \frac{2^k}{k}$ (d) $\displaystyle\sum_{n=0}^{4} (-1)^n 3^{n-1}$

Solutions (a) $\displaystyle\sum_{j=1}^{4} 2j$ This sum consists of four terms, where j is 1, 2, 3, and 4. That is,

$$\begin{aligned}\sum_{j=1}^{4} 2j &= 2 \cdot 1 + 2 \cdot 2 + 2 \cdot 3 + 2 \cdot 4\\ &= 2 + 4 + 6 + 8\\ &= 20\end{aligned}$$

(b) $\sum_{j=3}^{6} (2 + j)$ The lower limit does not have to be 1. Notice that this sum starts with $j = 3$.

$$\sum_{j=3}^{6} (2 + j) = (2 + 3) + (2 + 4) + (2 + 5) + (2 + 6)$$

$$= 5 + 6 + 7 + 8 = 26$$

(c) $\sum_{k=1}^{4} \frac{2^k}{k} = \frac{2^1}{1} + \frac{2^2}{2} + \frac{2^3}{3} + \frac{2^4}{4} = 2 + 2 + \frac{8}{3} + 4 = \frac{32}{3}$

In this sum, notice that the *index of summation* is k instead of j. The index can be any letter as long as it is not used in some other role in the expression.

(d) $\sum_{n=0}^{4} (-1)^n 3^{n-1} = (-1)^0 3^{0-1} + (-1)^1 3^0 + (-1)^2 3^1 + (-1)^3 3^2 + (-1)^4 3^3$

$$= 1 \cdot 3^{-1} + (-1) \cdot 1 + 1 \cdot 3 + (-1) \cdot 9 + 1 \cdot 27$$

$$= \frac{1}{3} - 1 + 3 - 9 + 27 = \frac{1}{3} + 20 = \frac{61}{3}$$

Notice in this sum how the factor $(-1)^n$ served to make the signs *alternate*. □

Sums may contain literal constants or variables other than the index of summation, and they may contain functions. However, the summation notation is unraveled in the same manner as illustrated in Example 1.

EXAMPLE 2 Expand each sum.

(a) $\sum_{j=-1}^{5} jx^j$ (b) $\sum_{j=1}^{4} e^{j-1} \ln j$ (c) $\sum_{n=1}^{4} (-1)^{n-1} 2^n x^{2n}$

Solutions (a) $\sum_{j=-1}^{5} jx^j = -1x^{-1} + 0 \cdot x^0 + 1x^1 + 2x^2 + 3x^3 + 4x^4 + 5x^5$

$$= \frac{-1}{x} + x + 2x^2 + 3x^3 + 4x^4 + 5x^5$$

(b) $\sum_{j=1}^{4} e^{j-1} \ln j = e^0 \ln 1 + e^1 \ln 2 + e^2 \ln 3 + e^3 \ln 4$

$$= e \ln 2 + e^2 \ln 3 + e^3 \ln 4$$

(c) $\sum_{n=1}^{4} (-1)^{n-1}2^n x^{2n} = (-1)^0 2^1 x^2 + (-1)^1 2^2 x^4 + (-1)^2 2^3 x^6 + (-1)^3 2^4 x^8$

$= 2x^2 - 4x^4 + 8x^6 - 16x^8$ □

Writing a sum in the compact summation notation is a little more difficult than expanding an existing sum that is in summation notation. A general term or *pattern* is needed. This pattern must represent all the terms in the sum. For example, consider the sum

$$e + \frac{e^2}{3} + \frac{e^3}{5} + \frac{e^4}{7} + \frac{e^5}{9} + \frac{e^6}{11} + \frac{e^7}{13}$$

We notice that there are seven terms and that each contains a power of e. If we write the first term in the form $e = \frac{e^1}{1}$, we see a definite pattern in the numerators. Our general term will look like $\frac{e^j}{?}$, where j starts at 1 and continues to 7. But what about the denominator? We notice the denominators are consecutive *odd* numbers starting with 1. A little fiddling on scratch paper shows that $2j - 1$ will do the trick. The sum can be written $\sum_{j=1}^{7} \frac{e^j}{2j-1}$.

EXAMPLE 3 Express each sum using summation notation.

(a) $2 + 4 + 6 + 8 + 10$

(b) $1 + 3z^2 + 9z^4 + 27z^6 + 81z^8 + 243z^{10}$

(c) $1 - \frac{1}{8} + \frac{1}{27} - \frac{1}{64} + \frac{1}{125} - \frac{1}{216}$

(d) $c_0 + c_1 \log_2 3 + 2c_2 + c_3 \log_2 5 + c_4 \log_2 6$

Solutions (a) $2 + 4 + 6 + 8 + 10$

This is the sum of the consecutive even integers from 2 to 10. The term $2j$ results in even integers starting with 2.

$$2 + 4 + 6 + 8 + 10 = \sum_{j=1}^{5} 2j$$

(b) $1 + 3z^2 + 9z^4 + 27z^6 + 81z^8 + 243z^{10}$

Since $1 = 3^0 z^0$, the powers of z are consecutive even integers starting with 0, while the coefficients are powers of 3 starting with 0.

$$1 + 3z^2 + 9z^4 + 27z^6 + 81z^8 + 243z^{10} = \sum_{j=0}^{5} 3^j z^{2j}$$

(c) $1 - \frac{1}{8} + \frac{1}{27} - \frac{1}{64} + \frac{1}{125} - \frac{1}{216}$

The signs alternate, which we can handle with $(-1)^{j+1}$, and the denominators are successive cubes.

$$1 - \frac{1}{8} + \frac{1}{27} - \frac{1}{64} + \frac{1}{125} - \frac{1}{216} = \sum_{j=1}^{6} (-1)^{j+1}\frac{1}{j^3}$$

(d) $c_0 + c_1 \log_2 3 + 2c_2 + c_3 \log_2 5 + c_4 \log_2 6$

The c's are just successively numbered constants, but the log's are a puzzle. Remembering that $\log_b b = 1$ and $\log_b x^k = k \log_b x$, we can rewrite the sum as

$$c_0 \log_2 2 + c_1 \log_2 3 + 2c_2 \log_2 2 + c_3 \log_2 5 + c_4 \log_2 6 =$$

$$c_0 \log_2 2 + c_1 \log_2 3 + c_2 \log_2 4 + c_3 \log_2 5 + c_4 \log_2 6 = \sum_{j=0}^{4} c_j \log_2(j + 2)$$

□

Summation notation for a sum is not unique. For example, the sum in part (d) of Example 3 could have just as easily been written $\sum_{j=2}^{6} c_{j-2} \log_2 j$.

Properties of Summation Notation

Sums written in summation notation have some interesting properties. For example, consider the sum $\sum_{j=1}^{5} (A_j + B_j)$. If we expand this sum and rearrange the terms, we obtain a useful result.

$$\begin{aligned}\sum_{j=1}^{5} (A_j + B_j) &= (A_1 + B_1) + (A_2 + B_2) + (A_3 + B_3) + (A_4 + B_4) + (A_5 + B_5)\\ &= A_1 + A_2 + A_3 + A_4 + A_5 + B_1 + B_2 + B_3 + B_4 + B_5\\ &= \sum_{j=1}^{5} A_j + \sum_{j=1}^{5} B_j\end{aligned}$$

In other words, the summation of a sum is the sum of the summations! This illustrates the first of the following list of summation properties.

Some Summation Notation Properties

For m, n, and p integers with $m < p < n$ and k a real number,

1. $\sum_{j=m}^{n} (a_j + b_j) = \sum_{j=m}^{n} a_j + \sum_{j=m}^{n} b_j$

2. $\sum_{j=m}^{n} (a_j - b_j) = \sum_{j=m}^{n} a_j - \sum_{j=m}^{n} b_j$

3. $\sum_{j=m}^{n} ka_j = k \sum_{j=m}^{n} a_j$

4. $\sum_{j=m}^{n} a_j = a_m + \sum_{j=m+1}^{n} a_j$

5. $\sum_{j=m}^{n} a_j = \sum_{j=m}^{n-1} a_j + a_n$

6. $\sum_{j=m}^{n} a_j = \sum_{j=m}^{p-1} a_j + \sum_{j=p}^{n} a_j$

7. $\sum_{j=m}^{n} a_j = \sum_{q=m}^{n} a_q$

Properties 1, 2, and 3 are often called the **linear properties** of summation. Properties 4, 5, and 6 show how terms may be ''peeled off'' either end of a sum. Property 7 illustrates that the index of summation is a **dummy variable** and may be changed to another letter (throughout the expression).

EXAMPLE 4 Use the properties of summation as directed.

(a) Write $\sum_{j=1}^{25} (3x^j - 5y^j)$ as two sums with constants factored out.

(b) Peel the first three terms off the sum $\sum_{j=0}^{17} (-1)^j x^{2j}$.

(c) Write $\sum_{j=2}^{11} \frac{v^j}{2} + 3 \sum_{k=2}^{11} u^{k+1}$ as a single sum.

Solutions (a) We utilize the linear properties to write

$$\sum_{j=1}^{25} (3x^j - 5y^j) = \sum_{j=1}^{25} 3x^j - \sum_{j=1}^{25} 5y^j$$

$$= 3 \sum_{j=1}^{25} x^j - 5 \sum_{j=1}^{25} y^j$$

(b) $\sum_{j=0}^{17} (-1)^j x^{2j} = (-1)^0 x^{2\cdot 0} + (-1)^1 x^{2\cdot 1} + (-1)^2 x^{2\cdot 2} + \sum_{j=3}^{17} (-1)^j x^{2j}$

$$= 1 - x^2 + x^4 + \sum_{j=3}^{17} (-1)^j x^{2j}$$

(c) We will use the linear properties and the fact that k is a dummy variable.

$$\sum_{j=2}^{11} \frac{v^j}{2} + 3\sum_{k=2}^{11} u^{k+1} = \sum_{j=2}^{11} \frac{v^j}{2} + \sum_{k=2}^{11} 3u^{k+1}$$

$$= \sum_{j=2}^{11} \frac{v^j}{2} + \sum_{j=2}^{11} 3u^{j+1}$$

$$= \sum_{j=2}^{11} \left(\frac{v^j}{2} + 3u^{j+1}\right)$$ □

PROBLEM SET 9.1

Warm-ups

In Problems 1 through 4, evaluate each sum. See Example 1.

1. $\sum_{j=1}^{5} 2^j$ **2.** $\sum_{j=1}^{5} j^2$ **3.** $\sum_{j=0}^{6} (2j+1)$ **4.** $\sum_{k=1}^{5} \frac{k^2}{k+1}$

In Problems 5 through 8, expand each sum. See Example 2.

5. $\sum_{j=1}^{4} \frac{x^{j-1}}{2j-1}$ **6.** $\sum_{i=0}^{3} \frac{2i-1}{2i+1} V^i$ **7.** $\sum_{v=6}^{10} (-1)^v \log v$ **8.** $\sum_{j=2}^{5} \frac{e^{j-2}}{(j-1)(j+1)}$

In Problems 9 through 12, express each sum using summation notation. See Example 3.

9. $3 + 5 + 7 + 9 + 11$ **10.** $\frac{1}{2} + \frac{2}{3} + \frac{3}{4} + \frac{4}{5} + \frac{5}{6} + \frac{6}{7}$

11. $1 - \frac{1}{2}y + \frac{1}{4}y^2 - \frac{1}{8}y^3 + \frac{1}{16}y^4$ **12.** $A_1 \log_5 2 + 5A_2 \log_5 3 + 9A_3 \log_5 4 + 13A_4$

In Problems 13 through 16, use the properties of summation as directed. See Example 4.

13. Write $\sum_{j=1}^{75} \left(b_j + \frac{x^j}{2}\right)$ as two sums with constants factored out.

14. Peel the last two terms off $\sum_{j=-1}^{99} \frac{j+1}{2j-1} x^j$.

15. Peel the first and last terms off the sum $\sum_{p=1}^{5} (-1)^{p-1} \frac{2p}{p+1}$

16. Write $\sum_{j=1}^{5} (j+1) - 2\sum_{h=1}^{5} (h-1) + 3\sum_{q=1}^{5} q$ as a single sum.

Practice Exercises

In Problems 17 through 22, evaluate each sum.

17. $\sum_{j=1}^{4} 3^j$ **18.** $\sum_{k=1}^{3} 5^k$ **19.** $\sum_{j=0}^{3} 3j^3$ **20.** $\sum_{j=2}^{6} (1-j)^2$

21. $\sum_{j=1}^{5} (-1)^{j-1}\frac{j}{j+1}$ **22.** $\sum_{j=0}^{5} (-1)^j\frac{j+2}{j+1}$

In Problems 23 through 28, expand each sum.

23. $\sum_{j=1}^{5} (2x)^j$ **24.** $\sum_{j=0}^{3} (x+1)^j$ **25.** $\sum_{k=0}^{5} (-1)^k x^{2k}$ **26.** $\sum_{n=1}^{5} (-1)^{n-1}x^{2n-1}$

27. $\sum_{j=1}^{5} \left(\frac{2}{3}\right)^j y^{-j}$ **28.** $\sum_{j=1}^{5} \left(\frac{3}{4}\right)^{j-1} z^{1-j}$

In Problems 29 through 32, express each sum in summation notation.

29. $6 + 9 + 12 + 15 + 18$

30. $9 + 14 + 19 + 24 + 29$

31. $3 - 9x^2 + 27x^4 - 81x^6 + 243x^8$

32. $x - 2x^3 + 4x^5 - 8x^7 + 16x^9$

33. $\frac{1}{2} - \frac{2}{3}s + \frac{3}{4}s^2 - \frac{4}{5}s^3 + \frac{5}{6}s^4$

34. $\frac{1}{w} + \frac{2}{3}w + \frac{4}{9}w^3 + \frac{8}{27}w^5 + \frac{16}{81}w^7$

35. Write $\sum_{j=1}^{15} 2x^j + 3\sum_{j=1}^{15} x^j$ as a single sum.

36. Write $5\sum_{j=1}^{5} y^{j-1} - 2\sum_{j=1}^{5} y^{j-1}$ as a single sum.

37. Peel the last two terms off $\sum_{j=1}^{25} \frac{j}{j+2}x^j$.

38. Peel the first two terms off $\sum_{j=0}^{75} (-1)^j x^{j+1}$.

In Problems 39 through 46, write each sum using only one summation notation symbol.

39. $\sum_{j=1}^{100} A_j + \sum_{n=1}^{100} B_n$ **40.** $\sum_{k=0}^{35} C^k - \sum_{v=1}^{35} D^v$ **41.** $\sum_{j=1}^{100} A_j - \sum_{j=1}^{101} B_j$ **42.** $\sum_{j=0}^{35} C^j + \sum_{j=1}^{35} D^j$

43. $\sum_{j=0}^{99} A_j + \sum_{k=1}^{100} B_k$ **44.** $\sum_{p=1}^{33} C^p - \sum_{j=3}^{35} D^j$ **45.** $\sum_{j=1}^{5} A_j + \sum_{j=6}^{15} A_j$ **46.** $\sum_{j=-1}^{25} B_j + \sum_{j=26}^{99} B_j$

Challenge Problems

*In Problems 47 and 48, write without summation notation. (*Hint: *Expand the outside sum first.)*

47. $\sum_{j=1}^{3}\left(\sum_{k=1}^{j} c_k x^k\right)$

48. $\sum_{j=1}^{3}\left(\sum_{k=1}^{j} \frac{j}{k}x^{j+k-2}\right)$

IN YOUR OWN WORDS . . .

49. Explain summation notation without using an example.

50. Explain the procedure for expanding a sum written in summation notation.

9.2 MATHEMATICAL INDUCTION

In this section we will study an important property of the set of natural numbers. Recall that the set of natural numbers is the set of counting numbers, also called the set of positive integers.

$$N = \{1, 2, 3, \ldots\}$$

Let's consider a set of numbers that has two properties.

1. The set contains the number 1.
2. *If* the set contains the natural number k, *then* it also contains $k + 1$.

We know the set is not empty because it contains 1. But, as 1 is a natural number, it must also contain $1 + 1$, or 2. But if it contains 2, it must contain 3, then 4, and so on. It seems that this set contains all the natural numbers. This, in fact, is what the Principle of Mathematical Induction says.

The Principle of Mathematical Induction

If S is a subset of the natural numbers with the following two properties:

1. 1 belongs to S.
2. If k belongs to S, then $k + 1$ belongs to S.

then S is the set of natural numbers.

Now suppose we wish to prove a suspicion we have about some statement that depends on the natural numbers. For example, we have noticed that the sum of the first few odd numbers always seems to be a perfect square.

$$\begin{aligned}
&1 &&= 1 &&= 1^2\\
&1 + 3 &&= 4 &&= 2^2\\
&1 + 3 + 5 &&= 9 &&= 3^2\\
&1 + 3 + 5 + 7 &&= 16 &&= 4^2\\
&1 + 3 + 5 + 7 + 9 &&= 25 &&= 5^2\\
&1 + 3 + 5 + 7 + 9 + 11 &&= 36 &&= 6^2
\end{aligned}$$

It appears that the sum of the first two odd natural numbers is 2^2, the sum of the first three odd natural numbers is 3^2, the sum of first four is 4^2, and so on. Can we *prove* this? First, we need to write our conjecture as a statement that depends on the natural numbers.

To show: The sum of the first n odd natural numbers is n^2.

The truth of this statement depends on the natural number n. We have seen, from the display above, that the statement is true when n is 1, 2, 3, 4, 5, and 6. But is it *always* true?

To use the Principle of Mathematical Induction, we must show two ideas for a subset of natural numbers S:

1. 1 belongs to S.

2. If k belongs to S, then $k + 1$ belongs to S.

Let S be the subset of the natural numbers for which our statement is true. We have seen that the statement is true when n is 1, 2, 3, 4, 5, and 6. Thus, 1 belongs to S. Part 1 is satisfied.

We show part 2. Let k be any member of S. If k belongs to S, then the statement is true when n is k. That means that the sum of the first k odd natural numbers is k^2. That is,

$$1 + 3 + 5 + \cdots + (2k - 1) = k^2$$

(Note that $2k - 1$ is a way to represent the kth odd natural number.) We must show that the statement is true when n is $k + 1$. That is, we must show

$$1 + 3 + 5 + \cdots + (2k - 1) + \textit{next odd natural number} = (k + 1)^2$$

What is the next odd natural number after $2k - 1$? It is $(2k - 1) + 2$ or $2k + 1$. Taking what we assume to be true,

$$1 + 3 + 5 + \cdots + (2k - 1) = k^2$$

We add $2k + 1$ to both sides.

$$\begin{aligned} 1 + 3 + 5 + \cdots + (2k - 1) + (2k + 1) &= k^2 + 2k + 1 \\ &= (k + 1)^2 \end{aligned}$$

We see that the sum of the first $k + 1$ odd natural numbers is $(k + 1)^2$. Therefore, $k + 1$ belongs to the set S. What have we shown? We have shown that

1. 1 belongs to S.

2. If k belongs to S, then $k + 1$ belongs to S.

Therefore, by the Principle of Mathematical Induction, S is the set of natural numbers. Thus our statement is true *for all natural numbers.*

EXAMPLE 1 Prove that the sum of the first n natural numbers is $\dfrac{n(n + 1)}{2}$.

Solution We are to show that for all natural numbers n,

$$1 + 2 + 3 + \cdots + n = \frac{n(n + 1)}{2}.$$

We divide our induction proofs into two parts.

Part 1. Show the statement is true when n is 1.

The left side of the statement becomes 1 when n is 1.

The right side becomes $\frac{n(n+1)}{2} = \frac{1(1+1)}{2} = 1$ when n is 1.

Since both the left and right sides are the same when n is 1, the statement is true when n is 1.

Part 2. *Assume* the statement is true when n is k. Then *show* it is true when n is $k + 1$.

Assume $1 + 2 + 3 + \cdots + k = \frac{k(k+1)}{2}$.

We must show that the sum of the first $k + 1$ natural numbers is $\frac{(k+1)(k+1+1)}{2}$.

The sum of the first $k + 1$ natural numbers is $1 + 2 + 3 + \cdots + k + (k + 1)$.

$$1 + 2 + 3 + \cdots + k = \frac{k(k+1)}{2} \qquad \text{Assumed to be true}$$

If we add $k + 1$ to both sides, we have the sum of the first $k + 1$ natural numbers on the left side.

$$\begin{aligned} 1 + 2 + 3 + \cdots + k + (k+1) &= \frac{k(k+1)}{2} + (k+1) \\ &= \frac{k(k+1) + 2(k+1)}{2} && \text{Add fractions.} \\ &= \frac{(k+1)(k+2)}{2} && \text{Common factor} \\ &= \frac{(k+1)[(k+1)+1]}{2} \end{aligned}$$

Therefore, the truth of the statement when n is k led to the truth of the statement when n is $k + 1$. This completes the proof that the sum of the first n natural numbers is $\frac{n(n+1)}{2}$. □

EXAMPLE 2 Prove that $2^0 + 2^1 + 2^2 + \cdots + 2^{n-1} = 2^n - 1$.

Proof: (By induction)

We wish to prove the statement $2^0 + 2^1 + 2^2 + \cdots + 2^{n-1} = 2^n - 1$ is true for every natural number. We show the two parts of the Principle of Mathematical Induction.

Part 1. When n is 1, the left side is

$$2^{1-1} = 2^0 = 1$$

and the right side is

$$2^1 - 1 = 2 - 1 = 1$$

The statement is true when n is 1.

Part 2. Assume the statement is true when n is k. That is,

$$2^0 + 2^1 + 2^2 + \cdots + 2^{k-1} = 2^k - 1$$

Now we examine the statement when n is $k + 1$.

$$2^0 + 2^1 + 2^2 + \cdots + 2^{k-1} + 2^{k+1-1} = 2^0 + 2^1 + 2^2 + \cdots + 2^{k-1} + 2^k$$

Substitute $2^k - 1$ for $2^0 + 2^1 + \cdots + 2^{k-1}$ (using the assumption)

$$= 2^k - 1 + 2^k$$

$$= 2 \cdot 2^k - 1$$

$$= 2^{k+1} - 1$$

which is $2^n - 1$ when n is $k + 1$.

Thus, we have shown the formula is true for all natural numbers. □

Next, we will prove that constants can be factored out of sums written in summation notation. (Property 3 of summation notation.)

EXAMPLE 3 Prove that for all natural numbers n, $\sum_{j=1}^{n} cA_j = c \sum_{j=1}^{n} A_j$.

Proof: (By induction)

Part 1. When n is 1,

$$\sum_{j=1}^{1} cA_j = cA_1 = c \sum_{j=1}^{1} A_j$$

Part 2. Assume $\sum_{j=1}^{k} cA_j = c \sum_{j=1}^{k} A_j$

$$\sum_{j=1}^{k+1} cA_j = \sum_{j=1}^{k} cA_j + cA_{k+1}$$ Peel off the last term.

$$= c \sum_{j=1}^{k} A_j + cA_{k+1} \qquad \text{By the assumption}$$

$$= c\left(\sum_{j=1}^{k} A_j + A_{k+1}\right) \qquad \text{Common factor}$$

$$= c \sum_{j=1}^{k+1} A_j$$

Which is $c \sum_{j=1}^{n} a_j$ when n is $k + 1$. □

The remaining properties of summation can be proved by mathematical induction in a similar manner. (See exercises at the end of the section.)

Extended Mathematical Induction

Sometimes it is convenient to start a proof by induction at an integer other than 1. Mathematical induction can be extended in a natural way.

Extended Principle of Mathematical Induction

Let j be an integer and P_n be a statement that depends on integers *greater than or equal to* j. If it can be shown that

1. P_j is true.

2. *If* P_k is true for any integer $k \geq j$, then it follows that P_{k+1} is true.

Then P_n is true for all integers greater than or equal to j.

Extended induction involves the same two-part proof as regular induction. In part 1, show that the statement holds for some integer. Then, in part 2, show that if it holds for some unspecified integer k *greater than or equal to* the starting integer, then it follows that it is true for $k + 1$. If *both* parts can be shown, then the statement is true for all integers greater than or equal to the starting integer.

EXAMPLE 4 Prove that $2^n > 2n + 1$ for all $n \geq 3$.

Proof:

Part 1. When n is 3,

$2^n = 2^3 = 8$ and

$2n + 1 = 2 \cdot 3 + 1 = 7$, thus

$2^n > 2n + 1$ when n is 3.

Part 2. Assume $2^k > 2k + 1$ for some $k \geq 3$. Then,

$2^k \cdot 2 > (2k + 1)2$ Multiply both sides of $2^k > 2k + 1$ by 2.

$2^{k+1} > 4k + 2.$

We must show that $2^{k+1} > 2(k + 1) + 1$.

Since $k \geq 3$, $2k - 1 > 0$. Therefore,

$4k + 2 > 4k + 2 - (2k - 1) = 2k + 3 = 2k + 2 + 1 = 2(k + 1) + 1.$

We have shown that *if* $2^k > 2k + 1$ and $k \geq 3$ *then*

$2^{k+1} > 4k + 2 > 2(k + 1) + 1$, which is the statement when n is $k + 1$.

Thus, $2^n > 2n + 1$ is true for all integers $n \geq 3$. □

PROBLEM SET 9.2

Warm-ups

In Problems 1 through 4, use mathematical induction to prove the given formula. For Problems 1 and 2, see Example 1. Assume n is a natural number.

1. $2 + 4 + 6 + \cdots + 2n = n(n + 1)$

2. $1 + 4 + 9 + \cdots + n^2 = \dfrac{n(n + 1)(2n + 1)}{6}$

For Problems 3 and 4, see Example 2.

3. $1 + 3 + 9 + \cdots + 3^{n-1} = \dfrac{3^n - 1}{2}$

4. $\dfrac{1}{2} + \dfrac{1}{6} + \dfrac{1}{12} + \cdots + \dfrac{1}{n(n + 1)} = \dfrac{n}{n + 1}$

In Problems 5 and 6, show that the given summation notation properties hold. See Example 3. Assume n is a natural number.

5. Property 1: $\sum_{j=1}^{n} (A_j + B_j) = \sum_{j=1}^{n} A_j + \sum_{j=1}^{n} B_j$

6. Property 7: $\sum_{j=1}^{n} A_j = \sum_{q=1}^{n} A_q$

In Problems 7 and 8, prove each proposition for the given values of n. See Example 4.

7. Property 5 of summation notation: $\sum_{j=1}^{n} A_j = \sum_{j=1}^{n-1} A_j + A_n$ for $n \geq 2$.

8. $2^n > n^2$ for $n \geq 5$; $n \in N$

Practice Exercises

In Problems 9 through 18, use mathematical induction to show each statement is true for all natural numbers.

9. $3 + 4 + 5 + \cdots + (n + 2) = \dfrac{n(n + 5)}{2}$

10. $4 + 5 + 6 + \cdots + (n + 3) = \dfrac{n(n + 7)}{2}$

11. $3 + 5 + 7 + \cdots + (2n + 1) = n(n + 2)$

12. $2 + 5 + 8 + \cdots + (3n - 1) = \dfrac{n(3n + 1)}{2}$

13. $1 + 2 + 4 + \cdots + 2^n = 2^n + 1$

14. $1 + 3 + 9 + \cdots + 3^{n-1} = \frac{1}{2}(3^n - 1)$

15. $3 + 2 + 1 + \cdots + (4 - n) = \frac{n(7 - n)}{2}$

16. $4 + 2 + 0 + \cdots + (6 - 2n) = n(5 - n)$

17. $1 + \frac{1}{2} + \frac{1}{4} + \cdots + \frac{1}{2^{n-1}} = 2 - \frac{1}{2^{n-1}}$

18. $1 + \frac{1}{3} + \frac{1}{9} + \cdots + \frac{1}{3^{n-1}} = \frac{3^n - 1}{2 \cdot 3^{n-1}}$

In Problems 19 through 22, use mathematical induction to show each statement is true for all the specified integers.

19. $n^3 > (n + 1)^2$ for $n \geq 3$

20. $n^2 > 6n + 7$ for $n \geq 8$

21. $1 + 4 + \cdots + 4^n = \frac{1}{3}(4^{n+1} - 1)$ for $n \geq 0$

22. $\frac{1}{8} + \frac{1}{4} + \cdots + 2^n = \frac{2^{n+4} - 1}{8}$ for $n \geq -3$

Challenge Problems

23. Show that $\sum_{j=1}^{n} 5 \cdot 2^{j-1} = 5(2^n - 1)$.

24. Show that $\sum_{j=0}^{n} (6 + 2j) = n^2 + 7n + 6$.

25. Show that 2 is a factor of $3^n - 1$ for integers $n \geq 0$.

26. Show that 2 is a factor of $n^2 - 3n + 2$ for *all* integers. (*Hint:* First show it is true for $n \geq 0$. Then prove 2 is a factor of $(-n)^2 - 3(-n) + 2$ for $n > 0$.)

IN YOUR OWN WORDS . . .

27. Describe the Principle of Mathematical Induction.

28. How do we prove a statement is true by Mathematical Induction?

9.3 THE BINOMIAL THEOREM

Factorials

Just as x^4 is shorthand notation for the product $x \cdot x \cdot x \cdot x$, there is shorthand notation for products like $1 \cdot 2 \cdot 3 \cdot 4$ and $1 \cdot 2 \cdot 3 \cdot 4 \cdot 5 \cdot 6 \cdot 7$. Such products are called **factorials.** We write them

$$4! = 1 \cdot 2 \cdot 3 \cdot 4$$

$$7! = 1 \cdot 2 \cdot 3 \cdot 4 \cdot 5 \cdot 6 \cdot 7$$

Read 4! as "four factorial," $K!$ as "K factorial," and $(n - 1)!$ as "n minus 1 factorial."

Definition: Factorial

For n a natural number,

1. $n! = 1 \cdot 2 \cdot 3 \cdots (n - 1) \cdot n$
2. $0! = 1$

Sometimes it is convenient to write factorials the other way around,

$$n! = n(n-1)\cdots 3 \cdot 2 \cdot 1$$

Factorials grow quickly, as illustrated here.

$$\begin{aligned}
0! &= 1 && = 1\\
1! &= 1 && = 1\\
2! &= 1 \cdot 2 && = 2\\
3! &= 1 \cdot 2 \cdot 3 && = 6\\
4! &= 1 \cdot 2 \cdot 3 \cdot 4 && = 24\\
5! &= 1 \cdot 2 \cdot 3 \cdot 4 \cdot 5 && = 120\\
&\vdots && \vdots\\
9! &= 1 \cdot 2 \cdot 3 \cdot 4 \cdot 5 \cdot 6 \cdot 7 \cdot 8 \cdot 9 && = 362{,}880\\
10! &= 1 \cdot 2 \cdot 3 \cdot 4 \cdot 5 \cdot 6 \cdot 7 \cdot 8 \cdot 9 \cdot 10 && = 3{,}628{,}800
\end{aligned}$$

Notice from the display that $5!$ is $4! \cdot 5$ and that $10!$ is $9! \cdot 10$. This idea illustrates a useful property of factorials.

Property of Factorials

If n is a natural number,

$$n! = (n-1)! \cdot n$$

For example, we can calculate $11!$ from $10!$ (using the value from the display).

$$\begin{aligned}
11! &= 10! \cdot 11\\
&= (3{,}628{,}800) \cdot 11\\
&= 39{,}916{,}800
\end{aligned}$$

Quotients of factorials are interesting and quite common. They often simplify with many cancellations.

EXAMPLE 1 Evaluate each factorial expression.

(a) $\dfrac{16!}{17!}$ (b) $\dfrac{20!}{18!}$ (c) $\dfrac{10!}{5! \cdot 5!}$ (d) $\dfrac{8!}{8! \cdot 0!}$

Solutions (a) $\dfrac{16!}{17!} = \dfrac{16!}{16! \cdot 17} = \dfrac{1}{17}$

Note that 16 factors were divided out from the numerator and denominator!

(b) $\dfrac{20!}{18!} = \dfrac{18! \cdot 19 \cdot 20}{18!} = 19 \cdot 20 = 380$

(c) $$\frac{10!}{5! \cdot 5!} = \frac{1 \cdot 2 \cdot 3 \cdot 4 \cdot 5 \cdot 6 \cdot 7 \cdot 8 \cdot 9 \cdot 10}{1 \cdot 2 \cdot 3 \cdot 4 \cdot 5 \cdot 1 \cdot 2 \cdot 3 \cdot 4 \cdot 5}$$
$$= \frac{6 \cdot 7 \cdot 8 \cdot 9 \cdot 10}{1 \cdot 2 \cdot 3 \cdot 4 \cdot 5} = 7 \cdot 2 \cdot 9 \cdot 2 = 252$$

(d) $\dfrac{8!}{8! \cdot 0!} = \dfrac{1}{0!} = \dfrac{1}{1} = 1$ □

Binomial Coefficients

Factorial expressions of the form $\dfrac{7!}{4! \cdot 3!}$ and $\dfrac{10!}{8! \cdot 2!}$ are quite common in mathematics. They are called **binomial coefficients** and are written

$$\binom{7}{4} = \frac{7!}{4! \cdot 3!} \quad \text{and} \quad \binom{10}{8} = \frac{10!}{8! \cdot 2!}$$

Binomial Coefficients

For n and r nonnegative integers, with $n \geq r$,

$$\binom{n}{r} = \frac{n!}{r!(n-r)!}$$

EXAMPLE 2 Calculate the following binomial coefficients.

(a) $\binom{7}{5}$ (b) $\binom{7}{2}$ (c) $\binom{6}{6}$ (d) $\binom{11}{0}$

Solutions (a) $$\binom{7}{5} = \frac{7!}{5!(7-5)!} = \frac{7!}{5! \cdot 2!}$$
$$= \frac{1 \cdot 2 \cdot 3 \cdot 4 \cdot 5 \cdot 6 \cdot 7}{1 \cdot 2 \cdot 3 \cdot 4 \cdot 5 \cdot 1 \cdot 2}$$
$$= 3 \cdot 7 = 21$$

(b) $$\binom{7}{2} = \frac{7!}{2!(7-2)!} = \frac{7!}{2! \cdot 5!}$$

$$= \frac{1 \cdot 2 \cdot 3 \cdot 4 \cdot 5 \cdot 6 \cdot 7}{1 \cdot 2 \cdot 1 \cdot 2 \cdot 3 \cdot 4 \cdot 5}$$

$$= 3 \cdot 7 = 21$$

Notice that $\binom{7}{5} = \binom{7}{2}$

(c) $\binom{6}{6} = \dfrac{6!}{6!(6-6)!} = \dfrac{6!}{6! \cdot 0!} = 1$

Remember, $0! = 1$.

(d) $\binom{11}{0} = \dfrac{11!}{0!(11-0)!} = \dfrac{11!}{0! \cdot 11!} = 1$ □

Example 2 illustrates some properties of binomial coefficients.

Properties of Binomial Coefficients

For integers n and r such that $0 \le r \le n$,

1. $\binom{n}{0} = 1$
2. $\binom{n}{n} = 1$
3. $\binom{n}{r} = \binom{n}{n-r}$

The Binomial Theorem

Let's look at a few powers of the binomial $(a + b)$.

$$(a + b)^0 = 1$$

$$(a + b)^1 = a + b$$

$$(a + b)^2 = a^2 + 2ab + b^2$$

$$(a + b)^3 = a^3 + 3a^2b + 3ab^2 + b^3$$

$$(a + b)^4 = a^4 + 4a^3b + 6a^2b^2 + 4ab^3 + b^4$$

$$(a + b)^5 = a^5 + 5a^4b + 10a^3b^2 + 10a^2b^3 + 5ab^4 + b^5$$

Notice the pattern of coefficients in the powers of $a + b$ from 0 to 5 in the above display. If we remove all but the coefficients, we get the pattern

0th row	1
1st row	1 1
	1 2 1
	1 3 3 1
	1 4 6 4 1
5th row	1 5 10 10 5 1

This is called **Pascal's Triangle.** Note that every row starts and ends with 1 and that each of the interior numbers is the sum of the two numbers directly above it. Pascal's Triangle has many remarkable properties, one of which is that it furnishes the coefficients for the expansion of $(a + b)^n$. Notice that the fifth row gives the coefficients for $(a + b)^5$. Also note that the fifth row can be written

$$\binom{5}{0} \quad \binom{5}{1} \quad \binom{5}{2} \quad \binom{5}{3} \quad \binom{5}{4} \quad \binom{5}{5}$$

using binomial coefficients. This is how they got their name.

The Binomial Theorem provides a general formula for the expansion of $(a + b)^n$ for n a natural number. Part of the formula is quite easy to see. The powers of a, for example, start at n and decrease by 1 for each term until the power becomes zero. Similarly, b starts with an exponent of zero and increases by 1 until b^n is reached. Except for coefficients, $(a + b)^7$ can be written

$$\_a^7 + \_a^6b + \_a^5b^2 + \_a^4b^3 + \_a^3b^4 + \_a^2b^5 + \_ab^6 + \_b^7$$

As we have seen, we can find the coefficients by either one of two ways. We can extend Pascal's Triangle to the seventh row and read them from the display, or we can calculate the binomial coefficients.

0th row	1
1st row	1 1
	1 2 1
	1 3 3 1
	1 4 6 4 1
	1 5 10 10 5 1
	1 6 15 20 15 6 1
7th row	1 7 21 35 35 21 7 1

From the display we can write

$$(a + b)^7 = a^7 + 7a^6b + 21a^5b^2 + 35a^4b^3 + 35a^3b^4 + 21a^2b^5 + 7ab^6 + b^7.$$

However, for large values of n this is an awkward way to find these coefficients. Let's write $(a + b)^7$ using binomial coefficients.

$$(a+b)^7 = \binom{7}{0}a^7 + \binom{7}{1}a^6b + \binom{7}{2}a^5b^2 + \binom{7}{3}a^4b^3 + \binom{7}{4}a^3b^4 + \binom{7}{5}a^2b^5 + \binom{7}{6}ab^6 + \binom{7}{7}b^7$$

Now, if we insert a^0 and b^0 in their appropriate places, we have

$$(a+b)^7 = \binom{7}{0}a^7b^0 + \binom{7}{1}a^6b^1 + \binom{7}{2}a^5b^2 + \binom{7}{3}a^4b^3 + \binom{7}{4}a^3b^4 + \binom{7}{5}a^2b^5 + \binom{7}{6}a^1b^6 + \binom{7}{7}a^0b^7$$

Notice that this can be abbreviated, using summation notation, to read

$$(a+b)^7 = \sum_{j=0}^{7} \binom{7}{j} a^{7-j}b^j$$

This suggests a general formula for expanding $(a + b)^n$ called the **binomial theorem.**

The Binomial Theorem

For n a natural number,

$$(a+b)^n = \sum_{j=0}^{n} \binom{n}{j} a^{n-j}b^j$$

EXAMPLE 3 Write the first four terms in the expansion of $(x + y)^{11}$.

Solution By the binomial theorem,

$$(x+y)^{11} = \sum_{j=0}^{11} \binom{11}{j} x^{11-j}y^j$$

The first four terms are given when j is 0, 1, 2, and 3. That is,

$$\binom{11}{0}x^{11-0}y^0 + \binom{11}{1}x^{11-1}y^1 + \binom{11}{2}x^{11-2}y^2 + \binom{11}{3}x^{11-3}y^3$$

We calculate the binomial coefficients.

$$\binom{11}{0} = \frac{11!}{0! \cdot 11!} = 1$$

$$\binom{11}{1} = \frac{11!}{1! \cdot 10!} = \frac{11}{1} = 11$$

$$\binom{11}{2} = \frac{11!}{2! \cdot 9!} = \frac{10 \cdot 11}{1 \cdot 2} = 55$$

$$\binom{11}{3} = \frac{11!}{3! \cdot 8!} = \frac{9 \cdot 10 \cdot 11}{1 \cdot 2 \cdot 3} = 3 \cdot 5 \cdot 11 = 165$$

Therefore, the first four terms of the expansion of $(x + y)^{11}$ are

$$x^{11} + 11x^{10}y + 55x^9y^2 + 165x^8y^3$$

□

EXAMPLE 4 Find the fourteenth term in the expansion of $(2 + k)^{16}$.

Solution From the binomial theorem, $(2 + k)^{16} = \sum_{j=0}^{16} \binom{16}{j} 2^{16-j}k^j$.

The fourteenth term will occur when j is 13.

$$\binom{16}{13} 2^{16-13}k^{13} = \frac{16!}{13! \cdot 3!} 2^3k^{13}$$

$$= \frac{14 \cdot 15 \cdot 16}{1 \cdot 2 \cdot 3} \cdot 8k^{13}$$

$$= 7 \cdot 5 \cdot 16 \cdot 8k^{13} = 4480k^{13}$$

□

Suppose we expand $(a - b)^n$ with the binomial theorem.

$$(a - b)^n = [a + (-b)]^n = \sum_{j=0}^{n} \binom{n}{j} a^{n-j}(-b)^j$$

$$= \sum_{j=0}^{n} (-1)^j \binom{n}{j} a^{n-j}b^j$$

We notice that the terms alternate in sign, starting with +.

EXAMPLE 5 Find the last three terms of $(3x - 2)^9$.

Solution

$$(3x - 2)^9 = \sum_{j=0}^{9} (-1)^j \binom{9}{j} (3x)^{9-j}2^j$$

The last three terms are given when j is 7, 8, 9.

$$(-1)^7\binom{9}{7}(3x)^2 2^7 + (-1)^8\binom{9}{8}(3x)^1 2^8 + (-1)^9\binom{9}{9}(3x)^0 2^9$$

$$= -1 \cdot \frac{9!}{7! \cdot 2!} \cdot 3^2x^2 2^7 + 1 \cdot \frac{9!}{8! \cdot 1!} \cdot 3x \cdot 2^8 + (-1)\frac{9!}{9! \cdot 0!} \cdot 1 \cdot 2^9$$

$$= -41472x^2 + 6912x - 512 \qquad \square$$

Proof of the Binomial Theorem (Optional)

The proof of the binomial theorem is an interesting exercise in summation notation, binomial coefficients, and proof by induction. Although not difficult, it is a little messy. Our approach is to prove two lemmas (little theorems) first, which will clean up some of the mess.

The first lemma deals with the reindexing of a sum in summation notation.

LEMMA 1: For k a natural number,

$$\sum_{j=0}^{k}\binom{k}{j}a^{k-j}b^{j+1} = \sum_{j=1}^{k+1}\binom{k}{j-1}a^{k+1-j}b^{j}$$

Proof:

$$\sum_{j=0}^{k}\binom{k}{j}a^{k-j}b^{j+1} = \binom{k}{0}a^{k}b^{1} + \binom{k}{1}a^{k-1}b^{2} + \cdots + \binom{k}{k}a^{0}b^{k+1}$$

$$= \sum_{j=1}^{k+1}\binom{k}{j-1}a^{k+1-j}b^{j} \qquad \square$$

The second lemma deals with a property of binomial coefficients.

LEMMA 2: For natural numbers j and k, with $j \le k$,

$$\binom{k}{j} + \binom{k}{j-1} = \binom{k+1}{j}$$

Proof:

$$\binom{k}{j} + \binom{k}{j-1} = \frac{k!}{j!(k-j)!} + \frac{k!}{(j-1)!(k-j+1)!}$$

First, we find a common denominator by multiplying the numerator and the denominator of the first fraction by $k - j + 1$ and the second fraction by j.

$$= \frac{k!(k-j+1)}{j!(k-j)!(k-j+1)} + \frac{j \cdot k!}{j(j-1)!(k-j+1)!}$$

Because $j \cdot (j-1)! = j!$ and $(k-j)! \cdot (k-j+1) = (k-j+1)!$, we have

$$= \frac{k!(k-j+1)}{j!(k-j+1)!} + \frac{j \cdot k!}{j!(k-j+1)!}$$

Now we add the fractions and then rewrite $k - j + 1$ as $k + 1 - j$.

$$= \frac{k!(k-j+1) + j \cdot k!}{j!(k-j+1)!} = \frac{k!(k+1-j) + j \cdot k!}{j!(k+1-j)!}$$

We distribute $k!$ and simplify the numerator.

$$= \frac{k!(k+1) - j \cdot k! + j \cdot k!}{j!(k+1-j)!} = \frac{k!(k+1)}{j!(k+1-j)!}$$

$$= \frac{(k+1)!}{j!(k+1-j)!} = \binom{k+1}{j}$$ □

The Binomial Theorem

For n a natural number,

$$(a+b)^n = \sum_{j=0}^{n} \binom{n}{j} a^{n-j} b^j$$

***Proof* (By induction):** Part 1. When n is 1,

$$\sum_{j=0}^{n} \binom{n}{j} a^{n-j} b^j = \sum_{j=0}^{1} \binom{1}{j} a^{1-j} b^j = 1 \cdot a^1 b^0 + 1 \cdot a^0 b^1 = a + b = (a+b)^n$$

This is true when n is 1.

Part 2. Assume $(a+b)^k = \sum_{j=0}^{k} \binom{k}{j} a^{k-j} b^j$. Now examine $(a+b)^{k+1}$.

$$(a+b)^{k+1} = (a+b)(a+b)^k$$

$$= (a+b) \sum_{j=0}^{k} \binom{k}{j} a^{k-j} b^j \qquad \text{Assumption}$$

$$= a \sum_{j=0}^{k} \binom{k}{j} a^{k-j} b^j + b \sum_{j=0}^{k} \binom{k}{j} a^{k-j} b^j \qquad \text{Distributive Property}$$

$$= \sum_{j=0}^{k} \binom{k}{j} a^{k-j+1} b^j + \sum_{j=0}^{k} \binom{k}{j} a^{k-j} b^{j+1} \qquad \text{Summation Property}$$

$$= \sum_{j=0}^{k} \binom{k}{j} a^{k-j+1} b^j + \sum_{j=1}^{k+1} \binom{k}{j-1} a^{k-j+1} b^j \qquad \text{Lemma 1}$$

Next we peel one term off each sum.

$$= \binom{k}{0} a^{k+1} + \sum_{j=1}^{k} \binom{k}{j} a^{k-j+1} b^j + \sum_{j=1}^{k} \binom{k}{j-1} a^{k-j+1} b^j + \binom{k}{k} b^{k+1}$$

Then we put the two sums together.

$$= \binom{k}{0} a^{k+1} + \sum_{j=1}^{k} \left\{ \binom{k}{j} a^{k-j+1} b^j + \binom{k}{j-1} a^{k-j+1} b^j \right\} + \binom{k}{k} b^{k+1}$$

$$= \binom{k}{0} a^{k+1} + \sum_{j=1}^{k} \left\{ \binom{k}{j} + \binom{k}{j-1} \right\} a^{k-j+1} b^j + \binom{k}{k} b^{k+1}$$

$$= 1 \cdot a^{k+1} + \sum_{j=1}^{k} \binom{k+1}{j} a^{k-j+1} b^j + 1 \cdot b^{k+1} \qquad \text{Lemma 2}$$

$$= \binom{k+1}{0} a^{k+1} + \sum_{j=1}^{k} \binom{k+1}{j} a^{k-j+1} b^j + \binom{k+1}{k+1} b^{k+1}$$

Last, we tuck the ends back in.

$$= \sum_{j=0}^{k+1} \binom{k+1}{j} a^{k-j+1} b^j$$

Note that this is $\sum_{j=0}^{n} \binom{n}{j} a^{n-j} b^j$ when n is $k + 1$. This completes the proof. □

GRAPHING CALCULATOR BOX

Factorials and Binomial Coefficients

Let's compute 8! which is 40,320 and $\binom{8}{4}$ which is 70.

TI-81

To compute 8!, enter 8 [MATH] [5] [ENTER].

Binomial coefficients are found in the Probability Menu which as accessed by pressing [MATH] and using the arrows to move the cursor so that PRB is darkened.

To compute $\binom{8}{4}$, select [3] from the PRB menu.

[8] [MATH] [→] [→] [→] [3] [4] [ENTER].

CASIO fx-7700G

Factorials and binomial coefficients are in the probability menu.
[MATH] [F2] selects PRB, the probability menu.
To compute 8!, enter [MATH] [F2] followed by 8 [F1] [EXE].

To compute $\binom{8}{4}$, enter [MATH] [F2] then press [8] [F3] [4] [EXE].

PROBLEM SET 9.3

Warm-ups

In Problems 1 through 4, evaluate each factorial expression. See Example 1.

1. $\dfrac{12!}{11!}$ **2.** $\dfrac{15!}{17!}$ 3. $\dfrac{14!}{7! \cdot 7!}$ **4.** $\dfrac{7!}{0! \cdot 7!}$

In Problems 5 through 8, calculate each binomial coefficient. See Example 2.

5. $\binom{8}{6}$ **6.** $\binom{9}{3}$ 7. $\binom{5}{0}$ **8.** $\binom{8}{4}$

In Problems 9 through 12, write the first four terms of each expansion. See Example 3.

9. $(s + t)^9$ **10.** $(x + 2)^8$ **11.** $(3x + y)^7$ **12.** $(2x + 1)^{11}$

In Problems 13 and 14, find the indicated term of each expansion. See Example 4.

13. Find the twelfth term of $(x + y)^{18}$. **14.** Find the middle term of $(2x + y)^{10}$.

In Problems 15 and 16, find the last three terms of each expansion. See Example 5.

15. $(x - y)^{11}$ **16.** $(5v - 1)^{22}$

Practice Exercises

In Problems 17 through 24, find the first three terms of the binomial expansion.

17. $(a + b)^{12}$ **18.** $(x + y)^{20}$ **19.** $(s + 2)^{15}$ **20.** $(a - b)^{13}$
21. $(2 + t)^9$ **22.** $(2a + 3b)^6$ **23.** $(3x - 1)^7$ **24.** $(z^3 - x^2)^{11}$

In Problems 25 through 32, find the last three terms of the binomial expansion.

25. $(a + b)^{13}$ **26.** $(x + y)^{21}$ **27.** $(s + 3)^5$ **28.** $(3 + t)^{17}$
29. $(3x + 2y)^7$ **30.** $(x^2 + y^2)^{30}$ **31.** $(a - b)^{12}$ **32.** $(x - y)^{24}$

33. Find the fifth term of $(a + b)^{13}$.

34. Find the ninth term of $(x + 1)^{12}$.

35. Find the fourth term of $\left(3x + \frac{1}{2}\right)^{11}$.

36. Find the middle term of $\left(\frac{x}{2} + \frac{1}{3}\right)^{10}$.

In Problems 37 through 48, use the Binomial Theorem to expand the given binomial.

37. $(a + b)^6$ **38.** $(x + y)^8$ **39.** $(x + 1)^9$ **40.** $(A + 2)^7$

41. $(2x + y)^6$ **42.** $(3x + 2y)^5$ **43.** $(a + b)^9$ **44.** $(x + 1)^8$

45. $(x + 2)^9$ **46.** $(3A + 1)^7$ **47.** $(x + 2y)^6$ **48.** $(2x + 3t)^5$

In Problems 49 through 52, find the first two terms of the binomial expansion.

49. $(a + b)^{50}$ **50.** $(x + 1)^{100}$ **51.** $(s + 3)^{500}$ **52.** $(a - b)^{150}$

Challenge Problems

53. Notice that $1 = 1^n = \left(\frac{1}{2} + \frac{1}{2}\right)^n$. Use the Binomial Theorem to expand $\left(\frac{1}{2} + \frac{1}{2}\right)^6$ and then show that it equals 1.

54. As $1.02 = 1 + 0.02$, powers of 1.02 can be found with the Binomial Theorem. Use this technique to find the exact value of $(1.02)^5$.

55. The expansion of $(a - b)^n$ is the same as the expansion of $(a + b)^n$, except that the signs alternate, starting with positive. Use this result with the technique of Problem 54 to find the exact value of $(0.98)^5$.

Proofs containing an ellipsis (. . .) are not considered correct by some purists. In Problems 56 and 57, prove Lemma 1 without using an ellipsis.

56. Prove Lemma 1 by induction.

57. Prove Lemma 1 making the substitution $j = q - 1$, simplifying, and then changing the q's back to j's.

IN YOUR OWN WORDS . . .

58. State the Binomial Theorem without using a formula.

9.4 SEQUENCES

An ordered list of numbers such as 1, 3, 5, 7, 9, . . . is called a **sequence.** The expressions separated by commas are called the **terms** of the sequence. If the sequence has a last member, it is called a **finite sequence.** If the sequence has no last member, it is called an **infinite sequence** or simply a sequence. Often, we can determine the general term that expresses every term of the sequence. For example,

Sequence	*Type*	*General Term*
1, 3, 5, 7, 9, . . .	Infinite	$2n - 1$
2, 8, 18, 32, 50	Finite	$2n^2$
$\frac{1}{2}, \frac{2}{3}, \frac{3}{4}, \frac{4}{5}, \ldots$	Infinite	$\frac{n}{n+1}$

Notice that replacing the variable in the general term by 1 gives the first term of the sequence. Replacing it by 2 gives the second term, and so on. In other words, the general term of the sequence is a *function* that yields the sequence when evaluated with the natural numbers. This leads to the definition of sequence.

Definition of Sequence

A sequence is a function whose domain is the set of natural numbers.

Since the domain of a sequence is the set of natural numbers, we use the general term rather than the usual functional notation. Instead of writing

$$f(x) = 2x - 1; \ x = 1, 2, 3, \ldots$$

for the sequence, 1, 3, 5, . . . , we use n as an element of the domain and a_n instead of $f(n)$, with the understanding that n is a natural number. Therefore, we can write the sequence as

$$a_n = 2n - 1$$

EXAMPLE 1 Write the first five terms of each sequence.

(a) $a_n = n^3$ (b) $A_n = 4n - 3$ (c) $b_n = n!$

Solutions (a) $a_n = n^3$ General term

$a_1 = 1^3 = 1$ First term, $n = 1$

$a_2 = 2^3 = 8$ Second term, $n = 2$

$a_3 = 3^3 = 27$

$a_4 = 4^3 = 64$

$a_5 = 5^3 = 125$

Therefore, the first five terms of the sequence are 1, 8, 27, 64, 125.

(b) $A_n = 4n - 3$

The first five terms are A_1, A_2, A_3, A_4, and A_5, or

$4 \cdot 1 - 3, 4 \cdot 2 - 3, 4 \cdot 3 - 3, 4 \cdot 4 - 3, 4 \cdot 5 - 3$ or

1, 5, 9, 13, 17

(c) $b_n = n!$

1!, 2!, 3!, 4!, 5! or

$1, 1 \cdot 2, 1 \cdot 2 \cdot 3, 1 \cdot 2 \cdot 3 \cdot 4, 1 \cdot 2 \cdot 3 \cdot 4 \cdot 5$ or

1, 2, 6, 24, 120 □

Often we are given some of the terms of a sequence and are asked to provide a suitable general term.

EXAMPLE 2 Find a suitable general term for each of the following sequences.

(a) 2, 6, 18, 54, 162, . . .

(b) $\sqrt{2}, 2, \sqrt{6}, 2\sqrt{2}, \sqrt{10}, \ldots$

(c) $1, -\frac{1}{2}, \frac{1}{4}, -\frac{1}{8}, \frac{1}{16}, \ldots$

Solutions (a) 2, 6, 18, 54, 162, . . .

Notice that each term, after the first, is three times the previous term. Thus

$a_2 = a_1 \cdot 3 = 2 \cdot 3$

$a_3 = a_2 \cdot 3 = 2 \cdot 3 \cdot 3$

$a_4 = a_3 \cdot 3 = 2 \cdot 3^3$

and

$a_5 = a_4 \cdot 3 = 2 \cdot 3^4$

Now we see the pattern, $a_n = 2 \cdot 3^{n-1}$.

(b) $\sqrt{2}, 2, \sqrt{6}, 2\sqrt{2}, \sqrt{10}, \ldots$

Rewriting this sequence as unsimplified principal square roots,

$\sqrt{2}, \sqrt{4}, \sqrt{6}, \sqrt{8}, \sqrt{10}, \ldots$

we see that each term is the square root of $2n$.

$a_n = \sqrt{2n}$

(c) $1, -\frac{1}{2}, \frac{1}{4}, -\frac{1}{8}, \frac{1}{16}, \ldots$

Since the terms alternate in sign, we think of a negative number raised to the nth power. Since powers of 2 are in the denominators, notice that

$$a_1 = 1 = \left(-\frac{1}{2}\right)^0$$

$$a_2 = -\frac{1}{2} = \left(-\frac{1}{2}\right)^1$$

$$a_3 = \frac{1}{4} = \left(-\frac{1}{2}\right)^2$$

$$a_4 = -\frac{1}{8} = \left(-\frac{1}{2}\right)^3$$

$$a_5 = \frac{1}{16} = \left(-\frac{1}{2}\right)^4$$

Thus, a suitable nth term is $a_n = \left(-\frac{1}{2}\right)^{n-1}$. □

A sequence such as

$$1, 4, 7, 10, 13, \ldots$$

where every term is *greater* than the previous term, is called an **increasing sequence.** A sequence such as

$$1, \frac{2}{3}, \frac{4}{9}, \frac{8}{27}, \frac{16}{81}, \ldots$$

where every term is *less* than the previous term, is called a **decreasing sequence.**

As an aid to our intuitive understanding of a function we often examine its graph. Since sequences are functions, we can study their graphs. The graphs will look a little unusual because the domain of a sequence is limited to the natural numbers. Thus the graph will be a collection of separated points on the plane. For example, the graph of the sequence $A_n = \dfrac{1}{n}$ follows.

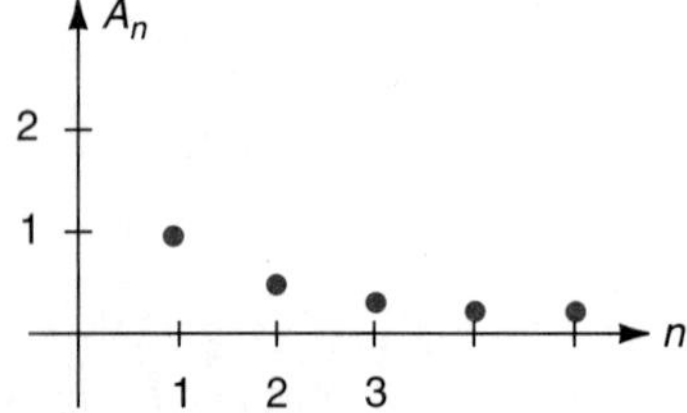

There are two common types of sequences. They are called arithmetic sequences and geometric sequences. An **arithmetic sequence** or **arithmetic progression** is a sequence obtained by adding a fixed number to each previous term. For example,

$$1, 5, 9, 13, 17, \ldots$$

is an arithmetic sequence as each term is 4 more than the previous term; 4 is called the **common difference** for this sequence. Thus arithmetic sequences have the form

$$\begin{aligned}
a_2 &= a_1 + d \\
a_3 &= a_2 + d = a_1 + 2d \\
a_4 &= a_3 + d = a_1 + 3d \\
&\vdots \\
a_n &= a_{n-1} + d = a_1 + d(n-1)
\end{aligned}$$

where d is the common difference.

An expression that gives a_n in terms of the previous term a_{n-1} is called a **recursive** formula for the sequence, while a formula that gives a_n in terms of a_1 and other constants is called an **explicit** formula. Although recursive formulas have their uses, we usually find an explicit formula more useful.

Arithmetic Sequences

The nth term of an arithmetic sequence is given by the explicit formula

$$a_n = a_1 + d(n-1)$$

EXAMPLE 3 For each of the following arithmetic sequences, find the common difference and write each in the form $a_n = a_1 + d(n - 1)$.

(a) $2, 11, 20, 29, \ldots$ (b) $-2, 5, 12, 19, \ldots$

(c) $3, \frac{7}{2}, 4, \frac{9}{2}, 5, \ldots$ (d) $8, 2, -4, -10, \ldots$

Solutions (a) $2, 11, 20, 29, \ldots$

If the sequence is an arithmetic sequence, we can find d by subtracting any term from the following term. That is, $d = 20 - 11 = 9$. Therefore, the common difference is 9, and we can write the sequence $a_n = 2 + 9(n - 1)$.

(b) $-2, 5, 12, 19, \ldots$

Since this is an arithmetic sequence, $d = 12 - 5 = 7$, and we may write

$$a_n = -2 + 7(n - 1)$$

(c) $3, \frac{7}{2}, 4, \frac{9}{2}, 5, \ldots$ By subtraction, $d = \frac{7}{2} - 3 = \frac{1}{2}$ so,

$$a_n = 3 + \frac{1}{2}(n - 1)$$

(d) $8, 2, -4, -10, \ldots$ By subtraction, $d = 2 - 8 = -6$, so

$a_n = 8 + (-6)(n - 1)$ or $a_n = 8 - 6(n - 1)$ □

EXAMPLE 4 Find an arithmetic sequence whose third term is 4 and seventh term is 24.

Solution We are to find an *arithmetic* sequence such that $a_3 = 4$ and $a_7 = 24$. Using the formula for the general term of an arithmetic sequence, we get two equations: one when n is 3 and the other when n is 7.

$$a_n = a_1 + d(n - 1) \qquad \text{General term}$$

$$4 = a_1 + d(3 - 1) \qquad a_3 = 4.$$

$$24 = a_1 + d(7 - 1) \qquad a_7 = 24.$$

We have two linear equations in the variables a_1 and d.

$$\begin{cases} a_1 + 2d = 4 \\ a_1 + 6d = 24 \end{cases}$$

We subtract the first equation from the second to obtain d and then a_1.

$$4d = 20$$

$$d = 5$$

$$a_1 = -6$$

Therefore, the sequence is

$$a_n = -6 + 5(n - 1) \qquad \square$$

A **geometric sequence** (or **geometric progression**) is a sequence obtained by multiplying a fixed number by each previous term. For example,

$$1, 2, 4, 8, 16, \ldots$$

is a geometric sequence as each term is the previous term multiplied by 2. Thus 2 is called the **common ratio** for this sequence. Thus geometric sequences have the form

$$\begin{aligned} a_2 &= a_1 \cdot r \\ a_3 &= a_2 \cdot r = a_1 \cdot r^2 \\ a_4 &= a_3 r = a_1 r^3 \\ &\vdots \\ a_n &= a_{n-1} r = a_1 r^{n-1} \end{aligned}$$

where r is the common ratio.

Geometric Sequences

The nth term of an geometric sequence is given by the explicit formula

$$a_n = a_1 r^{n-1}$$

Some examples of geometric sequences are

$$2, 10, 50, 250, \ldots \qquad \text{Common ratio is 5.}$$

$$4, -2, 1, -\frac{1}{2}, \frac{1}{4}, \ldots \qquad \text{Common ratio is } -\frac{1}{2}.$$

Writing these sequences in the form $a_n = a_1 r^{n-1}$, we have

$$a_n = 2 \cdot 5^{n-1}$$

$$b_n = 4\left(-\frac{1}{2}\right)^{n-1}$$

If a sequence is a geometric sequence, we can find the common ratio by dividing any term into the term following it. That is,

$$\frac{a_{k+1}}{a_k} = \frac{a_1 r^k}{a_1 r^{k-1}} = r^{k-(k-1)} = r$$

EXAMPLE 5 For each of the following geometric sequences, find the common ratio and write each in the form $a_n = a_1 r^{n-1}$.

(a) 4, 12, 36, 108, . . . (b) $3, -2, \frac{4}{3}, -\frac{8}{9}, \frac{16}{27}, \ldots$

Solutions (a) 4, 12, 36, 108, . . .

The common ratio is $r = \frac{36}{12} = 3.$

Thus, $a_n = 4 \cdot 3^{n-1}$.

(b) $3, -2, \frac{4}{3}, -\frac{8}{9}, \frac{16}{27}, \ldots$

The common ratio is $r = \frac{-2}{3} = -\frac{2}{3}$

Thus, $a_n = 3 \cdot \left(-\frac{2}{3}\right)^{n-1}$. □

EXAMPLE 6 Find a geometric sequence whose fourth term is $-\frac{3}{8}$ and seventh term is $\frac{3}{64}$.

Solution

$$a_n = a_1 r^{n-1} \qquad \text{General term}$$

$$-\frac{3}{8} = a_1 r^3 \qquad \text{Fourth term}$$

$$\frac{3}{64} = a_1 r^6 \qquad \text{Seventh term}$$

We have two *nonlinear* equations in two variables a_1 and r.

$$\begin{cases} a_1 r^3 = -\frac{3}{8} \\ a_1 r^6 = \frac{3}{64} \end{cases}$$

We use the method of substitution. We solve the first equation for a_1,

$$a_1 = -\frac{3}{8r^3}$$

and substitute for a_1 in the second equation.

$$-\frac{3}{8r^3} r^6 = \frac{3}{64}$$

$$r^3 = \frac{3}{64}\left(-\frac{8}{3}\right) = -\frac{1}{8}$$

$$r = -\frac{1}{2}$$

Now we can find a_1 by substitution.

$$a_1 = -\frac{3}{8\left(-\frac{1}{8}\right)} = 3$$

A geometric sequence with the desired properties is $a_n = 3\left(-\frac{1}{2}\right)^{n-1}$. □

PROBLEM SET 9.4

Warm-ups

In Problems 1 through 6, write the first five terms of the given sequence. See Example 1.

1. $a_n = (2n + 1)!$

2. $b_n = 14 - 3n$

3. $c_n = 2(n - 1)^3$

4. $A_n = \dfrac{3n}{2n - 1}$

5. $z_n = (-1)^{n-1}(1 - n)$

6. $k_n = (-1)^n \dfrac{n^2}{n + 1}$

In Problems 7 through 12, find a suitable general term for the given sequence. See Example 2.

7. 2, 4, 6, 8, . . .

8. 1, 4, 9, 16, . . .

9. 2, 5, 8, 11, . . .

10. −1, 2, −3, 4, . . .

11. 1, 9, 25, 49, . . .

12. 2, −4, 8, −16, . . .

Problems 13 through 18 are arithmetic sequences. In each, find the common difference and write in the form $a_n = a_1 + d(n - 1)$. See Example 3.

13. 3, 5, 7, . . .

14. $2, \frac{3}{2}, 1, \ldots$

15. 1, 5, 9, . . .

16. 0, −2, −4, . . .

17. $11, \frac{48}{5}, \frac{41}{5}, \ldots$

18. 12, 0, −12, . . .

For Problems 19 and 20, see Example 4.

19. Find an arithmetic sequence whose second term is 5 and whose eighth term is 17.

20. Find an arithmetic sequence whose fourth term is 1 and whose tenth term is 5.

Problems 21 through 26 are geometric sequences. In each, find the common ratio and write in the form $a_n = a_1 r^{n-1}$. See Example 5.

21. 3, 6, 12, . . .

22. 1, 5, 25, . . .

23. $\frac{1}{3}, \frac{1}{9}, \frac{1}{27}, \ldots$

24. 2, −4, 8, . . .

25. 4, 2, 1, . . .

26. $1, -\frac{1}{4}, \frac{1}{16}, \ldots$

For Problems 27 and 28, see Example 6.

27. Find a geometric sequence whose fourth term is 40 and whose sixth term is 160.

28. Find a geometric sequence whose third term is $\frac{8}{9}$ and whose sixth term is $-\frac{64}{243}$.

Practice Exercises

In Problems 29 through 34, label each sequence as increasing, decreasing, or neither increasing nor decreasing.

29. $a_n = 3 + 2n$

30. $a_n = 3 - 2n$

31. $b_n = n^{-2}$

32. $A_n = (-n)^{-1}$

33. $v_n = (-n)^{-n}$

34. $k_n = 1 - n^{-1}$

In Problems 35 through 40, write the first five terms of the given sequence.

35. $a_n = n + 1$

36. $b_n = 3n - 1$

37. $c_n = 8 - 2^n$

38. $A_n = n^2 - 1$

39. $B_n = n^2 - n + 1$

40. $C_n = (-1)^{n-1}(n - n^2)$

In Problems 41 through 47, find a suitable general term for the given sequence.

41. $4, 6, 8, 10, \ldots$

42. $0, 3, 8, 15, \ldots$

43. $-1, 3, 7, 11, \ldots$

44. $1, -2, 3, -4, \ldots$

45. $0, 1, 8, 27, \ldots$

46. $-1, 2, -4, 8, \ldots$

Problems 47 through 52 are arithmetic sequences. In each, find the common difference and write in the form $a_n = a_1 + d(n - 1)$.

47. $1, 4, 7, \ldots$

48. $-1, 4, 9, \ldots$

49. $10, 8, 6, \ldots$

50. $0, 11, 22, \ldots$

51. $-\frac{1}{2}, 0, \frac{1}{2}, \ldots$

52. $\frac{2}{3}, \frac{1}{3}, 0, \ldots$

Problems 53 through 58 are geometric sequences. In each, find the common ratio and write in the form $a_n = a_1r^{n-1}$.

53. $1, 3, 9, \ldots$

54. $2, 4, 8, \ldots$

55. $2, -6, 18, \ldots$

56. $5, 25, 125, \ldots$

57. $27, 9, 3, \ldots$

58. $128, -64, 32, \ldots$

Challenge Problems

59. Use mathematical induction to prove that the nth term of an arithmetic sequence is given by

$$a_n = a_1 + d(n - 1)$$

60. Use mathematical induction to prove that the nth term of a geometric sequence is given by

$$a_n = a_1r^{n-1}$$

61. Find two different geometric sequences, each with a third term of 12 and a fifth term of 48.

62. Find two different geometric sequences, each with a third term of -18 and a fifth term of -162.

IN YOUR OWN WORDS . . .

63. What is a sequence?

64. Explain the relationship between the sequence $a_n = 2n$ and the function $f(x) = 2x$.

9.5 SERIES

In Section 9.2 we proved by mathematical induction that the sum of the first n odd natural numbers is n^2. In other words, the sum of the terms of the finite sequence 1, 3, 5, 7, . . . , $(2n - 1)$ is n^2. The sums of the terms of a finite sequence have an important role in mathematics.

Notice how we can write the sums of finite sequences by using the general term of a sequence and summation notation. For example, the sum above can be written

$$1 + 3 + 5 + 7 + \cdots + (2n - 1) = \sum_{j=1}^{n} (2j - 1)$$

EXAMPLE 1 Write the sum of each finite sequence in summation notation.

(a) 3, 6, 12, 24, 48 (b) − 1, 4, 9, 14, 19, 24, 29, 34

Solutions (a) 3, 6, 12, 24, 48

Since each term of this finite sequence is twice the previous term, it is a geometric sequence with ratio 2. Therefore, its general term is given by $3 \cdot 2^{n-1}$. In summation notation the sum is

$$3 + 6 + 12 + 24 + 48 = \sum_{j=1}^{5} 3 \cdot 2^{j-1}.$$

(b) −1, 4, 9, 14, 19, 24, 29, 34

In this finite sequence, each term is 5 more than the previous term. It is an arithmetic sequence with 5 as its common difference. Its general term is $-1 + 5(n - 1)$. Therefore,

$$-1 + 4 + 9 + 14 + 19 + 24 + 29 + 34 = \sum_{j=1}^{8} [-1 + 5(j - 1)].$$ □

Of particular importance is the *sum* of the terms of a geometric sequence. Consider the *finite* geometric sequence 2, 6, 18, 54, 162. Let's develop a formula for its sum. Notice that the common ratio is 3. Let S be the *sum* of its terms. Then,

$$S = 2 + 6 + 18 + 54 + 162$$

Now, if we multiply both sides of this equation by the common ratio 3, we get a second equation.

$$3S = 6 + 18 + 54 + 162 + 486$$

If we subtract the first equation from the second, notice what occurs.

$$\begin{aligned} 3S &= \quad 6 + 18 + 54 + 162 + 486 \\ -S &= -2 - 6 - 18 - 54 - 162 \\ \hline 2S &= -2 \qquad\qquad\qquad\qquad\quad + 486 \end{aligned}$$

Therefore,

$$2S = 484$$

$$S = 242$$

and we have found the desired sum. Note how all the middle terms drop out when we subtract the sum from the common ratio times the sum. Let's try that in general.

The general term of a geometric sequence is given by ar^{n-1}, where a is the first term and r is the common ratio. Therefore, the sum of the first n terms of a geometric sequence is

$$S = \sum_{j=1}^{n} ar^{j-1}$$

Multiply by the common ratio and simplify.

$$rS = r\sum_{j=1}^{n} ar^{j-1}$$

$$= \sum_{j=1}^{n} ar \cdot r^{j-1} \qquad \text{Property of summation notation}$$

$$= \sum_{j=1}^{n} ar^{j} \qquad \text{Property of exponents}$$

Change the index so the exponent is $j - 1$.

$$= \sum_{j=2}^{n+1} ar^{j-1} \qquad \text{Property of summation notation}$$

Now we subtract S from rS.

$$rS - S = \sum_{j=2}^{n+1} ar^{j-1} - \sum_{j=1}^{n} ar^{j-1}$$

Next, we peel the last term off the first sum and the first term off the last sum.

$$= \sum_{j=2}^{n} ar^{j-1} + ar^{n+1-1} - \left(ar^{1-1} + \sum_{j=2}^{n} ar^{j-1}\right)$$

$$= \sum_{j=2}^{n} ar^{j-1} + ar^{n} - a - \sum_{j=2}^{n} ar^{j-1}$$

$$= ar^{n} - a$$

Factor S from the left side and a from the right and solve for S.

$$S(r-1) = a(r^{n}-1)$$

$$S = \frac{a(r^{n}-1)}{r-1}$$

The Sum of the Terms of a Finite Geometric Sequence

$$\sum_{j=1}^{n} ar^{j-1} = \frac{a(r^n - 1)}{r - 1}$$

The sum of the first n terms of a geometric sequence can be found by knowing the common ratio r and the first term a.

EXAMPLE 2 Find each sum using the formula.

(a) The sum of the terms of 2, 8, 32, 128, 512, 1024.

(b) The sum of the terms of the finite geometric sequence 24, 12, . . . , $\frac{3}{4}$.

(c) $\sum_{j=1}^{11} 2 \cdot 4^{j-1}$

Solutions (a) The sum of the terms 2, 8, 32, 128, 512, 1024

From the sequence we see that a is 2, r is $\frac{8}{2} = 4$, and n is 6. Thus the sum is

$$\frac{2(4^6 - 1)}{4 - 1} = \frac{2(4096 - 1)}{3} = 2730$$

Check by adding the terms!

(b) The sum of the terms of the finite geometric sequence 24, 12, . . . , $\frac{3}{4}$

We see that a is 24 and r is $\frac{1}{2}$, but we need n to use the formula. Since the general term is given by ar^{n-1} and we know the last term is $\frac{3}{4}$, we can find n.

$$\frac{3}{4} = 24\left(\frac{1}{2}\right)^{n-1}$$

$$\frac{3}{4 \cdot 24} = \left(\frac{1}{2}\right)^{n-1} \qquad \text{Divide by 24.}$$

$$\frac{1}{32} = \left(\frac{1}{2}\right)^{n-1}$$

$$\left(\frac{1}{2}\right)^5 = \left(\frac{1}{2}\right)^{n-1}$$

$$5 = n - 1 \qquad \text{One-to-One Property of exponentials}$$

$$6 = n$$

Now we can find the sum.

$$\frac{24\left(\left(\frac{1}{2}\right)^6 - 1\right)}{\frac{1}{2} - 1} = \frac{24\left(\frac{1}{64} - 1\right)}{-\frac{1}{2}} = \frac{24\left(-\frac{63}{64}\right)}{-\frac{1}{2}} = \frac{-\frac{3 \cdot 63}{8}}{-\frac{1}{2}} = \frac{189}{4}$$

(c) $\sum_{j=1}^{11} 2 \cdot 4^{j-1} = \frac{2(4^{11} - 1)}{4 - 1} = 2796202$ □

We now know how to sum the terms of a *finite* geometric sequence. However, we would like to attach meaning to the sum of *all* the terms of a sequence.

Series

The sum

$$\sum_{j=1}^{+\infty} A_j$$

is called a **series.**

In the geometric case that we have been discussing, we will find meaning for the sum

$$\sum_{j=1}^{+\infty} ar^{j-1}$$

Since the terms are those of a geometric sequence, this sum is called a **geometric series.** To try to make sense out of the geometric series, we examine the sum of a *finite* geometric sequence when n is very large. That is, we look at the formula

$$\sum_{j=1}^{n} ar^{j-1} = \frac{a(r^n - 1)}{r - 1}$$

when n is a large number. Suppose r is $\frac{1}{2}$. Let's look at some values of r^n when n starts to get large.

$$\left(\frac{1}{2}\right)^2 = 0.25$$

$$\left(\frac{1}{2}\right)^9 = 0.001953125$$

$$\left(\frac{1}{2}\right)^{25} \approx 0.000000029802$$

$$\left(\frac{1}{2}\right)^{100} \approx \underset{(30 \text{ zeros})}{0.00\ldots00007889}$$

We see that $\left(\frac{1}{2}\right)^n$ gets *very* small as n grows larger. 100 is not a very large number, but a decimal that starts with 30 zeros is vanishingly small. For n really large, $\left(\frac{1}{2}\right)^n$ is, for any practical use, equivalent to zero. We write

$$\lim_{n\to+\infty} \left(\frac{1}{2}\right)^n = 0$$

and we read this as "the limit of $\left(\frac{1}{2}\right)^n$ as n goes to positive infinity is zero." In calculus, an exact meaning is given to this notation. It turns out that if r is between -1 and 1, the same result holds.

Limit of r^n

If $|r| < 1$, then

$$\lim_{n\to+\infty} r^n = 0$$

Let's return to the sum of the first n terms of a geometric series.

$$\sum_{j=1}^{n} ar^{j-1} = \frac{a(r^n - 1)}{r - 1}$$

We rewrite the right side.

$$= \frac{ar^n - a}{r - 1}$$

$$= \frac{ar^n}{r - 1} - \frac{a}{r - 1}$$

$$= r^n \cdot \frac{a}{r - 1} + \frac{a}{1 - r}$$

Notice that the second term on the right side is the same for *all values of n,* but the first term changes as n changes. Now suppose $|r| < 1$. Since $r^n \to 0$ as $n \to +\infty$, the first term on the right side also goes to zero. Therefore, we have a sensible definition for the sum of a geometric series whenever $|r| < 1$.

The Sum of a Geometric Series

If $|r| < 1$, the sum of the series $a + ar + ar^2 + \cdots + ar^n + \cdots$ is given by

$$\sum_{j=1}^{+\infty} ar^{j-1} = \frac{a}{1-r}$$

If $|r| \geq 1$, the sum $\sum_{j=1}^{n} ar^{j-1}$ *does not* approach a fixed number as n grows large but grows larger and larger itself. We say the geometric series **diverges** when $|r| \geq 1$ and attach very little meaning to it.

EXAMPLE 3 Find the sum of each series if it exists.

(a) $\sum_{j=1}^{+\infty} 2\left(\frac{3}{5}\right)^{j-1}$

(b) $1 + \frac{1}{2} + \frac{1}{4} + \cdots + \frac{1}{2^{n-1}} + \cdots$

(c) $1 - \frac{1}{2} + \frac{1}{4} - \cdots + \left(-\frac{1}{2}\right)^{n-1} + \cdots$

(d) $\sum_{j=1}^{+\infty} 3\left(\frac{7}{6}\right)^{j-1}$

Solutions (a) $\sum_{j=1}^{+\infty} 2\left(\frac{3}{5}\right)^{j-1}$

This is a geometric series with $r = \frac{3}{5}$. Since $|r| = \frac{3}{5} < 1$,

$$\sum_{j=1}^{+\infty} 2\left(\frac{3}{5}\right)^{j-1} = \frac{2}{1 - \frac{3}{5}} = \frac{10}{5-3} = 5$$

(b) $1 + \frac{1}{2} + \frac{1}{4} + \cdots + \frac{1}{2^{n-1}} + \cdots$

This is a geometric series with $r = \frac{\frac{1}{2}}{1} = \frac{1}{2}$. As $|r| < 1$,

$$\sum_{j=1}^{+\infty} 1\left(\frac{1}{2}\right)^{j-1} = \frac{1}{1 - \frac{1}{2}} = 2$$

(c) $1 - \frac{1}{2} + \frac{1}{4} - \cdots + \left(-\frac{1}{2}\right)^{n-1} + \cdots$

This is a geometric series with $r = -\frac{1}{2}$. But, since $\left|-\frac{1}{2}\right| = \frac{1}{2}$, we have

$$\sum_{j=1}^{+\infty} 1\left(-\frac{1}{2}\right)^{j-1} = \frac{1}{1 - \left(-\frac{1}{2}\right)} = \frac{1}{\frac{3}{2}} = \frac{2}{3}$$

(d) $\sum_{j=1}^{+\infty} 3\left(\frac{7}{6}\right)^{j-1}$

This is a geometric series with $r = \frac{7}{6}$. Since $|r| = \left|\frac{7}{6}\right| > 1$, the series *diverges.*

The sum does not exist. □

Now let's develop a formula for the sum of the terms of a finite *arithmetic* sequence. Consider the sum of a 10-term arithmetic sequence with first term 5 and a common difference of 3.

$$S = 5 + 8 + 11 + 14 + 17 + 20 + 23 + 26 + 29 + 32$$

The sum is the same if we add the terms in the *opposite* order. So,

$$S = 32 + 29 + 26 + 23 + 20 + 17 + 14 + 11 + 8 + 5$$

Now if we add these sums together, we find an interesting result.

$$\begin{aligned} S &= 5 + 8 + 11 + 14 + 17 + 20 + 23 + 26 + 29 + 32 \\ S &= 32 + 29 + 26 + 23 + 20 + 17 + 14 + 11 + 8 + 5 \\ \hline 2S &= 37 + 37 + 37 + 37 + 37 + 37 + 37 + 37 + 37 + 37 \end{aligned}$$

$$\begin{aligned} 2S &= 10 \cdot 37 \\ &= 10(5 + 32) \end{aligned}$$

or

$$S = \frac{10(5 + 32)}{2}$$

Notice that the sum is one-half the number of terms times the sum of the first and last terms. Thus we have a formula for the sum of the terms of a *finite arithmetic sequence.*

Sum of the Terms of a Finite Arithmetic Sequence

If a_n is a finite arithmetic sequence with first term a_1 and last term a_k, then the sum

$$S = a_1 + a_2 + \cdots + a_k$$

is given by

$$S = \frac{k(a_1 + a_k)}{2}$$

EXAMPLE 4 Find each of the following sums using the formula.

(a) $6 + 11 + 16 + 21 + 26 + 31 + 36 + 41 + 46$

(b) $3 + \frac{5}{2} + 2 + \frac{3}{2} + 1 + \frac{1}{2} + 0 - \frac{1}{2}$

Solutions (a) $6 + 11 + 16 + 21 + 26 + 31 + 36 + 41 + 46$

As each term is 5 more than the previous term, this is an arithmetic sum with nine terms.

$$S = \frac{9(6 + 46)}{2} = 234$$

Check by adding the terms.

(b) $3 + \frac{5}{2} + 2 + \frac{3}{2} + 1 + \frac{1}{2} + 0 - \frac{1}{2}$

As each term is obtained by adding $-\frac{1}{2}$ to the previous term, this is an arithmetic sum.

$$S = \frac{8\left[3 + \left(-\frac{1}{2}\right)\right]}{2} = 4\left(3 - \frac{1}{2}\right) = 10$$

□

A *series* whose terms are an arithmetic sequence diverges.

Arithmetic Series

The arithmetic series $\sum_{j=1}^{+\infty} [a + d(j - 1)]$ diverges.

PROBLEM SET 9.5

Warm-ups

In Problems 1 and 2, write the sum of each finite sequence in summation notation. See Example 1.

1. $3, 5, 7, 9, 11, 13$

2. $4, -2, 1, -\frac{1}{2}, \frac{1}{4}, -\frac{1}{8}$

In Problems 3 through 6, find each sum using an appropriate formula. See Example 2.

3. $1 + 3 + 9 + 27 + 81 + 243$

4. $18 + 12 + 8 + \frac{16}{3} + \frac{32}{9}$

5. $\sum_{j=1}^{5} 3 \cdot 5^{j-1}$

6. $\sum_{j=1}^{10} 2^{j-1}$

In Problems 7 through 10, find the sum of each series, if it exists. See Example 3.

7. $1 + \frac{1}{3} + \frac{1}{9} + \cdots + \frac{1}{3^{n-1}} + \cdots$

8. $16 - 8 + 4 - \cdots + \left(-\frac{1}{2}\right)^{n-5} + \cdots$

9. $\sum_{j=1}^{+\infty} 2\left(\frac{3}{2}\right)^{j-1}$

10. $\sum_{n=1}^{+\infty} 4\left(\frac{3}{4}\right)^{n-1}$

In Problems 11 through 14, find each sum using an appropriate formula. See Example 4.

11. $-12 - 1 + 10 + 21 + 32 + 43$

12. $47 + 53 + 59 + 65 + 71$

13. $\sum_{j=1}^{5} [11 + 5(j - 1)]$

14. $\sum_{n=1}^{8} \left[-14 + \frac{1}{2}(n - 1)\right]$

Practice Exercises

In Problems 15 through 26, find each sum using appropriate formulas.

15. $3 + 5 + 7 + 9 + 11 + 13$

16. $7 + 10 + 13 + 16 + 19 + 22$

17. $32 + 16 + 8 + 4 + 2 + 1 + \frac{1}{2}$

18. $9 + 3 + 1 + \frac{1}{3} + \frac{1}{9} + \frac{1}{27}$

19. $\sum_{j=1}^{5} \frac{\pi}{2}\left(\frac{2}{3}\right)^{j-1}$

20. $\sum_{j=1}^{15} 6\left(-\frac{1}{3}\right)^{j-1}$

21. $\sum_{n=1}^{6} (3n + 2)$

22. $\sum_{j=1}^{5} \left(11 - \frac{j}{2}\right)$

23. $\sum_{k=-1}^{500} k$

24. $\sum_{j=-5}^{1000} (j - 11)$

25. $\sum_{j=0}^{12} 5\left(\frac{4}{3}\right)^{j}$

26. $\sum_{p=0}^{19} 2\left(\frac{3}{2}\right)^{p}$

In Problems 27 through 34, find the sum of each series, if it exists.

27. $\sum_{n=1}^{+\infty} (0.1)^{n-1}$

28. $\sum_{n=1}^{+\infty} (0.9)^{n-1}$

29. $\sum_{j=1}^{+\infty} (1.001)^{j-1}$

30. $\sum_{n=1}^{+\infty} (0.999)^{n-1}$

31. $\frac{9}{8} + \frac{3}{4} + \frac{1}{2} + \cdots + \frac{2^{n-4}}{3^{n-3}} + \cdots$

32. $5 + \frac{10}{3} + \frac{20}{9} + \cdots + 5\left(\frac{2}{3}\right)^{n-1} + \cdots$

33. $\sum_{n=1}^{+\infty} (-1)^{n-1}5^{n}6^{-(n-1)}$

34. $\sum_{q=1}^{+\infty} (-1)^{q-1}6^{q}5^{-(q-1)}$

35. A certain Maxi-flight golf ball is dropped from a height of 10 feet. Each time the ball strikes the ground it bounces to four-fifths the height of the previous bounce. How high will the ball bounce after it strikes the ground the sixth time? What total distance will it have traveled when it strikes the ground the sixth time? If left to bounce forever, how far will the ball travel?

36. A poorly inflated basketball will recover only 24% of its height when dropped. If it rolls off the rim and falls 10 feet, how high will the ball bounce after it strikes the ground the fifth time? What total distance will it have traveled when it strikes the ground the fifth time? If left to bounce forever, how far will the ball travel?

37. Martin's starting salary is \$24,000 a year. If he gets a 5% salary increase at the end of each year, what will his salary be when he begins his eleventh year? How much total salary will he have received at the end of 10 years?

38. Beverly's starting salary is \$20,000 a year. If she gets an 8% salary increase at the end of each year, what will her salary be when she begins her eleventh year? How much total salary will she have received at the end of 10 years?

Challenge Problems

39. Evaluate the series $\sum_{j=1}^{3}\left(\sum_{n=1}^{+\infty}\left(\frac{j}{4}\right)^{n-1}\right)$.

40. Suppose the following points are plotted on a rectangular coordinate system;

$$(1, 1), \left(\frac{1}{2}, \frac{1}{3}\right), \left(\frac{1}{4}, \frac{1}{9}\right), \ldots, \left(\frac{1}{2^{n-1}}, \frac{1}{3^{n-1}}\right), \ldots$$

How far is the tenth point from the origin?

IN YOUR OWN WORDS . . .

41. What is a series?

9.6 PERMUTATIONS AND COMBINATIONS, COUNTING

Counting the number of different ways that something can happen is basic to the studies of probability, statistics, and computer science. In this section we will examine methods of counting the number of ways that various events can occur.

Suppose a college student wishes to take math at 8:00 and history at 9:00. If there are 2 sections of her math course offered at 8:00 and 3 sections of history at 9:00, how many different ways can she do this? She has two choices for her 8:00 math class and, *for each one of these,* she has three choices for history. A diagram showing the possibilities is helpful.

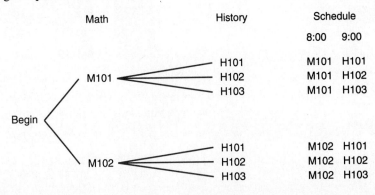

	Math	History	Schedule 8:00	Schedule 9:00
Begin	M101	H101	M101	H101
		H102	M101	H102
		H103	M101	H103
	M102	H101	M102	H101
		H102	M102	H102
		H103	M102	H103

A diagram, such as the one on the previous page, that shows every possibility is called a **tree diagram.** Notice that for each of the 2 different possibilities at 8:00 there are 3 possibilities for 9:00 or a total of $2 \cdot 3 = 6$ different schedules. This illustrates our basic counting principle.

Fundamental Principle of Counting

If *event 1* can be done in m ways and *event 2* can be done in n ways, then *event 1 followed by event 2* can be done a total of $m \cdot n$ different ways.

EXAMPLE 1 Many states use three letters in combination with numbers on their license plates. How many different three-letter arrangements can be made?

Solution By the fundamental principle there are $26 \cdot 26$ different two-letter arrangements. Thus, applying the fundamental principle again, there are $(26 \cdot 26) \cdot 26$ different three-letter arrangements. Since $26^3 = 17576$, there are 17,576 different three-letter arrangements. □

An arrangement of a collection of objects is called a **permutation.** Suppose 3 books are to be arranged on a bookshelf. How many different ways can this be done? If the books are labeled A, B and C, the possibilities are

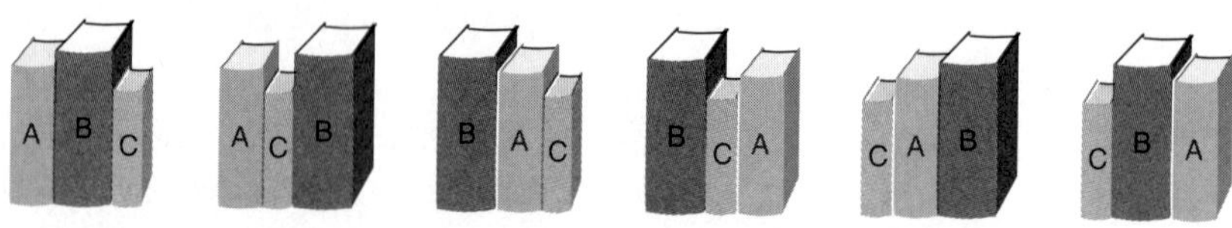

As there are no other possibilities, we conclude there are 6 different ways to arrange the 3 books. Now, suppose there is room for only 2 books on our bookshelf. Again, we list the various ways to place the books.

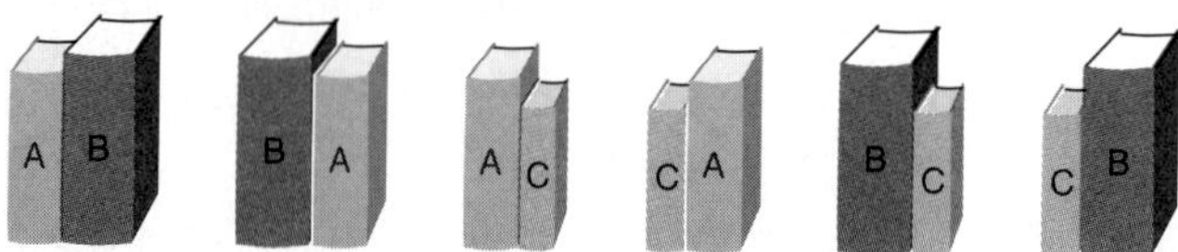

There are 6 ways to do this task. We say that there are "six permutations of three things taken two at a time." A common notation used for this is $P(3, 2)$. As there are also 6 permutations of 3 things taken 3 at a time (the first arrangement), we see that

$$P(3, 3) = 6$$

and

$$P(3, 2) = 6$$

If there were only 1 space on our shelf, we could put either book A, book B, or book C in that space. Thus,

$$P(3, 1) = 3$$

Now suppose there are 12 books and room for only 8. If we start listing the possibilities, we will soon tire of the task (or run out of paper). Let's consider the 8 spaces.

$$\square\ \square\ \square\ \square\ \square\ \square\ \square\ \square$$

Any one of the 12 books can be placed in the first space, so there are 12 ways to fill it. Suppose we put book G there.

$$\boxed{\text{G}}\ \square\ \square\ \square\ \square\ \square\ \square\ \square$$

Now there are only 11 books left to fill the next space. Thus, by the Fundamental Principle of Counting, there are $12 \cdot 11$ ways to fill the first two spaces. Suppose book D is chosen for the second space.

$$\boxed{\text{G}}\ \boxed{\text{D}}\ \square\ \square\ \square\ \square\ \square\ \square$$

Now there are only 10 books left to fill the third space. So there are $12 \cdot 11 \cdot 10$ ways to fill the first three spaces. If we continue placing books until we have one space left,

$$\boxed{\text{G}}\ \boxed{\text{D}}\ \boxed{\text{F}}\ \boxed{\text{A}}\ \boxed{\text{K}}\ \boxed{\text{J}}\ \boxed{\text{H}}\ \square$$

we find that we have 5 books left to fill the last space, so we can fill it 5 different ways. Therefore, there are

$$12 \cdot 11 \cdot 10 \cdot 9 \cdot 8 \cdot 7 \cdot 6 \cdot 5 = 19958400$$

different ways to place 12 books in 8 spaces. That is,

$$P(12, 8) = 19958400$$

There are 19,958,400 permutations of 12 things taken 8 at a time.

We extend these ideas to a formula for the number of permutations of n things taken r at a time.

$$P(n, r) = n \cdot (n - 1) \cdot (n - 2) \cdots [n - (r - 1)];\ 0 \leq r \leq n,\ n \neq 0$$

The formula for $P(n, r)$ can now be written using factorials. To see how, we reexamine $P(12, 8)$.

$$P(12, 8) = 12 \cdot 11 \cdot 10 \cdot 9 \cdot 8 \cdot 7 \cdot 6 \cdot 5$$

If we multiply and divide by $4 \cdot 3 \cdot 2 \cdot 1$, we get

$$P(12, 8) = \frac{12 \cdot 11 \cdot 10 \cdot 9 \cdot 8 \cdot 7 \cdot 6 \cdot 5 \cdot 4 \cdot 3 \cdot 2 \cdot 1}{4 \cdot 3 \cdot 2 \cdot 1}$$

Therefore,

$$P(12, 8) = \frac{12!}{4!}$$

$$P(12, 8) = \frac{12!}{(12-8)!}$$

This suggests a general formula for permutations.

Number of Permutations of n Things Taken r at a Time

For integers n and r; $0 \leq r \leq n$, $n \neq 0$

$$P(n, r) = \frac{n!}{(n-r)!}$$

EXAMPLE 2 Find each value.

(a) $P(6, 2)$ (b) $P(5, 3)$ (c) $P(8, 0)$ (d) $P(7, 7)$

Solutions (a) $P(6, 2) = \frac{6!}{(6-2)!} = \frac{6!}{4!} = \frac{6 \cdot 5 \cdot 4 \cdot 3 \cdot 2 \cdot 1}{4 \cdot 3 \cdot 2 \cdot 1} = 6 \cdot 5 = 30$

(b) $P(5, 3) = \frac{5!}{(5-3)!} = \frac{5!}{2!} = \frac{5 \cdot 4 \cdot 3 \cdot 2 \cdot 1}{2 \cdot 1} = 5 \cdot 4 \cdot 3 = 60$

(c) $P(8, 0) = \frac{8!}{(8-0)!} = \frac{8!}{8!} = 1$

(d) $P(7, 7) = \frac{7!}{(7-7)!} = \frac{7!}{0!} = \frac{7!}{1} = 7! = 5040$ Remember, $0! = 1$ □

EXAMPLE 3 If 9 horses are entered in the Motor City Handicap, how many different win, place, and show results are possible? (Win is first, place is second, and show is third.)

Solutions As each possible selection of first, second and third-place horses is a permutation of 9 horses taken 3 at a time, we must compute $P(9, 3)$.

$$P(9, 3) = \frac{9!}{(9-3)!} = \frac{9!}{6!} = \frac{9 \cdot 8 \cdot 7 \cdot 6 \cdot 5 \cdot 4 \cdot 3 \cdot 2 \cdot 1}{6 \cdot 5 \cdot 4 \cdot 3 \cdot 2 \cdot 1} = 9 \cdot 8 \cdot 7 = 504$$

There are 504 different win, place, and show possibilities. □

Suppose we have 5 books and only have room for 3 of them on the bookshelf, and it doesn't matter in what order they are placed. This is not a permutation problem, because the order of an arrangement is part of a permutation. In this problem, the collection of books A C D is the same as the collection D A C.

We call such collections **combinations** and use the notation $C(5, 3)$ for the number of combinations of 5 things taken 3 at a time. We list the different combinations of the 5 books, A, B, C, D, and E, taken 3 at a time.

A B C	A B D	A B E	A C D	A C E
A D E	B C D	B C E	B D E	C D E

Any other combination is an arrangement of one of the 10 combinations listed. Therefore, $C(5, 3) = 10$. Notice that there are 6 *permutations* of each of the listed combinations. Therefore, $6 \cdot C(5, 3)$ will be equal to $P(5, 3)$.

$$C(5, 3) = \frac{P(5, 3)}{6}$$

$$= \frac{1}{6} \cdot \frac{5!}{(5 - 3)!}$$

$$= \frac{5!}{6(5 - 3)!}$$

But $6 = 3!$

$$= \frac{5!}{3!(5 - 3)!}$$

This idea suggests the general formula for combinations.

Number of Combinations of n Things Taken r at a Time

For integers n and r; $0 \le r \le n$, $n \ne 0$

$$C(n, r) = \frac{n!}{r!(n - r)!}$$

Notice that this is exactly the formula for the binomial coefficients that was developed in Section 9.3.

$$C(n, r) = \binom{n}{r}$$

EXAMPLE 4 Find each value.

(a) $C(6, 2)$ (b) $C(6, 4)$ (c) $C(8, 0)$ (d) $C(7, 7)$

Solutions (a) $C(6, 2) = \dfrac{6!}{2!(6 - 2)!} = \dfrac{6!}{2! \cdot 4!} = \dfrac{6 \cdot 5 \cdot 4 \cdot 3 \cdot 2 \cdot 1}{2 \cdot 1 \cdot 4 \cdot 3 \cdot 2 \cdot 1} = 3 \cdot 5 = 15$

(b) $C(6, 4) = \dfrac{6!}{4!(6-4)!} = \dfrac{6!}{4! \cdot 2!} = \dfrac{6 \cdot 5 \cdot 4 \cdot 3 \cdot 2 \cdot 1}{4 \cdot 3 \cdot 2 \cdot 1 \cdot 2 \cdot 1} = 3 \cdot 5 = 15$

(c) $C(8, 0) = \dfrac{8!}{0!(8-0)!} = \dfrac{8!}{0! \cdot 8!} = \dfrac{8!}{1 \cdot 8!} = 1$

(d) $C(7, 7) = \dfrac{7!}{7!(7-7)!} = \dfrac{7!}{7! \cdot 0!} = \dfrac{7!}{7! \cdot 1} = 1$ □

Example 4 illustrates two interesting properties of combinations.

Combination Properties

For integers r and n; $0 \leq r \leq n$, $n \neq 0$

1. $C(n, r) = C(n, n-r)$
2. $C(n, 0) = C(n, n) = 1$

EXAMPLE 5 The Clarkston Women's Club has 16 members. A steering committee made up of 4 members is to be formed. How many different ways can this be done?

Solution Since there is no particular order to this committee, the number of different committees is the number of combinations of 16 things taken 4 at a time.

$$C(16, 4) = \frac{16!}{4! \cdot 12!} = \frac{16 \cdot 15 \cdot 14 \cdot 13 \cdot 12!}{4 \cdot 3 \cdot 2 \cdot 1 \cdot 12!} = 4 \cdot 5 \cdot 7 \cdot 13 = 1820$$

There are 1820 different ways to select the steering committee. □

EXAMPLE 6 How many different 5-card poker hands can be dealt from an ordinary deck of 52 cards?

Solution Since the order in which the cards are dealt does not matter, we need to find the number of combinations of 52 things taken 5 at a time.

$$C(52, 5) = \frac{52!}{5! \cdot 47!} = \frac{52 \cdot 51 \cdot 50 \cdot 49 \cdot 48 \cdot 47!}{5 \cdot 4 \cdot 3 \cdot 2 \cdot 1 \cdot 47!}$$
$$= 13 \cdot 17 \cdot 10 \cdot 49 \cdot 24 = 2598960$$

There are 2,598,960 different 5-card poker hands. □

There are other notations for permutations and combinations in common usage. The permutation $P(n, r)$ is sometimes written ${}_nP_r$. The combination $C(n, r)$ is also written ${}_nC_r$, as well as $\binom{n}{r}$.

GRAPHING CALCULATOR BOX

Permutations and Combinations

Let's see how to compute $P(8, 4)$ and $C(8, 4)$. $P(8, 4) = 1680$ and $C(8, 4) = 70$.

TI-81

Permutations and combinations are found in the Probability Menu which is accessed by pressing [MATH] and using the arrows to move the cursor so that PRB is darkened. To compute $P(8, 4)$, press [8] [MATH] [→] [→] [→] [2] [4] [ENTER].

To compute $C(8, 4)$, select [3] from the PRB menu.

[8] [MATH] [→] [→] [→] [3] [4] [ENTER].

CASIO fx-7700G

[MATH] [F2] selects PRB, the probability menu.

To compute $P(8, 4)$, enter [MATH] [F2] then press [8] [F2] [4] [EXE]

To compute $C(8, 4)$, use [8] [F3] [4] [EXE].

PROBLEM SET 9.6

Warm-ups

For Problems 1 and 2, see Example 1.

1. How many different three-digit numbers can be formed using only the odd digits 1, 3, 5, 7, and 9?

2. How many different three-letter "words" can be made with the letters X, Y, and Z? (Letters may be used more than once.)

In Problems 3 through 10, find each value. See Example 2.

3. $P(4, 2)$
4. $P(7, 4)$
5. $P(5, 0)$
6. $P(5, 1)$
7. $P(5, 2)$
8. $P(5, 3)$
9. $P(5, 4)$
10. $P(5, 5)$

For Problems 11 and 12, see Example 3.

11. How many different ways can a president, vice-president, and secretary be selected from a club of 12 eligible members?

12. How many different ways can first through fourth place be determined in the Kentucky Derby if 11 horses run?

In Problems 13 through 20, find each value. See Example 4.

13. $C(4, 2)$
14. $C(9, 0)$
15. $C(9, 1)$
16. $C(9, 2)$
17. $C(9, 3)$
18. $C(9, 5)$
19. $C(9, 8)$
20. $C(9, 9)$

For Problems 21 and 22, see Examples 5 and 6.

21. How many different 5-person committees can be made up from a club that has 12 members?

22. How many different 7-card poker hands can be dealt from an ordinary deck of 52 cards?

Practice Exercises

In Problems 23 through 46, evaluate each expression.

23. $P(8, 4)$
24. $P(7, 5)$
25. $P(6, 0)$
26. $P(6, 1)$
27. $P(6, 2)$
28. $P(6, 3)$
29. $P(6, 4)$
30. $P(6, 5)$
31. $P(6, 6)$
32. $C(8, 4)$
33. $C(7, 5)$
34. $C(8, 5)$
35. $C(8, 0)$
36. $C(8, 1)$
37. $C(8, 2)$
38. $C(8, 3)$
39. $C(8, 8)$
40. $C(8, 7)$
41. $C(8, 6)$
42. $C(8, 5)$
43. $P(14, 4)$
44. $P(14, 10)$
45. $C(14, 4)$
46. $C(14, 10)$

47. Seven players are to be chosen from a roster of 13. In how many ways can this be done?

48. The local Ace Hardware carries 14 different shades of interior latex paint. How many different combinations of 4 different shades can be selected?

49. How many different four-letter "words" can be made with the letters of the word OUTFIELD?

50. The officers of the Spin-around Dance Club are president, vice-president, and secretary-treasurer. No person is allowed to hold more than one office. If there are 28 members, how many ways can a slate of officers be selected?

51. Nine players are to be chosen from a roster of 16. In how many ways can this be done?

52. In how many ways can 6 algebra books be taken from a pile of 13 algebra books?

Challenge Problems

Suppose N is an integer greater than 1. Simplify each of the following.

53. $P(N, N)$
54. $P(N, 0)$
55. $P(N, N - 2)$
56. $C(N, N)$
57. $C(N, 0)$
58. $C(N, N - 2)$
59. $C(N + 2, N)$
60. $P(N + 1, N - 1)$

61. Restate the Binomial Theorem using the notation of combinations.

IN YOUR OWN WORDS . . .

62. Describe permutations without using a formula.

63. Describe combinations without using a formula.

9.7 INTRODUCTION TO PROBABILITY

Suppose a machine makes bolts that hold critical parts of an airplane wing together. Several questions may be asked about this machine. How likely is it that a bolt produced by this machine will be faulty? How likely is it that there will be a faulty bolt in a sample of 500 bolts made by this machine? How likely is it that there will be no faulty bolts in a sample of 1000? The answers to such questions are the domain of probability theory.

A simple case will explain some of the terms used in the study of probability.

Suppose a bag contains a black marble, a green marble, and a white marble. One marble is drawn from the bag sight unseen. This is called an **experiment.** There are three possible **outcomes:**

1. The black marble will be drawn.
2. The green marble will be drawn.
3. The white marble will be drawn.

The collection of all the outcomes is called the **sample space** of the experiment. If we call the outcomes B, G, and W, the sample space is the set {B, G, W}. An **event** is a subset of the sample space. For example, the event, *the marble is not black,* is the subset {G, W}, while the event, *the marble is green,* is the subset {G}. Notice that the outcomes are **equally likely.** (If there are *two* black marbles, one green, and one white in the bag, the outcomes are *not* equally likely.) We will consider only experiments with equally likely outcomes.

Probability

If the outcomes in the sample space of an experiment are equally likely, the **probability** of an event E is given by

$$P(E) = \frac{n(E)}{n(S)}$$

where $n(S)$ is the number of elements in the sample space and $n(E)$ is the number of elements in the event.

In other words, the probability of an event with equally likely outcomes is

$$\text{Probability} = \frac{\text{number of } \textit{favorable} \text{ outcomes}}{\text{number of } \textit{possible} \text{ outcomes}}$$

EXAMPLE 1 In the above experiment, what is the probability that the marble selected will not be black?

Solution The sample space {B, G, W} contains three elements, and the event $E = \{G, W\}$ contains two.

Thus the probability is given by,

$$P(E) = \frac{n(\{G, W\})}{n(\{B, G, W\})} = \frac{2}{3}$$

□

EXAMPLE 2 An experiment consists of throwing a fair 6-sided die and noting the number of dots on the upper face.

(a) What is the sample space for this experiment?

(b) What is the probability that the number of dots will be either 2 or odd?

Solutions (a) The sample space is the set of all possible outcomes $S = \{1, 2, 3, 4, 5, 6\}$.

(b) The event of 2 or odd is the subset $E = \{1, 2, 3, 5\}$. Thus

$$P(E) = \frac{n(E)}{n(S)} = \frac{4}{6} = \frac{2}{3}$$ □

EXAMPLE 3 Suppose an ordinary pair of fair dice are thrown.

(a) What is the probability of obtaining a total of 7 or 11 on the top faces of the dice?

(b) What is the probability of obtaining a total of 2, 3, or 12?

(c) What is the probability of *not* obtaining a total of 2, 3, 7, 11, or 12?

Solutions (a) What is the probability of obtaining a total of 7 or 11 on the top faces of the dice?

Since we want the outcomes of this experiment to be *equally likely,* we analyze this experiment as if one of the dice is white and the other blue. The total 7 can be obtained the following ways.

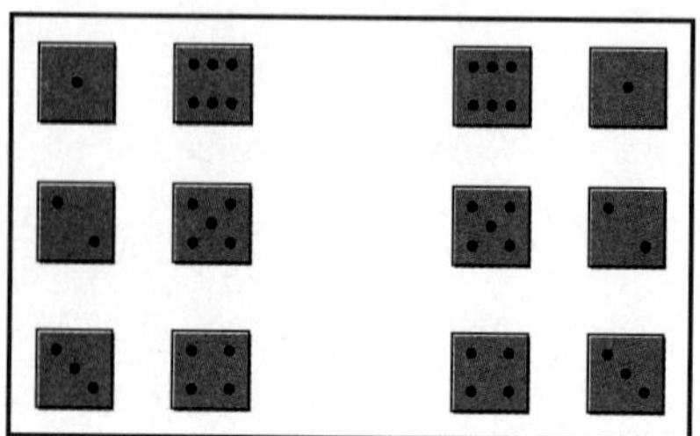

So there are 6 favorable outcomes that give the total 7. There are 2 ways to get 11.

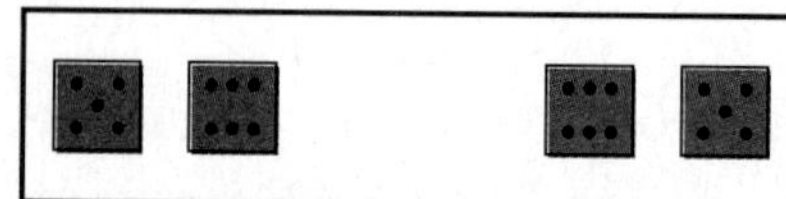

Therefore there are 8 favorable outcomes for this experiment. We now need the total number of outcomes. As the white die can have 6 different numbers and the blue die can have 6 different numbers and one cannot influence the other, there are $6 \cdot 6 = 36$ total outcomes by the Fundamental Principle of Counting (Section 9.6). Thus the probability is

$$P(7 \text{ or } 11) = \frac{6+2}{36} = \frac{2}{9}$$

(b) What is the probability of obtaining a total of 2, 3, or 12?
The only ways to get 2 or 12 are

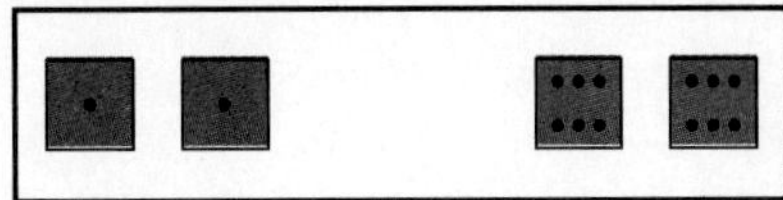

And 3 can be obtained with

Therefore, the probability is

$$P(2, 3, \text{ or } 12) = \frac{1+2+1}{36} = \frac{1}{9}$$

(c) What is the probability of *not* obtaining a total of 2, 3, 7, 11, or 12?
Since there are 8 outcomes that give 7 or 11 and 4 outcomes that give 2, 3, or 12, there are 12 outcomes that give a total of 2, 3, 7, 11, or 12. As there are 36 total outcomes, there must be $36 - 12 = 24$ outcomes that *are not* 2, 3, 7, 11, or 12.

$$P(\text{not } 2, 3, 7, 11, \text{ or } 12) = \frac{24}{36} = \frac{2}{3}$$

□

Notice in Example 3 that the probability of an event *occurring* plus the probability of the same event *not occurring* is 1. Also note that probability cannot be negative or greater than 1.

Two Properties of Probability

1. $0 \le P(E) \le 1$

2. $P(E) + P(\text{not } E) = 1$

However, probability *can* be zero or 1.

EXAMPLE 4 Suppose an ordinary pair of fair dice is thrown.

(a) What is the probability that a total of 17 will appear on the top faces of the dice?
(b) What is the probability that the total will be less than 17?

Solutions (a) Since there is no way to make 17 with two ordinary dice, there are 0 favorable outcomes.

$$P(17) = \frac{0}{36} = 0$$

(b) As *every* sum from two ordinary dice is less than 17, there are 36 favorable outcomes.

$$P(<17) = \frac{36}{36} = 1$$

□

If two events from the same sample space have no outcomes in common, they are said to be **mutually exclusive.** Example 3 illustrates that rolling a 7 with a pair of ordinary dice and rolling an 11 are mutually exclusive and that $P(7 \text{ or } 11) = P(7) + P(11)$.

Probability of Mutually Exclusive Events

Suppose A and B are mutually exclusive events from the same sample space. Then,

$$P(A \text{ or } B) = P(A) + P(B)$$

EXAMPLE 5 Suppose one card is drawn from an ordinary deck of playing cards. What is the probability that it will be an ace or a red 10?

Solution Since the one card drawn cannot be *both* a red 10 and an ace, the events are mutually exclusive. As there are two red 10s and 52 cards, $P(\text{red } 10) = \frac{2}{52}$. There are four aces, so $P(\text{ace}) = \frac{4}{52}$. Since the events are mutually exclusive,

$$P(\text{red } 10 \text{ or ace}) = \frac{2}{52} + \frac{4}{52} = \frac{3}{26}$$

□

Many interesting events *are not* mutually exclusive. As in Example 5, suppose one card is drawn from a deck of cards. What is the probability that it will be a spade or an ace? These events (drawing a spade, drawing an ace) *are not* mutually exclusive events, as there is a card that is both a spade and an ace. The number of favorable outcomes of a spade is 13, and of an ace is 4. However, if we add them together, we will have counted the ace of spades *twice.* That is, the number of favorable outcomes is 13 + 4 − 1.

$$P(\text{spade } or \text{ ace}) = \frac{13 + 4 - 1}{52} = \frac{13}{52} + \frac{4}{52} - \frac{1}{52}$$

$$P(\text{space } or \text{ ace}) = P(\text{spade}) + P(\text{ace}) - P(\text{spade and ace})$$

This suggests the following rule.

Addition Property of Probability

Suppose A and B are events from a sample space S, then

$$P(A \text{ or } B) = P(A) + P(B) - P(A \text{ and } B)$$

If two events occur in such a manner that one event has no effect on the other, they are said to be **independent.** The probability of independent events A and B *both* occurring is the product of their probabilities. $P(A \text{ and } B) = P(A) \cdot P(B)$

Probability of Independent Events

Suppose A and B are *independent* events from a sample space S, then

$$P(A \text{ and } B) = P(A) \cdot P(B)$$

EXAMPLE 6 A tool and die machine performs two independent operations on bar stock to produce a certain bolt. Out of every 100 bolts processed, the machine, in random fashion, makes a faulty thread on one bolt in operation 1. Out of every 50 bolts processed, again in random fashion, it misaligns the head of one bolt in operation 2. If this machine makes a bolt, what is the probability that it will be faulty?

Solution The probability that it will produce a faulty thread on a bolt is $P(T) = \frac{1}{100}$, while the probability that it will misalign the head is $P(H) = \frac{1}{50}$. Therefore, by the addition property of probability,

$$P(T \text{ or } H) = P(T) + P(H) - P(T \text{ and } H)$$

However, it may make a faulty thread and misalign the head *on the same bolt.* As the two operations are independent, the probability of this event (T and H) is the product of $P(T)$ and $P(H)$.

$$= P(T) + P(H) - P(T)P(H)$$

$$= \frac{1}{100} + \frac{1}{50} - \frac{1}{100} \cdot \frac{1}{50}$$

$$= \frac{149}{5000} = 0.0298$$

The probability that a faulty bolt is produced is 0.0298. □

PROBLEM SET 9.7

Warm-ups

For Problems 1 and 2, see Examples 1 and 2.

1. A bag contains four slips of paper of equal size numbered 1, 2, 3, and 4. A slip of paper is drawn from the bag sight unseen. What is the probability that its number will be less than 4?

2. An experiment consists of throwing a fair 12-sided die and noting the number of dots on the upper face.
(a) What is the sample space for this experiment?
(b) What is the probability that the die will show a 2 or a 4 or be odd?

In Problems 3 through 6, suppose an ordinary pair of fair dice are thrown. See Example 3.

3. What is the probability that an 8 will be thrown?

4. What is the probability that a "hard" 8 will be thrown (two 4s)?

5. What is the probability that a 6 or an 8 will be thrown?

6. What is the probability that a hard 6 or a hard 8 will be thrown? (A hard 6 is two 3s.)

In Problems 7 through 10, suppose one card is drawn from an ordinary deck of playing cards. See Examples 4 and 5.

7. What is the probability that it will be a red ace or a 7?

8. What is the probability that it will be a 10 or a face card? (Face cards are jacks, queens, and kings.)

9. What is the probability that it will be a club, a spade, or a red card?

10. What is the probability that it will be a 4 *and* a face card?

In Problems 11 and 12, see Example 6.

11. An experiment consists of rolling an ordinary die and flipping a coin. What is the probability of obtaining a 6 on the die or a head on the coin?

12. In the experiment in Problem 11, what is the probability of obtaining a number on the die less than 5 or a tail on the coin?

Practice Exercises

13. A bag contains 5 marbles, 3 black and 2 white. The black marbles are marked 1, 2, and 3, and the white marbles are marked 1 and 2. One marble is drawn from the bag, sight unseen.
(a) What is the sample space for this experiment?
(b) What is the event, *the marble drawn is black?*
(c) What is the probability that the marble drawn will be marked 2?

14. A hat contains seven slips of paper marked 1 through 7. One slip is drawn from the hat, sight unseen.
(a) What is the sample space for this experiment?
(b) What is the event, *the slip drawn is marked with an odd number?*
(c) What is the probability that the slip drawn will be less than 3?

15. A fair die is rolled. What is the probability of each event?
(a) A 5 shows on the die.
(b) A number greater than 2 shows on the die.
(c) An 8 shows on the die.
(d) A number less than 7 shows on the die.

16. A card is drawn from an ordinary deck. What is the probability of each event?
(a) The card is an 8.
(b) The value of the card is less than 10. (2 is the lowest card.)
(c) The card is a red spade.
(d) The card is a spade, diamond, heart, or club.

In Problems 17 through 28, consider the following experiment. Three dice, colored red, white, and blue, are rolled. Let A, B, C, D and E be the following events.

A. *Two dice show 3s.*
B. *All dice show even numbers.*
C. *All dice show odd numbers.*
D. *No die shows less than 5.*
E. *The blue die shows a 3.*

Compute each probability.

17. $P(A)$ **18.** $P(B)$ **19.** $P(C)$ **20.** $P(D)$
21. $P(A \text{ or } B)$ **22.** $P(B \text{ or } E)$ **23.** $P(B \text{ or } C)$ **24.** $P(A \text{ or } D)$
25. $P(A \text{ or } E)$ **26.** $P(C \text{ and } E)$ **27.** $P(B \text{ and } C)$ **28.** $P(D \text{ and } E)$

29. A certain experiment has several outcomes; among them, O_1 and O_2. Suppose $P(O_1) = \frac{1}{3}$, $P(O_2) = \frac{2}{5}$.
(a) What is $P(O_1 \text{ or } O_2)$ if O_1 and O_2 are mutually exclusive?
(b) What is $P(O_1 \text{ or } O_2)$ if $P(O_1 \text{ and } O_2) = \frac{1}{10}$
(c) What is $P(O_1 \text{ or } O_2)$ if O_1 and O_2 are independent events.

30. An economist whose hobby is studying weather patterns predicts that the probability of a recession this year is $\frac{3}{5}$ and that the probability of increased rainfall is $\frac{7}{15}$. If she is correct, what is the probability of a recession *or* increased rainfall this year?

Challenge Problems

31. Karen Smith assembles computers from parts furnished by other vendors. Suppose 0.5% of the memory modules contain a faulty chip and 0.03% of the power supplies are incorrectly wired. Assume each computer shipped by Karen contains one memory module and one power supply and all other components operate correctly. What is the probability there will be at least one faulty computer in a shipment of 100 of Karen's computers?

IN YOUR OWN WORDS . . .

32. What are mutually exclusive events?
33. What are independent events?
34. What is meant by the probability of an event?

CHAPTER SUMMARY

GLOSSARY

Arithmetic sequence: A sequence obtained by adding a fixed number to each previous term.

Combination: A set of objects, regardless of arrangement.

Event: A subset of a sample space.

Experiment: An occurrence with an observable outcome.

Explicit formula for a sequence: Formula for a_n given in terms of a_1 and other constants.

Geometric sequence: A sequence obtained by multiplying a fixed number by each previous term.

Permutation: An arrangement of a set of objects.

Probability: The study of the likelihood of events.

Recursive formula for a sequence: Formula for a_n given in terms of a_{n-1} and other constants.

Sample space: The set of all outcomes of an experiment.

Sequence: A function whose domain is the set of natural numbers.

Summation notation: A compact way of writing sums.

SUMMATION NOTATION

For integers m and n, with $m \leq n$,

$$\sum_{j=m}^{n} A_j = A_m + A_{m+1} + A_{m+2} + \cdots + A_n$$

j is called the **index of summation,** m is the **lower limit,** and n is called the **upper limit.**

PROPERTIES OF SUMMATION NOTATION

For m, n, and p integers, with $m < p < n$ and k a real number,

1. $\sum_{j=m}^{n} (a_j + b_j) = \sum_{j=m}^{n} a_j + \sum_{j=m}^{n} b_j$
2. $\sum_{j=m}^{n} (a_j - b_j) = \sum_{j=m}^{n} a_j - \sum_{j=m}^{n} b_j$
3. $\sum_{j=m}^{n} ka_j = k \sum_{j=m}^{n} a_j$
4. $\sum_{j=m}^{n} a_j = a_m + \sum_{j=m+1}^{n} a_j$
5. $\sum_{j=m}^{n} a_j = \sum_{j=m}^{n-1} a_j + a_n$
6. $\sum_{j=m}^{n} a_j = \sum_{j=m}^{p-1} a_j + \sum_{j=p}^{n} a_j$
7. $\sum_{j=m}^{n} a_j = \sum_{q=m}^{n} a_q$

THE PRINCIPLE OF MATHEMATICAL INDUCTION

If S is a subset of the natural numbers with the following two properties,

1. 1 belongs to S.

2. If k belongs to S, then $k + 1$ belongs to S

then S is the set of natural numbers.

HOW TO WRITE A PROOF BY MATHEMATICAL INDUCTION

1. Write the conjecture as a statement that depends on the natural number n.

2. Show that the statement is true when n is 1.

3. Assume the statement is true for some unspecified natural number k.

4. Show that it follows from the assumption in step 3 that the statement is true when n is $k + 1$.

EXTENDED PRINCIPLE OF MATHEMATICAL INDUCTION

Let j be an integer and P_n be a statement that depends on integers *greater than or equal to* j. If it can be shown that

1. P_j is true

2. *If* P_k is true for any integer $k \geq j$, then it follows that P_{k+1} is true

then P_n is true for all integers greater than or equal to j.

FACTORIALS

Definitions: For n a nonnegative integer,

1. $n! = 1 \cdot 2 \cdot 3 \cdots (n - 1) \cdot n$

2. $0! = 1$

Property: If n is a natural number,

$$n! = n \cdot (n - 1)!$$

THE BINOMIAL COEFFICIENTS

For n and r nonnegative integers with $n \geq r$,
Definition:

$$\binom{n}{r} = \frac{n!}{r!(n - r)!}$$

Properties:

1. $\binom{n}{0} = \binom{n}{n} = 1$

2. $\binom{n}{r} = \binom{n}{n - r}$

THE BINOMIAL THEOREM

For n a natural number, $(a + b)^n = \sum_{j=0}^{n} \binom{n}{j} a^{n-j}b^j$

FINITE SUMS

Arithmetic sum: $a_1 + a_2 + a_3 + \cdots + a_n = \dfrac{n(a_1 + a_n)}{2}$

Geometric sum: $\sum_{j=1}^{n} a \cdot r^{j-1} = \dfrac{a(r^n - 1)}{r - 1}$

SUM OF A GEOMETRIC SERIES

If $|r| < 1$,

$$\sum_{j=1}^{+\infty} a \cdot r^{j-1} = \frac{a}{1 - r}$$

FUNDAMENTAL PRINCIPLE OF COUNTING

If event 1 can be done in m ways and event 2 can be done in n ways, then event 1 followed by event 2 can be done a total of $m \cdot n$ different ways.

NUMBER OF PERMUTATIONS OF *n* THINGS TAKEN *r* AT A TIME

For integers n and r, where $0 \le r \le n$, $P(n, r) = \dfrac{n!}{(n - r)!}$

NUMBER OF COMBINATIONS OF *n* THINGS TAKEN *r* AT A TIME

For integers n and r, where $0 \le r \le n$, $C(n, r) = \dfrac{n!}{r!(n - r)!}$

Some properties of combinations:

1. $C(n, 0) = C(n, n) = 1$

2. $C(n, r) = C(n, n - r)$

3. $C(n, r) = \binom{n}{r}$

PROBABILITY

Definition:
If the outcomes in the sample space of an experiment are equally likely, the probability of an event E is given by

$$P(E) = \frac{n(E)}{n(S)}$$

where $n(E)$ is the number of elements in the event and $n(S)$ is the number of elements in the sample space.

Three properties of probability:

1. $0 \leq P(E) \leq 1$
2. $P(E) + P(\text{not } E) = 1$
3. $P(A \text{ or } B) = P(A) + P(B) - P(A \text{ and } B)$

Probability of mutually exclusive events:
If A and B are mutually exclusive events from the same sample space, then

$$P(A \text{ or } B) = P(A) + P(B)$$

Probability of independent events:
If A and B are independent events, then

$$P(A \text{ and } B) = P(A) \cdot P(B)$$

REVIEW PROBLEMS

In Problems 1 through 4, expand each sum.

1. $\sum_{j=1}^{5} 2^{j-1}x^j$ **2.** $\sum_{i=-2}^{2} \frac{i+4}{i+3}y^{-i}$ **3.** $\sum_{j=1}^{5} (-1)^{j-1}j^2$ **4.** $\sum_{n=0}^{4} (-1)^n \frac{x^{n+1}}{n!}$

In Problems 5 and 6, write each sum using summation notation.

5. $\frac{1}{x} + 4 + 9x + 16x^2 + 25x^3 + 36x^4$ **6.** $\frac{2}{5} - \frac{4}{7}x^2 + \frac{2}{3}x^4 - \frac{8}{11}x^6 + \frac{10}{13}x^8$

7. Write $3\sum_{j=1}^{k} \frac{x^j}{2} + 2\sum_{j=1}^{k} \frac{x^j}{3}$ as a single sum in summation notation.

8. Write $\sum_{j=1}^{500} A_j + \sum_{n=0}^{499} A_n$ as a sum using only one summation notation symbol.

In Problems 9 and 10, use mathematical induction to prove each statement for any natural number n.

9. $\frac{1}{1 \cdot 2} + \frac{1}{2 \cdot 3} + \frac{1}{3 \cdot 4} + \cdots + \frac{1}{n(n+1)} = \frac{n}{n+1}$ **10.** If $K \geq 1$, then $K^n \geq 1$

In Problems 11 through 14, use the Binomial Theorem to expand the given binomial.

11. $(x + k)^7$ **12.** $(2x + 3)^5$ **13.** $(x - 2y)^6$ **14.** $\left(\frac{2}{x} - \frac{y}{2}\right)^8$

In Problems 15 and 16, write the first five terms of the given sequence.

15. $a_n = 2^{n-1}x^{n+1}$ **16.** $b_n = (-1)^{n-1}(2x)^{2n-1}$

Problems 17 through 20 are either arithmetic or geometric sequences. Identify each and find the common difference or common ratio.

17. 2, 5, 8, . . .
18. 1, 3, 9, . . .
19. 36, −24, 16, . . .
20. 20, 5, −10, . . .

In Problems 21 through 24, find the sum of each series, if it exists.

21. $\sum_{n=1}^{+\infty} 5\left(\frac{7}{9}\right)^{n-1}$
22. $\sum_{n=1}^{+\infty} \left(-\frac{1}{10}\right)^{n-1}$
23. $\sum_{k=1}^{+\infty} \frac{11}{15}\left(\frac{9}{7}\right)^{k-1}$
24. $\sum_{n=1}^{+\infty} (-1)^n \frac{2^{n+1}}{3^n}$

In Problems 25 through 36, evaluate each expression.

25. $P(7, 2)$
26. $P(8, 4)$
27. $P(4, 4)$
28. $P(11, 0)$
29. $C(7, 2)$
30. $C(8, 4)$
31. $C(4, 4)$
32. $C(11, 0)$
33. $P(8, 5)$
34. $C(16, 13)$
35. $P(24, 3)$
36. $C(52, 3)$

In Problems 37 through 40, an ordinary pair of fair dice are thrown. Find each probability.

37. What is the probability that a total of 4 will be thrown?
38. What is the probability that the total will be more than 10?
39. What is the probability that a total of 4 or an odd total will be thrown?
40. What is the probability that a total of 14 will be thrown?
41. An experiment consists of drawing a card from an ordinary deck of cards and flipping a quarter. What is the probability of an ace being drawn and a tail on the coin?
42. The probability that a flu shot will prevent the flu is 0.68. If the shot is given to 750 people, how many people can we expect to have the flu?

In Problems 43 through 46, use mathematical induction to show that each statement is true for all natural numbers.

43. $1^3 + 2^3 + 3^3 + \cdots + n^3 = \frac{n^2(n+1)^2}{4}$
44. $\frac{1}{2} + \frac{1}{4} + \frac{1}{8} + \cdots + \frac{1}{2^n} < 1$
45. $1 + 5 + 9 + \cdots + (4n - 3) = n(2n - 1)$
46. $4 + 8 + 12 + \cdots + 4n = 2n(n + 1)$
47. Sam earns \$1 on the first day in June, \$2 on the second day, \$4 on the third day, and the pay doubles each day of the month. How much does Sam earn in June?
48. Enrollment in statistics is growing at a rate of 1 percent each year. If the enrollment this year is 28, what will the enrollment be in 10 years?

LET'S NOT FORGET . . .

49. Which form is more useful? Pascal's Triangle or the binomial coefficients when expanding a binomial to a power.
50. Watch the role of the negative signs. Write the first five terms of the sequence $(-1)^k x^{-k}$.
51. From memory. Expand $(2x + t)^7$.
52. With a calculator. Evaluate:
(a) 8!
(b) $C(6, 2)$
(c) $P(6, 2)$

ANSWERS

PROBLEM SET 1.1

1. $S \cap T = \{0\}$; $S \cup T = \{-4, -2, -1, 0, 1, 2, 3\}$ **3.** $X \cap Y = \varnothing$; $X \cup Y = Z$, the set of integers **5.** $M \cap \varnothing = \varnothing$; $M \cup \varnothing = M$ **7.** 10 **9.** 62 **11.** Estimate: -1.5 Approximation: -1.538 **13.** a **15.** $-b$ **17.** $3 - \sqrt{5}$ **19.** $\sqrt{11} - \sqrt{7}$ **21.** $\frac{8}{|x|}$ **23.** $\frac{\pi}{7}$ **27.** true **29.** true **31.** $-8\frac{7}{12}$ **33.** -16 **35.** -8 **37.** 0 **39.** $10\frac{1}{2}$ **41.** 26 **43.** $3a - 3b$ **45.** x **47.** 0 **49.** $2 + (3 + 4)$ **51.** $(x + y)5$ **53.** $-y$ **55.** x^2 **57.** $2z^2$ **59.** $\frac{y^2}{2}$ **61.** $x - y$ **63.** x **65.** $\sqrt{5} - 1$ **67.** $1 - y$ **69.** Assuming that only packages listed can be bought, he should buy packages labeled 2.23, 2.14, 1.37. **71.** It will cost Ray \$4,671.20 to fence his field. **73.** Beth, June, Ken, Harold, Jim **75.** $C(2s + 2t - 3n) + n(244)$ **77.** $y, |1 + y|; -y, |1 - y|$ **79.** $|p + q| \le |p| + |q|$. **81.** Consider all possible cases of p and q being positive or negative.

PROBLEM SET 1.2

1. base is 3; -81 **3.** base is $2x$; $8x^3$ **5.** base is 5; -1 **7.** $-3d^2$ is the base for 0, d is the base for 2; q is the base for 0; $-\frac{1}{2}$ **9.** 5 is the base; $\frac{1}{25}$ **11.** -5 is the base; $\frac{1}{25}$ **13.** x^{10} **15.** $\frac{t^3}{27}$ **17.** $2^6x^9y^{12}$ **19.** $k^6 - 4k^3s^2 + 4s^4$ **21.** $\frac{1}{8x^6y^3}$ **23.** $-\frac{d^3z^9}{27h^6k^3x^3y^6}$ **25.** $\frac{1}{x^4 - 6x^2 + 9}$ **27.** 2.23179 **29.** 0.00249183 **31.** 1.55×10^5 **33.** -2.54×10^{-6} **35.** 56,473,000,000 **37.** -0.000000039022 **39.** 0.0200 **41.** 1.20×10^{25} **43.** x^4 **45.** $-\frac{1}{y^5}$ **47.** $-\frac{9}{2}$ **49.** -16 **51.** -2 **53.** $-\frac{x^4}{y^8}$ **55.** $\left(\frac{1}{s^2} + t\right)^4$ **57.** $-\frac{1}{z^2}$ **59.** $\frac{4q - 4}{27p}$ **61.** $\frac{z^6}{4k^4p^4}$ **63.** $\frac{12}{7}$ **65.** $2x - 1$ **69.** The U.S. consumes 26.4% of the energy consumed by the human race. **71.** The distance is approximately 1.27×10^{-10} inches. **73.** The distance traveled by light in a year is $5.87452608 \times 10^{12}$ or 5,874,526,080,000 miles. **75.** Claude should expect to lose about \$.89. **77.** For $x > 2$, $(5^x)^2 < 5^{(x^2)}$ and for $x < 0$, $(5^x)^2 < 5^{(x^2)}$. **79.** 5.60187682 billion

PROBLEM SET 1.3

1. -6 **3.** -4 **5.** 56.8745 **7.** $-10a^2b^3$ **9.** $6x^2 - 21xy - 2xy^2 + 7y^3$ **11.** $2v^5 - 7v^3w - 15vw^2 - v^2w^3 + 5w^4$ **13.** $49x^2 - 42xy + 9y^2$ **15.** $27 + 54g + 36g^2 + 8g^3$ **17.** $3s - 6 + \frac{3}{s}$ **19.** $x^2 + x - 1 + \frac{-1}{2x - 3}$ **21.** $2x + 2$ **23.** $9t^4 - 12t^3 + 11t^2 + 2t$ **25.** $x^2 - 5x + 13$ **27.** $4x^2 + 12x + 9$ **29.** $6x - 18$ **31.** $3x^2 - x - 5$ **33.** $2y^2 + 9y - 56$ **35.** $36z^2 - 169w^2$ **37.** $27x^3 - 18ax^2 + 12a^2x - 8a^3$ **39.** $y^2 - 2y + 4 + \frac{-10}{y + 7}$ **41.** $8x^3 + 12x^2y + 6xy^2 + y^3$

43. $15y^2 - 6x^3 - 45xy + 22x^4 + 167x^2y$ **45.** $\frac{q^5}{p} - \frac{q^4}{p} - p^2 + \frac{p^2}{q}$ **47.** $100t^4 - 340t^2 + 289$ **49.** $256 - 625x^8$ **51.** $64x^6 + 125$ **53.** $x - 3y + \frac{4y^2}{x+y}$ **55.** $-2y^2 + 2xy$ **57.** $2s + 2 + \frac{7s-1}{3s^2 - s - 2}$ **59.** $x^3 + x^2 + 2x + 2 + \frac{5}{x^2 - 2}$ **61.** $x^2 - x + 3 + \frac{-1}{2x+1}$ **63.** -601.733 **65.** 0.380 **67.** $8x^2 - 12x + 4$; If x is 5 units, the perimeter is 144 units. **69.** Jim should not close Tuesday through Thursday. He should close either Sunday through Tuesday or Monday through Wednesday. **73.** $x^2 + y^2 = (p^2 - q^2)^2 + (2pq)^2 = p^4 + 2p^2q^2 + q^4 = (p^2 + q^2)^2$; Pythagorean triples: 3, 4, 5; 5, 12, 13; 7, 24, 25; 40, 42, 58 **75.** $x^{2n} + 1 = (-1)^{2n} + 1 = 1 + 1 = 2$ **77.** $x^{2n-1} + 1 = (-1)^{2n-1} + 1 = -1 + 1 = 0$ **79.** $z^{2n} - 10z^n + 25$ **81.** $\frac{1}{x^{2n+2}} - x^{2n+2}$ **83.** 20

PROBLEM SET 1.4

1. $3abc(a - 2bc + 3)$ **3.** $(z + t)(y - x)$ **5.** $(5 - 2x)(5 + 2x)$ **7.** $(2y - 1)(4y^2 + 2y + 1)$ **9.** prime **11.** $(x + 2)(x + 1)$ **13.** $(z - 5)(z + 1)$ **15.** $(5x + 3)(2x + 3)$ **17.** $(3x + 1)(2x - 1)$ **19.** $(x^2 + 7)(x^2 - 6)$ **21.** $(x + 4)(x^2 + 1)$ **23.** $(x - 1)(x^3 + 4)$ **25.** $(x^2 - 2)(x^2 + 2)(x^4 + 4)$ **27.** $-6(t - 1)(t + 1)$ **29.** $18x^2y^3z^2(5y - 3xz)$ **31.** $(3 - t)(9 + 3t + t^2)$ **33.** $(x - 1)(r - s)(r + s)$ **35.** $(y - 1)(y + 1)(3y - 2)$ **37.** $b(7 - 4a)(7 + 4a)$ **39.** $(x - 1)(x + 1)(x^2 + 1)(x^4 + 1)$ **41.** $x(x^2 + x + 2)$ **43.** $3(2a - 1)(2a + 1)(2x - 1)$ **45.** prime **47.** $(5w - 1)(25w^2 + 5w + 1)$ **49.** prime **51.** factored completely **53.** $9x^2(x^2 + 1)$ **55.** $y(xy - 1)(xy + 1)$ **57.** $(2t - 1)(a + b)$ **59.** $x(x - 1)(x + 1)(x + y)$ **61.** $a(a - 5)(a + 1)$ **63.** $(x^3 - 2)(x + 1)(x^2 - x + 1)$ **65.** $(3x^2 + 1)(2x^2 - 1)$ **67.** $(2 - b)(a - r)(a + r)$ **69.** $(a + 3b)(a^2 - 3ab + 9b^2)$ **71.** $8c^2d^2(4c - 9d)$ **73.** $(2x - a - b)(2x + a + b)$ **75.** n and $n + 1$ are consecutive integers. One must be even. $n = 2k$, where k is an integer. $\frac{n(n+1)}{2} = \frac{2k(n+1)}{2} = k(n + 1)$ **77.** $(x^n - 2)(x^n + 1)$ **79.** $(x^n - 1)(x^{2n} + x^n + 1)$ **81.** Difference of squares: $(x - 1)(x + 1)(x^2 + x + 1)(x^2 - x + 1)$; Difference of cubes: $(x - 1)(x + 1)(x^4 + x^2 + 1)$

PROBLEM SET 1.5

1. 0, 1 **3.** 4 **5.** $\frac{2-x}{x+1}$ **7.** $\frac{s^2 + 3r^2 - 5rs}{r^3s^4}$ **9.** $\frac{4x}{(x-2)(x+2)}$ **11.** $\frac{w^2 + 4w - 4}{(w-2)^2(w+2)}$ **13.** x; $x \neq 0, 3$ **15.** $-\frac{1}{s+t}$; $s \neq t$ **17.** $\frac{-4(x-2)}{(x-4)}$ **19.** $\frac{x(x-7)}{2(x+1)(x-1)}$; $x \neq -1, -3$ **21.** $\frac{2}{15}$; $s \neq -t$ **23.** $\frac{1}{p-q}$; $p \neq -\frac{1}{2}q$ **25.** $\frac{5x+5}{(x+5)(x-5)}$ **27.** $\frac{x(K^2+2)}{4(x+2)}$; $x \neq 2$, $k \neq 0$ **29.** $\frac{y^2}{x(x+1)}$; $x \neq 0, 1$ **31.** $\frac{-p}{q}$; $p \neq q$ **33.** $\frac{(x-y)^2}{(s-t)(s+t)}$; $x \neq -y$ and $s \neq t$ **35.** $\frac{x^2 + 2x + 3}{(x-1)(x+1)^2}$ **37.** $\frac{9 - 6y + 4y^2}{x^3}$; $y \neq -\frac{3}{2}$ **39.** $\frac{(x+2)^2}{2(x-1)(x-2)}$; $x \neq -1, 3$ **41.** $\frac{y-5}{y+2}$ **43.** $\frac{7}{(z+1)}$; $z \neq 3$ **45.** $\frac{3-q}{(q-1)(q+1)}$ **47.** $\frac{1}{k-5}$ **49.** $\frac{2(2x+3)}{(x+1)(x+2)(x+3)}$ **51.** xy; $x \neq -5, 0$ **53.** $\frac{y+2x}{3y-4x}$ **55.** t; $t \neq -\frac{5}{3}$ **57.** $\frac{-1}{x(x+h)}$; $h \neq 0$ **59.** $\frac{-4}{(x+h-1)(x-1)}$; $h \neq 0$ **61.** $\frac{(x^n-3)(x^p+1)}{(x^p+3)(x^n+2)}$; $x^n \neq -2$ and $x^p \neq 1$ **63.** $\frac{1}{x^2}$ gets larger and larger as x gets closer and closer to zero. $\frac{1}{x^2}$ does not have a largest value. $\frac{1}{x^2}$ gets closer and closer to zero when x is a very large positive number. $\frac{1}{x^2}$ gets closer and closer to zero when x is a very small negative number. $\frac{1}{x^2}$ does not have a smallest value. **65.** As x gets closer and closer to 2 (from values larger than 2 and from values smaller than 2) the value of $\frac{x^2-4}{x-2}$ gets closer and closer to 4, the value of $x+2$ when $x=2$.

67. $\frac{7}{4}$ **69.** $\frac{x^4+2x^3+x^2+x}{x^3+x^2+x}$ **71.** Add (and subtract) 1 to both sides of $\frac{A}{B}=\frac{C}{D}$.

PROBLEM SET 1.6

1. 7 **3.** $\sqrt{-25}$ is not a real number. **5.** $\frac{2}{3}$ **7.** $\frac{|t|}{2}$ **9.** $-2s$ **11.** $4\sqrt{3}$ **13.** $4x^2\sqrt[3]{2}$ **15.** $|x-y|$ **17.** $\left|\frac{x}{y}\right|$ **19.** $7\sqrt[3]{2}$ **21.** $4\sqrt{5y}$ **23.** 75 **25.** $11-4\sqrt{6}$ **27.** 4 **29.** $\frac{5}{\sqrt{15}}$ **31.** $\frac{\sqrt{10}}{6}$ **33.** $\frac{7\sqrt[4]{4}}{2}$ **35.** $\frac{-47}{16+11\sqrt{6}}$ **37.** 2 **39.** -4 **41.** -8 **43.** $\frac{1}{4}$ **45.** $\frac{1}{4}$ **47.** $\frac{1}{x^{1/6}}+1$ **49.** $\frac{xy^{3/2}}{z^6}$ **51.** $(x^{1/5}-5)(x^{1/5}+2)$ **53.** 58.8 **55.** 2.04 **57.** $\sqrt{-100}$ is not a real number. **59.** $2t^2|s|\sqrt[4]{2s^3}$ **61.** $-4|x|\sqrt{3}$ **63.** $\frac{x\sqrt[3]{x}}{2y^2}$ **65.** cannot be simplified **67.** $|x-y|$ **69.** $2|x|$ **71.** 5 **73.** $-\frac{1}{5}$ **75.** not a real number **77.** $\frac{1}{25}$ **79.** $-\frac{1}{25}$ **81.** $12\sqrt{2}$ **83.** $-5\sqrt{5x}$ **85.** $8x^2\sqrt[3]{xy}$ **87.** $\sqrt{6xy}$ **89.** $5x^2$ **91.** $3x^2$ **93.** $6-4\sqrt{3}$ **95.** $78+9\sqrt{2}$ **97.** 1 **99.** $17+4\sqrt{15}$ **101.** 5 **103.** $x^{5/6}$ **105.** $\frac{1}{x^{2/3}}-1$ **107.** $x+2+\frac{1}{x}$ **109.** $\frac{z^2}{x^{12}y^3}$ **111.** $\frac{1}{x^4y}$ **113.** $\frac{\sqrt{2}}{4}$ **115.** $\frac{5\sqrt[3]{9}}{3}$ **117.** $\frac{1}{\sqrt{7}+1}$ **119.** $-3\sqrt{2}+3\sqrt{3}$ **121.** $\frac{1}{\sqrt{2+h}+\sqrt{2}}$; $h\neq 0$ **123.** $\frac{1}{2+\sqrt{x_1}}$ where $x_1\neq 4$ **125.** $2^{-1/3}$ is larger. **127.** $\sqrt{\frac{1}{7}}$ is larger. **129.** $\left(\frac{22}{7}\right)^{1/2}$ is larger. **131.** $\frac{5}{\sqrt{7}-\sqrt{2}}\approx 4.059964873$; $\sqrt{7}+\sqrt{2}\approx 4.059964873$. **133.** There are approximately 37,188 people who have an income of over \$25,000. **135.** $|s|^2=(\sqrt{s^2})^2=s^2$ **137.** $\sqrt[n]{pq}=(pq)^{1/n}=p^{1/n}q^{1/n}=\sqrt[n]{p}\cdot\sqrt[n]{q}$ **139.** $\frac{2-\sqrt{x_1}}{4-x_1}=\frac{2-\sqrt{x_1}}{(2-\sqrt{x_1})(2+\sqrt{x_1})}=\frac{1}{2+\sqrt{x_1}}$; $\frac{\sqrt{7}-\sqrt{t_1}}{7-t_1}=\frac{\sqrt{7}-\sqrt{t_1}}{(\sqrt{7}-\sqrt{t_1})(\sqrt{7}+\sqrt{t_1})}=\frac{1}{(\sqrt{7}+\sqrt{t_1})}$

PROBLEM SET 1.7

1. $4+2i$; real part: 4; imaginary part: 2 **3.** $1+0i$; real part: 1; imaginary part: 0 **5.** $-2\sqrt{3}-2\sqrt{3}i$; real part: $-2\sqrt{3}$; imaginary part: $-2\sqrt{3}$ **7.** $3+0i$; real part: 3; imaginary part: 0 **9.** -1 **11.** $-i$ **13.** $-1+8i$ **15.** $-27-11i$ **17.** $-4+0i$ **19.** $-48-14i$ **21.** $3+11i$ **23.** $-36-30i$ **25.** $\frac{21}{29}-\frac{20}{29}i$ **27.** $-i$ **29.** $\frac{1}{2}-\frac{1}{2}i$ **31.** $15-3i$ **33.** $1+7i$ **35.** $16-12i$ **37.** $-3\sqrt{5}+3i$ **39.** $\frac{23}{13}-\frac{54}{13}i$ **41.** $3+4i$ **43.** $-3-4i$ **45.** $21+20i$ **47.** $-2+6i$ **49.** $-1-2\sqrt{2}i$ **51.** $-2-2\sqrt{2}i$ **53.** $-4+3i$ **55.** $-\frac{3}{25}+\frac{4}{25}i$ **57.** $\sqrt{2}i$ **59.** $-\frac{11}{13}+\frac{4\sqrt{3}}{13}i$ **61.** $12-5i$ **63.** 13 **65.** $10+24i$ **67.** $10+24i$

69. $-24 + 10i$ **71.** $\frac{1}{2} + \frac{5}{2}i$ **73.** $-9 + 46i$ **75.** Let $Z = a + bi$. $Z\bar{Z} = (a + bi)(a - bi) = a^2 - b^2i^2 = a^2 + b^2$
77. Let $Z = a + bi$. $Y = c + di$. $\bar{Z}\bar{Y}(a - bi)(c - di) = (ac - bd) - (ad + bc)i = \overline{ZY}$ **79.** $(x + \sqrt{7}i)(x - \sqrt{7}i)$
81. -1 **83.** i or $-i$ **85.** 1

CHAPTER 1 REVIEW PROBLEMS

1. $\frac{-2y^3}{x}$ **3.** -1 **5.** $3x$ **7.** -4 **9.** -27 **11.** $9x^8$ **13.** $-\frac{1}{y}$ **15.** $4t^2 - 20t + 25$
17. $x^3 + 2x^2 - 3x - 6$ **19.** $-\sqrt{6}$ **21.** $\frac{1}{2} + \frac{3}{2}i$ **23.** 2 **25.** $24\sqrt{6}$ **27.** $3x^{1/2} - 1$ **29.** $y^{1/2} + 2y^{1/4} + 1$

31. $8r^3 + 12r^2 + 6r + 1$ **33.** 3 **35.** -1 **37.** $\frac{21}{4}$ **39.** $2x + 7 + \frac{12x + 4}{2x^2 - 3x + 1}$ **41.** $r^2 - \frac{3}{4} + \frac{1}{4r}$
43. $-\frac{2}{9}$ **45.** 9 **47.** $(x - 1)^2(x + 1)^2$ **49.** $(x - 4)(x + 1)$ **51.** $2x(x - 1)(x^2 + 4)$ **53.** $\frac{xy^2 - 2x^2}{y^2 + x^2y}$
55. $x + 1, x \neq 6$ **57.** $\frac{1}{\sqrt{2}}$ **59.** $\frac{-1}{6 - 6\sqrt{2}}$ **61.** $\frac{\sqrt{14} + 3\sqrt{7}}{-7}$ **63.** $-4 + \sqrt{15}$ **65.** 2.665448 **67.** 0.448901

69. (a) $-\frac{1}{16}$ (b) $-\frac{1}{4}$ **71.** (a) 46,568,203,100,000 (b) -0.1481 (c) 6269.2893

PROBLEM SET 2.1

1. $\{-2\}$ **3.** R **5.** $\{5.73\}$ **7.** His initial average speed was 60 mph. **9.** It is 137.5 miles from Janet's home to Bad Axe. **11.** Dr. Hope should use $\frac{2}{5}$ ounce of the 40% alcohol syrup and $\frac{3}{5}$ ounce of the 15% alcohol syrup. **13.** Miguel should invest \$40,000 at $8\frac{1}{2}$% and \$80,000 at $10\frac{3}{4}$%. **15.** Early invested \$76,000 in the deal with 18% profit and \$124,000 in the deal with 2% loss. **17.** It would take Joan 36 minutes to wash the dishes working alone. **19.** She must sell 100 shirts to break even. She will make a profit after she has sold 100 shirts. To make a \$1000 profit, Virginia must sell 300 shirts. **21.** $h = \frac{3V}{\pi r^2}$ **23.** $C = \frac{5F - 160}{9}$ **25.** $\{1\}$ **27.** $\{0\}$ **29.** $\{-23\}$ **31.** $\left\{\frac{-5}{18}\right\}$ **33.** $\left\{\frac{2}{3}\right\}$
35. R **37.** $\left\{\frac{84}{65}\right\}$ **39.** $m_1 = \frac{R^2F}{Gm_2}$ **41.** $t = \frac{v - v_0}{-32}$ **43.** $R_2 = \frac{RR_1}{R_1 - R}$ **45.** $r = \frac{S - a}{S}$ **47.** There were 2321 videos rented last month when the profit was \$2367.15. To break even, the Video Store must rent 1220 videos each month. **49.** The CD for the Ford Taurus changes from 0.33 to 0.297 when the drag force is decreased by 10%. **51.** The simple interest rate is $16\frac{2}{3}$%. **53.** To the nearest cent, \$987.65 was deposited originally. **55.** To the nearest cent, Dr. Krebs' payments were \$1,723.06. **57.** Karen should drain and add $2\frac{1}{4}$ gallons. **59.** **OPTION 1:** Invest \$20,000 in the stock that yields 7.25% and \$100,000 in the stock that yields 8.55%. **OPTION 2:** Invest \$16,250 in the stock that yields 6.95% and \$103,750 in the stock that yields 8.55%. **61.** Val's average speed is 6 mph and Allison's average speed is 8 mph. The distance to the Rib Shack is 2.4 miles. **63.** He should eat 9 of his wife's cookies so that his total

amount of saturated fatty acid is approximately 30%. **65.** He should eat 13 of his mother-in-law's biscuits and 11 of his wife's muffins so that his total amount of saturated fatty acid is approximately 28%. **67.** The time recorded by the Spanish team to the nearest tenth of a second was 39.5 seconds **69.** It would take them $3\frac{12}{13}$ hours to paint the large room working together. **71.** Since the rainstorm started at noon, the cistern will be full at 3:45 p.m.

PROBLEM SET 2.2

1. $\{x \mid x < 2\}$ **3.** $\{x \mid x \geq 2\}$ **5.** R **7.** $\left\{x \mid x > -\frac{1}{12}\right\}$; [number line: open at $-\frac{1}{12}$, arrow right] **9.**  [number line: open at 3, closed at 5] **11.** $\{x \mid -1 < x < 2\}$ **13.** $\left\{w \mid w \leq \frac{3}{4}\right\}$; $\left(-\infty, \frac{3}{4}\right]$; [number line: arrow left, closed at $\frac{3}{4}$] **15.** R; [number line: arrows both directions] **17.** The maximum efficiency is 57%. **19.** Dr. Jackson consumed between $246\frac{2}{3}$ mg and $346\frac{2}{3}$ mg of caffeine. **21.** $(-\infty, -1]$ **23.** $(-12, 12)$ **25.** $(-\infty, 4] \cup [9, +\infty)$ **27.** $\left(-\infty, -\frac{5}{2}\right) \cup \left(-\frac{1}{2}, \frac{7}{2}\right)$ **29.** $\{t \mid t > 5\}$; $(5, +\infty)$; [number line: open at 5, arrow right] **31.** $\left\{x \mid x \leq \frac{8}{5}\right\}$; $\left(-\infty, \frac{8}{5}\right]$; [number line: arrow left, closed at $\frac{8}{5}$] **33.** $\left\{x \mid x \leq \frac{1}{3}\right\}$; $\left(-\infty, \frac{1}{3}\right]$; [number line: arrow left, closed at $\frac{1}{3}$] **35.** $\left\{p \mid p > -\frac{14}{61}\right\}$; $\left(-\frac{14}{61}, +\infty\right)$; [number line: open at $-\frac{14}{61}$, arrow right] **37.** $\left\{y \mid -4 < y \leq \frac{5}{2}\right\}$; $\left(-4, \frac{5}{2}\right]$; [number line: open at -4, closed at $\frac{5}{2}$] **39.** $\{x \mid x < -3$ or $x \geq 50\}$; $(-\infty, -3) \cup [50, +\infty)$; [number line: arrow left to open at -3, closed at 50 arrow right] **41.** $\{x \mid 0 < x \leq 4\}$; $(0, 4]$; [number line: open at 0, closed at 4] **43.** $\left(-\infty, \frac{49}{67}\right)$ **45.** $\varnothing$; No interval **47.** $(-\infty, 18)$ **49.** $\left[-\frac{3}{2}, +\infty\right)$ **51.** $[20, +\infty)$ **53.** $\varnothing$; No interval **55.** $(-\infty, +\infty)$ **57.** The original temperature of the mug should be between 35°C and 65°C. **59.** The dimensions of the scarf must be such that the width is at most $6\frac{2}{3}$ inches and the length is at most $13\frac{1}{3}$ inches. **69.** $\left\{x \mid x < \frac{b}{c}\right\}$ **71.** $\left\{x \mid x > -\frac{a}{c}\right\}$ **73.** $\left\{x \mid x \geq -\frac{bc}{a^2 + 1}\right\}$ **75.** $\left\{x \mid \frac{3c - a}{b} \leq x < \frac{a^2 - a}{b}\right\}$

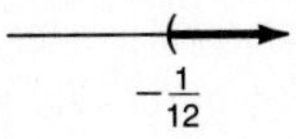

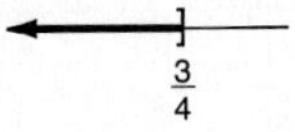

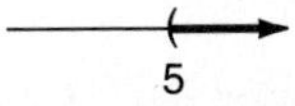

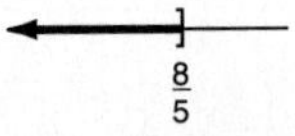

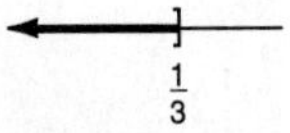

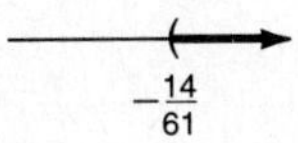

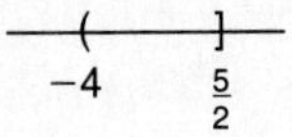

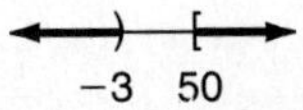

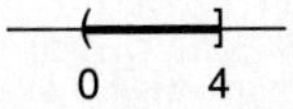

PROBLEM SET 2.3

1. $\{-9, -3\}$ **3.** $\{2\}$ **5.** $\{\ \}$ **7.** $\left\{-\frac{3}{2}, 5\right\}$ **9.** $\left\{-1, -\frac{1}{3}\right\}$ **11.** $(-\infty, -2) \cup \left(\frac{2}{3}, +\infty\right)$ **13.** $\left(-\infty, \frac{2}{3}\right] \cup [2, +\infty)$ **15.** $\left[-\frac{3}{2}, 5\right]$ **17.** $\left(-\infty, -\frac{5}{9}\right) \cup (5, +\infty)$ **19.** $(-\infty, 13)$ **21.** $[-1, 1]$ **23.** $\{\ \}$ **25.** $-6 \leq x \leq 1$ **27.** $|x| = 17$ **29.** $\left\{\frac{1}{8}, \frac{9}{8}\right\}$ **31.** $\{\ \}$ **33.** $\left\{\frac{7}{2}\right\}$ **35.** $\left\{-\frac{50}{3}, -\frac{46}{3}\right\}$ **37.** $\left\{\frac{5}{3}, \frac{2}{3}\right\}$ **39.** $\left\{-3, \frac{1}{3}\right\}$ **41.** $\left\{-3, -\frac{3}{2}\right\}$ **43.** $\{\ \}$ **45.** $\left\{\frac{1}{2}\right\}$ **47.** R **49.** $(5.999, 6.001)$

51. $\left(-\frac{1}{2}, \frac{7}{2}\right)$ **53.** $[-8, 0]$ **55.** R **57.** $\{\ \}$ **59.** $\left\{-\frac{3}{2}\right\}$ **61.** $\left(-4, -\frac{3}{2}\right)$ **63.** $[4, +\infty)$ **65.** $\left(\frac{1}{12}, +\infty\right)$ **67.** $(-\infty, -3) \cup \left(-\frac{18}{7}, +\infty\right)$ **69.** $\{\ \}$ **71.** $(46, +\infty)$ **73.** $50 < T < 130$ **75.** The range of temperatures in which both drugs can be stored is $10 < T < 15$, where T is in Celsius degrees. **77.** To hundredths of a percent, the range of percent of the client's body fat is $15.51\% < BF < 20.58\%$ **83.** $\left\{\frac{1-k}{2}, \frac{-1-k}{2}\right\}$ **85.** $\left\{\frac{k+3}{7}, \frac{-k+3}{7}\right\}$ **87.** $(-p - m, p - m)$ **89.** $\{\ \}$ **91.** $\{q\}$ **93.** $\{x \mid x \neq m\}$ **95.** $\{\ \}$

PROBLEM SET 2.4

1. $\left\{0, \frac{2}{3}\right\}$ **3.** $\left\{\frac{4}{3}\right\}$ **5.** $\{-3, 3\}$ **7.** The dimensions are 13 inches by 5 inches. **9.** $\{\pm 3\sqrt{3}\}$ **11.** $\{\pm 2\sqrt{3} - 8\}$ **13.** $\{\pm 2\sqrt{2} + 3\}$ **15.** $\left\{-\frac{1}{2} \pm \frac{\sqrt{3}}{2}i\right\}$ **17.** $\{1 \pm \sqrt{5}i\}$ **19.** $y \approx -8.1266$ and $y \approx -0.035522$ **21.** It will take 8 seconds for the object to strike the ground. **23.** Two complex roots. **25.** $(-\infty, -5) \cup (5, +\infty)$ **27.** $\left[-\frac{3}{2}, \frac{2}{3}\right]$ **29.** R **31.** The time for one period must be less than 2.721 seconds if the length of the pendulum is less than 6 inches. **33.** $\{\pm 3\sqrt{3}\}$ **35.** $\left\{0, \frac{3}{16}\right\}$ **37.** $\{3\}$ **39.** $\left\{-\frac{1}{4}, 2\right\}$ **41.** $\left\{-\frac{3}{4}, \frac{2}{5}\right\}$ **43.** $\{3 \pm 2\sqrt{2}\}$ **45.** $\{2 \pm 2\sqrt{2}\}$ **47.** $\{-5, 8\}$ **49.** $\left\{-\frac{1}{2} \pm \frac{\sqrt{11}}{2}i\right\}$ **51.** $\left\{-\frac{7}{2}, \frac{4}{3}\right\}$ **53.** $\left\{-\frac{2}{3}, 7\right\}$ **55.** $\left\{\frac{1}{2} \pm \frac{1}{2}i\right\}$ **57.** $(-\infty, -1) \cup (7, +\infty)$ **59.** $\left[-\frac{3}{2}, 4\right]$ **61.** $\left(-\frac{2}{3}, \frac{5}{3}\right)$ **63.** $(-\infty, -1 - \sqrt{7}] \cup [-1 + \sqrt{7}, +\infty)$ **65.** $\{\ \}$ **67.** $\left(-\infty, -\frac{3}{2}\right] \cup \left[\frac{1}{3}, +\infty\right)$ **69.** The rate of interest is 9%. **71.** The width of the border must be no more than 2 feet. **73.** (a) When p is \$100 or p is \$160 (b) When p is between \$100 and \$160 (c) \$22,500 **75.** About 4,854 miles. **81.** About 0.754 inches; $5\frac{11}{16}$ inches.

PROBLEM SET 2.5

1. $\{3\}$ **3.** $\{4\}$ **5.** $\left\{-\frac{1}{6}, \frac{2}{3}\right\}$ **7.** Larry would row 10 mph in still water. **9.** It would take the cold water tap $1 + \sqrt{11}$ hours to fill the tank alone. **11.** $\left(-\infty, \frac{5}{2}\right) \cup (4, +\infty)$ **13.** $(-\infty, -7) \cup (-2, 3)$ **15.** The concentration will be less than 0.04 mg/cm^3 after 1 minute. **17.** $\left\{\frac{1}{2}\right\}$ **19.** $\{-1, -2\}$ **21.** $\{-1\}$ **23.** $[-4, 0)$ **25.** $\left\{\frac{-4 \pm \sqrt{385}}{9}\right\}$ **27.** $\{\ \}$ **29.** $\{-1\}$ **31.** $\left\{-\frac{1}{6}, 1\right\}$ **33.** $(-5, 0)$ **35.** $\left\{\frac{1}{3}, \frac{4}{9}\right\}$ **37.** $\{-1\}$ **39.** $(-\infty, 0) \cup \left(0, \frac{1}{2}\right) \cup \left(\frac{1}{2}, +\infty\right)$ **41.** $\left\{-\frac{1}{3}\right\}$ **43.** $[-3, -1) \cup [3, +\infty)$ **45.** $(-\infty, -3)$ **47.** $(-\infty, -2) \cup [0, 2) \cup [6, +\infty)$ **49.** The train's original average speed was 65 miles per hour. **51.** It takes the fill pipe 6 hours to fill the pool working alone. **53.** The current's rate is $2\frac{2}{3}$ miles per hour. **55.** At least 22,000 can be produced. **59.** $\{-a \pm a\sqrt{2}\}$

PROBLEM SET 2.6

1. $\{0, \pm 2, \pm 2i\}$ **3.** $\left\{\frac{1}{3}, 2, -1 \pm \sqrt{3}i\right\}$ **5.** $(-2, 0) \cup (2, +\infty)$ **7.** $\{12\}$ **9.** $\{0, 1\}$ **11.** $\{\pm 3, \pm 3i\}$
13. $\{0, 2, -1 \pm \sqrt{3}i\}$ **15.** $\{\pm\sqrt{2}, \pm\sqrt{3}\}$ **17.** $\{1, 2, -2\}$ **19.** $\{2\}$ **21.** $\left\{\frac{2}{3}, -\frac{1}{3} \pm \frac{1}{3}\sqrt{3}i\right\}$ **23.** $\{\pm 8\}$
25. $\{-2, 0, 3\}$ **27.** $\{6\}$ **29.** $\{4\}$ **31.** $\{1, \pm i\}$ **33.** $\left\{\pm\frac{1}{2}\sqrt{6}i, \pm\sqrt{5}i\right\}$ **35.** $\left\{\frac{2}{3}, 2, -1 \pm \sqrt{3}i\right\}$ **37.** $\{-3\}$
39. $\{25\}$ **41.** $\{-1, 8\}$ **43.** $\{0, 4\}$ **45.** $\left\{\pm\frac{\sqrt{2}}{2}\right\}$ **47.** $\{\pm 1\}$ **49.** $(-\infty, -3) \cup (-1, 0) \cup (0, +\infty)$
51. $[1, +\infty)$ **53.** $(-2, 2)$ **55.** $\sqrt{2}$ is not a solution to the equation. **57.** $x \approx 1.9129$ **59.** $t \approx \pm 1.2247, \pm 2.2361$
61. The speed of the boat is 5 mph. **63.** When the radius of the cone is less that one-half the height of the cone, the niece gets the most yogurt and when the radius of the cone is more than one-half the height of the cone, Anne gets the most yogurt.

CHAPTER 2 REVIEW PROBLEMS

1. $\left\{\frac{20}{9}\right\}$ **3.** $\left\{-4, \frac{16}{3}\right\}$ **5.** $\left\{0, \frac{3}{7}\right\}$ **7.** $\{-7\}$ **9.** $\{\pm 2\sqrt{2}, \pm 2\sqrt{2}i\}$ **11.** $\{5\}$ **13.** $\{\ \}$ **15.** $\{0, \pm 3, \pm 3i\}$
17. $\{-3, 2\}$ **19.** $\{\pm 1, \pm 2, \pm i\}$ **21.** $\left\{\frac{20}{11}, 8\right\}$ **23.** $\left(-\infty, \frac{29}{2}\right)$ **25.** $(-\infty, 4) \cup (8, +\infty)$ **27.** $\left(-\frac{2}{3}, \frac{3}{2}\right)$
29. $\left[-\frac{1}{2}, +\infty\right)$ **31.** $\left\{\frac{6}{5}\right\}$ **33.** $[-4, +\infty)$ **35.** $\left\{-\frac{3}{2} \pm \frac{\sqrt{11}}{2}i\right\}$ **37.** $(-2, 0) \cup (2, +\infty)$ **39.** $\{\pm 2, \pm 4\}$
41. $R_1 = \dfrac{R_T R_2 R_3}{R_2 R_3 - R_T R_3 - R_T R_2}$ **43.** $x \approx -1.535$ or 0.8685 **45.** $z \approx -2.831$ or 0.1850
47. $z \approx 0.5000 + 1.936i$ or $0.5000 - 1.936i$ **49.** 261; Two distinct real solutions. **51.** -8; Two complex solutions.
53. $\frac{16}{81} - \frac{44}{17} < 0$; Two complex solutions. **55.** 10 liters of a 15% brine solution must be added.
57. It would take them $1\frac{1}{5}$ days to do the paving working together.
59. It will take Ly Tan $1\frac{1}{3}$ hours to drive from work to the Civic Center. **61.** It will take one of the valves $\frac{2}{3}$ hours working alone to fill the vat. **63.** The tolerance level is between -0.996 and 1.004. **65.** The Celsius reading will be greater than $37\frac{7}{9}$°. **67.** (a) Either can be entered with one series of calculator entries for each number.
(b) $(x + 2)(x - 1) = 0$ is more useful because now we can set each factor equal to zero and solve.
69. (b) $x + 27 + 10\sqrt{x + 2}$ (c) $8x^3 + 1 = (2x + 1)(4x^2 - 2x + 1)$

PROBLEM SET 3.1

1. 5 **3.** 10 **5.** $2\sqrt{2}$ **7.** 5.9036 **9.** 10.5372 **11.** $\left(-\frac{1}{10}, -\frac{3}{4}\right)$ **13.** (2, 8)

15.

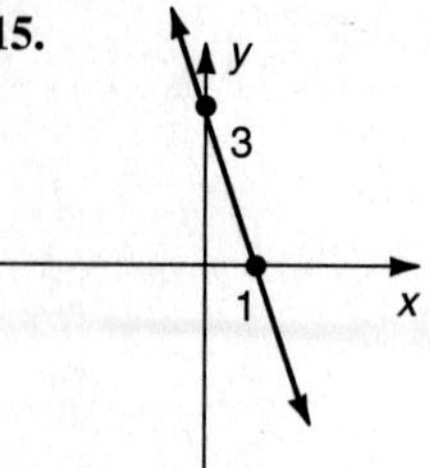

17.

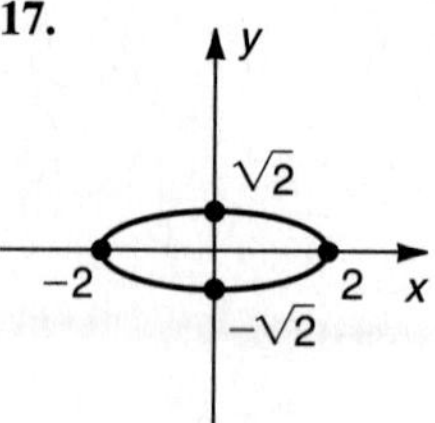

19. x: ± 2; y: none **21.** x: none; y: 2

23.

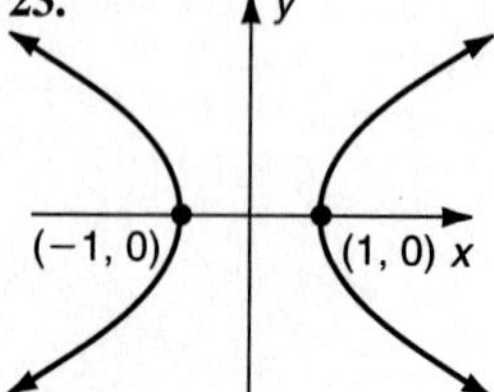

25.

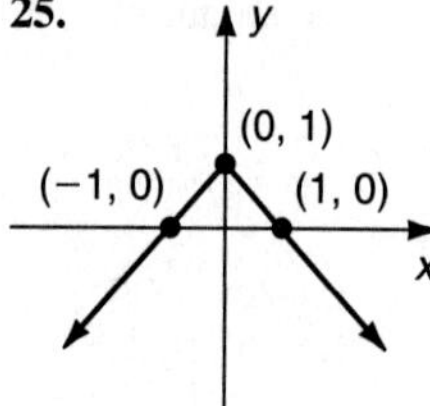

27. yes **29.** yes

31. $7\sqrt{2}$ **33.** $\frac{\sqrt{106}}{6}$ **35.** $\sqrt{21}$ **37.** right **39.** isosceles **41.** right **43.** 1 **45.** (5, 4), (−3, 4)

47. ± 8 **49.** 4 or 6 **51.** 11.32, −1.325 **53.** (−4,11)

55.

57.

59.

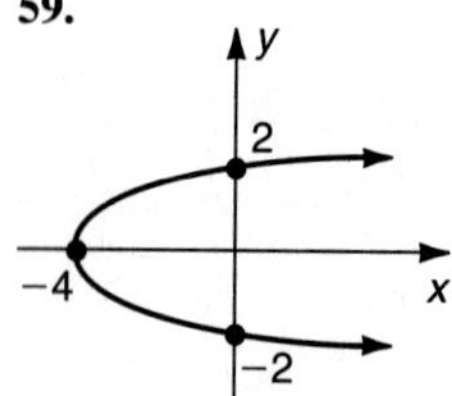

61.

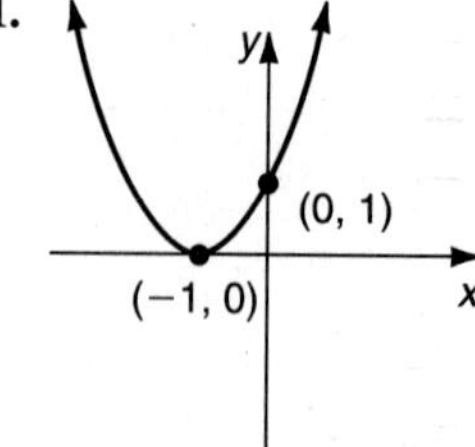

63.

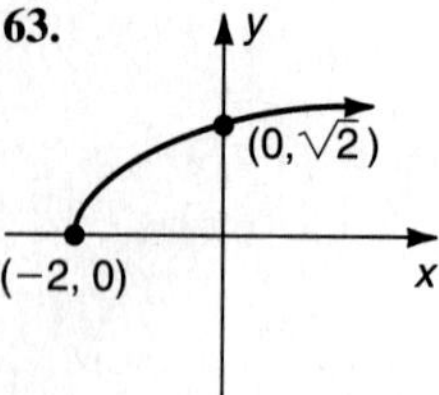

65.

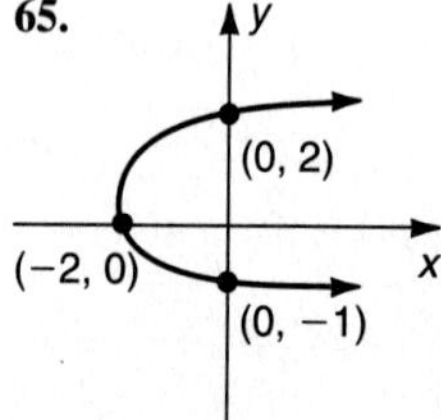

67.

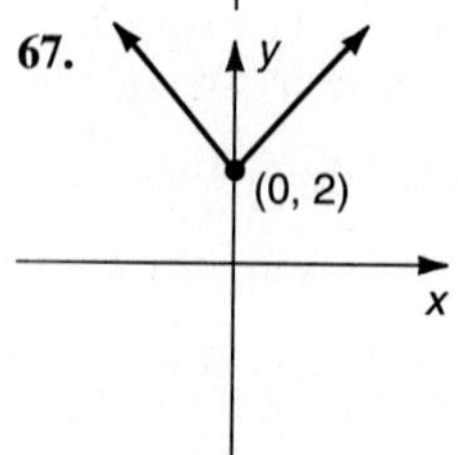

69.

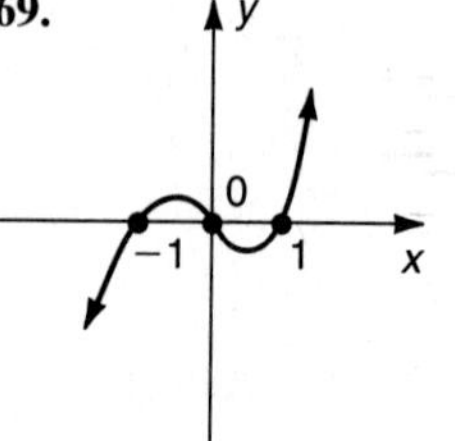

71. (a) the time it takes the ball to strike the ground. (b) 16 ft (c) The graph is symmetric with respect to $x = 1$. The ball is at a given height going up and also going down. **73.** $x_2 = x_1 + 2p$; $y_2 = y_1 + 2q$

75.

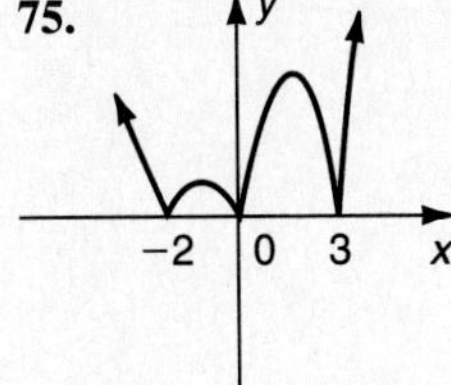

77. (a)

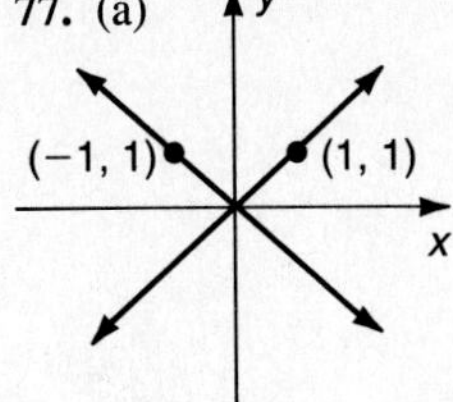

77. (b)

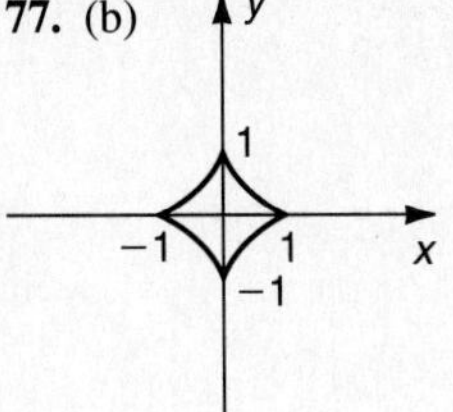

PROBLEM SET 3.2

1.

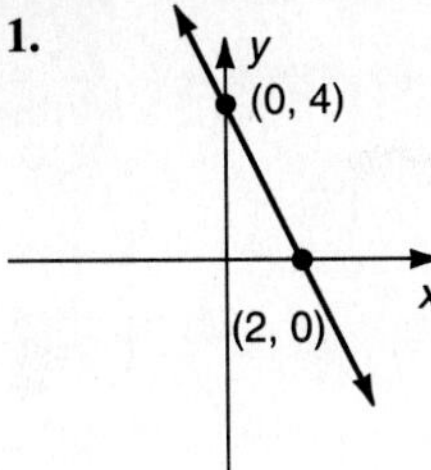

3.

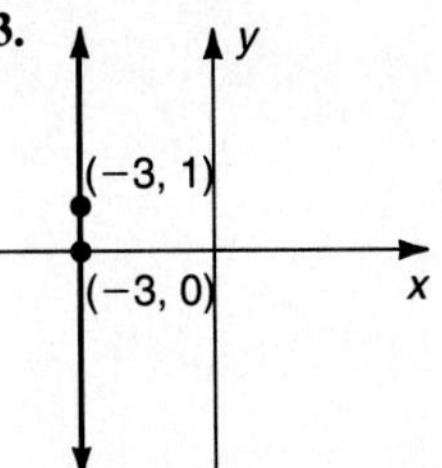

5. (a)

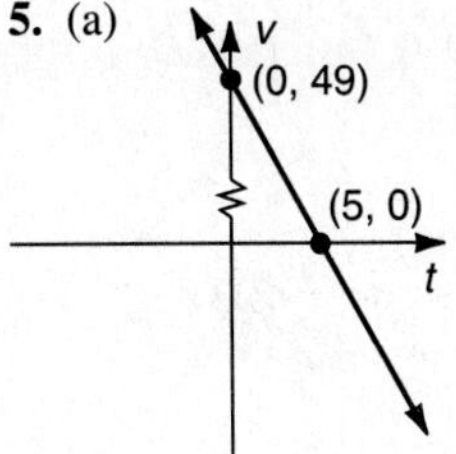

(b) 29.4 m/sec
(c) after 5 sec
(d) About 4 sec
(e) decreases

7. $-\frac{1}{3}$ 9. -1 11. 2 13. undefined

15.

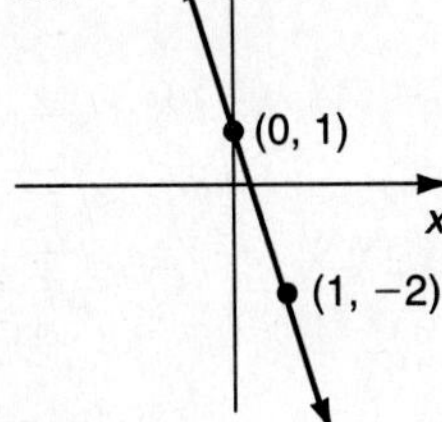

17.

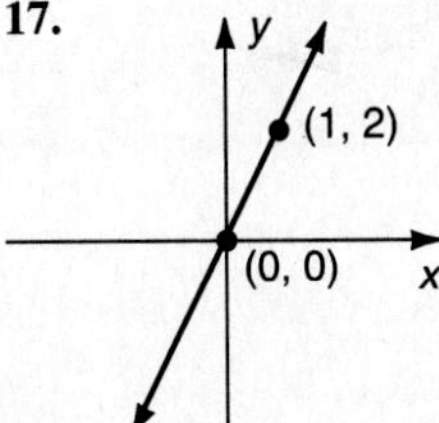

19. $3x - y = 1$ 21. $x + y = -3$

23. $x - y = -1$ 25. $x - 2y = -4$ 27. $6x - y = 11$ 29. $x = 2$ 31. $-\frac{2}{3}$ 33. $\frac{3}{2}$ 35. $\frac{1}{4}$ 37. $-\frac{1}{2}$

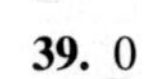

39. 0 41. $\frac{1}{2}$ 43. $-\frac{7}{5}$

45.

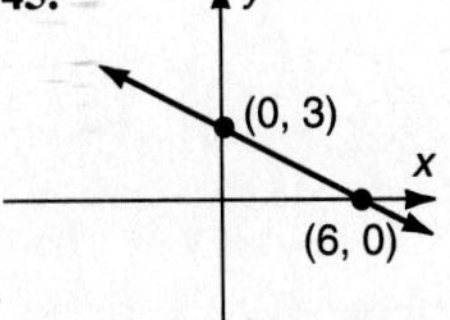

47.

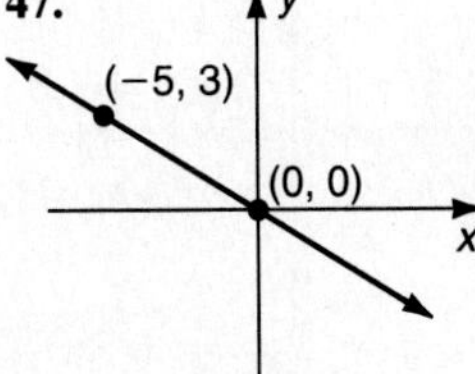

49.

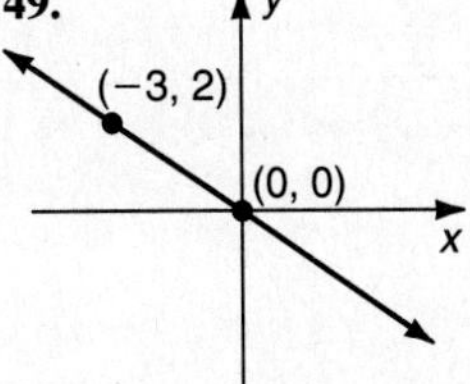

51.

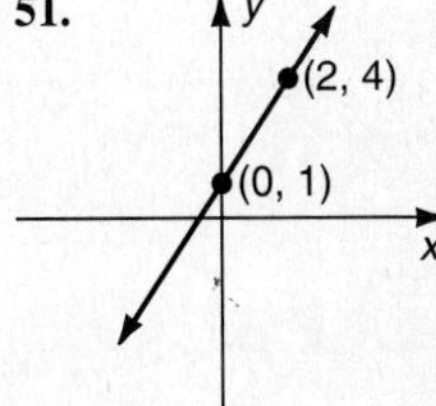

53.

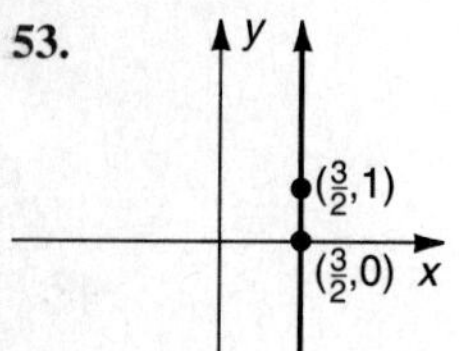

55.

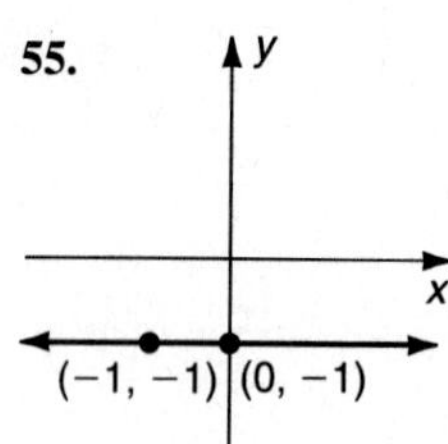

57. $x + 2y = 0$ **59.** $y = 7$

61. $x = 2$ **63.** $3x - y = 2$ **65.** $4x - y = 3$ **67.** $7y = -1$ **69.** $x = 7$ **71.** $x + y = 3$

73. (a)

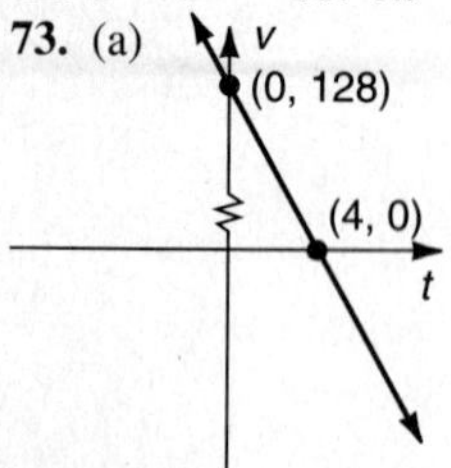

v-int is initial velocity;
t-int is time it takes to strike ground

(b) -32 (c) As time increases, velocity decreases.

75. (a)

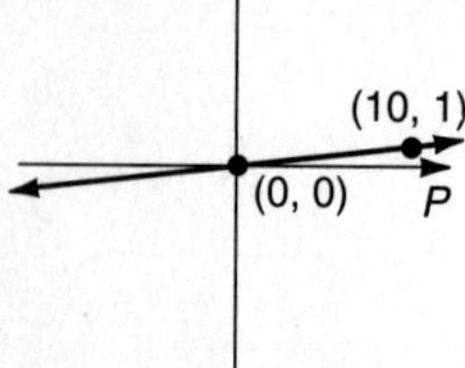

(b) .10 is the slope. (c) No meaning for problem; investment ≥ 0.

77. (a) no (b) $5q + 4p = 350$ (c) Demand is 70 when price is 0. Price is \$87.50 when demand is 0. (d) 22

79. All are parallel **81.** $2x - y = 5$; $x + 3y = 13$; $x - 4y = -1$ **83.** $y = -\frac{A}{B}x + \frac{C}{B}$; $B \neq 0$ **85.** $\frac{x}{2} + \frac{y}{4} = 1$

PROBLEM SET 3.3

1.

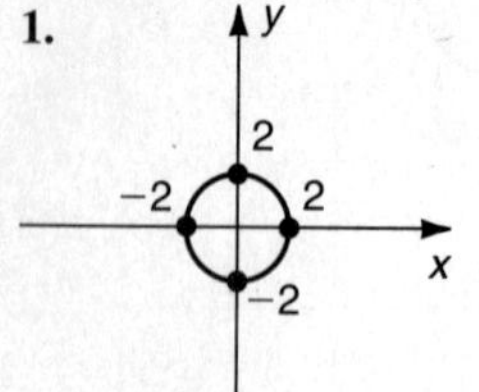

3.

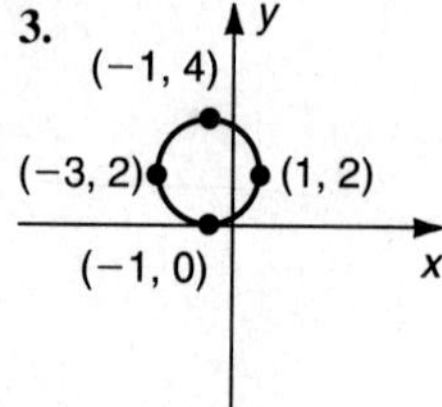

5.

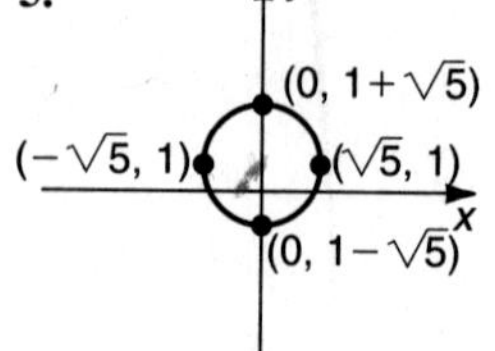

7.

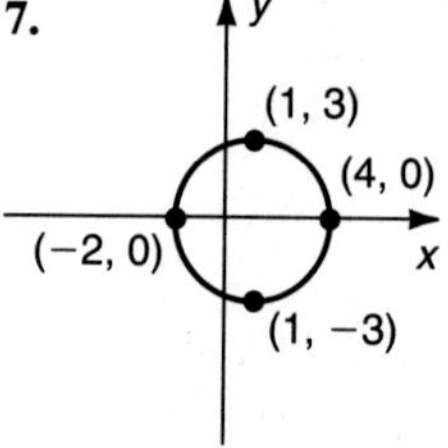

9.

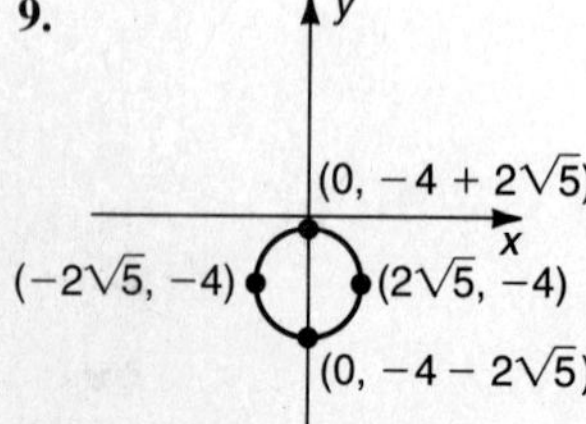

11.

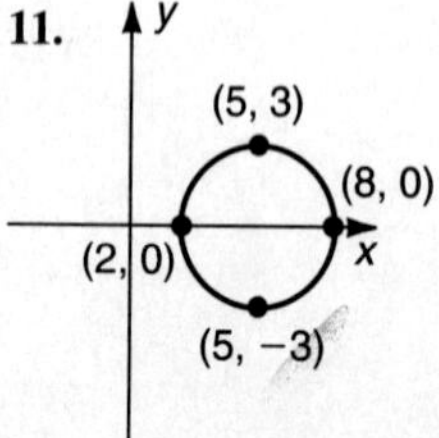

13.

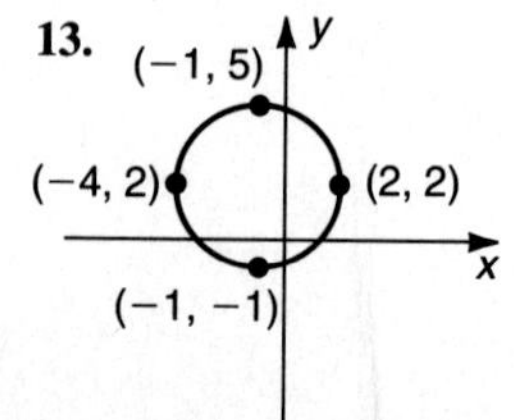

15.

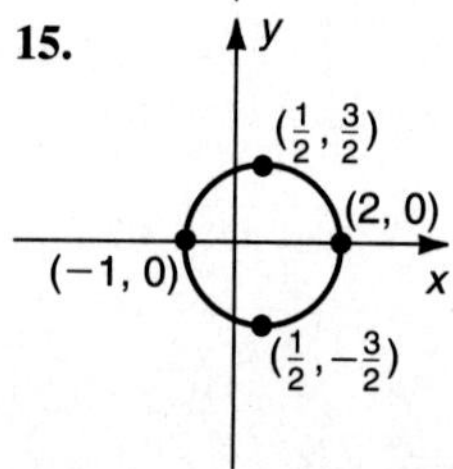

17. no graph **19.** $(x - 7)^2 + (y - 5)^2 = 5$ **21.** x-int: 1,5; y-int: none **23.**

25. C: (0, 0); r: 3; x-int: ± 3; y-int: ± 3 **27.** C: (0, 2); r: 4; x-int: $\pm 2\sqrt{3}$; y-int: -2 and 6 **29.** C: $(-3, -2)$; r: 7; x-int: $-3 \pm 3\sqrt{5}$; y-int: $-2 \pm 2\sqrt{10}$ **31.** none **33.** C: (0, 4); r: 4; x-int: 0; y-int: 0 and 8 **35.** C: $(-4, 0)$; r: 3; x-int: -1 and -7; y-int: none **37.** C: $(0, -1)$; r: $2\sqrt{2}$; x-int: $\pm\sqrt{7}$; y-int: $-1 \pm 2\sqrt{2}$ **39.** C: $(-6, 0)$; r: 7; x-int: 1 and -13; y-int: $\pm\sqrt{13}$ **41.** C: (2, 1); r: 4; x-int: $2 \pm \sqrt{15}$; y-int: $1 \pm 2\sqrt{3}$ **43.** none **45.** C: $(-3, 3)$; r: 0; x-int: : none; y-int: none **47.** C: $(-2, -3)$; r: 5; x-int: -6 and 2; y-int: $-3 \pm \sqrt{21}$ **49.** C: $(-2, 0)$; r: 2; x-int: -4 and 0; y-int: 0 **51.** C: $(-6, -2)$; r: $2\sqrt{10}$; x-int: -12 and 0; y-int: -4 and 0 **53.** $(x + 5)^2 + (y + 7)^2 = 0$ **55.** $(x - 4)^2 + (y + 7)^2 = 16$ **57.** $x^2 + (y + 1)^2 = 20$ **59.** $x^2 + y^2 = 26$

61.

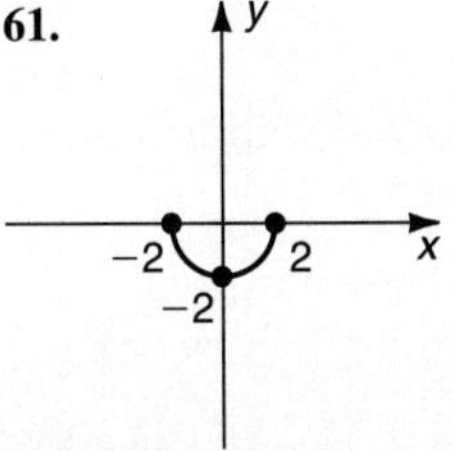

63.

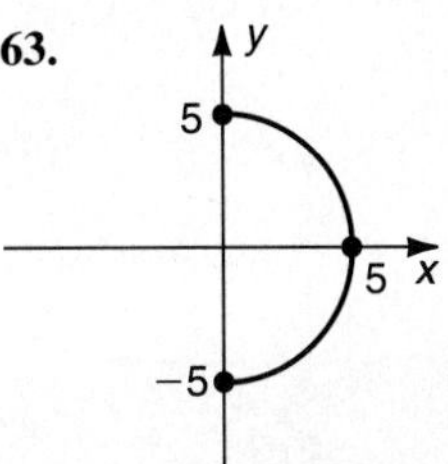

65.

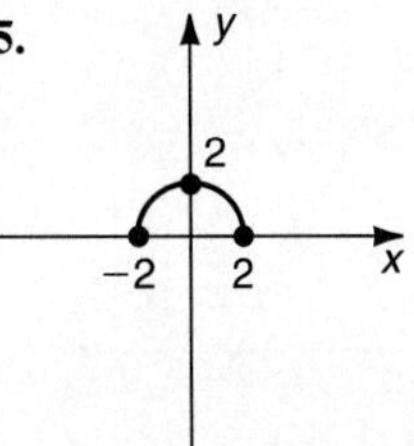

67.

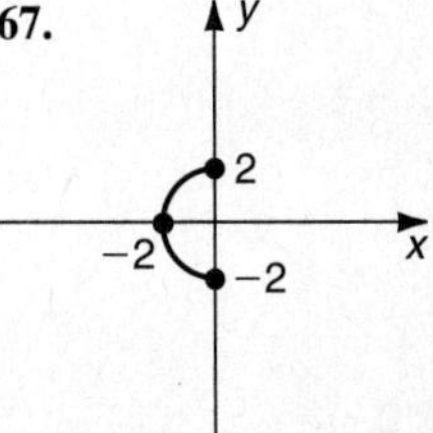

69. (3, 4) and $(-3, -4)$

PROBLEM SET 3.4

1.

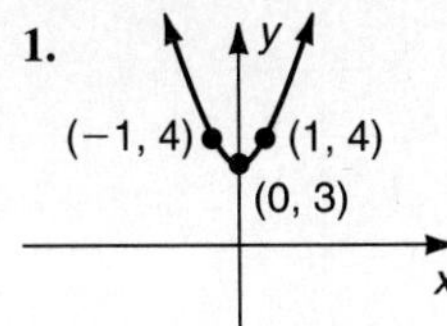

Axis of Symmetry: $x = 0$

3.

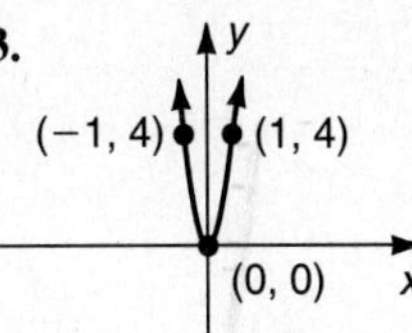

Axis of Symmetry: $x = 0$

5. $D_1 = \frac{7}{4}$; $D_2 = 3$

7.

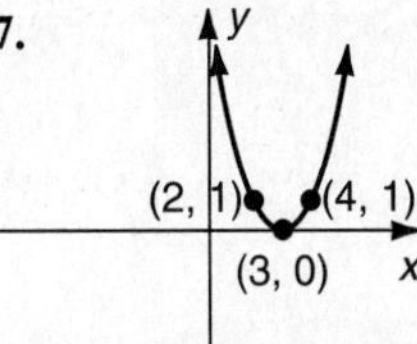

Axis of Symmetry: $x = 3$

9.

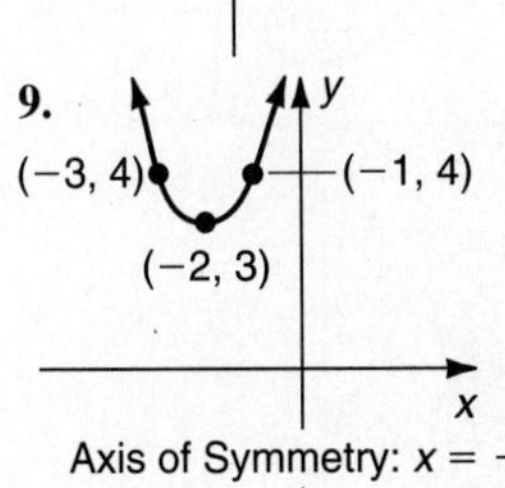

Axis of Symmetry: $x = -2$

11.

Axis of Symmetry: $x = -2$

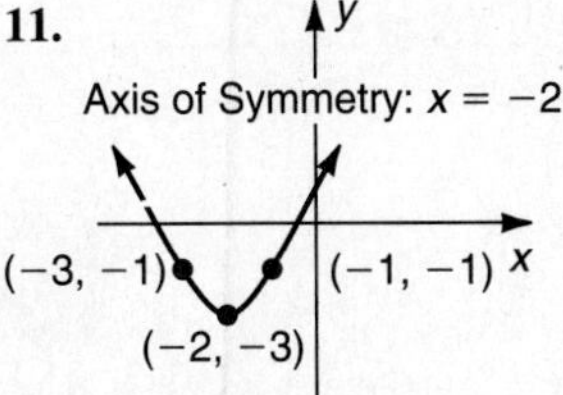

13.

Axis of Symmetry: $x = 4$

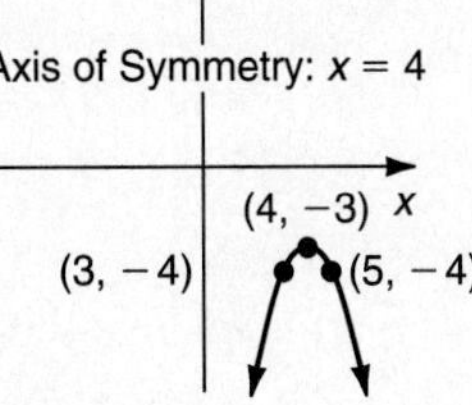

15.

Axis of Symmetry: $x = -2$

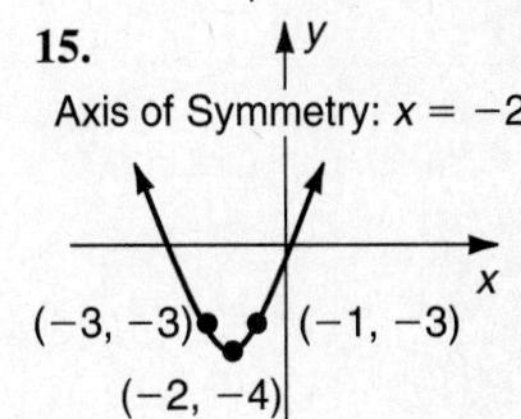

17.

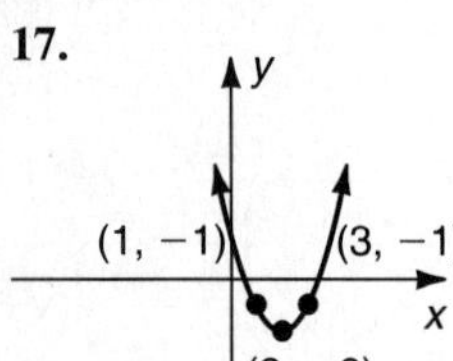

Axis of Symmetry: $x = 2$

19.

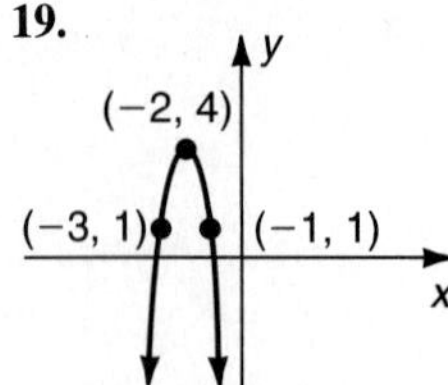

Axis of Symmetry: $x = -2$

21.

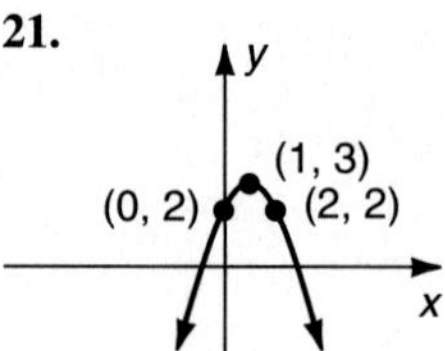

Axis of Symmetry: $x = 1$

23. Bill should make 600 bagels per day.

25.

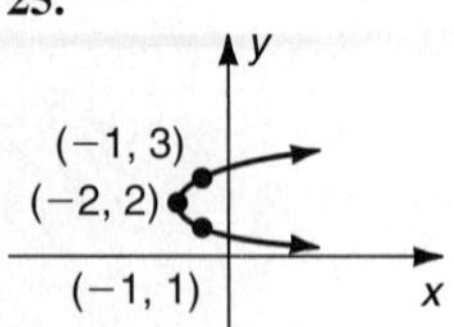

Axis of Symmetry: $y = 2$

27.

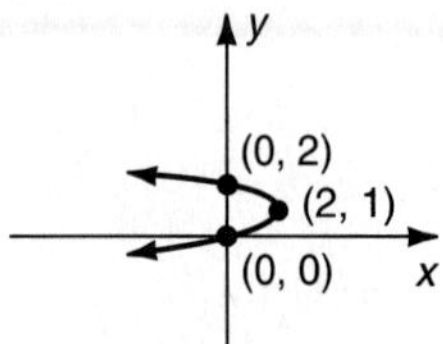

Axis of Symmetry: $y = 1$

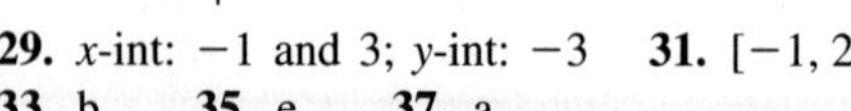

29. x-int: -1 and 3; y-int: -3 **31.** $[-1, 2]$

33. b **35.** e **37.** a

39.

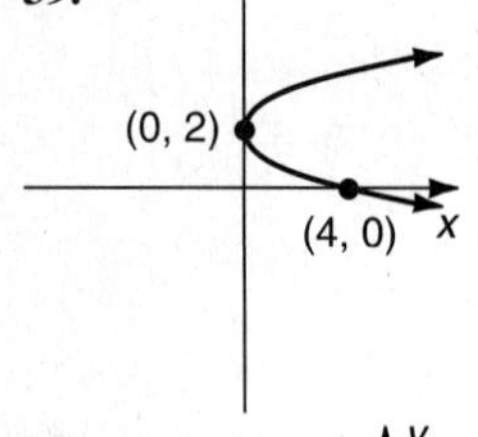

41.

(0, 3) (1, 2)

43.

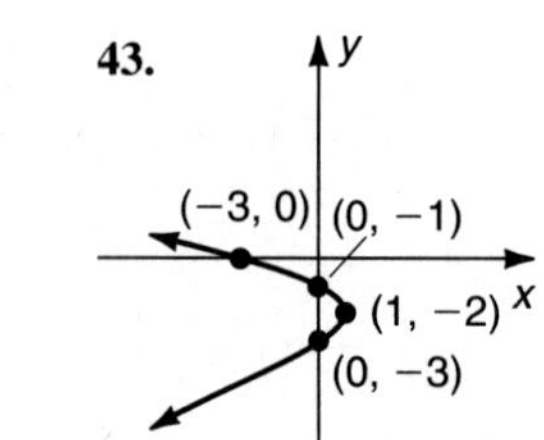

45.

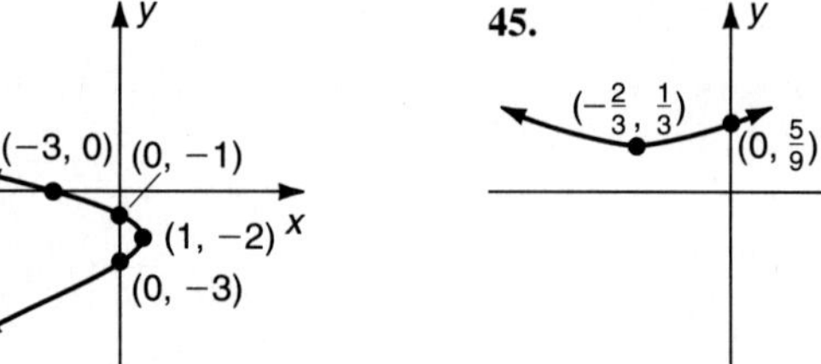

47.

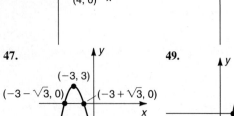

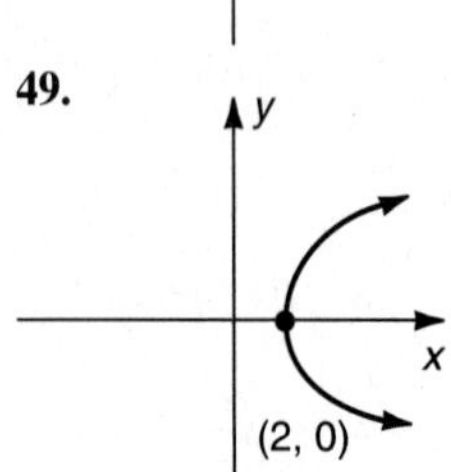

49.

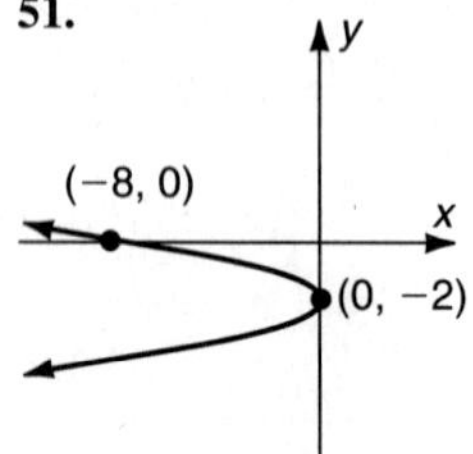

51.

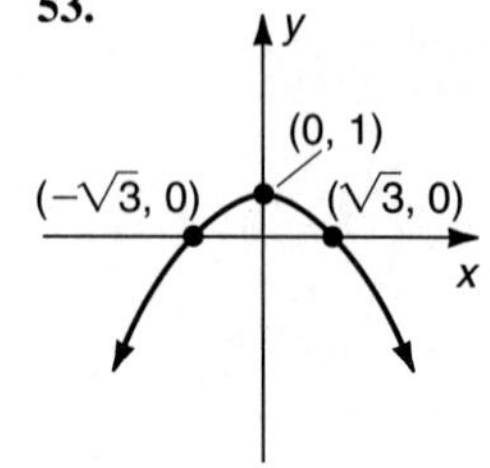

53.

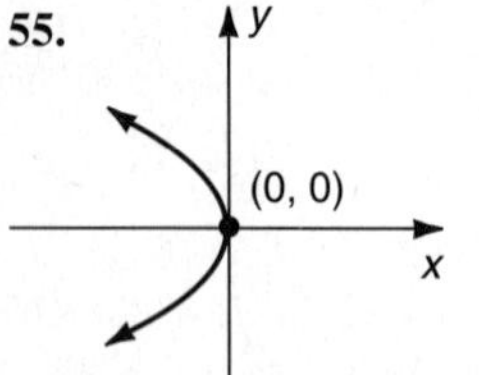

55.

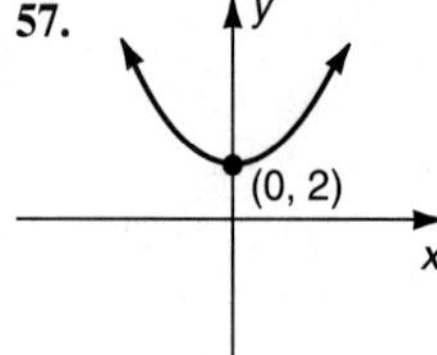

57.

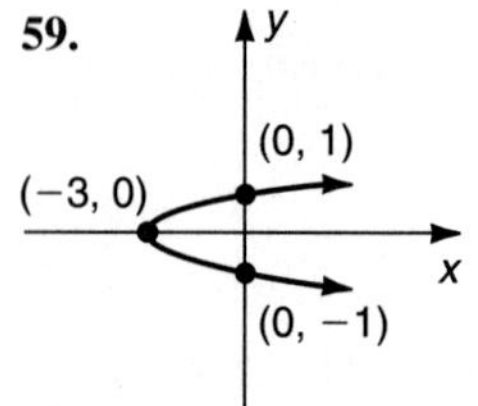

59.

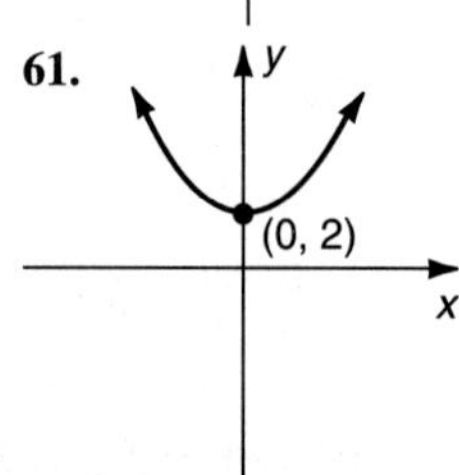

61.

(0, 2)

63.

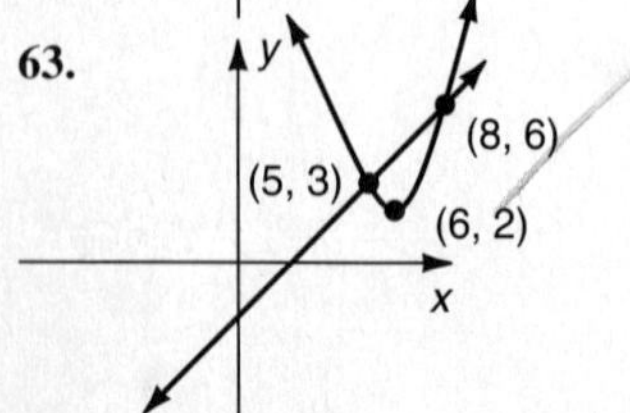

65.

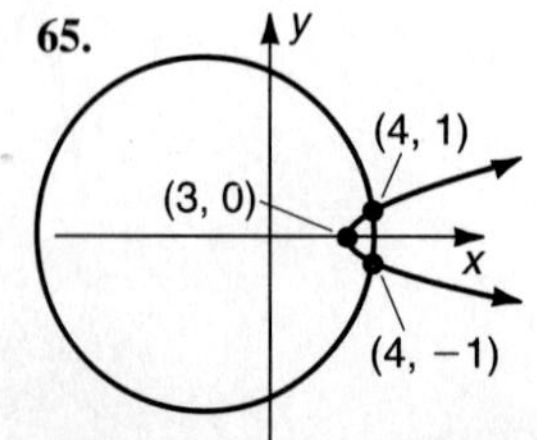

67.

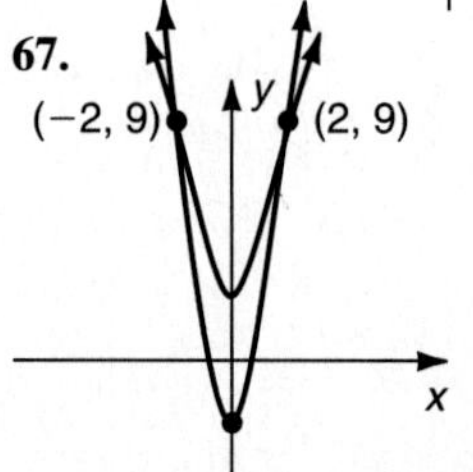

69. $-0.804, 1.554$ **71.** $x = -2(y-1)^2 + 2,\ y = -\frac{1}{4}(x-2)^2 + 1$ **73.** $y = 2(x-4)^2 - 2$

75. (a) (b) 16 ft above ground (c) $\frac{\sqrt{5}}{2}$ sec (d) Approx. 0.56 sec

20 ft is height of binoculars above ground. $\sqrt{5}/2$ sec is number of sec to hit ground.

77. (a)

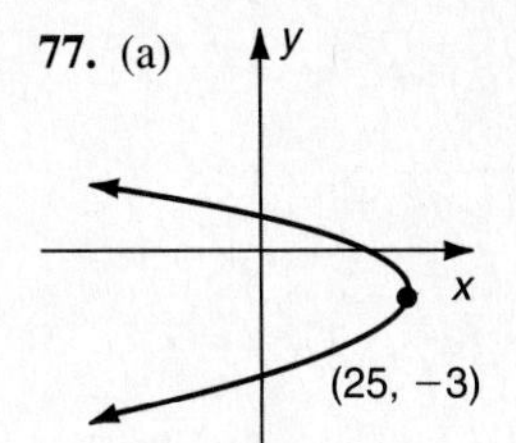

(b) 16 tons (c) 2 tons

79. $[-1, 1]$ **81.** $(-\infty, 1-\sqrt{3}) \cup (1+\sqrt{3}, +\infty)$ **83.** $(-\infty, -1] \cup [3, +\infty)$ **85.** $y - 4 = -\frac{1}{9}(x-3)$

87. v: $(1.25, -2.125)$ **89.** $-1.75, 0.26$ **91.** $-2.2, 1.43$

PROBLEM SET 3.5

1. Function: (3, 4), (2, 5), (1, 5) **3.** Not a function; (6, −1), (6, 0), (6, 1) **5.** 14 **7.** 1 **9.** $4 + 3\sqrt{2}$

11. $2t^2 + 3t$ **13.** -2 **15.** $\frac{1}{5}$ **17.** 1 **19.** $|a+1|$ **21.** $2(x+h)^2 + 3(x+h)$ **23.** $4x + 2h + 3$ **25.** R

27. $(-\infty, -4) \cup (-4, +\infty)$ **29.** $[-3, +\infty)$ **31.** R **33.** R **35.** $P(s) = 4s$; $A(s) = s^2$ **37.** $x(y) = \frac{7-2y^2}{3}$

39. not possible for either variable **41.** $\{-6, -3, 0, 4, 7\}$ **43.** $\{3, 7, 11, 15, \ldots\}$ **45.** $\{11\}$ **47.** R **49.** R
51. $(-\infty, -8) \cup (-8, +\infty)$ **53.** $(-\infty, -3) \cup (-3, 2) \cup (2, +\infty)$ **55.** R **57.** $[-1/3, +\infty)$ **59.** $(-\infty, -3] \cup [3, +\infty)$

61. R **63.** 1 **65.** undefined **67.** $\frac{3}{8}$ **69.** $\frac{1}{6}$ **71.** 17 **73.** 72 **75.** 59 **77.** $x^2 + 5$ **79.** 2

81. $2t + h + 2$ **83.** $\frac{-1}{t(t+h)}$ **85.** $3t^2 + 3th + h^2$ **87.** \$1050 **89.** (a) 800 sq ft (b) 1200 sq ft (c) 1200 sq ft (d) 25 ft **91.** 475 miles **93.** $50\sqrt{34}$ mi **95.** $V(r) = \frac{10}{3}\pi r^2$ **97.** $V(h) = 25\pi h$ **99.** $(-\infty, -2) \cup (2, +\infty)$

101. $(2, +\infty)$

PROBLEM SET 3.6

1. (a) Domain: $[-3, +\infty)$; Range: $[0, +\infty)$ (b) 2 (c) -3 (d) none (e) yes **3.** (a) Domain: R; Range: $[-3, +\infty)$ (b) -2 (c) $\pm\sqrt{2}$ (d) $(-\sqrt{2}, \sqrt{2})$ (e) yes **5.** yes **7.** no **9.** increasing on $[2, 3]$; decreasing on $[-2, 2]$; constant on $[-5, -2]$ **11.** even **13.** odd **15.** Domain: $[-\pi, \pi]$; Range: $[0, 2]$ **17.** Domain: $(-2, 2]$; Range: $\{1, 2, 3, 4\}$
19. $[-5, -3]$, $[0, +\infty)$ **21.** $-4, -2, 2$ **23.** neither **25.** no **27.** 2 **29.** none **31.** $x = \pm 1$ **33.** odd

35. no **37.** $\frac{4}{5}$ **39.** s **41.** negative **43.** positive **45.** $[p, q]$ **47.** yes **49.** even **51.** even

53. neither **55.** even **57.** odd **59.** even **61.** increasing on $[500, 1500]$; decreasing on $[0, 500]$, $[1500, 2000]$
63. 34 sq units **65.** 38.5 sq units; 44.625 sq units **73.** $(1.1347, 1.4610)$ **75.** $(-0.0053976, 1.0054)$, $(1.2185, 1.2663)$

CHAPTER 3 REVIEW PROBLEMS

1. $\sqrt{53}, \left(-\frac{1}{2}, 6\right)$ **3.** $\sqrt{181}, \left(\frac{9}{2}, 0\right)$ **5.** $5\sqrt{5}, \left(\frac{5}{2}, -6\right)$ **7.** $\sqrt{61}, \left(3, \frac{3}{2}\right)$ **9.** $\sqrt{130}, \left(\frac{7}{2}, -\frac{9}{2}\right)$

11. $\sqrt{170}, \left(-\frac{5}{2}, -\frac{1}{2}\right)$ **13.** $\frac{2}{3}$ **15.** $\frac{4}{3}$ **17.** $-\frac{1}{2}$ **19.** undefined **21.** undefined **23.** $y - 5 = \frac{1}{3}(x + 2)$

25. $x - 4y = 4$ **27.** $3x + 2y = -12$ **29.** $x = 2$ **31.** $y = -\frac{4}{3}(x - 5)$

33. (a)

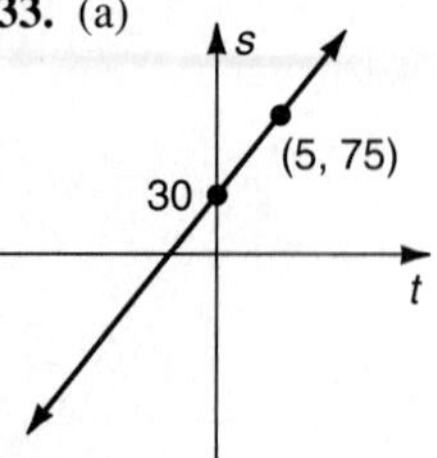

(b) 75 knots, 300 knots (c) 10 sec

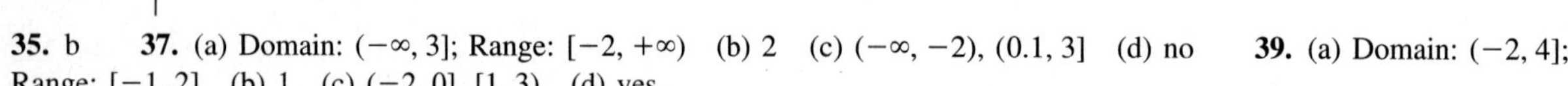

35. b **37.** (a) Domain: $(-\infty, 3]$; Range: $[-2, +\infty)$ (b) 2 (c) $(-\infty, -2)$, $(0.1, 3]$ (d) no **39.** (a) Domain: $(-2, 4]$; Range: $[-1, 2]$ (b) 1 (c) $(-2, 0]$, $[1, 3)$ (d) yes

41.

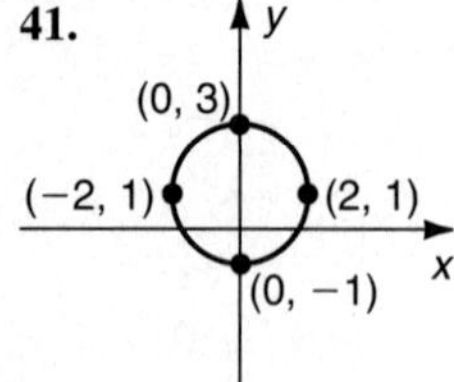

43.

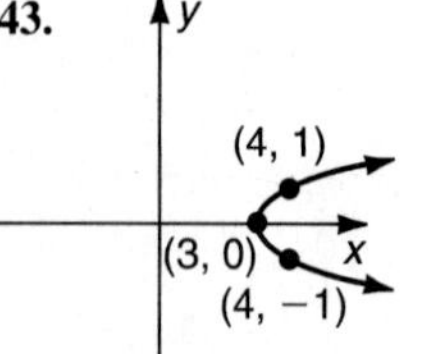

45.

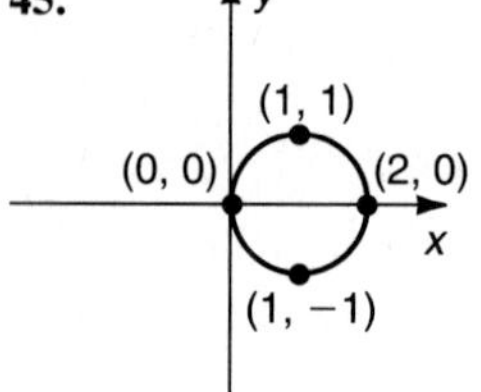

47.

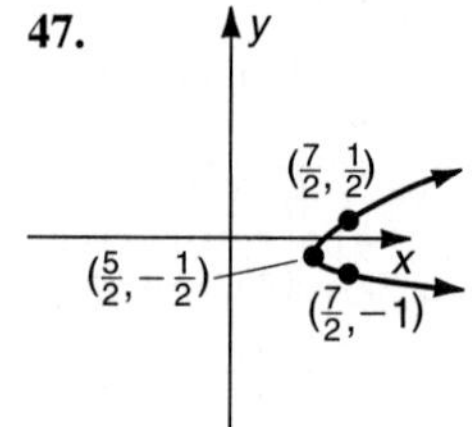

49.

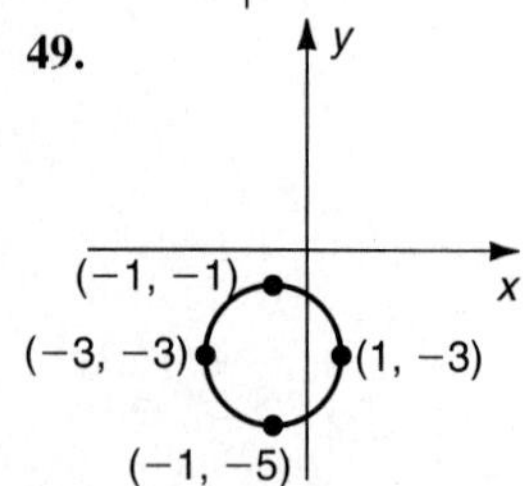

51.

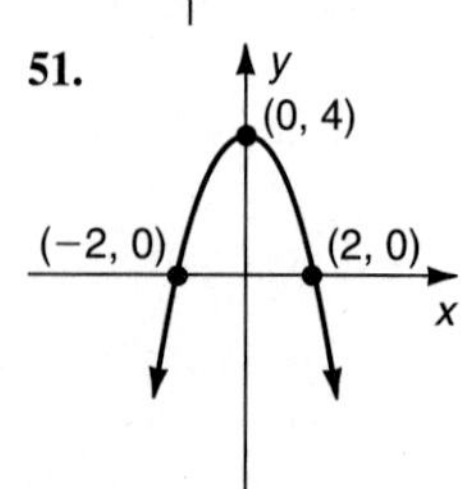

53.

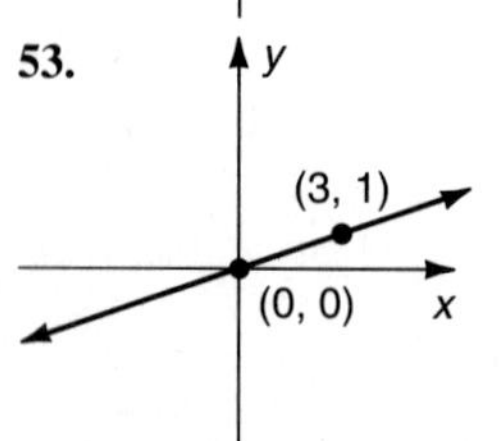

55.

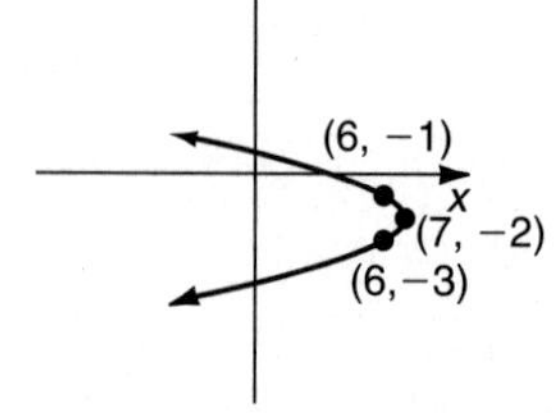

57. R **59.** $(-\infty, -2) \cup (-2, 2) \cup (2, +\infty)$ **61.** R **63.** $(-\infty, 4]$ **65.** R **67.** $-\frac{1}{2}$ **69.** $\frac{1}{(x-1)^2}$ **71.** -2

73. $\frac{-1}{(x-1)(x+h-1)}$ **75.** $[-1, 2]$, $(2, 3]$ **77.** $[-1, 0)$, $(2, 3)$, $(3, +\infty)$ **79.** Domain: $[-1, +\infty)$; Range: $[-3, +\infty)$

81. 0 **83.** 3 **85.** odd **87.** neither **89.** odd **91.** Maximum profit is \$860. 40 flags must be sold. Increasing on $(0, 40]$; Decreasing on $[40, 40 + 10\sqrt{86}]$ **93.** 7 **95.** $1 - 2x - h$ **97.** $\frac{-2}{(1 + x + h)(1 + x)}$

99. $\dfrac{-1}{\sqrt{x(x+h)}\,(\sqrt{x}+\sqrt{x+h})}$ **101.** (a) factored (b) $\left(x-\dfrac{1}{2}\right)^2-\dfrac{25}{4}$ **103.**

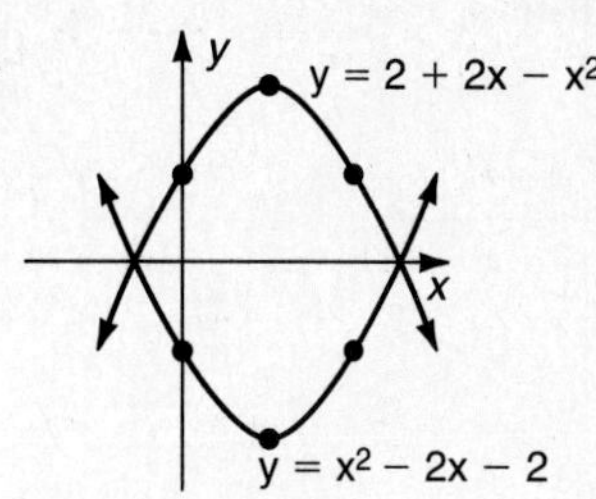

105. −0.8792 and 0.3792

PROBLEM SET 4.1

1.

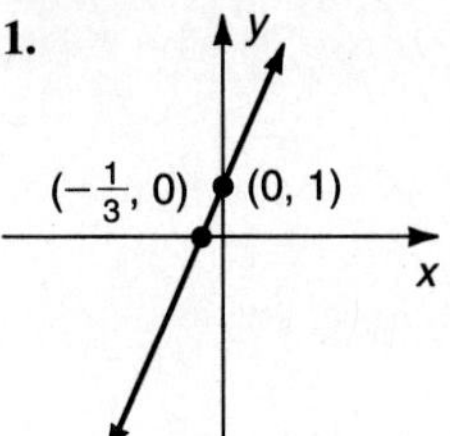

3.

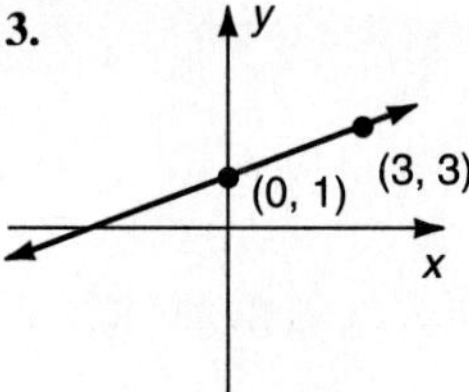

5. $A(x) = 12x - 20$

7.

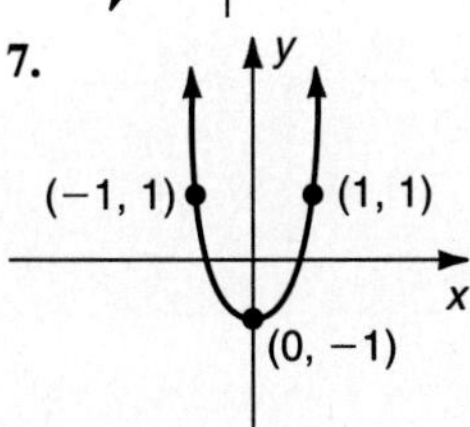

9.

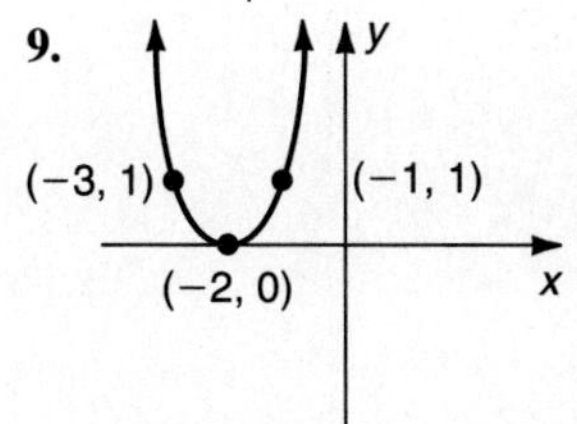

11.

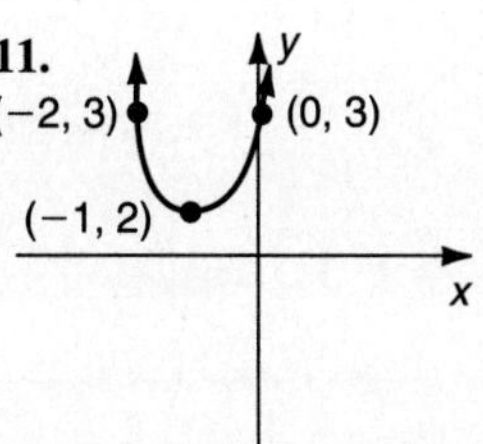

13.

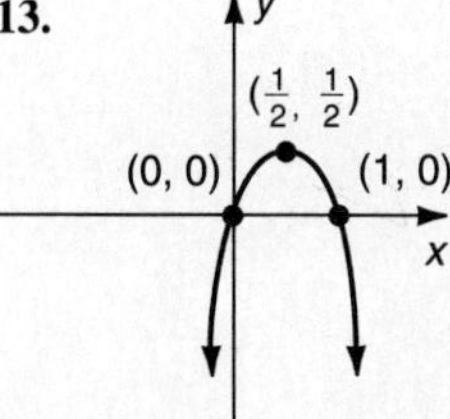

15. 120; $8,500 **17.** 7

19.

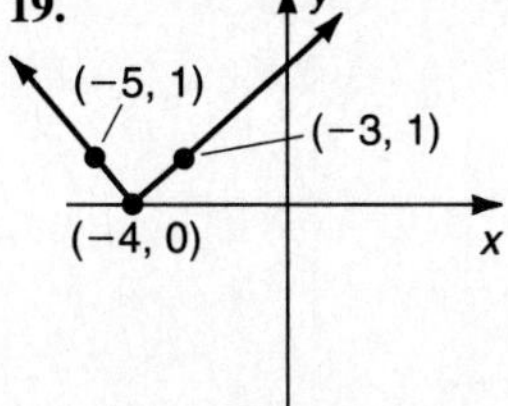

21.

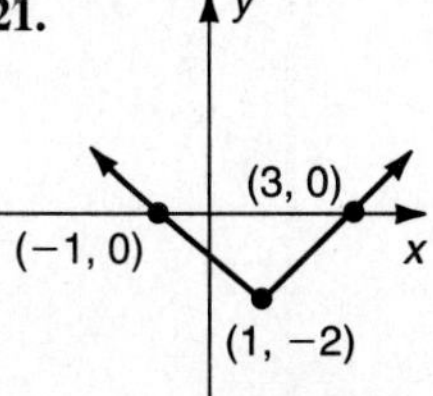

23.

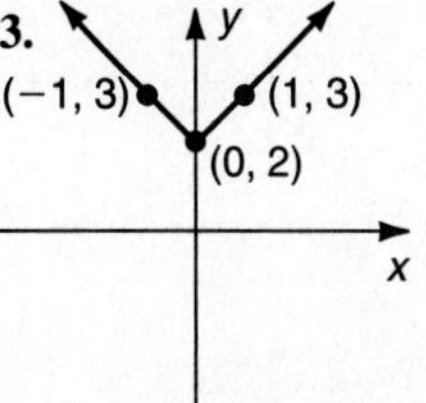

25.

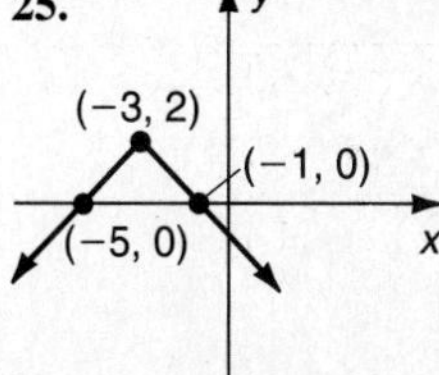

27.

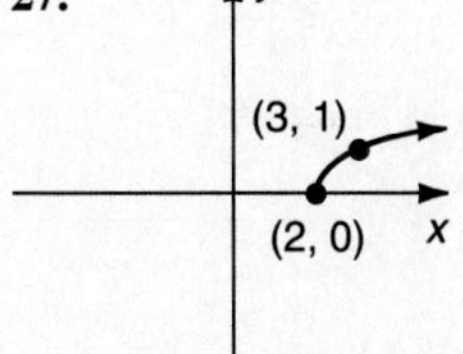

29.

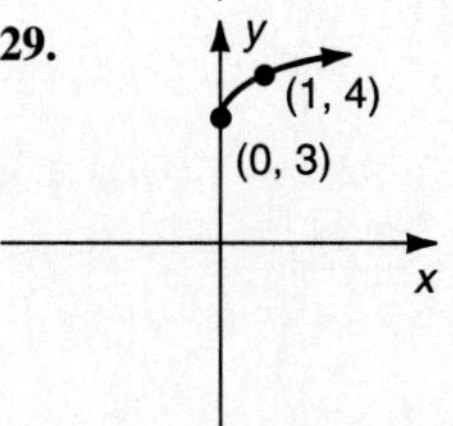

31.

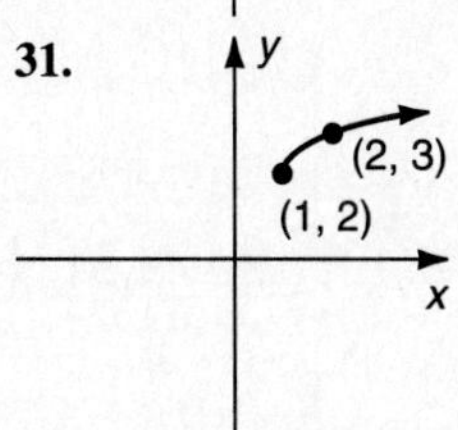

33.

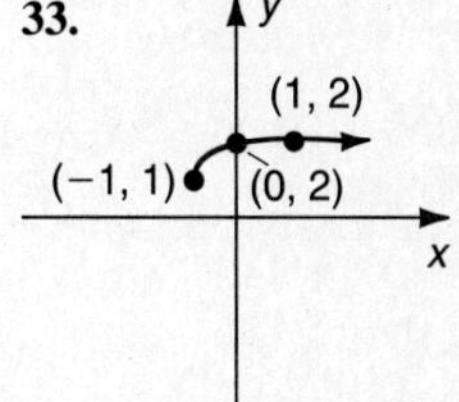

35.

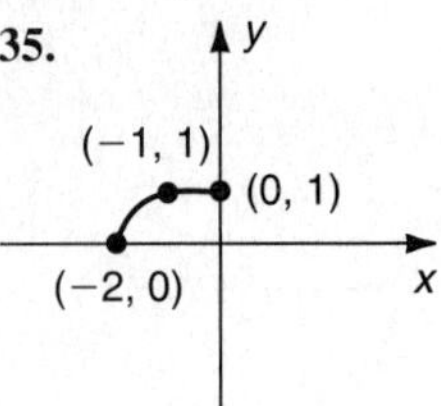

37.

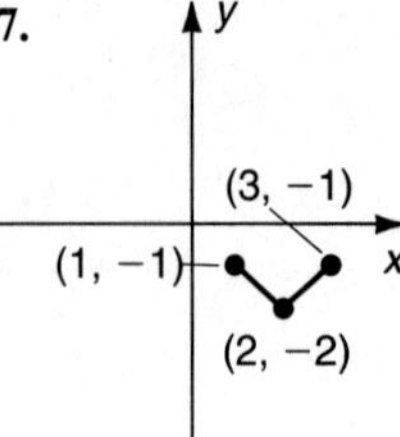

39.

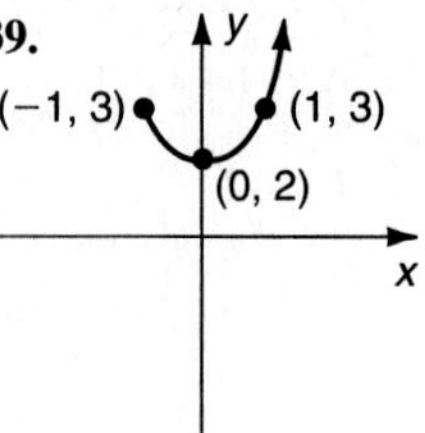

41. 0

43.

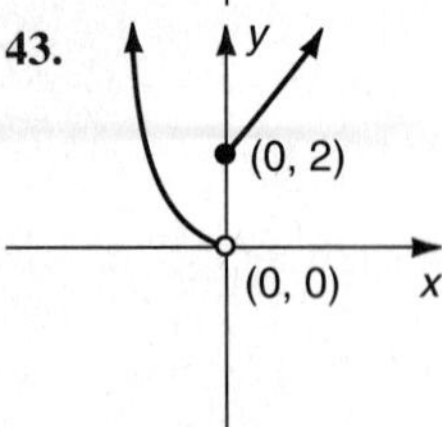

45.

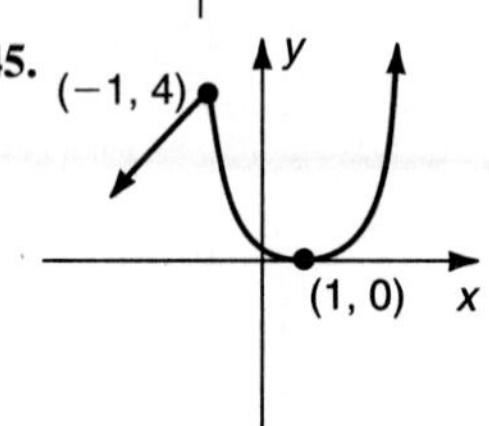

47. $\begin{cases} 6.4 + 2x & \text{if} \quad 0 < x \leq 8 \\ 12 + 1.3x & \text{if} \quad 8 < x \leq 25 \\ 22 + 0.9x & \text{if} \quad 25 < x \leq 40 \end{cases}$

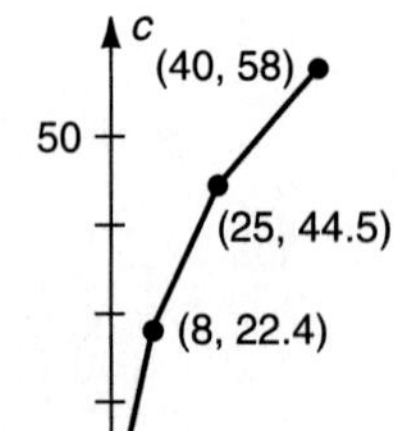

49.

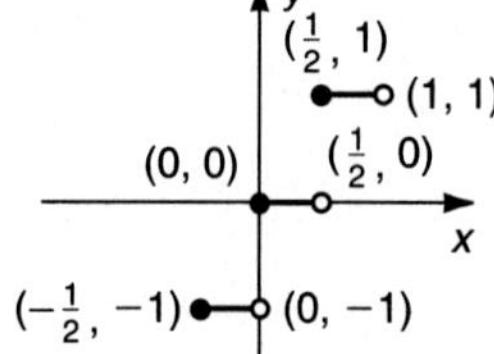

51.

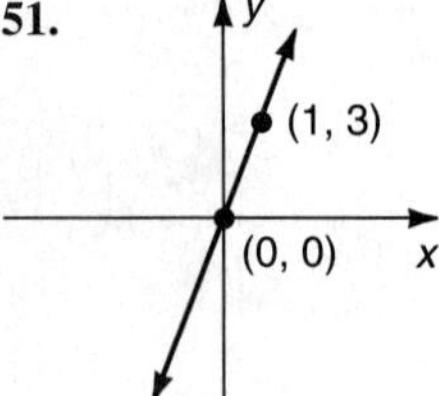

53.

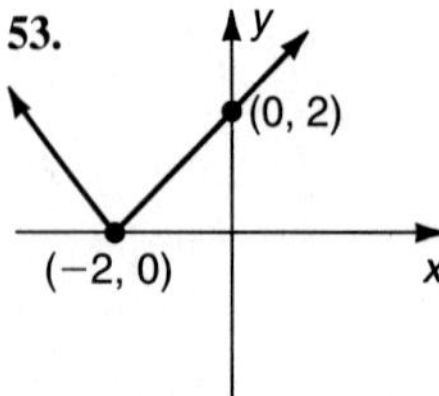

55.

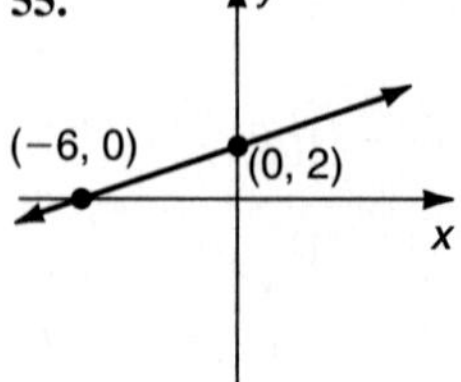

57.

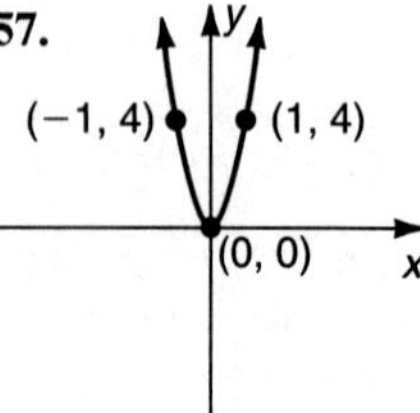

59.

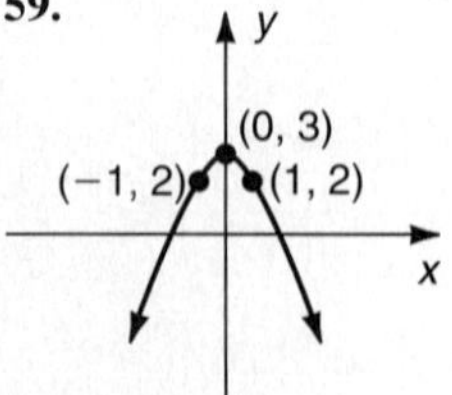

61.

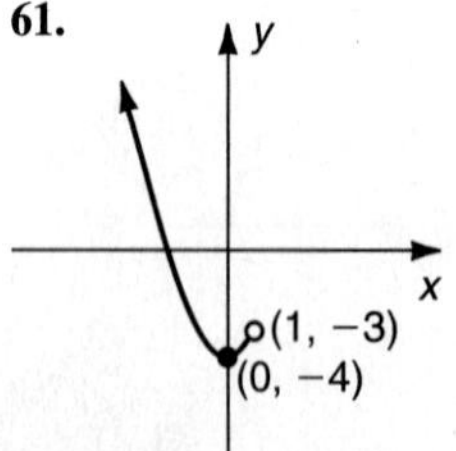

63.

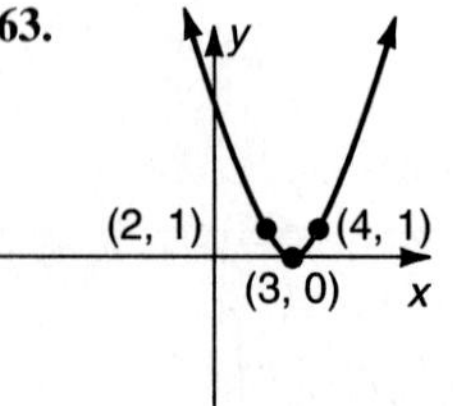

65.

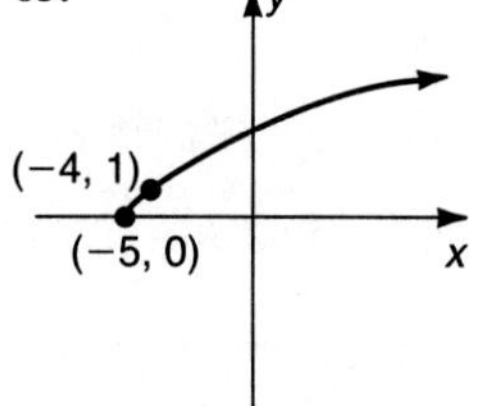

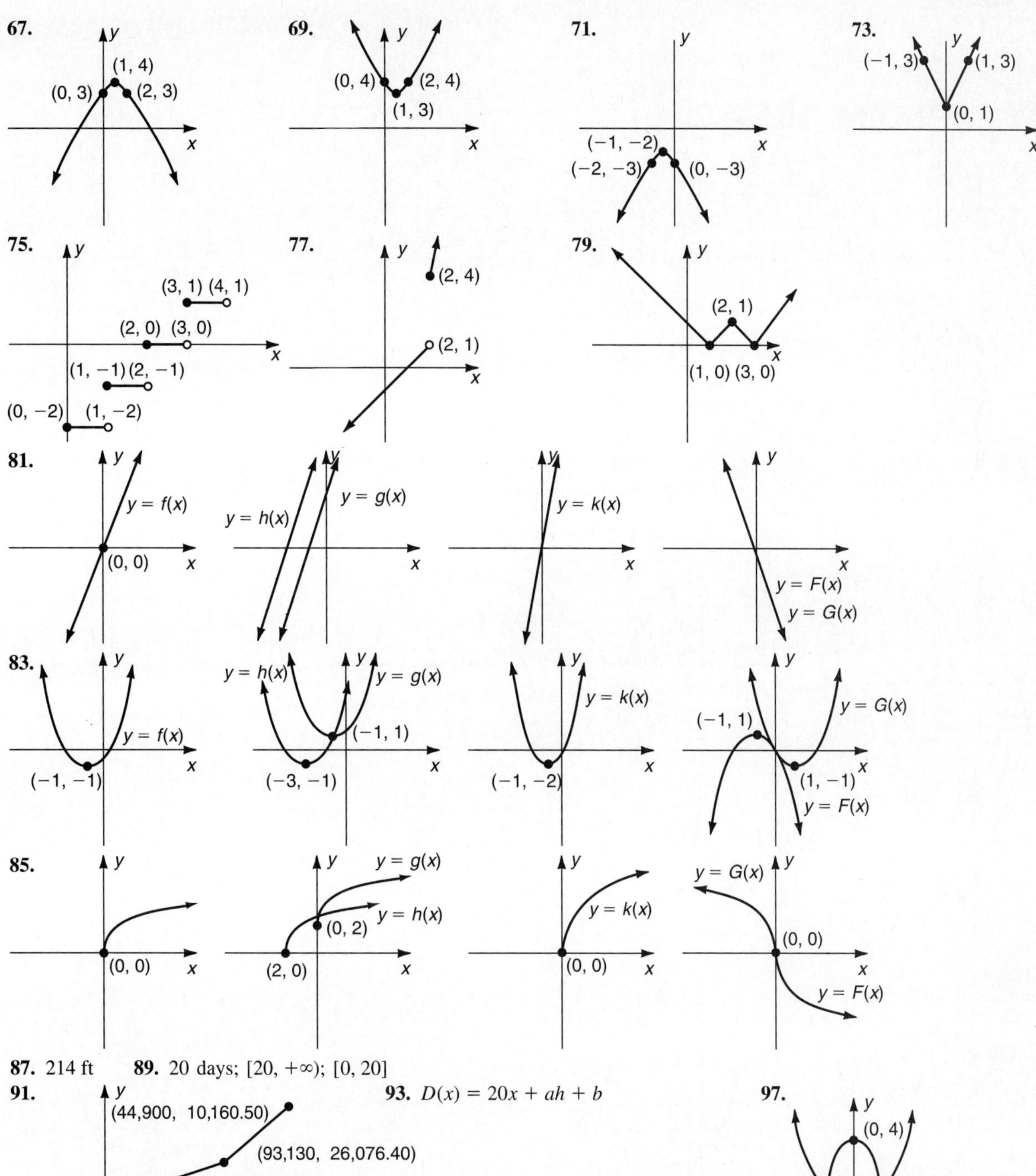

87. 214 ft **89.** 20 days; $[20, +\infty)$; $[0, 20]$

91.

(44,900, 10,160.50)
(93,130, 26,076.40)
(18,550, 2,782.50)

93. $D(x) = 20x + ah + b$

97.

(0, 4)
(−2, 0) (2, 0)

99. $D(x) = \begin{cases} -p & \text{if } x < a \\ p & \text{if } x \geq a \end{cases}$

PROBLEM SET 4.2

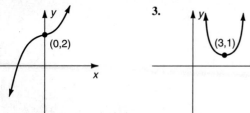

1. **3.**

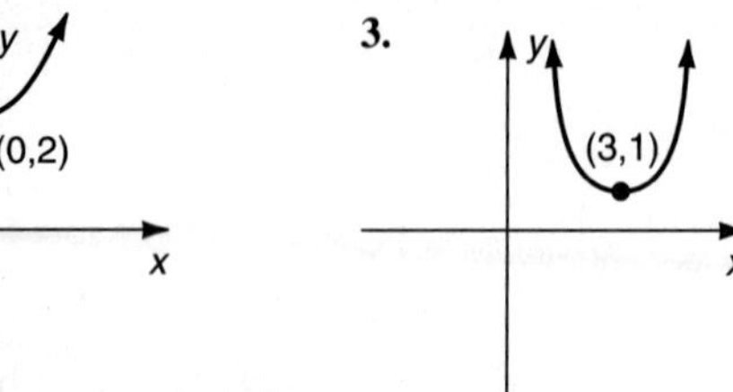

5.

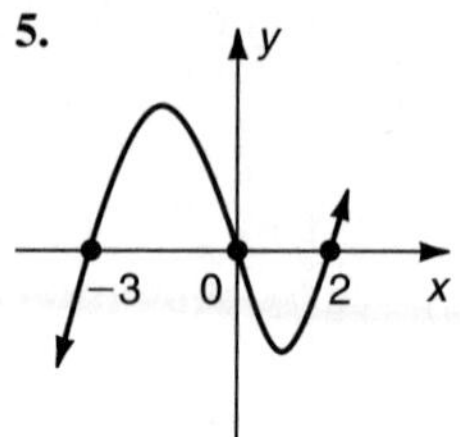

7.

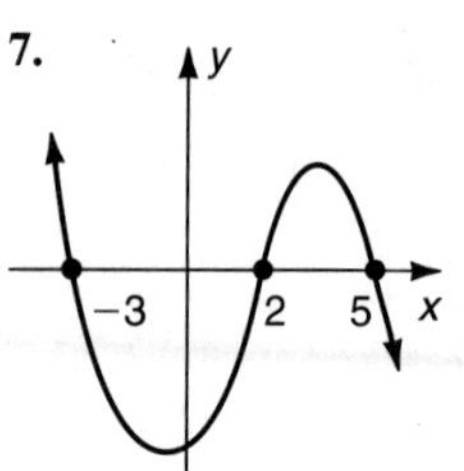

9.

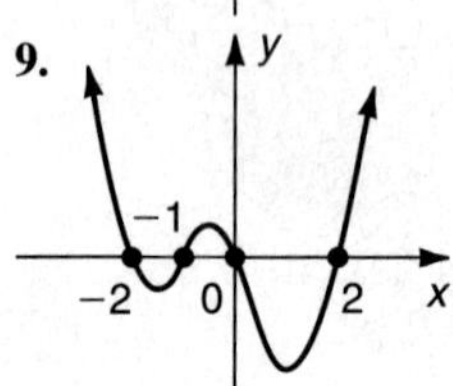

11.

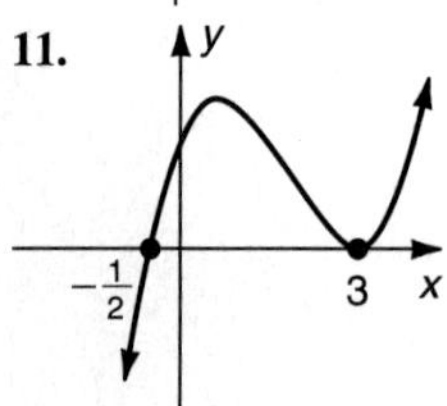

13.

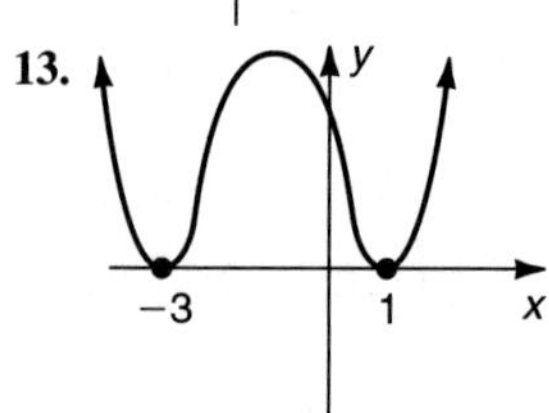

15.

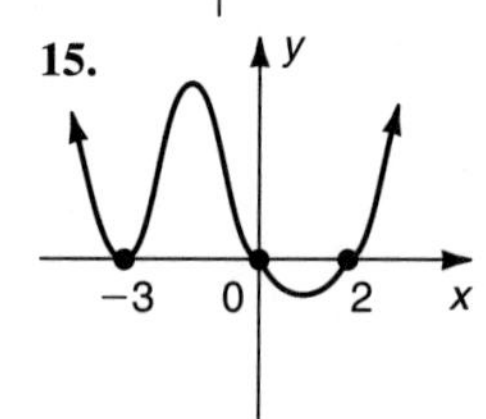

17.

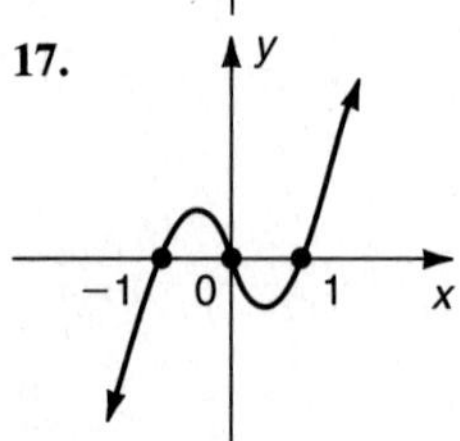

19.

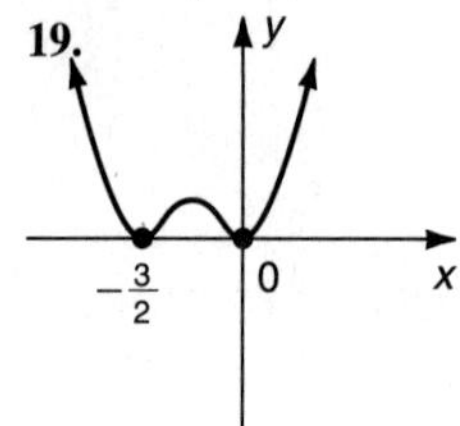

21.

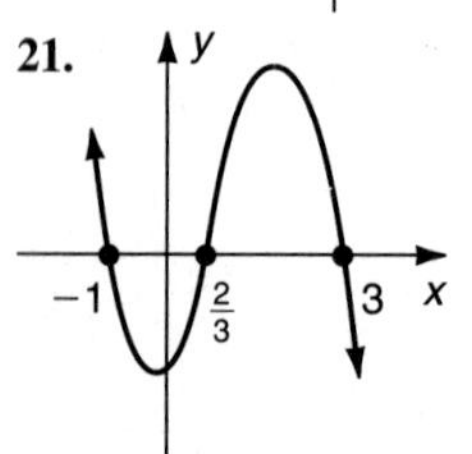

23.

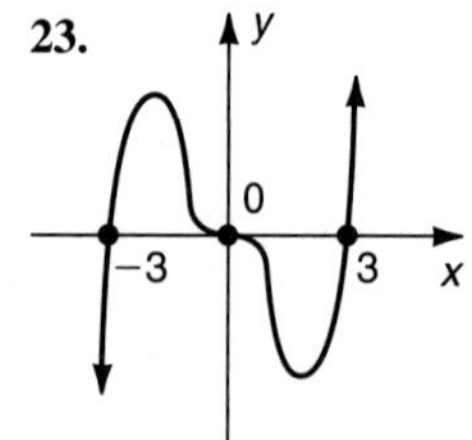

25.

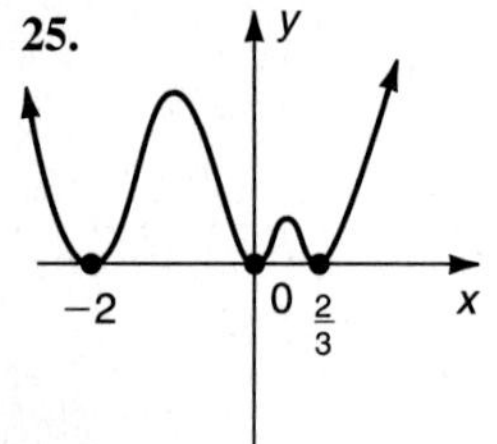

27.

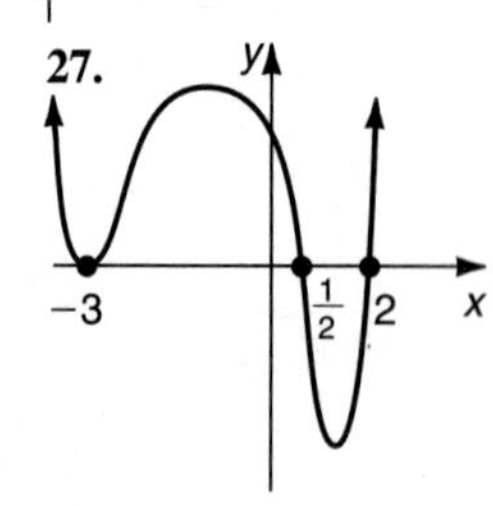

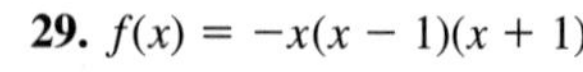

29. $f(x) = -x(x - 1)(x + 1)$

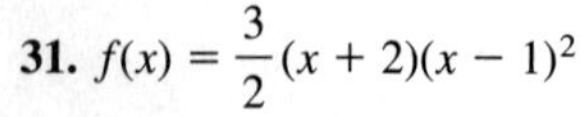

31. $f(x) = \dfrac{3}{2}(x + 2)(x - 1)^2$

33. $f(x) = (x - 2)(x - 3)(x - 4)$

35. $f(x) = (x + 1)(2x + 1)(2x - 1)$

37.

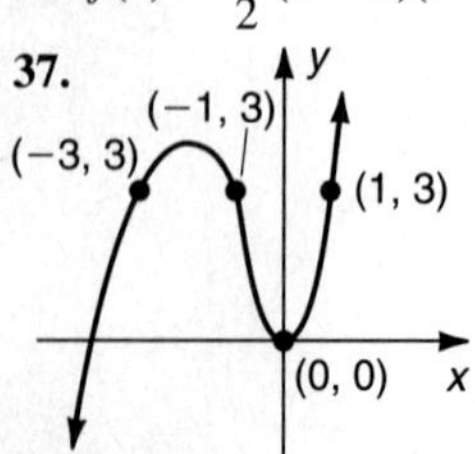

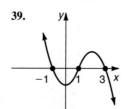

39.

41. $2x^3 - x$

43.

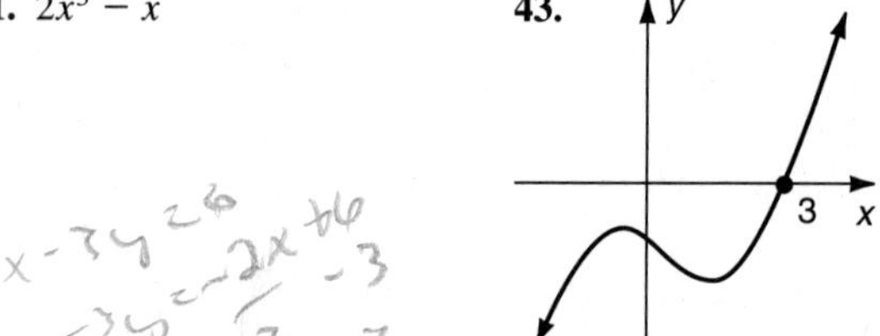

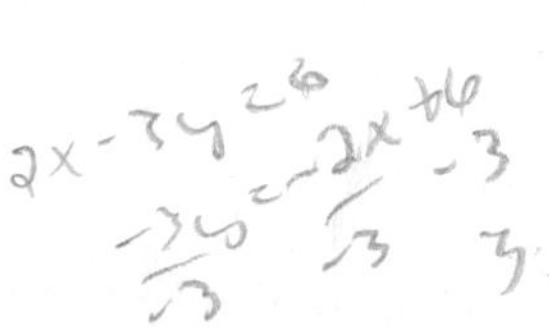

45.

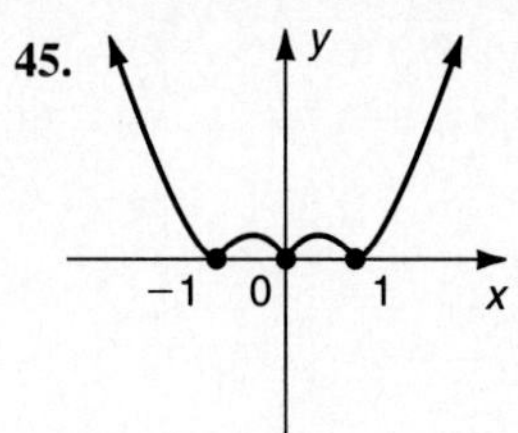

47. Rel. max (0.334, 1.926)
Rel. min (1.500, −1.250)

49. Rel. min (−0.626, −0.146)
Rel. max (0.312, 1.348)
Rel. min (2.564, −14.369)

51. 22,402; 48

PROBLEM SET 4.3

1. $(-\infty, -3) \cup (-3, +\infty)$ **3.** $(-\infty, -2) \cup (-2, 2) \cup (2, +\infty)$ **5.** $y = 0$, $t = 0$

7. $y = -5$, $x = -3$, $x = 1$

9. **11.**

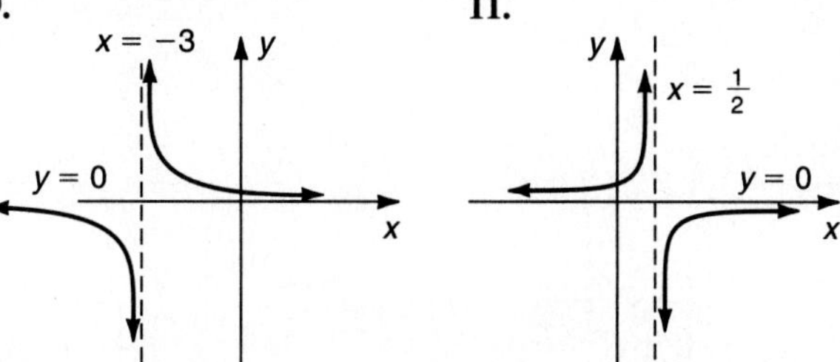

13.

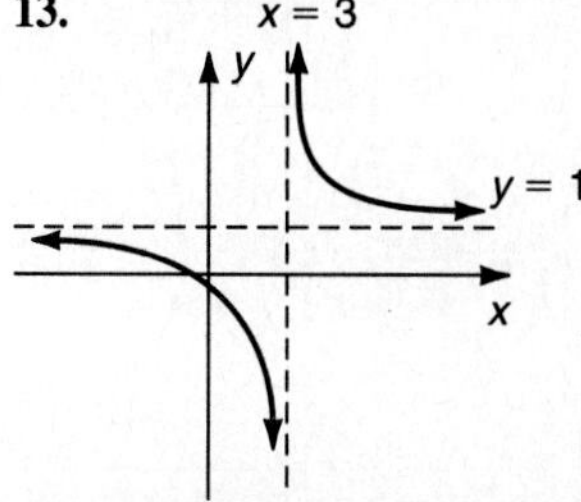

15.

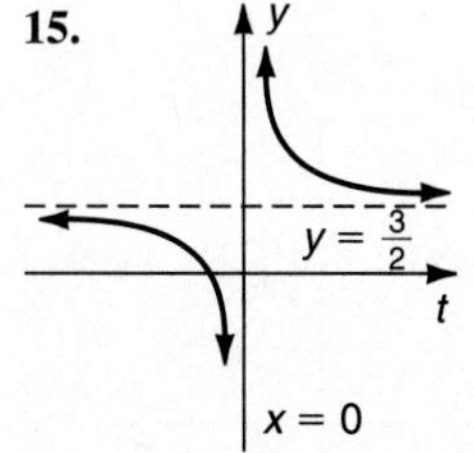

17.

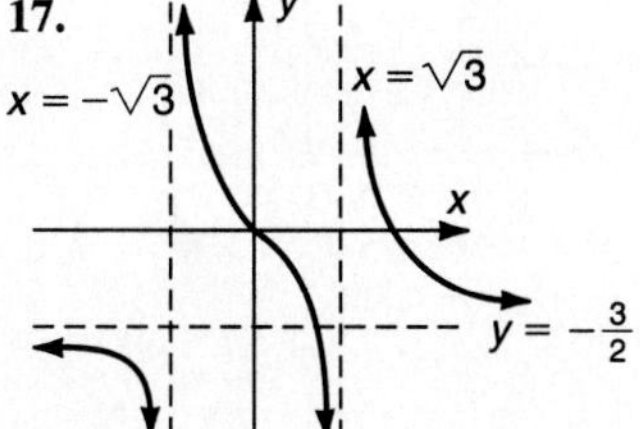

19.

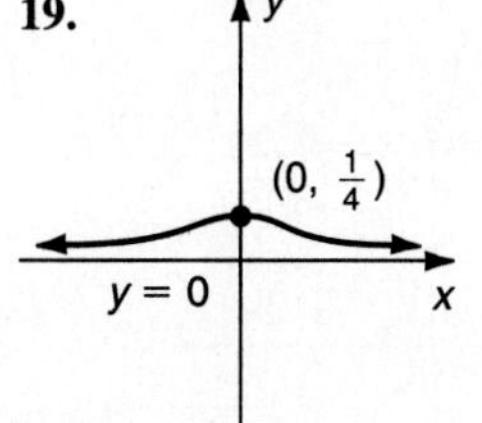

21.

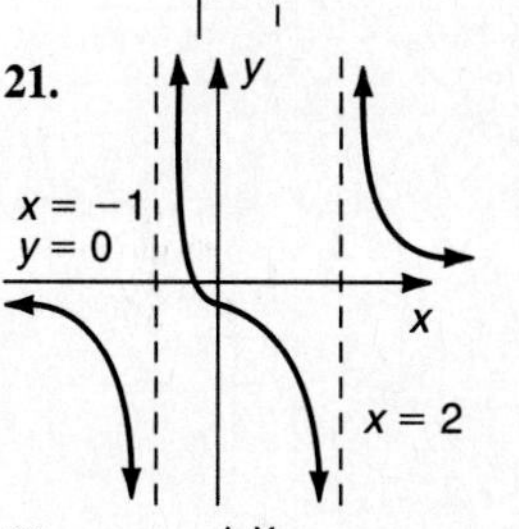

23.

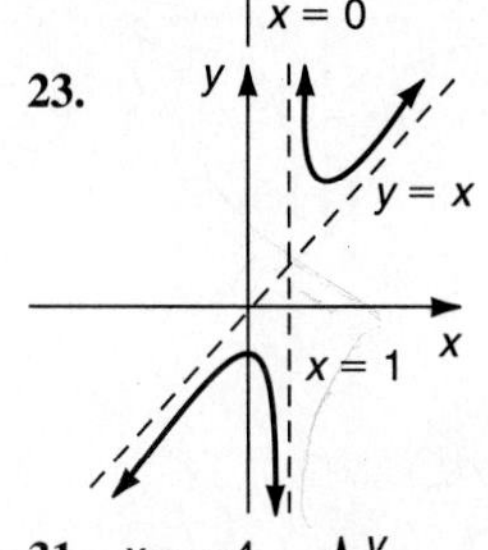

25.

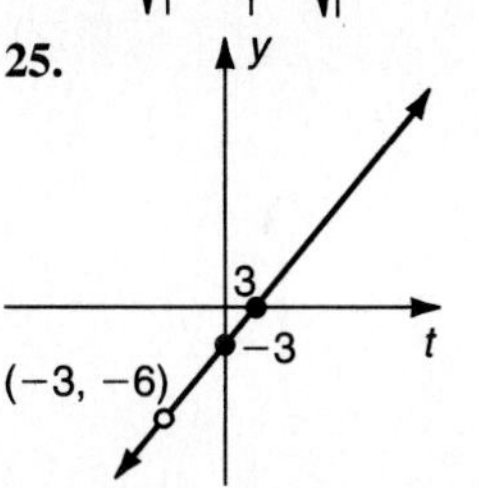

27.

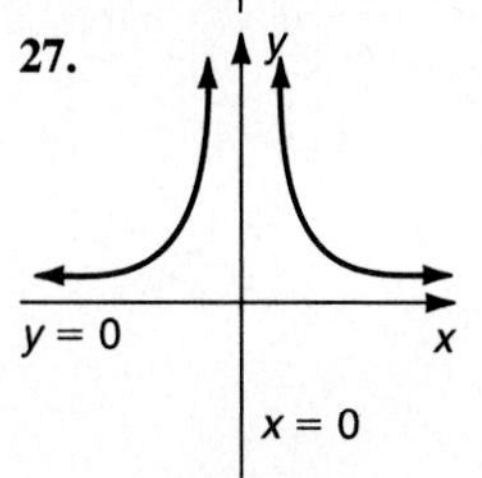

29.

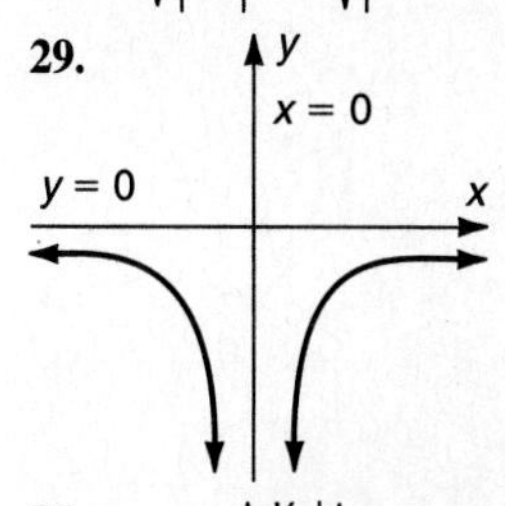

31.

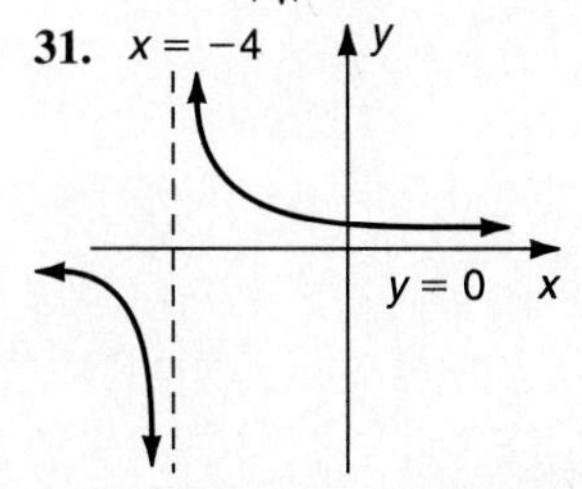

33.

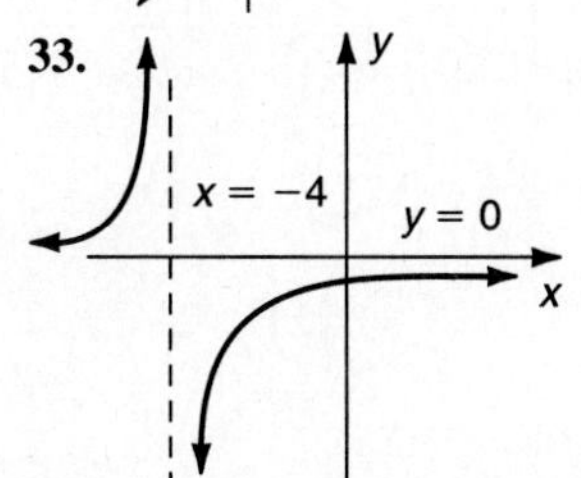

35.

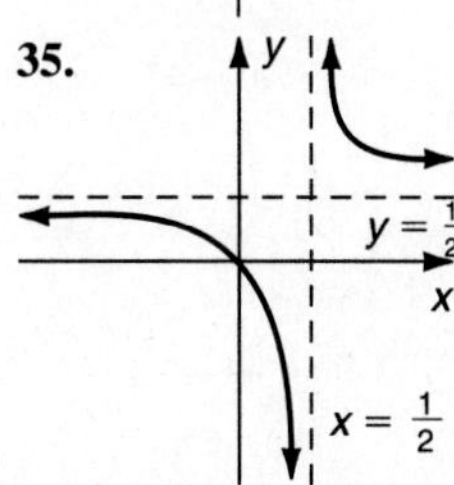

37.

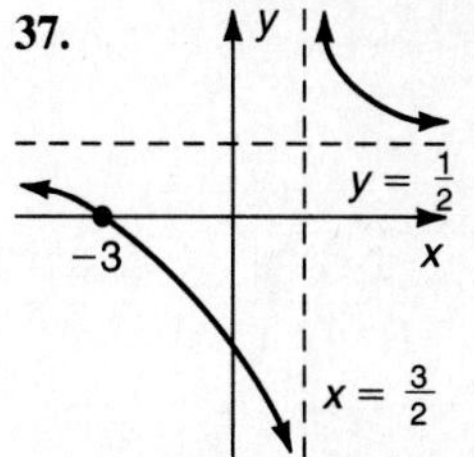

39.

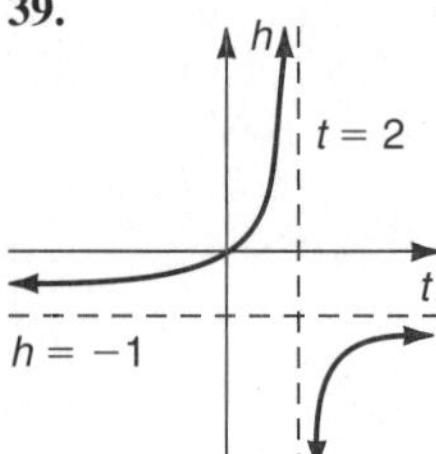

41.

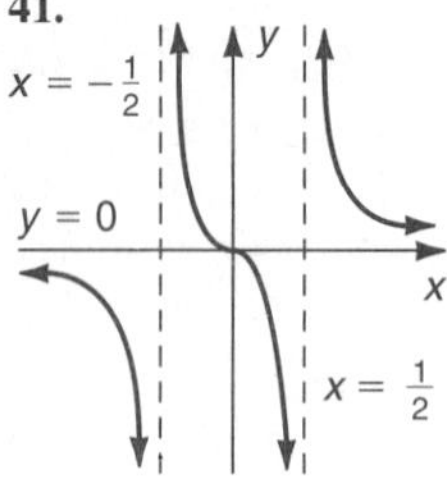

43.

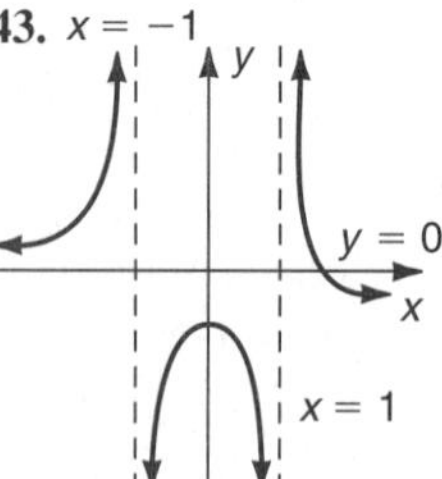

45.

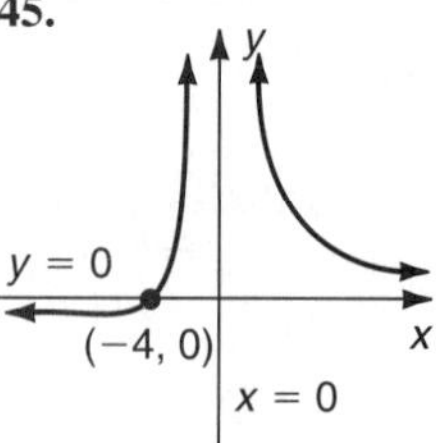

47. $x = -\frac{1}{2}$ $x = \frac{1}{2}$ y $y = \frac{1}{4}$ x

49.

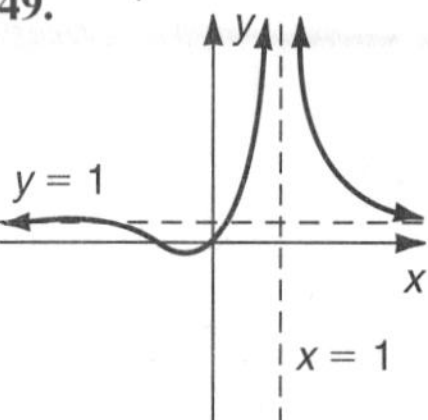

51.

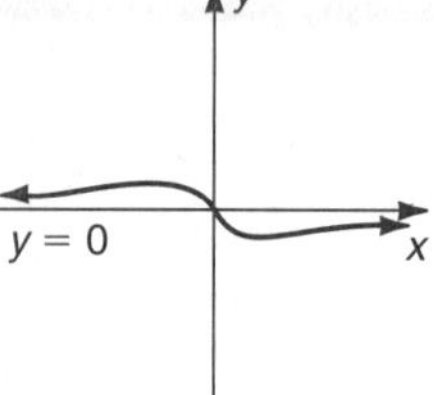

53.

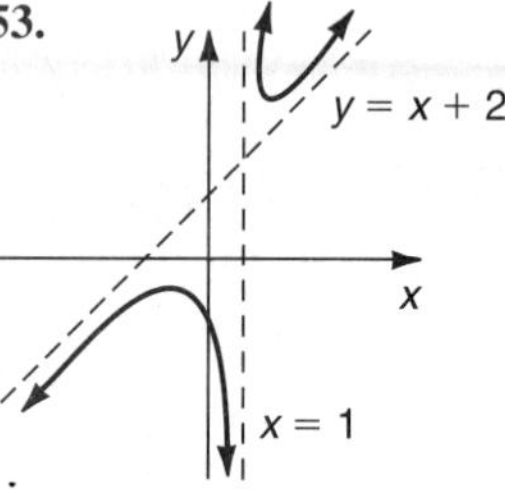

55.

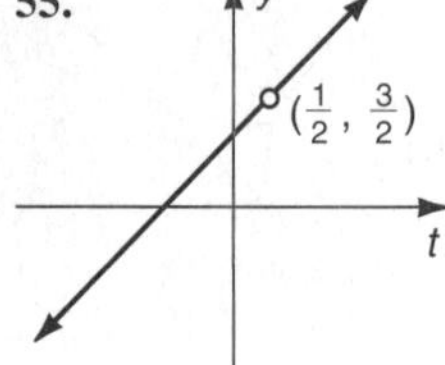

57.

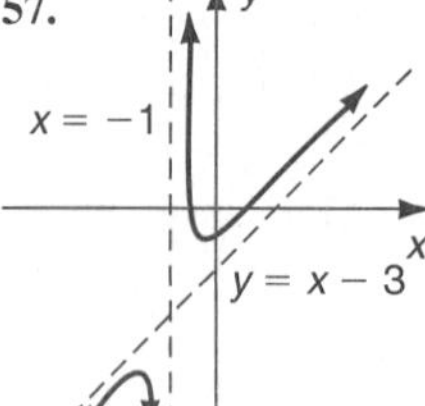

59.

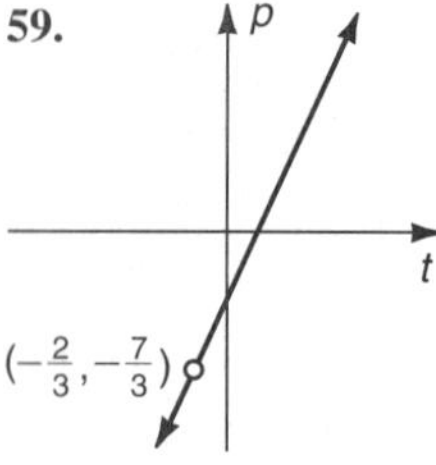

61.

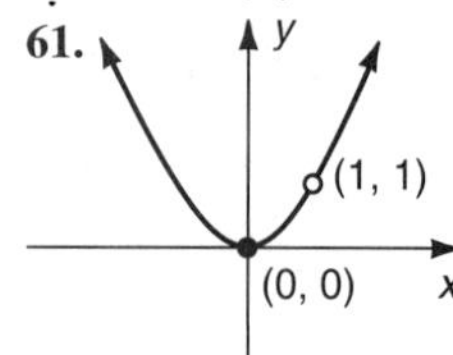

63.

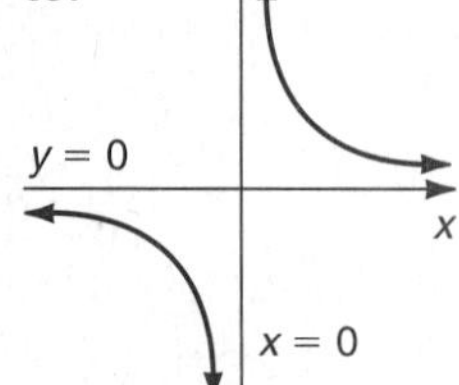

65.

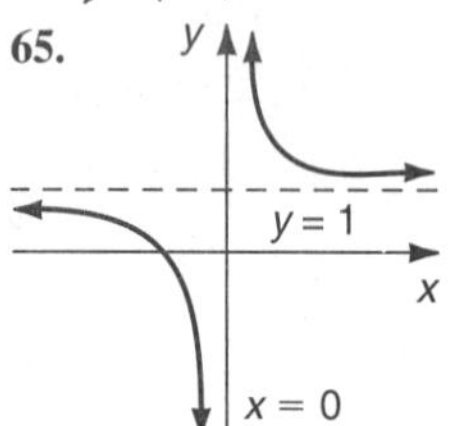

67.

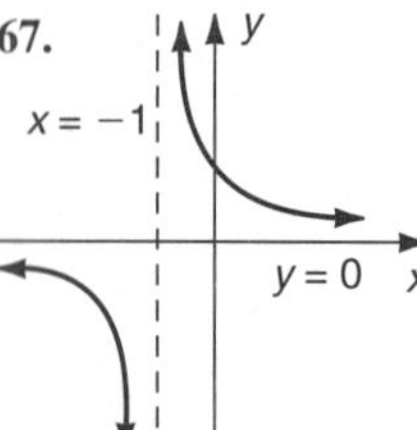

69.

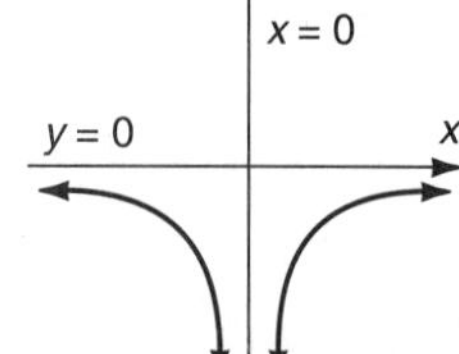

71. y $(-1, 0)$ $(1, 0)$ x $y = -1$ $x = 0$

73. $f(x) = \dfrac{3x}{x - 2}$

75. $f(x) = \dfrac{3x^2}{x^2 - 4}$

77.

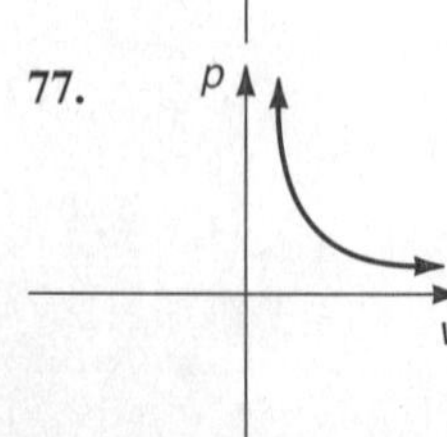

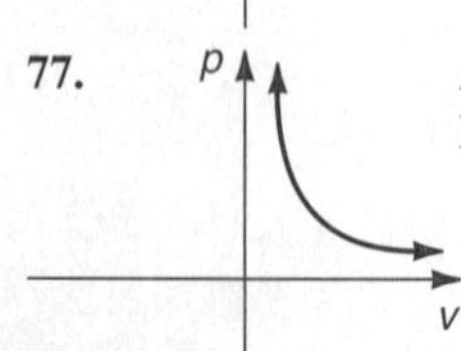

As V increases, P decreases.
If P increases, V decreases.

79.

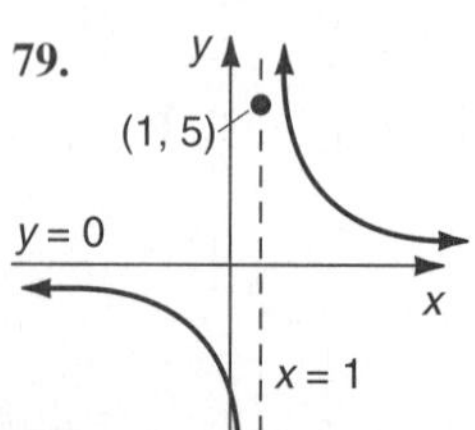

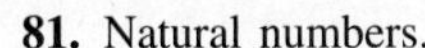
81. Natural numbers.

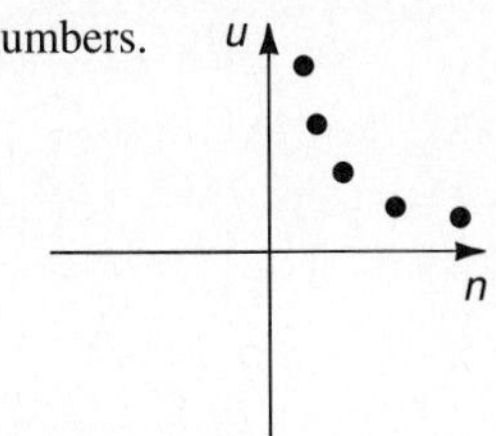

83.

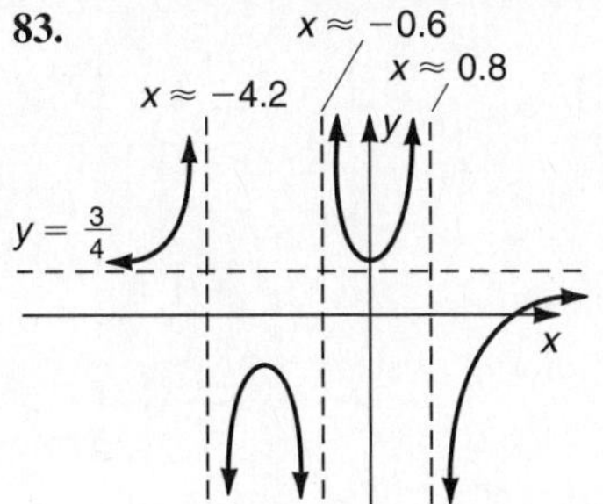

85.

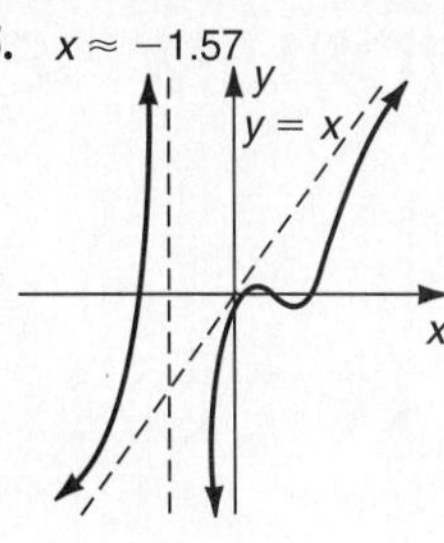

PROBLEM SET 4.4

1. $(f+g)(x) = 5x + 5$; $(f-g)(x) = -x + 5$; $(fg)(x) = 6x^2 + 10x$; $\left(\frac{f}{g}\right)(x) = \frac{2x}{3x+5}$ **3.** $(f+g)(x) = 3 - x^2 + \sqrt{x+1}$; $(f-g)(x) = 3 - x^2 - \sqrt{x+1}$; $(fg)(x) = (3-x^2)\sqrt{x+1}$; $\left(\frac{f}{g}\right)(x) = \frac{3-x^2}{\sqrt{x+1}}$ **5.** 0 **7.** $-\frac{5}{3}$ **9.** $[-2, +\infty)$

11. $[-2, 3) \cup (3, +\infty)$ **13.** $(f \circ g)(x) = 35x - 17$; $(g \circ f)(x) = 35x + 51$ **15.** $(f \circ g)(x) = \frac{1}{x^2 + 6x + 11}$; $(g \circ f)(x) = \frac{1}{x^2+2} + 3$ **17.** -1 **19.** -1 **21.** R; $(-\infty, -2) \cup (-2, +\infty)$ **23.** $\sqrt{x}$ and $3x + 2$; (Other possibilities exist)

25. $\frac{-2}{x-1} + \frac{3}{x+1}$ **27.** $3 + \frac{2}{x-3} - \frac{2}{x+1}$ **29.** $\frac{2}{x} + \frac{3}{x^2} - \frac{1}{x+2}$ **31.** 4 **33.** 1 **35.** 1 **37.** 1 **39.** 1

41. $[-3, -1) \cup (1, +\infty)$ **43.** $4 - \sqrt{5}$ **45.** -480 **47.** 2 is not in the domain. **49.** $\sqrt{11}$ **51.** $\sqrt{2\sqrt{7} + 5}$

53. $(f+g)(x) = \frac{13}{2}x - 1$; $(f-g)(x) = \frac{11}{2}x - 1$; $(fg)(x) = 3x^2 - \frac{1}{2}x$; $\left(\frac{f}{g}\right)(x) = \frac{12x-2}{x}$ **55.** $(f+g)(x) = x^2 - 4$; $(f-g)(x) = x^2 - 2x + 6$; $(fg)(x) = x^3 - 6x^2 + 6x - 5$; $\left(\frac{f}{g}\right)(x) = \frac{x^2 - x + 1}{x - 5}$ **57.** $(f+g)(x) = \sqrt{x+7} + \sqrt{x+4}$; $(f-g)(x) = \sqrt{x+7} - \sqrt{x+4}$; $(fg)(x) = \sqrt{(x+7)(x+4)}$; $\left(\frac{f}{g}\right)(x) = \sqrt{\frac{x+7}{x+4}}$ **59.** $(f \circ x)(x) = x$; Domain R; $(g \circ f)(x) = x$; Domain R **61.** $(f \circ g)(x) = \sqrt{x^2 + 2}$; Domain R; $(g \circ f)(x) = x + 2$; Domain $[-2, +\infty)$

63. $(f \circ g)(x) = \frac{2}{x^2+1}$; Domain R; $(g \circ f)(x) = \frac{4}{x^2} + 1$; Domain $(-\infty, 0) \cup (0, +\infty)$ **65.** $(f \circ g)(x) = \frac{x}{2 + 2x}$; Domain $(-\infty, -1) \cup (-1, 0) \cup (0, +\infty)$; $(g \circ f)(x) = 2x + 4$, $(-\infty, -2) \cup (-2, +\infty)$ **67.** $f(x) = x^3$; $g(x) = x + 2$ (Others exist) **69.** $f(x) = \frac{3}{x}$; $g(x) = x^2 - 5$ (Others exist) **71.** $f(x) = \sqrt{x}$; $g(x) = x^2 - 1$ (Others exist) **73.** $b + c$

75. c **77.** $\frac{-2}{x+1} - \frac{3}{x-3}$ **79.** $\frac{1}{2x+1} + \frac{2}{x-3}$ **81.** $\frac{3}{t+1} - \frac{1}{(t-1)^2}$ **83.** $2 + \frac{-8}{5(3x+2)} + \frac{1}{5(x-1)}$

85. (a) $[-\sqrt{2}, -1] \cup [1, \sqrt{2}]$; $[0, 1]$ (b) $\{0\}$; $\{0\}$ **87.** (a) $\left(0, \frac{1}{2}\right) \cup \left(\frac{1}{2}, 1\right]$; $(-\infty, -1] \cup (0, +\infty)$ (b) $(-\infty, -\sqrt{2}] \cup [\sqrt{2}, +\infty)$; $[0, +\infty)$ **89.** $\frac{-3}{(x-1)^2} + \frac{-1}{x+1} + \frac{5}{(x+1)^2}$

PROBLEM SET 4.5

1. Yes **3.** Yes **5.** Yes **7.** No **9.** Yes **11.**

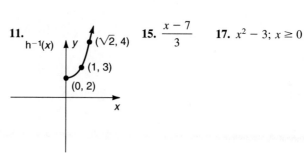

$h^{-1}(x)$ y ($\sqrt{2}$, 4) (1, 3) (0, 2) x

15. $\dfrac{x-7}{3}$ **17.** $x^2 - 3$; $x \geq 0$

19. 1 to 1

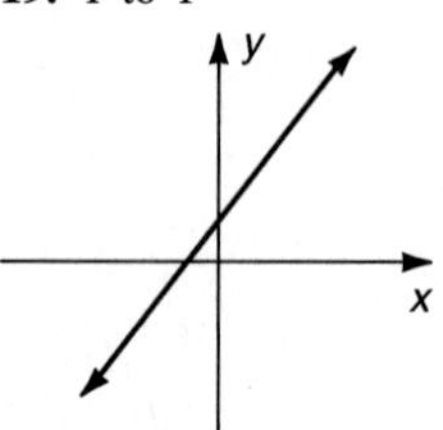

21. Not 1 to 1

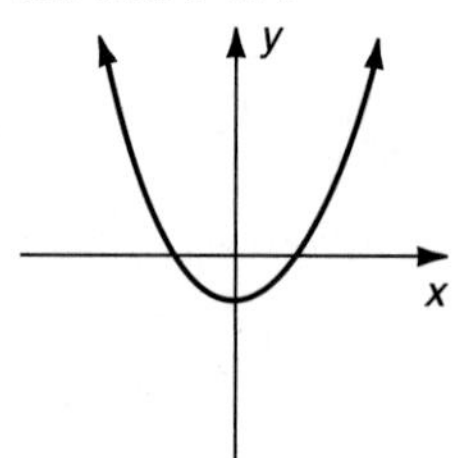

23. Not 1 to 1

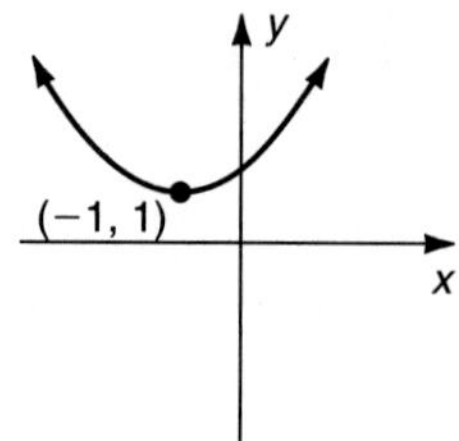

25. Not 1 to 1

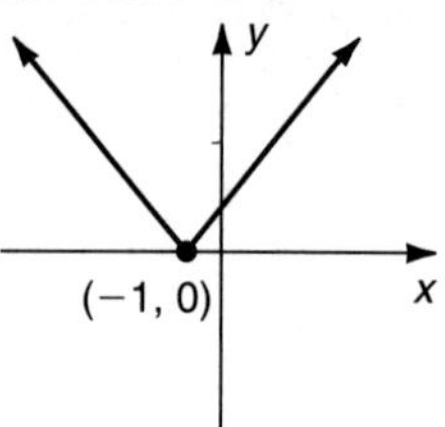

27. 1 to 1

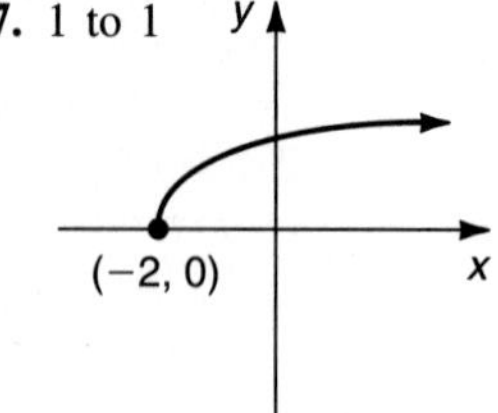

29. Not 1 to 1

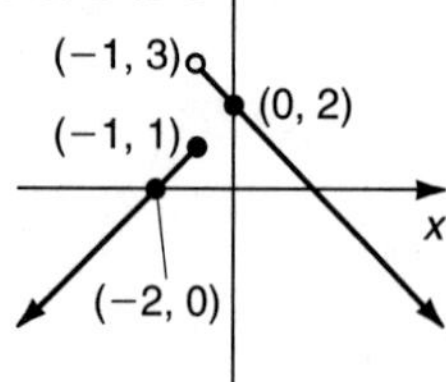

31. $f^{-1}(x) = x - 3$
Dom f: R
Ran f: R
Dom f^{-1}: R
Ran f^{-1}: R

y y = f(x) (0, 3) (3, 0) (−3, 0) x y = $f^{-1}(x)$ (0, −3)

33. $f^{-1}(x) = \dfrac{1}{x-1}$
Dom f: $x \neq 0$
Ran f: $x \neq 1$
Dom f^{-1}: $x \neq 1$
Ran f^{-1}: $x \neq 0$

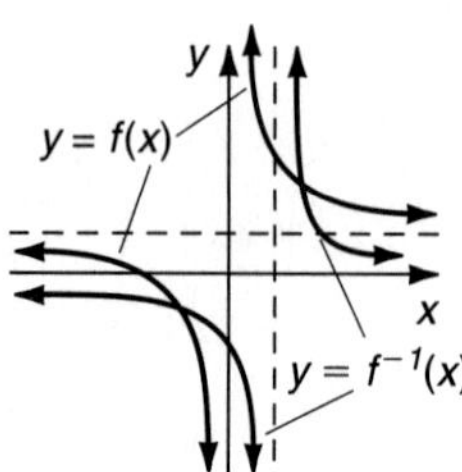

35. $f^{-1}(x) = x^2 - 7$; $x \geq 0$
Dom f: $[-7, +\infty)$
Ran f: $[0, +\infty)$
Dom f^{-1}: $[0, +\infty)$
Ran f^{-1}: $[-7, +\infty)$

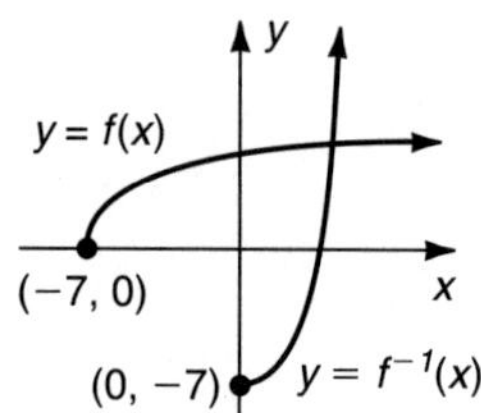

37. $f^{-1}(x) = \sqrt{x-1}$
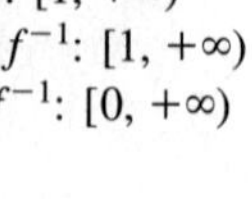
Dom f: $[0, +\infty)$
Ran f: $[1, +\infty)$
Dom f^{-1}: $[1, +\infty)$
Ran f^{-1}: $[0, +\infty)$

y y = f(x) (0, 1) x (1, 0) y = $f^{-1}(x)$

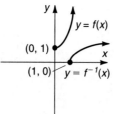

39. Yes; $\dfrac{x+9}{5}$ **41.** Yes; $\dfrac{1+t}{1-t}$

43. No **45.** No **47.** No **49.** Yes; $\sqrt{x} + 2$ **51.** Yes; $\theta^{2/3}$, $\theta \geq 0$

53. Yes

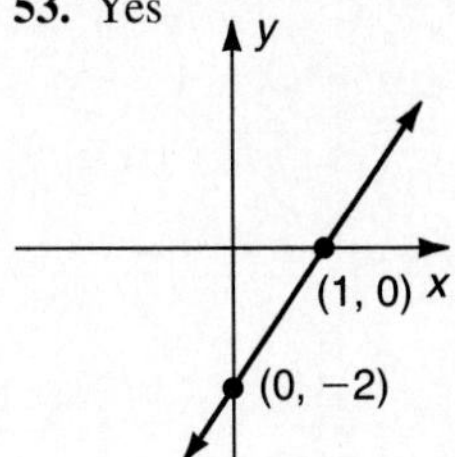

55. No

57. Yes

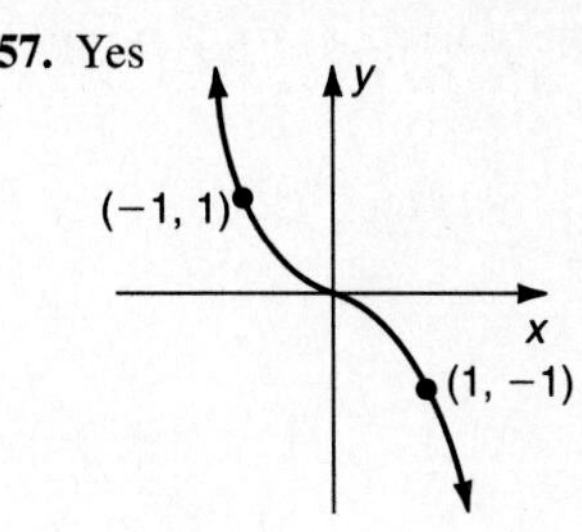

63. $(f \circ g)^{-1}(x) = \dfrac{x+9}{2}$; $(f^{-1} \circ g^{-1})(x) = \dfrac{x+9}{2}$

PROBLEM SET 4.6

1. 144 **3.** $\dfrac{55}{2}$ **5.** 56 **7.** Multiplied by 4. **9.** 16.1 ft/sec^2; 6.84 m/sec^2; 27.4 m; 64.4 ft **11.** $M = kt^3$
13. $U = \dfrac{k}{L^2}$ **15.** $x = \dfrac{kU}{s^2}$ **17.** $I = \dfrac{kw_1w_2}{\sqrt{t}}$ **19.** $z = kd^2$; 5 **21.** $R = \dfrac{k}{n^3}$; $\dfrac{81}{40}$ **23.** $\theta = \dfrac{kt^2}{w}$; $\dfrac{1}{2}$
25. $A = \dfrac{ka_1c^4}{a_2}$; 0.048 **27.** $\sqrt{2}$ sec **29.** 3,600 rpm **31.** 134 lbs **33.** 92.16 watts
35. The resistance increases by a factor of 16. **37.** 174,879 lbs

CHAPTER 4 REVIEW PROBLEMS

1. 1 to 1

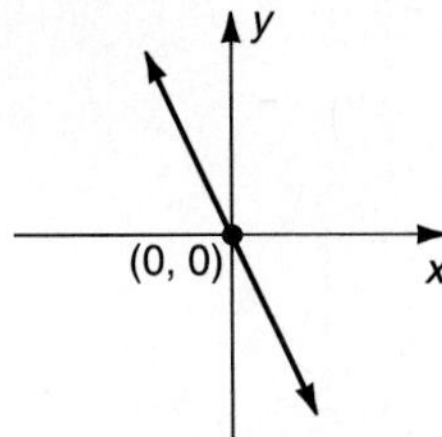

3. Not 1 to 1

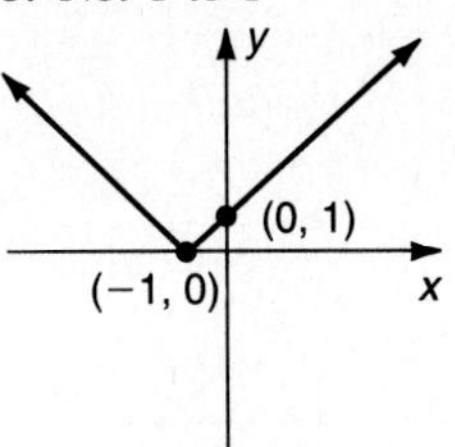

5. Not 1 to 1

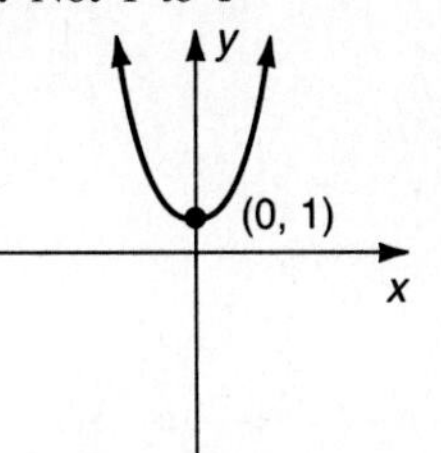

7. 1 to 1

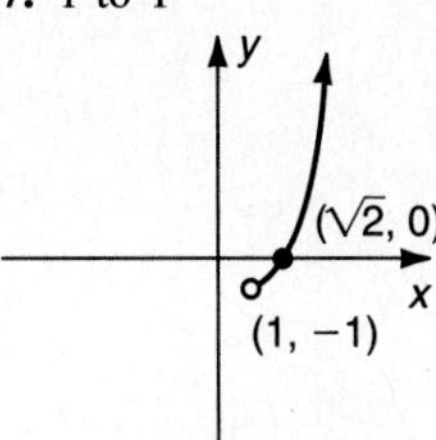

9. Not 1 to 1

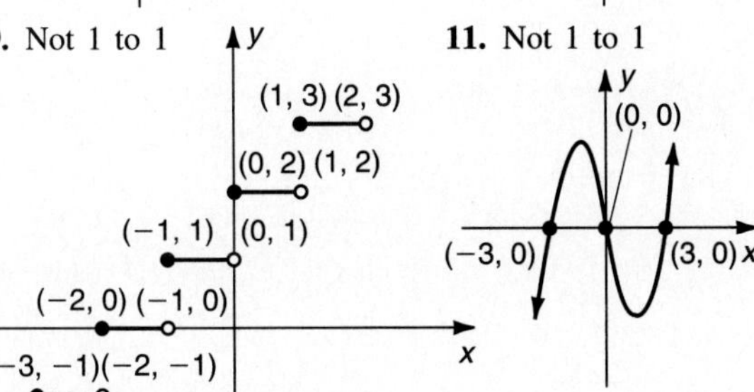

11. Not 1 to 1

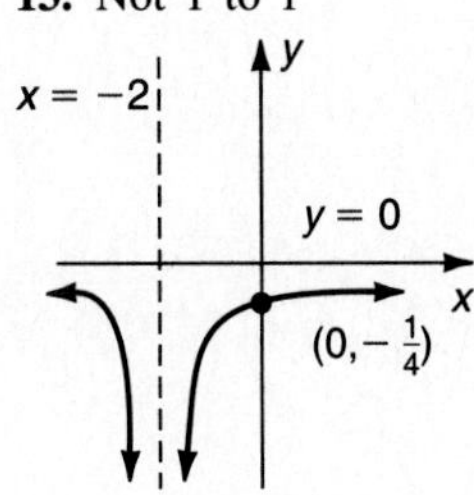

13. Not 1 to 1

15. Not 1 to 1

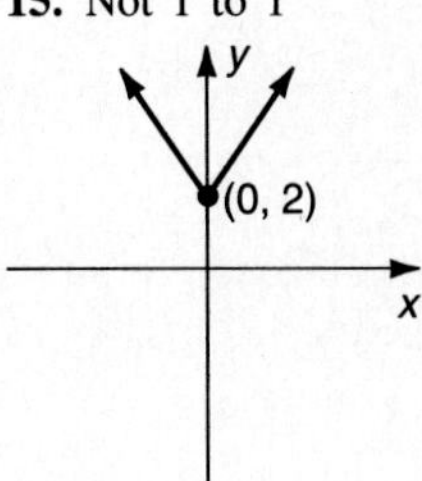

17. 1 to 1

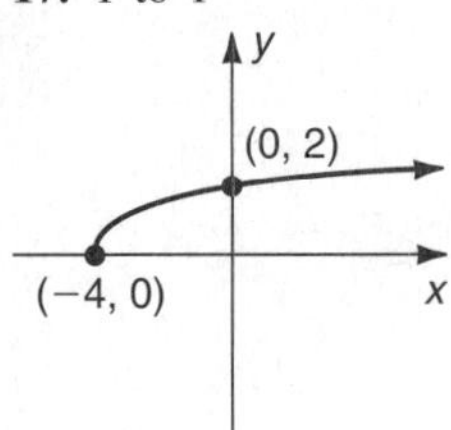

19. Not 1 to 1

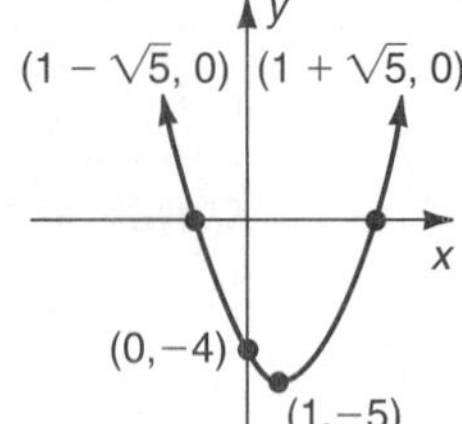

21. Not 1 to 1

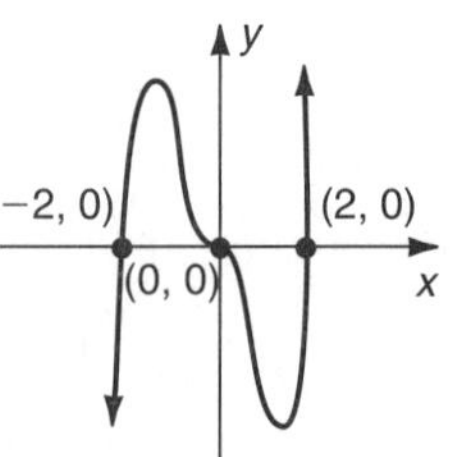

23. 1 to 1

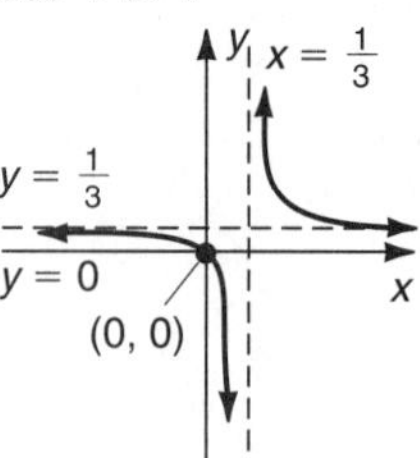

25. Not 1 to 1

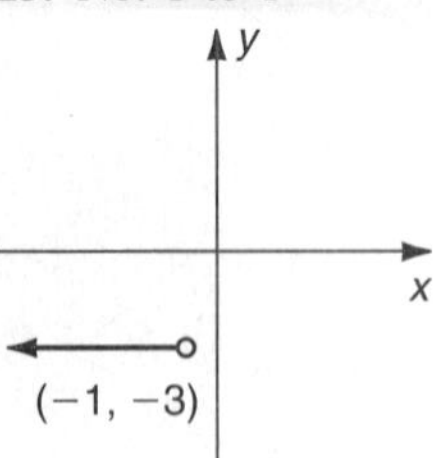

27. Not 1 to 1

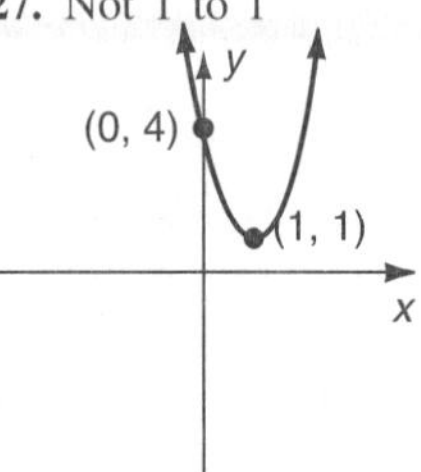

29. Not 1 to 1

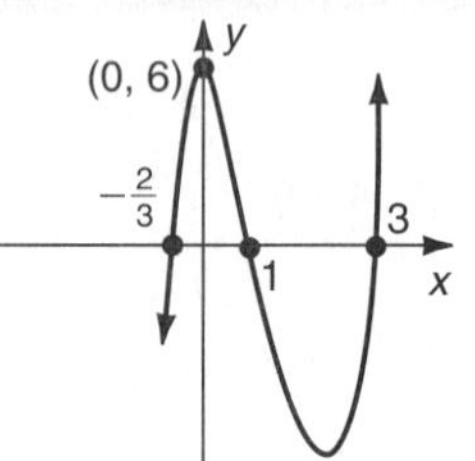

31. Not 1 to 1

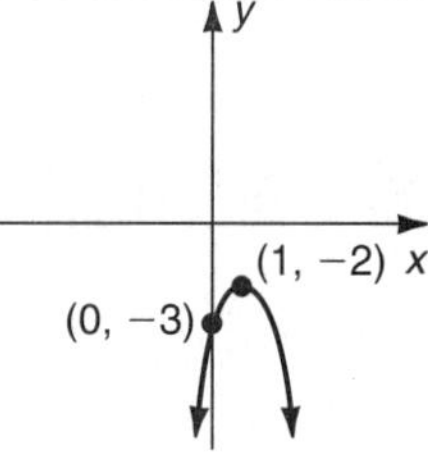

33. 1 to 1

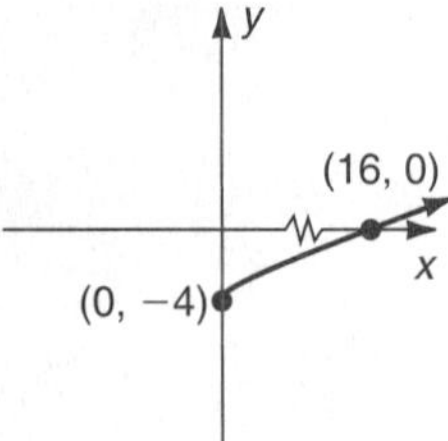

35. 1 to 1

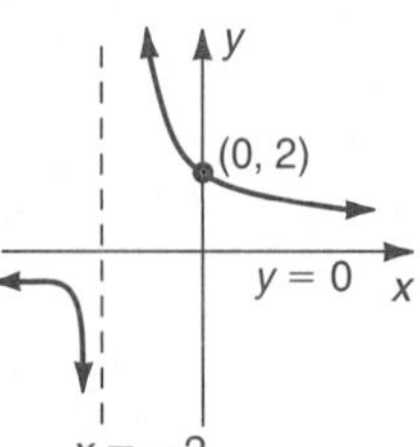

37. Not 1 to 1

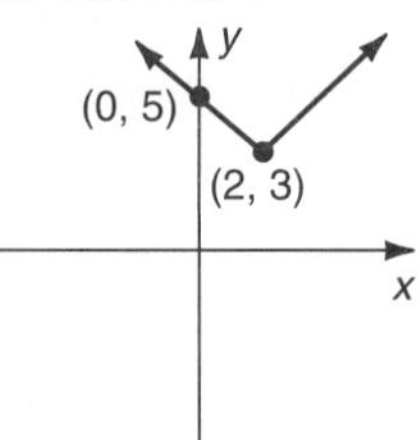

39. Not 1 to 1

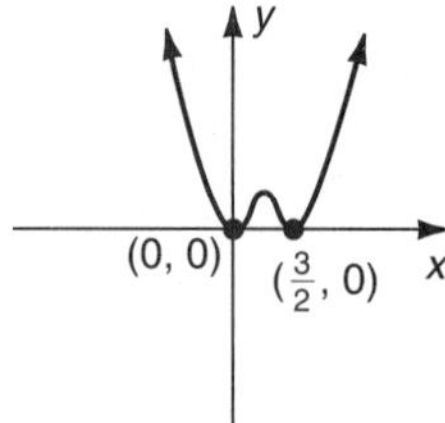

41. −4 **43.** Undefined **45.** 3 **47.** −1

49.

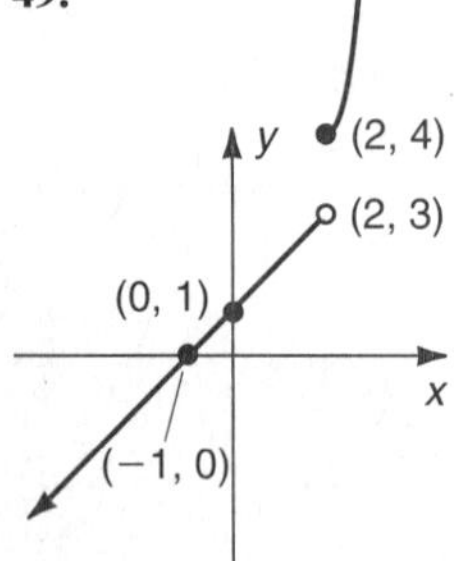

51.

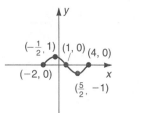

53.

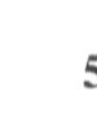
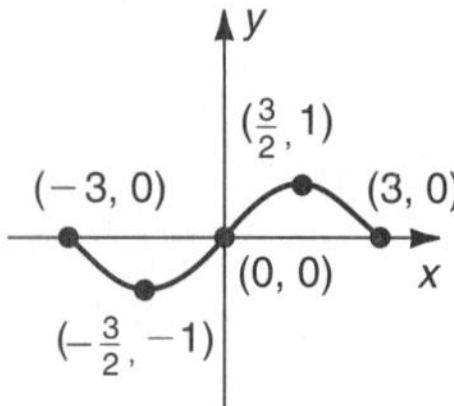

55.

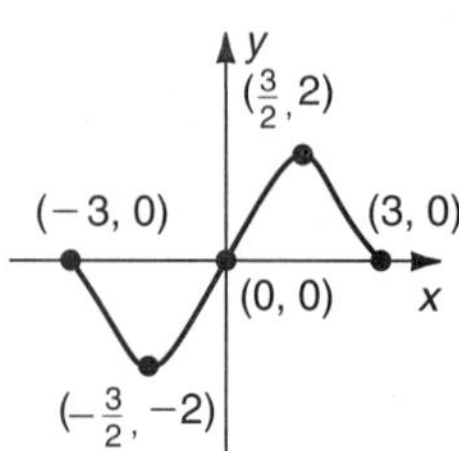

57. $(f+g)(x) = \frac{9}{2}x - 2$; $(f-g)(x) = \frac{3}{2}x - 2$; $(fg)(x) = \frac{9}{2}x^2 - 3x$; $\left(\frac{f}{g}\right)(x) = \frac{6x-4}{3x}$ **59.** 0 **61.** 1

63. $\sqrt{2\sqrt{3}+1}$ **65.** $(f \circ g)(x) = \frac{x^2+4}{3}$; $(g \circ f)(x) = \frac{x^2}{9} + 4$ **67.** $h(x) = (f \circ g)(x)$ where $f(x) = x^4$ and $g(x) = x - 2$

69. $\frac{3}{t-1} - \frac{2}{t+3}$

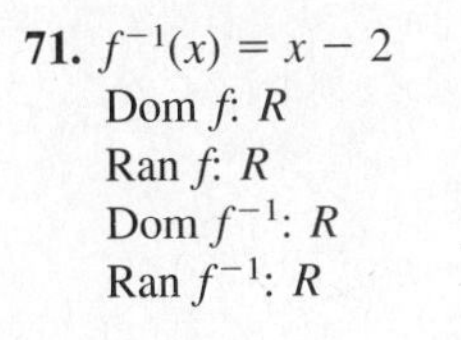

71. $f^{-1}(x) = x - 2$
Dom f: R
Ran f: R
Dom f^{-1}: R
Ran f^{-1}: R

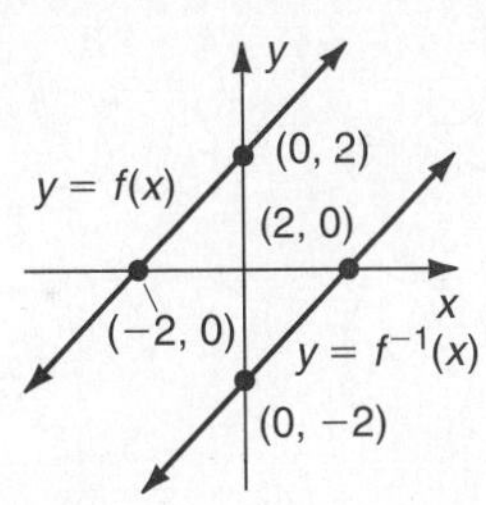

73. $f^{-1}(x) = x^2 + 3$; $x \geq 0$
Dom f: $[3, +\infty)$
Ran f: $[0, +\infty)$
Dom f^{-1}: $[0, +\infty)$
Ran f^{-1}: $[3, +\infty)$

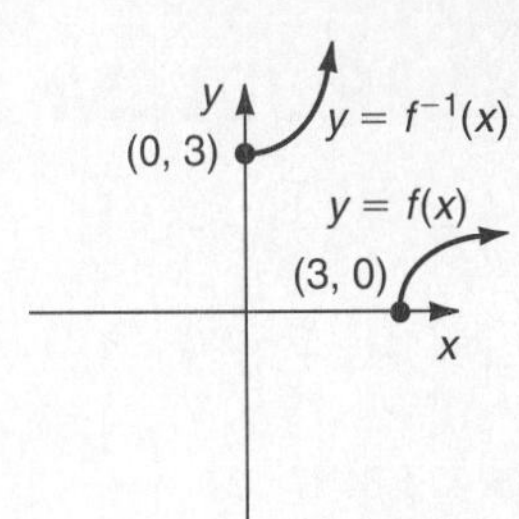

75. $\frac{81}{4}\pi^2$; It's the area of the circular ripple; 324π **79.** 1-*c*, 2-*d*, 3-*b*, 4-*e*, 5-*a*

PROBLEM SET 5.1

1. $2x^2 + 3x + 3 + \frac{2}{x-2}$ **3.** $x^3 + 2x^2 + 2x - 1 + \frac{1}{x-2}$ **5.** 65 **7.** −11 **9.** $x^3 + ix^2 - 3x - (1 - 3i) + \frac{1+i}{x-i}$
11. no **13.** $x^3 + 3x^2 + 3x + 1 = (x^2 + 4x + 7)(x - 1) + 8$ **15.** $q^3 + 3q^2 - 4q - 12 = (q^2 + 5q + 6)(q - 2)$
17. $t^3 - 5t^2 + 3t + 10 = (t^2 - 2t - 3)(t - 3) + 1$ **19.** $2x^3 + 2x^2 + 5x + 5 = (2x^2 + 5)(x + 1)$
21. $3k^3 + 7k^2 - 25k + 10 = (3k^2 - 5k - 5)(k + 4) + 30$ **23.** $x^6 - 2^6 = (x^5 + 2x^4 + 4x^3 + 8x^2 + 16x + 32)(x - 2)$
25. $x^4 + 2x^2 + 1 = (x^3 + ix^2 + x + i)(x - i)$ **27.** 4 **29.** 8 **31.** 0 **33.** 1 **35.** 1052 **37.** 9 **39.** 0
41. no **43.** no **45.** no **47.** yes **49.** $(x + 1)(x^4 - x^3 + x^2 - x + 1)$ **51.** $2x^3 + 4x^2 - 4x - 2$
53. $3z^3 + 6z^2 - 2z + \frac{11}{3} + \frac{7}{9z+3}$ **55.** $x^2 - (2 + \sqrt{2})x + 1 + \sqrt{2}$ **57.** −16 **59.** $x^3 + x^2 - 2x + 1$
61. $x^3 - kx^2 + k^2x - k^3$ **63.** $x^2 + kx + k^2 + \frac{k^3}{x-k}$ **65.** (a) $2x + 1$ (b) $x^2 + x + 1$

PROBLEM SET 5.2

1.

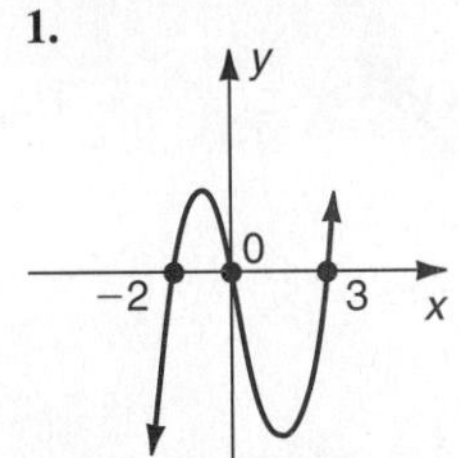

Zeros: −2, 0, 3

3.

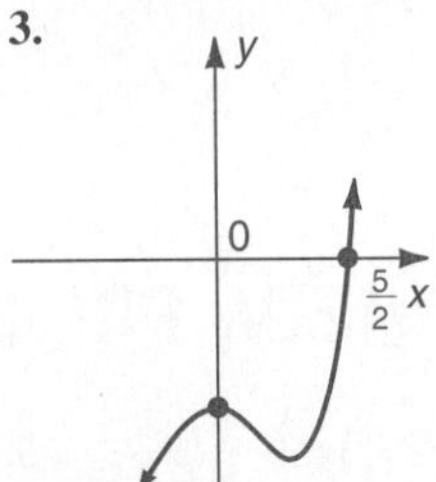

Zeros: 5/2, $\pm i$

5. −91 **7.** 24 **9.** $P(4) = 0$ **11.** $x(x + 3)(x - 2)^2 = 0$ or $x^2(x + 3)(x - 2) = 0$ or $x(x + 3)^2(x - 2) = 0$
13. $f(x) = (2x - 1)^2$ **15.** $\{1, -1 \pm \sqrt{2}\}$ **17.** −4, −2, 1, 3 **19.** $(z + 3)(z + 1)^2$ **21.** $(x + 2)(x - 3)$
$\left(x - \frac{3 + \sqrt{13}}{2}\right)\left(x - \frac{3 - \sqrt{13}}{2}\right)$ **23.** $5, -\frac{1}{2} \pm \frac{\sqrt{3}}{2}i$ **25.** $-5, -\sqrt{2}, \sqrt{2}, \pm i$ **27.** $\{4, 1 \pm \sqrt{2}\}$
29. $\{-1, 0, 1, 11\}$ **31.** $\{1, \pm\sqrt{2}, -1 \pm i\}$ **33.** $\left\{\frac{2}{3}, \pm i\right\}$ **35.** a) 3 real of multiplicity 1 b) 3 real zeros, r_1 multiplicity 1, r_2 multiplicity 2 c) 2 real both of multiplicity 2 **37.** At most 2; no real zeros when discriminant is negative.
39. A is $\left(\frac{1}{3}, \frac{52}{27}\right)$; B is $\left(\frac{3}{2}, -\frac{5}{4}\right)$ **41.** 45.0625 **43.** $2i$

PROBLEM SET 5.3

1. $6\left(x-\left(-\frac{2}{3}\right)\right)(x-5)\left(x-\frac{3}{2}\right)$ **3.** $(x-1-i)(x-1+i)$ **5.** $f(x)=\frac{5}{2}(x-1)^3(x-2)$
7. $F(x)=3(x+2)^2(x+3)^2$ **9.** $\frac{1}{3}, 3\pm\sqrt{2}$ **11.** $\left\{-1,-\frac{1}{3},\frac{1}{2}\right\}$ **13.** $\left\{-1,-\frac{1}{2},2\right\}$ **15.** none **17.** none
19. $\{-1,2,3\}$ **21.** $\left\{-\frac{1}{2},\frac{1}{3}\right\}$ **23.** $\left\{-2,-1,\frac{3}{2}\right\}$ **25.** $\left\{-\frac{1}{2},2,3\right\}$ **27.** $\left\{-2,1,-\frac{3\pm\sqrt{17}}{4}\right\}$
29. $\left\{-\frac{1}{2},\frac{1}{3},-\frac{1}{2}\pm\frac{\sqrt{3}}{2}\right\}$ **31.** $\left\{-2,\frac{3}{2},\frac{1}{6}\pm\frac{\sqrt{11}}{6}i\right\}$ **33.** $\left\{\pm\frac{2}{3},-\frac{1}{2},1\right\}$ **35.** $\{1\}$ **37.** $-\frac{1}{2},\frac{2}{3},3,\pm i$
39. The dimensions are 11 in by 6 in by 3 in. **41.** The lengths of the sides are 6 ft, 8 ft and 10 ft. **43.** 385
45. 24,000 souvenirs should be sold. **47.** $(-\infty,-2]\cup\left[\frac{1}{2},1\right]$ **49.** $(-\infty,-1)\cup(-1,+\infty)$

PROBLEM SET 5.4

1. 2 or 0 positive; 2 or 0 negative **3.** 3 or 1 positive; 1 negative **5.** 1.74 **7.** −0.78 **9.** no positive, 1 negative **11.** 1 positive, 1 negative **13.** 3 or 1 positive; 0 negative **15.** 2 or 0 positive; 1 negative **17.** 0 positive; 3 or 1 negative **19.** 1.72 **21.** 1.88 **23.** −1.27, 0.19, 2.08 **25.** −1.80 **27.** −1.17 **29.** 0.75 **31.** The inner radius should be about 3.37 cm. **33.** −1.3 **35.** 0.3473

PROBLEM SET 5.5

1. $-3, 1\pm i$ **3.** $-3,\frac{1}{2},2\pm 3i$ **5.** $x^3-7x^2+15x-25=0$ **7.** $(x^2+1)(x^2-4x+2)$ **9.** $\left\{\pm 2i,-1\pm\frac{\sqrt{2}}{2}\right\}$
11. $\left\{1\pm 2i,\frac{-3\pm\sqrt{5}}{2}\right\}$ **13.** $\left\{\frac{3}{2},\pm 2i,\frac{3\pm\sqrt{5}}{2}\right\}$ **15.** $\{\pm\sqrt{3}i, 2, \pm\sqrt{3}\}$ **17.** $\left\{1\pm\sqrt{2}i,-\frac{2}{3},\pm\frac{\sqrt{2}}{2}\right\}$
19. $\left\{1\pm\sqrt{5}i,\frac{1}{4},\pm\sqrt{7}\right\}$ **21.** $(x-\sqrt{5})(x+\sqrt{5})(x^2-4x+7)$ **23.** $(z+1)(2z-1)(z^2+6z+11)$

CHAPTER 5 REVIEW PROBLEMS

1. $3s^2+s-2+\frac{9}{s+1}$ **3.** $x^3-x^2+3x+\frac{1}{x+3}$ **5.** 5 **7.** 1 **9.** 2 **11.** 224 **13.** 2 or 0 positive; 2 or 0 negative **15.** 1 positive; 0 negative **17.** $3, 1\pm\sqrt{2}$ **19.** $-1, 2, -2\pm\sqrt{5}$ **21.** $\{2, 1\pm i\}$
23. $\{\pm i, -2\pm\sqrt{2}\}$ **25.** $\left\{-2,-\frac{2}{3},\frac{1}{2}\right\}$ **27.** $\{\pm 2, 3, \pm 2i\}$ **29.** $\left\{1\pm i,\frac{3}{2},-1\pm\sqrt{2}\right\}$ **31.** −1.48
33. $(x^2+2x+3)(x-2)(x+2)$ **35.** $\{2,\pm i\}$ **37.** $f(x)=(4x-1)(x^2+1)$ **39.** Its dimensions are 20 in by 10 in by 9 in. **41.** s is 6 ft. **43.** Either form is useful. **45.** 2 or 0 **47.** 1.52

PROBLEM SET 6.1

1. 4.7288 **3.** 2.4573 **5.** Dom: R
Ran: $(0, +\infty)$

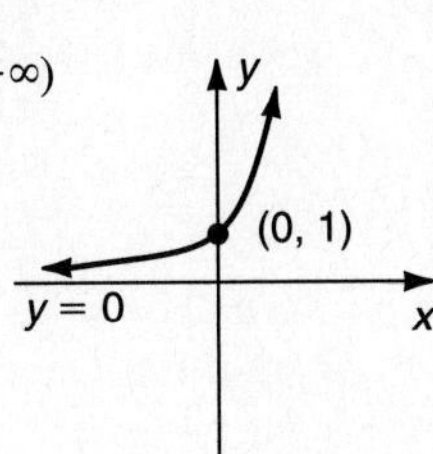

7. 0.13534 **9.**

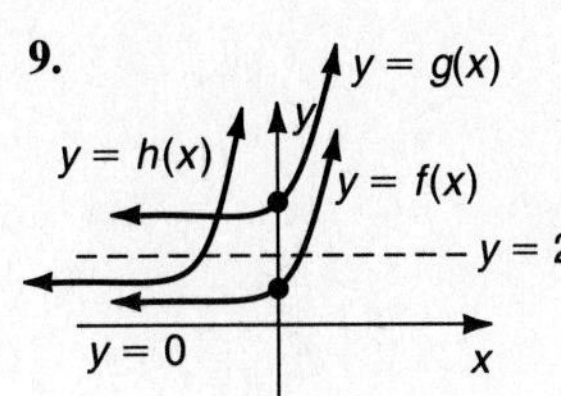

11. $\{7\}$ **13.** $\{11\}$ **15.** $\left(\frac{4}{3}, +\infty\right)$ **17.** $(-\infty, 0]$ **19.** $5^{0.7}$ **21.** $26^{\sqrt{5}}$ **23.** $(\sqrt[3]{47})^{\sqrt{7}}$ **25.** e **27.** $\{5\}$

29. $\{-3\}$ **31.** $\left\{\frac{5}{2}\right\}$ **33.** $\left\{\frac{5}{3}\right\}$ **35.** $\left(-\infty, \frac{7}{2}\right)$ **37.** $\left\{\frac{3}{7}\right\}$ **39.** $\{2\}$ **41.** $(-\infty, 1]$ **43.** $\left\{\frac{5}{3}\right\}$ **45.** $\{-23\}$

47. $\left\{\frac{7}{6}\right\}$ **49.** $\varnothing$ **51.** $\{2, -2\}$ **53.** $\{0\}$ **55.** $\{1, 2\}$ **57.** $\{\pm 1\}$

59. Dom: R
Ran: $(0, +\infty)$

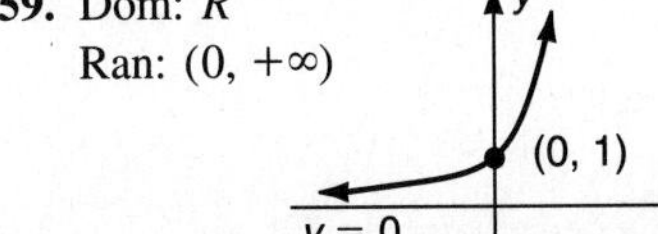

61. Dom: R
Ran: $(0, +\infty)$

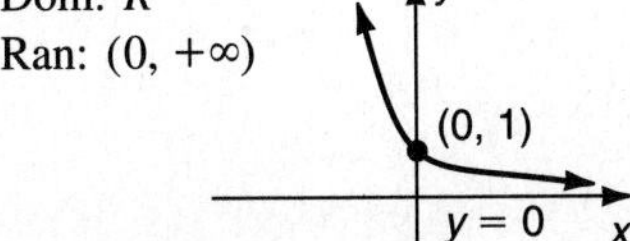

63. Dom: R
Ran: $(1, +\infty)$

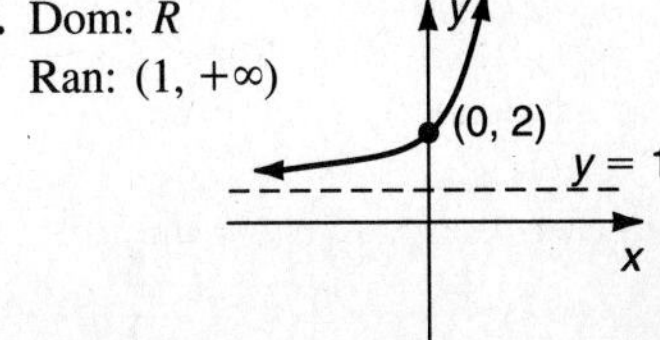

65. Dom: R
Ran: $(0, +\infty)$

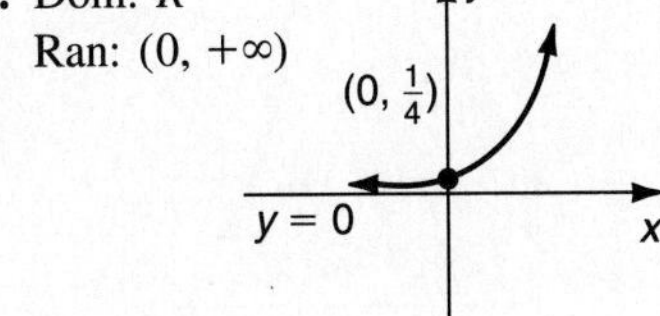

67. Dom: R
Ran: $(-2, +\infty)$

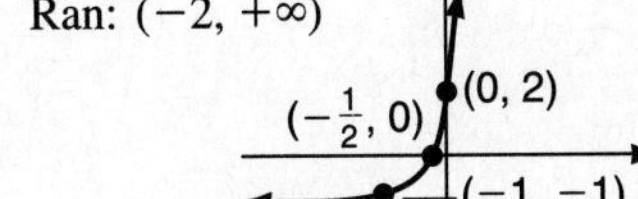

69. Dom: R
Ran: $(-\infty, 0)$

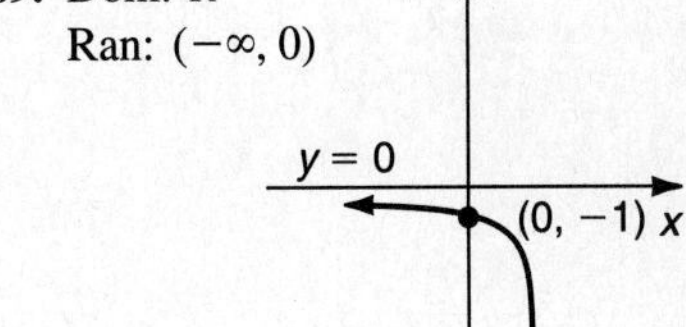

71. Dom: R
Ran: $(2, +\infty)$

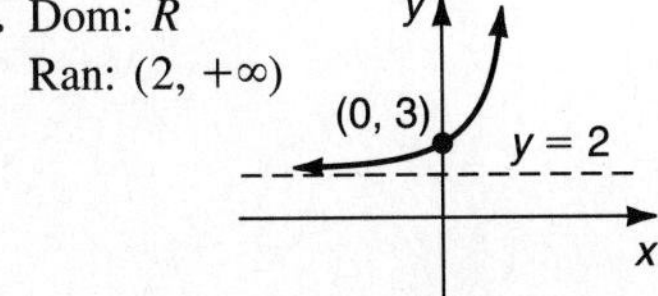

73. Dom: R
Ran: $(0, +\infty)$

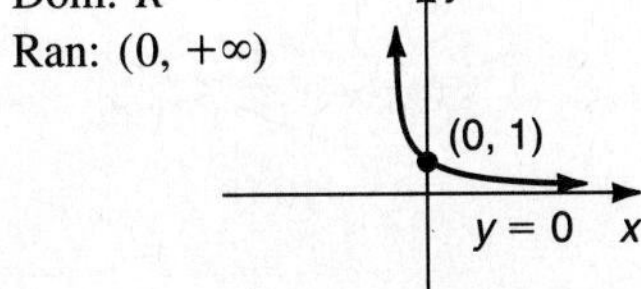

75. Dom: R
Ran: $[1, +\infty)$

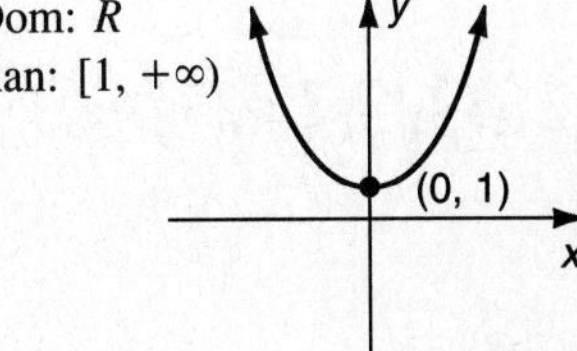

77. Dom: R
Ran: $[1, +\infty)$

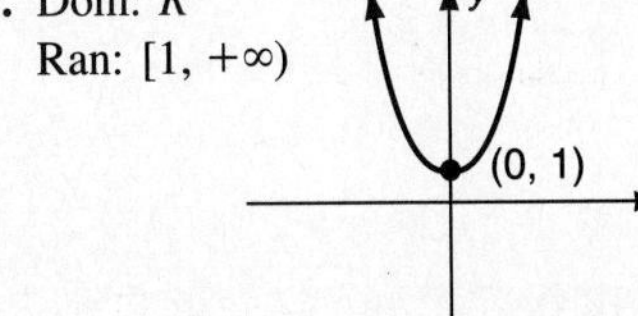

79. Dom: R
Ran: $(0, +\infty)$

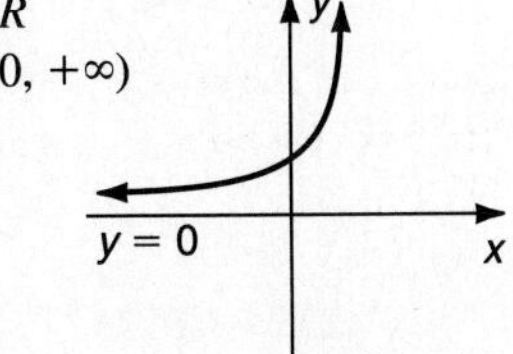

81. $[1, +\infty)$ **83.** $(-\infty, -2)$

85. (a) 0, (b) 1, (c) $\frac{1}{2}\left(e - \frac{1}{e}\right)$ (d) $\frac{1}{2}\left(e + \frac{1}{e}\right)$ **87.** $\cosh^2 x - \sinh^2 x = 1$ **89.**

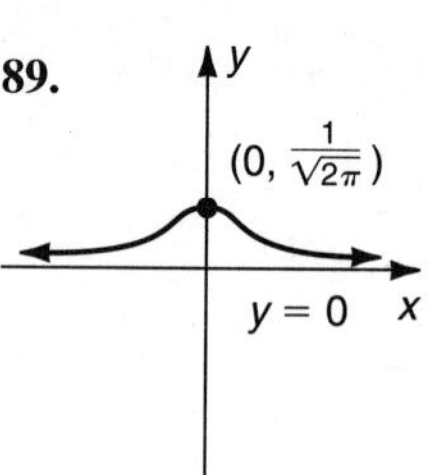

PROBLEM SET 6.2

1.

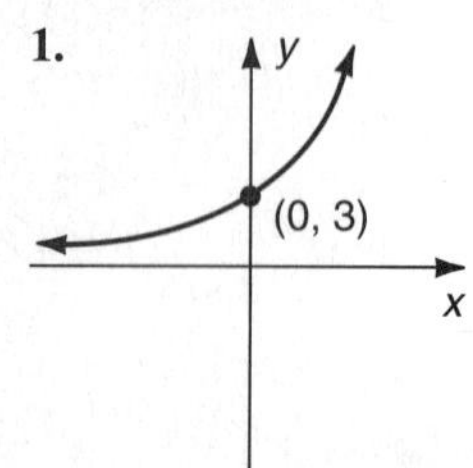

3.

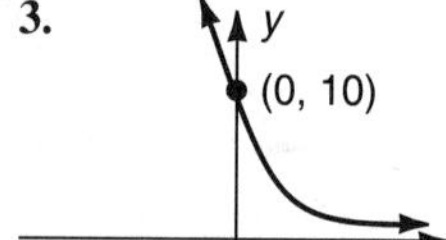

5. $f(t) = 250(1.12)^t$ **7.** 41,472

9. 952,831 **11.** 0.029 mg **13.** $104,817.75; $104,935.69 **15.** Invest at 6.6% compounded quarterly. $20.64
17. $3.48 **19.** Approx. 192 mg **21.** 2,484 **23.** $f(t) = 5000 \cdot 2^{t/20}$; 280,000; 17,920,000
25. Approx. 8.287 lb/in^2 **27.**

1	1.08
10	1.082942308
100	1.083252422
1,000	1.083283601
10,000	1.083286721
1,000,000	1.083287064

29.

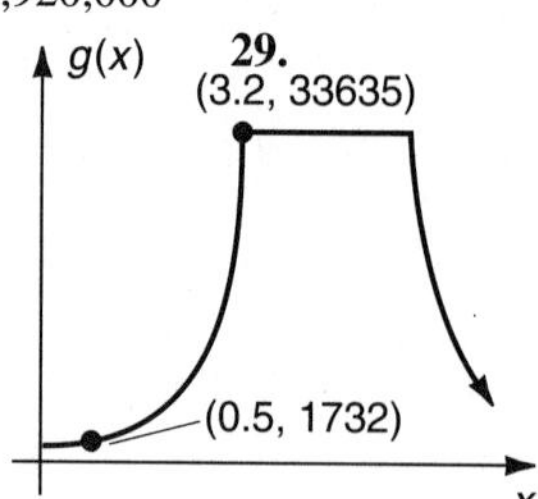

31. $f(m) = 1000(1.01)^{m-1}$ $g(m) = 750(1.015)^{m-1}$
$f(36) = 1417$ $g(36) = 1263$
$f(60) = 1799$ $g(60) = 1805$

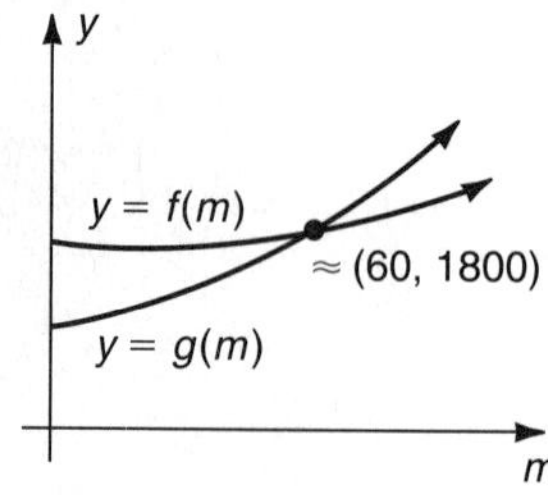

PROBLEM SET 6.3

1. $5^3 = 125$ **3.** $\left(\frac{1}{2}\right)^4 = \frac{1}{16}$ **5.** $6^{-1} = \frac{1}{6}$ **7.** $\log_2 8 = 3$ **9.** $\log_{10} 10000 = 4$ **11.** $\log_{1/4} 64 = -3$ **13.** 2
15. Undefined **17.** -3 **19.** -1 **21.** 4 **23.** -1.0573 **25.** 3.1570 **27.** -14.886 **29.** ≈ 0.08703
31. ≈ 15.15 **33.** ≈ 1.002 **35.** $\{4\}$ **37.** $\{15\}$ **39.** $\left\{\frac{4}{3}\right\}$

41.

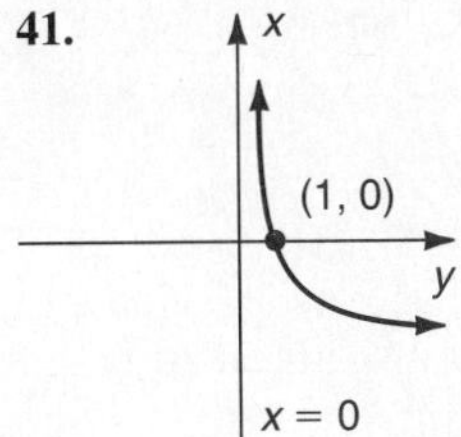

43.

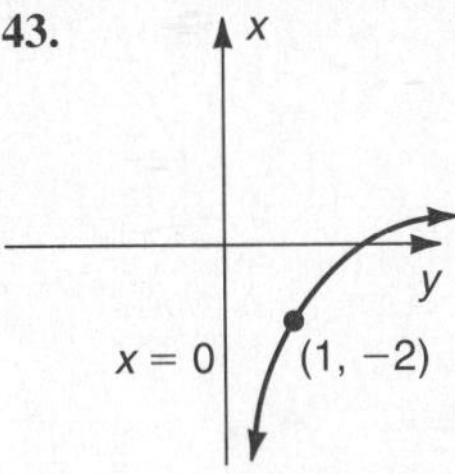

45.

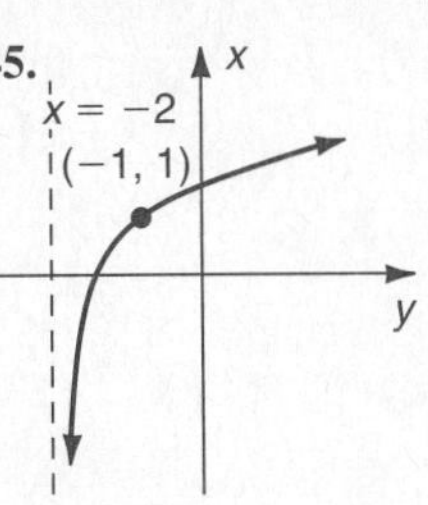

47.

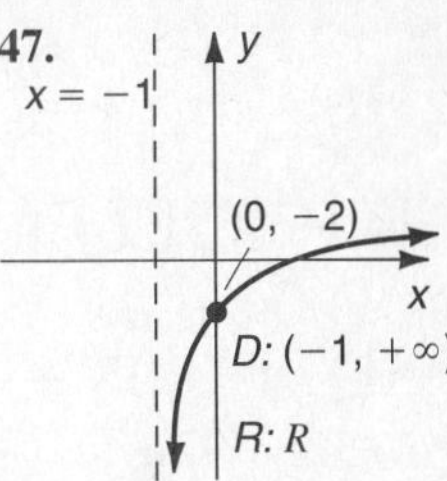

49. 100 dB **51.** 1×10^{-7} **53.** -2 **55.** -3 **57.** 1.3222 **59.** -0.43078 **61.** 5.4161 **63.** 7.3335
65. 5.4161 **67.** 1.1761 **69.** 0.90309 **71.** 1.1761 **73.** $\{64\}$ **75.** $\{7\}$ **77.** ≈ 20.086 **79.** ≈ 221.06
81. $\{-15\}$ **83.** $\{\pm 3\}$

85.

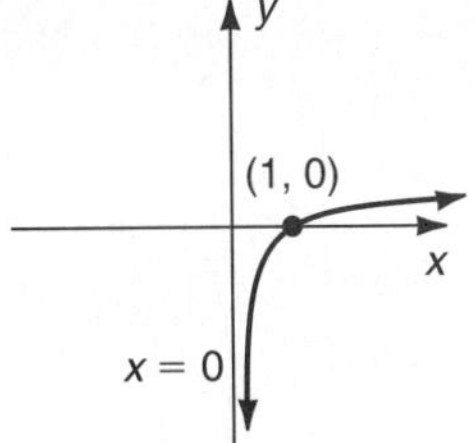

87. Dom: $(0, +\infty)$
Ran: R

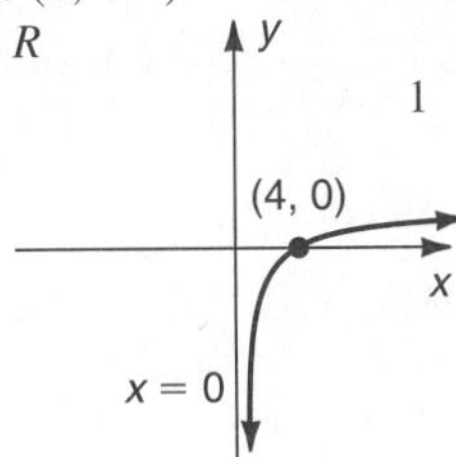

1

89. Dom: $(2, +\infty)$
Ran: R

y
x = 2
x
(3, 0)

91. Dom: $(-1, +\infty)$
Ran: R

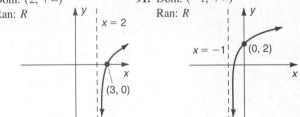

y
x = −1
(0, 2)
x

93. 120 dB **95.** 7.9×10^{-6} **97.** $(0, 1)$ **99.** $\left(0, \frac{1}{2}\right)$ **101.** 9.3×10^{-6} **103.** 98 times as bright.

PROBLEM SET 6.4

1. $\log_b p + \log_b q + 2\log_b r$ **3.** $\ln p + \frac{1}{2}\ln q$ **5.** $\frac{2}{3}\log_b p$ **7.** $\log_b 90$ **9.** $\ln \frac{32}{3x^3}$ **11.** $\log_b 48$ **13.** 2.4

15. -0.3 **17.** 1.95 **19.** $\frac{1}{81}$ **21.** 10 **23.** $\frac{-3}{\log 7}$ **25.** 2.3370 **27.** -0.54827 **29.** 0 **31.** $\frac{1}{3}$ **33.** n

35. $-\frac{1}{2}$ **37.** $\sqrt{2}$ **39.** $\frac{1}{2}$ **41.** $-0.5; \frac{4}{9}$ **43.** 0.2; 0.6325 **45.** 0.9; $\frac{10}{9}$ **47.** $\log_b \frac{b^2}{8}$ **49.** $\log \frac{8t^2}{s}$

51. $\ln 24$ **53.** $\ln \frac{y^6 z^3}{x^4}$ **55.** $\log \frac{(t-1)^2}{t+1}$ **57.** $\ln \frac{t+1}{(t-1)(t+2)}$ **59.** 0.8271 **61.** 0.9318 **63.** 1.0566

69.

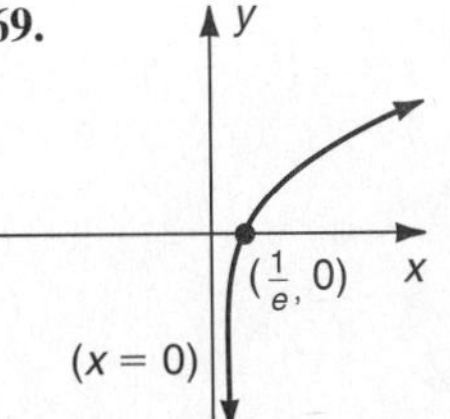

PROBLEM SET 6.5

1. $\{\pm 9\}$ **3.** $\left\{\pm\frac{27}{2}\right\}$ **5.** $\left\{1000, \frac{1}{1000}\right\}$ **7.** $\{3\}$ **9.** $\{2\}$ **11.** 0.84510 **13.** 2.79176 **15.** 3.90709

17. $\{\pm 8\}$ **19.** $\left\{\frac{25}{2}\right\}$ **21.** $\{-9\}$ **23.** 1.325 **25.** 0.2747 **27.** $\{7\}$ **29.** $\varnothing$ **31.** $\{1\}$ **33.** $\{2\}$

35. $\{\pm 1\}$ **37.** -1.900 **39.** $\left\{-\frac{35}{2}, \frac{37}{2}\right\}$ **41.** $\{5\}$ **43.** -15.33 **45.** $\left\{\frac{5}{2}\right\}$ **47.** $\{6\}$ **49.** -0.6931

51. {0} **53.** {3} **55.** ∅ **57.** {e} **59.** {1, 100} **61.** {1} **63.** ∅ **65.** $\{\ln(2 + \sqrt{5})\}$ **67.** {1}
69. 1.937 **71.** 5.669 **73.** 1.335

PROBLEM SET 6.6

1. About 14.69 years **3.** Approx. 48.60 mg **5.** 2812 **7.** Five times as powerful. **9.** About 7.4 years **11.** In about 6.8 years **13.** 46 **15.** About 98%; 81% **17.** About 7:15 pm **19.** 5272 **21.** 6.61 ft; 14.61 ft; 78.61 ft **23.** $t_2 = \dfrac{\ln 2}{r}$ **25.** $r \approx 11\%$ **27.** 5,700 years **29.** $t_2 = \dfrac{\ln 2}{\ln(1 + r)}$

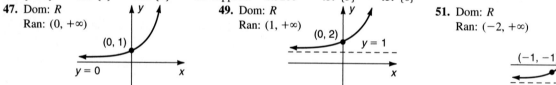

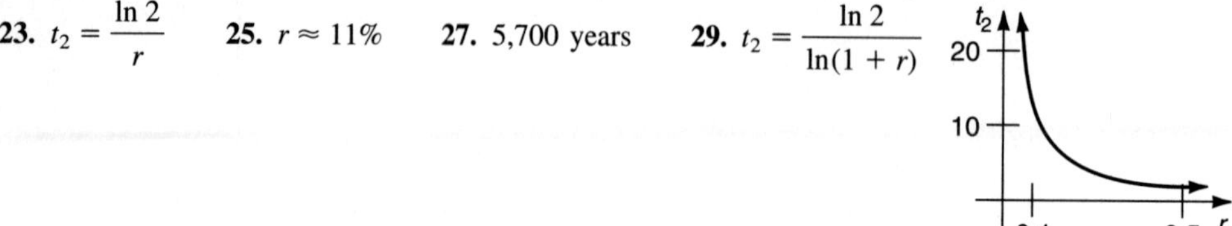

CHAPTER 6 REVIEW PROBLEMS

1. 4.2567 **3.** 9.3565 **5.** 1.4472 **7.** −0.15082 **9.** 1.2091 **11.** {4} **13.** $\left\{\dfrac{3}{2}\right\}$ **15.** $(-\infty, 5)$ **17.** $\left\{\dfrac{4}{11}\right\}$
19. $\log_b \dfrac{23}{64}$ **21.** $\log \dfrac{xy^5}{36}$ **23.** $\ln \dfrac{t^4V^6}{U^3}$ **25.** $\ln \dfrac{x^3(x + 1)}{x + 2}$ **27.** {81} **29.** {±27} **31.** $\left\{-\dfrac{9}{2}\right\}$ **33.** $\{10^{-8}\}$
35. {−33} **37.** {2} **39.** {6} **41.** Approx. −2.3260 **43.** {5} **45.** {0}

47. Dom: R
Ran: $(0, +\infty)$

49. Dom: R
Ran: $(1, +\infty)$

51. Dom: R
Ran: $(-2, +\infty)$

53. Dom: $(0, +\infty)$
Ran: R

55. Dom: $(2, +\infty)$
Ran: R

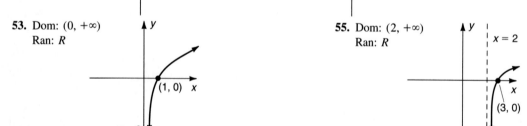

57. $(2, e + 2)$ **59.** $[e, +\infty)$ **61.** 25 years **63.** 100; 900; 144 **65.** 3.23 years **67.** 6.6 miles
69. \$77.84; \$181.12 **71.** a) $\log_3 5 = x$ b) $3^x = 9$ **73.** $(-\infty, 0)$ **75.** 1.771 **77.** 12.18

PROBLEM SET 7.1

1. (0, 2); $y = -2$ **3.** $\left(2, -\dfrac{5}{2}\right)$; $y = \dfrac{1}{2}$ **5.** $y^2 = 8(x - 1)$ **7.** $x^2 = 20y$ **9.** $\left(1, -\dfrac{15}{8}\right)$; $y = -\dfrac{17}{8}$
11. $\left(-\dfrac{4}{3}, \dfrac{1}{2}\right)$; $x = -\dfrac{7}{6}$ **13.** The diameter is 24 inches. **15.** $\left(\dfrac{1}{2}, 0\right)$; $x = -\dfrac{1}{2}$ **17.** $\left(2, \dfrac{5}{4}\right)$; $y = \dfrac{3}{4}$
19. $\left(\dfrac{15}{8}, 1\right)$; $x = \dfrac{17}{8}$ **21.** $\dfrac{1}{2}y = x^2$ **23.** $(y - 1)^2 = 4(x + 2)$ **25.** $(y - 5)^2 = -\dfrac{25}{4}(x - 2)$; $(x - 2)^2 = -\dfrac{16}{5}(y - 5)$
27. $\left(1, -\dfrac{15}{4}\right)$, $y = -\dfrac{17}{4}$; $\left(1, \dfrac{15}{4}\right)$, $y = -\dfrac{17}{4}$ **29.** The height of the cable is 13.75 meters.
31. The height is $8.\overline{8}$ meters.

PROBLEM SET 7.2

1.
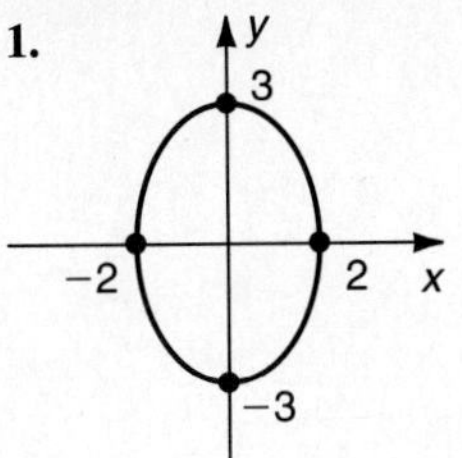

3.
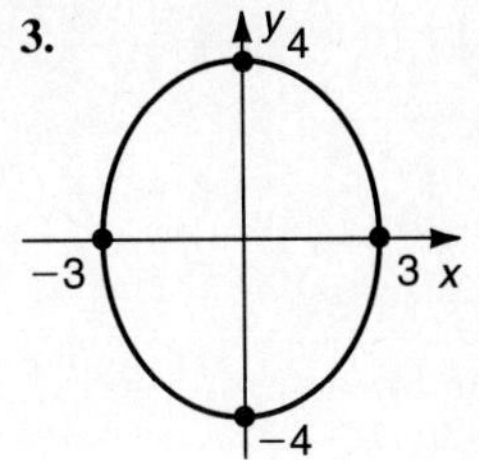

5.
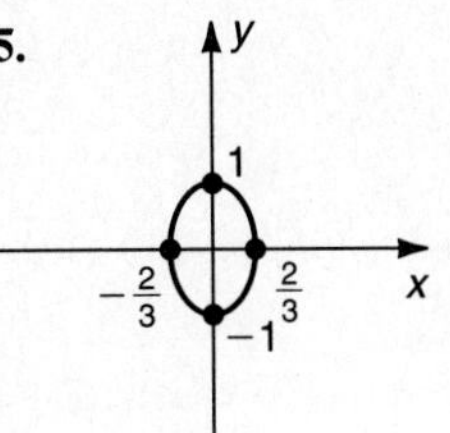

7.
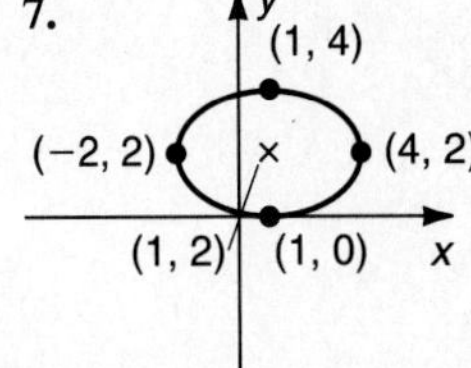

9.
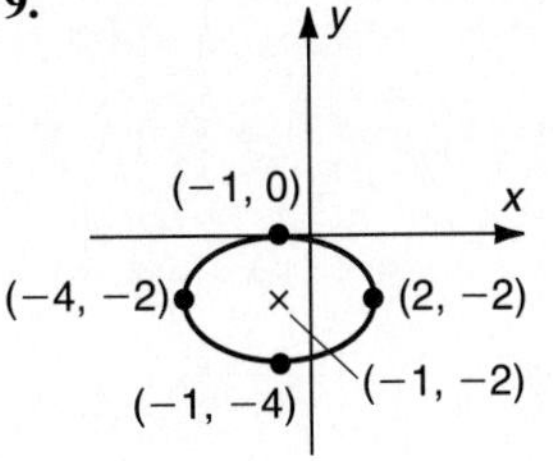

11.
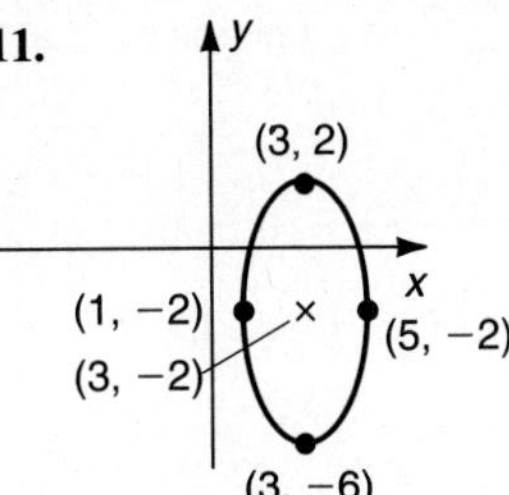

13.
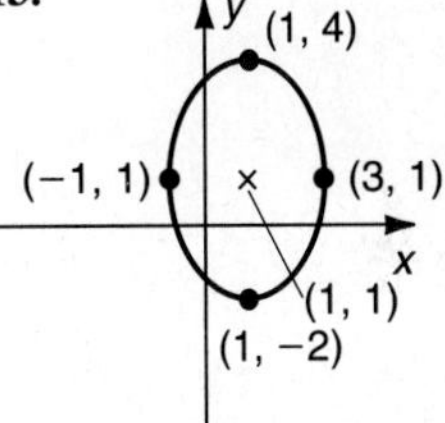

15. $\dfrac{x^2}{25} + \dfrac{y^2}{9} = 1$

17. $(\pm\sqrt{5}, 0)$

19.
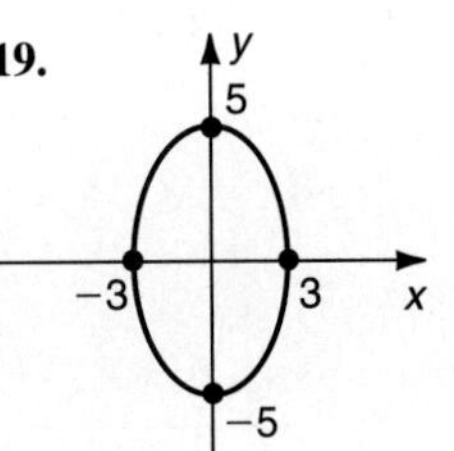

21.
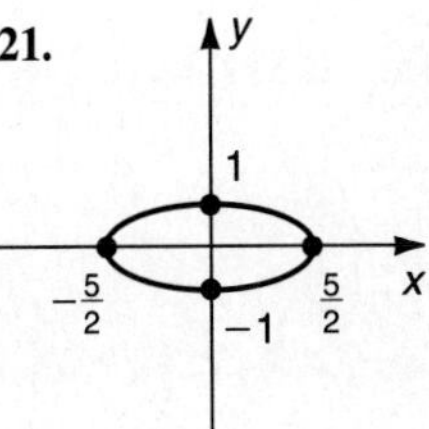

23.
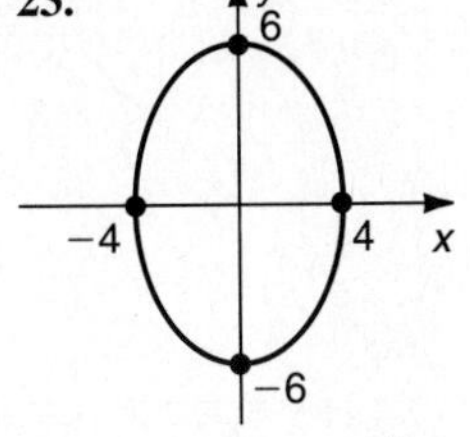

25.
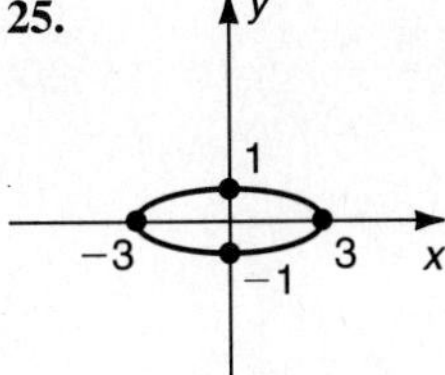

27.
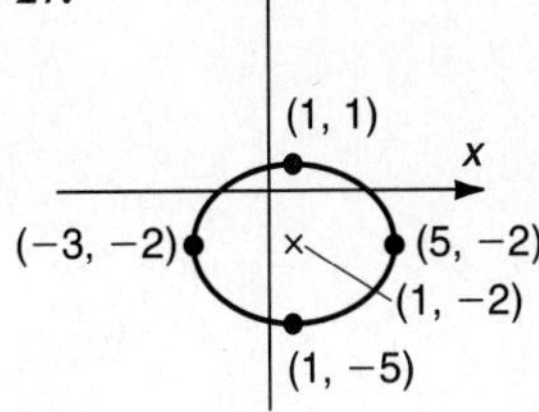

29.
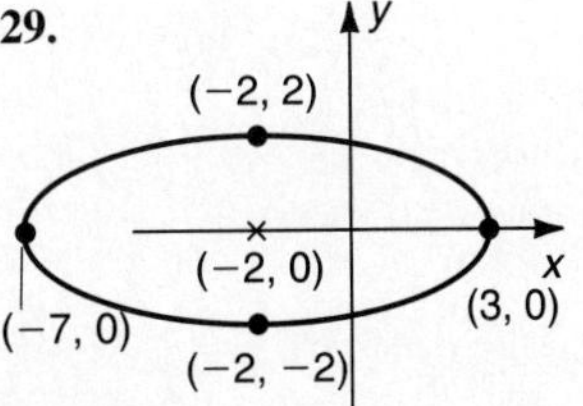

31.
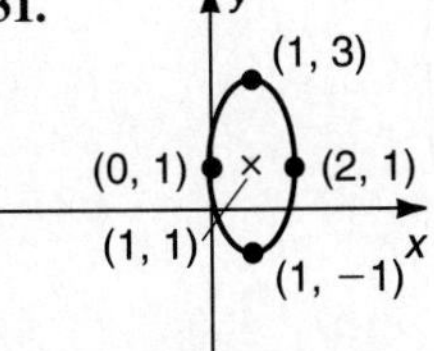

33.
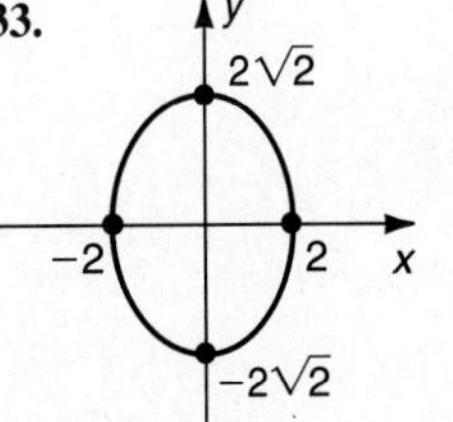

35.
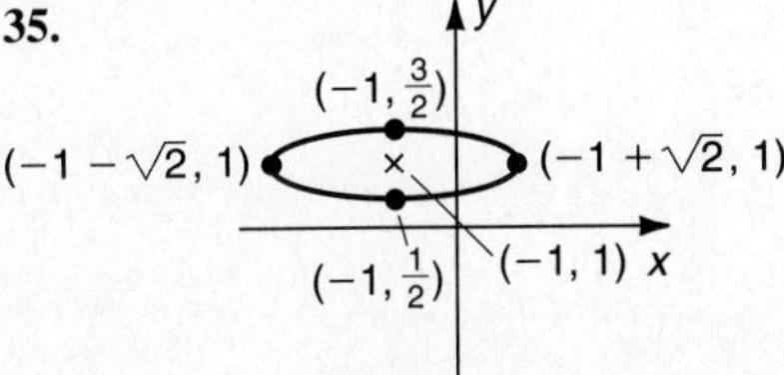

37.
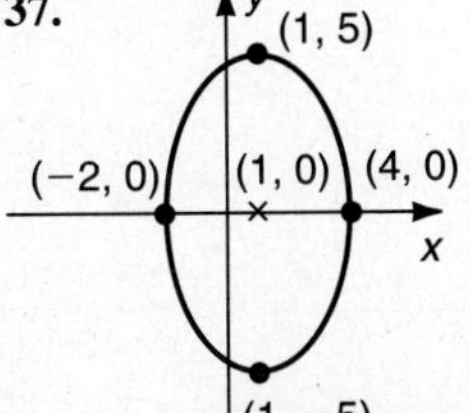

39.
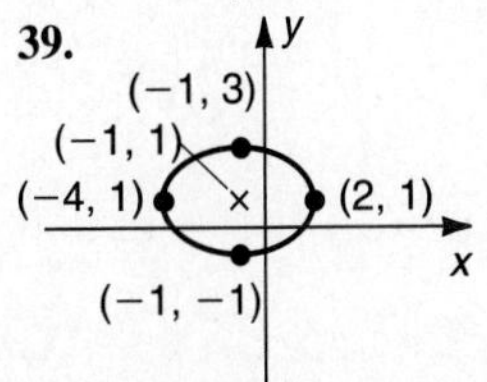

41.

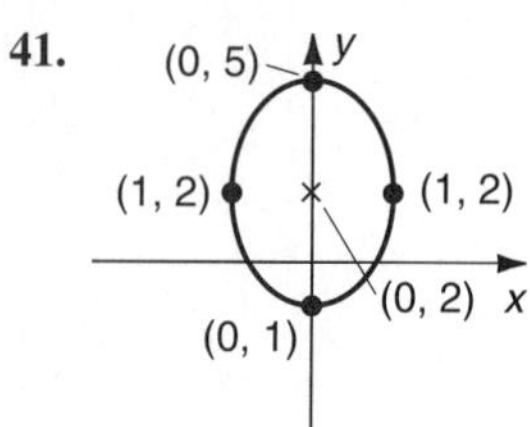

43.

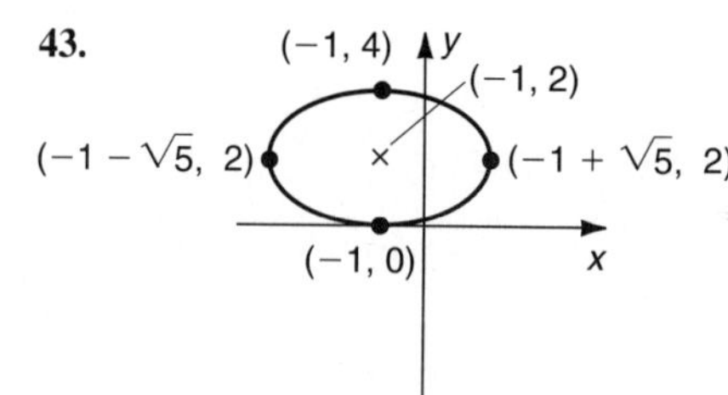

45. perihelion = 92,960,000 mi; aphelion = 94,960,000

47. The arch is $5\sqrt{3}$ feet high 12 feet from the center. The height at the foci is less than $5\sqrt{3}$.

53.

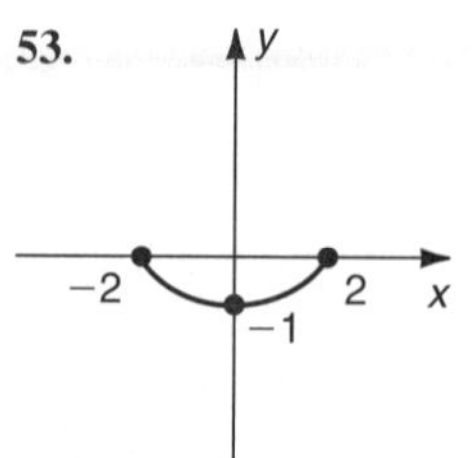

55.

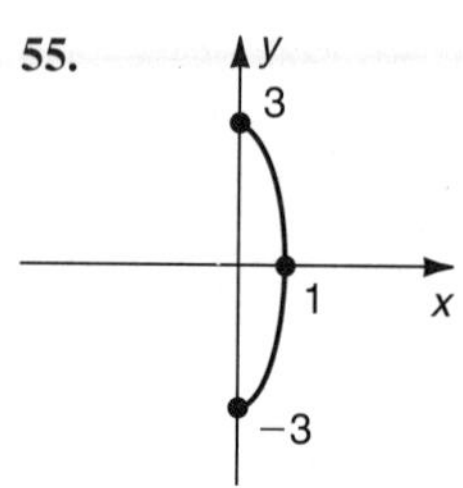

57. $x = 2\sqrt{1 - \frac{(y-1)^2}{2}}$

PROBLEM SET 7.3

1.

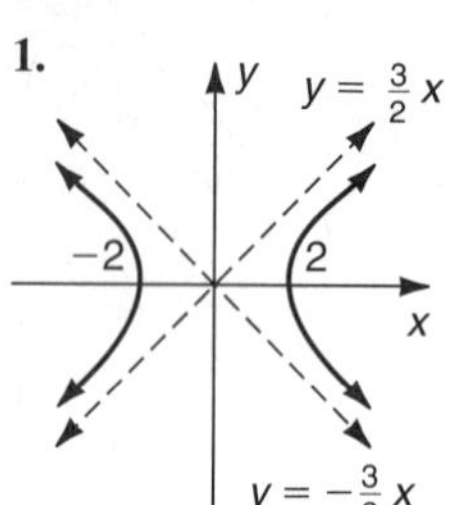

3.

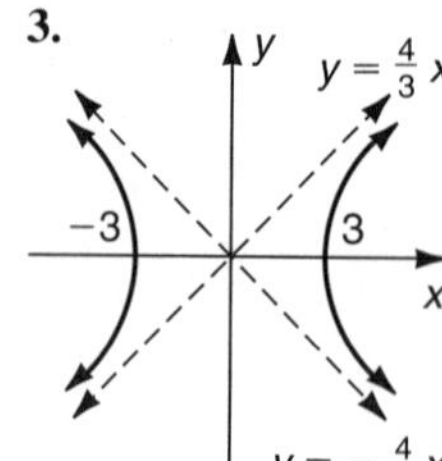

5.

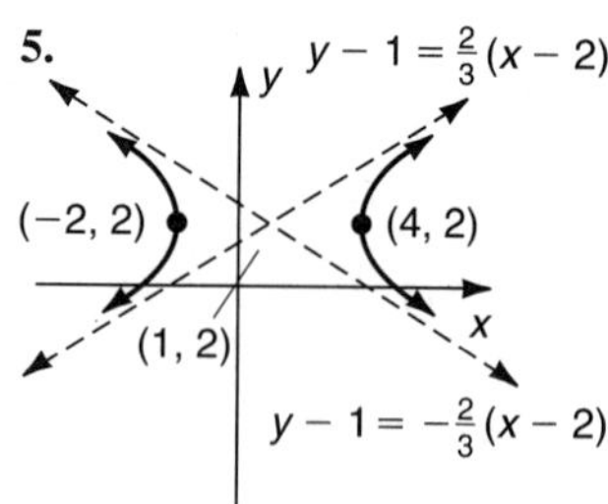

7.

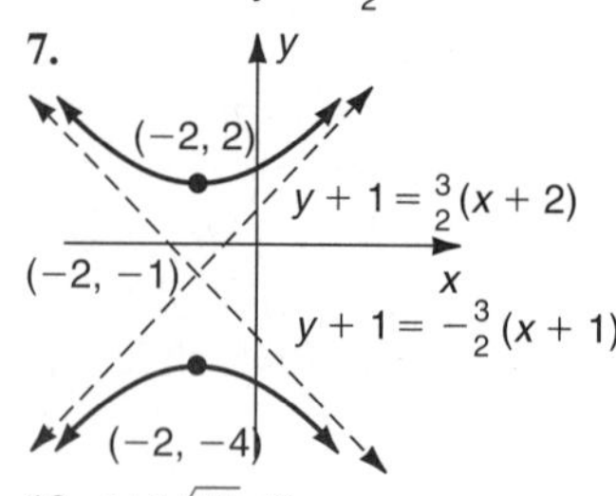

9.

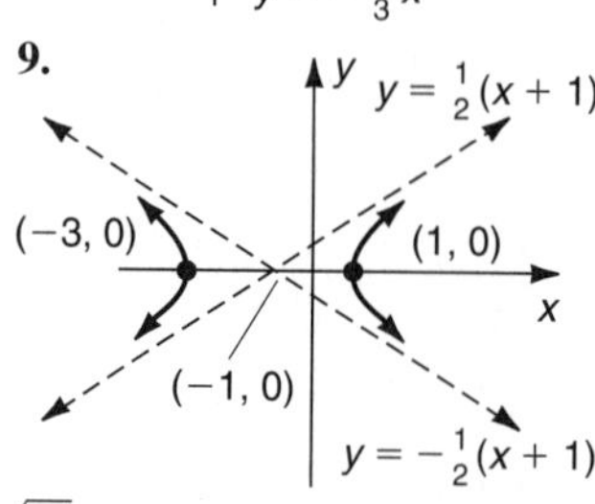

11.

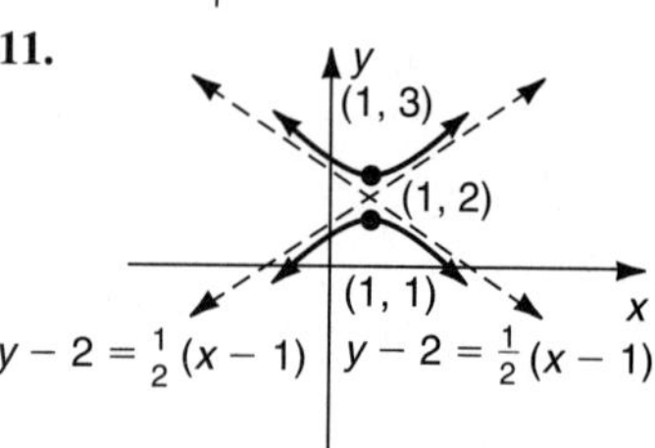

13. $(\pm\sqrt{13}, 0)$

15. $(0, \pm\sqrt{13})$

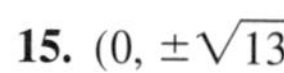

17.

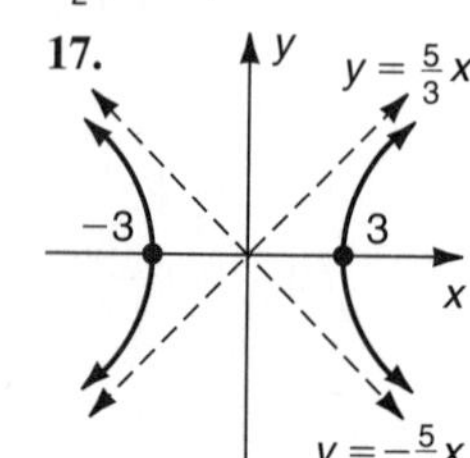

19.

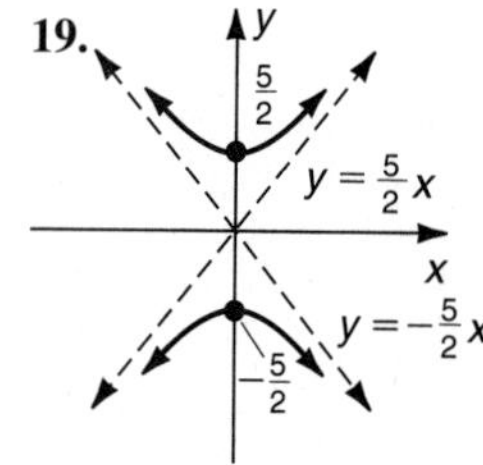

21.

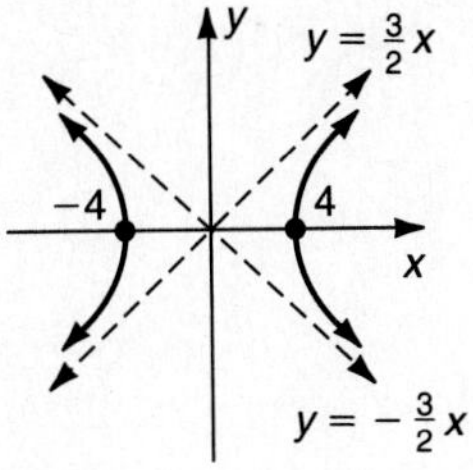

$y = \frac{3}{2}x$, $y = -\frac{3}{2}x$, -4, 4, x, y

23.

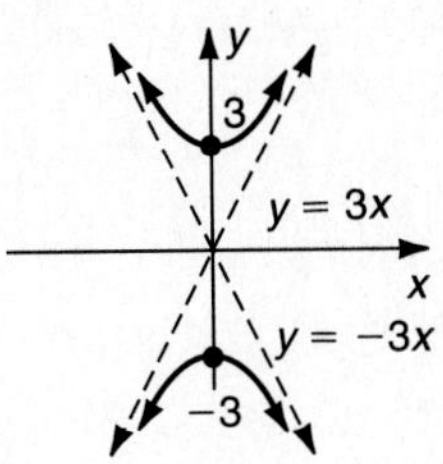

3, -3, $y = 3x$, $y = -3x$, x, y

25.

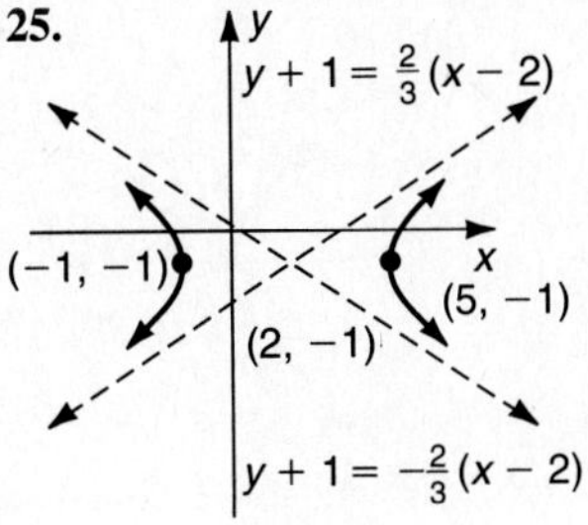

$y + 1 = \frac{2}{3}(x - 2)$, $y + 1 = -\frac{2}{3}(x - 2)$, $(-1, -1)$, $(5, -1)$, $(2, -1)$, x, y

27.

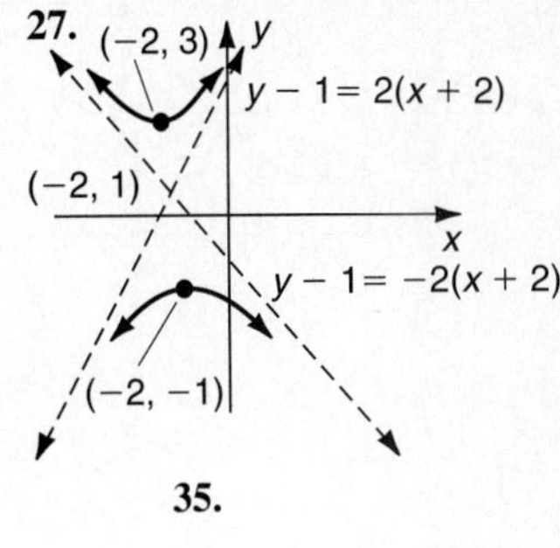

29.

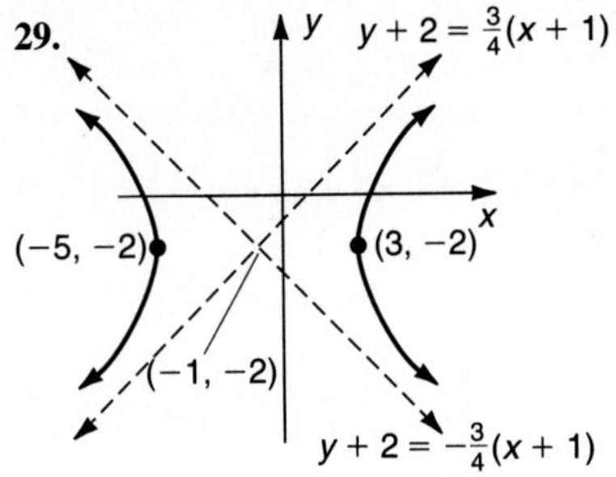

$y + 2 = \frac{3}{4}(x + 1)$, $y + 2 = -\frac{3}{4}(x + 1)$, $(-5, -2)$, $(3, -2)$, $(-1, -2)$, x, y

31.

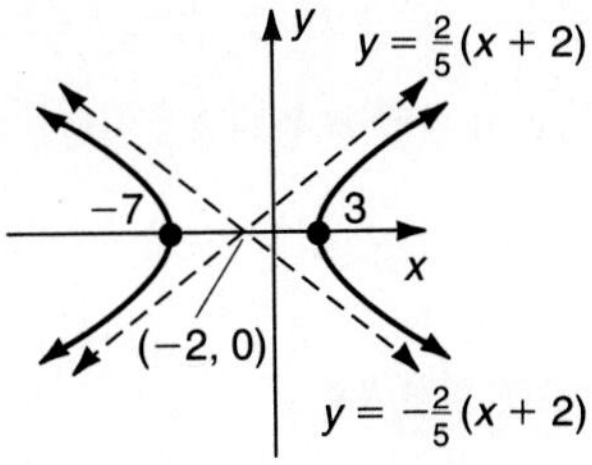

$y = \frac{2}{5}(x + 2)$, $y = -\frac{2}{5}(x + 2)$, -7, 3, $(-2, 0)$, x, y

33.

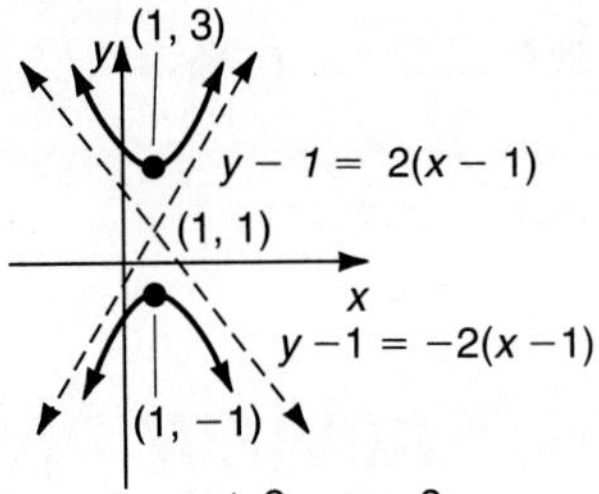

35.

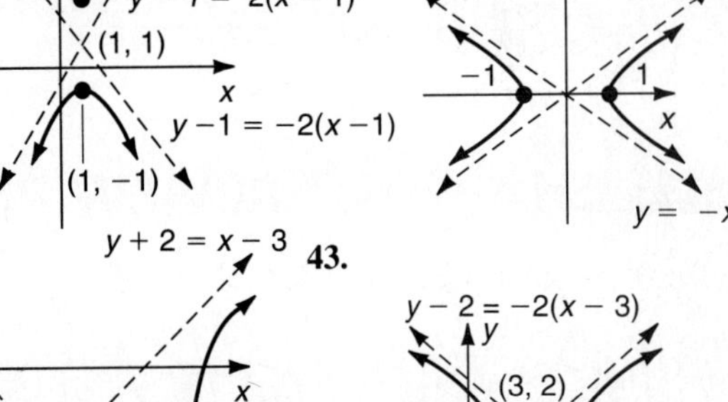

37.

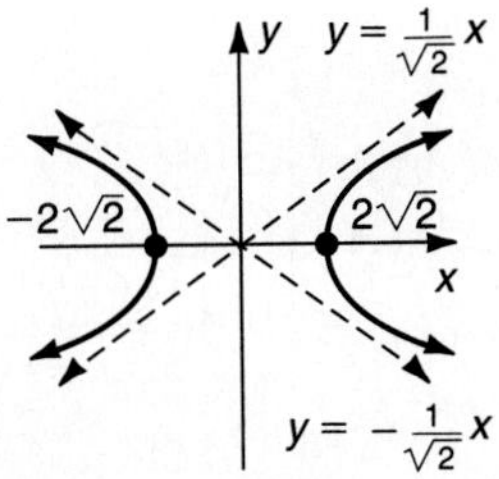

$y = \frac{1}{\sqrt{2}}x$, $y = -\frac{1}{\sqrt{2}}x$, $-2\sqrt{2}$, $2\sqrt{2}$, x, y

39.

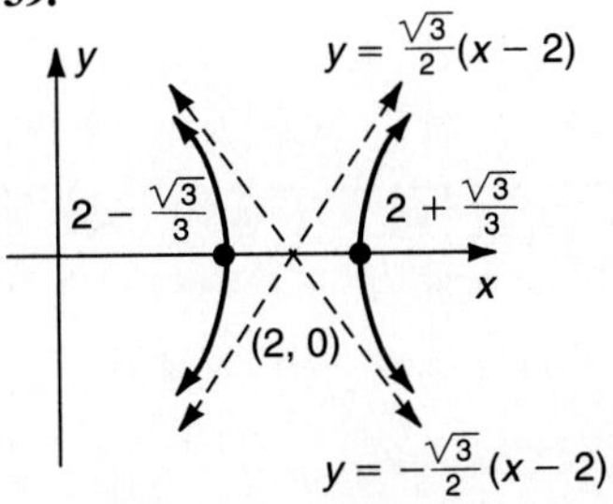

$y = \frac{\sqrt{3}}{2}(x - 2)$, $y = -\frac{\sqrt{3}}{2}(x - 2)$, $2 - \frac{\sqrt{3}}{3}$, $2 + \frac{\sqrt{3}}{3}$, $(2, 0)$, x, y

41.

43.

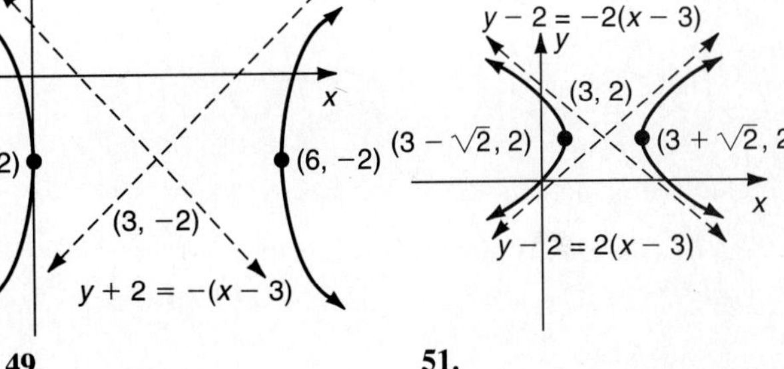

45.

47.

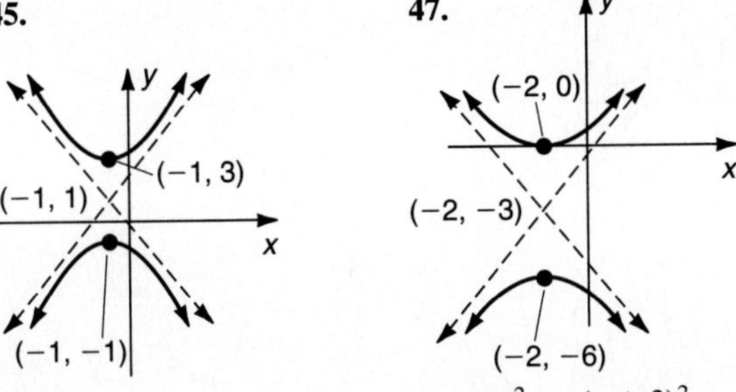

49.

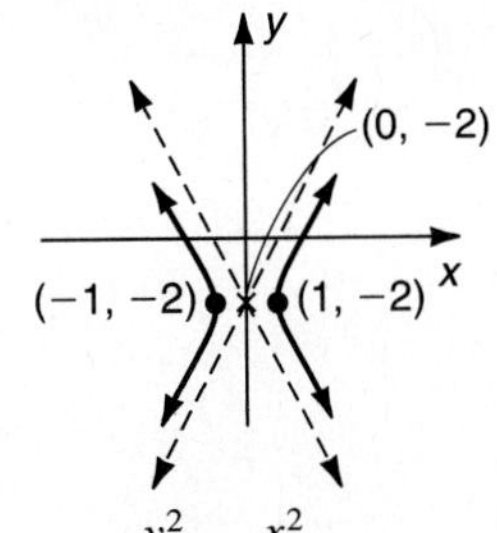

51.

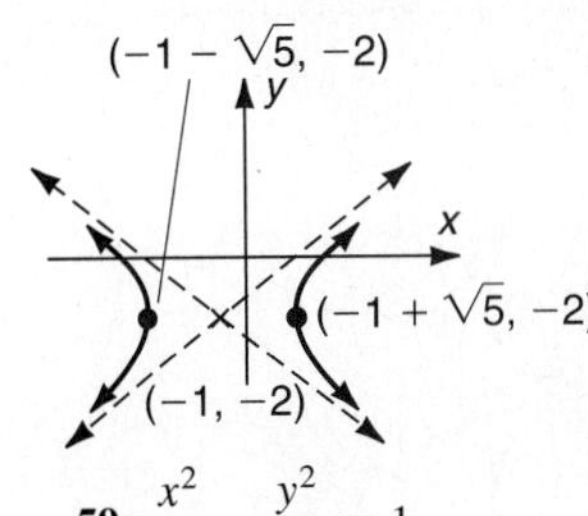

53.

$\left(\frac{1}{2}, -\frac{1}{3}\right)$, $\left(-\frac{5}{2}, -\frac{1}{3}\right)$, $\left(\frac{7}{2}, -\frac{1}{3}\right)$, x, y

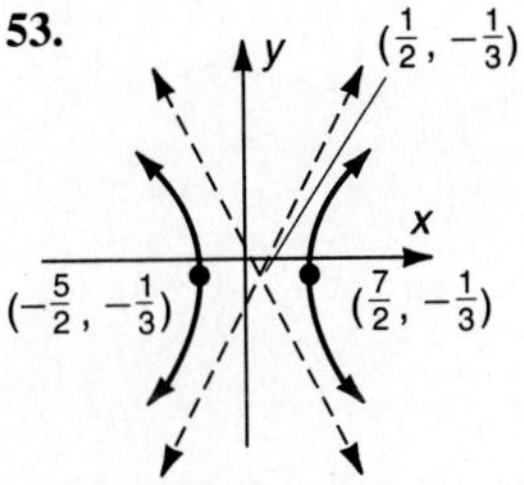

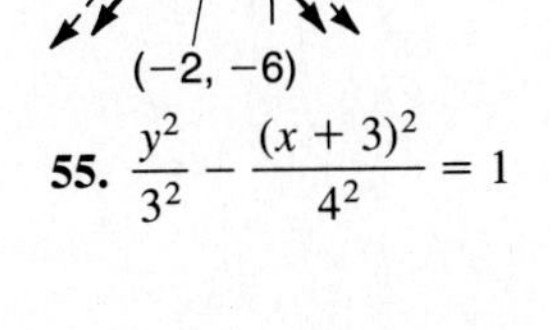

55. $\dfrac{y^2}{3^2} - \dfrac{(x + 3)^2}{4^2} = 1$

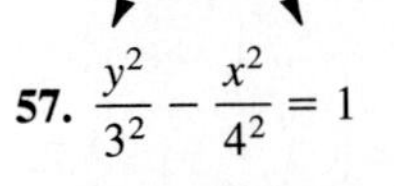

57. $\dfrac{y^2}{3^2} - \dfrac{x^2}{4^2} = 1$

59. $\dfrac{x^2}{4} - \dfrac{y^2}{12} = 1$

63.

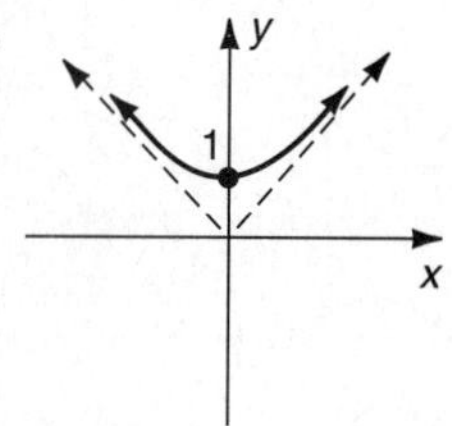

65.

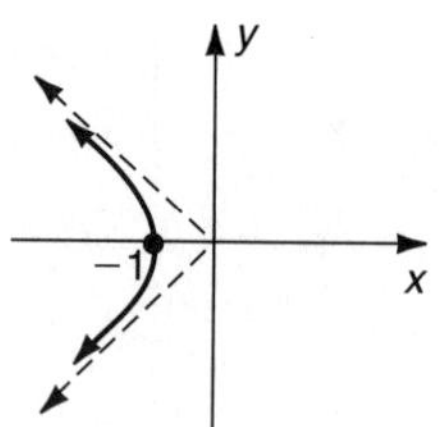

67. $y = -1 + 3\sqrt{\dfrac{(x-1)^2}{4} - 1}$

PROBLEM SET 7.4

1. parabola 3. circle 5. hyperbola 7. circle 9. parabola; y-axis 11. not a conic 13. hyperbola 15. ellipse 17. not a conic 19. hyperbola

CHAPTER 7 REVIEW PROBLEMS

1.

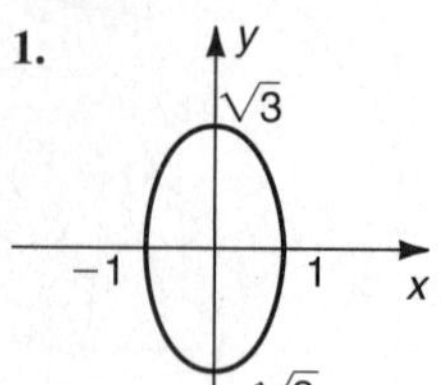

3.

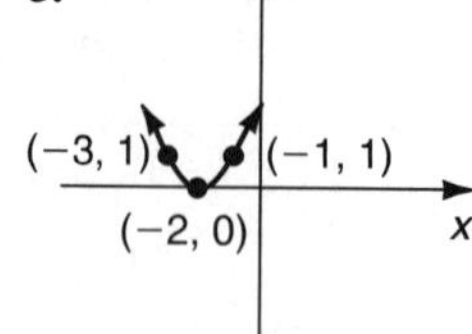

5.

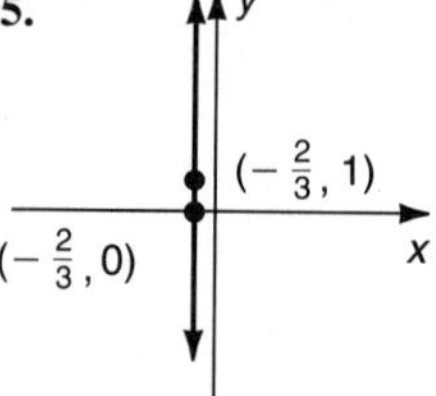

7.

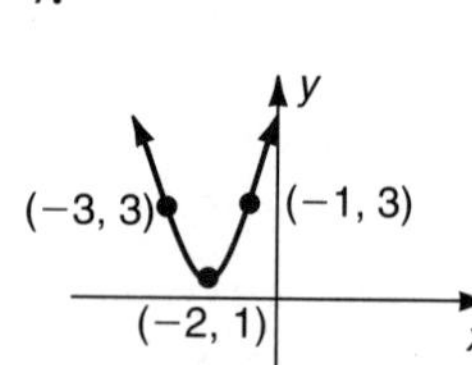

9.

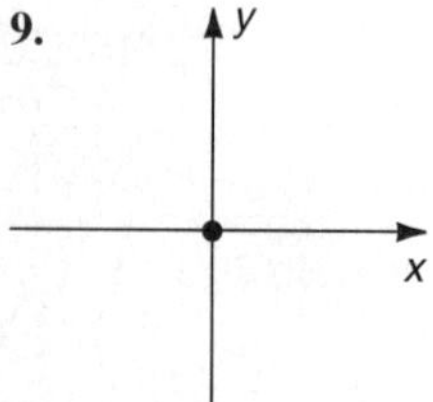

11.

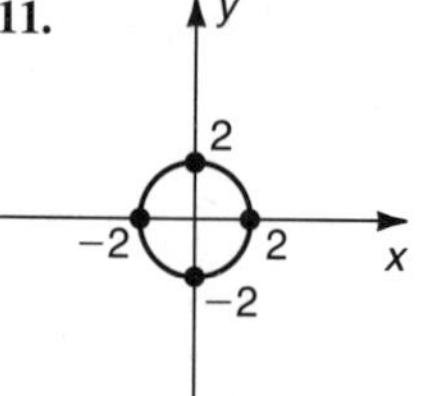

13.

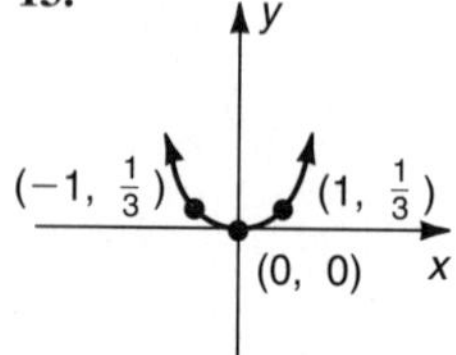

15.

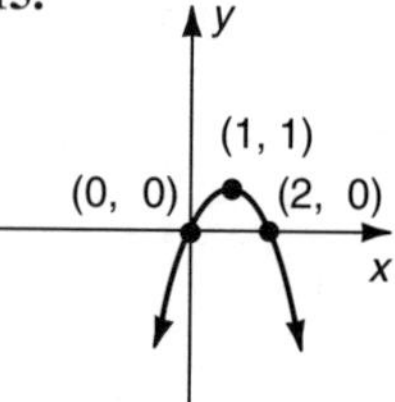

17.

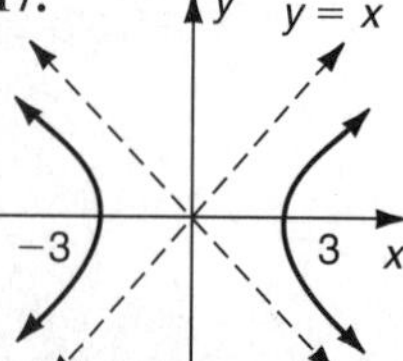

19.

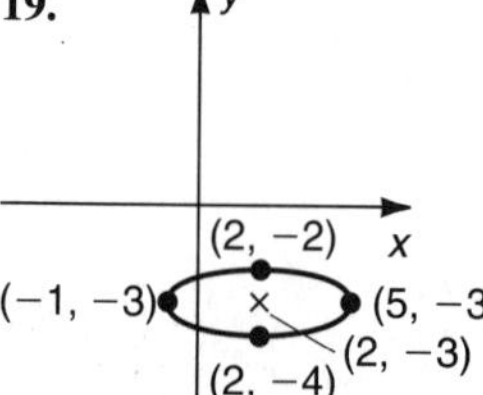

21.

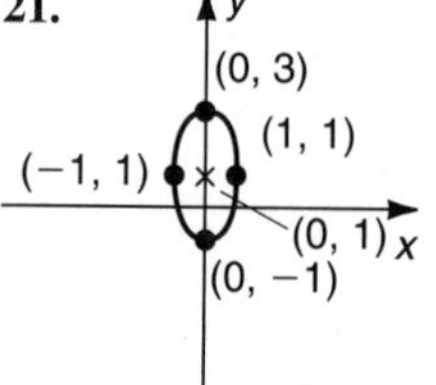

23.

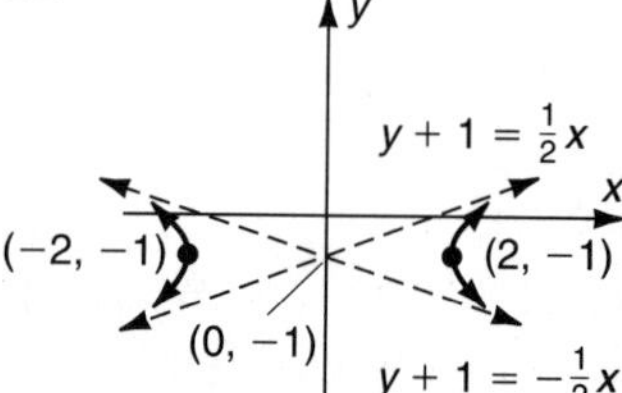

25. $\dfrac{x^2}{9} + \dfrac{y^2}{25} = 1$ 27. $\dfrac{(y+1)^2}{25} - \dfrac{(x-2)^2}{24} = 1$ 29. $8(y+1) = (x-1)^2$ 31. Focus is 18 meters above the vertex of the parabola. 33. $\dfrac{(x-1)^2}{4} + \dfrac{y^2}{9} = 1$

PROBLEM SET 8.1

1. Nonlinear **3.** Nonlinear **5.** (1, 1)

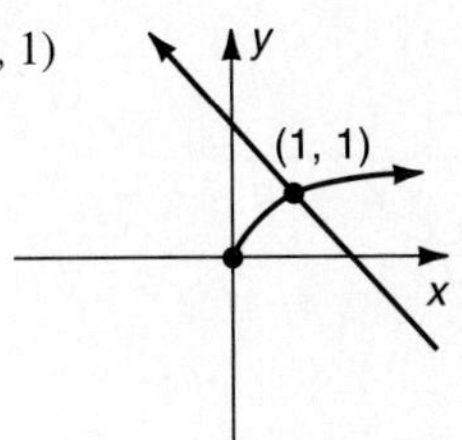

7. No intersection

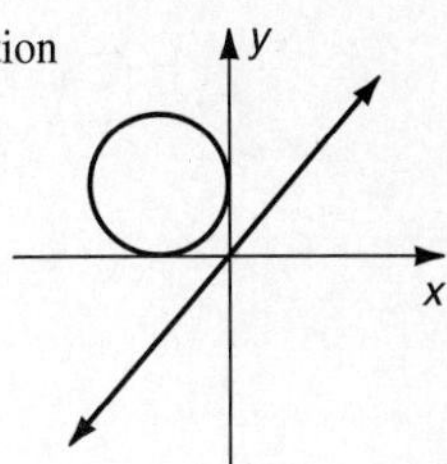

9. $\{(0, 3), (-3, 0)\}$

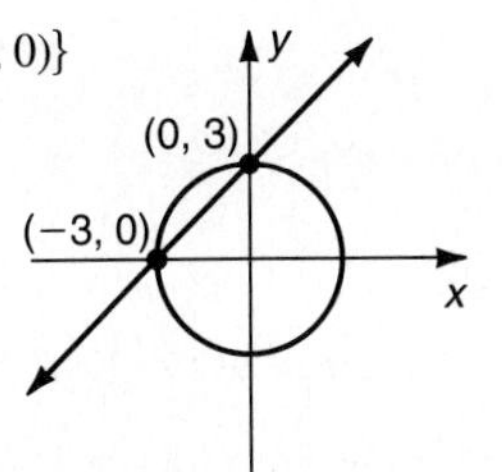

11. $\varnothing$

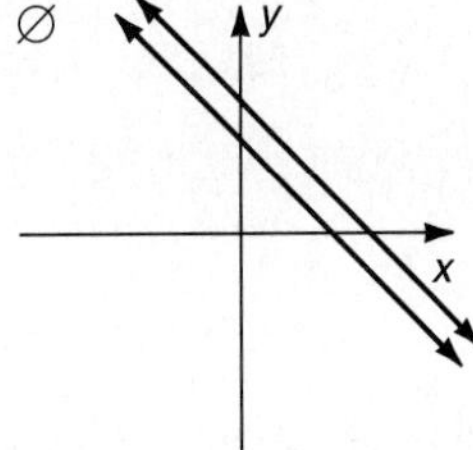

13. $\{(2\sqrt{2}, 0), (-2\sqrt{2}, 0)\}$ **15.** $\{(-2, 0)\}$ **17.** $\{(s, 3s + 6) \mid s \in R\}$ **19.** $\{(1, -1)\}$ **21.** Substitution **23.** Substitution **25.** 4 by 24 miles **27.** 69¢ a head and 99¢ a pound **29.** Boat: 10 mph; current: 6 mph **31.** 12 hrs and 24 hrs **33.** $\{(3, -4)\}$ **35.** $\varnothing$ **37.** $\{(t, 3t - 4) \mid t \in R\}$ **39.** $\{(2, 0), (-2, 0)\}$ **41.** $\varnothing$ **43.** $\left\{\left(-3, -\frac{4}{3}\right)\right\}$ **45.** $\left\{\left(-\frac{24}{23}, \frac{24}{13}\right)\right\}$ **47.** $\{(2, 2)\}$

49.

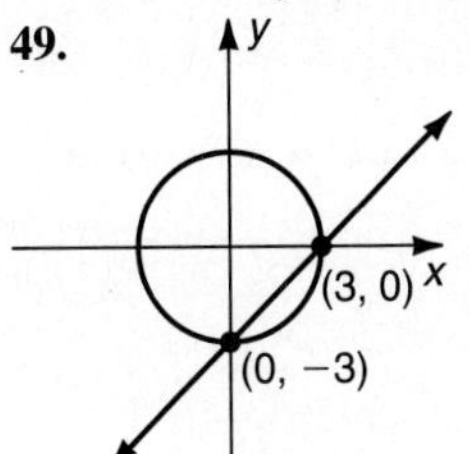

51.

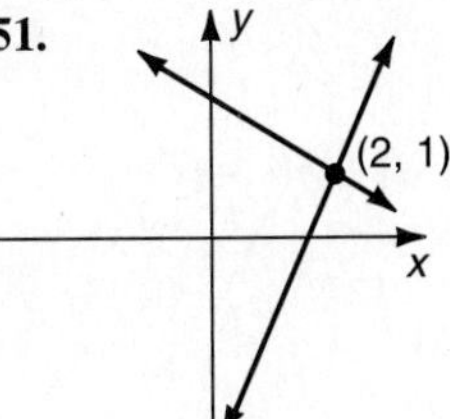

53.

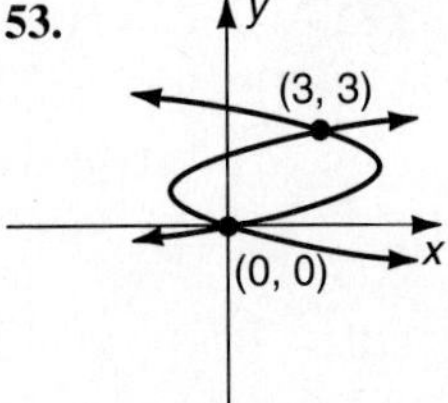

55.

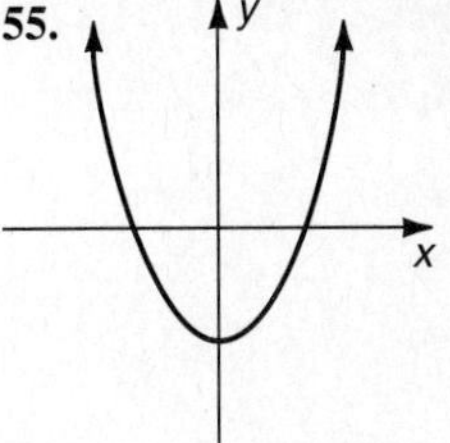

57.

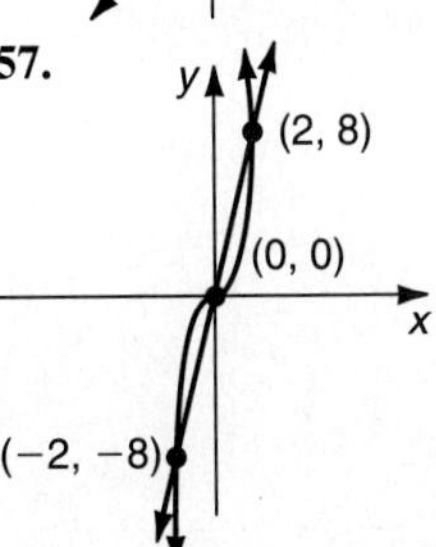

59. They are 0. −3 or 3, 0. **61.** 56 mph and 64 mph **63.** Penn: \$3.15; Wilson: \$2.70

65. 12 L of 17% solution and 14 L of the 24% solution. **67.** 10 by 17 ft **69.** All real numbers **73.** Exact: (1, 0) and (0, 1); approximate: (0.5437, 0.7044) **75.** Exact: (0, 1); approximate: (−2.6079, 0.0737) and (−1.4438, 0.2360) **77.** After approximately 14 minutes, 15 seconds.

PROBLEM SET 8.2

1. 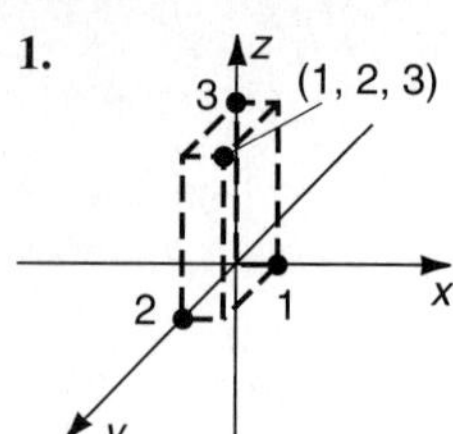

3. 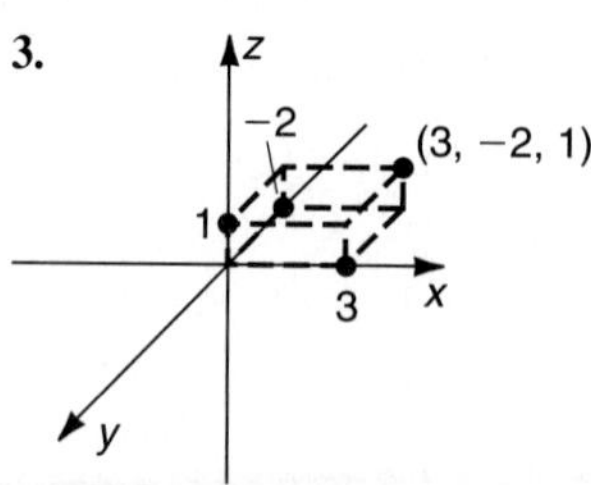

5. $\{(-2, 0, -1)\}$

7. $\left\{\left(-\frac{7}{5}, \frac{2}{5}, -\frac{1}{5}\right)\right\}$ 9. $\{(4, -1, 2)\}$ 11. $\varnothing$ 13. $\{(4t + 5, -t - 1, t) \mid t \in R\}$ 15. $y = x^2 - 4x + 4$

17. $\{(-7t, 3t + 2, t) \mid t \in R\}$ 19. $\{(1 - t, t - 1, -t) \mid t \in R\}$ 21. $\{(-1, 1)\}$ 23. $\left\{\left(\frac{13}{7}, \frac{-11}{7}\right)\right\}$ 25. $\{(0, 0)\}$ 27. $\varnothing$

29. $\{(1 - 6t, t) \mid t \in R\}$ 31. $\{(2, -1, 3)\}$ 33. $\varnothing$ 35. $\{(10, 4, 3)\}$ 37. $\{(42, 25, -7)\}$ 39. $\varnothing$
41. $\{(3t - 4, 2 - t, t) \mid t \in R\}$ 43. $\{(-1, 2, 1, 2)\}$ 45. They are 32, 19 and 13. 47. Small drinks, 65¢; large, \$1.00; hot dog \$1.50. 49. 40 minutes 51. No 53. Yes; $x = -\frac{1}{2}y^2 + \frac{11}{2}y - 14$ 55. $(x - 3)^2 + (y + 1)^2 = 4$

57. $k \neq 0$; $\left(-4, 8 - \frac{10}{k}, \frac{10}{k}\right)$

PROBLEM SET 8.3

1. $\begin{pmatrix} 4 & 1 & 0 & 7 \\ 1 & 2 & 3 & -8 \\ 2 & 0 & 3 & -1 \end{pmatrix}$ 3. $\begin{cases} x \qquad = -3 \\ -4x + 5y = 8 \end{cases}$ 5. $\{(1, 2, 3)\}$ 7. $\{(7, -11, 10)\}$ 9. $\{(0, 0)\}$

11. $\{(3 - t, t - 1, t \mid t \in R\}$ 13. $\left\{\left(\frac{5 - 2s}{3}, 1 - s, s\right) \mid s \in R\right\}$ 15. $\varnothing$ 17. $\left\{\left(-\frac{13}{2}, 3\right)\right\}$ 19. $\{(-7, 4, 0)\}$

21. $\{(-26, 10)\}$ 23. $\left\{\left(\frac{18}{13}, -\frac{14}{13}\right)\right\}$ 25. $\varnothing$ 27. $\left\{\left(-\frac{7}{3}t, t\right) \mid t \in R\right\}$ 29. $\left\{\left(-2, 1, \frac{1}{3}\right)\right\}$ 31. $\{(1, -2, 3)\}$

33. $\varnothing$ 35. $\left\{\left(\frac{2t - 1}{11}, \frac{3t - 7}{11}, t \mid t \in R\right\}$ 37. $\{(0, 0, 0)\}$ 39. $\{(-2t, t, t) \mid t \in R\}$ 41. $\{(2, -1, 3, 4)\}$ 43. They are −3, 4, 7, and 11. 45. Add −2 times row 1 to row 2, then exchange rows 1 and 2.

PROBLEM SET 8.4

1. $x = \frac{5}{2}$; $y = \frac{3}{2}$ 3. $\begin{pmatrix} -2 & 20 \\ -10 & 12 \end{pmatrix}$ 5. Not compatible 7. $\begin{pmatrix} 8a & 7a \\ -6a & 16a \end{pmatrix}$ 9. $\begin{pmatrix} 20 & 11 \\ -6 & 5 \\ -45 & -4 \end{pmatrix}$

11. Not compatible 13. $\begin{pmatrix} 13 & -14 \\ -13 & -5 \end{pmatrix}$ 15. $\begin{pmatrix} 8 & -1 & 0 \\ 5 & -1 & -1 \end{pmatrix}$ 17. Not compatible 19. $\begin{pmatrix} 8 & -3 \\ 3 & -1 \end{pmatrix}$

21. $\begin{pmatrix} 11 & -1 & 1 \\ 3 & 0 & 1 \end{pmatrix}$ 23. $\begin{pmatrix} 1 & -3 & 2 \\ 4 & -12 & 8 \\ 5 & -15 & 10 \end{pmatrix}$ 25. $\begin{pmatrix} 7 \\ -1 \\ 7 \\ 9 \end{pmatrix}$ 27. Not compatible 29. $\begin{pmatrix} -26 & 7 \\ 11 & -21 \end{pmatrix}$

31. $\begin{pmatrix} -17 & -1 \\ 30 & -30 \end{pmatrix}$ **33.** $\begin{pmatrix} 5 & -2 \\ 2 & -1 \end{pmatrix}$ **35.** $x_1 = 24$; $x_2 = 11$ **37.** $x_1 = 144$; $x_2 = 66$ **39.** $x_1 = \frac{3}{2}$; $x_2 = \frac{7}{2}$

41. $x_1 = -3$; $x_2 = 2$ **43.** $x_1 = 0$; $x_2 = 0$ **49.** $\begin{pmatrix} a & b & c \\ d & e & f \\ g & h & i \end{pmatrix}$ (Both parts) **51.** $A = \begin{pmatrix} 1 & 0 \\ 5 & 0 \end{pmatrix}$

53. $A = \begin{pmatrix} 1 & 0 \\ 0 & 2 \end{pmatrix}$ $B = \begin{pmatrix} 3 & 0 \\ 0 & 4 \end{pmatrix}$ **55.** $\begin{pmatrix} 94.5 & 90 & 93 & 90.9 \\ 1076.5 & 898 & 1017 & 933.7 \\ 7.2 & 19.2 & 11.2 & 16.8 \end{pmatrix}$ Grams of protein, carbohydrates, and fat in each blend. (b) 94.5 gm (c) 16.8 gms

PROBLEM SET 8.5

1. $\begin{pmatrix} \frac{3}{7} & -\frac{1}{7} \\ -\frac{4}{7} & -\frac{1}{7} \end{pmatrix}$ **3.** $\begin{pmatrix} -\frac{1}{6} & -\frac{1}{3} \\ -\frac{2}{3} & -\frac{1}{3} \end{pmatrix}$ **5.** $\begin{pmatrix} 1 & 1 & 1 \\ 3 & 5 & 4 \\ 3 & 6 & 5 \end{pmatrix}$ **7.** No inverse **9.** $\begin{pmatrix} 5 & -7 \\ 3 & 8 \end{pmatrix}\begin{pmatrix} x \\ y \end{pmatrix} = \begin{pmatrix} 1 \\ 0 \end{pmatrix}$

11. $\begin{pmatrix} 1 & 2 & -1 \\ 3 & 0 & 2 \\ 0 & 1 & -5 \end{pmatrix}\begin{pmatrix} x \\ y \\ z \end{pmatrix} = \begin{pmatrix} 11 \\ 5 \\ 0 \end{pmatrix}$ **13.** $\left\{\left(-\frac{6}{7}, \frac{1}{7}\right)\right\}$ **15.** $\{(1, 1)\}$ **17.** $\{(1, 2, 3)\}$ **19.** $\begin{pmatrix} \frac{1}{4} & 0 \\ 0 & \frac{1}{2} \end{pmatrix}$

21. $\begin{pmatrix} 5 & -1 \\ -4 & 1 \end{pmatrix}$ **23.** $\begin{pmatrix} -2 & 1 \\ \frac{5}{3} & -\frac{2}{3} \end{pmatrix}$ **25.** $\begin{pmatrix} 1 & 0 & 0 \\ 0 & 1 & 0 \\ 0 & 0 & 1 \end{pmatrix}$ **27.** $\begin{pmatrix} \frac{1}{3} & 0 & 0 \\ 0 & \frac{1}{4} & 0 \\ 0 & 0 & \frac{1}{5} \end{pmatrix}$ **29.** $\begin{pmatrix} -24 & 7 & 1 & -2 \\ -10 & 3 & 0 & -1 \\ -29 & 7 & 3 & -2 \\ 12 & -3 & -1 & 1 \end{pmatrix}$

31. $\{(9, -6)\}$ **33.** $\{(5, -4)\}$ **35.** $\{(-18, 79, -27)\}$ **37.** $\{(-2a + 2b + c, 9a - 8b - 3c, -3a + 3b + c)\}$

39. $a_{ij} = 1$ all i, j **41.**

	#1	#2	#3
X	872	347	775
Y	442	330	506
Z	250	511	225

PROBLEM SET 8.6

1. 5 **3.** 17 **5.** $-\begin{vmatrix} b & c \\ h & i \end{vmatrix}$ **7.** $\begin{vmatrix} e & f \\ h & i \end{vmatrix}$ **9.** 51 **11.** 30 **13.** −12 **15.** Yes **17.** No **19.** $\left\{\left(\frac{1}{9}, \frac{7}{3}\right)\right\}$

21. $\{(0, 0, 0)\}$ **23.** (a) B: Interchange rows; C: Multiply row 1 by 5; D: Multiply row 1 by −3 and add to row 2. (b) $\det A = -2$; $\det B = 2$; $\det C = -10$; $\det D = -2$ **25.** adf **27.** 0 **29.** 24 **31.** 0 **33.** 18 **35.** −30

37. $\{(-12, 11)\}$ **39.** $\left\{\left(\frac{2}{5}, -\frac{12}{5}, \frac{4}{5}\right)\right\}$ **41.** $\{(1, -2, 3)\}$ **43.** $\{(1, 2, -1)\}$ **45.** $\{(1, 2, -3, 4)\}$ **47.** -10

PROBLEM SET 8.7

1.

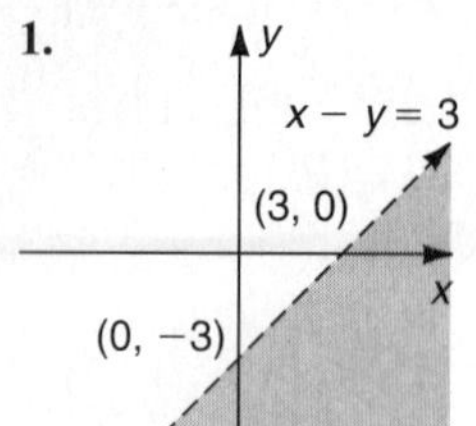

3.

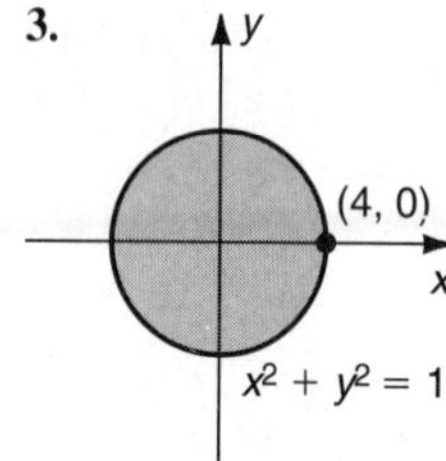

5.

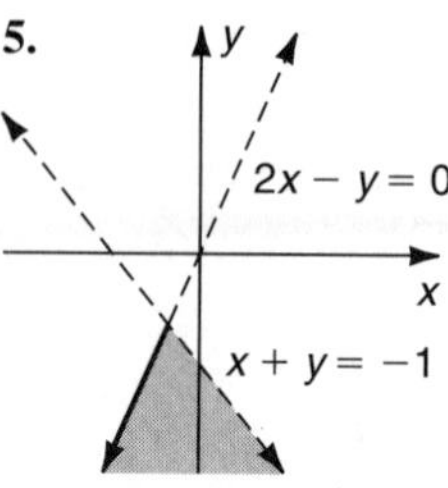

7.

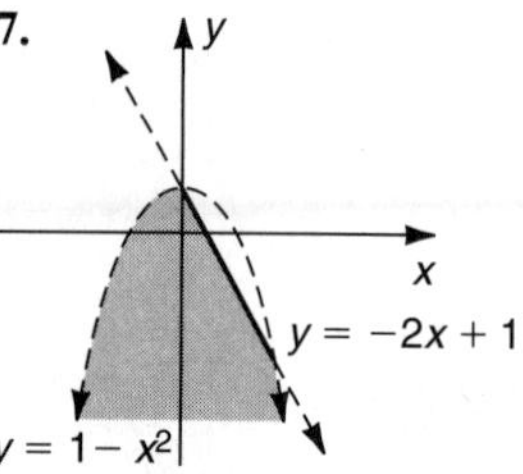

9. 0 **11.** 210 **13.** 10 oz dry and 20 oz canned.

15.

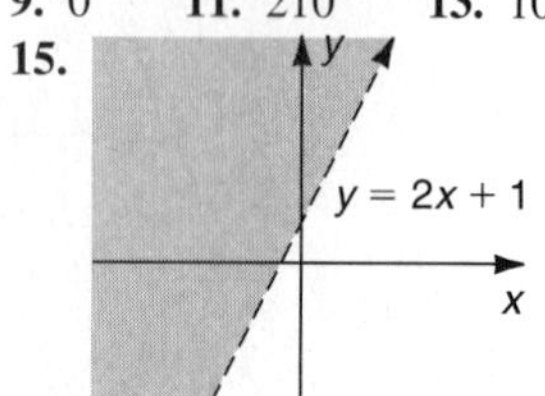

17.

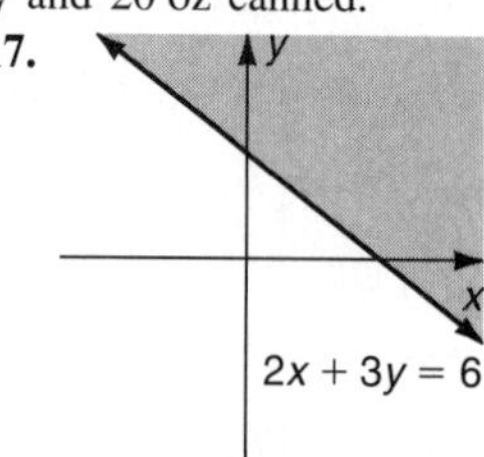

19.

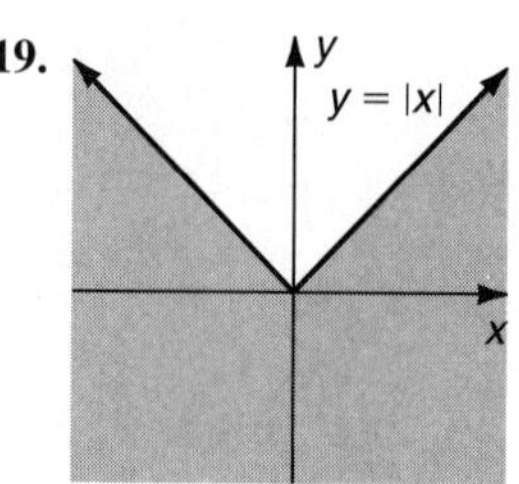

21.

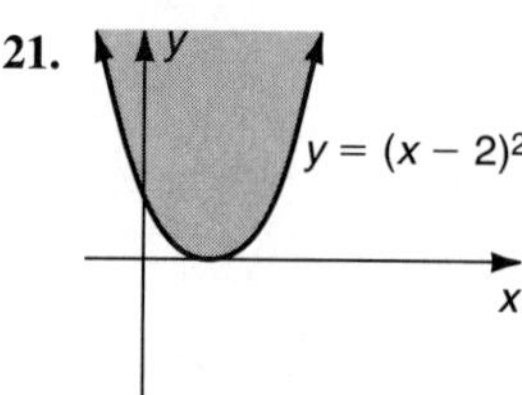

23.

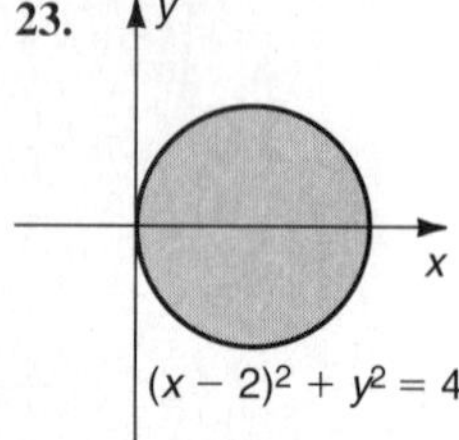

25.

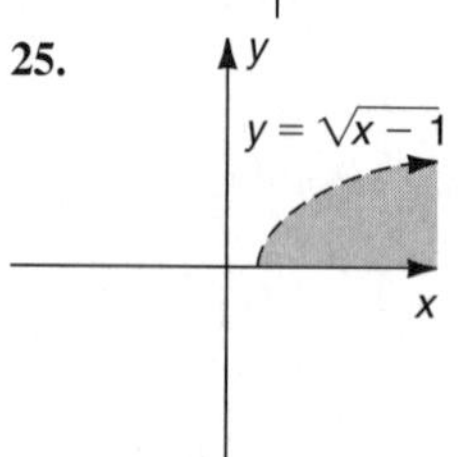

27.

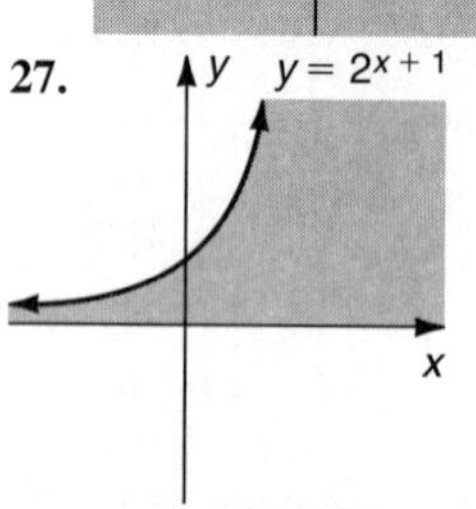

29.

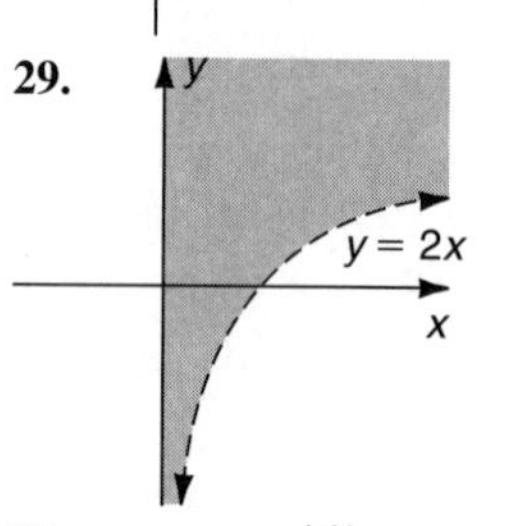

31.

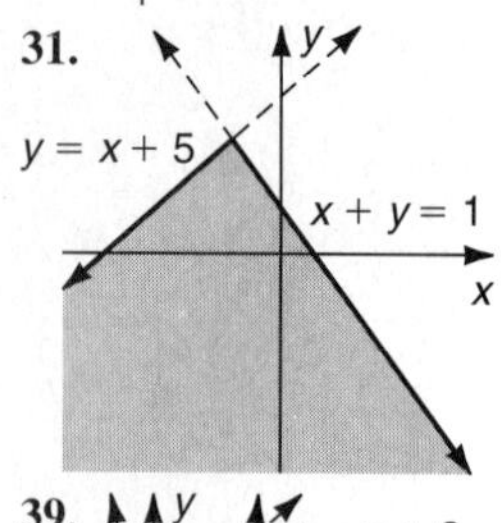

33.

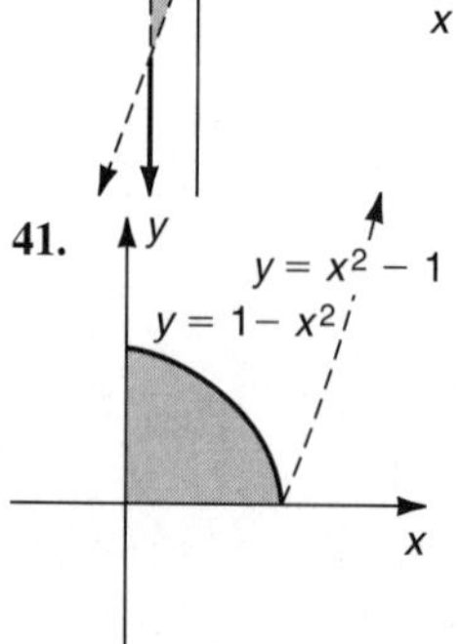

35.

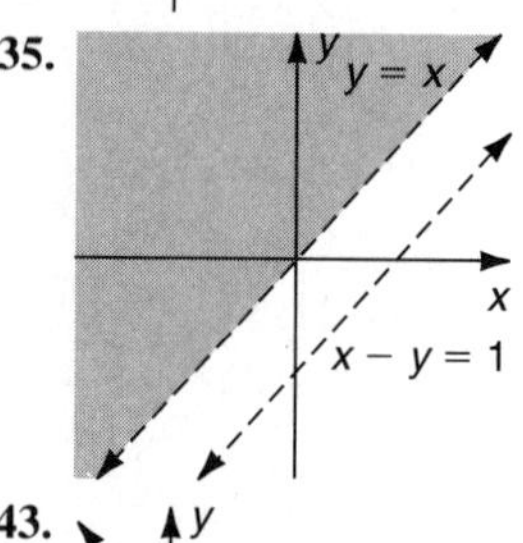

37.

y
x + 2y = 4
x
x + 2y = 0

39.

y
y = x + 3
y = (x − 1)² + 2
x

41.

y
y = x² − 1
y = 1 − x²
x

43.

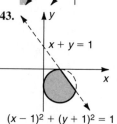

y
x + y = 1
x
(x − 1)² + (y + 1)² = 1

45.

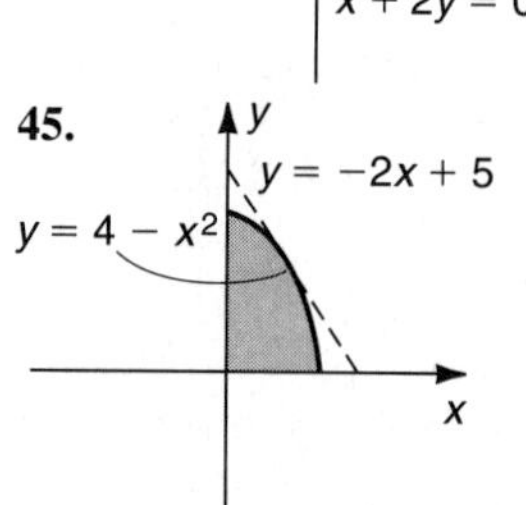

47.

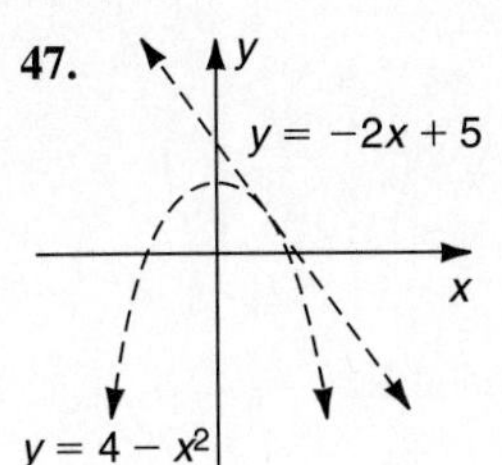

49.

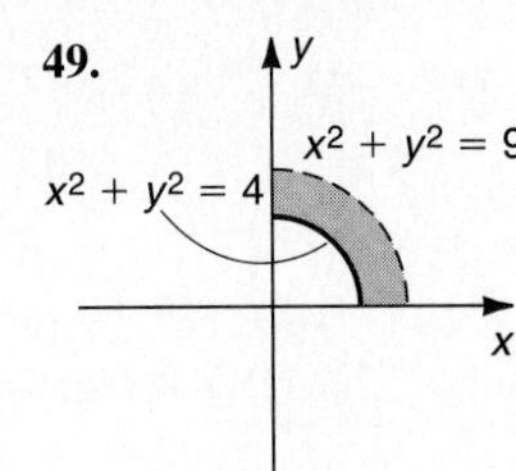

51.

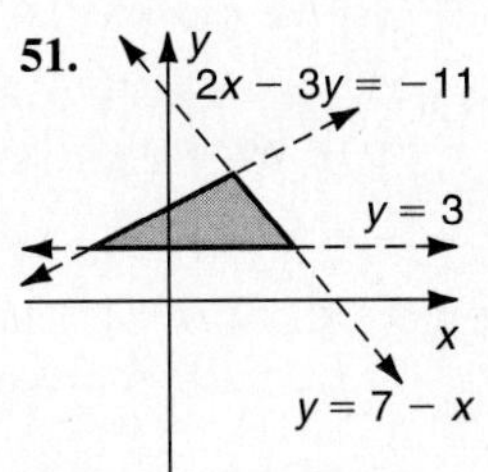

53. Max: 108; min: 27

55. Max: 96; min: 30 **57.** Max: $73\frac{1}{3}$; min: 36 **59.** She should invest \$30,000 in AAA and \$15,000 in AA.

PROBLEM SET 8.8

1. 3 **3.** 1 **5.** 0, ±2 **7.** −1, 0 **9.** 0.7 **11.** 1, 12.3 **13.** −2, 0, 2, 3 **15.** −2, 0, 2 **17.** (−3, 0), (3, 3)
19. $-3, -1, \frac{3}{2}$ **21.** [−3, 3] **23.** 0, 1 **25.** −1, 1 **27.** ≈ −1.6, 1 **29.** ≈0.7 **31.** (a)

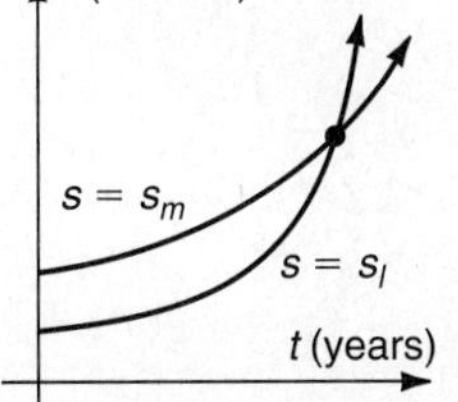

(c) Approx. 18.5 years. (d) Approx. 12.6 years
33. −1.2, −0.4, 0, 1.9

CHAPTER 8 REVIEW PROBLEMS

1. $\{(3, \pm\sqrt{5}), (-4, \pm 2\sqrt{3})\}$ **3.** {(2, 7), (4, 5)} **5.** {(1, 1), (−2, 2)} **7.** $\left\{\left(-\frac{10}{9}, -\frac{5}{3}\right)\right\}$ **9.** $\{(t, -3t) \mid t \in R\}$

11. ∅ **13.** $\left\{\left(\frac{7}{5}, -\frac{8}{5}, \frac{6}{5}\right)\right\}$ **15.** {(0, 0, 0)} **17.** $\left\{\left(\frac{41}{9}, 0, -\frac{4}{3}\right)\right\}$ **19.** $\left\{\left(\frac{11}{4}, \frac{1}{2}, \frac{11}{4}\right)\right\}$ **21.** {(0, 0, −5)}

23. $\{(t, -1 - t, t, -1 - t) \mid t \in R\}$

25.

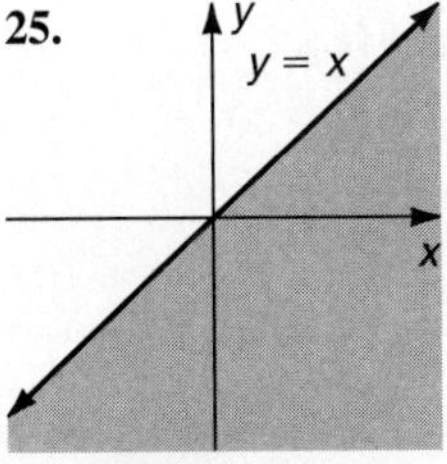

27.

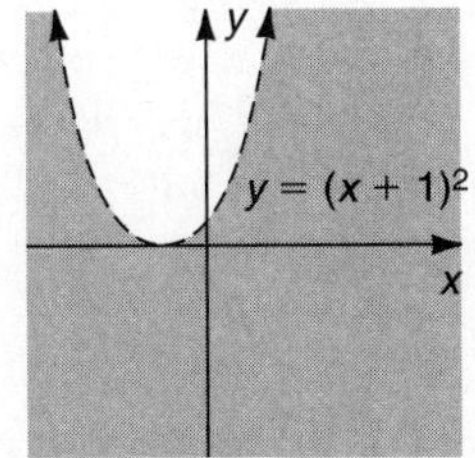

29.

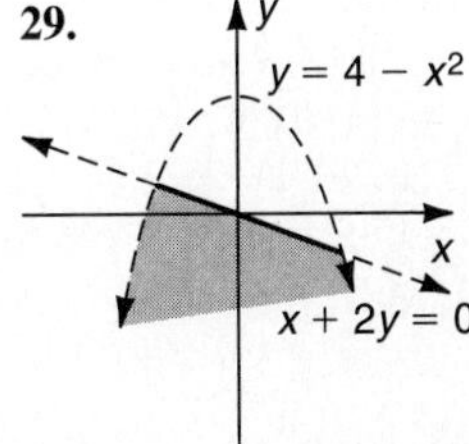

31.

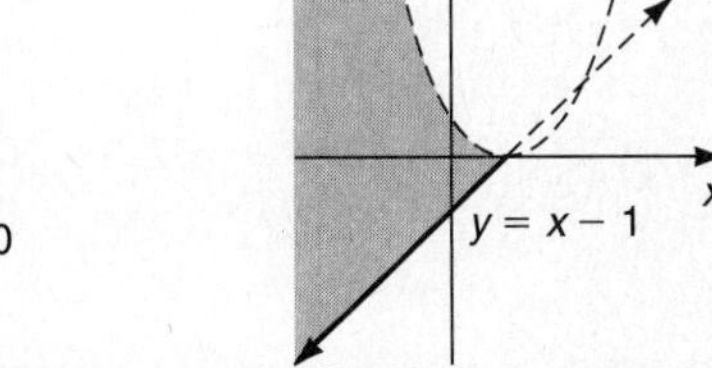

33.

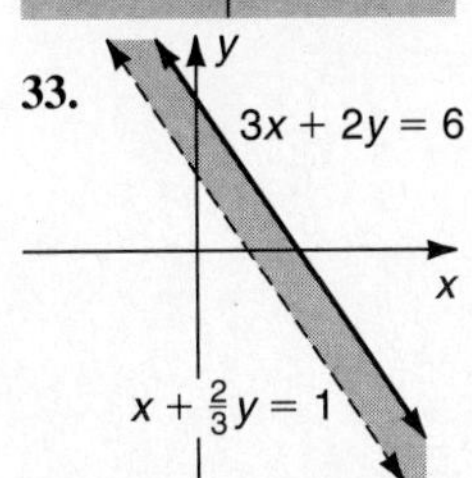

35. {(2, 3)} **37.** {4, −1, 6)} **39.** {(0, −5, 3)} **41.** $\begin{pmatrix} 3 & 2 \\ 1 & 1 \end{pmatrix}\begin{pmatrix} x \\ y \end{pmatrix} = \begin{pmatrix} 4 \\ 0 \end{pmatrix}$

43. $\{(-3, 2)\}$ **45.** -1 **47.** $\begin{pmatrix} -15 \\ -38 \end{pmatrix}$ **49.** $\begin{pmatrix} 3 & 0 & -2 \\ -6 & 1 & 4 \\ 1 & 0 & -1 \end{pmatrix}$ **51.** $\begin{pmatrix} \frac{36}{13} \\ -\frac{11}{13} \end{pmatrix}$ **53.** 13 **55.** $\begin{pmatrix} 10 & 4 & 9 \\ 6 & 3 & 6 \end{pmatrix}$

57. No inverse **59.** Max: 6; min: 0 **61.** Max: 84; min: 35 **63.** (1, 3) and approx. (−1.9, 0.1)
65. 8 ft by 48 ft **67.** 1,416 **69.** The matrix in row echelon form. **71.** (a) Not singular (b) Singular

PROBLEM SET 9.1

1. 62 **3.** 49 **5.** $1 + \frac{x^1}{3} + \frac{x^2}{5} + \frac{x^3}{7}$ **7.** $\log 6 - \log 7 + \log 8 - \log 9 + \log 10$ **9.** $\sum_{j=1}^{5} (2j + 1)$

11. $\sum_{j=0}^{4} \frac{(-1)^j}{2^j} y^j$ **13.** $\sum_{j=1}^{75} b_j + \frac{1}{2}\sum_{j=1}^{75} x^j$ **15.** $1 + \sum_{p=2}^{4} (-1)^{p-1}\frac{2p}{p-1} + \frac{5}{3}$ **17.** 120 **19.** 108 **21.** $\frac{67}{60}$

23. $2x + 4x^2 + 8x^3 + 16x^4 + 32x^5$ **25.** $1 - x^2 + x^4 - x^6 + x^8 - x^{10}$ **27.** $\frac{2}{3y} + \frac{4}{9y^2} + \frac{8}{27y^3} + \frac{16}{81y^4} + \frac{32}{243y^5}$

29. $\sum_{i=0}^{4} (6 + 3i)$ **31.** $\sum_{i=0}^{4} (-1)^i 3^{i+1} x^{2i}$ **33.** $\sum_{i=1}^{5} (-1)^{i+1}\frac{i}{i+1} s^{i-1}$ **35.** $\sum_{j=1}^{15} 5x^j$ **37.** $\sum_{j=1}^{23} \frac{j}{j+2}x^j + \frac{24}{26}x^{24} + \frac{25}{27}x^{25}$

39. $\sum_{j=1}^{100} (A_j + B_j)$ **41.** $\sum_{j=1}^{100} (A_j - B_j) - B_{101}$ **43.** $\sum_{j=0}^{99} (A_j + B_{j+1})$ **45.** $\sum_{j=1}^{15} A_j$ **47.** $3c_1x^1 + 2c_2x^2 + c_3x^3$

PROBLEM SET 9.3

1. 12 **3.** 3432 **5.** 28 **7.** 1 **9.** $s^9 + 9s^8t + 36s^7t^2 + 84s^6t^3$ **11.** $2187x^7 + 5103x^6y + 5103x^5y^2 + 2853x^4y^3$
13. $31{,}824x^7y^{11}$ **15.** $-55x^2y^7 + 11xy^{10} - y^{11}$ **17.** $a^{12} + 12a^{11}b + 66a^{10}b^2$ **19.** $s^{15} + 30s^{14} + 420s^{13}$
21. $512 + 2304t + 4608t^2$ **23.** $2187x^7 - 5103x^6 + 5103x^5$ **25.** $78a^2b^{11} + 13ab^{12} + b^{13}$ **27.** $270s^2 + 405s + 243$
29. $6048x^2y^5 + 1344xy^6 + 128y^7$ **31.** $66a^2b^{10} - 12ab^{11} + b^{12}$ **33.** $715a^9b^4$ **35.** $\frac{1{,}082{,}565}{8}x^8$
37. $a^6 + 6a^5b + 15a^4b^2 + 20a^3b^3 + 15a^2b^4 + 6ab^5 + b^6$
39. $x^9 + 9x^8 + 36x^7 + 84x^6 + 126x^5 + 126x^4 + 84x^3 + 36x^2 + 9x + 1$
41. $64x^6 + 192x^5y + 240x^4y^2 + 160x^3y^3 + 60x^2y^4 + 12xy^5 + y^6$
43. $a^9 + 9a^8b + 36a^7b^2 + 84a^6b^3 + 126a^5b^4 + 126a^4b^5 + 84a^3b^6 + 36a^2b^7 + 9ab^8 + b^9$
45. $x^9 + 18x^8 + 144x^7 + 672x^6 + 2016x^5 + 4032x^4 + 5376x^3 + 4608x^2 + 2304x + 512$
47. $x^6 + 12x^5y + 60x^4y^2 + 160x^3y^3 + 240x^2y^4 + 192xy^5 + 64y^6$ **49.** $a^{50} + 50a^{49}b$ **51.** $s^{500} + 1500s^{499}$

PROBLEM SET 9.4

1. 6, 120, 5040, 362880, 39916800 **3.** 0, 2, 16, 54, 128 **5.** 0, −1, 2, −3, 4 **7.** $a_n = (-1)^n n$ **9.** $a_n = 3n - 1$
11. $a_n = (2n - 1)^2$ **13.** $a_n = 3 + 2(n - 1)$ **15.** $a_n = 1 + 4(n - 1)$ **17.** $a_n = 11 - \frac{7}{5}(n - 1)$
19. $a_n = 3 + 2(n - 1)$ **21.** $a_n = 3(2)^{n-1}$ **23.** $a_n = \frac{1}{3}\left(\frac{1}{3}\right)^{n-1}$ **25.** $a_n = 4\left(\frac{1}{2}\right)^{n-1}$ **27.** $a_n = 5(2)^{n-1}$
29. increasing **31.** decreasing **33.** neither **35.** 2, 3, 4, 5, 6 **37.** 6, 4, 0, −8, −24 **39.** 1, 3, 7, 13, 21

41. $a_n = 2n + 2$ **43.** $a_n = 4n - 5$ **45.** $a_n = (n - 1)^3$ **47.** $a_n = 1 + 3(n - 1)$ **49.** $a_n = 10 - 2(n - 1)$
51. $a_n = -\frac{1}{2} + \frac{1}{2}(n - 1)$ **53.** $a_n = 1(3)^{n-1}$ **55.** $a_n = 2(-3)^{n-1}$ **57.** $a_n = 27\left(\frac{1}{3}\right)^{n-1}$

PROBLEM SET 9.5

1. $\sum_{n=1}^{6} (2n + 1)$ **3.** 364 **5.** 2343 **7.** $\frac{3}{2}$ **9.** diverges **11.** 93 **13.** 105 **15.** 48 **17.** $63\frac{1}{2}$
19. $\frac{211\pi}{162}$ **21.** $\frac{75}{2}$ **23.** 125249 **25.** $15\left(\frac{4}{3}\right)^{13} - 3$ **27.** $\frac{10}{9}$ **29.** diverges **31.** $\frac{27}{8}$ **33.** $\frac{30}{11}$
35. a) $10\left(\frac{4}{5}\right)^6$ b) 63.7856 c) 90 **37.** a) $39,093.47 b) $301,869.42 **39.** $\frac{22}{3}$

PROBLEM SET 9.6

1. 125 **3.** 12 **5.** 1 **7.** 20 **9.** 120 **11.** 1,320 **13.** 6 **15.** 9 **17.** 84 **19.** 9 **21.** 792
23. 1680 **25.** 1 **27.** 30 **29.** 360 **31.** 720 **33.** 21 **35.** 1 **37.** 28 **39.** 1 **41.** 28 **43.** 24,024
45. 1001 **47.** 1716 **49.** 1680 **51.** 11,440 **53.** N! **55.** $\frac{N!}{2}$ **57.** 1 **59.** $\frac{(N + 1)(N + 2)}{2}$

61. $(a + b)^n = \sum_{j=0}^{n} C(n, j)a^{n-j}b^j$

PROBLEM SET 9.7

1. 3/4 **3.** 5/36 **5.** 5/18 **7.** 3/26 **9.** 1 **11.** 7/12 **13.** {B, W}; {B}; 2/5 **15.** (a) 1/6 (b) 2/3 (c) 0
(d) 1 **17.** 1/72 **19.** 1/8 **21.** 35/216 **23.** 1/4 **25.** 11/216 **27.** 1/64 **29.** (a) 11/15 (b) 19/30 (c) 2/15
31. 0.007985

CHAPTER 9 REVIEW PROBLEMS

1. $x + 2x^2 + 4x^3 + 8x^4 + 16x^5$ **3.** $1 - 4 + 9 - 16 + 25$ **5.** $\sum_{j=1}^{4} (j + 2)^2 x^j$ **7.** $\sum_{j=1}^{k} \frac{13x^j}{6}$
11. $x^7 + 7x^6k + 21x^5k^2 + 35x^4k^3 + 35x^3k^4 + 21x^2k^5 + 7xk^6 + k^7$
13. $x^6 - 12x^5y + 60x^4y^2 - 160x^3y^3 + 240x^2y^4 - 192xy^5 + 64y^6$ **15.** $x^2, 2x^3, 4x^4, 8x^5, 16x^6$ **17.** $d = 3$
19. $r = -2/3$ **21.** 45/2 **23.** diverges **25.** 42 **27.** 24 **29.** 21 **31.** 1 **33.** 6720
35. 12144 **37.** 1/12 **39.** 7/12 **41.** 1/26 **47.** $1,073,741,823
49. Binomial coefficients with large powers.
51. $128x^7 + 448x^6t + 672x^5t^2 + 560x^4t^3 + 280x^3t^4 + 84x^2t^5 + 14xt^6 + t^7$

Index

A

B

C

D

E

I

J

L

M

Q

R

S